Table of Atomic Weights and Numbers

Element	Symbol	Atomic Number	Atomic Weight*
Actinium	Ac	89	(227)
Aluminum	Al	13	26.982
Americium	Am	95	(243)
Antimony	Sb	51	121.75
Argon	Ar	18	39.948
Arsenic	As	33	74.9216
Astatine	At	85	(210)
Barium	Ba	56	137.33
Berkelium	Bk	97	(247)
Beryllium	Be	4	9.012
Bismuth	Bi	83	208.980
Boron	B	5	10.81
Bromine	Br	35	79.904
Cadmium	Cd	48	112.41
Calcium	Ca	20	40.08
Californium	Cf	98	(251)
Carbon	C	6	12.011
Cerium	Ce	58	140.12
Cesium	Cs	55	132.905
Chlorine	Cl	17	35.453
Chromium	Cr	24	51.996
Cobalt	Co	27	58.9332
Copper	Cu	29	63.546
Curium	Cm	96	(247)
Dysprosium	Dy	66	162.50
Einsteinium	Es	99	(254)
Erbium	Er	68	167.26
Europium	Eu	63	151.96
Fermium	Fm	100	(257)
Fluorine	F	9	18.998
Francium	Fr	87	(223)
Gadolinium	Gd	64	157.25
Gallium	Ga	31	69.72
Germanium	Ge	32	72.59
Gold	Au	79	196.967
Hafnium	Hf	72	178.49
Hahnium	Ha	105	(260)
Helium	He	2	4.003
Holmium	Ho	67	164.930
Hydrogen	H	1	1.008
Indium	In	49	114.82
Iodine	I	53	126.9045
Iridium	Ir	77	192.2
Iron	Fe	26	55.847
Krypton	Kr	36	83.80
Kurchatovium	Ku	104	(260)
Lanthanum	La	57	138.91
Lawrencium	Lr	103	(260)
Lead	Pb	82	207.2
Lithium	Li	3	6.941
Lutetium	Lu	71	174.97
Magnesium	Mg	12	24.305
Manganese	Mn	25	54.9380
Mendelevium	Md	101	(258)
Mercury	Hg	80	200.59
Molybdenum	Mo	42	95.94
Neodymium	Nd	60	144.24
Neon	Ne	10	20.179
Neptunium	Np	93	237.0482
Nickel	Ni	28	58.70
Niobium	Nb	41	92.906
Nitrogen	N	7	14.007
Nobelium	No	102	(259)
Osmium	Os	76	190.2
Oxygen	O	8	15.999
Palladium	Pd	46	106.4
Phosphorus	P	15	30.9738
Platinum	Pt	78	195.09
Plutonium	Pu	94	(244)
Polonium	Po	84	(209)
Potassium	K	19	39.098
Praseodymium	Pr	59	140.908
Promethium	Pm	61	(145) (147)
Protactinium	Pa	91	231
Radium	Ra	88	226.0254
Radon	Rn	86	(222)
Rhenium	Re	75	186.2
Rhodium	Rh	45	102.906
Rubidium	Rb	37	85.47
Ruthenium	Ru	44	101.07
Samarium	Sm	62	150.4
Scandium	Sc	21	44.956
Selenium	Se	34	78.96
Silicon	Si	14	28.086
Silver	Ag	47	107.868
Sodium	Na	11	22.990
Strontium	Sr	38	87.62
Sulfur	S	16	32.06
Tantalum	Ta	73	180.948
Technetium	Tc	43	(97)
Tellurium	Te	52	127.60
Terbium	Tb	65	158.925
Thallium	Tl	81	204.37
Thorium	Th	90	232.038
Thulium	Tm	69	168.934
Tin	Sn	50	118.69
Titanium	Ti	22	47.90
Tungsten	W	74	183.85
Uranium	U	92	238.03
Vanadium	V	23	50.941
Xenon	Xe	54	131.30
Ytterbium	Yb	70	173.04
Yttrium	Y	39	88.906
Zinc	Zn	30	65.38
Zirconium	Zr	40	91.22

*Based on $^{12}_{6}C = 12.00 \ldots$

CONCEPTS OF GENERAL, ORGANIC, AND BIOLOGICAL CHEMISTRY

CONCEPTS OF GENERAL, ORGANIC, AND BIOLOGICAL CHEMISTRY

Robert D. Whitaker

Jack E. Fernandez

Janice O. Tsokos

University of South Florida

Contributor: *Carol Swezey, St. Elizabeth Hospital Medical Center, Lafayette, Indiana*

Houghton Mifflin Company *Boston*

Dallas *Geneva, Illinois* *Hopewell, New Jersey* *Palo Alto* *London*

This text is accompanied by the following publications which may be used along with it:
Study Guide with Solutions
Instructor's Manual
Laboratory Exercises for General, Organic, and Biological Chemistry, with Instructor's Manual, by Sandra M. Kotin of Sullivan County Community College, New York

Cover: The photograph shows a computer-generated image of a molecule of thyroid hormone (thyroxine) bound to its transport protein, forming a complex that is probably circulating in your blood stream right now. Thyroxine is a hormone that regulates the rate of all chemical reactions that proceed in your body. This hormone is manufactured by the thyroid gland and then released into the blood stream to be carried to the other tissues in the body. Hormones such as thyroxine are carried in the blood stream by specific transport proteins—thyroxine's transport protein is called prealbumin.

In the photograph, the pink dots represent the molecular surface of the hormone thyroxine, and the green dots represent the binding site of the protein prealbumin. (Photo used with the permission of Dr. Robert Langridge, Computer Graphics Laboratory, University of California, San Francisco. Color separations copyright 1980, *Chemical & Engineering News.*)

Printed in the U.S.A.
Library of Congress Catalog Card Number: 80-82738
ISBN: 0-395-29273-5

To our students

LIST OF CHAPTERS

CONTENTS

PREFACE

This text is written for a one-year course in general, organic, and biological chemistry to prepare students to enter nursing and allied health sciences, as well as nutrition, physical education, agriculture, home economics, and the environmental and biological sciences. Our overriding goal has been to present the chemistry needed in these fields and to make it readily accessible and understandable to the beginning student. We have therefore emphasized topics that lead to an understanding of life and life processes.

THOROUGH COVERAGE OF BASIC CHEMICAL CONCEPTS AND PROBLEM-SOLVING SKILLS

We have treated fundamental ideas of general, organic, and biological chemistry as thoroughly as possible, always keeping in mind the needs of the beginning student. Ideas are presented conceptually rather than quantitatively. Essential mathematical manipulations are covered in a careful, step-by-step fashion. To build the student's problem-solving skills, we have provided abundant worked-out examples throughout the text and more than 800 class-tested problems for the student. Approximately 350 of these problems are within the chapters, giving students the opportunity to test and reinforce their understanding as they read through a chapter. An additional 400 end-of-chapter problems encourage further review. An end-of-book self-test containing more than 90 problems helps students review for final exams. The mathematical techniques that are needed to understand certain topics (such as scientific notation in Chapter 1 and pH in Chapter 7) are presented right in the text with plenty of problems for students to develop the skills they need. The appendixes offer additional flexibility: Appendix 1 deals with unit analysis, Appendix 2 covers SI units, Appendix 3 contains a diagnostic math test and a review of basic mathematical skills, Appendix 4 provides a glossary defining all important terms, and Appendix 5 gives the answers to all in-chapter problems and half of the end of-chapter problems.

APPLICATIONS OF CHEMISTRY TO HEALTH AND EVERYDAY LIFE

To show students the relevance of chemistry and to reduce the abstractness of chemical concepts, we have highlighted the role of chemistry in everyday life, in matters of health, and in the health sciences. Chapter-opening discussions present examples of the role of chemistry in health and disease to emphasize the importance of the basic chemistry the student is about to learn. For example, Chapter 7, "Acids and Bases," is preceded by a brief discussion of diabetic acidosis; Chapter 16, "Carbohydrates," begins with a few paragraphs about glucose tolerance testing; and Chapter 20, "Lipids," opens with information about the measurement of lipids in amniocentesis. Throughout the narrative, we weave in as many brief examples of chemistry in health and everyday life as possible. Optional Focus essays located between chapters point to the reality of chemistry in everyday life with discussions of such topics as soft and hard water, food additives, and atherosclerosis and dietary lipids.

FLEXIBLE ORGANIZATION BUILT AROUND A THEME

The text begins with simple ideas in chemistry and gradually builds into the complex area of biochemistry. Chapters 1–7 present those principles of general and inorganic chemistry that are needed to understand later chapters on organic and biological chemistry. Chapters 8–16 contain the organic chemistry needed for biochemistry. Finally, Chapters 17–27 on biochemistry focus on the chemistry of life and its sustaining processes. Because courses in general, organic, and biological chemistry may differ in emphasis and selection of topics, we have used a unifying theme and extensive cross-referencing to make our coverage as adaptable and flexible as possible.

We organize the learning experience around the theme of the relationship between chemical structure on the one hand and chemical properties and biological functions on the other. We meet the theme in the early paragraphs of most chapters. Through an exploration of the theme in individual chapters, we deal with such questions as: How does sodium chloride, a requirement for life, differ from the very poisonous elements sodium and chlorine? Why does water have higher melting and boiling points than we might expect? What structural features are responsible for the catalytic activity of enzyme proteins? Why are starch and cellulose so different even though they both consist of units of the same compound, glucose? How does DNA direct the transmission of genetic information?

Certain concepts are referred to throughout the book. Each time we mention a particular concept we develop it further so the idea can grow in the student's mind. An example is hydrogen bonding. Hydrogen bonding is introduced in Chapter 5 to explain the unusual properties of water. Later, it is used in Chapters 10, 13, 14, and 15 to explain the properties of alcohols, ethers, amines, and other simple organic substances. In Chapters 17–20, hydrogen bonding serves to explain the properties of large biomolecules. Other chemical concepts that recur throughout the text include energetics, electronic structure, covalent bonding, equilibrium, reaction rates, and the Bronsted-Lowry acid-base theory.

Within each chapter, we employ a format that makes it easy for individual instructors to pick and choose from the material to fit their own courses. Wherever possible, we have limited each section (designated by a heading) to a single idea, method, or concept. Not only does this breakdown permit instructors to assign only the material that is relevant to their own courses, but it also enables students to master small, manageable units of subject matter. In our outline of "Alternative Sequences" we suggest possible selections of material for courses of several different emphases, lengths, and sequences.

CHAPTERS STRUCTURED FOR EASY READING AND EFFECTIVE REVIEW

Each chapter of the text has been carefully structured to be easy for the student to read and later to review and study. The various features of the format work together as follows:

Chapter-openers (illustrations and paragraphs) present a health-related topic, such as diabetic acidosis in Chapter 7, and point to the relevance of the chemistry the student is about to encounter in the chapter.

Chapter outlines appear at the beginning of each chapter to give the student an overview.

Chapter introductions offer a preview of the ideas to be covered in the chapter and bring out the book's theme of the relationship between chemical structure and life processes.

Sections (designated by headings) present a single idea, method, or concept so that ideas are carefully delineated for the reader.

Questions in the margins of the text introduce many sections. These questions and comments are designed to place students in an inquisitive attitude as they read the section. Our own students tell us that knowing what to look for is a great help in approaching a new technical subject.

Worked-out examples demonstrate key methods and concepts that students must master.

Approximately 350 in-chapter problems at the ends of sections encourage students to test mastery of the major idea presented in each section before moving to the next one. Answers to all in-chapter problems are given at the back of the book. Step-by-step worked-out solutions appear in a separate *Study Guide.*

Illustrations and photographs are used liberally throughout the text. They reinforce and supplement visually the concepts that the text presents in words.

Formulas of several different types are used along with drawings of models to help students make the transition from the two-dimensional page to the three-dimensional reality of molecules. We have found in our own classrooms that this transition is one of the most difficult that the student faces in introductory chemistry.

New terms appear in boldface type in the text and are carefully defined.

Key terms are listed at the end of each chapter along with references to the specific sections in which they first occur.

Summaries conclude the chapters and review the important ideas in them.

Approximately 400 additional problems at the ends of chapters give students a chance to review the chapter after going through it. All of the problems have been class tested, and the more challenging problems are marked with an asterisk. Answers to the problems that are numbered in color are given at the back of the book. Worked-out solutions are presented in the separate *Study Guide.*

Hints for approaching problems, given at the end of the problem set, refer students back to relevant sections of the text if they have trouble solving a problem.

Focus essays between chapters discuss interesting, practical applications of the chemistry the student has learned or introduce somewhat more challenging topics. The Focus essays are intended to enrich the course and may be omitted without discontinuity.

A glossary defining all important terms is located at the end of the text. The terms are also referenced to the sections of the text in which they are defined and discussed.

Appendixes on unit analysis, SI Units, and basic mathematical operations appear at the end of the book as resources for instructors and students.

A self-test at the end of the book helps students review for final exams.

COMPLETE LEARNING PACKAGE

This text is accompanied by the following publications which may be used along with it:

A Study Guide with Solutions by the authors contains a self-test for each chapter and the step-by-step worked-out solutions for all of the problems in the text.

The Instructor's Manual to accompany the text includes a description of each chapter, a set of test questions for each chapter, and the answers to all the problems in the book.

Laboratory Exercises in General, Organic, and Biological Chemistry by Sandra M. Kotin of Sullivan County Community College, New York, provides a wide range of experiments suitable for the abilities and interests of beginning students. It begins with simple laboratory procedures and techniques and builds to more complex operations as the organic and biological experiments are introduced. The exercises are complete and independent of one another to provide maximum flexibility in assignments. Some exercises include graphing and interpretation of data to provide students with background in an area that is frequently desired by employers of graduates from allied health career programs. A separate *Instructor's Manual* lists equipment, chemicals, quantities, and special notes and instructions for setting up the exercises.

ACKNOWLEDGMENTS

It is impossible to thank all the people who helped us individually and collectively in writing this book. There were many, and they helped us in many ways. Some of those to whom we owe thanks at the University of South Florida are Jeff C. Davis, George Jurch, Graham Solomons, Jon Weinzierl, and George Wenzinger. Each of these kind persons contributed by making suggestions, taking part in discussions, and, often, just by listening to what must have seemed endless talk about our book.

Questionnaire Survey Respondents

We are grateful to the more than 100 instructors who responded to Houghton Mifflin Company's lengthy questionnaire survey in the spring of 1979. The needs they expressed and the information they provided were invaluable to us in developing a text that we hope will make courses in general, organic, and biological chemistry an enjoyable experience for instructors and students alike.

Reviewers

Many chemistry professors around the country added their insights and suggestions through their meticulous and thoughtful reviewing of the manuscript. We especially wish to thank:

Douglas J. Bauer — *Mohawk Valley Community College*, Utica, New York

Thomas D. Berke — *Brookdale Community College*, Lincroft, New Jersey

W. Robert Coley — *Montgomery College*, Germantown, Maryland

Ruth M. Doherty — *University of Maryland*, College Park, Maryland

Duane W. Fish — *West Valley College*, Saratoga, California

Henry Heikkinen — *University of Maryland*, College Park, Maryland

Sandra Migdalof Kotin — *Sullivan County Community College*, South Fallsburg, New York

Nicholas Lembo — *Boston State College*, Boston, Massachusetts

Henry Marriani — *Boston State College*, Boston, Massachusetts

M. O. Naumann	*City College of San Francisco* San Francisco, California
Ambrose Pendergrast	*Georgia State University* Atlanta, Georgia
Alan R. Price	*University of Michigan* Ann Arbor, Michigan
Jerry Swanson	*Daytona Beach Community College* Daytona Beach, Florida
Carol B. Swezey	*St. Elizabeth Hospital Medical Center* Lafayette, Indiana
Albert Zabady	*Montclair State College* Upper Montclair, New Jersey

Contributor

We would like to give special acknowledgment to Carol B. Swezey for her role in creating the chapter-opening discussions of health-related applications of chemistry. As an instructor of general, organic, and biological chemistry at Purdue who subsequently became a developmental research chemist at St. Elizabeth Hospital Medical Center in Lafayette, Indiana, she is uniquely qualified to generate applications capable of arousing the interest of students preparing for health-related careers.

Our Students

Finally, we wish to thank our students who, as is usually the case, have taught us at least as much as we have taught them. We especially thank the students enrolled in our course at the University of South Florida during the years that we class-tested a preliminary version of this manuscript.

ALTERNATIVE SEQUENCES

Below we have suggested several possible alternative selections and sequences of topics that might be followed in courses of somewhat different lengths and emphases. Many other selections and adaptations are also possible with this text.

A. **Full year course** with approximately equal coverage of general, organic, and biological chemistry: Do all chapters in the text with Focus essays and starred sections optional.

B. **Full year course** with the first half devoted to general chemistry and the second half divided equally between organic chemistry and biological chemistry.

First Half

Self-test on math in Appendix III
Appendix I on unit analysis
Chapters 1–7 and first five "Focus" essays

Second Half

Chapter	*Recommended Sections*
8	8.1–8.12
9	9.1–9.7
10	10.1–10.2, 10.4–10.6, 10.12, 10.15, 10.16
12	12.1–12.8, 12.10, 12.11, 12.14
13	13.1–13.6, 13.8, 13.11
14	14.1–14.5, 14.8, 14.9, 14.11, 14.12
15	15.1–15.7, 15.12, 15.13, 15.15, 15.17
16	16.1–16.11
17	17.1–17.15
18	18.1–18.6
19	19.1–19.5
20	20.1–20.9
21	21.1–21.4, 21.8–21.13
22	22.1–22.9, 22.12–22.17
23	23.1–23.11
24	24.1–24.6, 24.8, 24.10, 24.11, 24.15–24.17
25	25.1–25.6, 25.8, 25.9, 25.16–25.20, 25.22

C. **Full year course** with the first half devoted to general chemistry and the second half emphasizing organic chemistry with light coverage of biological chemistry.

First Half: Same as B

Second Half

Chapter	*Recommended Sections*
8	8.1–8.12
9	9.1–9.9
10	10.1–10.6, 10.12–10.17
12	12.1–12.11, 12.14, 12.15
13	13.1–13.6, 13.8–13.11
14	14.1–14.12
15	15.1–15.17
16	16.1–16.11
17	17.1–17.15
18	18.1–18.6
19	19.1–19.5
20	20.1–20.9
21	21.1–21.4, 21.8–21.13

D. **Full year course** with the first half devoted to general chemistry and the second half emphasizing biological chemistry after light coverage of organic chemistry.

First Half: Same as B

Second Half

Chapter	*Recommended Sections*
8	8.1–8.12
9	9.1–9.4, 9.6
10	10.1, 10.2, 10.4, 10.6, 10.12, 10.16
12	12.1, 12.3, 12.8, 12.10, 12.11, 12.14, 12.15, 12.19
13	13.1, 13.2, 13.4–13.6, 13.9, 13.11
14	14.1–14.3, 14.8, 14.11, 14.12
15	15.1, 15.2, 15.7, 15.10, 15.11, 15.12, 15.13, 15.15–15.17
16	16.1–16.3, 16.8, 16.9, 16.11
17	17.1–17.15
18	18.1–18.6
19	19.1–19.5
20	20.1–20.9
21	21.1–21.4, 21.8–21.13
22	22.1–22.9, 22.12–22.17
23	23.1–23.11
24	24.1–24.18
25	25.1–25.22
26	26.1–26.14

E. **One semester** with equal coverage of general, organic, and biological chemistry.

Chapter	*Recommended Sections*
1	1.16
2	2.1–2.20
3	3.1–3.14
4	4.1–4.3, 4.6–4.10
5	5.1–5.11
6	6.1–6.9
7	7.1–7.10
8	8.1–8.12
9	9.1–9.4, 9.6
10	10.1, 10.2, 10.4, 10.6, 10.12, 10.16
12	12.1, 12.3, 12.8, 12.10, 12.11, 12.14, 12.15, 12.19
13	13.1, 13.2, 13.4–13.6, 13.11
14	14.1–14.3, 14.8, 14.11, 14.12
15	15.1, 15.2, 15.7, 15.12, 15.13, 15.15, 15.17
16	16.1–16.3, 16.8, 16.11
17	17.1–17.15
18	18.1–18.6
19	19.1–19.5
20	20.1–20.9
21	21.1–21.4, 21.8–21.13

F. **One semester** with emphasis on organic chemistry and light coverage of biological chemistry.

Chapter	*Recommended Sections*
1	1.16
2	2.1–2.20
3	3.1–3.14
4	4.1–4.3, 4.6–4.10
5	5.1–5.11
6	6.1–6.9
7	7.1–7.10
8	8.1 8.12
9	9.1–9.7
10	10.1–10.2, 10.4–10.6, 10.12, 10.15, 10.16
12	12.1–12.8, 12.10, 12.11, 12.14
13	13.1–13.6, 13.8, 13.11
14	14.1–14.5, 14.8, 14.9, 14.11, 14.12
15	15.1–15.7, 15.12, 15.13, 15.15, 15.17
16	16.1–16.11
17	17.1–17.15
18	18.1–18.6
19	19.1–19.5
20	20.1–20.9

G. **One semester** with emphasis on biological chemistry and light coverage of organic chemistry.

Chapter	*Recommended Sections*
1	1.16
2	2.1–2.20
3	3.1–3.14
4	4.1–4.3, 4.6–4.10
5	5.1–5.11
6	6.1–6.9
7	7.1–7.10
8	8.1–8.12
9	9.1, 9.2, 9.4
10	10.1, 10.4, 10.12
12	12.1, 12.3
13	13.1, 13.2, 13.4–13.6, 13.11
15	15.1, 15.2, 15.4, 15.7, 15.13, 15.15
16	16.1–16.3, 16.11
17	17.1–17.15
18	18.1–18.6
19	19.1–19.5
20	20.1–20.9
21	21.1–21.4, 21.8–21.13
22	22.1–22.9, 22.12–22.17
23	23.1 23.11
24	24.1–24.6, 24.8, 24.10, 24.11, 24.15–24.17
25	25.1–25.6, 25.8, 25.9, 25.16–25.20, 25.22

A QUICK GUIDE TO TOPICS RELEVANT TO THE CHEMISTRY OF LIFE

Why study chemistry? Many students take a course in general, organic, and biological chemistry because they are interested in health-related careers, the biological sciences, or how chemistry relates to themselves as human beings. For easy reference, we have provided the following list of topics relevant to the chemistry of life.

Adipose tissue	Section 25.1
Adrenaline (epinephrine)	Section 23.3
Air pollution	Section 4.5
Alcohols	Opener, Chapter 12; Section 12.10
Aldehydes	Section 14.12
Alkaloids	Section 13.10
Alkaptonuria	Section 25.21
Allergy	Section 20.8
Amides	Sections 15.7, 15.15–15.17
Amino acids	Opener, Chapter 13; Section 13.11, Chapter 17
Amniocentesis	Opener, Chapter 20
Anesthesia	Opener, Chapter 4
Anion gap	Opener, Chapter 27
Antibiotics	Section 26.14
Antinuclear antibodies	Opener, Chapter 18
Artificial radioisotopes	Section 2.18
Aspirin	Section 20.8
Atherosclerosis	Focus 12
Biodegradability	Section 25.7
Black lung disease	Opener, Chapter 9
Blood-brain barrier	Section 25.3
Blood clotting	Section 27.2
Blood components	Section 27.1
Blood pressure	Section 4.3
Blood storage	Opener, Chapter 1
Blood sugar (blood glucose)	Section 24.16
Body fluids	Section 27.1
Brownian motion	Section 5.10
Buffers	Sections 7.10, 27.4
Cancer chemotherapy	Section 26.5
Carbon-14 dating	Focus 2
Carcinogenic compounds	Focus 15
Catalysts	Section 6.5
CAT scanner	Opener, Chapter 2
Chlorinated hydrocarbons	Sections 11.1–11.5

TO THE STUDENT

STUDYING CHEMISTRY FROM THIS BOOK

Chemistry is most easily mastered through regular reading, review, and working of problems. Cramming for a test usually does not work well in chemistry. We have written this book to make chemistry accessible to you. The chapters are meant to be interesting and straightforward when you first read them, and they are set up to be easy to review and study afterward. We have included many examples of chemistry in everyday life, in health, and in the health sciences to show you how chemistry affects us in so many memorable ways. We hope you will develop the habit of studying chemistry and referring to this book regularly. Your own efforts will bring you success.

We suggest that you approach each chapter in this way. First, read the chapter opener that shows health-related applications of the chemistry you are about to learn. Then read the chapter outline and the introductory paragraphs to get an overview of the chapter. Next, read the first section after thinking about the question posed in the margin. Work the problems in the section. If you have trouble with a problem, reread the section. Check your answer at the back of the book only if you can work the problem. Then go on to the next section assigned by your instructor. At the end of the chapter, review the entire chapter by reading the summary. If any concepts are not clear, review the relevant sections. Reviewing the key terms also helps to review the chapter.

At this point, you should work as many of the additional problems at the end of the chapter as you can or as your instructor assigns. If you cannot figure out how to answer a problem or a question, see the *Problem-Solving Hints* given at the end of the problem set. The hints refer you back to the relevant sections of the book. Once you have worked a problem, check your answer at the back of the book. If you have a *Study Guide*, do the self-test in the *Study Guide*. Go over the step-by-step solutions for any questions or problems you missed.

When you are preparing for a major examination in the course, you can use the comprehensive self-test at the end of the book to help complete your review of the material.

CONCEPTS OF GENERAL, ORGANIC, AND BIOLOGICAL CHEMISTRY

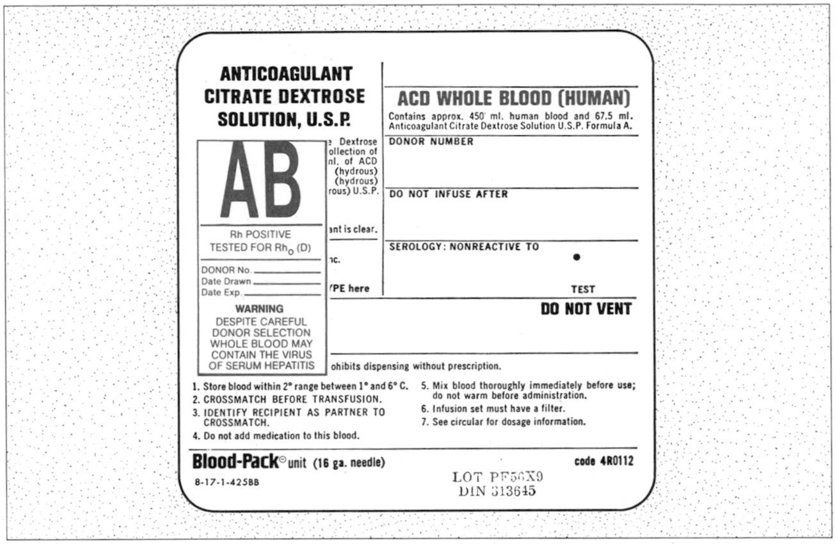

ANTICOAGULANT
CITRATE DEXTROSE
SOLUTION, U.S.P.

ACD WHOLE BLOOD (HUMAN)
Contains approx. 450 ml. human blood and 67.5 ml. Anticoagulant Citrate Dextrose Solution U.S.P. Formula A.

AB
Rh POSITIVE
TESTED FOR Rh_o (D)
DONOR No.
Date Drawn
Date Exp.

WARNING
DESPITE CAREFUL DONOR SELECTION WHOLE BLOOD MAY CONTAIN THE VIRUS OF SERUM HEPATITIS

DONOR NUMBER
DO NOT INFUSE AFTER
SEROLOGY: NONREACTIVE TO
TEST
DO NOT VENT

ohibits dispensing without prescription.

1. Store blood within 2° range between 1° and 6° C.
2. CROSSMATCH BEFORE TRANSFUSION.
3. IDENTIFY RECIPIENT AS PARTNER TO CROSSMATCH.
4. Do not add medication to this blood.
5. Mix blood thoroughly immediately before use; do not warm before administration.
6. Infusion set must have a filter.
7. See circular for dosage information.

Blood-Pack© unit (16 ga. needle) code 4R0112
8-17-1-425BB
LOT PF56X9
DIN 313645

We will introduce the idea of temperature in this chapter. Temperature is a critical factor in many basic chemical reactions, including many life-sustaining biochemical reactions. The important role that temperature plays becomes evident when we look at blood storage and see how sensitive blood is to temperature.

The primary function of the human blood cell is to deliver oxygen and other necessary nutrients to the human tissues. The patient who has lost a significant amount of blood lacks the means of delivering oxygen to the tissues in sufficient amounts. Such a patient is likely to require a blood transfusion, to provide a new source of red blood cells that can be used to nourish the tissues.

The transfusion procedure depends on the successful collection and storage of blood. Blood is collected from a donor into a sterile solution, which prevents the blood from clotting and provides nutrients for the blood cells to continue their metabolism while outside of the body. To keep the blood cells alive and to maintain their ability to function, collected blood must be stored at a temperature of from 1° to 6° C. Within this temperature range, blood can be transported and stored for up to twenty-one days. When blood is transported, it is subjected to special cooling conditions so that the temperature of the blood stays within acceptable limits. When blood is stored in blood banks, it is placed in special refrigerators that are monitored constantly

1

INTRODUCTION

for any variation in temperature. Beyond twenty-one days, certain bacteriologic and biochemical changes take place that affect the composition of the blood.

The importance of temperature control becomes evident when we look at what happens to stored blood when temperature conditions are not carefully maintained. The stored blood undergoes a chemical reaction in which the blood cells are broken down. The blood is said to be hemolyzed and, in this condition, is no longer able to carry oxygen.

[Illustration: The storage of blood is a critical factor in the transfusion of blood from donor to recipient.]

INTRODUCTION

The primary goal of this book is to help us understand how the structure of molecules accounts for the properties of substances. An appreciation of the structural basis for the properties of molecules is essential for an understanding of life processes.

For example, consider water, something with which we are all familiar. Water is absolutely essential to nearly all forms of life. Roughly two-thirds of our body weight is due to water. It constitutes up to 90% of the weight of some organisms.

The smallest particle of water is called a water molecule. No one has ever seen a water molecule because it is far too tiny for even the most powerful microscope to make it visible. Yet we know that each water molecule is made of three even tinier particles—two hydrogen atoms and one oxygen atom. The two hydrogen atoms are joined to the oxygen atom.

The three atoms that make up the water molecule do not lie in a straight line like this: o—o—o. Instead, they are bent like this: o/o\o This shape of the water molecule is responsible for the fact that table salt dissolves in water. It also partly explains why gasoline and water do not mix. (In a later chapter, we will see exactly why this shape is so important.) If water molecules were straight, the reverse would likely be true: gasoline and water would probably mix, and salt would not dissolve to any extent in water. Certainly, life, if it existed, would be totally different from what it is.

To realize our goal, we must learn about the things that determine molecular structure: properties of atoms; the combinations of atoms to form molecules; the way changes can occur to transform certain substances into different ones.

This first chapter will provide some fundamental ideas about matter and energy and their measurement that we will use throughout our venture into chemistry.

MATTER AND ENERGY 1.1

Matter is the physical stuff that makes up the universe. The most obvious properties of matter are that it has **mass** and takes up space. "Mass," in fact, is just another way of saying "quantity of matter." Mass is related to **weight,** and we often use the words interchangeably, but there is a difference between them. Weight is the force of gravity acting on a mass. Without gravity, an object still has mass, but it has no weight. Anywhere on the earth's surface, the force of gravity on a given mass is practically the same. Therefore, as long as we do not venture into outer space, beyond the pull of gravity, the distinction between mass and weight is not critical.

Matter can be altered, and the capacity to alter it is called **energy.** Every change that matter undergoes involves a transfer of energy. The measurement of matter and energy is fundamental to the study of chemistry, and we will discuss it at some length later on in this chapter.

PHYSICAL CHANGES 1.2

Every day we see a great many physical changes. At breakfast, we spread butter on hot toast. The form of the stick of butter is changed by the spreading. But the composition of the butter has not changed: the thin

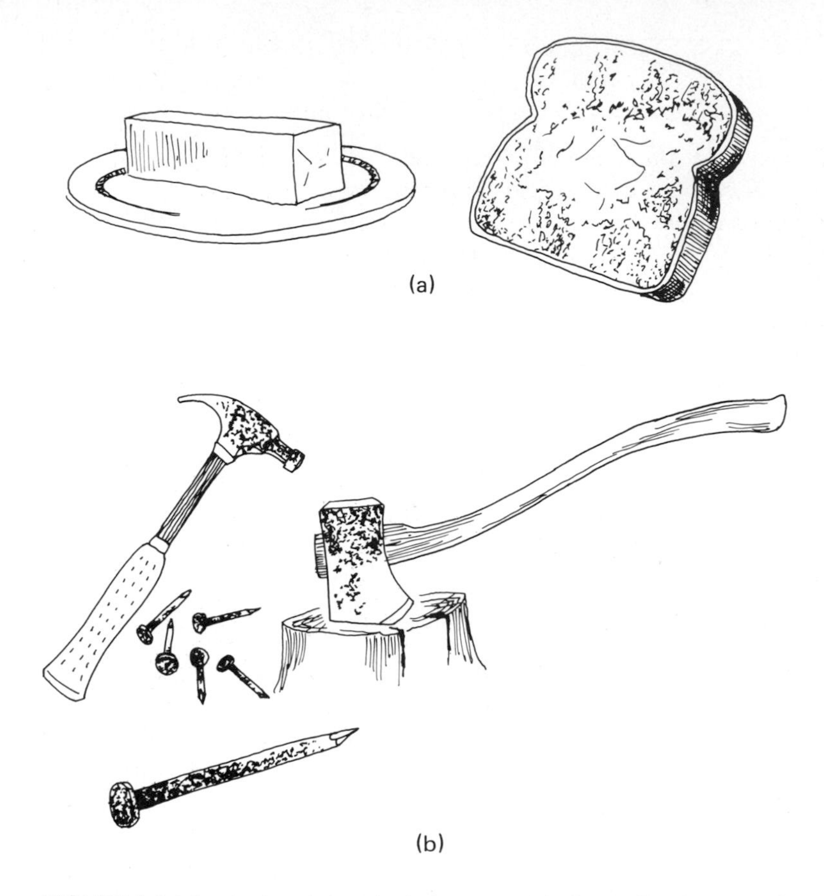

FIGURE 1.1 Physical and chemical changes. (a) Spreading butter on hot toast is an example of a physical change. The form of the butter changes from solid stick to a thin melted layer but its composition does not change. (b) Unprotected metal objects often become coated with rust. The composition of rust is different from that of iron or steel, so this is an example of a chemical change.

layer of melted butter on the toast is still butter (Figure 1.1a). *Form* refers to the appearance of something; *composition* refers to the nature of its constituent parts. When we dissolve sugar in coffee or tea, the form of the sugar is certainly changed, but the composition is not. Whether in the sugar bowl or dissolved in a liquid, sugar is still sugar. Changes such as these, which affect the form but not the composition of a substance, are called **physical changes.**

CHEMICAL CHANGES 1.3

What is the difference between a physical and a chemical change?

A change that alters not only the form of matter but also its composition is called a **chemical change.** Iron rusts, and the rust that is produced is something quite different from iron (Figure 1.1b). Green leaves turn yellow, red, or brown in the fall; the changing colors signal a complete alteration in the composition of the leaves. We eat and digest food, and the food is drastically altered. In all of these changes, substances with certain compositions and properties are transformed into different substances with quite different compositions and properties. In other words, all of these changes are chemical ones.

Problem 1.1 Classify the following changes as physical or chemical.

(a) freezing liquid water to ice
(b) sawing a board in two
(c) the fermentation of sugar to alcohol
(d) burning a piece of wood
(e) boiling water

We also refer to chemical changes as **chemical reactions.** We can write equations to show what happens in a reaction. For example, when iron rusts, it combines with oxygen in the air to form iron oxide, better known as rust. We can express this reaction in an equation:

$$\text{iron} + \text{oxygen} \rightarrow \text{iron oxide}$$

The arrow means "produces" or "yields."

ELEMENTS AND COMPOUNDS 1.4

How do elements and compounds differ?

Matter is made up of simple substances called elements. An **element** is a substance that cannot be broken down into a simpler chemical substance. Water is not an element because it can be broken down by an electrical current into the simpler chemical substances hydrogen and oxygen (see Figure 1.2). Hydrogen and oxygen are elements because they cannot be broken down into simpler chemical substances. Water is a compound. A **compound** is composed of certain elements combined in a fixed and definite proportion by weight.

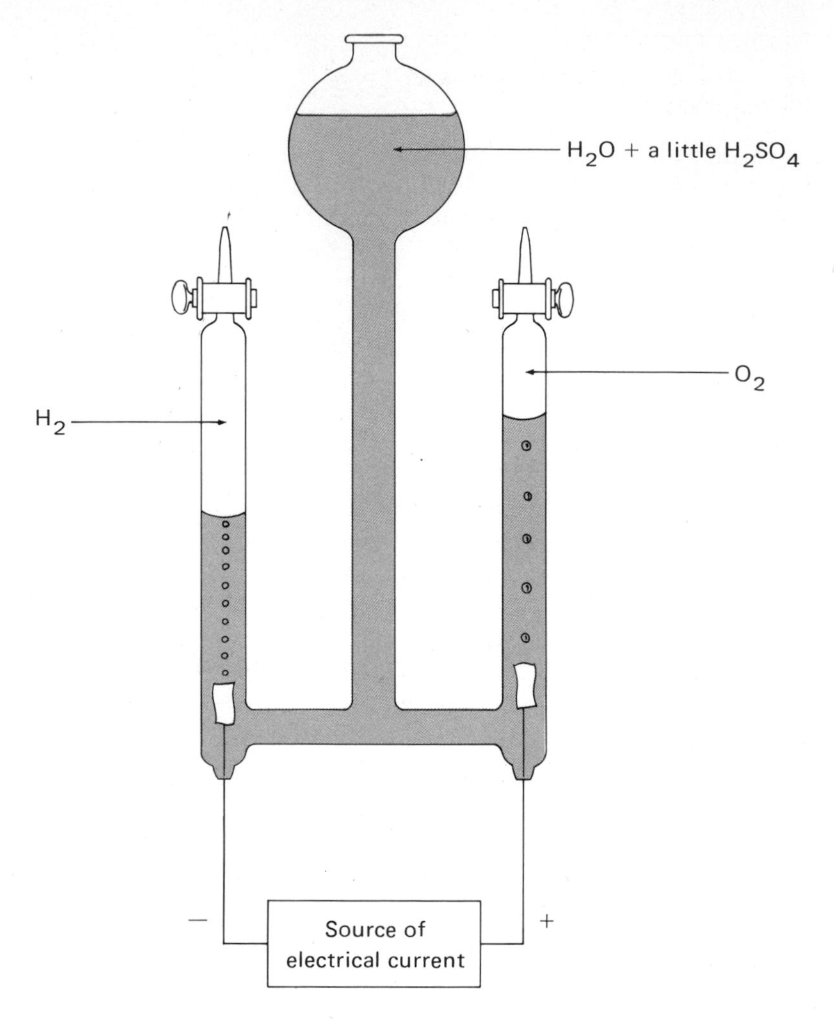

FIGURE 1.2 Electrolysis of water (H_2O) produces pure hydrogen (H_2) and oxygen (O_2). Pure water will not conduct an electrical current, so a small amount of some substance such as sulfuric acid (H_2SO_4) is added to make the solution conducting. The sulfuric acid is unchanged by the electrolysis.

1.5 CHEMICAL SYMBOLS

Chemists commonly use symbols to stand for the chemical elements. There is one **chemical symbol** for each element. Most of these

TABLE 1.1 Twenty-five Common Elements and Their Symbols

Aluminum	Al	Manganese	Mn
Calcium	Ca	Nickel	Ni
Carbon	C	Nitrogen	N
Chlorine	Cl	Oxygen	O
Chromium	Cr	Phosphorus	P
Cobalt	Co	Potassium	K
Copper	Cu	Silicon	Si
Fluorine	F	Sodium	Na
Helium	He	Sulfur	S
Hydrogen	H	Titanium	Ti
Iodine	I	Vanadium	V
Iron	Fe	Zinc	Zn
Magnesium	Mg		

symbols are simply the first letter, or first two letters, of the English name of the element: O is the symbol for oxygen, H for hydrogen, and He for helium. Some symbols are derived from non-English names. For example, Na, the symbol for sodium, comes from *natrium*, the Latin word for sodium.

Table 1.1 lists twenty-five common chemical elements and their symbols. These should be memorized. It is very important that symbols be written correctly—changing a small letter to a capital can alter the meaning completely. Co is the symbol for the metal cobalt, but if we change the o to O we get CO, the formula for the poisonous gas carbon monoxide.

Problem 1.2 What is wrong with using SI as the symbol for silicon? What is the correct symbol?

SCIENTIFIC UNITS 1.6

Why do we use metric units instead of British units in chemistry?

In the United States, we have traditionally used British Imperial units of measurement: feet, pounds, and seconds. In most other countries, the **metric system** is used, which is based on meters, kilograms, and seconds. The International System of Units (abbreviated SI) is a modification of the metric system. The fundamental **SI units** are given in Appendix 2.

Someday everyone will use the metric system. Even the United States now has a federal law that encourages the eventual conversion from the British Imperial to the metric system. Table 1.2 shows some equivalent British and metric units. There are many metric units besides those listed in Table 1.2, but we will not be concerned with most of them. The units we will use often in this book are listed in Table 1.3.

One big advantage of metric units over British units is that one metric unit can be converted into another by multiplying or dividing

TABLE 1.2 British Imperial and Metric Units: Equivalent Units

Quantity expressed	Metric unit		Imperial unit
Length	One kilometer (km)	=	0.6214 mi
	One meter (m)	=	39.37 in
	0.3048 m	=	One foot
	One centimeter (cm)	=	0.3937 in
	2.54 cm	=	One inch
	30.48 cm	=	One foot
Mass	One kilogram (kg)	=	2.20 lb
	One gram (g)	=	0.00220 lb
	453.59 g	=	One pound
Volume*	One liter (L)	=	1.06 U.S. qt
	One milliliter† (mL)	=	0.00106 U.S. qt
	3.785 L	=	One U.S. gallon

*The U.S. system differs from the British system in the size of its units of volume. The U.S. pint holds 16 fluid ounces, the British, 20 fluid ounces. The relationships of pint, quart, and gallon are the same in both systems, so the British gallon is the equivalent of 1.201 U.S. gallons.
†One milliliter equals one cubic centimeter (cm^3 or cc).

TABLE 1.3 Metric Units Often Used in Chemistry

Length	10 millimeters (mm)	=	1 centimeter (cm)
	100 cm	=	1 meter (m)
Mass	1,000 milligrams (mg)	=	1 gram (g)
	1,000 g	=	1 kilogram (kg)
Volume	1,000 milliliters (mL)	=	1,000 cubic centimeters (cm^3)
		=	1 liter (L)

TABLE 1.4 Metric Prefixes Used in Chemistry

Prefix	Symbol	Meaning
mega-	M	million
kilo-	k	thousand
deci-	d	one-tenth
centi-	c	one-hundredth
milli-	m	one-thousandth
micro-	μ	one-millionth
nano-	n	one-billionth

by ten, a hundred, a thousand, or some other multiple of ten. There is no uniform relationship of this sort among British units. A yard has 3 feet, but a foot has 12 inches. By contrast, a meter contains 100 centimeters, and a centimeter contains 10 millimeters. Metric units of different sizes are related to each other by prefixes. The prefixes tell you whether one unit is ten times another, a hundred times, or whatever. For example, the prefix *kilo-* means a thousand, so we know that a kilometer is 1,000 meters and a kilogram is 1,000 grams. The prefix *milli-* means one-thousandth, so we know a millimeter is one-thousandth of a meter, and a milligram is one-thousandth of a gram. The metric prefixes used most often in chemistry are shown in Table 1.4.

Here are some examples showing how to convert from one metric unit to another and from metric to British units. These examples employ the method of unit analysis, which we will use throughout this book to solve numerical problems. Appendix 1 explains this method in detail.

Example 1.1 Niagara Falls has a height of 50.9 m. Express this height in kilometers.

Solution

$$50.9\ \cancel{\text{m}} \times \frac{1\ \text{km}}{1000\ \cancel{\text{m}}} = 0.0509\ \text{km}$$

Example 1.2 What is the mass (in grams) of a 0.468-kg steak?

Solution

$$0.468\ \cancel{\text{kg}} \times \frac{1000\ \text{g}}{1\ \cancel{\text{kg}}} = 468\ \text{g}$$

Example 1.3 A cup holds 0.235 L of water. How many milliliters of water does it hold?
Solution

$$0.235\ \cancel{\text{L}} \times \frac{1000\ \text{mL}}{1\ \cancel{\text{L}}} = 235\ \text{mL}$$

Example 1.4 Lake Tanganyika is 1.43 km deep. What is its depth in miles?
Solution

$$1.43\ \cancel{\text{km}} \times \frac{0.6214\ \text{mi}}{1\ \cancel{\text{km}}} = 0.889\ \text{mi}$$

Example 1.5 How many pounds of tomatoes is 1.5 kg of tomatoes?
Solution

$$1.5\ \cancel{\text{kg}} \times \frac{2.20\ \text{lb}}{1\ \cancel{\text{kg}}} = 3.3\ \text{lb}$$

Example 1.6 How many milliliters of milk are in 1.00 quart of milk?
Solution

$$1.00\ \cancel{\text{qt}} \times \frac{1\ \text{mL}}{0.00106\ \cancel{\text{qt}}} = 943\ \text{mL}$$

Problem 1.3 Some antihistamine tablets each contain 0.00600 g of active ingredient. How many milligrams of active ingredient are contained in each tablet?
Problem 1.4 What is the volume in liters of 200 mL of a solution?
Problem 1.5 What is the length in centimeters of a 3.50-in. nail?
Problem 1.6 What is the mass in kilograms of 11.0 lb of potatoes?

1.7 DENSITY

What is the difference between the mass of an object and its density?

The mass of a specific volume of a substance is a measure of how compact the material is. **Density** is the scientific name for this compactness. It is defined as the mass of a substance per unit of volume.

$$\text{Density} = \frac{\text{mass}}{\text{volume}}$$

The word "per" can always be read as "divided by."

It makes no sense at all to say, "concrete is much heavier than water," since 10 g of water obviously weigh ten times as much as 1 g of concrete. But if we say, "concrete is much more dense than water," we express a meaningful idea. If we know the mass of a definite volume of any substance, we can calculate its density.

Example 1.7 What is the density of ethyl alcohol if 6.0 mL of it weigh 4.8 g?

Solution

$$\text{volume} = 6.0\ \text{mL} \qquad \text{mass} = 4.8\ \text{g}$$

$$\text{density} = \frac{\text{mass}}{\text{volume}}$$

$$= \frac{4.8\ \text{g}}{6.0\ \text{mL}}$$

$$\text{density} = 0.80\ \text{g/mL}$$

Recall from Tables 1.2 and 1.3 that one milliliter (mL) and one cubic centimeter (cm^3 or cc) are identical. Consequently, we could just as well have expressed the density of ethyl alcohol as 0.80 g/cm^3.

Example 1.8 Use the result from Example 1.7 to calculate the mass of 10 mL of ethyl alcohol.

Solution

$$\text{density} = 0.80\ \text{g/mL} \qquad \text{volume} = 10\ \text{mL}$$

$$\frac{\text{mass}}{\text{volume}} = \text{density}$$

$$\frac{\text{mass}}{10\ \text{mL}} = 0.80\ \frac{\text{g}}{\text{mL}}$$

$$\text{mass} = 0.80\ \text{g/mL} \times 10\ \text{mL}$$

$$\text{mass} = 8.0\ \text{g}$$

Problem 1.7

(a) Calculate the density of gasoline, given that 100 mL weigh 70 g.

(b) Use the density from part (a) to calculate the volume occupied by 5.0 g of gasoline.

The density of a substance usually decreases when it is heated. In other words, the higher the temperature, the lower the density of a substance.* The change is often not great, however. The density of water is very nearly 1 g/mL at all ordinary temperatures. The density of water is exactly 1.000 g/mL at 4°C.

Why doesn't specific gravity have units?

The **specific gravity** of a substance is defined as follows:

$$\text{Specific gravity} = \frac{\text{density of substance}}{\text{density of water at same temperature}}$$

Because the density of liquid water is about 1 g/mL regardless of the temperature, the specific gravity of a substance is numerically about equal to its density. There are no units attached to specific gravity, since the units in the numerator (density of the substance) and, in the denominator (density of water) cancel out. Substances with specific gravities of less than 1 will float on water (provided they do not dissolve in it) and those with specific gravities greater than 1 will sink. The hydrometer is a simple instrument that measures specific gravities of liquids directly (Figure 1.3).

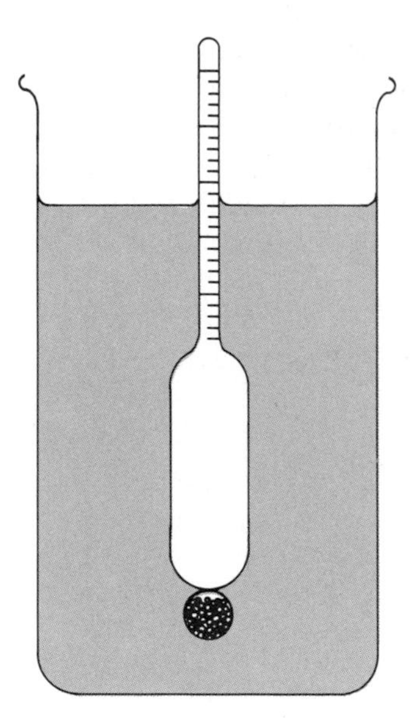

FIGURE 1.3 The hydrometer is a simple instrument to measure the specific gravity of liquids. The stem is calibrated to read specific gravity directly at the point where the surface of the liquid crosses the stem.

*Water is an interesting exception to this rule between 0°C and 4°C (32°–39°F), since its density increases slightly with increasing temperature.

The liquid inside an automobile storage battery is a mixture of sulfuric acid and water. As the battery discharges, the specific gravity of the mixture changes from about 1.28 to about 1.12. Simply measuring the specific gravity of the battery water thus serves to show how much charge remains in the battery.

SCIENTIFIC NOTATION 1.8

Some numbers are so big, and some so small, that it is awkward to write them out—4,500,000,000, for example, or 0.0000000045. We can write these numbers much more simply by using **scientific notation:** we can write 4,500,000,000 as 4.5×10^9, and we can write 0.0000000045 as 4.5×10^{-9}. In scientific notation, we express a number as a coefficient multiplied by 10 raised to some power. In the examples above, 4.5 is the coefficient. The phrase "10 raised to some power" simply means that 10 is multiplied by itself as many times as is shown by the exponent (the number written in the upper right-hand corner of the ten): 10^9 means $10 \times 10 \times 10 \times 10 \times 10 \times 10 \times 10 \times 10 \times 10$, which equals 1,000,000,000. So, when we write 4.5×10^9, we are really writing $4.5 \times 1{,}000{,}000{,}000$, which is just the same as 4,500,000,000. When we write a negative exponent, such as 10^{-9}, we mean 1 divided by 10^9, or $1/10^9$. So 4.5×10^{-9} equals $4.5/10^9$, which in turn equals 0.0000000045.

As you may have noticed, the 10, with its exponent, takes the place of a lot of zeros. To figure out how big an exponent you need, write out the whole number first, and then move the decimal point to the left or the right until you get a coefficient of the right size (usually what you want is a coefficient with only one number to the left of the decimal point). For each place you move the decimal point to the left, you add one number to the exponent. For each place you move the decimal point to the right, you subtract one number from the exponent. For example,

$$4_{\uparrow}5\ 0\ 0\ 0\ 0\ 0\ 0\ 0\ 0.$$

To reduce this number to 4.5, we must move the decimal point nine places to the left. This tells us that our 10 must have the exponent 9. So, to express 4,500,000,000 in scientific notation, we write 4.5×10^9. If for some reason we wanted a coefficient of 45, we would move the decimal point only eight places to the left, so we would get 45×10^8. The expression 45×10^8 has the same numerical value as 4.5×10^9, and they are each equal to 4,500,000,000. To write 0.0000000045 in scientific notation, we must move the decimal point nine places to the right:

$$0.000000004_{\wedge}5$$

Moving the decimal point to the right gives us a negative exponent, so we write this number as 4.5×10^{-9}.

Table 1.5 shows how numbers are written in scientific notation.

TABLE 1.5 Scientific Notation

Number	Equivalent in scientific notation	Mathematical meaning
1	1×10^{0}	1
10	1×10^{1}	10
100	1×10^{2}	10×10
1,000	1×10^{3}	$10 \times 10 \times 10$
10,000	1×10^{4}	$10 \times 10 \times 10 \times 10$
100,000	1×10^{5}	$10 \times 10 \times 10 \times 10 \times 10$
1,000,000	1×10^{6}	$10 \times 10 \times 10 \times 10 \times 10 \times 10$
0.1	1×10^{-1}	$\frac{1}{10}$
0.01	1×10^{-2}	$\frac{1}{10 \times 10}$
0.001	1×10^{-3}	$\frac{1}{10 \times 10 \times 10}$
0.0001	1×10^{-4}	$\frac{1}{10 \times 10 \times 10 \times 10}$
0.00001	1×10^{-5}	$\frac{1}{10 \times 10 \times 10 \times 10 \times 10}$
0.000001	1×10^{-6}	$\frac{1}{10 \times 10 \times 10 \times 10 \times 10 \times 10}$

1.9 SIGNIFICANT FIGURES

If you count five apples from a basket, you can be sure that you have exactly five apples. However, if you weigh the apples, you cannot be quite so sure that the weight you get is exactly correct. In measuring almost any quantity, there is a certain margin for uncertainty. How precisely a quantity has been measured is shown by the number of significant figures in the measurement. The **significant figures** include all the digits that are known with certainty, plus the first digit about which there is some uncertainty.

How do you tell the number of significant figures in a number?

For example, suppose we weigh an object using a balance that is accurate only to the nearest tenth of a gram. If we get a mass of 10.5 g, we can be sure that the 1 and the 0 are correct, but we cannot be sure about the 5. If we were to use a more precise balance, we might get a reading of 10.46 g. Rounded off, that would still be 10.5, but it does mean that there is some uncertainty about the 5. The 5 is the *first* uncertain figure, so it is the *last* significant figure. We say, then, that 10.5 has three significant figures.

If the balance gave us a reading of 10.5 g, and we wrote it down as 10.50 g, we would be making a mistake. That number would imply that we knew the mass to the nearest hundredth of a gram, when in fact we only know it to the nearest tenth of a gram.

A zero counts as a significant figure if there is a nonzero integer before it. For example, in 10.5, the zero is significant, because a 1 (an integer that is not zero) comes before it. But in 0.015, the zeros are not significant; they merely serve to locate the decimal point. The numbers 0.015, 0.0015, and 0.000000015 each have only two significant figures. We can see why this is so if we compare the measurements 0.15 mm and 0.015 cm. These measurements are equal (and equally precise), so they must have the same number of significant figures. Note carefully, though, that 0.0150 does contain three significant figures, as does 15.0. Can you see why?

Sometimes it is difficult to determine the significant figures in very large numbers. When we say the world population is 4,400,000,000, we obviously do not mean that there are exactly four billion, for hundred million people in the world. Our knowledge of world population is only approximate. Yet, every one of these zeros might appear to be significant, because they follow a nonzero integer. Actually, these zeros serve only to locate the decimal point, just as do the zeros in 0.00015, so they are not significant.

The best we can say is that, if a number is known exactly, the zeros at the end are significant, but if the number is an estimate and the zeros are intended only to locate the decimal point, then they are not significant. When we say there are 1,000 g in a kilogram, we mean exactly 1,000. Not only are those four digits significant, but we could write the number as 1,000.00000 . . . , and keep writing zeros forever, and all of them would be significant. We can say that the number 1,000 in this case has an infinite number of significant figures. But if we roughly calculate that there are 1,000 people in a crowd, that 1,000 has perhaps only one significant figure.

We can avoid the problem of showing how many significant figures there are in a large number if we consistently use scientific notation. If we write the world population as 4.4×10^9, it becomes obvious that there are only two significant figures.

Problem 1.8 How many significant figures are there in each of the following numbers?

(a) 15
(b) 1.5
(c) 104
(d) 560.3
(e) 560.30
(f) 0.105
(g) 0.01082
(h) 0.001
(i) 0.0010

Problem 1.9 Write the following numbers using scientific notation. In each case retain as many significant figures as the original number has.

(a) 6,351
(b) 0.00501
(c) 6128.503
(d) 0.00001040
(e) 0.002
(f) 102
(g) 5,654,940
(h) 0.034

1.10 RULES FOR SIGNIFICANT FIGURES IN CALCULATIONS

Significant figures are useful because they keep us from going too far wrong when we perform mathematical calculations with quantities we have measured. The answer must always have the correct number of significant figures. When we use defined numbers, such as

$$\frac{100 \text{ cm}}{1 \text{ m}}$$

we do not have to worry about how many significant figures there are. These numbers are exact. But when we carry out calculations using measured values, we must pay attention to significant figures. We can determine how many significant figures an answer should have by following a few simple rules.

Addition and Subtraction

In addition or subtraction, the answer should have as many places to the right of the decimal point as the number that has the fewest places to the right of the decimal point. When adding or subtracting numbers, you must first line them up on their decimal points. For example:

$$\begin{array}{r} 520.3 \\ 2.001 \\ 52.38 \\ 0.010 \\ \hline 574.691 \end{array}$$

But 574.691 is *not* the correct answer, because it has too many decimal places—that is, too many figures to the right of the decimal point. It must be rounded off so that it contains the same number of decimal places as the number with the fewest decimal places. Since 520.3 has only one decimal place, the answer can have only one decimal place. We must round off 574.691 to 574.7. In rounding off numbers, if the digit to the right of the last allowable significant figure is 4 or less, you simply drop it; if it is 5 or more, you add 1 to the last significant figure. In this case, the last allowable significant figure will fall in the slot occupied by the 6 in 574.691. The digit following 6 is 9, so we add 1 to the 6. The correct answer, then, is 574.7.

When we add or subtract numbers written in scientific notation, the exponent of 10 must be the same for each number. For example, if we want to subtract 4.02×10^{-4} from 5.612×10^{-2}, we are going to have to change the way we have written one of the numbers. We can rewrite 4.02×10^{-4} as 0.0402×10^{-2}, or we can rewrite 5.612×10^{-2} as 561.2×10^{-4}. When the two numbers are expressed to the same power of 10, we can subtract them.

$$\begin{array}{r} 5.6120 \times 10^{-2} \\ -\ 0.0402 \times 10^{-2} \\ \hline 5.5718 \times 10^{-2} \end{array}$$

To make the subtraction possible, we have had to add a nonsignificant zero to the end of 5.612. This is a clear sign that our answer has too many figures, that is, too many decimal places. We can have only three decimal places, because 5.612 (the number with the fewest decimal places) has only three. We must round off the answer to 5.572×10^{-2}.

Multiplication and Division

The final answer can have only as many significant figures as the number with the fewest significant figures. For example,

$$\frac{12 \times 4.683}{6420} = 0.00875327 = 0.0088 = 8.8 \times 10^{-3}$$

In this example, we assume that all the numbers come from measurements and are each expressed to the correct number of significant figures. We have to round the answer off to two significant figures because one of the numbers in the calculation, 12, has only two significant figures.

It is very easy to multiply and divide numbers written in scientific notation. For the 10s part of the number, you simply add the exponents if you are multiplying and subtract the exponents if you are dividing. You multiply and divide the coefficients just like ordinary numbers. The following examples illustrate these points.

$$10^2 \times 10^3 = 10^5$$

$$10^5 \times 10^{-8} = 10^{-3}$$

$$10^{-4} \times 10^{-6} = 10^{-10}$$

$$\frac{10^{12}}{10^{10}} = 10^2$$

$$\frac{10^8}{10^{-2}} = 10^{10}$$

$$\frac{10^{-4}}{10^5} = 10^{-9}$$

$$\frac{10^{-4}}{10^{-3}} = 10^{-1}$$

$$\frac{10^6 \times 10^{-3}}{10^{-8}} = 10^{11}$$

The coefficients are just multiplied or divided as usual.

$$\frac{(6.321 \times 10^{-2})(4.89 \times 10^3)}{5.32 \times 10^2} = 5.810 \times 10^{-1} = 5.81 \times 10^{-1}$$

Problem 1.10 Simplify the following and express all answers to the correct number of significant figures.
(a) $(5.14 \times 10^2) + 7.79 + (2.110 \times 10^3)$
(b) $(4.4 \times 10^{-5}) + (9.76 \times 10^{-4}) + (4.3 \times 10^{-6})$
(c) $6.5 \times 7.2 \times (2.03 \times 10^6)$
(d) $(1.00 \times 10^9)(3.0 \times 10^{-3})(6.000 \times 10^{-7})$
(e) $8 \times (6.4 \times 10^2)(4.0 \times 10^5)$
(f) $\dfrac{(6.002)(7.70 \times 10^{-2})}{405 \times 10^{-4}}$

ENERGY TRANSFORMATIONS 1.11

During every physical or chemical change, energy is absorbed or given off. The energy changes that occur during chemical reactions can be as important as the chemical changes or even more important. For example, when gasoline and oxygen react in an automobile engine, what is important and useful is not the products of the reaction—the exhaust fumes—but the energy that is produced.

Life depends on chemical reactions. In living organisms, complicated chemical reactions occur to supply the energy needed to carry out essential activities. Without energy, there would be no life, there would be no activity, there would be no change.

Energy can exist in various forms, and it is constantly being converted from one form to another. For example, in a flashlight battery, chemical energy is converted into electrical energy. The electrical energy, in turn, is changed into heat and light, which are yet other forms of energy.

Problem 1.11 An electric company burns coal to generate steam that drives generators that make electricity. Starting with the chemical energy of the coal, name some other forms of energy that appear during the burning of the coal and the production of electricity.

POTENTIAL AND KINETIC ENERGY 1.12

Potential energy is energy that is stored. **Kinetic energy** is the energy of motion. A waterfall offers an excellent example of potential

FIGURE 1.4 Niagara Falls is an example of mechanical potential and kinetic energies. Water at the top of the falls has potential energy, which becomes kinetic energy as it falls. Part of this energy is used to turn large turbines and produce electrical energy.

What are some examples of kinetic energy and potential energy?

and kinetic energies (Figure 1.4). Water at the top of the fall possesses potential energy because it can fall and "do work" (generate power, for example). Thus it has energy stored in it because of its position. As the water actually falls, the potential energy becomes kinetic energy, or energy of motion. Any object in motion has kinetic energy. An automobile in motion has kinetic energy. An atom in motion has kinetic energy.

Problem 1.12 Describe the kinetic energy and potential energy conversions that occur as a pole-vaulter runs, vaults the bar, and falls to the ground.

1.13 HEAT

Heat is another very common form of energy. **Temperature** is a measure of the intensity of heat. The natural direction of heat transfer is from hotter objects (higher temperature) to cooler objects (lower temperature). Chemical reactions often release or absorb heat. Those that release heat are called **exothermic reactions,** and those that absorb heat are called **endothermic reactions.**

Problem 1.13 A refrigerator reverses the natural direction of heat transfer (from hotter to cooler objects) by doing work. What happens to the heat that the refrigerator removes from its interior? Can you cool a kitchen by leaving the refrigerator open? Explain.

Problem 1.14 Water at the top of a waterfall has potential energy, which is converted into kinetic energy as the water falls. Into what form of energy is the kinetic energy of the water converted when it reaches the bottom of the falls?

TEMPERATURE SCALES 1.14

Everyone knows that temperature is a measure of relative hotness or coldness. Weather reports in the United States usually give the temperature highs and lows in terms of the **Fahrenheit scale** (degrees Fahrenheit, or **°F**). We know that an outdoor temperature of 85°F means a hot day, whereas 35°F means a cold day. We normally measure our body temperature in °F too. For example, 98.6°F is considered normal for humans, and a temperature of 104°F indicates a serious illness.

Scientists (and most people outside the United States) usually use another temperature scale, the **Celsius scale(°C)**. The normal body temperature of a human, which is 98.6° on the Fahrenheit scale, is 37.0°C. The relationship between these two temperature scales is shown in Figure 1.5.

Figure 1.5 shows that there are exactly 1.8 Fahrenheit degrees for every Celsius degree. For example, the difference between the boiling and freezing points of water is 180 Fahrenheit degrees (212 − 32), but only 100 Celsius degrees. Therefore, to convert a temperature reading on the Celsius scale to one on the Fahrenheit scale, we must adjust the Celsius degrees to Fahrenheit degrees by multiplying by 1.8. Since 0°C = 32°F, we must also add 32° to take care of the difference in zero points.

$$°F = (°C \times 1.8) + 32°$$

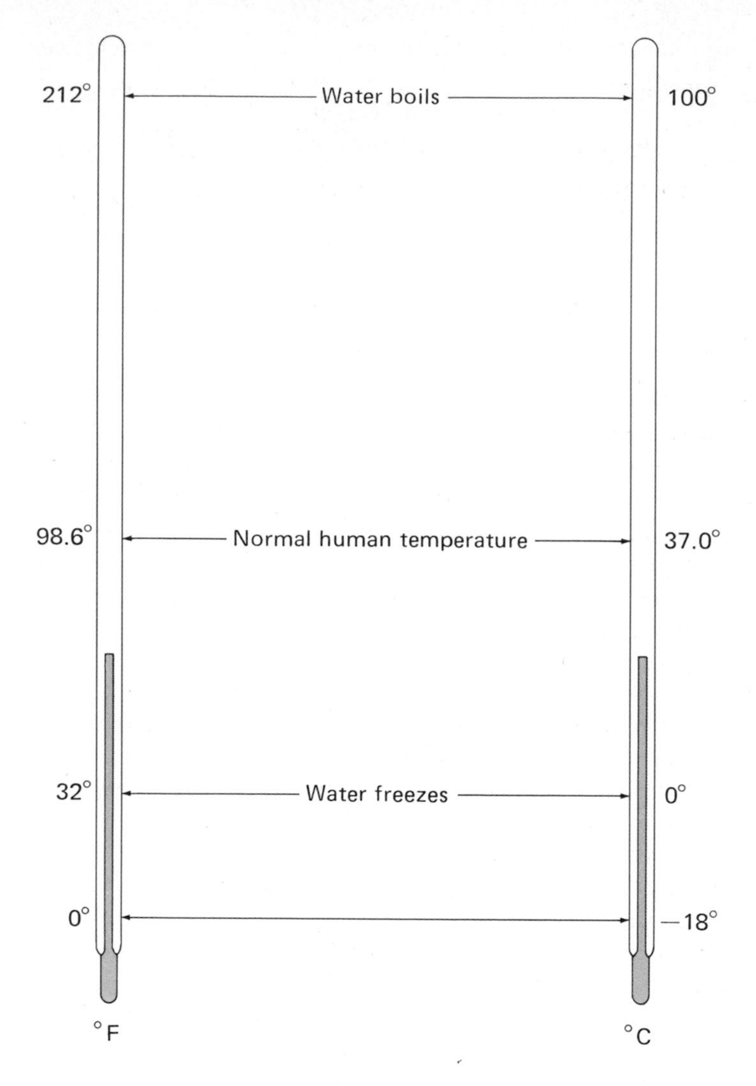

FIGURE 1.5 The Fahrenheit and Celsius temperature scales.

We can easily change °F into °C by solving this equation for °C.

$$°C \times 1.8 = °F - 32°$$

$$°C = \frac{(°F - 32°)}{1.8}$$

Problem 1.15
(a) What is 70°F in °C?
(b) −40°C corresponds to what temperature in °F?

Problem 1.16 Is a 1° change on the Celsius scale a larger or smaller temperature change than a 1° change on the Fahrenheit scale?

FREE ENERGY 1.15

Chemical energy is a special and important form of energy that is stored in substances. When a chemical reaction takes place with the liberation of energy, the chemical energy of the substances formed in the reaction (the products) is less than the chemical energy of the substances that react (the reactants). This difference in chemical energy is energy that can theoretically be obtained from a chemical reaction (at a given temperature and pressure). Some of this chemical energy that is liberated is available to do useful work for us. This useful energy is called **free energy.** The useful energy that we can get when a chemical reaction occurs, then, is what is known as the **change in free energy** for the reaction. Consider, for example, a flashlight battery: the change in free energy is equal to the electrical energy that can be obtained from the chemical reaction that occurs in the battery.

The symbol ΔG (read "delta gee") stands for the change in free energy. The Greek letter Δ (delta) is often used to indicate a change in some quantity. The G is in honor of the nineteenth-century American physicist J. W. Gibbs, who first used the concept of free energy.

The concept of free energy is very important in understanding energy transfer in chemical reactions. For example, when a muscle contracts, a chemical reaction occurs that provides energy to do work (Section 22.10). The substances involved in the reaction lose free energy: that is, some of the chemical energy originally stored in the reactants is converted into mechanical and heat energy. Such reactions, in which useful energy is given off and the change in free energy is negative, are called **spontaneous reactions.** Once started, they do not require an external source of energy in order to take place. Reactions for which the free energy of the products is greater than the free energy of the reactants have a positive value of ΔG. Such reactions are called **nonspontaneous** reactions and do not occur on their own. They can be forced to occur only if useful energy is supplied.

What does the algebraic sign of the change in free energy for a chemical reaction tell us?

An automobile storage battery offers a simple example of spontaneous and nonspontaneous reactions. As the battery discharges, a chemical reaction occurs that releases free energy in the form of electrical energy (Figure 1.6). This electrical energy can be used to start the car and operate the radio, horn, wipers, and so on. Since energy is given off, the ΔG for the reaction is negative, and the reaction is spontaneous. When the battery runs down, we recharge it by putting electrical energy back into it. This causes the chemical reaction to run in reverse and the battery to regain its free energy. The ΔG for this reaction is positive, so we say the reaction is nonspontaneous. The recharged battery once again has the ability to lose free energy and thus perform useful work.

We must be careful not to identify spontaneous and nonspontaneous with exothermic and endothermic. The change in free energy for a reaction is a measure of the net energy transfer. The terms exothermic and endothermic apply only to energy in the form of heat. Although it is true that many spontaneous reactions are exothermic, and many nonspontaneous reactions are endothermic, this correspondence is not always realized. For example, the reaction in a discharging automobile storage battery is spontaneous because electrical energy is provided by the

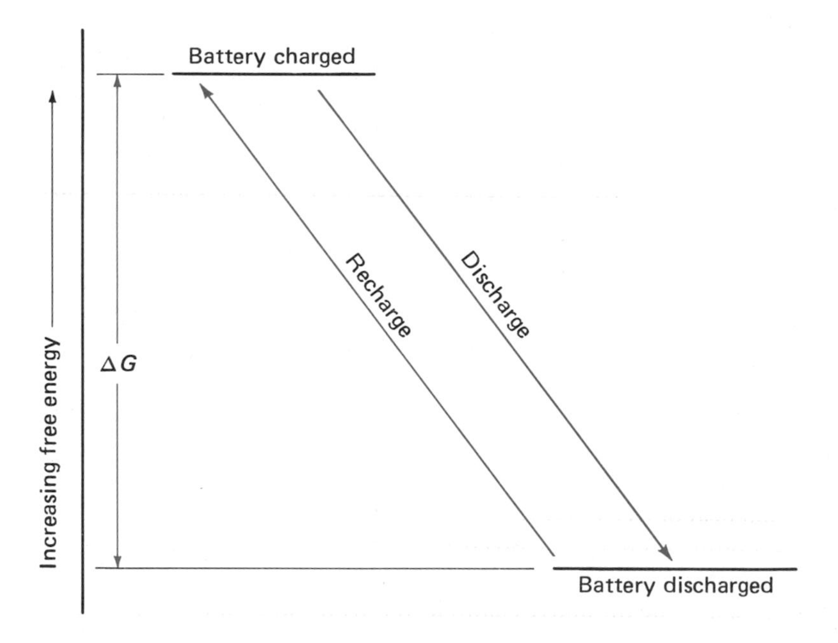

FIGURE 1.6 Chemical reactions in a storage battery illustrate the concept of free energy. ΔG for the battery reaction is negative during discharge and positive during recharge.

reaction. However, this reaction is endothermic, because the battery absorbs heat from its surroundings as it discharges. As we have already seen, the reaction when the battery is recharging is nonspontaneous; it is also exothermic because heat is given off.

Problem 1.17 A piece of metal undergoes chemical reaction by slowly rusting. Is there a loss or gain in free energy during this process? Briefly discuss.

ENERGY UNITS 1.16

We can express any form of energy in any energy unit. This statement is true because one form of energy can be converted into another form. The *total* energy available, however, is constant. In other words, energy is conserved.

There are many energy units in use today. However, we shall need to know just a few of these units. The **joule** (J) is the SI energy unit (see Appendix 2). A more traditional energy unit, one with which you may be familiar, is the **calorie** (cal). One calorie is equivalent to the energy required to increase the temperature of 1 g of water by 1°C. A still larger unit of energy, one we often use in chemistry, is the kilocalorie (kcal). A kilocalorie is, as you may have guessed, 1,000 calories. Nutritionists speak of energy values of foods in Calories (Cal), which are actually identical to kilocalories. People on a diet "count their Calories" to avoid an energy intake greater than that needed for essential bodily functions. An average apple (approximately 170 g) has a value of 85 kcal, but an equal weight of cheddar cheese has a value of 725 kcal. Dieters have to watch what they eat as well as how much they eat.

Why is the kilocalorie energy unit so often used in chemistry?

The kilocalorie* is commonly used in chemistry because the energy involved in most chemical reactions ranges from a few kilocalories to several thousand kilocalories. For example, the formation of 1 kg of water from the reaction of hydrogen with oxygen releases a little more than 3,000 kcal of free energy.

The amount of energy involved in a chemical reaction depends not only on what substances are reacting, but also on how much of them reacts. To continue the above example, if only 0.5 kg of water is produced

*The kilojoule is also sometimes used to express the energy associated with chemical reactions. A rough conversion of kilocalories to kilojoules can be estimated by multiplying kilocalories by 4.

by the reaction of hydrogen and oxygen, then only half as much energy is given off—about 1,500 kcal.

Problem 1.18 Hydrogen and oxygen combine to form water. Large amounts of energy must be expended to produce hydrogen and oxygen from water, however. If $\Delta G = -3 \times 10^3$ kcal for the formation of 1 kg of water from hydrogen and oxygen, explain:

(a) the meaning of the negative sign;

(b) the sign and value of ΔG for the conversion of 1 kg of water completely into hydrogen and oxygen.

Problem 1.19 Calculate ΔG for the formation of 2 kg of water from the reaction between hydrogen and oxygen. (See Problem 1.18.)

SUMMARY

Energy causes **matter** to undergo various **physical** and **chemical changes.** Matter is composed of **compounds** and the simpler chemical materials called **elements** that make up compounds. Each element has its own **chemical symbol.**

Density and **specific gravity** are important properties of matter. In the measurement of density and other properties, we use units from the **metric system. Scientific notation** is usually the most convenient way to express numbers. Any measured quantity must be expressed with the proper number of **significant figures.**

Potential energy, kinetic energy, heat, and **free energy** are different forms of energy. Any form of energy can be expressed in any energy unit, such as the **calorie** or kilocalorie. The **change in free energy** (ΔG) for a reaction tells whether the reaction takes place with the net liberation of energy (**spontaneous reaction**), or whether a continuous supply of energy is necessary (**nonspontaneous reaction**). ΔG is negative for a spontaneous reaction and positive for a nonspontaneous reaction. Reactions that give off energy in the form of heat are called **exothermic reactions** and those that absorb heat from the surroundings are called **endothermic reactions.** The liberation or absorption of heat does not determine whether or not a reaction can occur on its own. Only the sign of ΔG determines this.

Temperature is usually expressed on the **Celsius scale** (**°C**) for scientific work, but the **Fahrenheit scale** (**°F**) is used in the United States for some purposes, such as human temperature readings and weather reports.

Key Terms

Calorie (Section 1.16)
Celsius scale (°**C**) (Section 1.14)
Change in free energy (ΔG) (Section 1.15)
Chemical change (Section 1.3)
Chemical energy (Section 1.15)
Chemical reaction (Section 1.3)
Chemical symbol (Section 1.5)
Compound (Section 1.4)
Density (Section 1.7)
Element (Section 1.4)
Endothermic reaction (Section 1.13)
Energy (Section 1.1)
Exothermic reaction (Section 1.13)
Fahrenheit scale (°**F**) (Section 1.14)
Free energy (G) (Section 1.15)
Heat (Section 1.13)
Joule (**J**) (Section 1.16)
Kinetic energy (Section 1.12)
Mass (Section 1.1)
Matter (Section 1.1)
Metric system (Section 1.6)
Nonspontaneous reaction (1.15)
Physical change (Section 1.2)
Potential energy (Section 1.12)
Scientific notation (Section 1.8)
SI units (Section 1.6)
Significant figures (Section 1.9)
Specific gravity (Section 1.7)
Spontaneous reaction (Section 1.15)
Temperature (Sections 1.13, 1.14)
Weight (Section 1.1)

ADDITIONAL PROBLEMS

1.20 Classify each of the following as a physical or chemical change.
(a) the natural evaporation of water from the ocean
(b) cooking some bacon and eggs
(c) melting snow
(d) the growth of a tree
(e) the combustion of gasoline in an engine

1.21 Heating solid mercuric oxide produces mercury metal and oxygen gas. Is mercuric oxide a compound or an element? Explain.

1.22 Write the chemical symbols for the following elements.

(a) calcium (b) carbon
(c) chlorine (d) hydrogen
(e) iron (f) nickel
(g) nitrogen (h) oxygen
(i) potassium (j) sodium
(k) sulfur (l) titanium
(m) vanadium (n) zinc

1.23 Give the names of the following elements.

(a) Al (b) Cr
(c) Co (d) Cu
(e) F (f) He
(g) H (h) I
(i) Fe (j) Mg
(k) Mn (l) P
(m) K (n) Si
(o) Na (p) S

1.24 Convert the following British unit lengths to centimeters.

(a) 0.394 in (b) 1.00 ft
(c) 1.00 yd

1.25 Convert the following British units to meters.

(a) 6.5 in (b) 1.00 ft
(c) 1.00 yd

1.26 In football, a team is required to advance the ball at least 10 yd in four downs to maintain possession. If metric units were used in football and 10 m were set as the required distance, would it be easier or more difficult for a team to maintain possession of the ball?

1.27 If milk were sold by the liter rather than by the gallon, what should be the price per liter for milk selling at $2.15 per gallon?

1.28 The average glass holds about 250 cubic centimeters (cm^3). What is the volume in milliliters and in liters of such a glass?

1.29 How many significant figures does each of the following numbers have?

(a) 16 (b) 16.0
(c) 5.21×10^6 (d) 5.210×10^{-8}
(e) 3261 (f) 3.260×10^3
(g) 0.0001 (h) 0.0502

1.30 Make the following conversions.

(a) 53 g to kg (b) 4.5 mg to g
(c) 0.5 kg to g (d) 6.5×10^{-3} g to mg
(e) 4.9×10^4 g to kg

1.31 Add the following numbers: 901 + 11 + 2. Now multiply the same numbers: 901 × 11 × 2. Why are the results expressed to different significant figures? What are the correct answers?

1.32 Mercury is the only metal that is a liquid at all ordinary temperatures. If 10.0 mL of mercury weigh 135 g at 25°C, what is the density of mercury at this temperature? Lead has a density of 11.2 g/cm^3 at 25°C. Will lead float on mercury?

1.33 What are the specific gravities of mercury and of lead? (See Problem 1.32.) The density of water at 25°C is 0.997 g/mL.

1.34 Generally speaking, is the free energy of a substance a form of kinetic energy or a form of potential energy? Base your answer on the definition of kinetic and potential energies.

1.35 A 100-watt bulb uses 24 cal/s while lit. Calculate its usage of:

(a) calories per hour
(b) kilocalories per hour

1.36 Which has more calories, an average apple or 30 g of cheddar cheese? (See Section 1.16.)

1.37 In an ordinary automobile storage battery, lead metal reacts with lead oxide and sulfuric acid to produce lead sulfate and water. Electrical energy is produced while the reaction is taking place. For a 6-volt battery, $\Delta G = -2.7 \times 10^2$ kcal for every 2.0×10^2 g of lead that reacts.

(a) What is the change in free energy per gram of lead that reacts?
(b) What is the change in free energy per gram of lead produced when the battery is recharged?

1.38 Change the following temperatures to °C.

(a) 40°F
(b) 100°F
(c) 250°F
(d) −20°F
(e) −459°F

1.39 Change the following temperatures to °F.

(a) 0°C
(b) 22°C
(c) 100°C
(d) −10°C
(e) −273°C

***1.40** How many significant figures should be retained in the answer to: 654.014 − 653.990? Explain why we try to avoid laboratory procedures that involve taking the difference of two numbers of about the same magnitude.

***1.41** An analytical balance weighs to the nearest 0.1 mg. An ordinary triple-beam balance weighs only to the nearest 0.1 g but is usually faster to use than an analytical balance. If an object of about 10 g must be weighed and the result divided by another experimental value known only to two significant figures, which balance should be used? If the object weighed less than 1 g, would you use the other balance? Explain.

***1.42** A small vial when empty weighs 4.682 g. When the vial is filled with a urine specimen, it weighs 6.580 g, and when it is filled with water, it weighs 6.534 g. Calculate the specific gravity of the specimen.

***1.43** A dieter wishes to make a lunch of apple and cheddar cheese. The diet calls for a lunch of 400 kcal with an intake of 250 g. Use the information in Section 1.16 as a basis for calculating the grams of apple and grams of cheese the dieter should eat.

***1.44** Electrical power is measured in watts. If 1 watt is about 0.24 cal/s, how many days would it take to electrolyze 50 kg of water completely to hydrogen and oxygen using 1,000 watts? (See Problem 1.18.)

***1.45** Given that a volt is a joule per coulomb, and an ampere is a coulomb per second, and a watt is a volt $\times$ an ampere, use the information given in Problem 1.44 and calculate the number of joules in a calorie.

***1.46** A *yazoo* is a fictitious metric unit. How many kiloyazoos are in 6.58 yazoos? How many milliyazoos are in 6.58 yazoos?

Problem-Solving Hints

If you have trouble solving any of the additional problems, refer to the sections listed next to the problem numbers.

1.20 1.2, 1.3	**1.21** 1.4	**1.22** 1.5
1.23 1.5	**1.24** 1.6	**1.25** 1.6
1.26 1.6	**1.27** 1.6	**1.28** 1.6
1.29 1.8, 1.9	**1.30** 1.6, 1.8	**1.31** 1.10
1.32 1.7	**1.33** 1.7	**1.34** 1.12, 1.15
1.35 1.6	**1.36** 1.16	**1.37** 1.15, 1.16
1.38 1.14	**1.39** 1.14	**1.40** 1.9
1.41 1.9	**1.42** 1.7	**1.43** 1.16
1.44 1.16	**1.45** 1.16	**1.46** 1.6

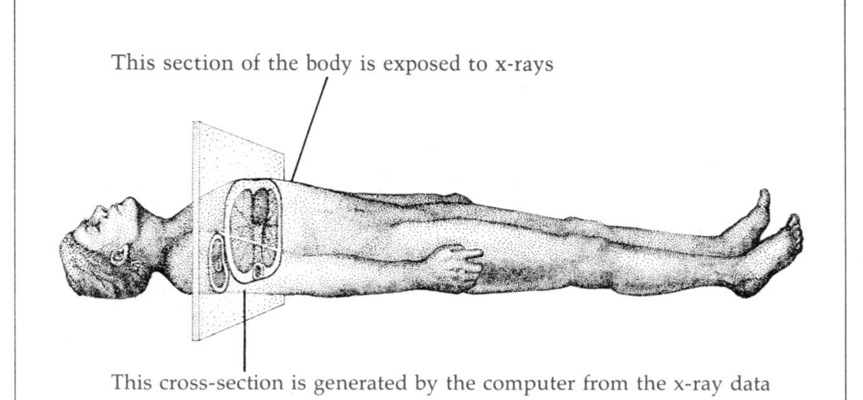
This section of the body is exposed to x-rays
This cross-section is generated by the computer from the x-ray data

In this chapter we will discuss such topics as the naturally occurring sources of radiation (alpha, beta, and gamma rays, for example) as well as the synthetic sources (the most notable being x rays). The medical application of radiation is familiar to many of us who have had x rays in the course of a routine physical or dental checkup. Until recently, radiological techniques were limited to the absorption of x-ray energy on film, which produced an image that was sometimes indistinct and made it difficult to locate the position of a tumor. In the mid-1970s, it became possible to quantify x-ray absorption more exactly with an instrument known as the CAT scanner. Today this instrument is used in radiology as part of a diagnostic method called "computerized axial tomography" (CAT).

The CAT scanner allows the radiologist to see many sections of the body in great detail at one time, rather than the simple front and side views previously available, which were often blocked by other images. Now the radiologist can get a close look at each of the overlapping layers of tissues on a selective basis. The instrument is known as a scanner *and produces a* scan *as the final result.*

The CAT scanner consists of four major parts, one of which is a tube that projects the x ray. The tube rotates around the patient and produces multiple x-ray pictures. A computerized method is incorporated to tabulate and store all of the data obtained from the multiple scans. The radiologist can then

2

ATOMS AND MOLECULES

choose to review a specific section of the body or to examine individually a collection of single pictures produced by the computer.

The CAT scanner has become an essential part of radiological diagnosis, and we have come to rely on it for precise diagnoses.

[Illustration: An x-ray "section" is produced by the CAT scanner.]

Our age is often called the Atomic Age, which might lead one to think that atoms are a recent discovery. But actually, the idea of the atom is over two thousand years old. In the fifth century B.C., a Greek named Democritus wrote, "The only existing things are the atoms and empty space; all else is mere opinion." Aristotle disagreed. He thought that if you started chopping up matter, you could keep chopping indefinitely, dividing it into finer and finer pieces. The atomists, on the other hand, argued that once you got down to the atom, you couldn't chop any further. The word *atom,* in fact, comes from a Greek word meaning "uncuttable" or "indivisible."

For a long time, science was based largely on what people like Aristotle had said. Still, the atomists persisted, so the debate went on. After twenty-three hundred years, the argument over atoms was still unsettled.

Then, in the early 1800s, an English scientist named John Dalton came up with the first theory of atoms that was scientifically meaningful. Dalton put an end to the ancient argument. Let's first see what he did and then look at some modern developments in atomic theory.

DALTON'S THEORY 2.1

According to John Dalton's atomic theory of matter, all matter is composed of atoms. Furthermore, atoms are indestructible and have their own unique properties. In the nearly 200 years since this theory was first proposed, it has been expanded to reflect many developments in atomic theory. Dalton's ideas may now be stated in the following modified form:

1. The atoms of one element are different from the atoms of every other element. An element is thus a substance made up of only one kind of atom.
2. Atoms of different elements can unite to form compound molecules.* Each molecule of a particular compound has a specific number of atoms of each element that it contains. For example, a molecule of water, H_2O, always consists of two hydrogen atoms and one oxygen atom.
3. Chemical reactions involve the breakdown of molecules and the rearrangement of the atoms into new molecules. The electrolysis of water (Figure 1.2) breaks down the H_2O molecules into molecules of the elements—hydrogen (H_2) and oxygen (O_2).
4. Atoms are neither created nor destroyed in chemical reactions, although they can be rearranged into different molecular combinations. There are always just as many atoms of each kind of element after a reaction as there were before the reaction. After the electrolysis of water, there are exactly the same number of hydrogen atoms and oxygen atoms in the H_2 and O_2 molecules as there were originally in the H_2O molecules.

*Dalton did not realize that atoms of a single element can also unite to form molecules. A molecule of an element is composed of two or more atoms of the same element; a molecule of a compound is composed of atoms of different elements.

2.2 CHEMICAL EQUATIONS

Dalton's theory makes the important point that atoms are conserved in chemical reactions, but molecules are not conserved. To take an everyday example of a chemical reaction, we can look at what happens when natural gas burns. Natural gas is mostly methane (CH_4). The molecules of oxygen in the air are composed of two oxygen atoms (O_2). When natural gas burns, methane molecules react with oxygen molecules. The products of this reaction are water (in the form of steam) and carbon dioxide (CO_2). The equation for this reaction is:

$$\text{Methane} + \text{oxygen} \rightarrow \text{water} + \text{carbon dioxide}$$

What must we do to balance a chemical equation?

We can express the burning of natural gas more precisely by using chemical symbols. (The equation is not balanced.)

$$CH_4 + O_2 \rightarrow H_2O + CO_2$$

The reactants (substances present at the start) of a chemical reaction appear to the left of the arrow and the products (substances present at the finish) of a reaction are written to the right of the arrow. This equation for the burning of natural gas shows the symbols for the reactants and products correctly, but it violates the rule that atoms cannot be created or destroyed in chemical reactions. To follow this rule we must balance the equation. There are three oxygen atoms on the right side of the equation, but only two on the left side. Only two hydrogen atoms appear on the right side, but there are four hydrogen atoms on the left side. We must not alter the subscripts in the formula of a compound, because to do so would turn it into a different compound. However, we may change the number of molecules on either side in order to obey the rule for conservation of atoms.

If we double the number of water molecules, we get

$$CH_4 + O_2 \rightarrow 2H_2O + CO_2$$

The hydrogen atoms are now balanced—four on each side of the equation—but the oxygen atoms are not yet balanced—there are four on the right side and only two on the left. If we double the number of oxygen molecules, then all the atoms are balanced.

$$CH_4 + 2O_2 \rightarrow 2H_2O + CO_2$$

The molecules are completely altered by the reaction, because methane molecules and oxygen molecules disappear and water molecules and carbon dioxide molecules appear. But we have the same number of the same kinds of atoms after the reaction as before. The above equation tells us that every molecule of methane that burns requires two molecules of oxygen. Likewise, for every molecule of carbon dioxide produced, two molecules of water are formed.

Problem 2.1 Balance the following equations.

(a) $H_2 + O_2 \rightarrow H_2O$

(b) $CaC_2 + H_2O \rightarrow C_2H_2 + Ca(OH)_2$

(c) $K + H_2S \rightarrow K_2S + H_2$

(d) $H_2 + N_2 \rightarrow NH_3$

(e) $PCl_3 + H_2O \rightarrow H_3PO_3 + HCl$

(f) $Al + O_2 \rightarrow Al_2O_3$

STATES OF MATTER 2.3

How does the distance between molecules change as a substance changes from gas to liquid to solid?

In general, substances can exist in three different physical states—gas, liquid, and solid. Water is the most common substance that we can observe in all three states—as steam, liquid water, and ice (Figure 2.1).

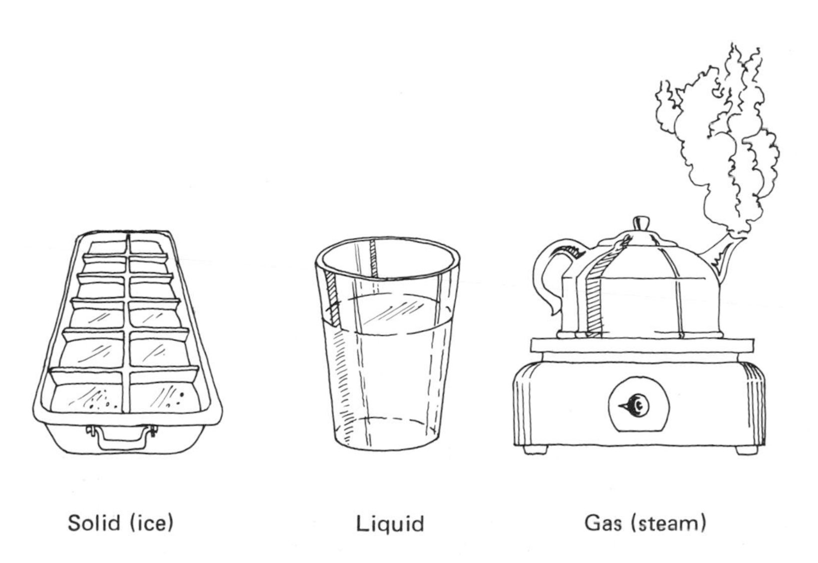

FIGURE 2.1 The three physical states of matter are readily observed with H_2O: (a) solid (ice); (b) liquid (water); (c) gas (steam).

TABLE 2.1 Densities of Substances at 20°C

Substance	Physical state	Density (g/mL)
Copper	Solid	8.9
Table salt (sodium chloride)	Solid	2.2
Sugar (sucrose)	Solid	1.6
Carbon tetrachloride	Liquid	1.6
Water	Liquid	1.0
Benzene	Liquid	0.9
Gasoline (isooctane)	Liquid	0.7
Oxygen	Gas	1.3×10^{-3}
Carbon dioxide	Gas	1.8×10^{-3}
Hydrogen	Gas	8.3×10^{-5}

A **gas** has no definite volume or shape and simply takes on the volume and shape of its container. The molecules of a gas are relatively far apart, and this fact accounts for the extremely low densities of gases (small mass in a large volume).

A **liquid** has a definite volume, but its shape is determined by its container. The molecules of a liquid are much closer together than those of a gas. They are not rigidly bound together, however, and can slide past one another. These facts account for the ability of liquids to flow and also for their high densities as compared with gases.

A **solid** has a definite shape and volume. The molecules are fairly closely packed, producing relatively high density and rigidity. Table 2.1 lists some densities of substances in various physical states at ordinary pressure and 20°C.

2.4 PURE SUBSTANCES AND MIXTURES

A **pure substance** consists of only one kind of molecule. Pure water, for example, is made up of H_2O molecules and no other molecules; pure carbon dioxide consists only of CO_2 molecules.

Mixtures usually contain two or more pure substances in the same or in different physical states. Sometimes a mixture is simply more than one physical state of a single pure substance. A mixture of pure liquid water and some pieces of pure ice is a pure substance because only H_2O

molecules are present, but it is also a mixture because it consists of two different physical states. Many mixtures, however, consist of several different pure substances: concrete, a birthday cake, and ocean water are everyday examples of this type of mixture (Figure 2.2).

We may further classify mixtures as heterogeneous or homogeneous. The components of a **heterogeneous mixture** are visually distinguishable from one another (although a magnifying glass may be necessary). Of the above examples, concrete, a birthday cake, and a mixture of pure ice and pure liquid water are all heterogeneous mixtures. But ocean water is a homogeneous mixture because it is visually uniform.

Are solutions always liquids?

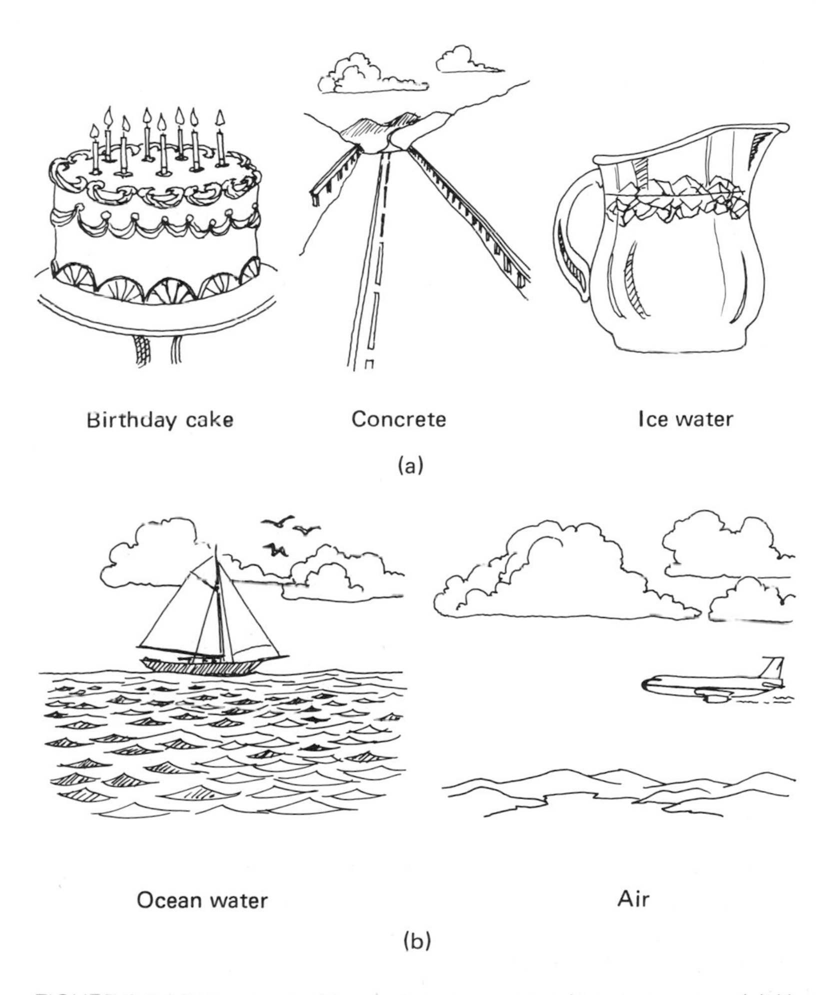

FIGURE 2.2 Mixtures are either heterogeneous or homogeneous. (a) Heterogeneous mixtures have visually distinct components. (b) Homogeneous mixtures are usually called solutions and are visually uniform throughout.

Homogeneous mixtures of two or more pure substances are usually called **solutions.** The physical state of a solution does not matter. What is essential is that the substances be uniformly intermingled. Air is a gaseous solution of oxygen, nitrogen, carbon dioxide, water vapor, and argon, plus several other minor constituents.

Problem 2.2 Classify each of the following as a pure substance, a heterogeneous mixture, or a homogeneous mixture (solution).

(a) 5 g of gold
(b) 1 g of sugar dissolved in 100 g of water
(c) a piece of granite
(d) some scrambled eggs
(e) 2 L of hydrogen gas
(f) some hydrogen and argon confined in a 2-L container

2.5 BOHR'S THEORY

Where is most of the mass of an atom found?

Dalton's simple atomic theory was very useful in advancing scientific understanding of chemical reactions and the properties of pure substances, mixtures, and solutions. By the end of the nineteenth century, however, a tremendous amount of evidence had accumulated to prove that atoms are not simply hard pieces of matter. We now know that atoms are not indivisible, but are composed of even smaller particles, some of which carry electrical charges.

Most of the mass of an atom is concentrated in a very small volume called the **nucleus.** The nucleus is positively charged. Nevertheless, atoms are electrically neutral because they also contain negative particles called **electrons** that exactly balance the positive charge of the nucleus. The electrons are not in the nucleus but are arranged in space around it.

The Danish physicist Niels Bohr developed the first successful modern theory of atoms. Bohr announced his theory in 1913, and although it has been substantially modified, the original theory offers a useful, simple model.

Figure 2.3 is a diagram of the simplest atom, hydrogen, according to Bohr's theory. The hydrogen atom has a nucleus that consists of only one positively charged particle, called a **proton.** Only one electron is necessary to balance the nuclear charge and, according to Bohr, the electron follows a specific path around the nucleus. Bohr imagined that there were many paths available to the electron, but only one path was

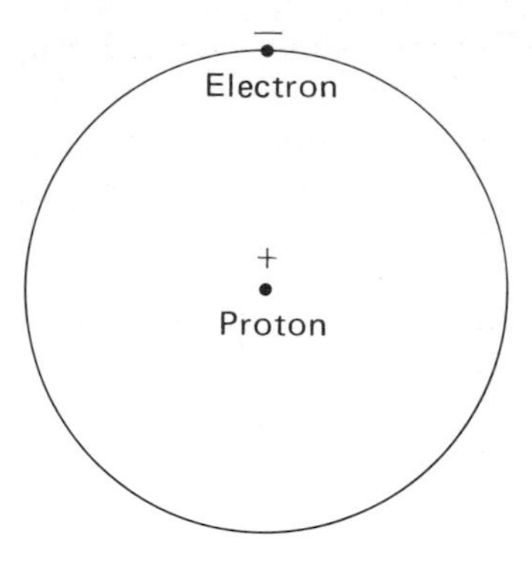

FIGURE 2.3 The hydrogen atom according to Bohr's theory.

followed at a time. The most stable situation occurred when the electron was in the path closest to the nucleus.

The distance between the electron and the proton is very great compared with the size of the proton. If we imagine the proton to be as large as a tennis ball, the electron would circle the proton at a radius of several miles. Most of an atom is empty space! Of course, its actual size is extremely small. The diameter of most atoms is approximately 10^{-8} cm.

Some years after Bohr's theory had been published, another nuclear particle was discovered—the **neutron.** A neutron is about the same mass as a proton, but it is electrically neutral—that is, it has no charge. Protons and neutrons together make up the nuclei of atoms and account for practically the entire mass of an atom. Hydrogen is the only atom whose nucleus does not contain any neutrons. Table 2.2 summarizes the properties of the three important subatomic particles—electrons, protons, and neutrons. The relative masses of these particles can be expressed in **unified atomic mass units (u).**

What are the three important subatomic particles?

TABLE 2.2 Properties of Subatomic Particles

Particle	Mass in g	Mass in u	Electrical charge
Proton	1.7×10^{-24}	1.0	+1
Neutron	1.7×10^{-24}	1.0	0
Electron	9.1×10^{-28}	5.4×10^{-4}	−1

2.6 ATOMIC NUMBER

What is the most fundamental property of an atom?

The number of protons in the nucleus of an atom is the **atomic number** of that atom. The atomic number also tells how many electrons there are in an atom, because the number of electrons equals the number of protons. The number and arrangement of electrons around the nucleus determine practically all the properties of an atom except its mass. Thus, the atomic number is the most important and fundamental property of an atom, and all the atoms of a given element have the same atomic number.

How many electrons can fit in a given shell?

We can use the Bohr theory to explain the structure of atoms. According to the theory, each electron follows a specific path. The various electron paths, in turn, can contain different numbers of electrons. We can think of these paths as concentric shells centered on and surrounding the nucleus. The shell closest to the nucleus can hold two electrons. The second shell can hold as many as eight electrons, and the third can hold up to eighteen. The larger shells can hold even more electrons. The maximum number of electrons that each shell can hold is given by the expression $2n^2$, where n is the shell number. For the first shell, $n = 1$ and $2n^2 = 2$; for the second shell, $n = 2$ and $2n^2 = 8$; and so on. Table 2.3 shows how many electrons each of the first five shells can hold. The electron structures of several atoms according to the Bohr theory are given in Figure 2.4.

Problem 2.3 What is the maximum number of electrons that the sixth Bohr shell ($n = 6$) can hold?

Problem 2.4 Draw Bohr electron structures for the following atoms.

(a) He (atomic number 2)
(b) O (atomic number 8)
(c) Na (atomic number 11)
(d) Cu (atomic number 29)

TABLE 2.3 Maximum Number of Electrons in First Five Shells

Shell number (n)	Maximum number of electrons ($2n^2$)
1	2
2	8
3	18
4	32
5	50

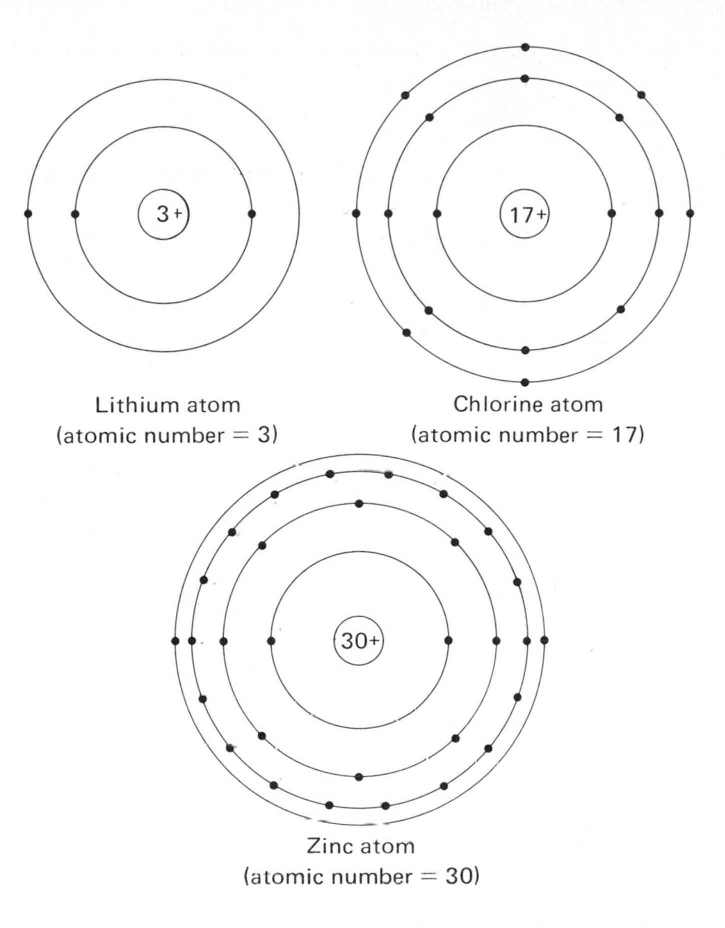

FIGURE 2.4 Electron structures of lithium, chlorine, and zinc according to the Bohr theory. The atomic number is the positive charge on the nucleus of an atom. This charge is equal to the total number of electrons in the atom.

MODERN ATOMIC THEORY 2.7

A lot of experimental work has been done since Bohr first proposed his theory, and the results show that electrons do not follow simple, well-defined paths around the nucleus of an atom. The situation is much more complicated.

The modern concept of atomic structure pictures general regions of space around the nucleus in place of the specific shells of the earlier Bohr theory. The exact positions of electrons in an atom are impossible to predict; all that can be specified is the chance or probability of an electron being in a certain region. The regions where electrons are likely to be are called **orbitals.** The details of orbital theory (sometimes called wave theory) are highly mathematical. Without getting into mathematics, we can use the results of the theory to get useful information.

How do s orbitals and p orbitals differ?

The different orbitals are designated by the letters *s, p, d,* and *f.* In our study, we will be mainly concerned with atoms where *s* and *p* orbitals are of primary importance. Figure 2.5 shows the spatial properties of these orbitals. Notice that there is only one kind of *s* orbital, but there are three different *p* orbitals. An electron in an *s* orbital is likely to be anywhere around the nucleus (at a certain distance), but the electrons in *p* orbitals are found along certain axes. Depending on which one of the three possible *p* orbitals an electron actually occupies, it is likely to be near the x-axis (p_x orbital), the *y*-axis (p_y orbital), or the *z*-axis (p_z orbital).

Electrons have a property called **spin.** To use a physical analogy, we may think of an electron as a top, which can spin either clockwise or counterclockwise. If there is only one electron in a given orbital, it makes

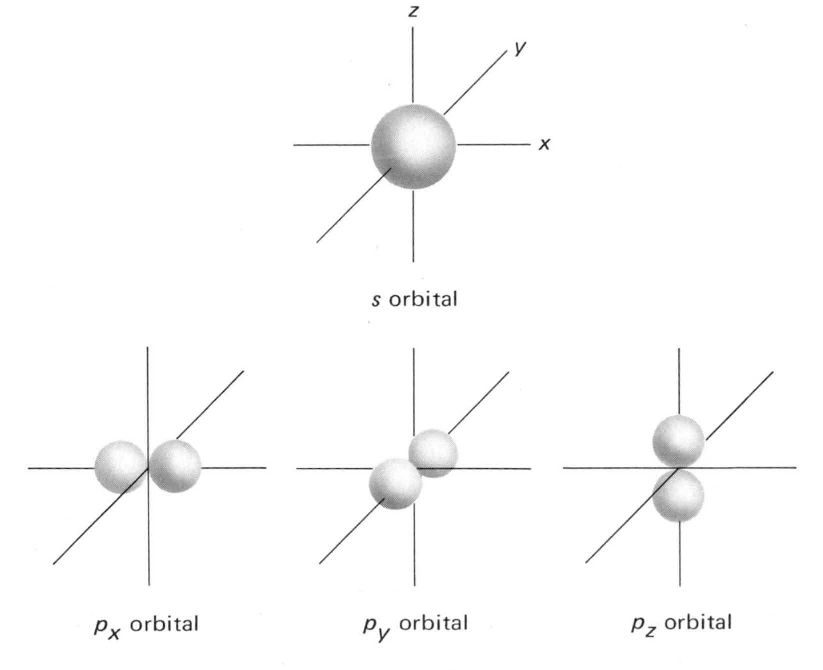

FIGURE 2.5 Shapes of *s*- and *p*-type orbitals. The nucleus is imagined to be at the point of intersection of the axes.

no difference in which direction it spins. However, if two electrons occupy the same orbital, they must spin in opposite directions. This is one of the most important facts of modern electronic theory. Electrons with opposite spins are said to be "paired." Thus, two electrons in the same orbital must be paired, which limits to two the number of electrons that can occupy a given orbital. Since there are only two possible spins for an electron, if more than two electrons occupied an orbital, at least two would have the same spin. But this would violate the rule that electrons must have opposite spins to occupy the same orbital.

How many electrons can occupy a given orbital? $2n^2$

The orbitals of an atom are grouped into various levels around the nucleus. These are called **quantum levels** and are roughly equivalent to the shells of the Bohr theory. (Today, the terms "shell" and "quantum level" are synonymous.) We assign numbers, similar to the shell numbers, to designate the different quantum levels. The first quantum level consists of a single *s* orbital. We call this the 1*s* orbital. The second quantum level consists of another *s* orbital and three *p* orbitals, which we call the 2*s* orbital and the three 2*p* orbitals, respectively. It is easy to see why the first quantum level can have only two electrons: the single 1*s* orbital can hold two electrons and no more. The second quantum level can hold as many as eight electrons, because this level consists of four orbitals (2*s*, $2p_x$, $2p_y$, and $2p_z$). The third level, in addition to the 3*s*, $3p_x$, $3p_y$, and $3p_z$ orbitals, has five *d*-type orbitals available. Since two electrons can fit into each orbital, a total of eighteen electrons can be contained in the nine third-level orbitals.

Problem 2.5 What is the maximum number of electrons that can fit in a 3*s* orbital? In a $4p_x$ orbital?

ELECTRON ARRANGEMENTS OF ATOMS 2.8

How do we write out the electron arrangements of atoms?

The way electrons are arranged around an atom is called the electron arrangement or electron configuration of that atom. Electrons always fill the orbitals closest to the nucleus before filling those further away. The most stable orbital, the 1*s*, fills first, then the 2*s* orbital, followed by the three 2*p* orbitals. The 3*s* orbital comes next, followed by the three 3*p* orbitals. The electrons in the 1*s* orbital are lowest in energy, followed by those in the 2*s*, 2*p*, 3*s*, and 3*p* orbitals, respectively. Figure 2.6 shows the relative energies of these orbitals. We will not deal here with atoms that use orbitals beyond the 3*p* levels.

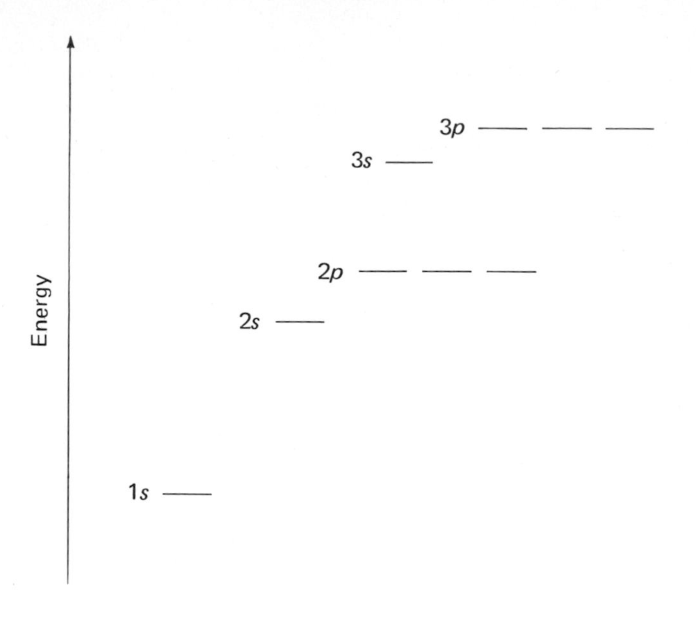

FIGURE 2.6 Relative energies of orbitals. The ls orbital is closest to the nucleus and thus has the lowest energy (so it is the most stable). The other orbitals have higher energies, as indicated.

In describing electron arrangements, we use superscripts to indicate the number of electrons in a particular orbital. The hydrogen atom has only one electron, and its electron arrangement is written $1s^1$. The helium atom has two electrons, and we express its arrangement as $1s^2$. The next atom, lithium, has three electrons and its arrangement is written $1s^2 2s^1$. Remember, there can be no more than two electrons in each orbital!

The next atom, beryllium, has the configuration $1s^2 2s^2$. Boron follows, with the arrangement $1s^2 2s^2 2p^1$. Carbon is next with $1s^2 2s^2 2p^2$. Nitrogen has seven electrons, and we write its configuration as $1s^2 2s^2 2p^3$. This does not violate the rule that an orbital can contain no more than two electrons, because there are three *p* orbitals. Together, the three *p* orbitals (p_x, p_y, p_z) can hold a maximum of six electrons. Table 2.4 gives the electron distributions for atoms of the first eighteen elements. These are the atoms with atomic numbers 1 through 18.

Theory predicts, and experiments confirm, that electrons in *p*-type orbitals do not pair up in any one of the three *p* orbitals if another *p* orbital is vacant. Thus, the arrangement for nitrogen (Table 2.4) is

TABLE 2.4 Electron Distribution for Atoms of the First Eighteen Elements

Element	Symbol	Atomic number	Electron distribution
Hydrogen	H	1	$1s^1$
Helium	He	2	$1s^2$
Lithium	Li	3	$1s^22s^1$
Beryllium	Be	4	$1s^22s^2$
Boron	B	5	$1s^22s^22p^1$
Carbon	C	6	$1s^22s^22p^2$
Nitrogen	N	7	$1s^22s^22p^3$
Oxygen	O	8	$1s^22s^22p^4$
Fluorine	F	9	$1s^22s^22p^5$
Neon	Ne	10	$1s^22s^22p^6$
Sodium	Na	11	$1s^22s^22p^63s^1$
Magnesium	Mg	12	$1s^22s^22p^63s^2$
Aluminum	Al	13	$1s^22s^22p^63s^23p^1$
Silicon	Si	14	$1s^22s^22p^63s^23p^2$
Phosphorus	P	15	$1s^22s^22p^63s^23p^3$
Sulfur	S	16	$1s^22s^22p^63s^23p^4$
Chlorine	Cl	17	$1s^22s^22p^63s^23p^5$
Argon	Ar	18	$1s^22s^22p^63s^23p^6$

usually written as $1s^22s^22p^3$ for simplicity, but the actual arrangement is $1s^22s^22p_x{}^12p_y{}^12p_z{}^1$.

A very detailed method of showing the electron configurations of atoms employs arrows to indicate electron spin. Thus,

$$\underline{\uparrow}$$

indicates one electron in one orbital; and

$$\underline{\uparrow\downarrow}$$

indicates a pair of electrons in one orbital. Several electron arrangements will illustrate this more detailed method.

Carbon (C) $1s\,\underline{\uparrow\downarrow}$ $2s\,\underline{\uparrow\downarrow}$ $2p_x\,\underline{\uparrow}\;2p_y\,\underline{\uparrow}\;2p_z\,\underline{}$

The orbitals are placed at different levels to emphasize the higher relative energies of electrons in the higher orbitals (see Figure 2.6).

Nitrogen (N)
$1s\ \underline{\uparrow\downarrow}$ $2s\ \underline{\uparrow\downarrow}$ $2p_x\ \underline{\uparrow}$ $2p_y\ \underline{\uparrow}$ $2p_z\ \underline{\uparrow}$

Oxygen (O)
$1s\ \underline{\uparrow\downarrow}$ $2s\ \underline{\uparrow\downarrow}$ $2p_x\ \underline{\uparrow\downarrow}$ $2p_y\ \underline{\uparrow}$ $2p_z\ \underline{\uparrow}$

Problem 2.6 Write electron structures for the following atoms by indicating the electrons in different orbitals ($1s^2 2s^2$ . . .).

(a) H (b) Li
(c) B (d) N
(e) O (f) Ne

Problem 2.7 Indicate how the electrons are arranged in the atoms in Problem 2.6 by using arrows to represent electron spin.

2.9 LEWIS DOT STRUCTURES

What does a Lewis dot structure show?

There is still another way to show the electron configurations of atoms, the use of **Lewis dot structures.** We will look at this method in some detail because it is probably the most useful in chemistry.

The outermost electron shell (quantum level) of an atom is the most important. The ability of an atom to combine with another atom depends on the electron arrangement in this shell, which is called the **valence shell.** A Lewis dot structure concentrates on the electron arrangement in the valence shell of an atom. The rules for writing Lewis dot structures are simple:

1. The chemical symbol for an atom is used to represent the nucleus and all electrons *except* the valence-shell ones.
2. Electrons in the valence shell are represented by small dots placed around the chemical symbol. Electron pairs are indicated by placing the dots close together.

Table 2.5 gives Lewis dot structures for atoms of the first eighteen elements.

Problem 2.8 Write Lewis dot structures for the atoms in Problem 2.6.

TABLE 2.5 Lewis Dot Structures for Atoms of the First Eighteen Elements

Element	Symbol	Electrons in valence shell	Lewis dot structure
Hydrogen	H	1	H·
Helium	He	2	He:
Lithium	Li	1	Li·
Beryllium	Be	2	Be:
Boron	B	3	·B:
Carbon	C	4	·Ċ:
Nitrogen	N	5	·N̈:
Oxygen	O	6	:Ö:
Fluorine	F	7	:F̈:
Neon	Ne	8	:N̈e:
Sodium	Na	1	Na·
Magnesium	Mg	2	Mg:
Aluminum	Al	3	·Al:
Silicon	Si	4	·Ṡi:
Phosphorus	P	5	·P̈:
Sulfur	S	6	:S̈:
Chlorine	Cl	7	:C̈l:
Argon	Ar	8	:Är:

PERIODICITY AND CHEMICAL FAMILIES 2.10

What primarily determines the properties of atoms?

Years before anything was known about electrons, chemists discovered that elements with similar properties could be grouped together in chemical families. The properties of the different elements are not haphazard, but instead recur in periodic fashion.

Today we know that the properties of atoms depend on how their electrons are arranged, especially in the valence shells. If we list the elements according to their atomic numbers, it is possible to arrange them into groups, or families, in which the atoms have similar valence-shell electron arrangements.

There are a few exceptions to this orderly arrangement. Hydrogen does not fit into any family, and two series of elements, with the atomic numbers 58 to 71 and 90 to 103, are difficult to place in families. We will say more about these interesting elements later.

Figure 2.7 is a listing of the elements by atomic number, and is one of many similar listings referred to as the **periodic table.** The vertical columns, indicated by Roman numerals, are the chemical families. Elements that fall in the same family have quite similar properties. The families are further classified as the **representative elements,** the **transition elements,** the **lanthanide series,** the **actinide series,** and the **noble gases.**

The representative elements are very important to the chemistry of life. Both metallic and nonmetallic elements are included among them.

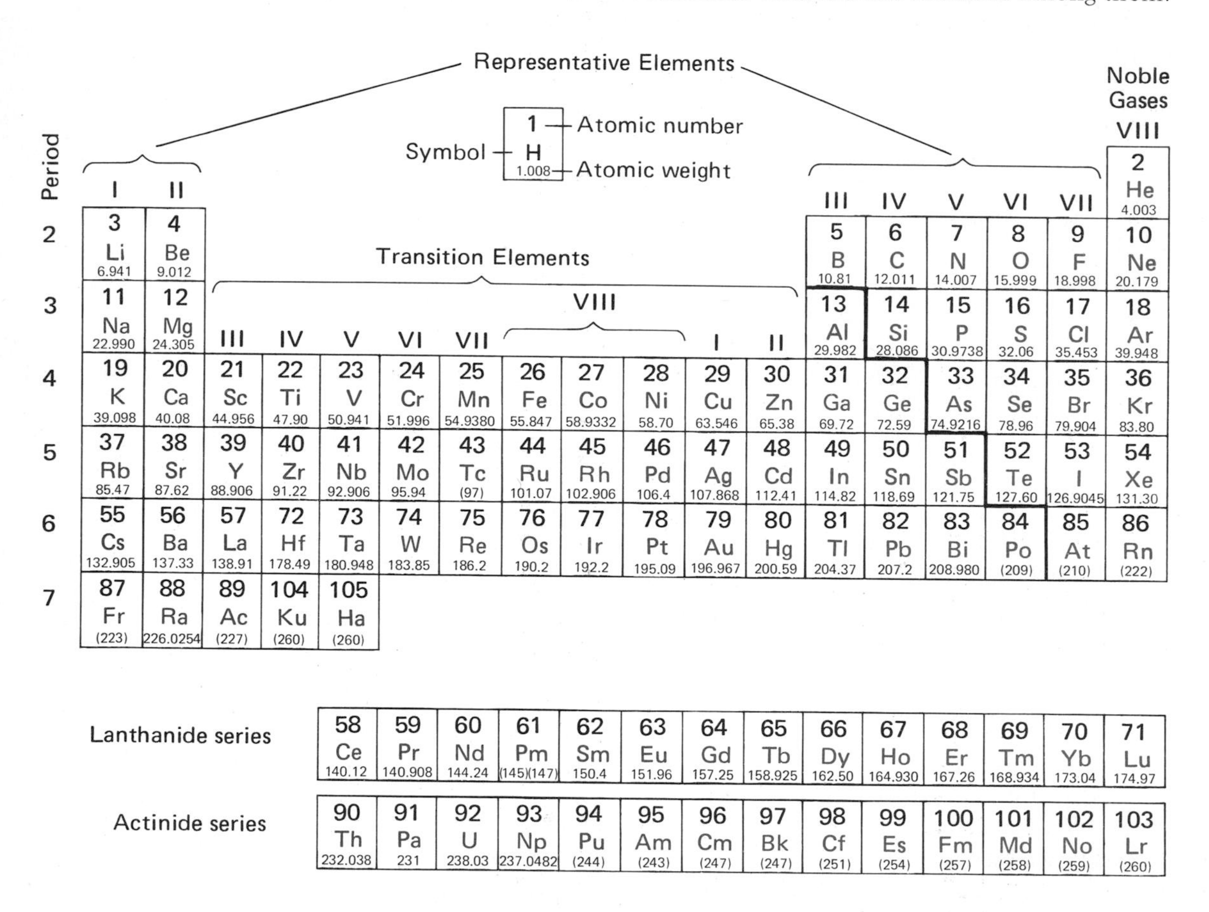

Period	I	II	III	IV	V	VI	VII	VIII	VIII	VIII	I	II	III	IV	V	VI	VII	VIII
																		2 He 4.003
2	3 Li 6.941	4 Be 9.012											5 B 10.81	6 C 12.011	7 N 14.007	8 O 15.999	9 F 18.998	10 Ne 20.179
3	11 Na 22.990	12 Mg 24.305											13 Al 29.982	14 Si 28.086	15 P 30.9738	16 S 32.06	17 Cl 35.453	18 Ar 39.948
4	19 K 39.098	20 Ca 40.08	21 Sc 44.956	22 Ti 47.90	23 V 50.941	24 Cr 51.996	25 Mn 54.9380	26 Fe 55.847	27 Co 58.9332	28 Ni 58.70	29 Cu 63.546	30 Zn 65.38	31 Ga 69.72	32 Ge 72.59	33 As 74.9216	34 Se 78.96	35 Br 79.904	36 Kr 83.80
5	37 Rb 85.47	38 Sr 87.62	39 Y 88.906	40 Zr 91.22	41 Nb 92.906	42 Mo 95.94	43 Tc (97)	44 Ru 101.07	45 Rh 102.906	46 Pd 106.4	47 Ag 107.868	48 Cd 112.41	49 In 114.82	50 Sn 118.69	51 Sb 121.75	52 Te 127.60	53 I 126.9045	54 Xe 131.30
6	55 Cs 132.905	56 Ba 137.33	57 La 138.91	72 Hf 178.49	73 Ta 180.948	74 W 183.85	75 Re 186.2	76 Os 190.2	77 Ir 192.2	78 Pt 195.09	79 Au 196.967	80 Hg 200.59	81 Tl 204.37	82 Pb 207.2	83 Bi 208.980	84 Po (209)	85 At (210)	86 Rn (222)
7	87 Fr (223)	88 Ra 226.0254	89 Ac (227)	104 Ku (260)	105 Ha (260)													

Lanthanide series	58 Ce 140.12	59 Pr 140.908	60 Nd 144.24	61 Pm (145)(147)	62 Sm 150.4	63 Eu 151.96	64 Gd 157.25	65 Tb 158.925	66 Dy 162.50	67 Ho 164.930	68 Er 167.26	69 Tm 168.934	70 Yb 173.04	71 Lu 174.97
Actinide series	90 Th 232.038	91 Pa 231	92 U 238.03	93 Np 237.0482	94 Pu (244)	95 Am (243)	96 Cm (247)	97 Bk (247)	98 Cf (251)	99 Es (254)	100 Fm (257)	101 Md (258)	102 No (259)	103 Lr (260)

FIGURE 2.7 Periodic table of the elements. The number above the chemical symbol is its atomic number. The Roman numerals designate the groups, or chemical families.

Metals are generally good conductors of heat and electricity, and they can be bent without breaking. Most **nonmetals** are poor conductors of heat and electricity. The solid nonmetals are brittle, and several are gases at ordinary temperatures. The heavy steplike line in Figure 2.7 marks the dividing line between the metals, to the left, and the nonmetals (including hydrogen), to the right. Notice that over three-fourths of the elements are metallic.

Most of the time, we will be concerned with the first eighteen elements, all of which are representative elements and noble gases. The period number tells which shell is the valence shell, and the group (or family) number tells the number of electrons in the valence shell. All the elements in the same chemical family thus have similar valence-shell electron arrangements, and the Lewis dot structures of their atoms are identical. For example, oxygen is a group VI element and so has six electrons in its valence shell.

$$\cdot\overset{\cdot\cdot}{\underset{\cdot}{\mathrm{O}}}:$$

Atoms of all the other elements in this group also have six electrons in their valence shells. For instance, we represent selenium (atomic number 34) as

$$\cdot\overset{\cdot\cdot}{\underset{\cdot}{\mathrm{Se}}}:$$

In similar fashion, we can easily write the dot structure for iodine (atomic number 53) by noting that it is in group VII.

$$\cdot\overset{\cdot\cdot}{\underset{\cdot\cdot}{\mathrm{I}}}:$$

Aluminum (atomic number 13) is in group III, so we know it has three valence-shell electrons.

$$\dot{\mathrm{Al}}:$$

Hydrogen and helium are exceptions, and it is necessary to remember that the hydrogen atom has only one electron (H·) and the helium atom has two electrons (He:).

Problem 2.9 Using *only* Figure 2.7, write Lewis dot structures for the following atoms.

(a) K
(b) Sn
(c) Ga
(d) atom with atomic number 15
(e) atom with atomic number 20
(f) atom with atomic number 35

2.11 IONIZATION ENERGY

How do metals and nonmetals differ in their ionization energies?

Opposite electrical charges attract. The electrons of an atom are thus attracted by the positive charge on the nucleus. Like charges repel. Therefore, electrons repel electrons, and protons repel protons.* The electrical repulsion between electrons is more than offset by the electrical attraction between each electron and the nucleus. Overall, the electron structure of an atom is stable. However, if we put enough energy into an atom, an electron can be removed. The minimum amount of energy necessary to remove an electron from a gaseous atom is called the **ionization energy** of the atom.

In the periodic table, ionization energies generally increase from left to right along a period and decrease from top to bottom in a family. Chemically speaking, metals have relatively low ionization energies, whereas nonmetals have high ionization energies. Figure 2.8 shows for the first eighteen elements, how ionization energy increases as one moves along the periodic table from metals to nonmetals.

Problem 2.10 Refer to Figure 2.8 to answer the following questions.

(a) Which of the first eighteen elements has the highest ionization energy? the lowest?

(b) Which atom has a higher ionization energy, C or F?

(c) Which atom has a lower ionization energy, O or S?

(d) Which atom has a higher ionization energy, N or Si?

(e) Which element is metallic, Be or B?

2.12 IONS

When an atom loses an electron, the atom is no longer electrically neutral because it now carries a positive charge. It has become a positive ion, called a **cation.** For example, if we supply the ionization energy

*The protons and neutrons are held together in the nucleus by the extremely strong nuclear force. We will not discuss this force, however, since it is apparently not involved in ordinary chemical reactions.

Increasing ionization energy →

						1 H	2 He
3 Li	4 Be	5 B	6 C	7 N	8 O	9 F	10 Ne
11 Na	12 Mg	13 Al	14 Si	15 P	16 S	17 Cl	18 Ar

↑ Increasing ionization energy

FIGURE 2.8 A segment of the periodic table showing the directions in which ionization energy increases as we move through the first eighteen elements. Metallic elements are shaded.

to a sodium atom and it loses an electron, the result is a sodium ion, which is a cation. The positive charge on the ion is indicated by a plus sign, the negative charge of the electron (e) by a minus.

What happens if an atom loses or gains electrons?

$$Na \rightarrow Na^+ + e^-$$

By adding more energy, it is possible to remove more than one electron from an atom. When metallic atoms participate in chemical reactions, they usually lose electrons to form cations. To indicate a positive charge greater than one, we place a superscript number before the plus, as in Mg^{2+} and Al^{3+}. In the next chapter, we will learn to predict how many electrons any particular metallic atom is likely to lose.

Nonmetallic atoms usually gain electrons in chemical reactions. A negative ion results in such a case. A negative ion is called an **anion.** If a chlorine atom gains an electron, a chloride ion forms, which is an anion.

$$Cl + e^- \rightarrow Cl^-$$

Nonmetallic atoms rarely form positive ions, because their very high ionization energies make it hard for them to give off electrons. They are much more likely to take on electrons, thereby gaining a negative charge. A negative charge greater than one is indicated by a superscript number before the minus sign, as in O^{2-} and N^{3-}. The general rule is that in chemical reactions where metallic atoms react with nonmetallic atoms, *metallic atoms lose electrons and nonmetallic atoms gain electrons.*

Problem 2.11 If the following atoms were to form ions in chemical reactions, which would be most likely to form cations and which most likely to form anions?

(a) Li
(b) N
(c) S
(d) Mg
(e) H

2.13 ATOMIC WEIGHT AND ISOTOPES

Can atoms of the same element have different weights?

Another important property of an atom is its **atomic weight.** In general, atoms of different elements have different masses. However, even atoms of the same element can have slightly different masses. That is, atoms with the same atomic number can have different atomic weights. Two atoms with the same atomic number but different atomic weights are called **isotopes.**

The mass of an atom depends on the sum of protons and neutrons in its nucleus, so isotopes are atoms that contain the same number of protons but different numbers of neutrons. Dalton and other chemists after him devised ingenious ways to measure atomic weights. Of course, an atom is far too small to see, much less weigh, so they had to use indirect methods. In all of their indirect methods, they were not able to detect the presence of isotopes of elements, and, in fact, the existence of isotopes was not suspected until early in the twentieth century. Today an instrument called the **mass spectrograph** is used to determine directly the relative masses of different atoms, including all the isotopes of a given element.

The standard atom used for calibrating the mass spectrograph is the $^{12}_{6}C$ isotope. The symbol has the following meaning:

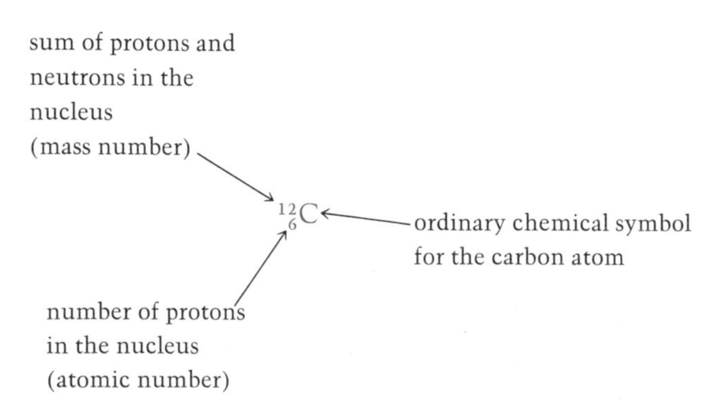

The **mass number** is essentially the mass of a particular isotope. Since the mass number is the sum of protons and neutrons, if you subtract the atomic number from the mass number, you get the number of neutrons.

$$(\text{Mass number}) - (\text{atomic number}) = (\text{number of neutrons})$$

This isotope of carbon has been assigned a mass of exactly 12 unified atomic mass units (u). Although some atomic weights are whole numbers, many of them are not. The main reason for this is that most elements contain fixed amounts of different isotopes, and the atomic weight of an element is an average of the masses of the isotopes.* Table 2.6 lists the atomic weights to two or three decimal places of the first eighteen elements. Many atomic weights are known even more precisely, and a complete listing of atomic weights can be found on the inside front cover of this book.

TABLE 2.6 Atomic Weights of Atoms of the First Eighteen Elements

Element	Symbol	Atomic weight
Hydrogen	H	1.008
Helium	He	4.003
Lithium	Li	6.941
Beryllium	Be	9.012
Boron	B	10.81
Carbon	C	12.011
Nitrogen	N	14.007
Oxygen	O	15.999
Fluorine	F	18.998
Neon	Ne	20.179
Sodium	Na	22.990
Magnesium	Mg	24.305
Aluminum	Al	26.982
Silicon	Si	28.086
Phosphorus	P	30.974
Sulfur	S	32.06
Chlorine	Cl	35.453
Argon	Ar	39.948

*There is another reason why atomic weights are not whole numbers. The nuclear force causes the nucleus to weigh slightly less than the sum of the masses of the neutrons and protons in the nucleus. However, this effect is less important than the presence of isotopes in explaining why atomic weights are not whole numbers.

Notice that carbon has an atomic weight slightly greater than 12. The reason is that ordinary elemental carbon, as we find it in nature, contains some $^{13}_{6}C$ atoms along with the $^{12}_{6}C$ atoms. Of every 100 carbon atoms in nature, one is a $^{13}_{6}C$ isotope and the other 99 are the $^{12}_{6}C$ isotopes. The atomic weight of carbon can thus be calculated as follows:

$$\begin{aligned} 1 \times 13.00 &= 13 \text{ u} \\ \underline{99 \times 12.00} &= \underline{1188} \text{ u} \\ \text{Total weight of 100 atoms} &= 1201 \text{ u} \\ \text{Average weight per atom} &= \frac{1201}{100} = 12.01 \text{ u} \end{aligned}$$

Problem 2.12 Define the following terms.
(a) element (b) isotope
(c) mass number

Problem 2.13 Ordinary neon consists of two isotopes, $^{20}_{10}Ne$ and $^{22}_{10}Ne$. Of every 100 ordinary neon atoms, 91 are $^{20}_{10}Ne$ and 9 are $^{22}_{10}Ne$. Calculate the atomic weight of neon.

2.14 RARE-EARTH ELEMENTS

The elements with atomic numbers 58 to 71 are referred to as the **rare-earth elements** or the lanthanide series (they follow lanthanum, atomic number 57), or simply as "the lanthanides." These elements are all metals and are chemically very similar to one another. Most of them are reasonably abundant in nature and are not really rare. However, their close chemical similarities made their identification and separation difficult.

2.15 RADIOACTIVE ELEMENTS

The elements with atomic numbers 90 to 103 are called the actinide series or just "the actinides" (they follow actinium, atomic number 89). These elements are all metals, but only the first three members of the

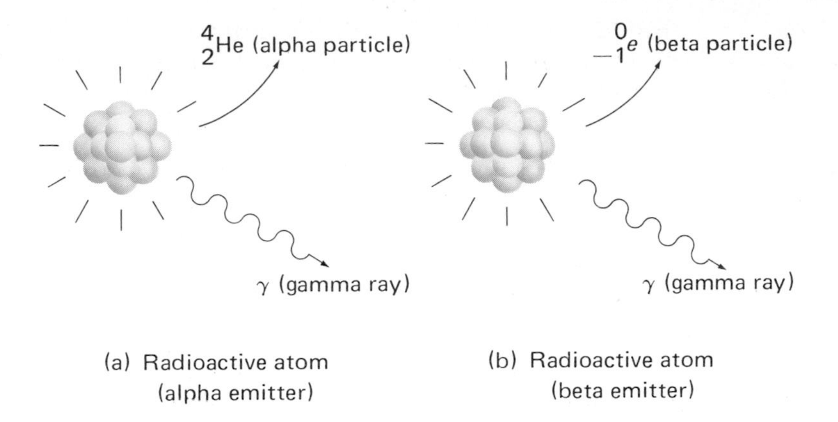

FIGURE 2.9 Some radioactive elements decay by emitting alpha (α) particles (a); others decay by emitting beta (β) particles (b). Gamma (γ) rays often accompany either type of decay.

series are found in nature. These three are thorium, protactinium, and uranium; along with radium (atomic number 88) and polonium (atomic number 84), they make up the bulk of the naturally occurring **radioactive elements.**

What emissions come from the nuclei of radioactive atoms?

The atoms of radioactive elements have unstable nuclei. Most atoms found in nature have nuclei that are quite stable and do not change even when the atoms combine with other atoms to form compounds. But radioactive atoms have nuclei that spontaneously emit particles.

The most common particles emitted by radioactive nuclei are identified by Greek letters—**alpha** (α) and **beta** (β) **particles. Gamma** (γ) **rays,** which are similar to x rays, are also sometimes released by radioactive atoms when their nuclei decay. An alpha particle consists of two neutrons and two protons—the same as the nucleus of a helium atom. A beta particle is a fast-moving electron, that is, an electron with a lot of kinetic energy. A gamma ray is a form of electromagnetic radiation, like an x ray or ordinary light (Figure 2.9). Gamma rays are even more penetrating than x rays and consequently can cause radiation damage to biological cells. For this reason, care must be exercised in handling radioactive materials. Beta particles are also fairly penetrating because of their high energies. Alpha particles are less penetrating. But, ingestion of a substance that emits either alpha or beta particles is extremely dangerous, especially if the emitting substance becomes incorporated into bone or other body tissue. Table 2.7 summarizes the properties of alpha, beta, and gamma emissions.

TABLE 2.7 Properties of Alpha (α), Beta (β), and Gamma (γ) Emissions

Emission	Mass in g	Mass in u	Electrical charge
α particle	6.6×10^{-24}	4.0	+2
β particle	9.1×10^{-28}	5.4×10^{-4}	−1
γ ray	0	0	0

2.16 RADIOISOTOPES

Does radioactive decay produce new elements?

Any isotope that is radioactive is called a **radioisotope.** All of the isotopes of elements with atomic numbers greater than 83 are radioactive. Only a very few radioisotopes are found in nature for elements with atomic numbers less than 83.

The fascinating thing about the decay of a radioisotope is that a new element is formed. For example, when a thorium isotope loses an alpha particle, an isotope of radium is formed.

$$^{232}_{90}\text{Th} \rightarrow {}^{4}_{2}\text{He} + {}^{228}_{88}\text{Ra}$$

The new atom is called a **daughter isotope.** This nuclear equation is balanced in terms of both the mass numbers and the atomic numbers of reactants and products. The alpha particle that is lost is expressed as a helium nucleus, with its atomic number and mass number explicitly shown to aid in balancing the equation.

Nuclear reactions are quite different from ordinary chemical reactions. They violate the principle, expressed in Dalton's atomic theory, that the same number and same kinds of atoms exist before and after a chemical reaction. However, the atomic numbers (which represent positive charges on the nuclei) and the mass numbers of the isotopes do have to balance.*

*In balancing nuclear reactions, we usually consider only *nuclear* charges. We do not consider the fact that ions are formed when nuclear charges are altered. Actually, there is always a lot of ionization when nuclear reactions occur. The energies involved in the nuclear changes are so much greater than ordinary ionization energies that we usually neglect the latter.

An example of a radioisotope that undergoes beta decay is the thorium isotope, $^{234}_{90}Th$.

$$^{234}_{90}Th \rightarrow ^{234}_{91}Pa + ^{0}_{-1}e$$

The daughter isotope has an atomic number—and thus a nuclear charge—one greater than the original thorium isotope. The reason is that the beta particle (which, you recall, is an electron) carries away a negative charge from the nucleus, leaving the daughter nucleus with a greater positive charge. We can view the loss of a beta particle by a nucleus as the conversion of a neutron into a proton. Since a beta particle has a mass number equal to zero, the mass numbers of parent and daughter atoms are identical. The daughter atom formed in the decay of a radioactive atom may also be unstable and therefore subject to decay. For example, both of the above-mentioned daughter atoms, $^{228}_{88}Ra$ and $^{234}_{91}Pa$, undergo beta decay.

Problem 2.14 Write nuclear reactions for the decay of (a) $^{228}_{88}Ra$ and (b) $^{234}_{91}Pa$.

Problem 2.15 Complete and balance the following nuclear decay reactions.

(a) $^{210}_{82}Pb \rightarrow ^{0}_{-1}e +$ (b) $^{212}_{84}Po \rightarrow ^{208}_{82}Pb +$

(c) $^{229}_{90}Th \rightarrow ^{4}_{2}He +$ (d) $^{235}_{92}U \rightarrow ^{4}_{2}He +$

Problem 2.16 Which of the reactions of Problem 2.15 involve alpha decay and which involve beta decay?

HALF-LIFE 2.17

The time required for half of the atoms of a given radioisotope to decay is called its **half-life.** In general, each radioisotope has its own half-life. Thorium-234 ($^{234}_{90}Th$), for example, has a half-life of 25 days. This means that if we start with 1 g of thorium-234, after 25 days only 0.50 g will remain undecayed. After 25 more days, one-half of the remaining atoms will decay, so that only 0.25 g of thorium-234 will be left, and so on. The decay sequence looks like this:

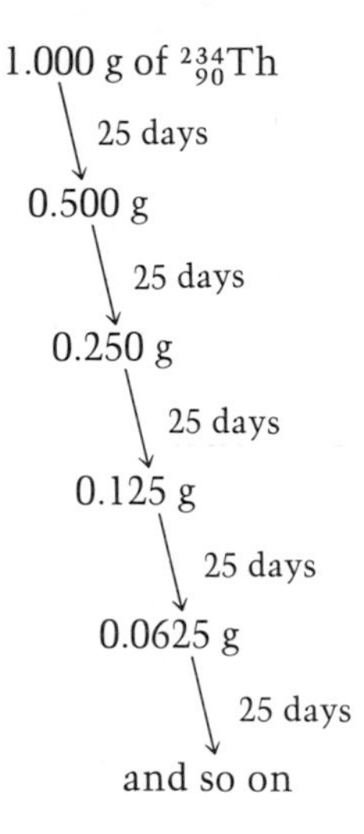

Problem 2.17 In the decay sequence above, 0.500 g of $^{234}_{90}Th$ disappears after 25 days. What happens to it?

Half-lives of radioisotopes vary enormously. For example, the half-life of $^{232}_{90}Th$ is 1.4×10^{10} years, whereas the half-life of $^{213}_{84}Po$ is 4.2×10^{-6} s!

Problem 2.18 The half-life of $^{214}_{83}Bi$ is 20 min. If we start with 12 mg of this radioisotope, how much of it will remain at the end of 1 h?

2.18 ARTIFICIAL RADIOISOTOPES

How do we use artificial radioisotopes?

Even more intriguing than the radioisotopes that occur naturally are the ones that are produced artificially. Almost any stable nucleus can be rendered unstable—and thus radioactive—by suitable treatment. The simplest way to produce an artificial radioisotope is to place a sample of an element in a nuclear reactor. The reactor produces a lot of neutrons, and when a stable nucleus absorbs one or more neutrons, a radioisotope is usually formed.

Cobalt-60 is a very important artificial radioisotope that is used in cancer therapy. It is formed when ordinary, stable cobalt-59 absorbs a neutron (n).

$$^{59}_{27}Co + ^{1}_{0}n \rightarrow ^{60}_{27}Co$$

The $^{60}_{27}Co$ is unstable and has a half-life of 5.3 years. When an atom of $^{60}_{27}Co$ decays, it releases a beta particle and gamma rays. It is the gamma radiation that destroys cancer cells. The decay of the cobalt-60 produces stable nickel-60 as a product.

$$^{60}_{27}Co \rightarrow ^{60}_{28}Ni + ^{0}_{-1}e + \gamma$$

Previously, naturally occurring radium was used in cancer therapy, but cobalt-60 has several advantages. It can be prepared as needed, and it is more effective, since it produces more gamma radiation than an equal weight of radium. And whereas the products of the decay of radium are also radioactive and dangerous, creating a disposal problem, nickel-60 is harmless.

Two other very useful artificial radioisotopes are $^{24}_{11}Na$ and $^{131}_{53}I$. Sodium-24 has a half-life of 15 h and is used to study the circulatory system. Some ordinary sodium chloride (table salt) that contains just a little sodium-24 is injected into a vein and is carried by the blood throughout the body. The beta particles and gamma rays produced as the sodium-24 decays can be detected at various points around the body. The speed at which they move from one point to another provides a measure of the rate of blood circulation.

Iodine-131 has a half-life of 8 days and is used to measure how rapidly the thyroid gland is making thyroid hormone. The rate of iodine uptake is an index of whether the gland is functioning normally, is overactive (hyperthyroid condition), or is underactive (hypothyroid condition). The patient ingests a small amount of sodium iodide that contains some iodine-131. The radioactivity that can be measured in the area of the neck tells how fast iodine is being concentrated by the thyroid gland. The radiation consists of beta particles and gamma rays emitted by iodine-131 atoms as they decay.

Problem 2.19 Write nuclear equations for the decay of $^{24}_{11}Na$ and $^{131}_{53}I$ by beta emission.

Radioisotopes have many other, nonmedical applications. The movement of fluids in pipes is easily monitored by the introduction of a small amount of a suitable radioactive material. Years ago, some scientists added a small amount of radioactive phosphorus-32 to fertilizers in order to measure the uptake of nutrients by plants. They discovered that it was more efficient to apply nutrients to the leaves of plants than to the roots.

The pathways and ultimate fate of insecticides and pesticides in biological systems can be studied by adding radioisotopes to these materials. Oceanographers use radioisotopes to study deep-water currents and the drift of sand and silt. These are but a few of the many practical uses for radioisotopes.

2.19 RADIATION UNITS

The particles emitted by the nuclei of radioactive isotopes can be detected by sensitive instruments such as Geiger-Müller counters, scintillation counters, and ionization chambers.

The amount of radioactivity being emitted by a sample is expressed in **curies.** One curie (Ci) of radioactivity is equal to 3.7×10^{10} nuclear disintegrations per second (dps). The millicurie (mCi) and the microcurie (μCi) are smaller units of radioactivity that are often more convenient to use: 1 mCi equals 3.7×10^7 dps and 1 μCi equals 3.7×10^4 dps.

Often we are more interested in the effects produced by radiation than in the total amount of radiation. Two units, the roentgen and the rad, are often used to express radiation effects.

One **roentgen** (R) is the quantity of x rays or gamma rays that will cause a gram of air to absorb 83.8 ergs of energy. (An *erg* is a very small unit of energy. One calorie is equal to 4.18×10^7 ergs.) A milliroentgen (mR) is a thousandth of a roentgen, so 1 mR results in the absorption of 8.38×10^{-2} ergs per gram of air.

The roentgen measures only gamma rays and x rays. A more general measure of radiation is the **rad** (*r*adiation *a*bsorbed *d*ose). One rad is the amount of any type of radiation that leads to the absorption of 100 ergs by 1 g of any material. In other words, when 1 g of substance has absorbed 100 ergs of energy through exposure to radiation, it has received 1 rad.

Still another unit of radiation is the **rem** (*r*oentgen *e*quivalent in *m*an), which is a measure of how much biological damage a dose of radiation will cause. One rem is defined as the amount of radiation that causes damage to tissue equivalent to the damage that 1 R would cause. The rem and the rad are roughly equivalent in most cases.

RADIATION HAZARDS 2.20

How does the average person receive radiation?

Radiation causes ionization and the breaking of chemical bonds. Living cells are fundamentally altered or destroyed by radiation. Consequently, we should all avoid exposing our bodies to any unnecessary radiation.

There is no general agreement on what level of radiation constitutes a safe exposure. Some scientists believe that a very small exposure, which has no obvious effect on an individual, can produce harmful genetic effects that may appear generations later.

We receive radiation from four sources: (1) natural radiation from cosmic rays and naturally occurring radioactive substances; (2) fallout from the atmospheric testing of nuclear weapons and other devices; (3) radioactive waste from industrial plants and research facilities; and (4) medical procedures that use x rays or nuclear particles. The average person in the United States receives 10 roentgens or less radiation over a lifetime from natural sources, and considerably less from other sources.

Most nations now refrain from atmospheric testing of nuclear devices. Fallout from such testing constitutes a serious problem because several dangerous radioisotopes are produced and spread over the earth. Large amounts of $^{131}_{53}I$ are produced in nuclear explosions. The body readily absorbs this radioisotope, but the short half-life of $^{131}_{53}I$ (8 days) reduces the danger to a few weeks. More dangerous are the radioisotopes strontium-90 ($^{90}_{38}Sr$, with a half-life of 28 years) and cesium-137 ($^{137}_{55}Cs$, with a half-life of 37 years). These isotopes are chemically similar to the essential elements calcium and potassium, respectively, so the body readily absorbs them along with the essential elements, and, they may remain in the body for many years.

SUMMARY

Dalton's atomic theory is adequate for balancing ordinary chemical equations and explaining the physical states of matter and the differences between mixtures and **pure substances.** Bohr's atomic theory gives us a simple model of the structure of atoms, which consist of a **nucleus** that contains **protons** and **neutrons,** with **electrons** occupying shells around the nucleus. Modern **orbital** theory provides the details of the electron structure. **Lewis dot structures** conveniently show the **valence-shell** electron arrangements. Elements whose atoms have identical valence-shell electron arrangements are grouped into families in the **periodic table.** The **ionization energy** of an element depends on whether it is a **metal** or **nonmetal.** The **atomic weight** of an element is the average of the masses of the **isotopes** that make it up. **Radioactive** atoms have unstable nuclei that emit alpha or beta particles along with gamma rays. Each **radioisotope** has a specific **half-life.** Artificial radioisotopes find many practical uses.

Key Terms

Actinide series (Sections 2.10, 2.15)
Alpha (α) **particle** (Section 2.15)
Anion (Section 2.12)
Atom (Section 2.1)
Atomic number (Section 2.6)
Atomic weight (Section 2.13)
Beta (β) **particle** (Section 2.15)
Cation (Section 2.12)
Curie (Section 2.19)
Daughter isotope (Section 2.16)
Electron (Section 2.5)
Erg (Section 2.19)
Gamma (γ) **ray** (Section 2.15)
Gas (Section 2.3)
Half-life (Section 2.17)
Heterogeneous mixture (Section 2.4)
Homogeneous mixture (Section 2.4)
Ionization energy (Section 2.11)
Isotopes (Section 2.13)
Lanthanide series (Sections 2.10, 2.14)
Lewis dot structure (Section 2.9)
Liquid (Section 2.3)
Mass number (Section 2.13)
Mass spectrograph (Section 2.13)
Metal (Section 2.10)
Molecule (Section 2.1)
Neutron (Section 2.5)
Noble gases (Section 2.10)

Nonmetal (Section 2.10)
Nucleus (Section 2.5)
Orbital (Section 2.7)
Periodic table (Section 2.10)
Proton (Section 2.5)
Pure substance (Section 2.4)
Quantum level (Section 2.7)
Rad (Section 2.19)
Radioactive elements (Section 2.15)
Radioisotope (Section 2.16)
Rare-earth elements (Section 2.14)
Rem (Section 2.19)
Representative elements (Section 2.10)
Roentgen (Section 2.19)
Solid (Section 2.3)
Solution (Section 2.4)
Spin (Section 2.7)
Transition elements (Section 2.10)
Unified atomic mass unit (u) (Sections 2.5, 2.13)
Valence shell (Section 2.9)

ADDITIONAL PROBLEMS

2.20 Balance the following equations.
(a) $C_2H_4 + O_2 \rightarrow CO_2 + H_2O$
(b) $Na + F_2 \rightarrow NaF$
(c) $CO + O_2 \rightarrow CO_2$
(d) $Fe + Cl_2 \rightarrow FeCl_3$
(e) $H_2O_2 \rightarrow H_2O + O_2$
(f) $Na + H_2O \rightarrow NaOH + H_2$

2.21 Distinguish between a heterogeneous mixture and a solution.

2.22 What is a pure substance?

2.23 Draw Bohr electron structures for the following atoms.
(a) Be
(b) C
(c) F
(d) Mg
(e) Al
(f) S
(g) Ar

2.24 Write out the electron arrangements for the atoms of Problem 2.23 by indicating the orbitals ($1s^2 2s^2$ . . .).

2.25 Indicate the electron structures of the atoms of Problem 2.23 by using arrows to represent electron spin.

2.26 Write Lewis dot structures for the atoms of Problem 2.23.

2.27 Write Lewis dot structures for the following atoms.

(a) Rb (b) Te
(c) I (d) In
(e) Ba (f) P
(g) Si (h) Xe

2.28 List the electrical charge and mass in u for each of the following.

(a) proton (b) neutron
(c) electron (d) alpha particle
(e) beta particle (f) gamma ray

2.29 Arrange the following atoms in order of increasing ionization energy.

(a) C (b) F
(c) Na (d) Mg
(e) O (f) Ne

2.30 Give the number of protons, the number of neutrons, and the mass number for each of the following atoms.

(a) $^{19}_{9}F$ (b) $^{12}_{6}C$
(c) $^{16}_{8}O$ (d) $^{32}_{16}S$
(e) $^{13}_{6}C$ (f) $^{17}_{8}O$
(g) $^{32}_{15}P$ (h) $^{18}_{8}O$
(i) $^{31}_{15}P$ (j) $^{14}_{6}C$

2.31 Group the list of atoms in Problem 2.30 into sets of isotopes.

2.32 Complete the following table.

Atom or ion	Atomic number	Neutrons	Mass number	Electrons
Si	______	______	28	______
Br^-	______	45	______	______
______	11	12	______	10
______	83	______	209	80
S^{2-}	______	16	______	______

2.33 The atomic weight of chlorine is 35.5. Ordinary chlorine consists of two isotopes, $^{35}_{17}Cl$ and $^{37}_{17}Cl$. Of every 100 chlorine atoms, how many of each isotope are there? [Hint: let Y = number of atoms of $^{37}_{17}Cl$ and $(100 - Y)$ = number of atoms of $^{35}_{17}Cl$.]

2.34 Balance the following nuclear reactions by adding the necessary symbols.

(a) $^{40}_{19}K \rightarrow {}^{0}_{-1}e +$ (b) $^{31}_{15}P + {}^{1}_{0}n \rightarrow$
(c) $^{32}_{15}P \rightarrow {}^{32}_{16}S +$ (d) $^{224}_{88}Ra \rightarrow {}^{4}_{2}He +$

2.35 What is the half-life of a radioisotope if three-fourths of it has decayed after 2 years?

2.36 Two radioisotopes are about equally useful for a particular medical diagnosis, except that one of them has a short half-life and the other a long half-life. Which one do you think would be preferred? Why?
2.37 Name several specific uses for radioisotopes.
2.38 How much energy is absorbed by 1.0 g of muscle tissue that is exposed to 1.0×10^4 rad?
2.39 Why are the radioisotopes $^{90}_{38}Sr$ and $^{137}_{55}Cs$ particularly dangerous?
***2.40** The element hydrogen exists in the form of diatomic molecules, H_2. With a boiling point of $-253°C$, hydrogen is almost always encountered as a gas. Some periodic tables list hydrogen in group I (representative elements) because the hydrogen atom has one electron in its valence shell. Criticize the inclusion of hydrogen in this chemical family.
***2.41** According to Figure 2.6, the difference in energy between a 1*s* and 2*s* orbital is much greater than the difference between a 2*s* and 2*p* orbital. Suggest an explanation.
***2.42** In nature, there exist three radioactive series in which a naturally occurring radioisotope decays into a daughter, which is itself radioactive and which decays to give another radioactive daughter, and so on until a stable nucleus is reached. One of these series begins with $^{238}_{92}U$ and ends with the stable lead isotope $^{206}_{82}Pb$. Calculate the total number of alpha and beta particles that are emitted in this series.
***2.43** As a general rule, the difference in mass between isotopes does not cause compounds of the isotopes to have any detectable differences in chemical or physical properties. However, there are detectable differences between the common hydrogen isotope, ^1_1H, and the deuterium isotope, ^2_1H (also symbolized as D). For example, the boiling point of H_2 is $-252.8°C$ and that of D_2 is $-249.5°C$. Why do hydrogen and deuterium have such different properties, when other isotopes do not? Estimate the boiling point of the compound HD.

Problem-Solving Hints

If you have trouble solving any of the additional problems, refer to the sections listed next to the problem numbers.

2.20 2.2	**2.21** 2.4	**2.22** 2.4
2.23 2.5	**2.24** 2.8	**2.25** 2.8
2.26 2.9	**2.27** 2.9	**2.28** 2.5, 2.15
2.29 2.11	**2.30** 2.13	**2.31** 2.13
2.32 2.6, 2.13	**2.33** 2.13	**2.34** 2.16
2.35 2.17	**2.36** 2.17, 2.18, 2.20	**2.37** 2.18
2.38 2.19	**2.39** 2.20	**2.40** 2.10
2.41 2.7, 2.8	**2.42** 2.16	**2.43** 2.13

FOCUS 1 Isotopes of Hydrogen

In nature, hydrogen is a mixture of three isotopes. The most abundant isotope is ^1_1H. Another isotope, ^2_1H, called deuterium, represents only one atom out of every 10^4. Neither hydrogen (^1_1H) nor deuterium (also written D) is radioactive. The third isotope, ^3_1H (also written T), called tritium, is radioactive and undergoes beta decay with a half-life of 12 years.

$$^3_1H \rightarrow {}^{0}_{-1}e + {}^3_2He$$

There is very little tritium in natural hydrogen. In ordinary water, less

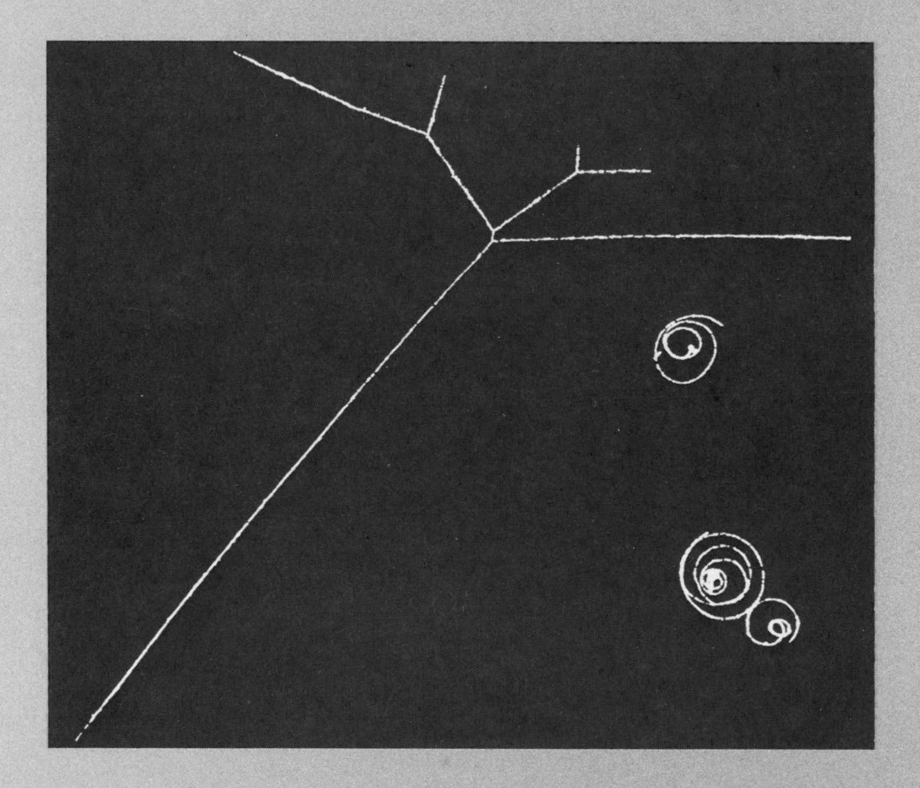

FIGURE 1 At the lower left, a high-energy proton enters a chamber filled with liquid hydrogen. The proton collides with other protons, producing trails that can be photographed. The small circular tracks are produced by relatively low-energy electrons. (Brookhaven National Laboratory)

than one hydrogen atom out of 10^{10} is tritium. However, tritium has been prepared artificially by nuclear reaction between deuterium atoms:

$$^{2}_{1}H + ^{2}_{1}H \rightarrow ^{1}_{1}H + ^{3}_{1}H$$

and also by the action of neutrons on lithium-6:

$$^{1}_{0}n + ^{6}_{3}Li \rightarrow ^{4}_{2}He + ^{3}_{1}H$$

A water molecule that contains deuterium atoms instead of hydrogen atoms is called heavy water. Ordinary water contains enough heavy water, D_2O, for the latter compound to be separated easily. When large amounts of water are electrolyzed to produce hydrogen and oxygen, the concentration of D_2O increases in the water that remains. The D_2O is not electrolyzed to D_2 and O_2 as readily as H_2O is to H_2 and O_2.

The deuterium atom is much heavier than the hydrogen atom. As a result, compounds of these isotopes have different properties (Table 1).

TABLE 1 Some Properties of H_2O and D_2O

	H_2O	D_2O
Melting point, °C	0.00	3.82
Boiling point, °C	100.00	101.42
Density, g/mL (20°C)	0.9982	1.1059

FOCUS 2 Radiocarbon Dating

About 99 out of every 100 carbon atoms are $^{12}_{6}C$ isotopes. One atom in 100 is a $^{13}_{6}C$ isotope. Both of these isotopes are stable. A tiny trace of an unstable isotope, $^{14}_{6}C$ (carbon-14), is found in natural carbon—only about one atom in 10^{12}. This trace amount is present because nitrogen in the upper atmosphere is bombarded by the neutrons that are a component of cosmic rays.

$$^{1}_{0}n + ^{14}_{7}N \rightarrow ^{14}_{6}C + ^{1}_{1}H$$

The small amount of $^{14}_{6}C$ produced by this naturally occurring nuclear reaction is incorporated into CO_2 in the atmosphere. Living plants and animals take in CO_2, including some that contains carbon-14, through respiration and photosynthesis. The amount of $^{14}_{6}C$ in plants and animals remains constant as long as the organism is alive. However, when a plant or animal dies, it stops taking in carbon, and the $^{14}_{6}C$ content per gram of ordinary carbon begins to decrease because of beta decay.

$$^{14}_{6}C \rightarrow ^{0}_{-1}e + ^{14}_{7}N$$

The half-life for this decay is 5,700 years. The amount of $^{14}_{6}C$ per gram of carbon thus decreases by half every 5,700 years after the plant or animal dies. The feeble radioactivity of the decaying carbon-14 can be

FIGURE 1 Radiocarbon dating can be used to establish the approximate age of fossils. Shown here is an extremely rare example of a prehistoric insect wing. (John H. Gerard from Monkmeyer Press Photo Service)

measured, compared to that found per gram of carbon in a living animal or plant, and used to determine how many half-lives the fossil remains or plant product (paper, for example) has passed through. This method, called radiocarbon dating, has proved very useful in dating archeological objects and in testing the authenticity of old paintings and manuscripts.

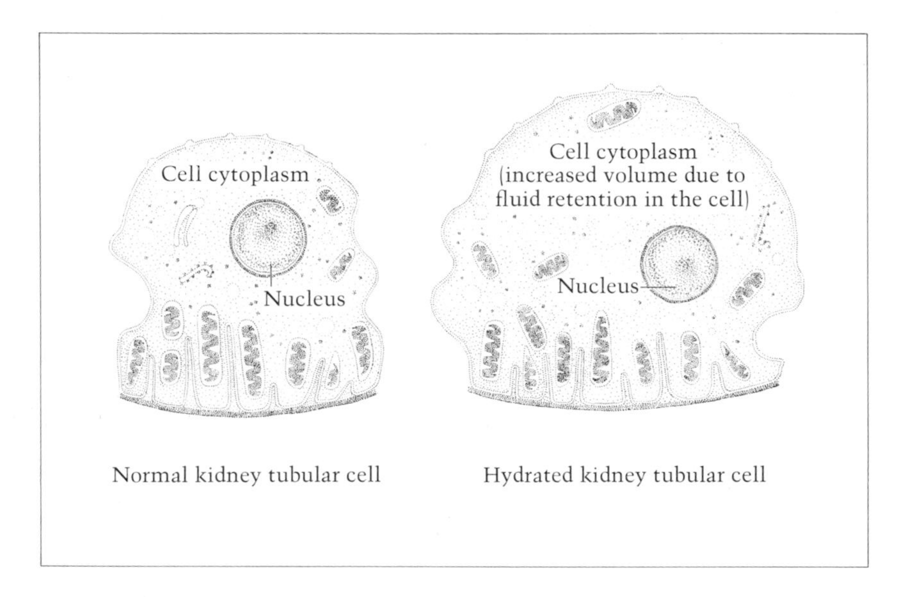

We will consider the formation and the nature of ions in this chapter. Certain ions function as electrolytes in the human body. Electrolytes are chemical substances that conduct an electrical current when dissolved in water. The electrolytes that are found in the protoplasm of the cell serve as a source of inorganic chemicals for various cellular reactions in the body. At the cell membrane, for example, electrolytes can assist in the transmission of electrochemical impulses in nerve and muscle fibers. Within the cell, electrolytes activate chemical reactions necessary for the life of the cell. Sodium, potassium, and chloride ions and the compound, carbon dioxide (which is an indicator of the level of bicarbonate ion), are among the common electrolytes at work in our bodies.

A diuretic is a substance that increases the rate at which urine is eliminated from the body. The accumulation of excess water in body cells can result in problems that range from general confusion to convulsions, leading, finally, to death. Such complications are avoided by using diuretics to reduce the total amount of fluid in the body to appropriate levels.

It is not pure coincidence that all valuable diuretics contain compounds that cause the loss of sodium ions along with the loss of water. If water alone is removed from the total volume of body fluids, the body's system of checks and balances responds by excreting a hormone that counteracts the effect of the diuretic, thus canceling the loss of water. However, if sodium ions are

3

CHEMICAL COMBINATION

lost along with the water, this system does not come into play, and the fluid loss is successful as long as the diuretic is administered.

The major side effect of the extended use of a diuretic drug is the loss of a critical amount of sodium ions. Nausea, weakness, and giddiness are the initial symptoms of sodium ion loss, followed by eventual circulatory collapse and death. Thus electrolyte balance is crucial to life processes. The therapeutic use of diuretic drugs can be life-saving, but only if there is constant monitoring of a patient's electrolyte levels.

[Illustration: Diuretic drug therapy can be useful in diminishing fluid retention in human body cells.]

We know that elements combine to form literally millions of compounds. We will be concerned with the properties of some of these compounds, particularly those that play a role in the chemistry of life.

Before we can understand the reasons for the different properties of compounds, we must see how atoms are able to unite chemically and form a compound. They can combine in two different ways. In an ionic compound, the atoms of one element donate electrons to the atoms of another element. In a covalent compound, the atoms share electrons. The fact that some atoms donate electrons, others accept electrons, and still others share them can be expressed fairly simply. Many atoms react to acquire eight electrons in their outermost shells.

An atom with eight electrons in the outermost shell is exceptionally stable, and evidence for this stability is found in the chemical family called the noble gases.

3.1 NOBLE GASES

When we consider all of the elements and their properties, one chemical family is outstanding by its *lack* of reactivity. The noble gases (Figure 2.7) are so unreactive that the first three members of the family—helium, neon, and argon—have thus far defied all attempts to make them form compounds with any other element. The heavier noble gases—krypton, xenon, and radon—do form compounds, but only with a few elements such as fluorine, chlorine, and oxygen.

The noble gases exist as single atoms (He, Ne, Ar, Kr, Xe, Rn), in contrast with other elemental gases such as hydrogen, nitrogen, oxygen, fluorine, and so on, which exist as diatomic molecules (H_2, N_2, O_2, F_2).

What is different about the noble gases?

ELECTRON CONFIGURATIONS OF THE NOBLE GASES 3.2

When atoms react, there is always some rearrangement of their valence-shell electrons. Since noble-gas atoms do not readily react, their valence-shell electrons must be very resistant to rearrangement. Table 3.1 shows Lewis dot structures for the noble gases. Notice that, except for helium, all the noble-gas atoms have eight valence-shell electrons. The helium atom has only two electrons, but these fill its valence shell.

We can gain some insight into the stability of the electron arrangements of the noble-gas atoms by remembering that ionization energy increases as we go from left to right across the periodic table. We find the noble gases on the right side of the periodic table, so their ionization energies are high, meaning that they do not readily give up electrons. Noble-gas atoms also show no tendency to add electrons. Consequently, the arrangement of eight valence-shell electrons (two in the case of He) is very stable.

TABLE 3.1 Lewis Dot Structures of the Noble-Gas Atoms

Noble gas	Atomic number	Lewis dot structure
He	2	He:
Ne	10	$:\ddot{\underset{..}{Ne}}:$
Ar	18	$:\ddot{\underset{..}{Ar}}:$
Kr	36	$:\ddot{\underset{..}{Kr}}:$
Xe	54	$:\ddot{\underset{..}{Xe}}:$
Rn	86	$:\ddot{\underset{..}{Rn}}:$

OCTET RULE 3.3

Early in the twentieth century, chemists realized that the noble-gas electron configurations serve as a useful guide in predicting the way

How is the valence-shell configuration of noble-gas atoms related to that of other atoms?

other atoms will react. We can state a general rule: atoms usually react to achieve a valence-shell electron configuration that is the same as that of a noble-gas atom. Except for helium, the valence shells of all the noble-gas atoms contain eight electrons. The general rule has thus been shortened to what is called the **octet rule:** *atoms usually react to achieve a valence shell of eight electrons.*

The atoms of hydrogen and lithium (Li) react to attain a valence shell of two electrons, and so technically they do not follow the octet rule. They do, however, follow the more general rule, because they achieve a valence shell that is the same as that of a noble-gas atom—helium. Sometimes atoms violate both the octet rule and the more general rule, but for most of the compounds that we will discuss in this book, the octet rule is valid and will be very useful.

Problem 3.1 Why is helium in group VIII of the periodic table, when it has only two valence-shell electrons and the other members of group VIII have eight electrons in their valence shells?

Problem 3.2 Consider the way ionization energies change within a chemical family and explain why the lighter noble gases (helium, neon, and argon) do not react even with fluorine.

3.4 IONIC COMPOUNDS

What happens when a metal and a nonmetal react?

A sodium atom has only one electron in its valence shell.

$$Na\cdot$$

The sodium atom cannot take on seven more electrons to fill its valence shell. That would produce an anion with a charge of -7. Such a high negative charge would make the ion unstable, because all those electrons would strongly repel one another. On the other hand, if the sodium atom loses its valence-shell electron, it becomes a cation with only a single positive charge.

$$Na\cdot \rightarrow Na^+ + e^-$$

The valence shell of this sodium cation has eight electrons, as a Bohr-theory representation shows.

A neutral sodium atom, however, is stable. It will not lose an electron to form a sodium ion unless some other atom is present to take the electron. We have already seen (Section 2.12) that nonmetallic atoms can take on electrons to form anions. For example, a chlorine atom will

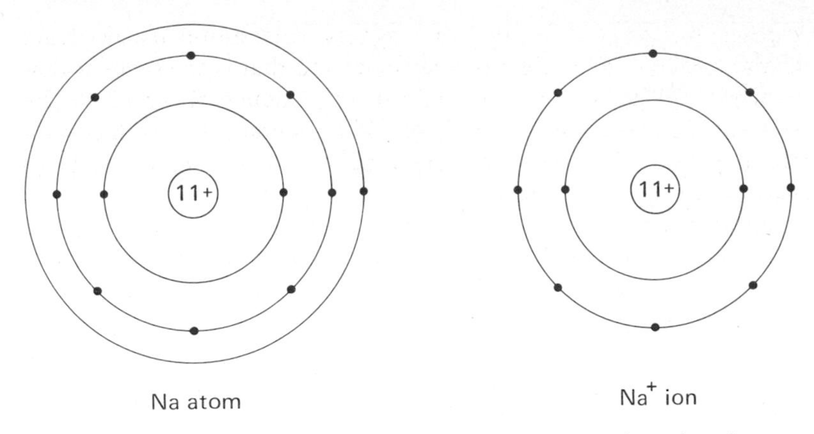

take on one electron, becoming a chloride ion with eight electrons in the valence shell.

$$:\ddot{\underset{..}{Cl}}\cdot + e^- \rightarrow :\ddot{\underset{..}{Cl}}:^-$$

Sodium, which can lose an electron, and chlorine, which can gain one, will thus react to give sodium chloride.

$$2Na + Cl_2 \rightarrow 2NaCl$$

Each sodium atom has transferred an electron to a chlorine atom. The resulting positive and negative ions strongly attract one another (opposite electrical charges attract) and arrange themselves in an orderly array of alternating sodium and chloride ions (Figure 3.1). Compounds of this type, which consist of positive and negative ions, are called **ionic compounds.** An ionic compound is commonly called a **salt.** Sodium

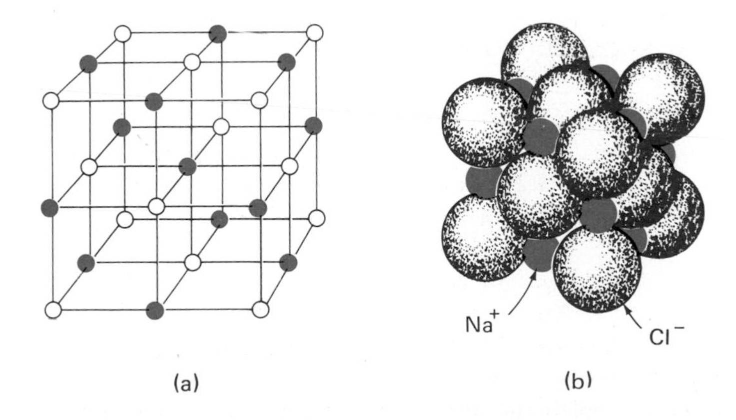

FIGURE 3.1 Sodium chloride consists of an orderly array of alternating sodium and chloride ions. (a) Relative positions of Na^+ and Cl^-. (b) Relative sizes of Na^+ and Cl^-.

TABLE 3.2 Melting and Boiling Points for Some Ionic and Covalent Compounds

Compound	Type	Melting point, °C	Boiling point, °C
Aluminum oxide, Al_2O_3	Ionic	2000	2,210
Potassium fluoride, KF	Ionic	880	1,500
Sodium chloride, NaCl	Ionic	801	1,413
Magnesium chloride, $MgCl_2$	Ionic	712	1,412
Lithium chloride, LiCl	Ionic	614	1,360
Water, H_2O	Covalent	0	100
Ammonia, NH_3	Covalent	−78	−33
Hydrogen bromide, HBr	Covalent	−86	−67
Chlorine, Cl_2	Covalent	−102	−35
Hydrogen chloride, HCl	Covalent	−111	−85

chloride is common table salt. We can state another general rule: *When a metal reacts with a nonmetal, they produce an ionic compound.*

The strong attraction between the positive and negative ions causes most ionic compounds to have high melting and boiling points. The melting or boiling of a compound requires separation of the individual particles. When the individual particles are oppositely charged ions, this takes a lot of energy. Table 3.2 gives melting and boiling points for some ionic (and covalent) compounds.

Problem 3.3 Figure 3.1(b) shows the relative sizes of Na^+ ions and Cl^- ions. Why is the Na^+ ion smaller than the Cl^- ion?

3.5 FORMULAS OF IONIC COMPOUNDS

The octet rule helps us predict how elements will combine to form compounds. For example, a magnesium atom has two electrons in its valence shell, so we know that it has to lose two electrons in order to form an octet.

$$Mg\colon \rightarrow Mg^{2+} + 2e^-$$

A chlorine atom has seven electrons in its valence shell, so it needs to add one electron to complete the octet. Consequently, when magnesium reacts with chlorine, two chlorine atoms are needed to accept the two electrons lost by the magnesium atom. The formula of magnesium chloride is therefore $MgCl_2$.

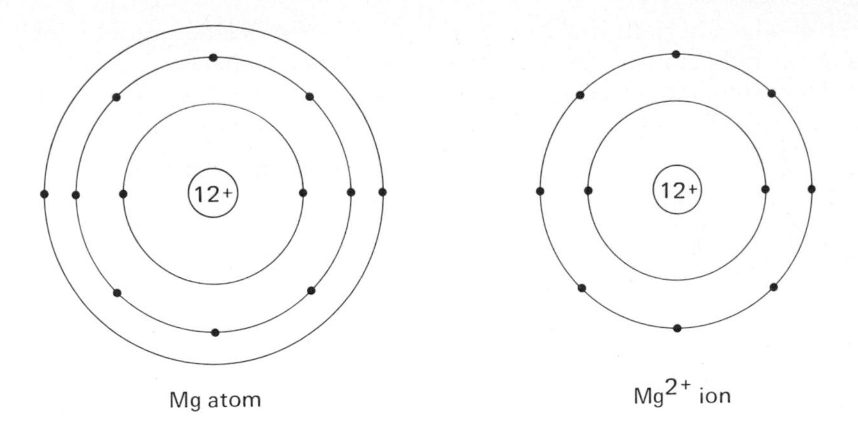

Another way of arriving at the correct formula is to consider the charges on the two ions and realize that the positive and negative charges must balance. The magnesium atom loses two electrons in becoming an ion, so it has two positive charges. The chlorine atom gains an electron, so the chloride ion has a single negative charge. Each doubly charged Mg^{2+} ion requires two singly charged Cl^- ions. This means that there must be twice as many Cl^- ions as Mg^{2+} ions, so the compound they form must have the formula $MgCl_2$. The equation for the reaction between magnesium and chlorine is

$$Mg + Cl_2 \rightarrow MgCl_2$$

Let us apply these ideas to some more examples. Consider sodium sulfide (the product of sodium and sulfur). Each sodium atom loses an electron to form an octet.

$$Na\cdot \rightarrow Na^+ + e^-$$

Each sulfur atom must gain two electrons to complete its octet.

$$:\ddot{S}\cdot + 2e^- \rightarrow :\ddot{S}:^{2-}$$

There must be two Na^+ ions for every S^{2-} ion in order for the charges to balance, so the formula of sodium sulfide is Na_2S. The equation for the reaction is thus

$$2Na + S \rightarrow Na_2S$$

Consider magnesium sulfide (the product of magnesium and sulfur). The Mg^{2+} ion and the S^{2-} ion balance when there are equal numbers of the ions. The formula of magnesium sulfide is therefore MgS.

Formulas for ionic compounds express only the relative number of positive and negative ions that are necessary for the charges to balance. There are no individual molecules of ionic compounds. A positive ion is surrounded by negative ions and the negative ions are surrounded by positive ions. We cannot say that a particular positive ion belongs to a particular negative ion.

Problem 3.4 Predict the formulas of the ionic compounds formed from the following elements.
(a) sodium and fluorine (b) magnesium and oxygen
(c) lithium and chlorine (d) aluminum and fluorine

Problem 3.5 Write balanced equations for the reactions between
(a) Al and O_2 (b) Li and N_2
(c) Na and F_2

3.6 NOMENCLATURE OF IONIC COMPOUNDS

Ionic compounds are simple to name. We give the name of the cation first, followed by the name of the anion. We have already considered several examples: sodium chloride, NaCl; magnesium chloride, $MgCl_2$; and sodium sulfide, Na_2S. Notice that the name of an ionic compound does not give the relative number of ions in the formula. We simply have to know how many of each ion are required to balance the charges.

Not all ions are single atoms. Some consist of several atoms and are called polyatomic ions. The sulfate ion is SO_4^{2-}. Sodium sulfate is, therefore, Na_2SO_4, and magnesium sulfate is $MgSO_4$. An important polyatomic cation is the ammonium ion, NH_4^+. Ammonium sulfate has the formula $(NH_4)_2SO_4$. We enclose the NH_4^+ in parenthesis to show

TABLE 3.3 Some Common Positive and Negative Ions

Positive ions				Negative ions			
NH_4^+	Ammonium	Li^+	Lithium	F^-	Fluoride	Cl^-	Chloride
K^+	Potassium	Na^+	Sodium	Br^-	Bromide	I^-	Iodide
Ca^{2+}	Calcium	Mg^{2+}	Magnesium	OH^-	Hydroxide	NO_3^-	Nitrate
Al^{3+}	Aluminum			O^{2-}	Oxide	S^{2-}	Sulfide
				CO_3^{2-}	Carbonate	SO_4^{2-}	Sulfate
				PO_4^{3-}	Phosphate		

that the subscript 2 applies to the entire ion—the nitrogen and all four hydrogens.

Table 3.3 lists some common cations and anions with their names. These should be memorized.

Problem 3.6 Name the following compounds.

(a) $CaSO_4$ (b) $(NH_4)_2CO_3$
(c) Al_2O_3 (d) NaBr
(e) $Mg(OH)_2$ (f) Li_2SO_4
(g) $Al(NO_3)_3$ (h) KI
(i) $MgBr_2$ (j) CaS
(k) NH_4NO_3 (l) Li_2O
(m) NaOH (n) Na_3PO_4

Some atoms can form ions with different charges, especially atoms of the transition elements (Figure 2.7). For example, iron has two cations, Fe^{2+} and Fe^{3+}. Consequently, iron can form more than one compound with a given anion. With chloride, for instance, it can form both $FeCl_2$ and $FeCl_3$. In naming these compounds, the charge on the cation is indicated by Roman numerals: $FeCl_2$ is iron(II) chloride and $FeCl_3$ is iron(III) chloride.

Problem 3.7 Name the oxides of iron, FeO and Fe_2O_3.
Problem 3.8 Write formulas for cobalt(II) bromide and cobalt(III) fluoride.

COVALENT COMPOUNDS 3.7

What happens when two nonmetals react?

When two nonmetallic atoms react, electrons are not transferred, and so ions are not formed. For example, two chlorine atoms cannot react by any transfer of electrons to complete the octets. There are two reasons. First, nonmetallic atoms have high ionization energies, so it is unlikely that a nonmetallic atom like chlorine will completely lose an electron. Furthermore, the loss of an electron by a chlorine atom would leave a positive ion with only six electrons in its valence shell.

However, there is a way that two chlorine atoms can combine and complete the octet for both atoms. The atoms can *share* a pair of electrons and form a chlorine molecule, as follows. Two chlorine atoms have seven electrons each in their valence shells.

$$:\ddot{\underset{..}{Cl}}\cdot \quad \cdot\ddot{\underset{..}{Cl}}:$$

The chlorine atoms share two electrons, and thereby each gets eight electrons in its valence shell.

$$:\underset{..}{\overset{..}{Cl}}:\underset{..}{\overset{..}{Cl}}:$$

The sharing of electrons by atoms is called **covalent bonding.**

Each atom in a chlorine molecule has eight electrons in its valence shell. The octet rule is thus satisfied for both atoms. Moreover, the two chlorine atoms that share the electrons belong to each other, in a way that ions in an ionic compound do not. The Cl_2 molecule is an individual entity, while there is really no such thing as an ionic molecule. We can state a general rule: *when nonmetallic atoms react with each other, they produce* ***covalent compounds.***

Chlorine atoms that share electrons are tightly bonded, but there is little interaction between individual Cl_2 molecules. It takes only a relatively small amount of energy to separate the molecules from each other. Therefore, covalent compounds usually have low melting points and low boiling points. (See Table 3.2.)

3.8 FORMULAS OF COVALENT COMPOUNDS

The octet rule is helpful in predicting the formulas of covalent compounds. Let us consider the combination of fluorine and nitrogen. The best way to start is to write their Lewis dot structures.

$$:\underset{..}{\overset{..}{F}}\cdot \qquad \cdot\underset{\cdot}{\overset{..}{N}}\cdot$$

The fluorine atom needs only one more electron to complete the octet, but the nitrogen atom needs three electrons. Both of these needs can be met if three fluorine atoms combine with one nitrogen atom.

$$\begin{array}{c} :\underset{..}{\overset{..}{F}}:\overset{..}{N}:\underset{..}{\overset{..}{F}}: \\ :\underset{..}{\overset{..}{F}}: \end{array}$$

The formula of this compound is thus NF_3.

Problem 3.9 Write a balanced chemical equation for the formation of NF_3 from nitrogen and fluorine.

Hydrogen is an important nonmetal that combines with most of the other nonmetals to form covalent compounds. The hydrogen atom, having only one electron, tends to share one with another nonmetal to complete its valence shell (which can take only two electrons). Thus hydrogen, H·, and chlorine, :Cl·, combine to give hydrogen chloride.

```
   ..
H:Cl:
   ..
```

Two H· and one oxygen, :Ö·, give water.

```
 ..
:O:H
 ..
 H
```

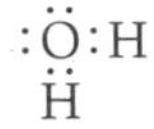

Three H· and one nitrogen, ·N·, give ammonia.

```
  ..
H:N:H
  ..
  H
```

Four H· and one carbon, ·C·, give methane.

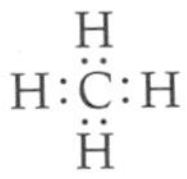

```
  H
  ..
H:C:H
  ..
  H
```

Notice that, in the above compounds, the hydrogen atom always has two electrons in its valence shell and the other atom always has eight.

Problem 3.10 Write Lewis dot structures for the compounds that would result from the reactions between the following pairs of elements.
(a) hydrogen and fluorine (b) nitrogen and chlorine
(c) hydrogen and sulfur

Problem 3.11 Write balanced chemical equations for reactions between the following pairs of elements.
(a) hydrogen and fluorine (b) nitrogen and chlorine
(c) nitrogen and hydrogen

ELECTRONEGATIVITY 3.9

Are the electrons shared equally by covalently bonded atoms?

If two identical nonmetallic atoms share electrons, we certainly would expect the sharing to be equal. We find that such is the case in the molecules H_2, Cl_2, and F_2, and in other molecules where identical atoms share electrons.

```
        ..  ..     .. ..
H:H    :Cl:Cl:    :F:F:
        ..  ..     .. ..
```

By the same token, a little thought makes us suspect that when two different nonmetallic atoms share electrons, the sharing probably will not be equal. One atom is likely to attract the electrons a little more than the other. Because of this uneven attraction, the bonding electrons will be a little closer on average to the atom with the greater attraction.

$H:\ddot{\underset{..}{O}}:$ H	$H:\ddot{\underset{..}{N}}:H$ H	$H:\ddot{\underset{..}{Cl}}:$	$:\ddot{\underset{..}{Cl}}:\ddot{\underset{..}{F}}:$
water	ammonia	hydrogen chloride	chlorine fluoride

This is the situation for molecules such as those shown above and, in fact, for most molecules where nonidentical atoms share electrons.

The **electronegativity** of an atom is a measure of how well it attracts electrons when it is covalently bonded with other atoms. If two atoms form a covalent bond, the atom of higher electronegativity will attract the electrons more strongly and thus will pull the bonding electrons closer to it.

Here is a list of some important elements arranged in order of increasing electronegativity. The symbol ~ means "approximately equal."

$$H \sim P < C \sim S \sim I < Br < N \sim Cl < O < F$$
$$\xrightarrow{\text{increasing electronegativity}}$$

This relative order will be sufficient for our purposes, although many attempts have been made to assign precise numerical values to the

		H 2.1				
Li 1.0	Be 1.5	B 2.0	C 2.5	N 3.0	O 3.5	F 4.0
Na 0.9	Mg 1.2	Al 1.5	Si 1.8	P 2.1	S 2.5	Cl 3.0
K 0.8	Ca 1.0	Ga 1.6	Ge 1.8	As 2.0	Se 2.4	Br 2.8
Rb 0.8	Sr 1.0	In 1.7	Sn 1.8	Sb 1.9	Te 2.1	I 2.5
Cs 0.7	Ba 0.9	Tl 1.8	Pb 1.8	Bi 1.9	Po 2.0	At 2.2

FIGURE 3.2 This chart shows Pauling's electronegativity values for most of the representative elements. The larger the number, the more electronegative the atom.

electronegativities of atoms. Linus Pauling made the first such attempt in 1932. Figure 3.2 gives Pauling's electronegativity values for most of the representative elements. The higher the electronegativity value, the more electronegative the atom.

POLAR COVALENT BONDS 3.10

What happens when atoms of differing electronegativity share electrons?

In a molecule such as water, atoms of different electronegativities share electrons. The oxygen atom is more electronegative than the hydrogen atoms. As a result, the oxygen atom has a greater share of the bonding electrons. Thus it has a little more negative charge than the hydrogen atoms. We can illustrate this situation by using arrows that point in the direction of greater electron attraction.

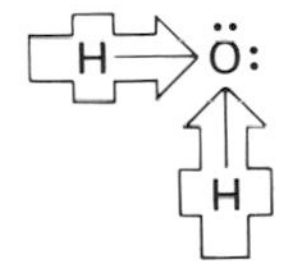

Another way to indicate that the oxygen atom is a little more negative than the hydrogen atoms is with the Greek letter δ (delta).

$$\begin{array}{l} \quad\;\; \delta^- \\ \mathrm{H:\ddot{O}:} \\ \delta^+\mathrm{\ddot{H}} \end{array}$$

The δ^- means a partial negative charge and the δ^+ means a partial positive charge.

Bear in mind that the water molecule is electrically neutral; there are no ions, just a small transfer of negative charge away from the hydrogen atoms toward the oxygen atom. Such a transfer always occurs when atoms with different electronegativities form a covalent bond. This is called a **polar covalent bond.**

Problem 3.12 For each of the following, predict whether or not the covalent bonds are polar. For those compounds with polar covalent bonds, indicate the direction of electron attraction by δ^- and δ^+.

(a) H:F: (b) H:N:H over H (c) H:H (d) :Br:Br: (e) :F:N:F: over :F:

3.11 NOMENCLATURE OF BINARY COVALENT COMPOUNDS

A binary compound contains atoms of only two elements. Binary covalent compounds are easy to name. More complicated covalent compounds are not so easy and require special rules. We will cover those rules later in the book when we come to organic chemistry.

In naming a binary covalent compound, the name of the less electronegative element is given first, followed by the name of the more electronegative element, which is named as if it were an anion (see Section 3.6).

We indicate the number of atoms of one or both elements by Greek prefixes: *mono-* (one), *di-* (two), *tri-* (three), *tetra-* (four), *penta-* (five), and *hexa-* (six).

Here are some examples of the use of these prefixes.

CO	carbon monoxide	NF_3	nitrogen trifluoride
CO_2	carbon dioxide	NO_2	nitrogen dioxide
SF_4	sulfur tetrafluoride	N_2O_4	dinitrogen tetroxide
SF_6	sulfur hexafluoride		

The prefix *mono-* is usually omitted, particularly when two elements can form only one compound. For example, HCl is called hydrogen chloride.

A few binary covalent compounds have special names that must simply be remembered. The two most common examples are water, H_2O, and ammonia, NH_3.

Problem 3.13 Name the following compounds.

(a) SO_2 (b) SO_3
(c) HBr (d) PCl_5
(e) NCl_3 (f) N_2O_3
(g) PF_3 (h) OF_2
(i) NH_3 (j) HI

3.12 OXIDATION-REDUCTION

How do we know if an atom has been oxidized or reduced?

The transfer of electrons from one atom to another in a chemical reaction is called oxidation–reduction, or **redox** for short. Atoms that lose electrons are said to undergo **oxidation.** Atoms that gain electrons are said to experience **reduction.**

We have seen how the reaction between a metallic atom and a nonmetallic atom results in complete electron transfer to form positive and negative ions. The metallic atoms lose electrons and are thus oxidized. The nonmetallic atoms gain electrons and are thus reduced. For example, when sodium, a metal, and chlorine, a nonmetal, react to form sodium chloride, the sodium atoms lose electrons and the chlorine atoms gain electrons. Sodium is oxidized and chlorine is reduced.

$$Na \rightarrow Na^+ + e^- \qquad \text{(oxidation)}$$

$$e^- + Cl \rightarrow Cl^- \qquad \text{(reduction)}$$

The overall reaction is a redox reaction in which sodium is oxidized and chlorine is reduced.

$$2Na + Cl_2 \rightarrow 2NaCl$$

Problem 3.14 Which atoms are oxidized and which are reduced in the following reactions?
(a) $Mg + Cl_2 \rightarrow MgCl_2$ (b) $Na + F_2 \rightarrow 2NaF$
(c) $4Li + O_2 \rightarrow 2Li_2O$

Reactions that produce covalent compounds can also be classified as redox reactions. Think about the reaction

$$2H_2 + O_2 \rightarrow 2H_2O$$

Electrons are not completely transferred because the hydrogen atoms share electrons with the oxygen atom. However, oxygen is more electronegative than hydrogen, so there is a *partial* transfer of electrons from the hydrogen atoms to the oxygen atom. We therefore say that the hydrogen atoms are oxidized and the oxygen atoms are reduced.

Oxygen is very electronegative (second only to fluorine), so when oxygen reacts with other substances it almost always results in the oxidation of the other element and the reduction of the oxygen.* Hydrogen, by contrast, has a very low electronegativity. When a hydrogen atom reacts with other substances, it usually loses part of its electronic charge and is oxidized, while the other atom gains the charge, to become reduced.

Redox reactions are very common. In later chapters, as you study organic reactions and biochemical processes, you will run into redox reactions time and time again.

---

*A clear exception would be the reaction between fluorine and oxygen, $2F_2 + O_2 \rightarrow 2OF_2$. In this case, oxygen is oxidized and fluorine is reduced.

Problem 3.15 Which atoms are oxidized and which are reduced in the following reactions?
(a) $H_2 + Cl_2 \rightarrow 2HCl$ (b) $4Cu + O_2 \rightarrow 2Cu_2O$
(c) $H_2 + F_2 \rightarrow 2HF$ (d) $S + O_2 \rightarrow SO_2$
(e) $C_2H_4 + H_2 \rightarrow C_2H_6$

3.13 QUANTITATIVE CONSIDERATIONS

What does a balanced chemical equation tell us?

When we write a balanced chemical equation, we can predict the amounts of the reactants and products that will be involved. Let us again look at a familiar reaction.

$$2Na + Cl_2 \rightarrow 2NaCl$$

We have already discussed what such an equation tells us about the relative numbers of atoms and molecules (Section 2.2). Specifically, we can say that two sodium atoms react with one chlorine molecule to produce two formula units of sodium chloride. A **formula unit** of sodium chloride is one sodium ion and one chloride ion. This term is used instead of molecule because there are no discrete molecules of an ionic compound (Section 3.5).

We could just as well say that two dozen sodium atoms and one dozen chlorine molecules give two dozen formula units of sodium chloride, or two million sodium atoms and one million chlorine molecules produce two million formula units of sodium chloride. As a matter of fact, the coefficients in front of the substances in any chemical equation tell the *relative* numbers of the different species that react and that are produced in the reaction. Of course, in any actual reaction between reasonable amounts of sodium and chlorine, literally billions upon billions of sodium atoms and chlorine molecules react. The point is that the above equation tells us that there will always be twice as many sodium atoms required as chlorine molecules, and that the number of formula units of sodium chloride produced will equal the number of sodium atoms that reacted.

Problem 3.16 In a reaction between sodium and chlorine, 6.0×10^{20} formula units of NaCl were produced. How many sodium atoms and chlorine molecules reacted?

THE MOLE CONCEPT 3.14

We cannot actually count the numbers of atoms and molecules in a sample of a substance. What we can do, though, is weigh the sample. Once we have the weight, a very useful concept called the **mole** will tell us how many units of that substance we have. (The units are atoms if the substance is an element, molecules if it is a covalent compound, and formula units if it is an ionic compound.)

Of what use is a mole?

A mole stands for a very large number of units—6.02×10^{23}. This is known as **Avogadro's number,** after the nineteenth-century Italian scientist Amedeo Avogadro. You may find it helpful to compare the word "mole" with some other words we use to stand for certain numbers. For example, when we say "a dozen eggs," we mean twelve eggs; "a gross of pajamas" means 144 pajamas; "a ream of paper" is 500 sheets of paper. In the same way, when we say "a mole of carbon" we mean 6.02×10^{23} *atoms* of carbon; a mole of water is 6.02×10^{23} *molecules* of H_2O; and a mole of sodium chloride consists of 6.02×10^{23} *formula units* of NaCl.

We know how many units there are in a mole, but how much does a mole weigh? The answer is, moles of different substances have different masses, just as a dozen large eggs weighs more than a dozen medium eggs. To find out how much a mole of a particular substance weighs, you simply take the formula of the substance, add up the atomic weights of all the atoms that appear in that formula, and change the total weight to grams. For example, carbon has the formula C. The elementary unit of carbon is a single carbon atom. The atomic weight of carbon is 12.0, so a mole of carbon weighs 12.0 g. If you have 12.0 g of carbon, you know you have a mole (6.02×10^{23}) of carbon atoms.

The formula for water is H_2O. Hydrogen has an atomic weight of 1.0, and oxygen has an atomic weight of 16.0. The **molecular weight** of a covalent compound like H_2O is the sum of the individual atomic weights. A molecule of H_2O, then, has a molecular weight of 18.0. So a mole of H_2O will weigh 18.0 g. We can thus say that 18.0 g of H_2O contain a mole (6.02×10^{23}) of H_2O molecules.

Sodium chloride, common table salt, has the formula NaCl. Na has an atomic weight of 23.0, and Cl an atomic weight of 35.5. Since NaCl is an ionic compound, it doesn't exist as molecules, so we speak of its **formula weight,** rather than its molecular weight. The formula weight of NaCl is 23.0 plus 35.5, which adds up to 58.5. So a mole of NaCl will weigh 58.5 g. If you have 58.5 g of sodium chloride, you have a mole (6.02×10^{23}) of formula units of NaCl (or you can think of it as a mole of Na^+ ions bonded to a mole of Cl^- ions).

Problem 3.17 What weight of the following substances will contain 1.00 mole of atoms (if elemental), 1.00 mole of molecules (if molecular), or 1.00 mole of formula units (if ionic)?

(a) HCl (b) Cu

(c) KF (d) H_2SO_4

(e) $Al_2(SO_4)_3$

Problem 3.18 What weight of each of the substances in Problem 3.17 will contain 0.10 mole of atoms (if elemental), 0.10 mole of molecules (if molecular), or 0.10 mole of formula units (if ionic)?

Now let's see how moles can help us understand a chemical reaction. Let's take the reaction of sodium and chlorine again.

$$2Na + Cl_2 \rightarrow 2NaCl$$

This equation tells us that 2 moles of sodium atoms will react with 1 mole of chlorine molecules to produce 2 moles of sodium chloride. Or 20 moles of sodium will react with 10 moles of chlorine to produce 20 moles of sodium chloride; or 1 mole of sodium will react with 0.5 mole of chlorine to give 1 mole of sodium chloride. Whenever sodium reacts with chlorine, the ratio of moles of sodium to moles of chlorine to moles of sodium chloride will be 2:1:2.

When we speak of a mole of a substance we are always referring to a specific weight of that substance. So when we say 1 mole of sodium reacts with 0.5 mole of chlorine to produce 1 mole of sodium chloride, it is the same as saying that 23.0 g of sodium (atomic weight 23.0) react with 35.5 g of chlorine (molecular weight 71.0) to produce 58.5 g of sodium chloride (formula weight 58.5).

Moles are useful in solving problems such as: What weight of chlorine is required to react completely with 46.0 g of sodium? Since the atomic weight of sodium is 23.0, 46.0 g of sodium is 2.00 moles of sodium!*

$$46.0\ \cancel{\text{g Na}} \times \frac{1\ \text{mol Na}}{23.0\ \cancel{\text{g Na}}} = 2.00\ \text{mol Na}$$

The equation says that 1 mole of chlorine reacts with 2 moles of sodium. We call this the **mole ratio.** We multiply the mole ratio by 2.00 moles Na, and we get 1.00 mole Cl_2.

$$2.00\ \cancel{\text{mol Na}} \times \frac{1\ \text{mol Cl}_2}{2\ \cancel{\text{mol Na}}} = 1.00\ \text{mole Cl}_2$$

*The abbreviation for mole is mol.

The 1 and 2 in the ratio (1 mol Cl_2)/(2 mol Na) are *exact ratios* given by the balanced equation. They do not count in determining how many significant figures we should have in the answer. Now, since the molecular weight of Cl_2 is 71.0, to calculate the weight of chlorine required, we simply multiply moles by molecular weight.

$$1.00\ \cancel{\text{mol Cl}_2} \times \frac{71.0\ \text{g Cl}_2}{\cancel{\text{mol Cl}_2}} = 71.0\ \text{g Cl}_2$$

So we need 71.0 g of chlorine to react completely with 46.0 g of sodium. Table 3.4 shows the amounts of Cl_2 required to react with various amounts of Na, and the amounts of NaCl that are produced. These relationships are called stoichiometric relationships.

Problem 3.19 Verify the last row in Table 3.4.

The stoichiometric relationships between the substances involved in a reaction can be calculated quite easily, by following a few simple steps (when we say *substance,* we mean any reactant or product):

1. Write a balanced chemical equation for the reaction.
2. Write the mass of one of the substances (usually given in grams) and convert this mass into moles.
3. To find the number of moles of any other substance, multiply the known moles from Step 2 by the proper mole ratio (we will see how to find that ratio in the example below).
4. Take the answer to Step 3 and multiply it by the atomic, molecular, or formula weight of that substance. This will tell you how many grams of the second substance will react with the given number of grams of the first substance.

These steps can be illustrated by using them with another reaction, the formation of water from hydrogen and oxygen.

TABLE 3.4 Stoichiometric Relationships for the Reaction $2Na + Cl_2 \rightarrow 2NaCl$

Mass Na, g	Mol Na	Mass Cl_2, g	Mol Cl_2	Mass NaCl, g	Mol NaCl
11.5	0.500	17.8	0.25	29.3	0.50
23.0	1.00	35.5	0.50	58.5	1.00
46.0	2.00	71.0	1.00	117	2.00
57.5	2.50	88.8	1.25	146	2.50
230	10.0	355	5.00	585	10.0

Example 3.1 What weight of water can be produced from 6.0 g of hydrogen? What weight of oxygen is required to react completely with that amount of hydrogen?
Solution

1. Write a balanced equation for the reaction.

$$2H_2 + O_2 \rightarrow 2H_2O$$

2. Convert the 6.0 g of hydrogen into moles.

$$6.0 \cancel{\text{g H}_2} \times \frac{1 \text{ mol H}_2}{2.0 \cancel{\text{g H}_2}} = 3.0 \text{ mol H}_2$$

3. Write the mole ratios. The equation tells us that 2 moles of H_2 will react to yield 2 moles of H_2O, so their ratio is (2 mol H_2O)/(2 mol H_2). Similarly, the ratio of O_2 to H_2 is (1 mol O_2)/(2 mol H_2). Multiply each mole ratio by the 3.0 moles of H_2 we obtained in step 2.

$$3.0 \cancel{\text{mol H}_2} \times \underbrace{\frac{2 \text{ mol H}_2\text{O}}{2 \cancel{\text{mol H}_2}}}_{\text{mole ratio}} = 3.0 \text{ mol H}_2\text{O}$$

$$3.0 \cancel{\text{mol H}_2} \times \overbrace{\frac{1 \text{ mol O}_2}{2 \cancel{\text{mol H}_2}}} = 1.5 \text{ mol O}_2$$

4. Multiply the moles of H_2O and O_2 by their respective molecular weights.

$$3.0 \cancel{\text{mol H}_2\text{O}} \times \frac{18 \text{ g H}_2\text{O}}{\cancel{\text{mol H}_2\text{O}}} = 54 \text{ g H}_2\text{O}$$

$$1.5 \cancel{\text{mol O}_2} \times \frac{32 \text{ g O}_2}{\cancel{\text{mol O}_2}} = 48 \text{ g O}_2$$

The answer to the problem is thus: 6.0 g of H_2 require 48 g of O_2 to react completely and produce 54 g of H_2O.

Problem 3.20 How many moles of H_2 are required to react completely with 10 moles of O_2, and how many moles of H_2O are produced?
Problem 3.21 How many grams of H_2O can be produced from 10 g of hydrogen? How many grams of oxygen would be required?
***Problem 3.22** How many grams of oxygen would remain unreacted if 10 g of hydrogen and 100 g of oxygen reacted as completely as possible to produce water?

SUMMARY

The **octet rule** states that atoms usually react to achieve a valence shell of eight electrons. This rule is helpful in predicting the formulas of both **ionic** and **covalent compounds.** In covalent compounds, atoms with different **electronegativities** share electrons unequally; this unequal sharing leads to a **polar covalent bond.** The naming of ionic and covalent compounds follows several simple rules. When atoms gain or lose electrons in reactions, the processes are called **reduction** or **oxidation,** respectively. Reactions of this type are called **redox** reactions. The amounts of reactants used or products formed in a chemical reaction can best be calculated by using the **mole** concept.

Key Terms

Avogadro's number (Section 3.14)
Covalent bonding (Section 3.7)
Covalent compound (Section 3.7)
Electronegativity (Section 3.9)
Formula unit (Section 3.13)
Formula weight (Section 3.14)
Ionic compound (Section 3.4)
Mole (Section 3.14)
Mole ratio (Section 3.14)
Molecular weight (Section 3.14)
Octet rule (Section 3.3)
Oxidation (Section 3.12)
Polar covalent bond (Section 3.10)
Reduction (Section 3.12)
Redox (Section 3.12)
Salt (Section 3.4)

ADDITIONAL PROBLEMS

3.23 Predict the formulas of the following ionic compounds.
(a) lithium oxide (b) magnesium fluoride
(c) sodium chloride

3.24 Write balanced chemical equations showing how the substances in Problem 3.23 can be formed from elements.

3.25 Predict the formulas of the following covalent compounds.
(a) chlorine fluoride
(b) chlorine oxide
(c) hydrogen sulfide

3.26 Give correct names for the substances shown in Problem 3.12.

3.27 Write balanced chemical equations showing how the substances in Problem 3.25 can be formed from the elements.

3.28 State which atoms are oxidized and which reduced in each of the redox reactions of Problems 3.24 and 3.27.

3.29 For each of the following pairs of substances, predict whether their reaction product will be ionic or covalent, and write a balanced chemical equation. State whether or not the covalent products are polar covalent compounds.
(a) Al and N_2
(b) Cl and Cl (chlorine atoms reacting)
(c) H_2 and N_2
(d) Na and H_2
(e) Li and N_2
(f) H_2 and P
(g) Mg and O_2

3.30 Name the following compounds.
(a) CaO
(b) CS_2
(c) $Mg(NO_3)_2$
(d) N_2O_5
(e) $Al_2(SO_4)_3$
(f) KOH
(g) HF
(h) IF

3.31 Give names that distinguish between the following compounds.
(a) Cu_2O and CuO
(b) $FeSO_4$ and $Fe_2(SO_4)_3$

3.32 Write the formulas for the following compounds.
(a) potassium bromide
(b) iron(II) bromide
(c) calcium iodide
(d) ammonium fluoride
(e) lithium carbonate
(f) phosphorus trichloride
(g) sodium hydroxide
(h) carbon tetrachloride
(i) magnesium hydroxide

3.33 Calculate the moles of molecules (if molecular), moles of atoms (if elemental), or moles of formula units (if ionic) represented by the following.
(a) 4 g of H_2 (calculate moles of H_2 molecules)
(b) 9 g of H_2O
(c) 7 g of Li
(d) 170 g of H_2S
(e) 10 g of MgO

3.34 Calculate the weight in grams of the following.
(a) 2.0 moles of NH_3
(b) 1.5 moles of HCl
(c) 0.50 mole of SCl_2
(d) 1.6 moles of F_2
(e) 4.0 moles of LiCl

3.35 The reaction of H_2 and Cl_2 produces 12.2 g of HCl. Calculate the following amounts.
(a) the moles of HCl produced
(b) the moles of H_2 and Cl_2 required

(c) the grams of H_2 and Cl_2 required
(d) the grams of excess Cl_2 if 12.0 g of Cl_2 react with 0.330 g of H_2

3.36 Methane (CH_4) burns to produce carbon dioxide and water.
(a) Write a balanced equation for the reaction of methane with oxygen.
(b) How many moles of carbon dioxide are produced when 80.0 g of methane burns?
(c) How many grams of water are produced when 80.0 g of methane burns?

3.37 Propane (C_3H_8) is sold commercially as a "bottled gas" fuel. It burns to produce carbon dioxide and water. Calculate the grams of propane that must burn to produce 220 g of carbon dioxide.

***3.38** Xenon forms a compound with oxygen, XeO_3. What is the name of this compound? Is it ionic or covalent? Draw a Lewis dot structure for the compound.

***3.39** If 100 g of H_2 react with 640 g of O_2 to produce the maximum possible amount of water, how many grams of each substance will be present after the reaction?

Problem-Solving Hints

If you have trouble solving any of the additional problems, refer to the sections listed next to the problem numbers.

3.23 3.5 **3.24** 3.5 **3.25** 3.8
3.26 3.11 **3.27** 3.8 **3.28** 3.12
3.29 3.4, 3.7, 3.10 **3.30** 3.6, 3.11 **3.31** 3.6
3.32 3.6 **3.33** 3.14 **3.34** 3.14
3.35 3.14 **3.36** 3.14 **3.37** 3.14
3.38 3.7, 3.11 **3.39** 3.14

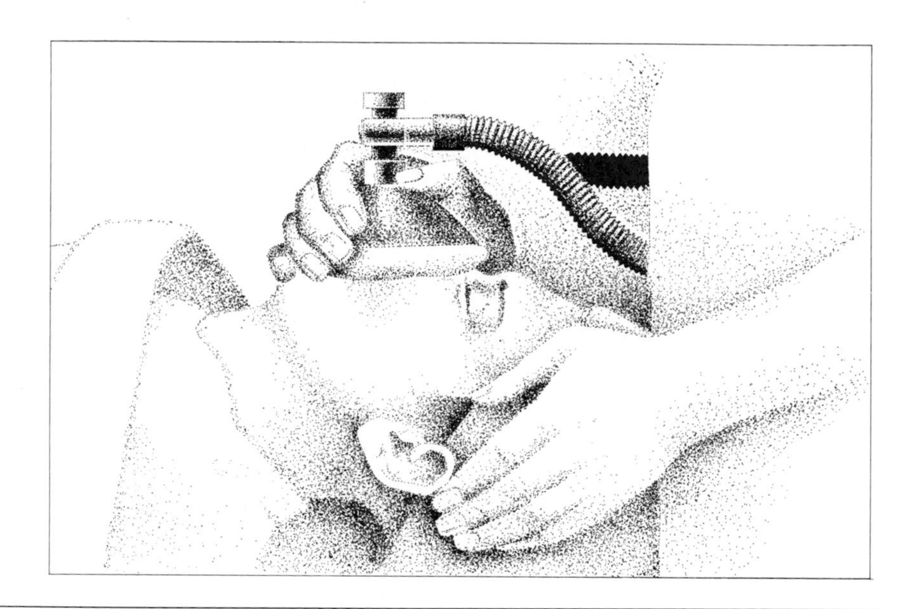

One of the important medical applications of gases is in their use as anesthetics. General anesthesia refers to the loss of pain sensation and consciousness that can be produced by various chemical agents. Among the agents available to today's anesthesiologists are a wide range of gaseous anesthetics. Each has its advantages and disadvantages, and choices must be made depending on individual patient needs.

The first surgical operations requiring anesthesia were amputations. At the time that they were first performed, over 100 years ago, opium and alcohol were primarily used to produce intoxication. Many strong assistants were necessary to restrain the patient, and the speed at which the surgeon could complete the operation was critical for all involved.

Nitrous oxide (N_2O) was introduced as an anesthetic agent by Dr. Horace Wells in 1844 for the extraction of teeth. This odorless gas has the advantage of being fast-acting and is neither flammable nor explosive. However, nitrous oxide has the disadvantage of producing very little muscle relaxation. Thus muscle-relaxing drugs must also be administered during the surgery. Nitrous oxide is frequently used as an anesthetic in present-day surgical procedures.

Diethyl ether $(C_2H_5)_2O$ was described by Valerius Cordus in 1540 and was used for anesthetic purposes in the early days of surgery. Ether is slow-acting and causes sickness by its irritating effect on the stomach. Unlike nitrous oxide, ether has a very pungent odor, is highly flammable, and may explode

4

GASES, LIQUIDS, AND SOLIDS

at fairly low temperatures. When stored, ether can react to form peroxides having much lower explosion points than pure ether. In view of these disadvantages, it is not surprising to find that the use of ether has dropped significantly.

Trichloroethylene and chloroform as well as some fluorinated hydrocarbons (such as halothane) are also used as anesthetics, but their toxic effects make them less favorable to use than some of the other anesthetics.

[*Illustration: General anesthesia frequently involves the use of certain gases that keep the patient asleep and unaware of pain during a surgical procedure.*]

GASES, LIQUIDS, AND SOLIDS

Although water always has the chemical formula H_2O, it does not always have the same properties. You can be burned by gaseous water, swim in liquid water, or walk on solid water. Water provides the most obvious everyday example of the three physical states of matter: gas, liquid, and solid.

As temperature increases, water molecules (or any molecules) gain energy. If the temperature exceeds 100°C, water molecules pick up so much energy that they overcome their attractions and fly apart. (They overcome the attractions between water molecules, not the chemical bonds that hold the atoms in each molecule together.) The liquid water changes to a gas, which we call steam. In this process, the molecules in steam gain so much energy that they can cause a bad burn.

As the temperature drops, water molecules lose energy. The attractions between molecules tighten their hold, the structure becomes rigid, and water enters the solid state. We have ice.

Pressure as well as temperature affects the state of matter. Even at a temperature well above 100°C, water can remain in a liquid state if the pressure is high enough. If the pressure is low enough, water can exist as a gas at temperatures below 0°C.

Temperature and pressure determine what physical state a particular substance will be in. (The chemical nature of the substance also plays a role, of course.) The reason why this is so is explained by one of the most important ideas in science, the kinetic molecular theory. In simplest terms, this theory says that molecules are in constant motion. For example, contact between the moving molecules of a gas and the walls of a container causes pressure. The more contact there is, the higher the pressure. If the temperature increases, the molecules become more energetic, move faster, and bump into the container walls more often and more violently. Thus, an increase in temperature results in an increase in pressure (assuming that everything else remains the same).

The complete kinetic molecular theory and how it affects the three states of matter is considerably more complicated. As we learn a little more about pressure, we will begin to develop a better understanding of this important theory.

PRESSURE 4.1

What is the difference between weight and pressure?

The term **pressure** means "force per unit of area." It is calculated by dividing the force by the area over which the force is exerted.

$$\text{Pressure} = \frac{\text{force}}{\text{area}}$$

Atmospheric pressure is the force or weight of the air pushing on the surface of the earth. To measure atmospheric pressure, we use an instrument known as a **barometer,** which consists of a column of mercury in a tube (Figure 4.1). The weight of the air presses down on a pool of mercury at the base of the tube, forcing a column of mercury up into the tube. There is a vacuum at the top of the tube, so there is no downward air pressure on the column. The column is pushed up until the pressure of the mercury is just enough to offset the pressure of the air. The width of the column doesn't matter, because what is important is not the total weight of the mercury, but the weight per area. A wider column of the same height will weigh more, but it will just spread its weight over a greater area. A higher column of the same width, however, will have *more* weight pushing down on the *same* area. So it is the height of the mercury column that serves as a measure of atmospheric pressure.

At sea level at a temperature of 0°C, the atmosphere typically supports a column of mercury about 760 mm (30 inches) high. Actual atmospheric pressures vary from time to time as meteorological conditions change. But the pressure that will support a column of mercury exactly 760 mm high is called an **atmosphere** (atm) and is used as a standard unit of pressure in scientific work. Another unit of pressure is the **torr,** which is 1/760 of an atmosphere. One torr is thus the pressure that will support a column of mercury 1 mm high. The following examples show how to express pressure readings in the various units.

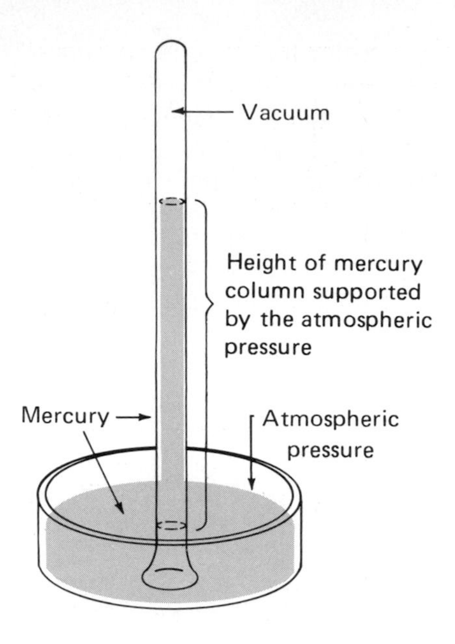

FIGURE 4.1 A simple barometer.

Example 4.1 Express a pressure of 1.50 atm in torr, millimeters of mercury, and centimeters of mercury.
Solution

$$1.50\ \cancel{\text{atm}} \times 760\ \frac{\text{torr}}{\cancel{\text{atm}}} = 1.14 \times 10^3\ \text{torr}$$

$$1.14 \times 10^3\ \cancel{\text{torr}} \times \frac{1\ \text{mm of mercury}}{\cancel{\text{torr}}} = 1.14 \times 10^3\ \text{mm of mercury}$$

$$1.14 \times 10^3\ \cancel{\text{mm}}\ \text{of mercury} \times \frac{1\ \text{cm}}{10\ \cancel{\text{mm}}} = 1.14 \times 10^2\ \text{cm of mercury}$$

Example 4.2 Express 1.61×10^3 torr in atm.
Solution

$$1.61 \times 10^3\ \cancel{\text{torr}} \times \frac{1\ \text{atm}}{760\ \cancel{\text{torr}}} = 2.12\ \text{atm}$$

Problem 4.1 Express a pressure of 2.0 atm in centimeters of mercury, millimeters of mercury, and torr.
Problem 4.2 Express 380 torr in atm.

KINETIC MOLECULAR THEORY 4.2

How do gas molecules behave?

The **kinetic molecular theory** is basic to every science because it helps us understand the behavior of all kinds of matter. We will mainly be interested in using kinetic molecular theory to explain the behavior of gases.

We can outline the main points of the kinetic molecular theory of gases as follows:

1. The molecules of a gas are in constant, rapid, random motion.
2. A molecule travels in a straight line until it strikes another molecule of the gas or the walls of the container (Figure 4.2).
3. A consequence of this motion is that gas molecules have kinetic energy. The average kinetic energy of the molecules is directly related to the **absolute temperature** of the gas.

Absolute temperatures are given in degrees **Kelvin** (**K**). The absolute temperature scale is related to the Celsius scale by the equation

$$K = °C + 273$$

The degree sign is not used with Kelvin degrees. Zero degrees Kelvin, for example—absolute zero—is represented as 0 K.

Problem 4.3 What is the absolute temperature of a substance at 25°C?
Problem 4.4 A substance has an absolute temperature of 373 K. What is its temperature on the Celsius scale?

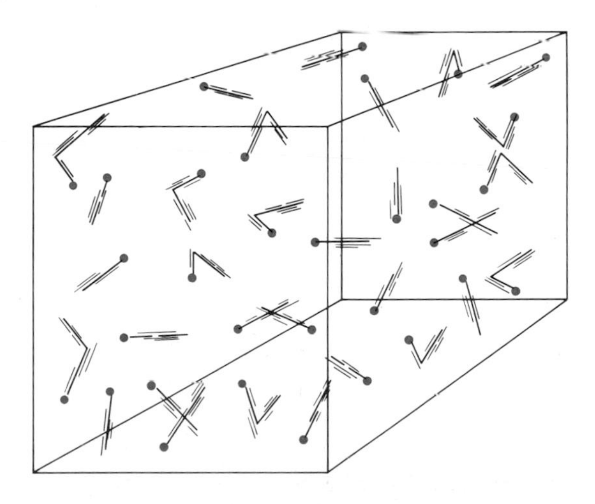

FIGURE 4.2 The molecules of a gas are in rapid motion and collide with each other and the walls of the container.

4.3 PRESSURE, VOLUME, AND TEMPERATURE BEHAVIOR OF GASES

How does a gas exert pressure?

The relationship between the pressure exerted by a gas and the volume of its container is well known. If the volume becomes smaller, the pressure of the gas increases; if the volume gets larger, the pressure drops. According to the kinetic molecular theory, the pressure of a gas is the result of countless collisions between the gas molecules and the sides of the container. The more collisions per second, the greater the pressure. (Collisions between the gas molecules themselves do not affect the pressure at all.) In a larger volume, the gas molecules have to travel farther between collisions with the walls, so there are fewer collisions per second and the pressure is lower. As volume decreases, the collisions occur more frequently and pressure increases (Figure 4.3).

If we put more gas into a fixed volume, the pressure also increases. The explanation is obvious: the more molecules of gas there are, the more frequent will be collisions with the walls of the container (Figure 4.4). The increase in pressure as air is pumped into an automobile or bicycle tire is due to the increase in the number of gas molecules per volume.

The principle that an increase in gas molecules per volume leads to an increase in pressure is employed in the **sphygmomanometer,** a device used to measure blood pressure in humans.

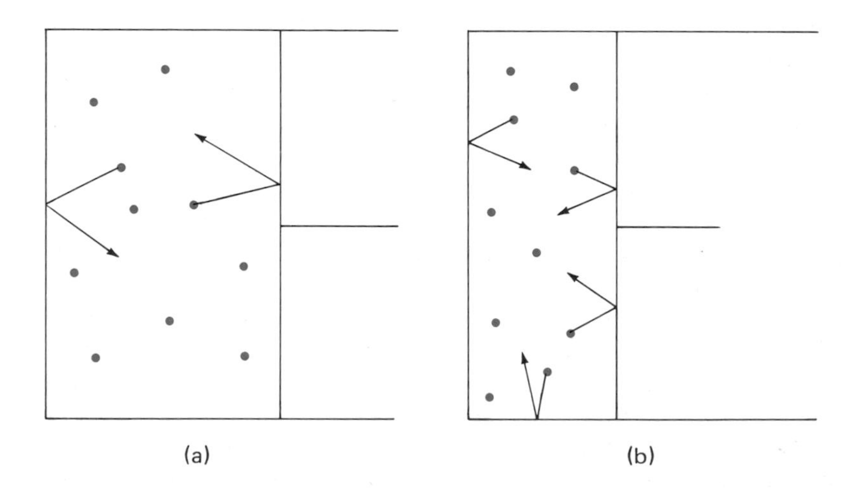

FIGURE 4.3 For a gas, as volume decreases, pressure increases. Gas molecules in a certain volume (a) strike the walls of the container at a given rate. In a smaller volume (b) the molecules strike the walls more frequently because they do not have to travel as far between wall collisions.

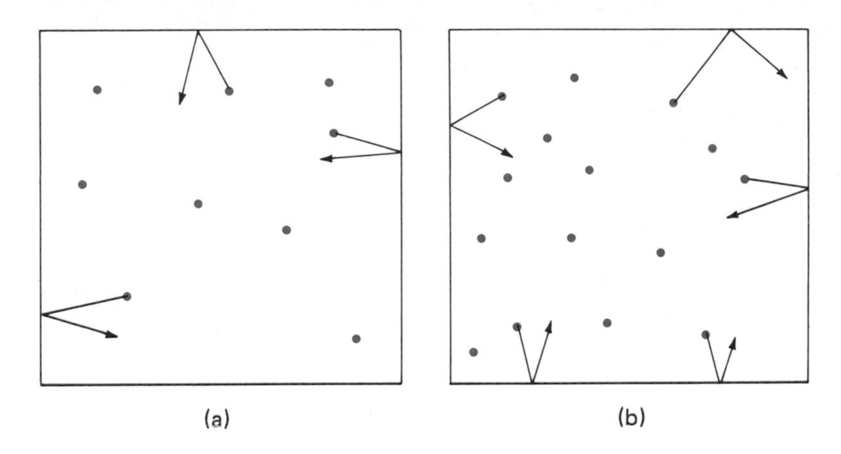

FIGURE 4.4 The pressure of a gas depends on the number of molecules in the container. A certain number of molecules (a) produce a certain number of collisions per second with the walls. If more gas is added (b), the number of molecules increases. This, in turn, causes the pressure to increase because there are more collisions per second between the molecules and the container walls.

This device consists of a cuff that can be pumped up with air and an instrument called a manometer that measures the pressure within the cuff. A manometer operates on the same principle as a barometer (Section 4.1). The cuff is wrapped around the upper arm and air is pumped into the cuff until the air pressure inside the cuff is great enough to cut off the flow of blood through the brachial artery in the arm. The technician places a stethoscope over the brachial artery just below the elbow and slowly allows air to escape from the cuff. As air escapes, the air pressure decreases and thus the pressure on the artery also decreases. When the technician first hears faint sounds corresponding to the heartbeat, he or she knows that the pressure of the air inside the cuff is equal to the blood pressure in the artery as the heart contracts. This pressure, which can be read directly on the manometer, is called the systolic blood pressure. As more air escapes from the cuff, and the pressure continues to decrease, the sounds become louder, then fainter, and finally disappear. The air pressure in the cuff at this point is equal to the blood pressure in the artery during the relaxation period of the heart, called the diastolic pressure. Blood pressure varies from individual to individual and from time to time for a person, depending on general health and physical activity. A normal range is a systolic pressure of 100–140 **millimeters of mercury** (mmHg) and a diastolic pressure of 60–100 mmHg. Blood pressure is usually written as a fraction, with the systolic pressure over the diastolic, for example, 120/80.

In discussing gas behavior up to now, we have not considered the effect of changes in **temperature.** However, it is well known that temperature greatly affects the pressure and volume of gases. If pressure remains constant, all gases expand when heated and contract when cooled. On the other hand, if a gas is confined in a rigid container while it is being heated, its pressure will increase as the temperature rises. If this same gas is cooled in the container, its pressure will decrease as the temperature goes down.

Kinetic molecular theory explains the temperature effect in terms of the kinetic energies of the gas molecules. The lower the temperature, the smaller the kinetic energy; the higher the temperature, the greater the kinetic energy. The speed of a molecule depends on its kinetic energy. Faster-moving molecules possess greater kinetic energy than that possessed by slower-moving molecules. Increasing the temperature increases kinetic energy, which means the molecules speed up so they strike the container walls more frequently and with greater force, increasing the pressure. Lowering the temperature decreases pressure, because the slower-moving molecules strike the container walls less frequently and with less force (Figure 4.5).

As a practical observation of this effect, the air pressure in an automobile tire increases on a long trip. The reason is that the temperature of the tire and that of the air in it go up when an automobile travels at high speed for a long time, but the volume of the tire is almost constant, so the air pressure increases. Tire manufacturers always warn drivers not to let air out of tires during a trip even if the pressure temporarily rises far above recommended values. After the car has been parked for

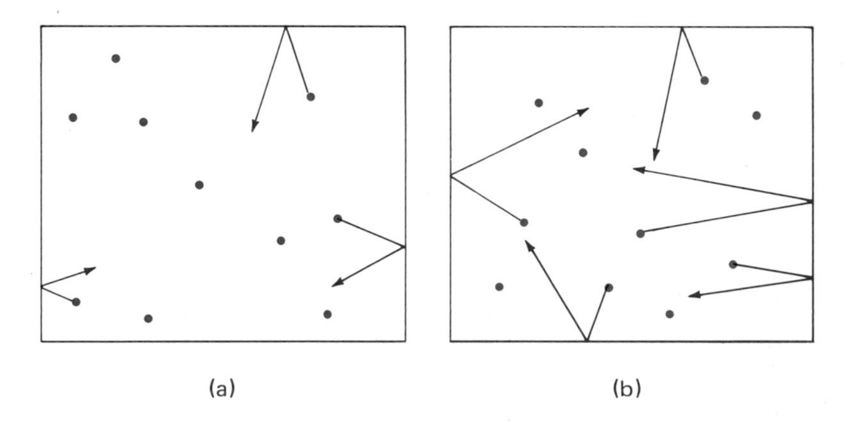

FIGURE 4.5 A higher temperature increases the pressure of a gas in a rigid container. At a certain temperature (a) the gas molecules strike the walls at a given rate and a certain average force. At a higher temperature (b) the molecules have more kinetic energy and hence move faster, so that they strike the walls more frequently and with greater force.

a time, the temperature of the tires drops, and the air pressure falls. However, if air has been let out of the tire while it was hot, the pressure will fall below normal levels as the tire cools. The tires will then be underinflated and may be damaged if the car is operated.

Problem 4.5 Predict what will happen to the volume of an inflated balloon if it is placed in a refrigerator for a while. Explain. [Hint: Remember that a balloon is not a rigid container of fixed volume.]

QUANTITATIVE TREATMENT OF GASES 4.4†

Several mathematical equations are available to express quantitatively the ideas presented in Section 4.3. These equations are often called the **gas laws.** The most familiar are Boyle's law and Charles's law.

Boyle's Law

Boyle's law describes the relationship between pressure (P) and volume (V) for a gas. This law says that for a fixed amount of gas at a constant temperature, volume is inversely proportional to pressure. In other words, the pressure of a gas multiplied by its volume is equal to a constant.

$$PV = \text{constant}$$

A more useful expression for making calculations is

$$P_1V_1 = P_2V_2$$

where the subscripts 1 and 2 refer to initial and final conditions, respectively. Boyle's law can be used to find the final pressure of a gas, when the volume is known, as the example shows.

Example 4.3 A sample of gas has a pressure of 2.0 atm and a volume of 10 L. What pressure would the same amount of gas exert in a volume of 12 L at the same temperature?

†Optional

Solution

$$P_1 = 2.0 \text{ atm} \qquad V_1 = 10 \text{ L} \qquad P_2 = ? \qquad V_2 = 12 \text{ L}$$

$$P_1V_1 = P_2V_2$$

$$2.0 \text{ atm} \times 10 \text{ L} = P_2 \times 12 \text{ L}$$

$$P_2 = \frac{2.0 \text{ atm} \times 10 \not{\text{L}}}{12 \not{\text{L}}}$$

$$P_2 = 1.7 \text{ atm}$$

Charles's Law

Charles's law expresses the relationship between the volume of a sample of gas and its temperature (T). This law states that, at a constant pressure, the volume of a gas is directly proportional to its absolute temperature (degrees Kelvin). In working with the gas laws, *absolute* temperature must *always* be used.

$$\frac{V}{T} = \text{constant}$$

In other words, as this equation shows, the volume of a gas divided by its absolute temperature is equal to a constant. A more useful expression for calculations is

$$\frac{V_1}{T_1} = \frac{V_2}{T_2}$$

Charles's law can be used to find the final volume of a gas, when the temperature is known, as the following example shows.

Example 4.4 A sample of gas occupies a volume of 250 mL at 25°C. What volume will it occupy at 80°C, if the pressure is unchanged?
Solution

$$V_1 = 250 \text{ mL} \qquad T_1 = 25 + 273 = 298 \text{ K}$$

$$V_2 = ? \qquad T_2 = 80 + 273 = 353 \text{ K}$$

$$\frac{V_1}{T_1} = \frac{V_2}{T_2}$$

$$\frac{250 \text{ mL}}{298 \text{ K}} = \frac{V_2}{353 \text{ K}}$$

$$V_2 = \frac{250 \text{ mL} \times 353 \cancel{\text{K}}}{298 \cancel{\text{K}}}$$

$$V_2 = 296 \text{ mL}$$

Combined Gas Law

Boyle's law and Charles's law are helpful in solving a variety of problems. They are simple and easy to use. However, many changes in the pressure, volume, and temperature of a gas take place simultaneously. In such cases, neither of these laws by itself is of much help. Boyle's law is true only if the temperature of the gas does not change, and Charles's law is true only if the pressure of the gas does not change. Fortunately, it is possible to combine these two laws into a single mathematical expression, called the **combined gas law.**

$$\frac{P_1V_1}{T_1} = \frac{P_2V_2}{T_2}$$

This expression combines and replaces the separate laws of Boyle and Charles. If $T_1 = T_2$ (constant temperature), then the combined expression becomes

$$P_1V_1 = P_2V_2$$

(that is, Boyle's law). If $P_1 = P_2$ (constant pressure), then the combined expression becomes

$$\frac{V_1}{T_1} = \frac{V_2}{T_2}$$

which is Charles's law. We can therefore use the combined gas law for all problems, rather than using Boyle's and Charles's laws separately.

Example 4.5 A sample of air in a balloon has a volume of 900 mL at a temperature of 25°C and a pressure of 850 torr. After the balloon has been in a refrigerator for a while, the pressure is 825 torr and the volume is 875 mL. What is the temperature inside the refrigerator?

Solution

$P_1 = 850$ torr $\quad V_1 = 900$ mL $\quad T_1 = 25 + 273 = 298$ K

$P_2 = 825$ torr $\quad V_2 = 875$ mL $\quad T_2 = ?$

$$\frac{P_1V_1}{T_1} = \frac{P_2V_2}{T_2}$$

$$\frac{850 \text{ torr} \times 900 \text{ mL}}{298 \text{ K}} = \frac{825 \text{ torr} \times 875 \text{ mL}}{T_2}$$

$$T_2 = \frac{298 \text{ K} \times 825 \text{ torr} \times 875 \text{ mL}}{850 \text{ torr} \times 900 \text{ mL}}$$

$$T_2 = 281 \text{ K}$$

or, temperature = 281 − 273 = 8°C

Example 4.6 A gas sample in a bulb of fixed volume has a pressure of 1.25 atm at 20°C. If we heat the gas until its temperature is 95°C, what will be its pressure?

Solution

$P_1 = 1.25$ atm $T_1 = 20 + 273 = 293$ K $V_1 = V_2$

$P_2 = ?$ $T_2 = 95 + 273 = 368$ K

$$\frac{P_1V_1}{T_1} = \frac{P_2V_2}{T_2}$$

$$\frac{1.25 \text{ atm} \times V_1}{293 \text{ K}} = \frac{P_2 \times V_1}{368 \text{ K}}$$

$$\frac{1.25 \text{ atm}}{293 \text{ K}} = \frac{P_2}{368 \text{ K}}$$

$$P_2 = \frac{368 \text{ K} \times 1.25 \text{ atm}}{293 \text{ K}}$$

$$P_2 = 1.57 \text{ atm}$$

Problem 4.6 A girl fills a bicycle tire with air to a pressure of 3.0 atm at 23°C. After she rides the bicycle for some time, the pressure is found to be 3.2 atm. What is the temperature of the tire? Assume that volume of the tire does not change.

Problem 4.7 A balloon is filled with helium to a pressure of 1.1 atm at 20°C. The volume of the balloon is 8.2×10^3 L. The balloon rises until the pressure has dropped to 0.85 atm and the temperature has fallen to 0°C. Calculate the volume of the balloon at this height.

Ideal Gas Law

Occasionally, we need to know the amount of gas in a sample. Neither the combined gas law, Boyle's law, nor Charles's law can be used, because these equations make no reference to the actual amount of gas—that is, the number of molecules. They simply assume that the amount is constant. An equation called the **ideal gas law** can be used when the exact amount of gas must be specified.

$$PV = nRT$$

The symbols P, V, and T have their usual meanings, namely, pressure, volume, and absolute temperature, respectively. The amount of gas, in moles, is given by n, and R is a constant. If we express this equation in the form

$$\frac{PV}{T} = nR$$

it is immediately obvious that whenever the amount of gas is fixed and unchanging, the right side of the equation is a constant.

$$\frac{PV}{T} = \text{constant}$$

This last expression can also be expressed as

$$\frac{P_1V_1}{T_1} = \frac{P_2V_2}{T_2}$$

which is simply the combined gas law. Thus, whenever the amount of gas is constant and the specific quantity need not be known, the ideal gas law reduces to the combined gas law.

In order to use the ideal gas law, we must know the value of the constant R. This value has been determined by careful measurements with a variety of different gases.

$$R = 0.08206 \frac{\text{liter} \times \text{atmosphere}}{\text{mole} \times \text{absolute degrees}}$$

The value of R is the same for all gases, though different units of pressure and volume can be used. We should just make sure that when the pressure of the gas is expressed in atmospheres and the volume in liters, we use a value of R that is also expressed in atmospheres and liters. Several examples will illustrate how we use the ideal gas law.

Example 4.7 How many moles of gas are there in a sample in a 200 mL container? The pressure of the gas is 500 mmHg, and its temperature is 22°C.

Solution

$$P = 500\ \text{mmHg} = 500\ \cancel{\text{mmHg}} \times \frac{1\ \text{atm}}{760\ \cancel{\text{mmHg}}} = 0.658\ \text{atm}$$

$$V = 200\ \text{mL} = 200\ \cancel{\text{mL}} \times \frac{1\ \text{L}}{1000\ \cancel{\text{mL}}} = 0.200\ \text{L}$$

$$T = 22 + 273 = 295\ \text{K}$$

$$R = 0.08206\ \frac{\text{L} \cdot \text{atm}}{\text{mol} \cdot \text{K}}$$

$$n = ?$$

$$PV = nRT$$

or,

$$n = \frac{PV}{RT}$$

$$= \frac{0.658\ \text{atm} \times 0.200\ \text{L}}{0.08206\ \frac{\text{L} \cdot \text{atm}}{\text{mol} \cdot \text{K}} \times 295\ \text{K}}$$

$$= 5.44 \times 10^{-3}\ \text{mol}$$

Example 4.8 How many grams of oxygen did a student collect if she filled a 150-mL bulb with O_2 at a pressure of 700 mmHg and a temperature of 27°C?

Solution

$$P = 700\ \text{mmHg} = 700\ \cancel{\text{mmHg}} \times \frac{1\ \text{atm}}{760\ \cancel{\text{mmHg}}} = 0.921\ \text{atm}$$

$$V = 150\ \text{mL} = 150\ \cancel{\text{mL}} \times \frac{1\ \text{L}}{1000\ \cancel{\text{mL}}} = 0.150\ \text{L}$$

$$T = 27 + 273 = 300\ \text{K}$$

$$R = 0.08206\ \frac{\text{L} \cdot \text{atm}}{\text{mol} \cdot \text{K}}$$

$$n = ?$$

$$PV = nRT$$

so,

$$n = \frac{PV}{RT}$$

$$= \frac{0.921 \text{ atm} \times 0.150 \text{ L}}{0.08206 \frac{\text{L} \cdot \text{atm}}{\text{mol} \cdot \text{K}} \times 300 \text{ K}}$$

$$= 5.61 \times 10^{-3} \text{ mol}$$

The molecular weight of O_2 is 32.0, so the weight of 1 mol of O_2 is 32.0 g.

$$5.61 \times 10^{-3} \cancel{\text{mol}} \times 32.0 \frac{\text{g}}{\cancel{\text{mol}}} = 0.180 \text{ g}$$

Example 4.9 A student found that the mass of an evacuated bulb was 132.651 g. The instructor filled this bulb with an unknown gas at a pressure of 1.10 atm and a temperature of 23°C. The student reweighed the bulb and found that it weighed 132.816 g. The volume of the bulb was known to be 130 mL. Calculate the molecular weight of the gas.

Solution

$$P = 1.10 \text{ atm}$$

$$V = 130 \text{ mL} = 130 \cancel{\text{mL}} \times \frac{1 \text{ L}}{1000 \cancel{\text{mL}}} = 0.130 \text{ L}$$

$$T = 23 + 273 = 296 \text{ K}$$

$$R = 0.08206 \frac{\text{L} \cdot \text{atm}}{\text{mol} \cdot \text{K}}$$

$$n = ?$$

$$PV = nRT$$

so,

$$n = \frac{PV}{RT}$$

$$= \frac{1.10 \text{ atm} \times 0.130 \text{ L}}{0.08206 \frac{\text{L} \cdot \text{atm}}{\text{mol} \cdot \text{K}} \times 296 \text{ K}}$$

$$= 5.89 \times 10^{-3} \text{ mol}$$

The mass of the gas is equal to the difference between the weights of the filled and evacuated bulb.

$$\text{Mass} = 132.816 - 132.651 = 0.165 \text{ g}$$

so

$$\frac{0.165\text{ g}}{5.89 \times 10^{-3}\text{ mol}} = 28.0\,\frac{\text{g}}{\text{mol}}$$

which is the molecular weight of the gas.

Example 4.10 What pressure will 4.91×10^{-3} mol of gas exert in a volume of 0.500 L at 80°C?
Solution

$$\text{P} = ?$$

$$\text{V} = 0.500\text{ L}$$

$$\text{T} = 80 + 273 = 353\text{ K}$$

$$R = 0.08206\,\frac{\text{L}\cdot\text{atm}}{\text{mol}\cdot\text{K}}$$

$$n = 4.91 \times 10^{-3}\text{ mol}$$

$$\text{PV} = nR\text{T}$$

or,

$$\text{P} = \frac{nR\text{T}}{\text{V}}$$

$$= \frac{4.91 \times 10^{-3}\text{ mol} \times 0.08206\,\frac{\text{L}\cdot\text{atm}}{\text{mol}\cdot\text{K}} \times 353\text{ K}}{0.500\text{ L}}$$

$$\text{P} = 0.284\text{ atm}$$

Problem 4.8 How many moles of gas are required to produce a pressure of 520 mmHg in a 750-mL container at 50°C?
Problem 4.9 What is the molecular weight of a gas if 0.250 g exerts a pressure of 0.900 atm in a 0.250-L bulb at 24°C?
Problem 4.10 Calculate the pressure exerted by 10.0 g of H_2 in a 50.0-L container at 20°C.

4.5 THE EARTH'S ATMOSPHERE

What makes up our atmosphere?

The most important gas to which we are daily exposed is, of course, the air. Actually, air is a mixture (gaseous solution) of various substances. Although the precise amounts of the substances in air vary somewhat

TABLE 4.1 Major Components of Dry Air

Substance	Formula	Concentration (percent by volume)*
Nitrogen	N_2	78.1
Oxygen	O_2	20.9
Argon	Ar	0.9
Carbon dioxide	CO_2	0.03
Neon	Ne	0.002

*Percent by volume can be imagined as the liters of a given component out of every 100 liters of air.

from place to place and from time to time, the average composition of air is fairly constant. Table 4.1 lists the major components of dry air. Water vapor is also a component of air, but its concentration varies so much over short periods of time that we cannot state a standard concentration of it.

As we see from Table 4.1, five substances account for the bulk of the composition of air. Three more noble gases (helium, krypton, and xenon), plus hydrogen, account for an additional 0.0007% by volume. The remaining fraction of a percentage consists of a number of components that vary considerably in concentration. Some of these are pollutants. Naturally, they occur mainly in industrial areas and in large cities with heavy flows of traffic. Table 4.2 lists the most serious pollutants found in the air, with their major and minor sources.

As Table 4.2 shows, transportation is a major or minor source of all serious air pollution. Automobiles account for more than half of this

TABLE 4.2 Sources of the Serious Air Pollutants

Pollutant	Major source	Minor source
Hydrocarbons	Transportation	Heating
	Industrial plants	Waste disposal
Nitrogen oxides (NO, NO_2)	Transportation	Heating
	Industrial plants	Waste disposal
	Power plants	
Sulfur oxides (SO_2, SO_3)	Power plants	Transportation
	Industrial plants	Waste disposal
	Heating	
Carbon monoxide (CO)	Transportation	Industrial plants
		Power plants
		Heating
		Waste disposal

pollution, and trucks, buses, and diesel trains account for the remainder. Industrial plants and power plants are also significant sources of pollution.

4.6 CONDENSATION OF GASES

What happens when a gas condenses to a liquid?

Think about what happens when a gas cools. We can use water in its gaseous state as an example. As the temperature of the gas drops, the kinetic energy of the molecules decreases. In other words, the molecules of the gas slow down. Eventually, the gas reaches a temperature at which the gas molecules begin to stick together when they collide. For water under a pressure of 1 atm, this temperature is 100°C. At this temperature, some of the molecules are not moving fast enough to overcome completely the attraction of other molecules. When two slow-moving molecules collide, they have insufficient energy to rebound. If another molecule strikes this molecular pair, it may also stick. Liquid begins to form. We call this process **condensation.** Now, a very interesting thing happens: as more heat is removed, more gas molecules stick together to form liquid water, but the temperature stays constant at 100°C. In fact, for every gram of gaseous water that condenses to liquid water, 540 cal have to be removed. But, once all the gas has condensed to water, the temperature of the liquid will drop below 100°C if more heat is removed.

4.7 GAS–LIQUID EQUILIBRIUM

When the liquid and gaseous states of a substance exist together at a fixed temperature and pressure, the substance is said to be in an **equilibrium state.** If heat is removed, more gas will condense to liquid. If heat is added, more liquid will turn to gas (**vaporization**). But whether heat is added or removed, the temperature will not change as long as there is even a little bit of both physical states present (Figure 4.6).

4.8 HEAT OF VAPORIZATION

The amount of heat per gram that must be added at the boiling point to vaporize a liquid, or removed to condense a vapor, is called the **heat**

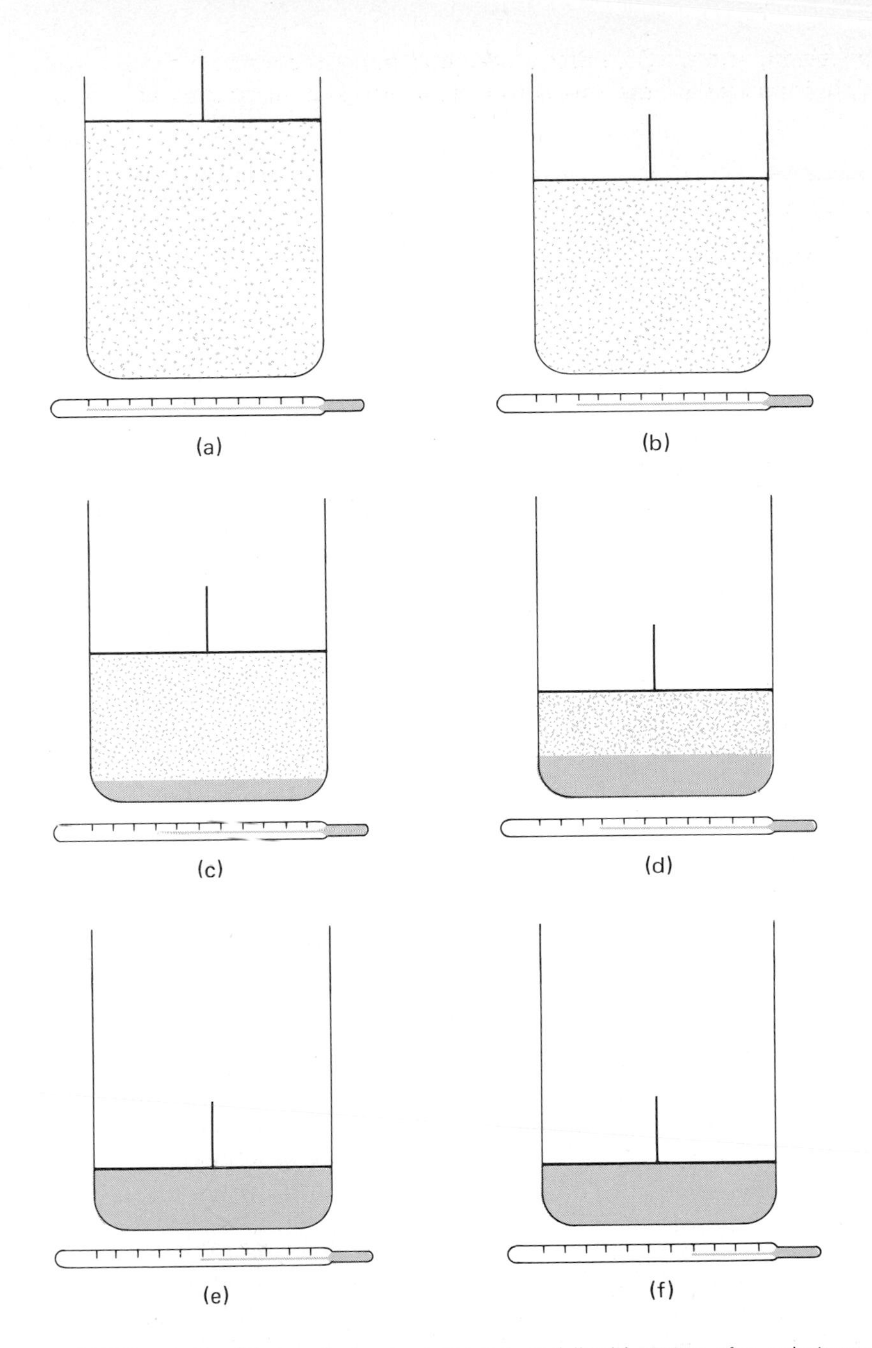

FIGURE 4.6 Equilibrium between gaseous and liquid states of a substance. Temperature of gas falls as heat is removed at constant pressure (a,b). Temperature remains constant as heat is removed and gas condenses to liquid (c,d). After all the gas has condensed, removal of more heat causes the temperature to start falling again (e,f).

TABLE 4.3 Heats of Vaporization

Substance	Boiling point, °C	Heat of vaporization, cal/g
Ammonia (NH_3)	−33	327
Benzene (C_6H_6)	80	94
Chloroform ($CHCl_3$)	62	59
Ethyl ether ($C_4H_{10}O$)	35	85
Water (H_2O)	100	540

of vaporization. For water, this is 540 cal/g. Every substance has a particular heat of vaporization. Table 4.3 lists heats of vaporization for different substances.

Why can steam at 100°C inflict a worse burn than water at 100°C?

At the boiling point of water, both liquid water and steam are at a temperature of 100°C, but the steam has a lot more energy than the liquid. In practical terms, this means that steam at 100°C will inflict a much worse burn than the same mass of liquid water at 100°C! The difference lies in the heat of vaporization that the steam possesses.

The heat of vaporization of a substance is the energy needed to pull apart the molecules in a liquid state, to produce the greater separation between molecules that exists in a gas. It can also be described as the energy necessary to overcome the attraction between molecules in a liquid state, in order to get them into a gaseous state. The same amount of energy is given off when a gas condenses to a liquid state. Molecules that have relatively strong attractions for each other in the liquid state—water molecules, for example—will have fairly high heats of vaporization (Table 4.3). We will see in the next chapter the reason for the fairly strong attractions between water molecules in the liquid state.

The energy involved in the heat of vaporization must be carefully distinguished from the kinetic energies of the molecules. The liquid and gas molecules in the equilibrium state each have the *same* temperature. Consequently, gaseous and liquid molecules have the same average kinetic energy when they are in equilibrium.

Problem 4.11 Given that 1 g of liquid benzene is in equilibrium with 1 g of gaseous benzene at 1 atm and 80°C, compare the following.

(a) the average kinetic energies of the benzene molecules in the liquid and gaseous states

(b) the total energies of the benzene molecules in the liquid and gaseous states

(c) the relative degrees of separation between the benzene molecules in the liquid and gaseous states.

MELTING AND FREEZING 4.9

When a solid is heated until it begins to form a liquid, we say it is melting. The temperature at which a solid melts is called the melting point. When a liquid is cooled until it begins to become solid, we say it is freezing. The temperature at which the solid begins to form is the freezing point.

The melting point and freezing point of a pure substance are the same. At this temperature, there is an equilibrium state between solid and liquid.

HEAT OF FUSION 4.10

The **heat of fusion** is the energy that must be supplied to change 1 g of a solid to a liquid at its melting point. The same amount of energy is given off when a liquid freezes to form a solid. For water, the heat of fusion is 80 cal/g. For every gram of ice that melts, 80 cal must be added.

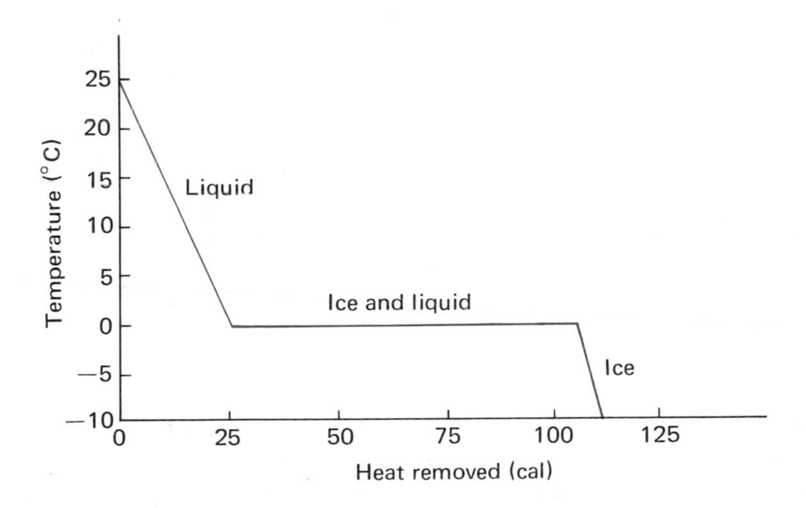

FIGURE 4.7 Temperature change as heat is removed from 1 g of water initially at 25°C. The temperature decreases as the liquid cools, until ice begins to form at 0°C. The temperature remains at 0°C as heat is removed, until all the liquid changes to ice. After all the liquid has frozen, further removal of heat causes the temperature of the ice to fall.

For every gram of liquid water that freezes, 80 cal must be removed. Water freezes or melts at 0°C, and again (as during vaporization or condensation), there is no temperature change during either the freezing or the melting process. Figure 4.7 shows the temperature change as heat is removed from water, initially at room temperature, until the water freezes.

Problem 4.12

(a) Is there a difference in the average kinetic energies of water molecules in the liquid and solid states at 0°C?

(b) What is the difference in energy, between 1 g of ice and 1 g of liquid water at 0°C?

4.11 SOLIDS

How do we classify solids?

Solids are divided into two major classes. **Amorphous solids** are composed of atoms or molecules that have no regular arrangement. Common examples of amorphous solids are soot and glass. In **crystalline solids** the particles (which may be atoms, molecules, or ions). are arranged in an orderly fashion in a three-dimensional pattern. We call such a pattern a **crystal lattice.**

All known crystal lattices fall into one of seven different systems (Figure 4.8). The classification is based on the angles between intersecting faces and the relative lengths of the sides of the crystal.

4.12 UNIT CELL

An external examination of a crystal is often sufficient to determine its system. Within each of the crystal systems, there are several ways to arrange the atoms, ions, or molecules to produce a **unit cell.** The unit cell consists of the minimum number of atoms, ions, or molecules necessary to display the overall pattern of the crystal structure.

Figure 4.9 shows the three unit-cell structures that fall within the cubic system. We cannot determine the unit-cell structure of a crystalline solid simply from an external examination of the solid. Such determinations require the use of x rays and a fairly involved mathematical analysis of the results.

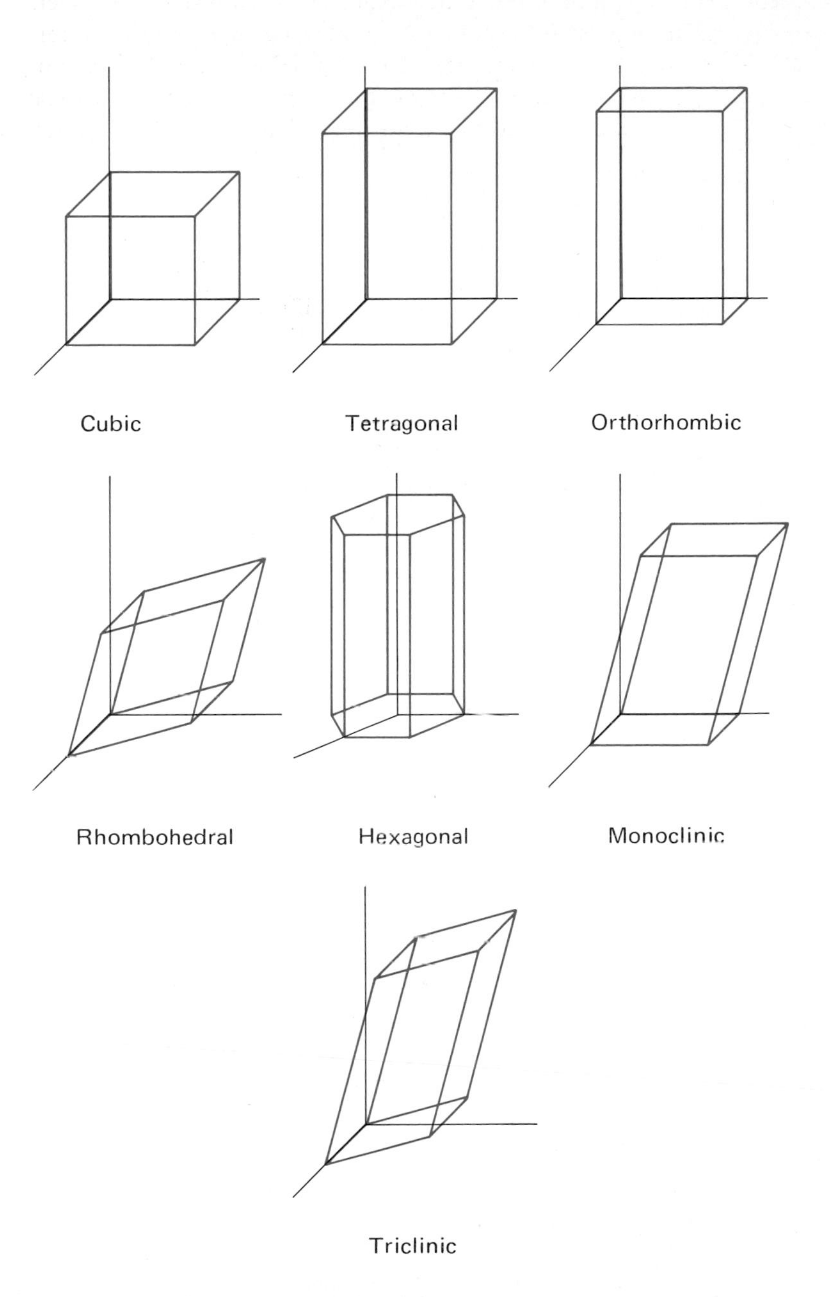

FIGURE 4.8 The seven crystal systems.

4.13 TYPES OF CRYSTAL LATTICES

How does the type of crystal lattice affect the melting point of a substance?

A crystal lattice is composed of atoms, ions, or molecules, depending on the chemical nature of the crystalline substance and the type of bonding.

The metal potassium has a body-centered cubic structure (Figure 4.9), in which the lattice is composed of potassium atoms. An ionic com-

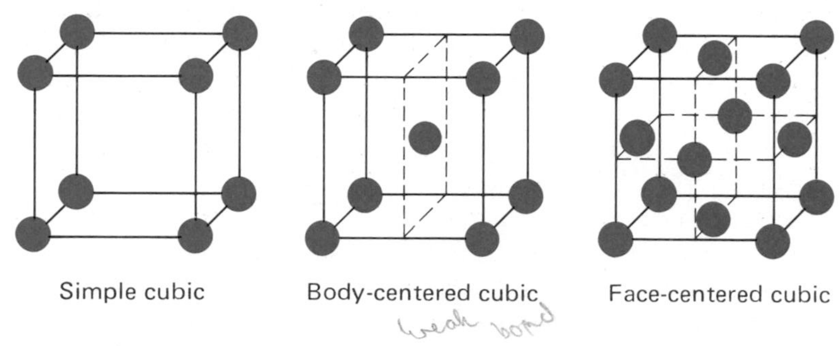

FIGURE 4.9 The three cubic unit cells.

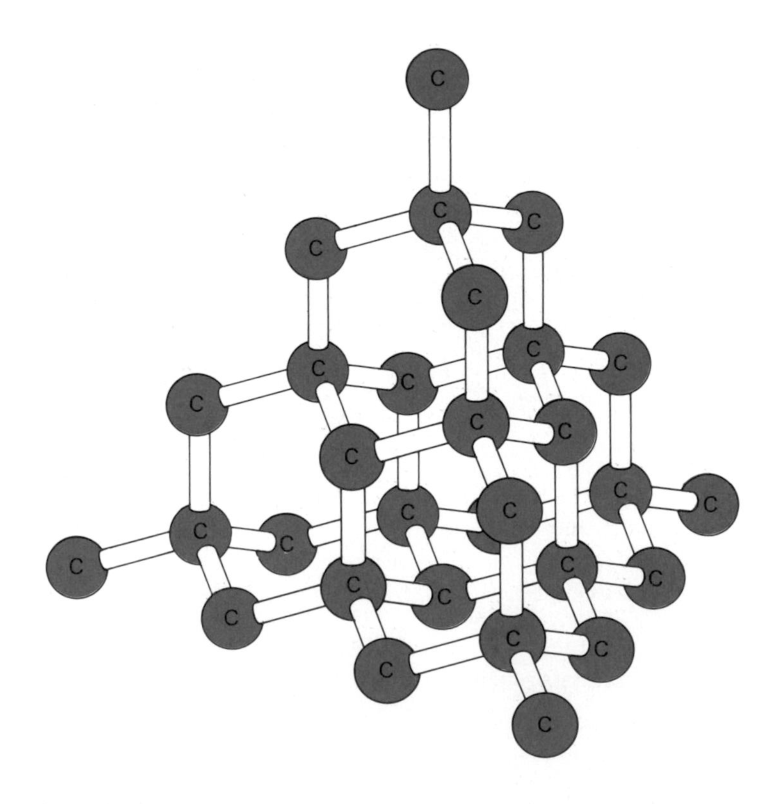

FIGURE 4.10 A portion of the structure of diamond.

pound, on the other hand, consists of positive and negative ions. The crystal lattice of a typical ionic compound, NaCl, was shown in Chapter 3 (Figure 3.1a).

The covalent compound CH_4 (methane) crystallizes in a face-centered cubic structure (Figure 4.9), in which the lattice consists of CH_4 molecules. The covalent C—H bonds in each molecule are very strong, but the forces of attraction between the molecules are very weak, so that crystals of solid methane are easily deformed. Methane thus has a very low melting point. Crystalline solids of this type are called molecular crystals.

Sometimes a crystal lattice consists of atoms that are covalently bonded to neighboring atoms. If the covalent bonds are strong, the crystals will be very strong and difficult to melt, because the covalent bonds themselves have to be broken to distort the crystal lattice. We call such substances network covalent crystals. Diamond consists of a lattice of carbon atoms covalently bonded to four other carbon atoms. The C—C bonds are very strong, making diamond a very hard, high-melting substance. A portion of the diamond structure is shown in Figure 4.10.

SUMMARY

The **pressure, volume,** and **temperature** behavior of gases can be explained in terms of **kinetic molecular theory.** Quantitative calculations require the use of the **gas laws.** Substances undergoing **condensation** and **vaporization** or melting and freezing exist in **equilibrium states** that are associated with **heats of vaporization** or **fusion,** respectively. Solids can be classified as either **amorphous** or **crystalline,** and the latter are characterized by a specific **crystal lattice** with a definite **unit-cell** structure.

Key Terms

Absolute temperature (Section 4.2)
Amorphous solid (Section 4.11)
Atmosphere (atm) (Section 4.1)
Atmospheric pressure (Section 4.1)
Barometer (Section 4.1)
Boyle's law (Section 4.4)
Charles's law (Section 4.4)
Combined gas law (Section 4.4)
Condensation (Section 4.6)
Crystal lattice (Section 4.11)
Crystalline solid (Section 4.11)

Equilibrium state (Section 4.7)
Gas laws (Section 4.4)
Heat of fusion (Section 4.10)
Heat of vaporization (Section 4.8)
Ideal gas law (Section 4.4)
Kelvin (K) (Section 4.2)
Kinetic molecular theory (Section 4.2)
millimeters of mercury (mmHg) (Section 4.1)
Pressure (Section 4.1)
Sphygmomanometer (Section 4.3)
Torr (Section 4.1)
Unit cell (Section 4.12)
Vaporization (Section 4.7)

ADDITIONAL PROBLEMS

4.13 Express the following pressures in millimeters of mercury (mmHg).
(a) 1.0 atm
(b) 0.50 atm
(c) 2.0 cmHg
(d) 700 torr

4.14 Express the following temperatures on the absolute scale (K).
(a) 20°C
(b) 50°C
(c) −15°C
(d) 0°F
(e) 200°F

4.15 In terms of kinetic molecular theory, explain why the pressure of a gas increases as its volume is decreased at constant temperature.

4.16 In terms of kinetic molecular theory, explain why the pressure of a gas decreases if the gas is cooled in a constant-volume container.

4.17 In terms of kinetic molecular theory, explain why the pressure of air in a tire decreases when air is released from the tire.

4.18 The blood pressure of a patient is 135/82. What do these numbers mean?

†**4.19** A gas sample occupies a volume of 1.0 L at 27°C and has a pressure of 800 torr. What volume would it occupy at 47°C with a pressure of 400 torr?

†**4.20** What is the pressure that a gas sample would exert in a 500-mL bulb, if the sample had a pressure of 0.90 atm in a 750-mL bulb at the same temperature?

†**4.21** A tire has an air pressure of 2.6 atm, when filled at 27°C. Predict the air pressure if the temperature increases to 62°C during a trip. (Assume that the volume of the tire is constant.)

†**4.22** What is the volume of the container needed to hold 0.160 g of O_2 so that the pressure of the oxygen is 0.800 atm at 25°C?

†Attempt these problems only if you have covered Section 4.4.

†**4.23** Calculate the temperature at which 0.100 mole of a gas occupies 2.24 L, when the pressure of the gas is 1.00 atm.
†**4.24** How many grams of methane (CH_4) are needed to create a pressure of 950 torr in a 250-mL bulb at 23°C?
4.25 An adult inhales and exhales about 7 L of air per minute. Use Table 4.1 to estimate the volume of each of the major components of dry air that an adult inhales and exhales in 1 min.
4.26 What are some of the serious air pollutants?
4.27 Would it be worse to be burned by a given mass of steam at 100°C, or by liquid water at 100°C, or would it be just the same? Explain.
4.28 According to Figure 4.7, ice begins to form after the removal of 25 cal from 1 g of water that starts at 25°C. What is the total amount of heat that must be removed to convert all the water in a 1-g sample from the initial liquid at 25°C to ice at 0°C?
4.29 Explain the difference between amorphous and crystalline solids.
4.30 Explain why ionic crystals have high melting points but molecular crystals have low melting points.
4.31 Explain the difference between covalent compounds that crystallize as molecular crystals and those that form network crystals.
*__4.32__ Why do you suppose mercury is the liquid chosen for barometers, rather than some other liquid such as water?
*__4.33__ Air is a mixture, but its composition does not vary greatly. Use Table 4.1 to calculate the "average molecular weight" of air. The percentage by volume is the same as the percentage by number of molecules. For purposes of this calculation, assume that the small fraction needed to make up 100% in Table 4.1 is nitrogen.
*__4.34__ The specific heat of a substance is the heat that must be removed or added to 1 g of the substance to change its temperature by 1°C. The specific heat of liquid water is 1 cal/(g°C). Use Figure 4.7 to estimate the specific heat of ice.
*__4.35__ Silicon carbide (SiC) is a covalent compound. Are crystals of this substance hard or soft? Do they have a high or low melting point? Base your answer on the type of crystal lattice you think the compound has in view of the periodic family relationship of carbon and silicon.

Problem-Solving Hints

If you have trouble solving any of the additional problems, refer to the sections listed next to the problem numbers.

4.13 4.1	**4.14** 4.2	**4.15** 4.3
4.16 4.3	**4.17** 4.3	**4.18** 4.3
4.19 4.4	**4.20** 4.4	**4.21** 4.4
4.22 4.4	**4.23** 4.4	**4.24** 4.4
4.25 4.5	**4.26** 4.5	**4.27** 4.8
4.28 4.10	**4.29** 4.11	**4.30** 4.13
4.31 4.13	**4.32** 4.1	**4.33** 4.5
4.34 4.10	**4.35** 4.13	

FOCUS 3 Ordinary and Photochemical Smog

Ordinary smog results simply from the concentration of certain pollutants in damp air. (The word "smog" is a combination of "smoke" and "fog.") The notorious smogs in London several decades ago resulted mainly from the burning of coal. Vast amounts of sulfur oxides, along with soot and other tiny particles, were produced. When the wind wasn't strong enough to blow these materials away, they collected to form a dense smog. Serious respiratory difficulties afflicted thousands and resulted in many deaths. The British have eliminated the London smogs by imposing severe controls on the burning of coal.

Photochemical smog results when certain materials released into the air undergo chemical reactions that require ultraviolet radiation from the sun. One example of photochemical smog is ozone, O_3, a very reactive substance that can cause irritation of the eyes, bronchia, and lungs. Ozone is produced by sunlight when oxides of nitrogen are present in the air.

Nitrogen oxides are produced when gasoline and air burn in the cylinders of automobiles at very high temperatures. The nitrogen and

FIGURE 1 Industrial pollution can be a major source of smog. (Environmental Protection Agency)

oxygen in the air react at these high temperatures to yield nitric oxide and nitrogen dioxide.

$$N_2 + O_2 \xrightarrow[\text{temperature}]{\text{high}} 2NO$$

$$2NO + O_2 \rightarrow 2NO_2$$

Under the influence of the ultraviolet (UV) radiation in sunlight, oxygen atoms are produced.

$$NO_2 \xrightarrow[\text{rays}]{\text{UV}} NO + O$$

The oxygen atoms can then react with ordinary oxygen molecules to produce ozone.

$$O + O_2 \rightarrow O_3$$

Ordinarily, most of the ozone reacts with the nitric oxide to regenerate nitrogen dioxide and oxygen.

$$O_3 + NO \rightarrow NO_2 + O_2$$

However, when hydrocarbons are present in the air, the oxygen atoms can react to produce certain oxidized hydrocarbon products that proceed to react with the nitric oxide.

$$O + \text{hydrocarbons} \rightarrow \text{oxidized hydrocarbons}$$

$$\text{oxidized hydrocarbons} + NO \rightarrow NO_2 + \text{hydrocarbons}$$

When the nitric oxide is removed by this last reaction, the reaction between nitric oxide and ozone cannot occur, so ozone levels slowly build up in the air.

Like the nitrogen oxides, hydrocarbons are given off when gasoline is burned in automobile engines. They become concentrated in the air when there is a lot of traffic and little wind. These conditions occur commonly in a number of large cities, and as a result photochemical smog is a continuing problem.

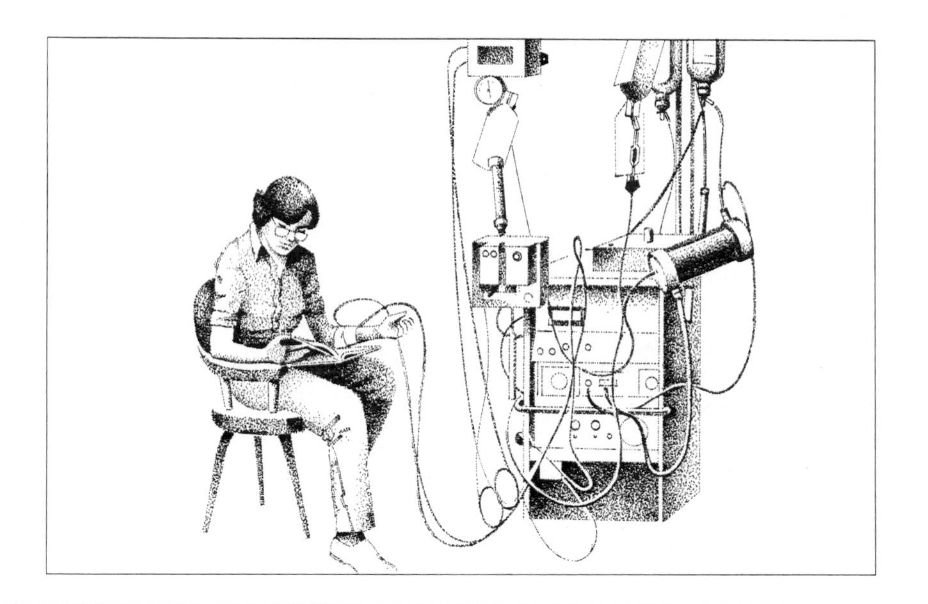

As we will learn later in this chapter, hemodialysis (artificial dialysis of the blood) is used to correct the chemical imbalances that occur when the kidney fails to function properly. The following case study will provide a valuable frame of reference for us.

R. G. is a 63-year-old white male who was found to have high blood pressure when examined three years ago for complaints of chest pain. At that time, his kidney function studies were normal, and treatment with medication resulted in a lowering of his blood pressure from 200/110 to within normal limits of 140/60. He was found to have protein in his urine (indicating abnormal kidney function) and small kidneys for his size, but his electrolytes and the rest of his routine laboratory tests were normal.

Approximately six months ago, he developed nausea and vomiting and became quite fatigued. He was readmitted by his family physician who found that the patient now had a hemoglobin level of 7 as compared to a normal hemoglobin of 14 two years ago. His carbon dioxide level was extremely low, indicating an acidotic (low blood pH) condition. Kidney studies showed a decrease in function. The patient's blood pressure had risen once again to 200/100. Urine cultures were taken, and no infection was revealed.

5

WATER AND ITS SOLUTIONS

The chest x ray showed an enlarged heart, but there was no evidence of any congestive heart failure. The patient was placed on a protein and sodium restricted diet. He was given bicarbonate tablets to correct his acidosis, but over the next 7 to 10 days, his renal (kidney) function did not improve. The patient did not respond to medical management and since no reversible causes of renal disfunction were identified, he was started on hemodialysis for control of his kidney abnormalities.

[Illustration: When normal kidney function is impaired and the body cannot filter the blood properly, kidney dialysis (purification of the blood by a machine) may be necessary.]

Since water is so common, we think of it as being ordinary. In fact, it is very peculiar stuff, and life depends on its uniqueness. If water had different properties, or didn't exist at all, the Earth would be unimaginably different.

An interesting thing about water is that it gets *less* dense as it freezes. Most substances become more dense when they change from liquid to solid. But since water gets less dense, ice floats. And since ice floats, the surfaces of lakes and rivers freeze first. The surface ice insulates the water below and helps to keep it from freezing. If ice were more dense than liquid water, it would sink to the bottom of lakes and rivers, allowing the water above to freeze until the whole body of water was ice. Fish couldn't survive a cold winter, and the whole ecology of the Earth would be different.

Although liquid water becomes less dense when it freezes, like most substances it becomes *more* dense as it cools (until it reaches its maximum density at 4°C). So we find that ice floats and cool water sinks. In the evening, the temperature drops at the surface of the water, and the cool surface water then sinks down through the warmer water below. In this way the water circulates, and oxygen and nutrients get spread around, which makes aquatic life possible.

But perhaps the most important property of water, to chemists if not to fish, is its remarkable ability to dissolve other substances. Water is sometimes called the universal solvent. This is an exaggeration, for, as you know, oil and water don't mix. Nonetheless, water can dissolve a great many compounds, and this ability is crucial to the operation of living organisms.

Why is water so exceptional? The answer lies in its molecular structure. We will look at the water molecule in detail and see just how the properties of water arise from its structure.

MOLECULAR STRUCTURE 5.1

What is the shape of the H_2O molecule?

We already know that the water molecule consists of two hydrogen atoms linked by polar covalent bonds to one oxygen atom (Section 3.10). Careful experimental measurements have shown that the H—O—H angle is 104.5°. The water molecule is therefore bent.

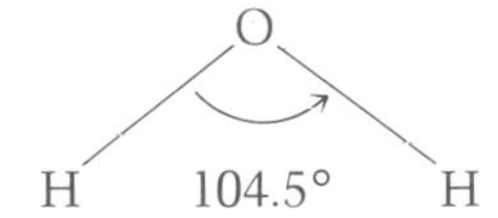

Oxygen is more electronegative than hydrogen, so the oxygen end of the molecule is slightly negative, and the hydrogen end is slightly positive. If the water molecule were linear (all three atoms in a straight line), the center of negative charge and the center of positive charge would coincide. Since the water molecule is bent, the centers of the charges are slightly separated. Such molecules, in which the positive and negative charges created by polar covalent bonds are a little separated, are called **dipoles.** They are sometimes pictured with a plus at one end and a minus at the other, to emphasize the separate positive and negative charges.

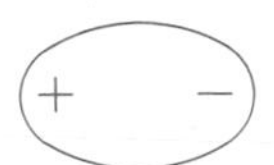

SOLUTIONS 5.2

Recall that solutions were defined in Section 2.4. The substance present in largest amount in a solution is called the **solvent,** and the substances dissolved in the solvent are called **solutes.** Water is the most common solvent in liquid solutions. When water or any other solvent has dissolved as much solute as possible, the result is a **saturated solution.**

5.3 WATER AS A SOLVENT FOR IONIC COMPOUNDS

Why is water a good solvent for salts?

Water is the best solvent for ionic substances (salts). The negative end of the dipolar water molecule attracts the positive ions, and the positive end attracts the negative ions (Figure 5.1). This sort of interaction—in which the poles of water molecules surround oppositely charged ions—is called **hydration of ions.** The hydration is sometimes so strong that the water molecules continue to surround the ions even in the solid state. The usual way to write the formula of a hydrated salt is to place a dot between the formula for the anhydrous salt and the number of

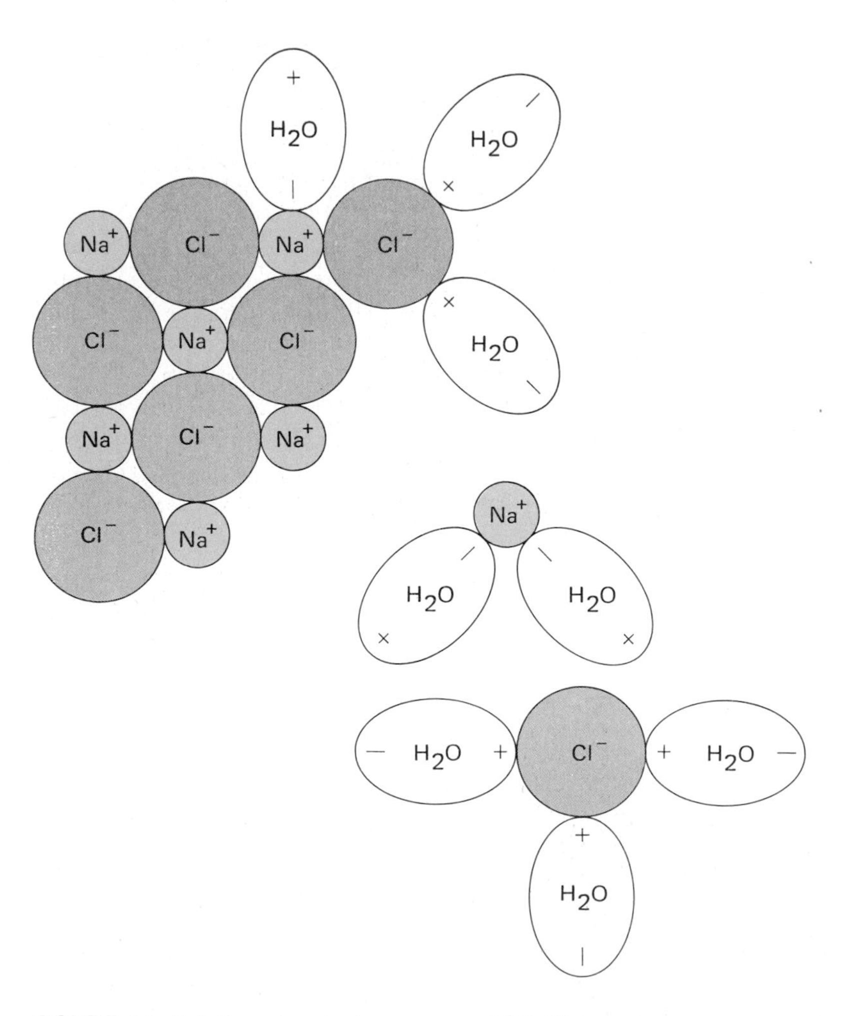

FIGURE 5.1 Solution of an ionic compound (NaCl) in water.

H_2O molecules combined with each formula unit. A few examples of solid hydrated salts are: $CuSO_4 \cdot 5H_2O$ (copper sulfate pentahydrate); $MgSO_4 \cdot 7H_2O$ (magnesium sulfate heptahydrate, also called epsom salts); $Na_2CO_3 \cdot 10H_2O$ (sodium carbonate decahydrate, also called washing soda). The water in hydrated salts is called water of hydration and can often be removed by heating. The unhydrated salt is then said to be **anhydrous** (without water). If the anhydrous salt is dissolved in water and recrystallized, the hydrated salt is formed once again.

Other dipolar solvents are known, but none is nearly as effective as water as a solvent for salts. As you might expect, molecules that have no separation of charge (that is, nonpolar molecules) make very poor solvents for ionic substances. Gasoline, benzene, and carbon tetrachloride are examples of nonpolar liquids in which ionic substances are insoluble.

HYDROGEN BONDING 5.4

When does hydrogen bonding occur?

Another result of the polarity of water molecules is **hydrogen bonding.** Whenever hydrogen shares a pair of electrons with one of the most electronegative atoms (N, O, or F), the partially positive hydrogen atom in one molecule is attracted to the partially negative atom (N, O, or F) in another molecule. This attraction creates a hydrogen bond (represented by a dotted line).

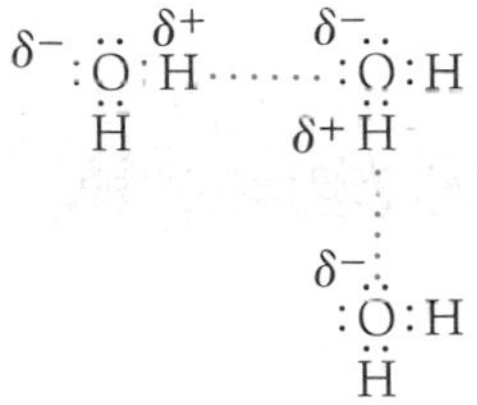

```
       ..  δ+      δ−..
  δ− :O:H·······:O:H
       ..          ..
       H        δ+ H
                   :
                   :
                δ−.:
                  :O:H
                   ..
                   H
```

This kind of attraction occurs only with hydrogen, because a hydrogen atom has only two electrons around it when it forms a covalent bond. So when a very electronegative atom such as oxygen shares these electrons, the proton (nucleus) of hydrogen is almost completely uncovered. The electrons spend most of their time with the other atom, and the hydrogen gets a partial positive charge. The positive charge will naturally attract the negative charge on the electronegative atom of an adjoining molecule.

Hydrogen bonds are only about one-tenth as strong as most covalent bonds. However, the hydrogen bond plays an important role in the properties of water. Hydrogen bonds also critically affect the molecular structure and hence the properties of important biological substances. These consequences of the hydrogen bond will be discussed later on in this book.

5.5 WATER AS A SOLVENT FOR COVALENT COMPOUNDS

Why do certain covalent substances dissolve in water?

Water is an excellent solvent for many covalent substances. Some covalent compounds, such as HCl, dissolve in water by reacting with it to produce ions.

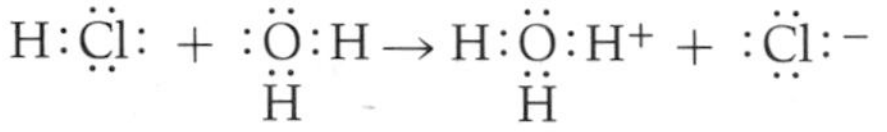

```
      ..      ..          ..          ..
    H:Cl: + :O:H  →  H:O:H⁺  +  :Cl:⁻
      ..      ..          ..          ..
              H           H
                  (hydronium ion)
```

In cases such as this, covalent bonds break and new ones form.

Many other covalent substances, such as alcohols, sugars (for example, glucose and sucrose), HF, and NH_3, readily dissolve in water because their molecules form hydrogen bonds with water molecules. Alcohols and sugars all have covalent —O:H bonds. As we learned in the last section, an —O:H bond will result in a partial positive charge on the hydrogen, which can then form a hydrogen bond with the negative end of a water molecule. Ammonia with its —N:H bonds and hydrogen fluoride with its F:H bond also meet the requirements for hydrogen bonding.

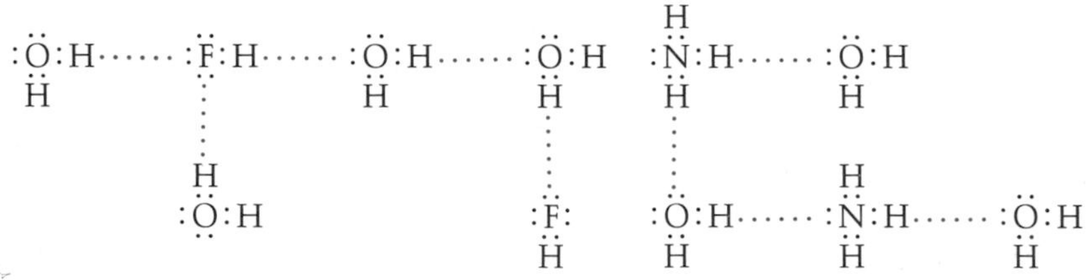

```
                                             H
 ..         ..          ..          ..       ..          ..
:O:H······:F:H······:O:H······:O:H   :N:H······:O:H
 ..         ..          ..          ..       ..          ..
 H          :           H           H        H           H
            :                       :        :
            H                       :        :           H
           ..                      ..       ..          ..          ..
          :O:H                     :F:      :O:H······:N:H······:O:H
           ..                      ..       ..          ..          ..
                                    H        H           H           H
```

The hydrogen bonds (dotted lines) provide the attractions necessary for solution in water.

5.6 MELTING AND BOILING POINTS OF WATER

We have seen that covalent compounds generally have low melting and boiling points (Section 3.7 and Table 3.2). In addition, substances with

TABLE 5.1 Melting and Boiling Points of Some Covalent Substances

Substance	Molecular weight	Melting point, °C	Boiling point, °C
Carbon tetrachloride, CCl_4	154	−23	77
Benzene, C_6H_6	78	6	80
Hydrogen chloride, HCl	37	−111	−85
Methyl alcohol (wood alcohol), CH_3OH	32	−98	65
Water, H_2O	18	0	100
Ammonia, NH_3	17	−73	−33
Methane, CH_4	16	−183	−162

lower molecular weights usually have lower melting and boiling points. Water is very unusual in this respect. Compared with covalent substances of similar or even higher molecular weight, water has peculiarly high melting and boiling points. Table 5.1 lists some melting and boiling points of covalent substances arranged in order of decreasing molecular weight. Notice how out of line water is.

In what respect are the melting and boiling points of water unusual?

Water also has high heats of vaporization and fusion (Sections 4.8 and 4.10). These high values fit in with its relatively high melting and boiling points. All these properties point to the same thing. Water molecules attract each other quite strongly. The attraction between water molecules is much stronger than the attraction between molecules of most other covalent compounds. The basis of these relatively strong **intermolecular attractions** is the dipolar nature of the water molecule and the ability of water molecules to hydrogen-bond. Both of these effects cause water molecules to cluster tightly together in the solid and liquid states.

SURFACE TENSION 5.7

Another manifestation of the intermolecular attraction between water molecules is the relatively high surface tension of water. **Surface tension** is a measure of the tendency of a liquid to resist the expansion of its surface area. This tendency causes drops of a liquid to be almost spherical, since a sphere has the least surface area for a given volume.

Why does water have a high surface tension?

An explanation for surface tension is illustrated in Figure 5.2. Molecules in the interor of a liquid are attracted equally in all directions, but those at the surface are only pulled downward. This effect causes the surface molecules to crowd together to produce a tight layer and

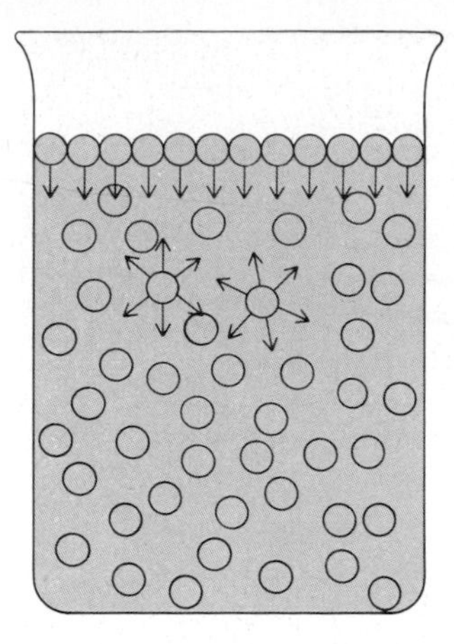

FIGURE 5.2 Surface tension is caused by surface molecules crowding together to produce a tight layer as they are pulled downward by the attraction of other molecules in the interior of the liquid.

gives a toughness to the surface. Lightweight insects can walk on the surface of ponds and streams because of the surface tension of water (Figure 5.3).

The greater the intermolecular attraction, the tighter the surface molecules will be pulled together. Water has a relatively high surface tension because of its dipolar interactions and extensive hydrogen bonding.

A **wetting agent** is a substance that lowers the surface tension of water. When a wetting agent is added, water spreads more readily and wets materials more effectively. Wetting agents act to disrupt the strong attractions between water molecules at the surface and those in the interior of the liquid. Because they operate on these surface forces, wetting agents are also called **surface-active agents** or **surfactants.**

5.8 TEMPERATURE AND PRESSURE EFFECTS ON SOLUBILITY

How is the water solubility of gases different from that of liquids and solids?

Most solids and liquids that dissolve in water become more soluble as the temperature rises. That is, the higher the temperature, the more solute is required to saturate a solution. Pressure has practically no effect on the solubilities of solids and liquids in water.

Temperature and pressure have entirely different effects on the solubility of gases. Most gases become less soluble in water as temperature

FIGURE 5.3 Lightweight insects can walk on water because of its high surface tension.

increases, whereas increasing pressure increases their solubility. Both of these effects have important biological implications.

Fish can survive only if there is enough dissolved oxygen (from the air) in the water. A relatively small increase in the temperature of a lake or bay can cause a significant decrease in the amount of dissolved oxygen. At the same time, a fish requires more oxygen as the temperature of the water rises. Discharge of waste heat from an industrial or power plant (thermal pollution) into natural water can thus affect fish life, by decreasing the dissolved oxygen available and increasing the amount needed.

Deep-sea divers often work at depths where the pressure is many times ordinary atmospheric pressure. Under high pressure, the solubility of air in the blood greatly increases. Air is mostly nitrogen. A diver who returns to the surface too rapidly experiences a sudden decrease in pressure. The solubility of nitrogen drops quickly, and the blood suddenly finds itself with more nitrogen than it can hold in solution. The excess nitrogen forms bubbles in the blood vessels, creating an extremely painful and dangerous condition called the bends. The bends can be avoided by a very slow ascent, so that the pressure is decreased over a longer time and the gas can escape from the blood without forming bubbles. Alternatively, divers can breathe a synthetic atmosphere of oxygen and helium instead of air. Helium is much less soluble than nitrogen, so much less of it dissolves in the blood.

5.9 Concentration

Often we need to know the exact amount of solute in a given amount of solution—in other words, the concentration of the solution. A number of different measures of concentration are used in scientific work. The three we will be concerned with are **weight–volume percent, volume–volume percent,** and **molarity.**

Weight–Volume Percent

Many routine clinical laboratory solutions are specified this way. To say a solution is 10% glucose means that there are 10 g of glucose in every 100 mL of the solution. This ratio of weight of solute (glucose) to volume of solution is the same no matter what the total volume of solution is. In 50 mL of 10% glucose solution there are 5 g of glucose.

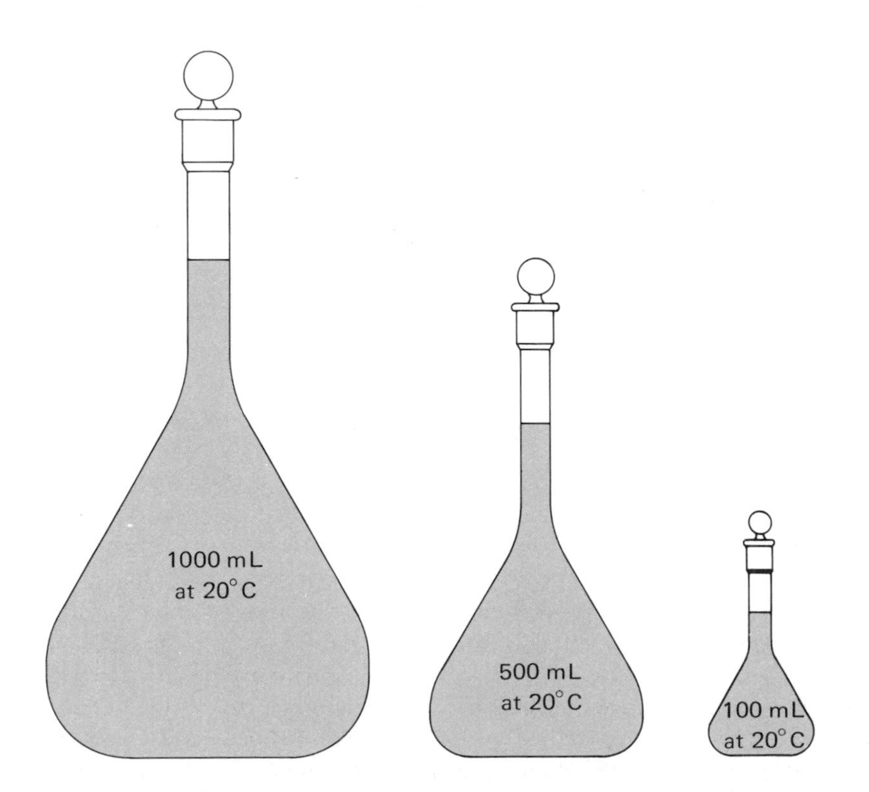

FIGURE 5.4 A volumetric flask is used to make up solutions of an exact volume. When solvent is added up to the mark on the stem, the volume at a specific temperature is equal to the value stamped on the flask.

In 500 mL of solution there are 50 g of glucose. In each case, the grams of glucose equal one-tenth (10%) of the volume of the solution. To prepare a 10% glucose solution, we must first decide how much solution we need. We can then use a simple equation.

$$\begin{pmatrix}\text{milliliters}\\\text{of solution}\end{pmatrix} \times \begin{pmatrix}\text{concentration}\\\text{of solution}\end{pmatrix} = \begin{pmatrix}\text{grams of}\\\text{solute}\end{pmatrix}$$

$$\cancel{\text{mL}} \times \frac{\text{g}}{100\ \cancel{\text{mL}}} = \text{g}$$

Example 5.1 How can we prepare 1.00 L of 10.0% glucose?
Solution

$$\text{mL} \times \frac{\text{g}}{100\ \text{mL}} = \text{g}$$

$$1.00\ \cancel{\text{L}} \times \frac{1000\ \cancel{\text{mL}}}{1\ \cancel{\text{L}}} \times \frac{10.0\ \text{g glucose}}{100\ \cancel{\text{mL}}} = 100\ \text{g glucose}$$

We weigh out 100 g of glucose on a balance, place it in a 1-L volumetric flask and add water to the mark (Figure 5.4).

Example 5.2 How many grams of potassium nitrate are contained in 250 mL of 4.8% KNO_3?
Solution

$$\text{mL} \times \frac{\text{g}}{100\ \text{mL}} = \text{g}$$

$$250\ \cancel{\text{mL}} \times \frac{4.8\ \text{g KNO}_3}{100\ \cancel{\text{mL}}} = 12\ \text{g KNO}_3$$

Volume–Volume Percent

This scale is commonly used to express concentrations of gaseous solutions. It gives the ratio of volume of solute to every 100 volumes of solution (see Table 4.1). This method of expressing concentration is never used for solutions in which a solid is dissolved in a liquid. Occasionally, it is used when one liquid is dissolved in another liquid. Rubbing alcohol is a 70% (vol/vol) solution of isopropyl alcohol in water. We could prepare 1 L of rubbing alcohol by pouring 700 mL of isopropyl alcohol into a 1,000-mL volumetric flask, adding water to the mark, and mixing.

Molarity

Molarity (M) is a measure of concentration in terms of the moles of solute per liter of solution.

$$\text{Molarity} = \frac{\text{moles of solute}}{\text{liter of solution}}$$

The term *molar concentration* is also used synonymously with molarity. In a 1 M solution, there is 1 mole (the molecular or formula weight in grams) of the solute in every liter of solution, 0.50 mole of solute in every 500 mL (half-liter) of solution, 0.10 mole of solute in every 100 mL of solution, and so on. You can see from the equation above that

$$\text{molarity} \times \text{liters of solution} = \text{moles of solute}$$

$$\frac{\text{mol}}{\cancel{\text{L}}} \times \cancel{\text{L}} = \text{mol}$$

Example 5.3 How can we prepare 300 mL of 0.50 M NaCl?
Solution

$$\frac{\text{mol}}{\text{L}} \times \text{L} = \text{mol}$$

$$0.50\,\frac{\text{mol NaCl}}{\cancel{\text{L solution}}} \times 300\,\cancel{\text{mL solution}} \times \frac{1\,\cancel{\text{L solution}}}{1000\,\cancel{\text{mL solution}}} = 0.15\text{ mol NaCl}$$

We need to take 0.15 mole of NaCl and add enough solvent to make a final volume of 300 mL. The weight of NaCl needed is easy to calculate.

$$0.15\,\cancel{\text{mol NaCl}} \times 58.5\,\frac{\text{g NaCl}}{\cancel{\text{mol NaCl}}} = 8.8\text{ g NaCl}$$

Example 5.4 Concentrated hydrochloric acid is a commercial product that is 12 M HCl in water. How can we prepare 500 mL of 1.0 M HCl from concentrated HCl?
Solution

$$\frac{\text{mol}}{\text{L}} \times \text{L} = \text{mol}$$

$$1.0\,\frac{\text{mol HCl}}{\cancel{\text{L}}} \times 500\,\cancel{\text{mL}} \times \frac{1\,\cancel{\text{L}}}{1000\,\cancel{\text{mL}}} = 0.50\text{ mol HCl}$$

Thus, 0.50 mole of HCl diluted with water to a volume of 500 mL gives the desired solution. The question now is how to get 0.50 mole of

HCl from 12 M HCl. One liter of 12 M solution will contain 12 moles of HCl. How many milliliters would contain 0.50 mole of HCl?

$$0.50 \text{ mol HCl} \times \frac{1.0 \text{ L}}{12 \text{ mol HCl}} = 0.042 \text{ L}$$

This calculation shows that 0.042 L (42 mL) of 12 M HCl contains 0.50 mole of HCl. We can now answer the original question: measure out exactly 42 mL of concentrated HCl, add it to sufficient water to make a total volume of 500 mL, and mix.

Problem 5.1 How many grams of silver nitrate are contained in 150 mL of 0.10% $AgNO_3$?

Problem 5.2 How would you prepare 250 mL of 15.0% potassium chloride solution?

Problem 5.3 How would you prepare 500 mL of 1.5 M NaCl?

Problem 5.4 How many moles of $MgSO_4$ are in 100 mL of 0.25 M $MgSO_4$? How many grams?

Problem 5.5 Commercial concentrated ammonia (sometimes called concentrated ammonium hydroxide) is 16 M NH_3. How can 800 mL of 0.10 M NH_3 be prepared from concentrated ammonia?

COLLIGATIVE PROPERTIES OF SOLUTIONS 5.10

What effect do solutes have on the properties of water?

A **colligative property** of a solution is a property that depends only on the number of particles of solute and not on the identity of the solute. We will look at three colligative properties of water solutions—**freezing-point lowering, boiling-point elevation,** and **osmotic pressure.**

As an example of the first of these properties, we use antifreeze in automobile radiators to lower the freezing point of the water and thus prevent damage to the cooling system from ice forming within it. In domestic ice-cream freezers, salt must be mixed with ice to lower the freezing point. This causes some of the ice to melt and remain liquid even at temperatures below 0°C.

Table 5.2 shows the effect on the freezing point of water of solutes that give different numbers of particles in solution. The freezing point of water drops as we increase the concentration of solute particles. Sucrose is a molecular substance. A 0.1 M solution of sucrose contains 0.1 M of solute particles. Sodium choride is ionic and each formula unit gives two ions. So 0.1 M sodium chloride yields 0.2 M of solute particles. Each formula unit of potassium sulfate gives three ions (two K^+

TABLE 5.2 Freezing Point of Water Solutions for Different Solutes

Solute	Total concentration of particles	Freezing point, °C
0.1 M $C_{12}H_{22}O_{11}$ (sucrose)	0.1 M	−0.18
0.1 M NaCl (sodium chloride)	0.2 M (0.1 M Na^+ + 0.1 M Cl^-)	−0.35
0.1 M K_2SO_4 (potassium sulfate)	0.3 M (0.2 M K^+ + 0.1 M SO_4^{2-})	−0.43

ions and one SO_4^{2-} ion), so 0.1 M yields 0.3 M of solute particles. Accordingly, sodium chloride depresses the freezing point of water roughly twice as much as sucrose does, and potassium sulfate depresses it even more.

A nonvolatile solute such as a salt will elevate the boiling point of water or any other solvent. A volatile solute may simply boil off and leave the solvent with its original boiling point. Antifreeze is a nonvolatile alcohol. Besides lowering the freezing point of water, it also elevates the boiling point, so it is routinely used in pressurized cooling systems.

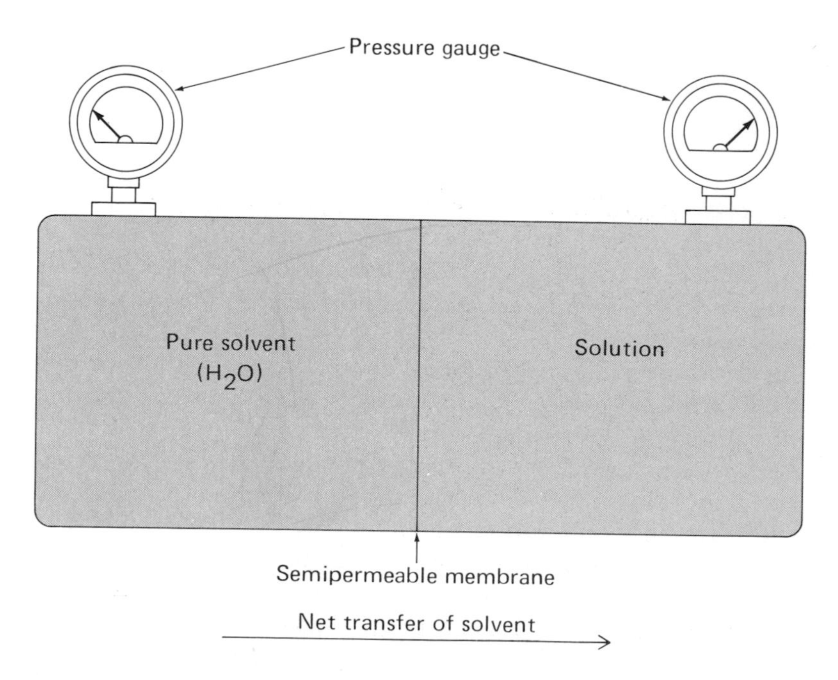

FIGURE 5.5 Measurement of osmotic pressure. The difference between the two pressure readings is the osmotic pressure of the solution.

Osmotic pressure is an interesting and important colligative property. Figure 5.5 shows how we can measure osmotic pressure. A **semipermeable membrane** allows water molecules to pass through it but does not allow solute particles to do so. Water molecules pass through from the solvent side and dilute the solution on the other side of the membrane. The increase in molecules on the solution side produces a pressure difference across the membrane. This pressure difference is called the osmotic pressure of the solution.

Problem 5.6 Which of the following water solutions would have the highest boiling point? Justify your answer.
(a) 0.1 *M* sucrose (b) 0.05 *M* NaF
(c) 0.05 *M* $MgCl_2$

COLLOIDAL DISPERSIONS 5.11

What happens when particles too large to dissolve are put into water?

Up to now we have talked about solutions in which the solute particles are molecules and ions. Sometimes a substance with particles much larger than individual molecules or ions can be briefly dispersed in a solvent. An example is a mixture of fine sand or clay in water. For a time, the sand or clay remains dispersed in the water, but soon the material will settle to the bottom of the container. Such a mixture is called a **suspension.** The solid material in a suspension can easily be filtered out. A particle of suspended material is usually 1,000 or more times larger than most molecules.

An interesting thing happens with particles of an intermediate size, between 10 and 1,000 times as big as most molecules. These particles do not settle out, nor can we remove them by ordinary filtration. They are too large to be considered in solution, but they remain dispersed for long periods. We call these particles **colloids** and such mixtures **colloidal dispersions.** Some protein molecules fall into the size range of colloids.

In a colloidal dispersion, the molecules of solvent are in rapid motion, constantly striking the colloidal particles and moving them about. This movement is called **Brownian motion.** It keeps the particles from settling out. There is another effect besides Brownian motion that plays a role in most stable colloidal dispersions: the colloidal particles carry the same electrical charge and thus repel each other.

A concentrated colloidal dispersion will often appear murky to the eye, but a very dilute dispersion may look like a solution. A solution can, however, easily be distinguished from a colloidal dispersion by

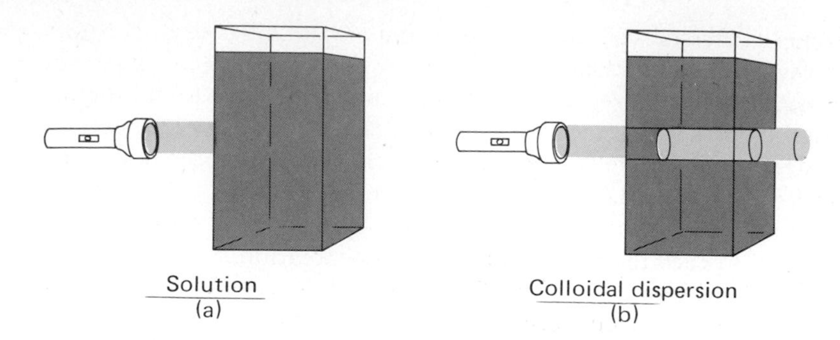

FIGURE 5.6 The Tyndall effect distinguishes a colloidal dispersion from a solution. Light passes through a solution (a) without scattering. Light is scattered by dispersed colloidal particles (b), so that the light beam becomes clearly visible.

means of the **Tyndall effect** (Figure 5.6). The large colloidal particles will reflect light and make the light rays visible as a shaft of light as they pass through a colloidal dispersion. A light beam is not visible as a shaft of light when it passes through a solution.

We see a very common example of the Tyndall effect when a beam of sunlight enters a darkened room. The beam is clearly visible as light is reflected by tiny dust particles. The dust particles are a colloidal dispersion in the air of the whole room, but are visible only in the beam of sunlight.

5.12 DIALYSIS

Why are cell membranes not semipermeable membranes?

The membranes that surround living cells are not of the osmotic type. Cells must be able to exchange food molecules and waste substances, as well as water molecules. A semipermeable membrane of the osmotic type would allow only solvent molecules to pass through. Cell membranes are more permeable and allow various ions and small molecules to pass. They prevent large molecules and colloids, however, from passing into or out of the cell.

Membranes that allow ions, water molecules, and other small molecules to pass, but hold back larger substances, are called **dialyzing membranes.** We can see that dialysis is similar to osmosis. The direction of migration of materials in dialysis is determined by the relative concentrations of substances on either side of the membrane. Materials tend to move in the direction that will dilute the more concentrated mixture.

When red blood cells are placed in pure water, they swell and ultimately rupture. This rupturing is called **hemolysis,** and it results from the passage of water through the cell wall into the more concentrated solution inside the cell (Figure 5.7c). Red blood cells that are placed in a concentrated salt solution will shrivel. This happens because water passes out of the cell through the cell wall to dilute the concentrated outside solution. The loss of water causes the cells to shrink. The term used to describe this process is **crenation** (Figure 5.7b). If red blood cells are placed into a 0.15 *M* NaCl solution, neither hemolysis nor crenation will occur, because the concentration of solute particles inside the cell is the same as in the outside solution and there is no net transfer of water (Figure 5.7a). Two solutions with equal concentrations of solute particles are said to be **isotonic.** Isotonic solutions will not transfer solvent across a semipermeable or dialyzing membrane. If one solution has a higher solute concentration than another, it is said to be **hypertonic.** The solution with the lower concentration is called **hypotonic.**

Problem 5.7 What concentration of sucrose will result in a solution isotonic with 0.05 *M* NaCl?

Problem 5.8 Will red cells undergo hemolysis or crenation when placed in a hypertonic solution? Explain.

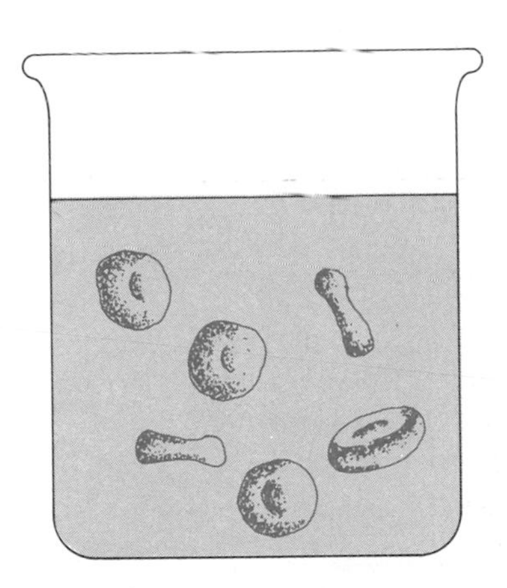

(a) Red cells in an isotonic solution

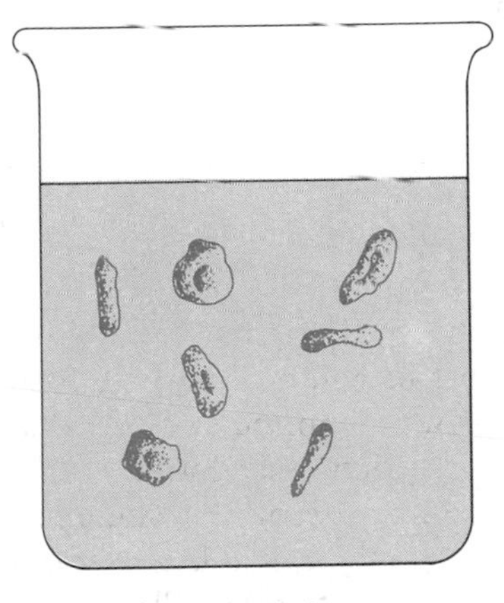

(b) Red cells in a hypertonic solution

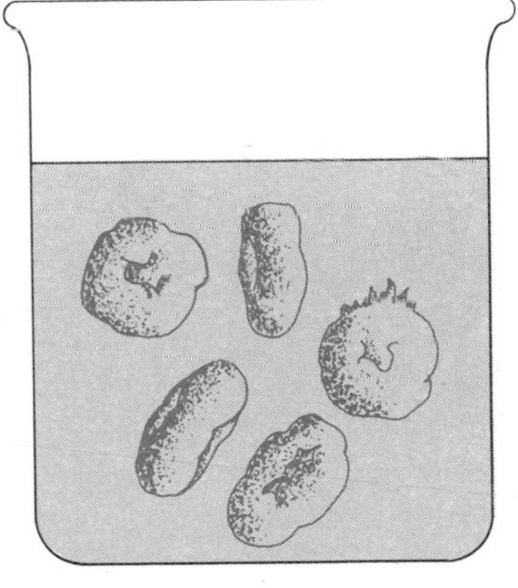

(c) Red cells in a hypotonic solution

FIGURE 5.7 Red blood cells may exchange water with the surrounding solution depending on the concentrations on either side of the cell membrane. (a) No net exchange occurs in an isotonic solution because there is no concentration difference. (b) Water passes out of the cells into a more concentrated (hypertonic) solution and the cells undergo crenation. (c) Water passes into the cells from a less concentrated (hypotonic) solution. The cells then expand and may undergo hemolysis (rupture).

The kidneys help to maintain the correct concentration of salts in the blood, and they remove wastes through a dialysis mechanism. If the kidneys do not work properly, fatal illness can result. Hemodialysis (artificial dialysis of the blood) is used in cases of chronic kidney disorder. Commonly known as a kidney machine, a hemodialyzer diverts the blood through a cellulose tube that serves as a dialyzing membrane (Figure 5.8). A solution containing the proper concentration of all the solutes that should be in the blood is circulated around the tube.

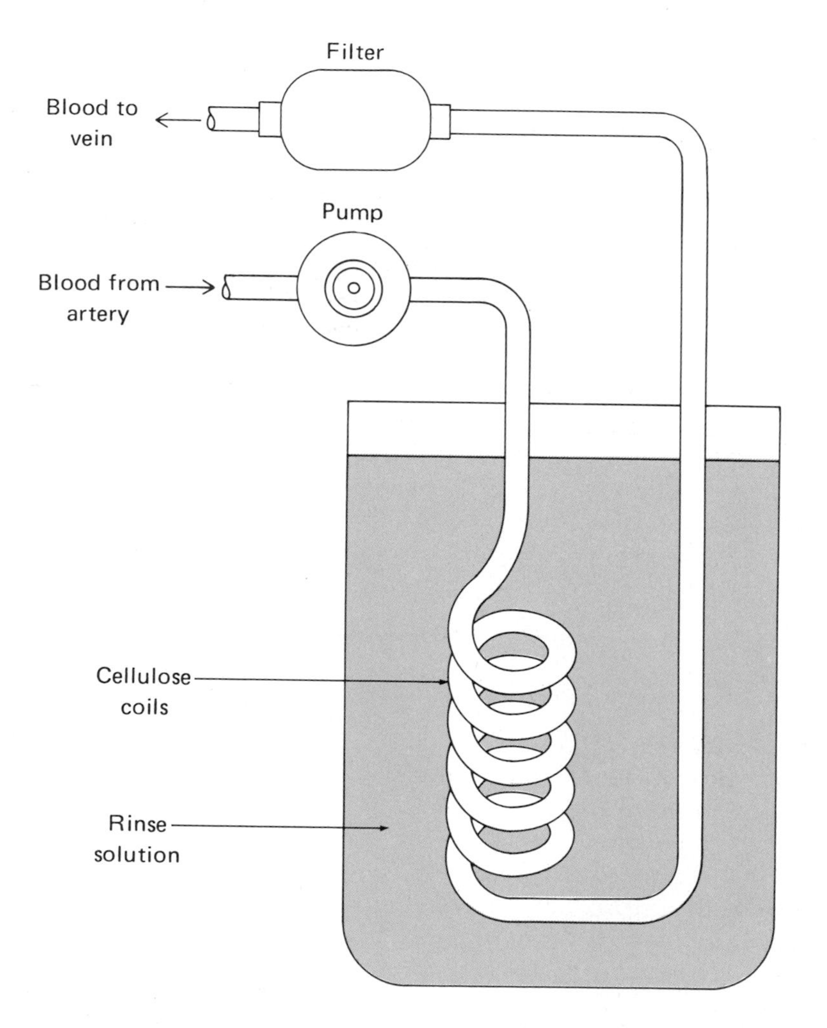

FIGURE 5.8 Hemodialysis takes place in an artificial kidney machine. Blood from an artery is pumped through coils of cellulose tubing. Surrounding these coils is a rinse solution containing the proper concentration of all the solutes that should be in the blood. This solution is circulated around the coils during operation. The waste materials in the blood pass through the cellulose into the solution. The cleansed blood is filtered and returned to the body through a vein.

Desirable substances are at equal concentrations in the blood and in this fluid, so they remain in the blood and do not pass through the cellulose tube. Waste products are at a higher concentration in the blood than in the fluid (there are, in fact, none in the fluid), so they pass through the tube, out of the blood and into the fluid, and are removed. In this way, the blood is cleansed and returned to the body.

SUMMARY

The polar covalent bond between hydrogen and oxygen and the bent structure of the water molecule causes water to be a **dipolar** substance capable of dissolving salts through **hydration of ions. Hydrogen bonding** accounts for water's ability to dissolve certain covalent substances and, together with the dipolar nature of water, it explains the high degree of **intermolecular attraction** between water molecules. This attraction accounts for the relatively high melting and boiling points of water, as well as for its high **surface tension.** The amount of **solute** in a solution can be expressed quantitatively by **weight–volume percent, volume–volume percent,** or **molarity.** The important **colligative properties** of water solutions are **freezing-point lowering, boiling-point elevation,** and **osmotic pressure.** Dialysis is similar to osmosis, except that small molecules in addition to water can cross a **dialyzing membrane.**

Key Terms

Anhydrous salt (Section 5.3)
Boiling-point elevation (Section 5.10)
Brownian motion (Section 5.11)
Colligative property (Section 5.10)
Colloid (Section 5.11)
Colloidal dispersion (Section 5.11)
Crenation (Section 5.12)
Dialyzing membrane (Section 5.12)
Dipole (Section 5.1)
Freezing-point lowering (Section 5.10)
Hemolysis (Section 5.12)
Hydration of ions (Section 5.3)
Hydrogen bonding (Section 5.4)
Hypertonic solution (Section 5.12)
Hypotonic solution (Section 5.12)

Intermolecular attraction (Section 5.6)
Isotonic solution (Section 5.12)
Molarity (Section 5.9)
Osmotic pressure (Section 5.10)
Saturated solution (Section 5.2)
Semipermeable membrane (Section 5.10)
Solute (Section 5.2)
Solvent (Section 5.2)
Surface-active agent (Section 5.7)
Surface tension (Section 5.7)
Surfactant (Section 5.7)
Suspension (Section 5.11)
Tyndall effect (Section 5.11)
Volume–volume percent (Section 5.9)
Weight–volume percent (Section 5.9)
Wetting agent (Section 5.7)

ADDITIONAL PROBLEMS

5.9 How does the fact that the H_2O molecule is bent rather than linear relate to its dipolar nature?
5.10 Why is water a good solvent for ionic compounds?
5.11 What types of covalent compounds dissolve in water?
5.12 What is hydrogen bonding?
5.13 Why does water have such high melting and boiling points for a substance of its molecular weight?
5.14 How do temperature and pressure affect the solubility of solids, liquids, and gases in water?
5.15 How does thermal pollution affect fish life?
5.16 What causes the bends?
5.17 How would you prepare 500 mL of 0.10% silver nitrate solution?
5.18 How many grams of potassium chloride are in 50 mL of 5.0% KCl solution?
5.19 What is the percent concentration of a solution prepared by mixing 6.00 g of lithium nitrate in enough water to make 175 mL of solution?
5.20 What is the molarity of a solution prepared by dissolving 28.6 g of washing soda ($Na_2CO_3 \cdot 10H_2O$) in enough water to make 500 mL of solution?
5.21 What is the molarity of a solution prepared by diluting 10 mL of 0.050 *M* HCl to a total volume of 50 mL?
5.22 How many moles of sucrose are contained in 25 mL of a 0.20 *M* solution?
5.23 What is a colligative property? Name some.

5.24 Which of the following solutions are isotonic with each other?
(a) 0.050 *M* KCl (b) 0.020 *M* $CaCl_2$
(c) 0.030 *M* NaCl (d) 0.020 *M* $Al_2(SO_4)_3$

5.25 What is the difference between an osmotic membrane and a dialyzing membrane?

***5.26** Certain anhydrous salts have such strong affinities for water that they are used as drying agents. One of these is anhydrous calcium chloride, $CaCl_2$. If 1.11 g of $CaCl_2$ combines with water to form 2.19 g of the hydrated salt, what is the formula of hydrated calcium chloride?

***5.27** Calculate the molarity of concentrated nitric acid (HNO_3), given that the concentrated acid is 69.5% HNO_3 and the solution has a density of 1.42 g/mL.

***5.28** Explain how the cell membranes in the roots of trees play a role in getting water to the branches and leaves at the top of tall trees.

Problem-Solving Hints

If you have trouble solving any of the additional problems, refer to the sections listed next to the problem numbers.

5.9 5.1
5.10 5.3
5.11 5.5
5.12 5.4
5.13 5.6
5.14 5.8
5.15 5.8
5.16 5.8
5.17 5.9
5.18 5.9
5.19 5.9
5.20 5.9
5.21 5.9
5.22 5.9
5.23 5.10
5.24 5.12
5.25 5.10, 5.12
5.26 5.3
5.27 5.9
5.28 5.10

FOCUS 4 Soft and Hard Water

If soap lathers easily in water, the water is said to be *soft;* if it is difficult to make the soap lather and a scum forms, the water is said to be *hard.* Hard water results from the presence of one or more of the ions Ca^{2+}, Mg^{2+}, and Fe^{2+}. These ions often occur in water that has passed through mineral deposits containing compounds of these metals ($MgCO_3$, $CaSO_4$, and so on).

Soap is made up of sodium or potassium salts of fatty acids. Fatty acids are moderately large molecules that are called "fatty" because they are found in animal fat (Chapter 20). A soap molecule is a long chain, and the two ends of the chain behave differently. The ionic end dissolves in water, whereas the hydrocarbon end dissolves in oils. Oil and water, as you know, do not mix. If you pour water on an oily stain, the water will simply pass over the oil. But if you mix soap with the water, the soap will combine with both the oil and the water. When you wash the water away, both the soap and the oil will go with it. It is this ability to combine with both water and oils that makes soap an effective cleanser.

The trouble with washing in hard water is that the calcium, magnesium, or iron ions react with the soap molecules to form an insoluble scum, so that a lot of soap is wasted in forming scum before an excess of soap is present and available for cleaning.

There are several ways to attack the problem of hard water. In some hard water, the main negative ion is the bicarbonate ion (HCO_3^-). Such water is said to be temporary hard water because it can be softened simply by heating.

$$Ca^{2+} + 2HCO_3^- \xrightarrow{\text{boiling}} CaCO_3 + H_2O + CO_2$$

The calcium carbonate, $CaCO_3$, is insoluble and precipitates from the solution. The carbon dioxide leaves the solution as a gas. A similar reaction occurs between magnesium or iron ions and the bicarbonate ion.

Permanent hard water contains other negative ions (Cl^-, NO_3^-, SO_4^{2-}), and heating will not soften it, because these anions cannot react to remove the metal ions the way HCO_3^- can. There are several ways to deal with permanent hard water.

1. Use a detergent that does not react with calcium, magnesium, or iron ions.
2. Add a precipitating agent such as sodium carbonate (washing soda) to remove the unwanted metal ions.

$$Mg^{2+} + CO_3^{2-} \rightarrow MgCO_3$$

The insoluble carbonate precipitates, leaving the water soft. The sodium ions introduced with the carbonate ions cause no trouble, since soap itself has sodium or potassium ions.

3. Remove the unwanted metal ions and replace them with sodium ions by means of ion-exchange techniques. Naturally occurring sodium aluminum silicates, known as zeolites, are used to soften water. The hard water runs through the zeolite, which retains the hard-water ions and releases sodium ions to the water. When the zeolite becomes saturated with calcium, magnesium, or iron, it can be regenerated by passing a concentrated sodium chloride solution (brine) through it; sodium ions displace the calcium, magnesium, or iron ions, and the zeolite is then ready to soften more hard water. Synthetic materials that can perform the ion-exchange function have also been developed.
4. Bind the calcium, magnesium, or iron ions in solution so that they cannot react with soap. Substances that strongly bind metal ions in solution are called *chelating agents*. One of the most effective chelating agents is the tripolyphosphate ion, $P_3O_{10}^{5-}$, which is also a good cleaner itself. This ion, in the form of its sodium salt, $Na_5P_3O_{10}$, is used in many phosphate detergents. Unfortunately, excessive phosphate levels in waste water that flows into lakes and rivers can cause pollution by stimulating the growth of algae. This problem is so troublesome that certain localities have limited or banned phosphates in detergents.

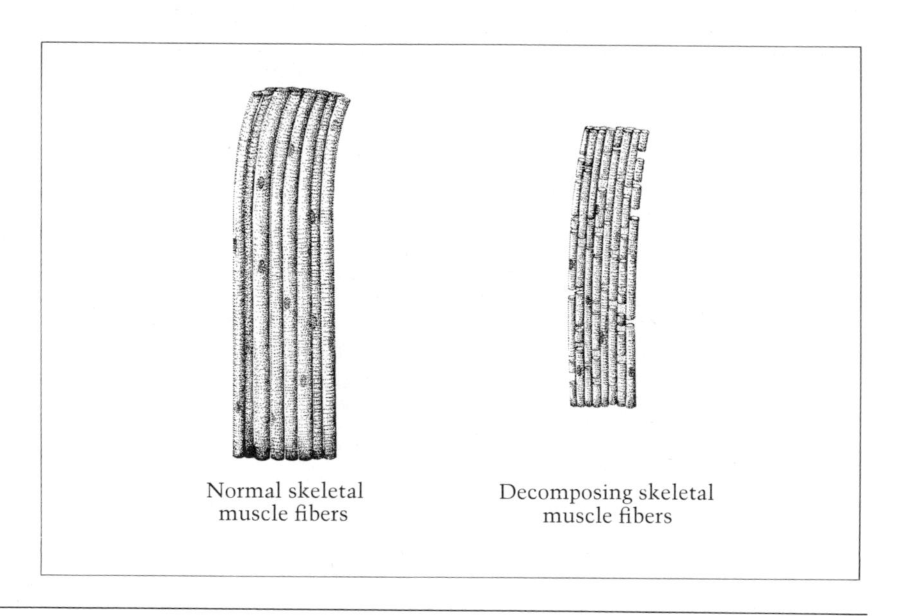

Normal skeletal muscle fibers

Decomposing skeletal muscle fibers

Death of a living animal organism results in tissue decomposition, which involves a series of chemical reactions. Some of these reactions are fast; others are slow. As is the case with any chemical reaction, the speed with which decomposition takes place depends on the environmental conditions present at the time. All activities of decomposition increase with an increase in temperature.

The fast-paced reactions begin with rigidity of the body of the animal—called rigor mortis—which commences shortly after death. Basically, lactic acid accumulates in the muscles and causes shortening of the muscle fibers that results in their becoming stiff. This process is accelerated by high temperatures and humidity and is delayed by freezing.

Enzymes originally present in the body of the animal play a very important role in decomposition. When respiration has stopped and contraction of the heart and circulation of the blood have ceased, certain enzyme activities still continue. These enzyme activities not only contribute to the clotting mechanism of the blood and the development of rigor mortis but also to the early changes of the cells after death. Various enzymes contribute to the decomposition of organs. This process is facilitated by weak acids and higher temperatures, is delayed by alkaline (base-oriented) reactions of the tissues and low temperatures, and occurs without bacterial influence (Tomio Watanabe, Atlas of Legal Medicine *(Philadelphia: Lippincott, 1968), p. 135).*

6

CHEMICAL REACTIONS

After the initial rapid chemical and enzymatic reactions take place, slower-paced bacterial changes assume responsibility for the continued decomposition of the body of the animal. These changes can be delayed by such processes as embalming (treatment with preservatives).

[Illustration: Normal skeletal muscle is smooth and continuous but is readily destroyed by enzymes in the chemical decomposition that occurs with the death of the organism.]

We know that some substances react with each other and some don't. We know that reactions fundamentally involve a rearrangement of electrons—either by a transfer from one atom to another, or by the sharing of electrons by atoms. But what determines how fast the reaction goes? It is obvious that some reactions take place much more swiftly than others. The rusting of iron occurs very slowly, but the burning of a match occurs very rapidly, with a burst of light.

The rate of a reaction is determined, first of all, by the nature of the substances that are reacting. Some substances won't react at all, or react very slowly. But assuming that the substances will react, we find that the rate at which they react depends on three additional factors: the concentrations of the substances, the temperature, and whether or not a catalyst is present. Catalysts are very important, and we will discuss them in detail later. For now, let's just say that they are substances that can speed up a reaction without being used up in the reaction.

Speed is not the only important thing in a chemical reaction; completeness is also very important. A reaction is said to be quantitative when all the reacting substances have been changed into the maximum possible amount of products. Many reactions, however, stop far short of converting reactants completely into products. The completeness of a reaction is affected by some of the same things that influence the rate of the reaction—the nature of the substances, their concentrations, and the temperature.

6.1 FAST AND SLOW REACTIONS

The rates of different reactions vary enormously. Sodium metal and chlorine gas react rapidly to form sodium chloride. At room temperature, hydrogen and iodine react very slowly to form hydrogen iodide, but at high temperatures, they form the compound much more rapidly.

At room temperature hydrogen and oxygen react so slowly to form water that a mixture of hydrogen gas and oxygen gas remains unchanged almost indefinitely. However, if some powdered platinum metal is introduced into the mixture, the gases react explosively to form water.

Clearly, there are several factors that affect the rates of chemical reactions. We will look at these separately.

NATURE AND CONCENTRATION OF REACTANTS 6.2

Do oppositely charged ions usually react faster than molecules?

Most reactions between oppositely charged ions in solution are very rapid. For example, insoluble silver chloride (AgCl) is formed as fast as Ag^+ ions and Cl^- ions are mixed in a solution. On the other hand, reactions between covalent molecules are usually slower, because covalent bonds have to be broken and new ones formed. This process is usually slower than the simple combination of ions of opposite charge. The chemical nature of reacting substances thus has an important effect on the speed of reaction.

Generally speaking, a higher concentration of reactants leads to a faster reaction. The quantitative relationship between the rate of a reaction and the concentration of reactants is not always a simple one, however. Although we will not analyze this relationship in detail, we can understand the overall effect of concentration on reaction rate if we think of reactant particles colliding with each other and reacting. The more particles there are in a given volume (the higher the concentration), the more frequent the collisions and the faster the reaction.

TEMPERATURE 6.3

Why haven't all possible reactions already taken place?

Temperature affects the rates of all chemical reactions in this way: *increasing temperature increases the rate of a reaction; decreasing temperature decreases the rate of a reaction.* We already know that the principal effect of increasing the temperature of a substance is to increase the kinetic energy of its molecules. Many reactions cannot occur until the reactant particles have a certain amount of energy. This is a very fortunate circumstance because otherwise all reactions that could have occurred would long since have done so.

The proteins of our skin could react with the oxygen in the air to form CO_2, H_2O, and oxides of nitrogen and sulfur, if the temperature were high enough. Oxygen molecules constantly collide with these proteins, giving ample opportunity for reaction. Yet this reaction does not take place, at least not at the temperatures in which we live. The collisions between the reactant molecules are not effective collisions—that is, they do not lead to a reaction. The protein molecules and the oxygen molecules simply do not have enough energy to react.

A molecule needs energy to react, just as a hiker needs energy to climb over a mountain range. The mountain range is a physical barrier to the hiker's progress. Similarly, many chemical reactions meet energy barriers that slow down their progress.

6.4 FREE ENERGY OF ACTIVATION

What is the difference between the ordinary change in free energy for a reaction and the free energy of activation?

We saw in Section 1.15 that free energy is lost in spontaneous reactions. We now see that there is also an energy barrier between reactants and products. The higher this energy barrier, the more energy the reactant molecules must have in order to react when they collide. We call this energy barrier the **free energy of activation,** $\Delta G^{\ddagger}$. Figure 6.1 illustrates the difference between $\Delta G^{\ddagger}$ and the ordinary change in free energy, ΔG.

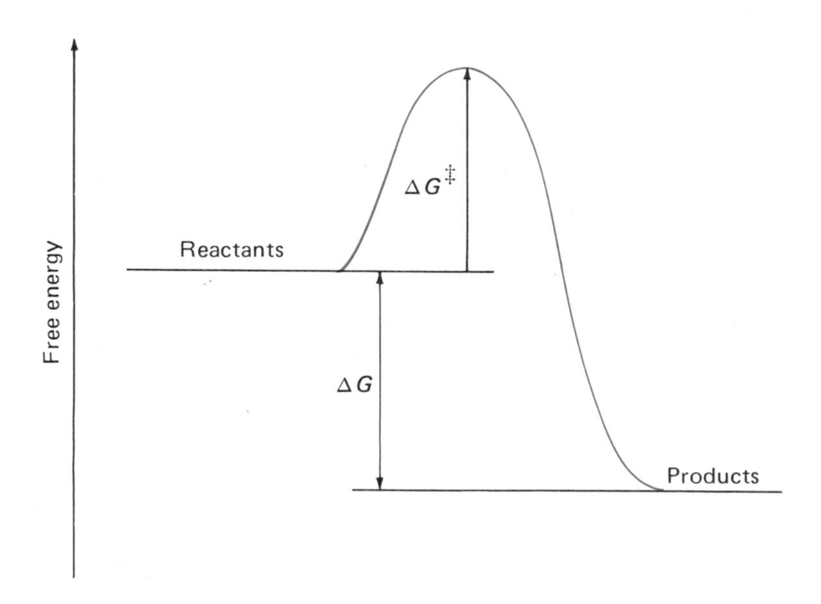

FIGURE 6.1 The free energy of activation, $\Delta G^{\ddagger}$, represents a barrier to reaction. The larger the $\Delta G^{\ddagger}$ for a reaction, the slower the reaction. The free energy of reaction, ΔG, represents the net change in free energy for the reaction.

There is no direct relationship between $\Delta G^{\ddagger}$ and ΔG. That is, both may be large or both small, or one may be large and the other small. The point to keep in mind is that the greater the $\Delta G^{\ddagger}$ for a reaction, the less likely it is that reactant molecules will actually react when they collide. Energy can be added by raising the temperature of the reactants or, in some cases, by introducing electrical or radiant energy. When molecules of higher energy collide, they are more likely to get over the free-energy-of-activation barrier. Reactions with small energies of activation are rapid, because almost all collisions between reactants are effective; even relatively low-energy molecules will have sufficient energy to get over a low barrier.

Problem 6.1 How does an increase in the concentrations of reactants affect the rate of a reaction? What effect does an increase in temperature have?

CATALYSTS 6.5

What are the two main types of catalyst?

Many essential biochemical reactions cannot take place in any reasonable length of time without the presence of **enzymes.** These are protein molecules that speed up biochemical reactions, but are not themselves consumed in the reactions. Enzymes are examples of catalysts (in Chapter 21 we will study enzyme catalysis in more detail). A **catalyst** in general terms is any substance that speeds up a reaction but remains unchanged at the end of the reaction.

Powdered platinum is a catalyst for the reaction between hydrogen and oxygen gases to form water (Section 6.1). In this case, the platinum acts as a **surface catalyst** and offers a suitable place for hydrogen and oxygen molecules to react. A surface catalyst is a solid that speeds up reactions between substances in the liquid or gaseous states. It is sometimes called a **heterogeneous catalyst,** because the reactants and the catalyst are in different physical states.

Catalysts that are in the same physical state as the reactants are called **homogeneous catalysts.** These catalysts actually combine chemically with one or more reactants, but are regenerated in a later step so that they are unchanged at the end of the reaction. An example is the catalysis by iodide ion of the decomposition of hydrogen peroxide (H_2O_2). The decomposition of H_2O_2 into water and oxygen at ordinary temperatures is fairly slow.

$$2H_2O_2 \rightarrow 2H_2O + O_2 \qquad (1)$$

If some I^- ions are added to the solution, however, the reaction speeds up, because the following reactions occur rapidly.

$$H_2O_2 + I^- \rightarrow H_2O + IO^- \qquad (2)$$

$$IO^- + H_2O_2 \rightarrow H_2O + O_2 + I^- \qquad (3)$$

If equations (2) and (3) are added together, the result is equation (1). The I^- ion is not used up, but its presence alters the reaction rate.

Problem 6.2 What is a catalyst? What is the difference between a surface catalyst and a homogeneous catalyst?
Problem 6.3 Verify that the addition of equations (2) and (3) results in equation (1) above.

A catalyst speeds up the rate of reaction by making it possible for the reaction to occur with a smaller free energy of activation. We can get an idea of what this means by returning to our analogy in Section 6.3 of a hiker confronted by a mountain range. A hiker who discovered a path that went around the mountain rather than over it would reach his or her destination sooner. A catalyst offers an easier way for substances to react with each other. The new reaction path has a lower $\Delta G^‡$

A critically important point to remember is that the net reactants and products are the same for both the catalyzed and uncatalyzed reactions. This means that although catalysts lower $\Delta G^‡$ for a reaction, they never affect the change of free energy of the reaction, ΔG. Figure 6.2 illustrates this crucial point.

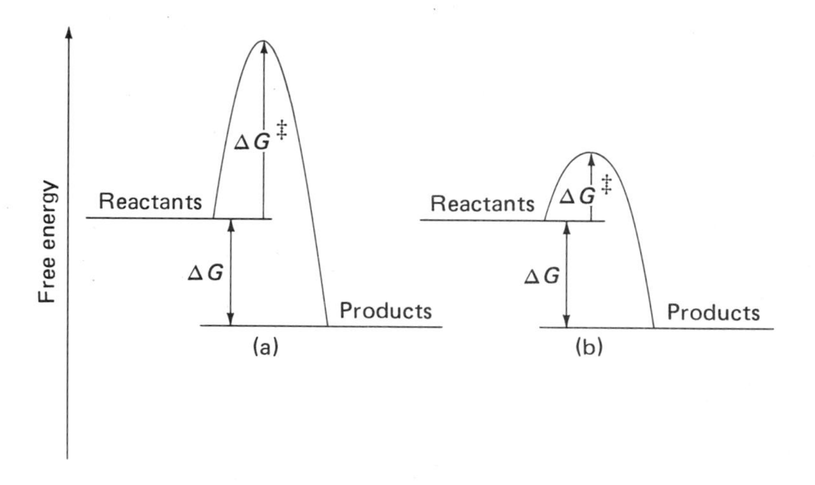

FIGURE 6.2 A catalyst lowers $\Delta G^‡$ and makes it easier for a reaction to take place. (a) Uncatalyzed reaction. (b) Same reaction in the presence of a catalyst. Note that ΔG is the same for both reactions.

Problem 6.4 How do the ΔG and $\Delta G^{\ddagger}$ of the uncatalyzed decomposition of hydrogen peroxide compare with the ΔG and $\Delta G^{\ddagger}$ of the I^--catalyzed decomposition?

REVERSIBLE REACTIONS 6.6

Why do some reactions seem to stop before all the reactants have been used up?

A **reversible chemical reaction** is one that can run in either direction. For example, if a mixture of H_2 and I_2 is heated in a closed container to about 400°C, HI is formed.

$$H_2 + I_2 \rightarrow 2HI$$

But not all of the H_2 and I_2 will be used up. After a while the reaction appears to stop, and a mixture of H_2, I_2, and HI remains. The reason why all the reactants are not used up is that the reverse reaction also takes place.

$$2HI \rightarrow H_2 + I_2$$

At first, when only H_2 and I_2 are present, there is no HI available and the reverse reaction cannot occur. As soon as some HI is formed, though, the reverse reaction can begin to operate. The forward and reverse reactions are then moving in opposite directions at the same time. As long as the concentrations of H_2 and I_2 are large, and the concentration of HI is very small, the forward reaction will be faster than the reverse reaction. But after a while, the concentration of HI will start increasing, and the concentrations of H_2 and I_2 will become smaller. So the forward reaction will slow down while the reverse reaction speeds up. Eventually, the rates of the forward and reverse reactions will become equal. When this happens, there can be no further changes in the concentrations of any of the substances, because as fast as H_2 and I_2 form HI, the HI molecules will react to produce H_2 and I_2 (Figure 6.3).

When the rates of forward and reverse reactions of a reversible chemical reaction become equal, the reaction appears to stop. The concentrations of the chemical substances no longer change. Actually, both forward and reverse reactions continue, but they cancel each other out. We call this situation a dynamic equilibrium or, simply, **chemical equilibrium.**

A double arrow is often used to emphasize the fact that a reversible reaction can take place in both directions and will eventually reach equilibrium.

$$H_2 + I_2 \rightleftarrows 2HI$$

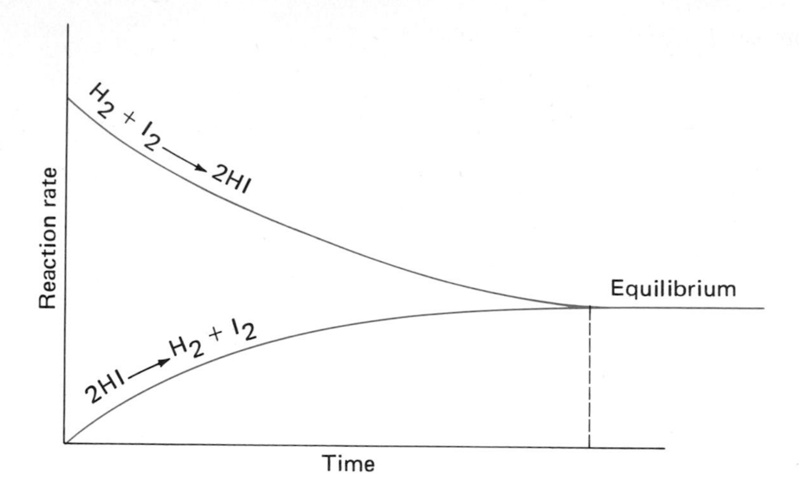

FIGURE 6.3 In a reversible chemical reaction, chemical equilibrium is established when the rate of the forward reaction equals the rate of the reverse reaction.

It is important to keep in mind that, although the *rates* of the forward and reverse reactions are equal at chemical equilibrium, the *concentrations* of the substances at equilibrium are usually not equal. The rate of a chemical reaction depends on the nature of the substances that react as well as on their concentrations. Thus, the rates of the two reactions—forward and reverse—can equalize under conditions in which the concentrations of the various substances are quite different.

Problem 6.5 How can the forward and reverse reactions of a reversible chemical reaction take place without causing a change in the concentrations of reactants and products?

6.7 EQUILIBRIUM CONSTANT

How do we express an equilibrium constant?

Every chemical reaction that reaches equilibrium has an **equilibrium constant** (K). This is the ratio of the concentration of products to the concentration of reactants. For the reaction we have been considering,

$$H_2 + I_2 \rightleftarrows 2HI$$

we express the equilibrium constant as

$$K = \frac{[HI][HI]}{[H_2][I_2]}$$

or, in slightly more compact form,

$$K = \frac{[HI]^2}{[H_2][I_2]}$$

We must pay attention to several important points concerning equilibrium constant expressions.

1. When square brackets are put around the formula of a substance, the expression stands for the molarity (or molar concentration) of the substance.
2. The molarities of the products are multiplied together in the numerator, and molarities of the reactants are multiplied together in the denominator.
3. The exact value of an equilibrium constant (K) depends on the particular chemical reaction and the temperature at which the equilibrium exists. For example, at 500°C, $K = 50$ for the above reaction. This means that, no matter how much H_2 and I_2 or HI we start out with, when equilibrium is reached at 500°C, the value of K equals 50. The individual molarities of HI, H_2, and I_2 at equilibrium will vary depending on the starting concentrations, but the *ratio* of the concentrations is a constant. If equilibrium is attained at 500°C, $[HI]^2$ will always be fifty times $[H_2][I_2]$. However, if equilibrium is reached at a different temperature, K will have a different value.

Once the equilibrium constant has been measured for a particular reaction at a particular temperature, it can be recorded and used when needed. Tables of various equilibrium constants are to be found in chemical reference books. Equilibrium constants allow us to calculate the equilibrium concentrations of substances without having to measure them directly.

Most of the time we will not be concerned with the actual concentrations of substances involved in an equilibrium reaction. So we will have little occasion to deal directly with equilibrium constants. However, the next chapter considers several special kinds of equilibrium constants.

Problem 6.6 What is the meaning of $[SO_2] = 0.25$?

Problem 6.7 Does the value of an equilibrium constant for a reaction depend on the concentrations of reactants at the beginning of the reaction? Briefly explain.

Problem 6.8 Does the value of an equilibrium constant for a reaction depend on the temperature at which equilibrium is achieved?
Problem 6.9 Write the equilibrium constant expression for the following reaction.

$$2SO_2 + O_2 \rightleftharpoons 2SO_3$$

6.8 CATALYSTS AND CHEMICAL EQUILIBRIUM

Can a catalyst affect the equilibrium constant, K?

Even if a chemical reaction is reversible and will eventually reach equilibrium, it may take a very long time to reach the equilibrium state. Introduction of a catalyst will speed up both the forward and the reverse reactions and cause equilibrium to be attained more rapidly. A catalyst, however, has no effect on the relative concentrations of reactants and products in the equilibrium state. In other words, a catalyst does not affect the value of the equilibrium constant. Catalysts can thus cause reversible chemical reactions to reach equilibrium more quickly, but the equilibrium state will be the same for the catalyzed reaction as for the uncatalyzed reaction.

6.9 LE CHATELIER'S PRINCIPLE

Once a reversible reaction has reached equilibrium, the concentrations of the substances will not change as long as the reaction is not disturbed. If we disturb the reaction, the equilibrium will be momentarily upset, though it will ultimately be restored. For example, if we disturb the equilibrium

$$H_2 + I_2 \rightleftharpoons 2HI$$

What happens if we add something to a reaction at equilibrium?

by adding more I_2 to the reaction mixture, what can we say about the concentrations of H_2 and HI after equilibrium is restored? The answer to this question is given by what is called **Le Chatelier's principle:** *an equilibrium reaction will shift so as to oppose a disturbance.*

In the example at hand, the disturbance is the addition of I_2. The only way the addition of more I_2 can be opposed is for some H_2 to react with some of the extra I_2 and use it up. In the process, some extra HI is formed, so we say the reaction has shifted to the right. If we add some extra HI, then the reaction would shift to the left, thereby using up some of the added HI and at the same time producing more H_2 and I_2.

Likewise, if we remove some HI, the reaction will oppose the loss of HI by causing more H_2 and I_2 to react to produce HI. In other words, the reaction will shift to the right.

Sometimes a product of a reversible reaction is deliberately and continuously removed in order to cause a permanent shift. Under such circumstances, equilibrium can never be established. Lime (CaO) is produced commercially by heating limestone ($CaCO_3$) to a very high temperature. In a closed container an equilibrium is established, which involves the two solids $CaCO_3$ and CaO and the gas CO_2.

$$CaCO_3 \rightleftarrows CaO + CO_2$$

Now, if we want to prepare as much lime as possible by reacting all of the $CaCO_3$, we need to avoid equilibrium. Simply by continuously withdrawing the gaseous CO_2, we shift the reaction continuously to the right so that all of the $CaCO_3$ decomposes.

Problem 6.10 Predict how the reversible reaction

$$3H_2 + N_2 \rightleftarrows 2NH_3$$

will shift for each of the following disturbances.

(a) H_2 is added
(b) NH_3 is removed
(c) NH_3 is added
(d) N_2 is removed
(e) H_2 and N_2 are added

SUMMARY

The rates of chemical reactions are affected by the nature of the reactants, their concentrations, the temperature, and the presence or absence of **catalysts.** The **free energy of activation** is an energy barrier that the reactants must overcome for any reaction to occur. Catalysts serve to lower the free energy of activation, but they do not affect the overall change of free energy of a reaction. **Reversible chemical reactions** can go in either direction, and they reach **chemical equilibrium** when the rate is the same in both directions. An **equilibrium constant** governs any reaction that has reached equilibrium. Catalysts cause reactions to reach equilibrium more rapidly, but they do not affect the equilibrium constant. **Le Chatelier's principle** predicts that when a reaction at equilibrium has been disturbed, it will shift to oppose the disturbance and re-establish equilibrium.

Key Terms

Catalyst (Section 6.5)
Chemical equilibrium (Section 6.6)
Enzymes (Section 6.5)
Equilibrium constant (Section 6.7)
Free energy of activation (Section 6.4)
Heterogeneous catalyst (Section 6.5)
Homogeneous catalyst (Section 6.5)
Le Chatelier's principle (Section 6.9)
Reversible chemical reaction (Section 6.6)
Surface catalyst (Section 6.5)

ADDITIONAL PROBLEMS

6.11 Explain the difference between $\Delta G^{\ddagger}$ and ΔG for a reaction.
6.12 What are enzymes?
6.13 Why are surface catalysts also called heterogeneous catalysts?
6.14 Can a catalyst change ΔG or $\Delta G^{\ddagger}$ for a reaction?
6.15 What is a reversible reaction?
6.16 What can be said about the rate of the forward and reverse reactions for a chemical reaction that is at equilibrium?
6.17 Write the equilibrium-constant expression for each of the following reactions.

(a) CH_3OH (methyl alcohol) $+$ HCO_2H (formic acid) $\rightleftarrows$ H_2O $+$ HCO_2CH_3 (methyl formate)

(b) $H_2 + Cl_2 \rightleftarrows 2HCl$

(c) $N_2O_4 \rightleftarrows 2NO_2$

(d) C_2H_5OH (ethyl alcohol) $+$ HBr $\rightleftarrows$ C_2H_5Br (ethyl bromide) $+$ H_2O

6.18 State Le Chatelier's principle and give some examples of its application.
6.19 Explain in terms of Le Chatelier's principle what happens to the concentrations of all substances in the following reaction when the equilibrium is changed in the ways described below. (All the substances are gases, and the mixture is in a closed container.)

$$2SO_2 + O_2 \rightleftarrows 2SO_3$$

(a) O_2 is added
(b) SO_2 is removed
(c) SO_3 is added
(d) SO_3 is removed

6.20 K for a certain reaction is 1.00×10^2 and K for a different reaction at the same temperature is 1.00×10^5 For which reaction will reactants be converted to products to a greater degree?

***6.21** At 500°C, K = 50 for

$$H_2 + I_2 \rightleftarrows 2HI$$

What is the value of K at 500°C for

$$2HI \rightleftarrows H_2 + I_2 ?$$

***6.22** Different amounts of H_2, I_2, and HI are placed in a closed container and held at 500°C until equilibrium is attained. Calculate [HI] at equilibrium, given that experiments show that $[H_2] = 1.5$ and $[I_2] = 0.33$ at equilibrium. (See Problem 6.21.)

Problem-Solving Hints

If you have trouble solving any of the additional problems, refer to the sections listed next to the problem numbers.

6.11 6.4	**6.12** 6.5	**6.13** 6.5
6.14 6.5	**6.15** 6.6	**6.16** 6.6
6.17 6.7	**6.18** 6.9	**6.19** 6.9
6.20 6.7	**6.21** 6.7	**6.22** 6.7

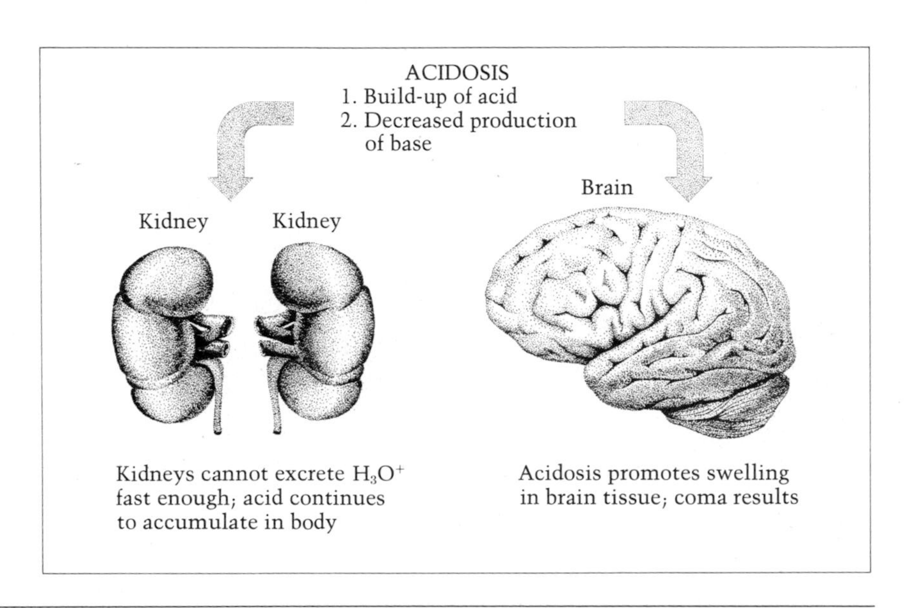
ACIDOSIS
1. Build-up of acid
2. Decreased production of base
Kidney
Kidney
Brain
Kidneys cannot excrete H_3O^+ fast enough; acid continues to accumulate in body
Acidosis promotes swelling in brain tissue; coma results

Acids and bases, which we will discuss in this chapter, are kept in a delicate balance in the human body. Unfortunately, if the balance is tipped significantly to one side or the other, death can result. The kidney is the primary organ responsible for maintaining this critical balance, and it does so by regulating the secretion of acids and bases in the urine.

Diabetes mellitus, a specific form of diabetes, frequently involves kidney complications that result in the body (1) retaining too many acidic compounds and (2) failing to produce a sufficient amount of basic (alkaline) compounds. The consequence of too much acid and too little base in the body is a condition known as acidosis. The blood of the patient with acidosis will show both a decrease in the normal level of carbon dioxide and a low pH (an abnormally high hydronium-ion concentration). In our discussion of pH later in this chapter, we'll see that the blood pH of the healthy person is between 7.35 and 7.45. On the other hand, the person suffering from acidosis will have a blood pH lower than 7.3.

The net result of acidosis is the inability of the red blood cells to take up sufficient oxygen for supply to the tissues. While steps are being taken to restore normal kidney function in a patient with acidosis, alkaline drugs can be administered to correct the acid-base balance. Thus the essential nutrients are again made available to the blood cells and the body tissues so that they can function normally.

7

ACIDS AND BASES

[*Illustration:* Acidosis can affect the normal function of the kidney as well as the brain.]

ACIDS AND BASES

Acids have a bad reputation. Some people think that all acids are dangerous. They imagine that any acid can eat its way through metal. But many everyday, harmless, and useful substances are acids. Vinegar is simply a dilute solution of acetic acid in water. Vitamin C is ascorbic acid. Lemon juice contains citric acid. Boric acid is so mild that a dilute solution is used as an eyewash.

A few strong acids really will dissolve some metals: for example, sulfuric acid, which is used in drain cleaners, and hydrochloric acid, which is used in your stomach to help digest food.

A number of bases also find practical uses. If your stomach puts out too much hydrochloric acid, you may swallow an antacid to quiet the troubled organ. All the popular antacids—Alka-Seltzer, Tums, Rolaids, or plain baking soda—contain mild bases that combine with the acid in your stomach. Strong bases are useful too. One such is sodium hydroxide, better known as lye. Lye can be used to clean drains, and although it won't dissolve steel, it can dissolve skin.

Traditionally, acids and bases have been distinguished on the basis of certain properties. Acids have a sour taste; bases are bitter. Acids and bases change the color of natural dyes such as litmus or cabbage juice. We will see that the most important difference between an acid and a base lies in their chemical natures. An acid can give up, and a base can combine with, a very tiny particle—an H^+ ion. This fact is the basis of modern definitions of an acid and a base.

DEFINITIONS 7.1

What is the difference between an acid and a base?

Acids are substances capable of giving up hydrogen ions. **Bases** are substances capable of combining with hydrogen ions. The hydrogen atom consists of one proton and one electron, and the hydrogen ion, H^+, is a hydrogen atom that has lost its electron. The hydrogen ion is thus simply a proton. The Brønsted-Lowry theory, in fact, defines an acid as a proton donor and a base as a proton acceptor.

An acid never gives up an H^+ ion unless there is some base present to combine with the H^+ ion. Hydrogen chloride, HCl, is an acid because it can lose an H^+ ion. Water, H_2O, is a base because it can combine with an H^+ ion and form the hydronium ion, H_3O^+. The reaction between hydrogen chloride and water is an example of an acid–base reaction.

$$HCl + H_2O \rightarrow H_3O^+ + Cl^-$$

When an acid loses its H^+ ion, the fragment left behind is called the **conjugate base** of the acid. The conjugate base of HCl is the Cl^- ion.

When a base combines with an H^+ ion, the new substance formed is called the **conjugate acid** of the base. The conjugate acid of H_2O is the H_3O^+ ion.

The interesting point about any acid–base reaction is that the products of such a reaction are another acid and base—namely, the conjugate acid and conjugate base of the reactant base and acid. We can label the reaction between HCl and H_2O as follows:

HCl	+	H_2O	$\rightarrow$	H_3O^+	+	Cl^-
acid		base		conjugate acid of H_2O		conjugate base of HCl

We can also call HCl the conjugate acid of the base Cl^- and H_2O the conjugate base of the acid H_3O^+.

AMPHOTERISM 7.2

How can a substance be both an acid and a base?

Some substances can either lose or gain an H^+ ion and thus can function either as an acid or as a base. We call them **amphoteric substances.** When sulfuric acid loses an H^+ ion, the conjugate base

HSO_4^- (hydrogen sulfate ion) results. For example, H_2SO_4 reacts with the base water.

$$H_2SO_4 + H_2O \rightarrow H_3O^+ + HSO_4^-$$

But the hydrogen sulfate ion, HSO_4^-, itself can lose an H^+ ion and act as an acid.

$$HSO_4^- + H_2O \rightarrow H_3O^+ + SO_4^{2-}$$

The HSO_4^- ion is a base because it can combine with an H^+ ion to form H_2SO_4. The HSO_4^- ion is also an acid because it can lose an H^+ ion. The HSO_4^- ion is an amphoteric substance.

Water is another example of an amphoteric substance. In all the above reactions, H_2O functions as a base. Many times, however, water molecules give up H^+ ions and thus serve as an acid. A common example is the reaction of the base ammonia with water.

NH_3	+	H_2O	$\rightarrow$	NH_4^+	+	OH^-
base		acid		ammonium ion (conjugate acid of NH_3)		hydroxide ion (conjugate base of H_2O)

Problem 7.1 Identify all acids and bases in the following reaction.

$$NH_3 + HSO_4^- \rightarrow NH_4^+ + SO_4^{2-}$$

7.3 STRONG ACIDS AND BASES

We will be concerned with water solutions of acids and bases. The acid or base concentration will usually be less than 1 *M*. Given these conditions, we can make some generalizations about strong acids and bases and weak acids and bases.

Why does 0.1 M HCl contain virtually no HCl molecules?

A **strong acid** reacts completely with water to form hydronium ion and the conjugate base of the acid. For example, HCl reacts completely with water.

$$HCl + H_2O \rightarrow H_3O^+ + Cl^-$$

A bottle marked 0.1 *M* HCl thus contains virtually no HCl molecules. What it does contain, in addition to water, is 0.1 *M* H_3O^+ ion and

TABLE 7.1 Some Common Strong Acids

Acid	Name of water solution
$HClO_4$	Perchloric acid
HI	Hydroiodic acid
HBr	Hydrobromic acid
HCl	Hydrochloric acid
HNO_3	Nitric acid
H_2SO_4	Sulfuric acid

0.1 *M* Cl^- ion. Solutions of strong acids do not contain molecules of the strong acid itself. They contain H_3O^+ and the conjugate base of the strong acid. Table 7.1 lists some common strong acids and the names given to their water solutions.

The thing to keep in mind is that the chemical formula of a strong acid does not represent what is actually in a solution of the acid. Solutions of 1 *M* or less of any strong acid contain the H_3O^+ ion and the conjugate base of the particular strong acid. Much higher concentrations of strong acids are required to give solutions that contain any un-ionized molecules of the acid.

The most important **strong base** is the hydroxide ion, OH^-. Soluble hydroxides such as sodium hydroxide (NaOH) and potassium hydroxide (KOH) simply dissolve in water and release the OH^- ion along with a positive ion, Na^+ or K^+.

WEAK ACIDS AND BASES 7.4

What is the difference between a weak acid and a strong acid?

A **weak acid** reacts only partially with water. Like a strong acid, it forms the H_3O^+ ion and a conjugate base, but it forms less of them. More of a weak acid remains in an un-ionized state. Weak acids undergo reversible reactions with water and establish an equilibrium.

Acetic acid (the acid in vinegar; CH_3CO_2H) is a typical weak acid.*

$$CH_3CO_2H + H_2O \rightleftharpoons H_3O^+ + \underset{\text{(acetate ion)}}{CH_3CO_2^-}$$

*The acetic acid molecule contains four hydrogen atoms, but only one of these can be lost as an H^+ ion. This fact is emphasized by writing one hydrogen atom separately from the other three in the formula.

Molecules of a weak acid (unlike molecules of a strong acid) can exist in solution because they only partially react with water. The ordinary equilibrium constant for this reaction would be written

$$K = \frac{[H_3O^+][CH_3CO_2^-]}{[CH_3CO_2H][H_2O]}$$

If we multiply both sides by $[H_2O]$, we get

$$[H_2O]K = \frac{[H_3O^+][CH_3CO_2^-]}{[CH_3CO_2H]}$$

As well as being a base, water is also the solvent in an acid solution and therefore present in a high concentration. When small concentrations of water react, the total concentration of water remains almost unchanged. The term $[H_2O]K$ is practically a constant and is replaced by the symbol K_a. The **acid constant, K_a,** for acetic acid is thus written

$$K_a = \frac{[H_3O^+][CH_3CO_2^-]}{[CH_3CO_2H]}$$

Equilibrium constants for the reaction of weak acids with water are always written this way.

The actual value for K_a must be determined by experiment. For acetic acid at 25°C, $K_a = 1.8 \times 10^{-5}$. Carbonic acid (H_2CO_3) is a weak acid with $K_a = 4.5 \times 10^{-7}$ at 25°C.

$$K_a = \frac{[H_3O^+][HCO_3^-]}{[H_2CO_3]} = 4.5 \times 10^{-7}$$

The larger value of K_a for acetic acid, as compared with the K_a for carbonic acid, means that acetic acid is a stronger acid than carbonic acid. Both acids are considered weak, but carbonic acid is weaker than acetic acid.

Ammonia is a typical **weak base.**

$$NH_3 + H_2O \rightleftarrows NH_4^+ + OH^-$$

Weak base molecules can exist in solution because they only partially react to produce OH^- ion and the conjugate acid of the weak base. The equilibrium constant for this reaction is expressed as the **base constant, K_b,** for ammonia.

$$K_b = \frac{[NH_4^+][OH^-]}{[NH_3]}$$

Again, $[H_2O]$ is virtually constant and so is omitted from the K_b expression. Table 7.2 lists some common weak acids and bases and their K_a and K_b values, respectively.

TABLE 7.2 Some Common Weak Acids and Bases

Acid	K_a(25°C)	Name
CH_3CO_2H	1.8×10^{-5}	Acetic acid
H_2CO_3	4.5×10^{-7}	Carbonic acid
H_3BO_3	8.0×10^{-10}	Boric acid
HCO_3^-	4.7×10^{-11}	Hydrogen carbonate ion (bicarbonate ion)
Base	K_b(25°C)	**Name**
CO_3^{2-}	2.1×10^{-4}	Carbonate ion
NH_3	1.8×10^{-5}	Ammonia
HCO_3^-	2.2×10^{-8}	Hydrogen carbonate ion (bicarbonate ion)
$CH_3CO_2^-$	5.6×10^{-10}	Acetate ion

Problem 7.2 Identify the following as acid, base, or amphoteric substance.

(a) HCl (b) CH_3CO_2H
(c) HSO_4^- (d) NH_3
(e) NH_4^+ (f) H_2O

Problem 7.3 For the substances listed in Problem 7.2, write the conjugate base of each acid, the conjugate acid of each base, and both the conjugate acid and conjugate base of amphoteric substances.

Problem 7.4 Write an equation for the equilibrium reaction between carbonic acid and water.

Problem 7.5 Write an equation for the equilibrium reaction between the carbonate ion and water.

Problem 7.6 Identify all acids and bases for the reactions of Problems 7.4 and 7.5.

Problem 7.7 Of the weak acids listed in Table 7.2, which one is the strongest? The weakest? Of the weak bases listed in Table 7.2, which one is the strongest? The weakest?

Problem 7.8 Write the algebraic expressions for the K_a's and K_b's listed in Table 7.2.

Problem 7.9 We do not speak of a K_a for a strong acid. Why? [Hint: What is the equilibrium concentration for a strong acid? Try to set up the expression for K_a for a strong acid such as HCl.]

7.5 RELATIVE ACID–BASE STRENGTH

If HCl is a strong acid, what can we say about the strength of Cl^- as a base?

A little reflection will show that the conjugate base of a strong acid is a very weak base. That is to say, a base like the Cl^- ion (the conjugate base of the strong acid HCl) is a very weak base, since the Cl^- ion has little ability to hold on to the H^+ ion in solution. The weakness of Cl^- as a base is what makes HCl such a strong hydrogen-ion donor. The conjugate bases of all strong acids have little ability to attract H^+ ions in solution, so they are very weak bases.

Table 7.1 lists some strong acids. The conjugate bases of all these acids are *very* weak bases. As a practical consequence, these ions, ClO_4^-, I^-, Br^-, Cl^-, and so on, have no effect on the acid–base properties of their solutions.

The conjugate base of a weak acid is a fairly strong base. This makes sense, because a weak acid is not a strong H^+-ion donor. This means that its conjugate base holds on to the H^+ ion fairly strongly. For example, the weak acid, acetic acid, has a conjugate base, acetate ion, that is strong enough to pull a proton away from water.

$$CH_3CO_2^- + H_2O \rightleftharpoons CH_3CO_2H + OH^-$$

Problem 7.10 Boric acid is a very weak acid. Write the formula for the conjugate base of boric acid, and write an equation for the reaction of this base with water.

7.6 NEUTRAL, ACIDIC, AND BASIC SOLUTIONS

What determines whether a solution is acidic, basic, or neutral?

A neutral water solution is one in which the concentration of the H_3O^+ ion equals the concentration of the OH^- ion. Remember that we indicate the molar concentration of a species by placing square brackets around it. We can express the condition for a **neutral solution** as

$$[H_3O^+] = [OH^-]$$

An **acidic solution** has $[H_3O^+]$ greater than $[OH^-]$ and a **basic solution** has $[OH^-]$ greater than $[H_3O^+]$.

Water solutions always contain some H_3O^+ ions and some OH^- ions, because water is both a weak acid and a weak base. The equilibrium reaction

$$H_2O + H_2O \rightleftarrows H_3O^+ + OH^-$$

is always taking place in water. In *pure* water, $[H_3O^+] = [OH^-]$, so we say that water is neutral. If an acid is added to pure water, the concentration of the H_3O^+ ion will increase and the equilibrium reaction will shift to the left. The $[OH^-]$ decreases, and the solution becomes acidic.

If a base is added to pure water, the concentration of OH^- ion will increase, and again the equilibrium reaction will shift to the left. This time, the $[OH^-]$ increases while the $[H_3O^+]$ decreases, and the solution becomes basic.

It is easy to see that any acid, strong or weak, will cause a solution to increase its H_3O^+-ion concentration and decrease its OH^--ion concentration, so that the solution becomes acidic. Likewise, any base, strong or weak, will cause an increase in the OH^--ion concentration and a decrease in the H_3O^+-ion concentration, so that the solution becomes basic.

Certain salts contain ions that function as acids or bases and hence affect the acidity of solutions. The explanation for these effects is quite simple, as the following examples illustrate.

Example 7.1 A 0.1 *M* NH_4Cl solution is slightly acidic. Why?
Solution We can see that the salt is composed of the ammonium ion (NH_4^+) and the chloride ion (Cl^-). We also see that NH_4^+ can lose an H^+ ion, that is, it is an acid. Of course, the Cl^- ion is a base, but it is such a weak base (the conjugate base of the strong acid HCl) that we can neglect it. An equilibrium reaction will occur between the weak acid NH_4^+ and the weak base H_2O.

$$NH_4^+ + H_2O \rightleftarrows H_3O^+ + NH_3$$

The increase in H_3O^+-ion concentration will thus cause the solution to be slightly acidic.

Example 7.2 A 0.1 *M* $Na(CH_3CO_2)$ solution is slightly basic. Why?
Solution We see that this salt, sodium acetate, is composed of sodium ions (Na^+) and acetate ions ($CH_3CO_2^-$). Now, the Na^+ ion has neither acid nor base properties (it can neither donate nor accept H^+ ions). However, the $CH_3CO_2^-$ ion can accept an H^+ ion. Water can serve as a weak acid, so we have the following equilibrium reaction.

$$CH_3CO_2^- + H_2O \rightleftarrows CH_3CO_2H + OH^-$$

The increase in OH^--ion concentration will cause the solution to become somewhat basic.

Problem 7.11 Predict whether the following solutions will be acidic, basic, or neutral.

(a) 0.1 *M* HCl
(b) 0.1 *M* NaOH
(c) pure water
(d) 0.1 *M* CH_3CO_2H
(e) 0.1 *M* NH_4NO_3
(f) 0.1 *M* NaCl
(g) 0.1 *M* $K(CH_3CO_2)$

Problem 7.12 Write equations for the acid–base equilibrium reactions for parts (c), (d), (e), (f), and (g) of Problem 7.11.

7.7 THE WATER CONSTANT, K_w

There is a simple relationship between the concentration of hydronium ions and the concentration of hydroxide ions in any water solution. It is always true that

$$[H_3O^+][OH^-] = K_w$$

How are $[H_3O^+]$ and $[OH^-]$ related in water?

The symbol $\mathbf{K_w}$ stands for a number that is called the **water constant.** K_w is just a special kind of equilibrium constant.

The value of K_w depends on the temperature of the solution. We will be interested in solutions at room temperature, about 25°C, for which

$$K_w = [H_3O^+][OH^-] = 1.0 \times 10^{-14}$$

This expression says that the hydronium-ion concentration multiplied by the hydroxide-ion concentration is always equal to 1.0×10^{-14} at 25°C. This is true no matter how many acids or bases may be present in a solution! If either the hydroxide-ion concentration or the hydronium-ion concentration is known, the other can be calculated by using the above expression. We will always assume a temperature of 25°C, unless otherwise stated.

Example 7.3 Find $[H_3O^+]$ and $[OH^-]$ in 0.10 *M* HCl.
Solution We know that HCl is a strong acid and is completely ionized in water to form H_3O^+ and Cl^-. Thus 0.10 *M* HCl means that

$$[H_3O^+] = 0.10$$

We must now use the water-constant expression to calculate the $[OH^-]$.

$$[H_3O^+][OH^-] = 1.0 \times 10^{-14}$$

$$[OH^-] = \frac{1.0 \times 10^{-14}}{[H_3O^+]}$$

$$= \frac{1.0 \times 10^{-14}}{0.10}$$

$$[OH^-] = 1.0 \times 10^{-13}$$

We see from this example that $[OH^-]$ is very small in an acidic solution. However, although $[OH^-]$ is small, it is never zero. No matter how high the hydronium-ion concentration becomes, there is always a very small hydroxide-ion concentration in any water solution. By the same token, no matter how basic a water solution is, there will always be a small hydronium-ion concentration. We discussed these ideas in Section 7.6, but now, with the water-constant expression, we can actually calculate the hydronium-ion or hydroxide-ion concentrations in any solution.

Problem 7.13 Find $[OH^-]$ and $[H_3O^+]$ in 0.010 *M* NaOH. [Hint: Remember that NaOH dissolves completely in water to give Na^+ and OH^- ions.]
Problem 7.14 A certain acetic acid solution has $[H_3O^+] = 1.0 \times 10^{-4}$. What is $[OH^-]$ in this solution?

A neutral solution is one in which $[H_3O^+] = [OH^-]$. So, for a neutral solution, the water-constant expression,

$$[H_3O^+][OH^-] = 1.0 \times 10^{-14}$$

can be rewritten as

$$[H_3O^+]^2 = 1.0 \times 10^{-14}$$

and then as

$$[H_3O^+] = 1.0 \times 10^{-7}$$

Since hydronium-ion and hydroxide-ion concentrations are the same in a neutral solution, we also get

$$[OH^-] = 1.0 \times 10^{-7}$$

A solution which is neither acidic nor basic will thus have hydronium-ion and hydroxide-ion concentrations of 10^{-7} *M*. Solutions with hydronium-ion concentrations greater than 10^{-7} are acidic, and those with hydronium-ion concentrations less than 10^{-7} are basic.

Problem 7.15 Predict whether solutions with the following concentrations would be acidic, basic, or neutral.
(a) $[H_3O^+] = 5 \times 10^{-5}$
(b) $[OH^-] = 1 \times 10^{-4}$
(c) $[H_3O^+] = 1 \times 10^{-10}$
(d) $[OH^-] = 1 \times 10^{-7}$
(e) $[OH^-] = 1$

7.8 pH

How is pH related to acidity?

We often speak of the acidity of a solution in terms of its **pH.** This is a measure of the hydronium-ion concentration in a solution. More precisely, the pH equals the negative logarithm of the hydronium-ion concentration.

$$pH = -\log [H_3O^+]$$

Another useful expression of pH is

$$[H_3O^+] = 10^{-pH}$$

To be able to go back and forth between pH and $[H_3O^+]$, we must have an understanding of logarithms. The logarithm of a number (written simply *log*) is just the power of 10 necessary to produce that number. For example, the logarithm of 100 (which is written log 100) is 2 because ten must be squared—multiplied by itself once—to produce 100.

$$10^2 = 100$$

so

$$\log 100 = 2$$

TABLE 7.3 Logarithms of Some Numbers

Number	Scientific notation	Logarithm
0.001	1×10^{-3}	−3.0
0.01	1×10^{-2}	−2.0
0.1	1×10^{-1}	−1.0
1.0	1×10^{0}	0.0
10	1×10^{1}	1.0

Table 7.3 lists the logarithms of some numbers. It is easy to see that a logarithm is simply a power of 10.

We can now use the definition of pH to express the acidity of solutions.

Example 7.4 What is the pH of a solution with $[H_3O^+] = 1.0$?
Solution

$$\begin{aligned} pH &= -\log[H_3O^+] \\ &= -\log(1.0) \\ &= -0 \\ pH &= 0 \end{aligned}$$

Example 7.5 What is $[H_3O^+]$ in a solution with a pH of 2.0?
Solution

$$\begin{aligned} [H_3O^+] &= 10^{-pH} \\ &= 10^{-2.0} \\ [H_3O^+] &= 1 \times 10^{-2} \end{aligned}$$

All of this is very easy so long as we deal with hydronium-ion concentrations that are whole powers of 10, that is, 0.01, 0.1, 1, and so on. But what about a solution with $[H_3O^+] = 6.2 \times 10^{-3}$? What is its pH? To answer this question, we must become familiar with an important property of logarithms.

$$\log(a \times b) = \log a + \log b$$

For a solution with $[H_3O^+] = 6.2 \times 10^{-3}$,

$$\begin{aligned} pH &= -\log[H_3O^+] \\ &= -\log(6.2 \times 10^{-3}) \\ &= -\log(6.2) - \log(10^{-3}) \end{aligned}$$

Now, the $\log(10^{-3})$ part is simply -3.00. The log of 6.2 is not so obvious; we have to refer to a table of logarithms to discover that $\log(6.2) = 0.79$. With this information, we can proceed.

$$\begin{aligned} pH &= -0.79 - (-3.00) \\ pH &= 2.21 \end{aligned}$$

Table 7.4 gives logarithms of numbers from 1 to 10 by tenths. The logarithms are given in the body of the table. The numbers are listed

TABLE 7.4 Logarithms of Numbers from 1 to 10 by Tenths

Number	.0	.1	.2	.3	.4	.5	.6	.7	.8	.9
1	.00	.04	.08	.11	.15	.18	.20	.23	.26	.28
2	.30	.32	.34	.36	.38	.40	.42	.43	.45	.46
3	.48	.49	.51	.52	.53	.54	.56	.57	.58	.59
4	.60	.61	.62	.63	.64	.65	.66	.67	.68	.69
5	.70	.71	.72	.72	.73	.74	.75	.76	.76	.77
6	.78	.79	.79	.80	.81	.81	.82	.83	.83	.84
7	.85	.85	.86	.86	.87	.88	.88	.89	.89	.90
8	.90	.91	.91	.92	.92	.93	.93	.94	.94	.95
9	.95	.96	.96	.97	.97	.98	.98	.99	.99	1.0

so that the part of the number to the left of the decimal point is in the left-hand column and the part to the right of the decimal point is across the top of the table. For example, to find the logarithm of 6.2, locate 6 in the left-hand column and .2 along the top. The logarithm is found at the point where row 6 and the column headed .2 intersect in the table, that is, 0.79.

Here are some more examples.

Example 7.6 What is the pH of a solution with $[H_3O^+] = 2.1 \times 10^{-9}$?
Solution

$$\begin{aligned} pH &= -\log[H_3O^+] \\ &= -\log(2.1 \times 10^{-9}) \\ &= -\log(2.1) - \log(10^{-9}) \\ &= -(0.32) - (-9.00) \\ pH &= 8.68 \end{aligned}$$

Example 7.7 What is the pH of a solution with $[OH^-] = 1.5 \times 10^{-4}$?
Solution

$$pH = -\log[H_3O^+]$$

We know that

$$[H_3O^+][OH^-] = 1.0 \times 10^{-14}$$

$$[H_3O^+] = \frac{1.0 \times 10^{-14}}{1.5 \times 10^{-4}}$$

$$[H_3O^+] = 6.7 \times 10^{-11}$$

$$\begin{aligned} pH &= -\log(6.7 \times 10^{-11}) \\ &= -\log(6.7) - \log(10^{-11}) \\ &= -0.83 - (-11.00) \\ pH &= 10.17 \end{aligned}$$

Example 7.8 What is the pH of a neutral solution?
Solution A neutral solution has $[H_3O^+] = 1.0 \times 10^{-7}$

$$\begin{aligned} pH &= -\log[H_3O^+] \\ &= -\log(1.0 \times 10^{-7}) \\ &= -\log(1.0) - \log(10^{-7}) \\ &= -0 - (-7.00) \\ pH &= 7.00 \end{aligned}$$

These examples illustrate a very important point. A neutral solution has a pH of 7. An acidic solution has a pH less than 7, and a basic solution has a pH greater than 7. Table 7.5 lists pH values for various solutions.

TABLE 7.5 pH Values for Various Solutions

Solution	pH
Gastric juice (before eating)	0.9–1.5
Lemon juice	2.0–2.5
Carbonated beverages	3–5
Saliva	6.8–7.8
Blood	7.4
Sea water	7.5–8.5
Soaps	9.5–10.5

We must now learn how to find $[H_3O^+]$ when we know the pH. If the pH is an exact whole number, such as, 6.00, 9.00, or 12,00, the $[H_3O^+]$ is obvious. Since $[H_3O^+] = 10^{-pH}$, if pH is 6.00, we know that $[H_3O^+] = 1.0 \times 10^{-6}$. For a pH of 9.00, $[H_3O^+] = 1.0 \times 10^{-9}$, and so forth. A slightly more complicated situation arises when the pH is not an exact whole number, which is handled as follows.

Example 7.9 A solution has pH = 7.38. Find $[H_3O^+]$.
Solution

$$[H_3O^+] = 10^{-pH}$$
$$= 10^{-7.38}$$

We can separate the power of 10 into an exact whole number and a decimal part.

$$[H_3O^+] = 10^{-7.38} = 10^{-7} \times 10^{-.38}$$

We can rewrite $10^{-.38}$ as

$$10^{-.38} = \frac{1}{10^{.38}}$$

Referring to Table 7.4, we see that 0.38 is the logarithm of 2.4, so that we get

$$10^{-.38} = \frac{1}{10^{.38}} = \frac{1}{2.4}$$

$$\frac{1}{2.4} = 0.42$$

Finally,

$$[H_3O^+] = 10^{-7} \times 0.42$$

or,

$$[H_3O^+] = 4.2 \times 10^{-8}$$

Example 7.10 A solution has pH = 3.08. Find $[OH^-]$ in this solution.
Solution

$$[H_3O^+] = 10^{-pH}$$
$$= 10^{-3.08}$$
$$= 10^{-3} \times 10^{-.08}$$
$$= 10^{-3} \times \frac{1}{10^{.08}}$$

$$= 10^{-3} \times \frac{1}{1.2}$$

$$= 10^{-3} \times 0.83$$

$$[H_3O^+] = 8.3 \times 10^{-4}$$

We know that

$$[H_3O^+][OH^-] = 1.0 \times 10^{-14}$$

so,

$$[OH^-] = \frac{1.0 \times 10^{-14}}{[H_3O^+]}$$

$$= \frac{1.0 \times 10^{-14}}{8.3 \times 10^{-4}}$$

$$[OH^-] = 1.2 \times 10^{-11}$$

Problem 7.16 Are solutions with the following pH values acidic, basic, or neutral?

(a) 7.61
(b) 10.92
(c) 7.00
(d) −0.30
(e) 4.53

Problem 7.17 Find the pH of each of the following solutions.

(a) 0.10 *M* HCl
(b) 0.10 *M* NaOH
(c) $[H_3O^+] = 5.0 \times 10^{-6}$
(d) $[OH^-] = 5.0 \times 10^{-6}$

Problem 7.18 Find $[H_3O^+]$ for each of the following solutions.

(a) 0.050 *M* $HClO_4$
(b) solution with a pH of 2.00
(c) solution with a pH of 3.75
(d) solution with a pH of 12.54

Problem 7.19 Find $[OH^-]$ for each of the solutions of Problem 7.18.

---

pK_a AND pK_b 7.9

Another way of expressing the K_a and K_b for weak acids and bases, respectively, makes use of logarithms. We have the following definitions for $\mathbf{pK_a}$ and $\mathbf{pK_b}$:

$$pK_a = -\log K_a$$

$$pK_b = -\log K_b$$

where K_a is an acid constant for some weak acid and K_b is a base constant for some weak base (Section 7.4). Suppose a certain weak acid has $K_a = 1.0 \times 10^{-4}$.

Then

$$pK_a = -\log(1.0 \times 10^{-4})$$

$$pK_a = 4.00$$

Another weak acid with $K_a = 2.5 \times 10^{-4}$ would have

$$pK_a = -\log(2.5 \times 10^{-4})$$

$$= -(0.40 - 4.00)$$

$$pK_a = 3.60$$

Notice that the *stronger* of the two weak acids (the one with $K_a = 2.5 \times 10^{-4}$) has the *lower* pK_a. The same is true for pK_b values; the lower the pK_b value for a base, the stronger base it is.

Problem 7.20 Calculate the pK_a and pK_b values for the weak acids and bases listed in Table 7.2

7.10 BUFFERS

How does blood maintain a constant pH?

The normal pH of blood is about 7.4. If the pH changes very much, the blood will not be able to transport oxygen, and a serious condition arises. Even though various acid–base reactions occur in the body, the pH of the blood under normal conditions remains nearly constant. We say that the blood contains a **buffer,** or that it is buffered. (Section 27.4 discusses the buffer system of the blood in more detail.)

A buffer system resists changes in the pH. A buffer contains both an acid and a base that can react with any added acids or bases and neutralize them. An acid and its conjugate base are especially useful as buffers, because when an acid reacts with its conjugate base, the products are identical with the reactants. For example, one of the buffers in the blood is the weak acid carbonic acid (H_2CO_3) and its conjugate base, hydrogen carbonate ion (HCO_3^-). If H_2CO_3 and HCO_3^- react, the products are simply HCO_3^- and H_2CO_3. In other words, there is no change in the concentrations of an acid and a base upon reaction if they are conjugates of each other.

H_2CO_3	+	HCO_3^-	→	HCO_3^-	+	H_2CO_3
acid		base		conjugate base of H_2CO_3		conjugate acid of HCO_3^-

For this reason, it is possible to have moderate concentrations of both the acid and base present to counteract the effects of any other acids or bases that may be added. The H_2CO_3 is available to react with any bases, and the HCO_3^- ion is available to react with any acids. For example, if some H_3O^+ ions are added to the solution, they will be removed through reaction with the basic HCO_3^- ions.

$$H_3O^+ + HCO_3^- \rightarrow H_2CO_3 + H_2O$$

By the same token, if some OH^- ions should be added, they would be removed through reaction with the acid H_2CO_3

$$OH^- + H_2CO_3 \rightarrow H_2O + HCO_3^-$$

The particular pH value that a buffer maintains depends on the strengths, as an acid and a base, of the acid–conjugate base pair and also on their relative concentrations. Buffers can resist changes in the pH only to a certain extent, depending on how much of the acid–conjugate base pair is present. Very large amounts of added acids or bases will eventually break down a buffer.

Problem 7.21 Another buffer system in the body involves the dihydrogen phosphate ion, $H_2PO_4^-$, a weak acid. What is the base in this buffer system?

Problem 7.22 A very common base used in making buffers in the laboratory is acetate ion. What acid would be used for such a buffer?

ACID–BASE INDICATORS 7.11

Why do indicators change color?

If a weak acid has one color and its conjugate base has a different color, the pair can be used as an **acid–base indicator.** The color of the solution tells something about the pH.

For example, the acid–base pair *methyl orange* is an indicator because the weak acid, $HC_{14}H_{14}N_2O_3S$, is red and its conjugate base, $C_{14}H_{14}N_2O_3S^-$ ion, is yellow. In solution, an equilibrium is established.

$$\underset{\text{acid (red)}}{HC_{14}H_{14}N_2O_3S} + H_2O \rightleftharpoons H_3O^+ + \underset{\text{base (yellow)}}{C_{14}H_{14}N_2O_3S^-}$$

The color of the solution will be red when the equilibrium is shifted to the left and yellow when the equilibrium is shifted to the right. The hydronium-ion concentration will determine the direction of shift for the reversible reaction. When the pH is 5 or higher, the solution is yellow. When the pH is 3 or lower, the solution is red. At a pH near 4, the solution contains roughly equal amounts of acid and conjugate base and so it has an orange color.

Each indicator has a certain pH range over which it changes colors. The specific pH color range for an indicator depends on the strength of the weak acid in the indicator pair. The stronger the acid, the lower the pH range at which the color changes. The weaker the acid, the higher the pH color range. Table 7.6 lists some indicators and their pH color ranges.

Problem 7.23 What $[H_3O^+]$ is required to cause the acid and conjugate base concentrations to be about equal in a solution containing methyl orange?

Problem 7.24 A certain solution turns yellow when either methyl orange or bromthymol blue is added to it. What is the approximate pH of the solution?

Litmus is a complex mixture of colored organic compounds found in lichen that grow in the Netherlands. Paper is impregnated with the dyestuff extracted from the lichen and is cut into strips after drying. These strips of litmus paper are used in chemistry laboratories as a convenient way to test the acidity of a solution. Litmus is red at any pH below 6 and blue at a pH of 8 or higher.

Mixtures of indicators known as *universal indicator* are available commercially. These show many more color changes at a wider range of pH's than single indicators can. Careful use of a universal indicator allows us to estimate the pH of any solution, whatever its acidity or

TABLE 7.6 Some Acid–Base Indicators

Indicator	pH	Color
Methyl orange	3 or lower	Red
	4	Orange
	5 or higher	Yellow
Bromthymol blue	6 or lower	Yellow
	7	Green
	8 or higher	Blue
Phenolphthalein	8 or lower	Colorless
	9	Pale pink
	10 or higher	Pink

basicity. The exact sequence of the color changes depends on the particular indicators used in the mixture. However, the general sequence is a reddish color at low pH's, which changes progressively through yellow, green, blue, and violet hues as the pH increases.

ACID–BASE TITRATIONS 7.12

Many times we wish to determine the concentration of an acid or a base. A convenient way to do so is to add measured amounts of another base or acid of known concentration until an acid–base reaction has taken place. We call this method **titration.** The solution of known concentration is called the *standard* solution.

How can we determine the concentration of an acid or base solution?

For example, suppose we wish to know the concentration of an HCl solution. We could find this by titrating a sample of the HCl solution with a standard NaOH solution. An NaOH solution contains Na^+ and OH^- ions, and an HCl solution contains H_3O^+ and Cl^- ions. The following acid–base reaction occurs when these solutions are mixed.

$$H_3O^+ + OH^- \rightarrow 2H_2O$$

Let us assume that we take 20.0 mL of our unknown HCl solution and titrate it with 0.100 *M* NaOH. As we add the NaOH solution, the above reaction takes place. We use a buret (Figure 7.1) to measure the volume of NaOH added. We can calculate exactly how many moles of OH^- ion have been added.

$$\text{Moles OH}^- \text{ added} = (\text{liters NaOH solution added}) \times \left(0.100\,\frac{\text{moles}}{\text{liter}}\right)$$

When we have added the same number of moles of OH^- ion as there were moles of H_3O^+ ion originally present in the 20-mL sample, the reaction will be complete. We know that since water itself is the product of the acid–base reaction, and water has a pH of 7, the pH at the end of the reaction (the end point) will be 7. Obviously, what we need is some way to tell when the pH of our reaction mixture reaches 7. One way to do this is to use an indicator.

Table 7.6 shows that bromthymol blue is green when the pH is 7, yellow when it is below 7, and blue above 7. Theoretically, bromthymol blue would be a good indicator for our titration. A few drops of a bromthymol blue solution would cause the HCl sample to be yellow. As we added our standard NaOH solution, the color would remain yellow until the pH of the solution approached 7. When the color of

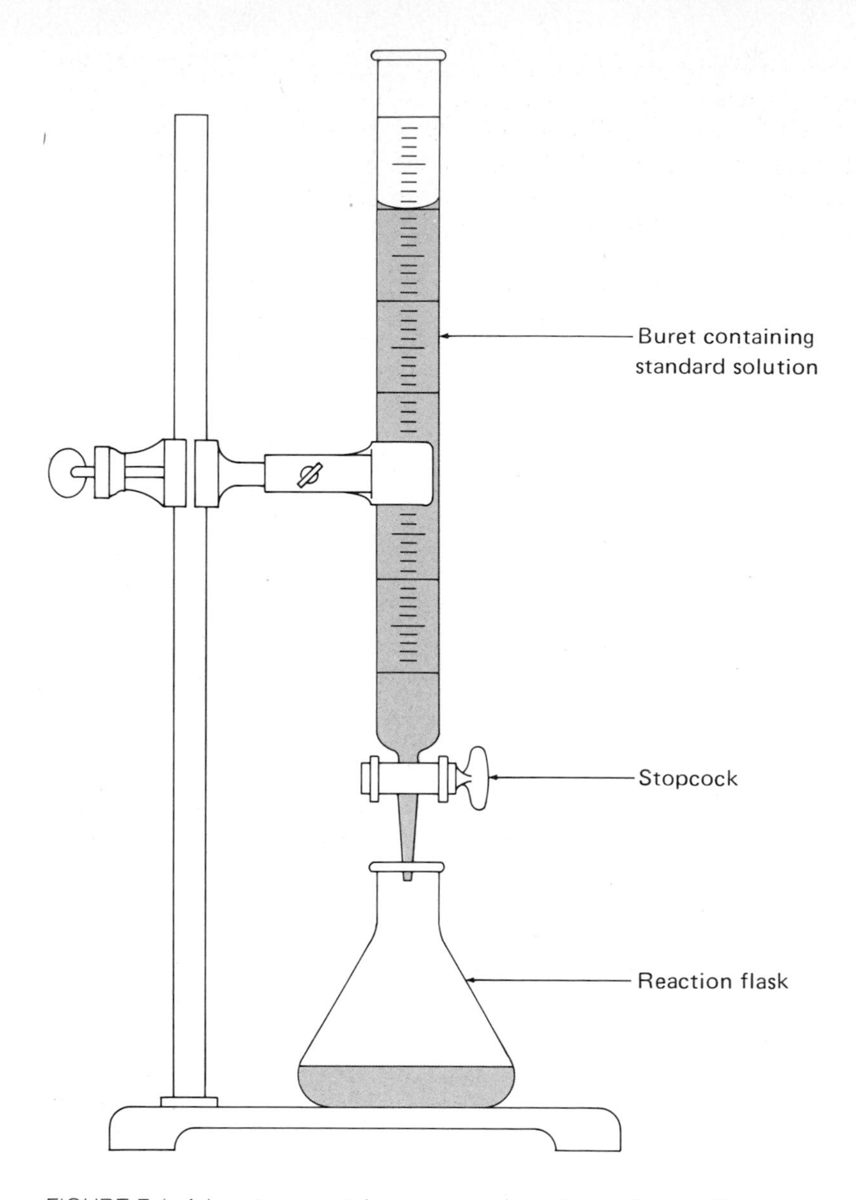

FIGURE 7.1 A buret accurately measures the volume of a solution.

the sample turned green, we would note how much standard NaOH we had added and proceed to calculate the concentration of the original HCl solution. All of this is good in theory, but not good in practice. The trouble is that our eyes cannot distinguish among subtle changes in hue from yellow to green to blue. Reproducible results cannot be obtained on successive titrations. Therefore, we use an indicator that undergoes a sharp color change.

The indicator most commonly used in acid–base titrations is phenolphthalein, which turns a pale pink at a pH of 9. It is easy to distinguish a pale-pink from a colorless solution. True, the end point of our reaction will occur at a pH of about 9 rather than the ideal value of 7. But the error introduced into our calculations is insignificant.

Now, let us run through the calculations for our example above.

Example 7.11 Twenty milliliters of an HCl solution require 22.8 mL of standard 0.100 *M* NaOH to reach the end point, with phenolphthalein used as the indicator. Find the concentration of the original HCl solution.

Solution At the end point, moles OH^- = moles H_3O^+

$$\text{moles } OH^- \text{ used} = 0.0228\ \not{L} \times 0.100\ \frac{\text{mol}}{\not{L}} = 0.00228 \text{ mol}$$

So,

$$\text{moles } H_3O^+ \text{ present in original 20.0 mL} = 0.00228 \text{ mol}$$

$$\text{Concentration of the HCl solution} = \frac{0.00228 \text{ mol}}{0.0200 \text{ L}} = 0.114\ M$$

Problem 7.25 It takes 18.0 mL of 0.100 *M* NaOH to titrate 20.0 mL of an HNO_3 solution to the end point, with phenolphthalein as indicator. What is the concentration of the HNO_3 solution?

Weak acids are also titrated with the strong base OH^- ion. The acid–base reaction during the titration of an acetic acid solution by standard NaOH solution can be expressed as

$$CH_3CO_2H + OH^- \rightarrow H_2O + CH_3CO_2^-$$

The hydroxide ion is such a strong base that it can remove all the H^+ ions from the weak acid.

The reaction is completed when the moles of OH^- ion that are added equal the moles of CH_3CO_2H molecules originally present. However, one of the products of this reaction is the acetate ion, a weak base. We have already seen that this ion reacts with water slightly to produce the OH^- ion.

$$CH_3CO_2^- + H_2O \rightleftarrows CH_3CO_2H + OH^-$$

The pH of the solution when the titration is over will thus be slightly greater than 7. Phenolphthalein is still the best indicator to use to detect the end point.

TABLE 7.7 Summary of Acid–Base Titrations

Titration type	Reactants	Products	pH of solution at completion
Strong acid–strong base	H_3O^+; OH^-	H_2O	7
Weak acid–strong base	Weak acid; OH^-	Conjugate base of weak acid; H_2O	Slightly greater than 7
Strong acid–weak base	H_3O^+; weak base	Conjugate acid of weak base; H_2O	Slightly less than 7

Sometimes we want to titrate a weak base with a standard solution of a strong acid. For example, if we titrate an unknown ammonia solution with standard HCl solution, the titration reaction is

$$H_3O^+ + NH_3 \rightarrow NH_4^+ + H_2O$$

One of the products of the titration is the ammonium ion, a weak acid. This ion reacts slightly with water to produce some H_3O^+ ions.

$$NH_4^+ + H_2O \rightleftharpoons H_3O^+ + NH_3$$

When the titration is finished, the pH of the solution will be slightly less than 7 because this equilibrium reaction will raise the concentration of hydronium ions.

Table 7.7 summarizes the types of acid–base titration reactions we have talked about.

Problem 7.26 It takes 17.0 mL of 0.0500 *M* HCl to titrate 14.5 mL of an ammonia solution to the end point, using phenolphthalein as indicator. Calculate the concentration of the ammonia solution.

7.13 pH TITRATION CURVES

At times, it is convenient to monitor the pH throughout a titration. When we do this, we get a **pH titration curve:** a plot of the pH of the solution against the volume of standard solution that has been added to the sample. In practice, a commercial instrument called a pH meter

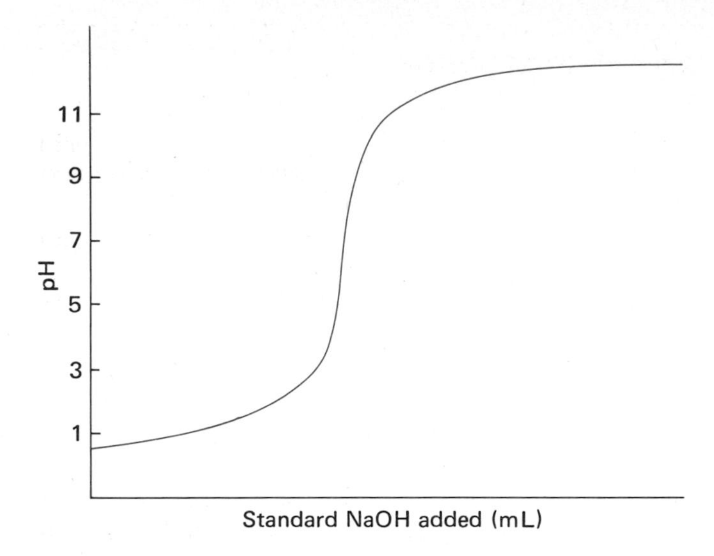

FIGURE 7.2 Typical strong acid–strong base titration curve. The titration is complete halfway up the steeply rising part of the curve, corresponding to pH of 7.

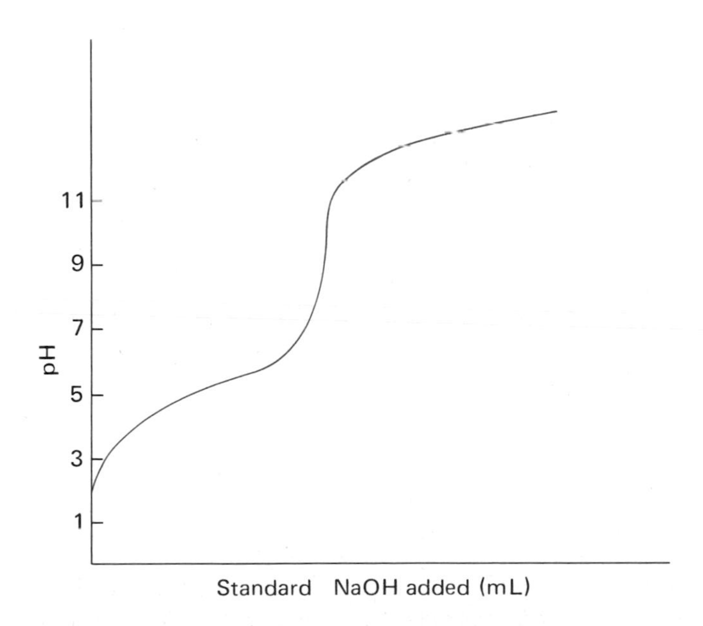

FIGURE 7.3 Typical weak acid–strong base titration curve. The titration is complete halfway up the steeply rising part of the curve, at a pH greater than 7.

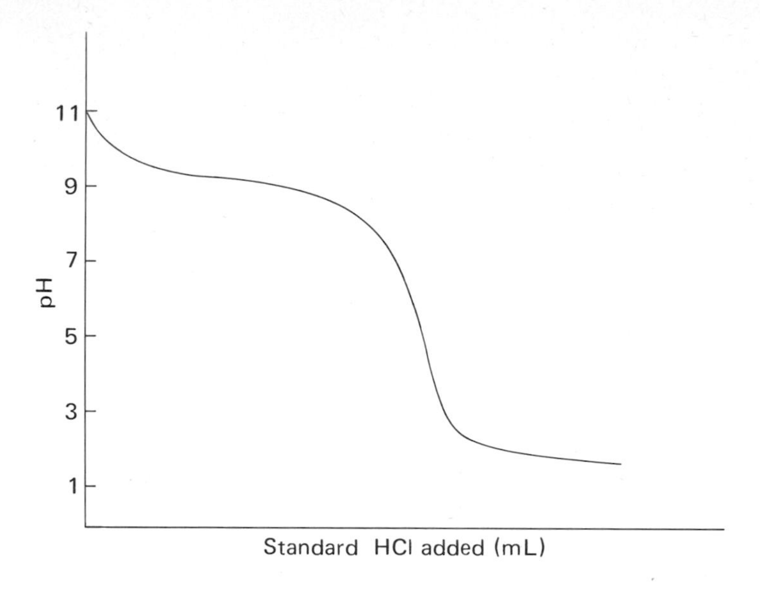

FIGURE 7.4 Typical strong acid–weak base titration curve. The titration is complete halfway down the steeply falling part of the curve, at a pH less than 7.

is used to follow the changes in the pH of a solution as the standard solution is added. We could not use indicators for this purpose. Figures 7.2, 7.3, and 7.4 show pH titration curves for the three different types of titration reactions that we discussed in Section 7.12.

The precise value of the pH when different volumes of standard solution have been added depends on the volume of the sample that is being titrated. In the case of a weak acid or weak base, the exact strength of the acid or base also affects the way the pH changes as the standard solution is added. Nevertheless, the general shapes of titration curves are the same as those we see in Figures 7.2, 7.3, and 7.4. The end points can be found by measuring half way up the steeply rising parts of the titration curves. The volume of standard solution that has to be added in order to reach the end point can be used to calculate the moles of acid or base contained in the original sample. We use the same calculations that we used in Section 7.12.

7.14 POLYPROTIC ACIDS AND BASES

An acid that can lose more than one H^+ ion, or a base that can combine with more than one H^+ ion, is said to be **polyprotic.** An acid like sulfuric acid, H_2SO_4, has two hydrogen atoms that can combine as

H^+ ions with a strong base. Likewise, the carbonate ion, CO_3^{2-}, can combine with two H^+ ions.

When we are titrating an acid or base that can give up or combine with more than one H^+, we must take this fact into account. For example, if we titrate a sulfuric acid solution with standard sodium hydroxide, the overall reaction is

$$H_2SO_4 + 2OH^- \rightarrow 2H_2O + SO_4^{2-}$$

We can use this reaction to find the concentration of a sulfuric acid solution.

Example 7.12 What is the concentration of a sulfuric acid solution, if 20.0 mL require 18.0 mL of 0.100 *M* NaOH for complete reaction?
Solution

$$\text{Moles OH}^- \text{ used} = 0.0180\ \cancel{\text{L}} \times 0.100\ \frac{\text{mol}}{\cancel{\text{L}}} = 0.00180\ \text{mol}$$

We have to take into account the equation for the reaction in order to calculate the moles of H_2SO_4.

$$0.00180\ \cancel{\text{mol OH}^-} \times \frac{1\ \text{mol H}_2\text{SO}_4}{2\ \cancel{\text{mol OH}^-}} = 0.00090\ \text{mol H}_2\text{SO}_4$$

$$\frac{0.00090\ \text{mol}}{0.0200\ \text{L}} = 0.045\ M\ \text{H}_2\text{SO}_4$$

Problem 7.27 What volume of 0.100 *M* NaOH would 15.0 mL of 0.100 *M* H_2SO_4 require for complete reaction?

SUMMARY

Acids and **bases** react to produce their **conjugate bases** and **conjugate acids,** respectively. **Strong acids** react completely in dilute water solutions, but **weak acids** react only slightly and establish an equilibrium that is governed by an acid constant, $\mathbf{K_a}$. **Weak bases** also establish an equilibrium in solution, one that is governed by a base constant, $\mathbf{K_b}$. The values of these constants are sometimes expressed in terms of $\mathbf{pK_a}$ and $\mathbf{pK_b}$. Strong acids have extremely weak conjugate bases, but weak acids and bases have moderately strong conjugate bases and acids. The relationship between [H_3O^+] and [OH^-] in solution is given by the water constant, $\mathbf{K_w}$. Solutions in which [H_3O^+] and [OH^-] are equal are called **neutral solutions,** those in which [H_3O^+] is greater are **acidic,** and those in which [OH^-] is greater are **basic.** An alternative way of expressing acidity is by means of **pH.** At 25°C, acidic solutions have a pH less than 7, basic solutions have a pH greater than 7, and neutral solutions have a pH of 7. A constant pH is achieved by the use of **buffers.** An **acid–base indicator** changes color as the pH of a solution changes. **Acid–base titrations** and **pH titrations** offer means of determining the concentrations of acid and base solutions.

Key Terms

Acid (Section 7.1)
Acid–base indicator (Section 7.11)
Acidic solution (Section 7.6)
Amphoteric substance (Section 7.2)
Base (Section 7.1)
Basic solution (Section 7.6)
Buffer (Section 7.10)
Conjugate acid (Section 7.1)
Conjugate base (Section 7.1)
K_a (acid constant) (Section 7.4)
K_b (base constant) (Section 7.4)
K_w (water constant) (Section 7.7)
Neutral solution (Section 7.6)
pH (Section 7.8)
pH titration curve (Section 7.13)
pK_a (Section 7.9)
pK_b (Section 7.9)
Polyprotic acid (Section 7.14)
Polyprotic base (Section 7.14)
Strong acid (Section 7.3)
Strong base (Section 7.3)
Titration (Section 7.12)
Weak acid (Section 7.4)
Weak base (Section 7.4)

ADDITIONAL PROBLEMS

7.28 Define the following terms.
(a) acid
(b) base
(c) amphoteric substance
(d) conjugate acid
(e) conjugate base

7.29 Write the acid–base reaction that takes place when solutions of HNO_3 and KOH are mixed.

7.30 Write formulas for the following.
(a) the conjugate acid of the Cl^- ion
(b) the conjugate base of the hydronium ion
(c) the conjugate base of water
(d) the conjugate base of the OH^- ion

7.31 Explain why 0.1 *M* $NaNO_3$ is neutral, whereas 0.1 *M* NH_4NO_3 is slightly acidic.

7.32 Calculate $[H_3O^+]$ in the following solutions.
(a) pure water
(b) 0.010 *M* $HClO_4$
(c) 0.10 *M* NaOH

7.33 Calculate $[OH^-]$ for parts (a), (b), and (c) of Problem 7.32.

7.34 Calculate the pH for parts (a), (b), and (c) of Problem 7.32.

7.35 Calculate $[H_3O^+]$ in solutions with the following pH values.
(a) 6.00
(b) 10.00
(c) 4.15
(d) 8.65
(e) 12.43

7.36 Calculate $[OH^-]$ for parts (a), (b), (c), (d), and (e) of Problem 7.35.

7.37 A certain weak acid, HX, in water solution has $[HX] = 0.10$, $[H_3O^+] = [X^-] = 2.0 \times 10^{-3}$. Calculate K_a and pK_a for HX.

7.38 A certain weak base B has a pK_b of 7.40. Another weak base C has a pK_b of 5.00. Is B or C the stronger of the two? Calculate the K_b's for these bases.

7.39 What acid would be used in preparing a buffer in which ammonia was the base? Name some compounds that would be a suitable source for this acid.

7.40 What properties must an acid–base indicator have?

7.41 A certain solution leaves phenolphthalein colorless but causes bromthymol blue to turn blue. What is the pH of the solution?

7.42 What is litmus paper?

7.43 What is a universal indicator? Would it be a good indicator for acid–base titrations?

7.44 What volume of 0.100 *M* NaOH is required for complete reaction with
(a) 20.5 mL of 0.120 *M* CH_3CO_2H?
(b) 16.8 mL of 0.0600 *M* H_2SO_4?

7.45 Sketch a pH titration curve for the titration of a weak base with a strong acid.

7.46 Sketch a pH titration curve for the titration of a weak acid with a strong base.

***7.47** Natural fresh waters (rivers and lakes) are slightly acidic (pH 5.5 to 6.0), but sea water is normally slightly basic (see Table 7.5). Suggest an explanation for this difference. (Part of the answer is given in Focus 5.)

***7.48** Prove that the K_a for any weak acid and the K_b for its conjugate base obey the relationship $K_a \times K_b = K_w$.

***7.49** A quantity called pOH is defined as $pOH = -\log[OH^-]$. Prove that $pH + pOH = 14$. [Hint: Use the K_w expression.]

***7.50** Calculate the pH of a solution prepared by mixing 25.0 mL of 0.100 *M* HCl, 25.0 mL of 0.100 *M* CH_3CO_2H, and 50.0 mL of 0.110 *M* NaOH.

Problem-Solving Hints

If you have trouble solving any of the additional problems, refer to the sections listed next to the problem numbers.

7.28 7.1
7.29 7.3
7.30 7.1
7.31 7.6
7.32 7.7
7.33 7.7
7.34 7.8
7.35 7.8
7.36 7.7, 7.8
7.37 7.4, 7.9
7.38 7.4, 7.9
7.39 7.10
7.40 7.11
7.41 7.11
7.42 7.11
7.43 7.11, 7.12
7.44 7.12, 7.14
7.45 7.13
7.46 7.13
7.47 7.8, Focus 5
7.48 7.4, 7.7
7.49 7.7, 7.8
7.50 7.8, 7.12

FOCUS 5

Limestone Caves: An Acid–Base Reaction in Nature

Natural caves form because of an acid–base reaction. Limestone ($CaCO_3$) contains the carbonate ion, CO_3^{2-}, a moderately strong base. Carbon dioxide from the atmosphere dissolves in all natural waters and forms carbonic acid (H_2CO_3), a weak acid.

$$CO_2 + H_2O \rightarrow H_2CO_3$$

Carbonic acid is present in any fresh water that has been exposed to air, and because of this the pH of fresh water is usually around 5.7, rather than the 7.0 value of pure water.

Mildly acidic water flowing through limestone reacts with the carbonate ion of $CaCO_3$ to form the hydrogen carbonate ion.

$$H_2CO_3 + CaCO_3 \rightarrow Ca^{2+} + 2HCO_3^-$$

FIGURE 1 A scientist explores a small limestone cave. (David Strickler from Monkmeyer Press Photo Service)

Calcium hydrogen carbonate, $Ca(HCO_3)_2$, is soluble, so the limestone slowly dissolves to make cracks, holes, and eventually caves.

Inside a cave, water carrying dissolved calcium hydrogen carbonate drips from ceiling to floor. As the water left on the ceiling evaporates, calcium carbonate is redeposited.

$$Ca^{2+} + 2HCO_3^- \rightarrow CaCO_3 + H_2O + CO_2$$

This deposit of $CaCO_3$ slowly builds into a stalactite hanging from the ceiling. The same reaction takes place where dripping water hits the floor, and a stalagmite begins to grow upwards.

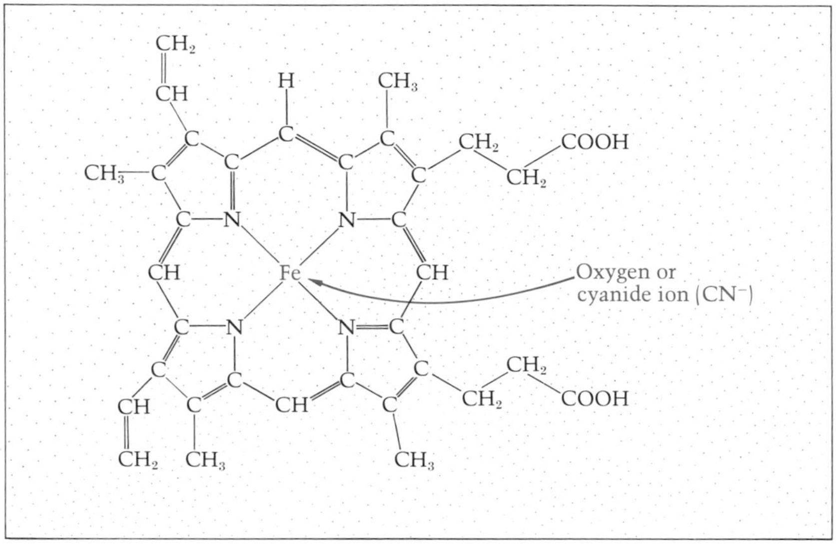

Carbon is the chemical element that provides the framework for the study of organic chemistry, which we will begin in this chapter. Carbon has the ability to combine with other elements to produce very different compounds that have varying effects on the body.

Carbon can combine with oxygen to form the compound carbon dioxide, CO_2*, which has a total of three atoms in its structure. In the body, carbon dioxide is a waste product excreted by the lungs. As oxygen is used by the tissues, carbon dioxide is given off. This carbon dioxide is the source of bicarbonate ion* HCO_3^-*, an important ion that is used to maintain electrical neutrality. Carbon dioxide serves as a blood buffer (that is, it helps maintain a steady pH). Therefore carbon dioxide must be present at a continually constant level.*

The combination of carbon with elements such as nitrogen and hydrogen can produce compounds with very different properties. An example is hydrogen cyanide HCN. This compound is metabolized in the body to yield the cyanide ion CN^-*, a lethal poison. The cyanide ion, which is frequently present in insecticides, reacts with the* Fe^{2+} *found in hemoglobin in the blood and makes it inactive toward oxygen. Since the* Fe^{2+} *normally acts to carry oxygen to the body cells, a sufficient amount of cyanide will deprive the cells of oxygen and cause death.*

8

ORGANIC CHEMISTRY, AN INTRODUCTION

Treatment of the patient with cyanide poisoning involves the administration of thiosulfate, a compound containing sulfur. The sulfur will react with the cyanide ion to form SCN^-, the thiocyanate ion, a relatively nontoxic substance that is excreted in the urine.

Carbon dioxide and hydrogen cyanide are just two very simple examples of the many interesting possibilities that exist for the combination of carbon with other elements.

[Illustration: The heme molecule contains a centralized ion group that binds oxygen, and the oxygen can be displaced by the lethal cyanide ion.]

ORGANIC CHEMISTRY, AN INTRODUCTION

By 1800 scientists had classified substances into two broad groups—organic and inorganic. Organic substances were the substances derived from plants and animals. Inorganic substances were the mineral substances of the inanimate earth. By the mid-nineteenth century, chemists had come to recognize two things about organic substances: (1) all organic compounds contain the element carbon, along with other elements; (2) organic compounds can be prepared outside of living organisms.

Most of us picture carbon as being black and smudgy—an unlikely candidate indeed for the leading role in the drama of life. Yet carbon has the ability to form an unequalled array of compounds with other elements. Organic chemists have been able to take advantage of this unique ability, and they can now tailor-make new compounds almost at will. Today, organic compounds pour from factories by the ton. Nylon, styrofoam, red dye number two, and aspirin are familiar examples of the thousands of synthetic organic compounds we produce.

In this chapter, we will begin our study of organic chemistry by looking at the several types of formulas that can be drawn to show precisely *how* atoms are bonded together in organic molecules. In our journey into organic chemistry we will try to explain the physical and chemical behavior of molecules in terms of their structures. We will begin with the simplest molecules, the hydrocarbons, and proceed to the more complex ones. This journey will take us first through the realm of simple organic substances and will lead us later on into biochemistry, the realm of life and living organisms.

THE NATURE OF ORGANIC COMPOUNDS 8.1

How do organic substances differ from inorganic substances?

Early chemists classified those compounds that were derived from plants and animals as **organic compounds.** Even in the early nineteenth century, chemists still believed that organic substances arose through the action of a "vital force" that was present only in living organisms.

In 1828, Friedrich Wöhler, a German chemist, overthrew the vital-force idea by a simple demonstration. He heated ammonium cyanate (NH_4OCN), an ionic, inorganic substance, and obtained urea (NH_2CONH_2), an organic substance found in the urine of animals. (In the equation below, the carbon atom shares more than one pair of electrons first with nitrogen and then with oxygen. This is explained in Section 8.7.)

$$NH_4^+ [:\ddot{O}:C:::N:]^- \xrightarrow{\text{heat}} H:\underset{H}{\ddot{N}}:\overset{\ddot{O}:}{\underset{}{C}}:\underset{H}{\ddot{N}}:H$$

ammonium cyanate — urea

Within a few years, similar experiments confirmed that organic substances could be prepared in the laboratory as readily as inorganic substances.

All organic compounds contain the element carbon. Most also contain hydrogen, and many contain other elements—oxygen, nitrogen, sulfur, phosphorus, and the halogens (fluorine, chlorine, bromine, and iodine).

By 1920 it was clear that most organic compounds are covalent compounds. However, we cannot use this fact to define them, because many inorganic compounds are also covalent—for example, water (H_2O) and ammonia (NH_3). The term *organic* started out as a simple, descriptive term, but it departed from its original meaning as chemists prepared more and more organic compounds in their laboratories. At present, about 90% of all known organic compounds are synthetic. How then do we define organic, and why do we continue to use the term? We define organic compounds as the hydrocarbons (compounds of carbon and hydrogen) and their derivatives, that is, compounds that have hydrocarbon parts. We continue to use the term *organic* because it allows us to classify systematically the chemical and physical properties of the over four million known organic compounds. The division of chemical compounds into two large groups—organic and inorganic—is a convenience that helps us organize a great many chemical facts.

Problem 8.1 Classify each of the following compounds as organic or inorganic.

(a) NaCl | (b) H_2O
(c) CH_4 | (d) CH_3OH
(e) C_6H_6 | (f) Na_3PO_4

8.2 PHYSICAL PROPERTIES

How does the bond type of a substance account for its physical properties?

Most organic compounds are covalently bonded and molecular in nature. Covalent molecules are attracted to one another by forces weaker than those that hold ions together (Table 3.2 and Figure 8.1). It is this difference between weak intermolecular forces and strong interionic forces that accounts for the fact that most organic compounds have lower melting and boiling points than most inorganic salts. In order for a substance to melt or boil, its molecules or ions must move apart; the more weakly they are attracted to each other, the more readily they move apart.

Many inorganic substances have melting and boiling points above 1,000°C. In contrast, few organic compounds melt or boil at temperatures above 350°C. In fact, the thermometers used in an organic chemistry laboratory will usually not measure a temperature higher than 360°C.

Organic compounds are usually less dense than inorganic salts. This distinction can also be explained by the difference between weak intermolecular attractions and strong interionic attractions: strong attractions result in tighter packing. Organic molecules, being weakly attracted to each other, are less tightly packed and therefore less dense.

8.3 CARBON—A UNIQUE ELEMENT

How is carbon unique?

Carbon can form millions of compounds, whereas all the remaining elements taken together form only a fraction of this number. Why isn't there a large-scale chemistry of any other element comparable to the organic chemistry of carbon? The answer is that carbon is the only element capable of forming covalent chains of unlimited length. Some **hydrocarbons**—compounds of carbon and hydrogen—have molecules in which literally thousands of carbon atoms are covalently bonded to one another. One example is the plastic called polyethylene.

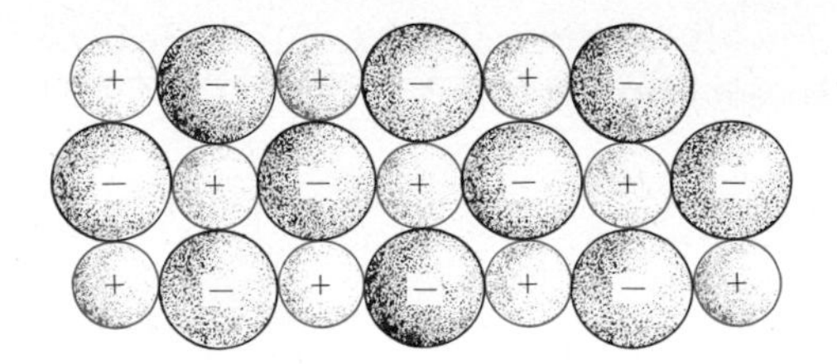

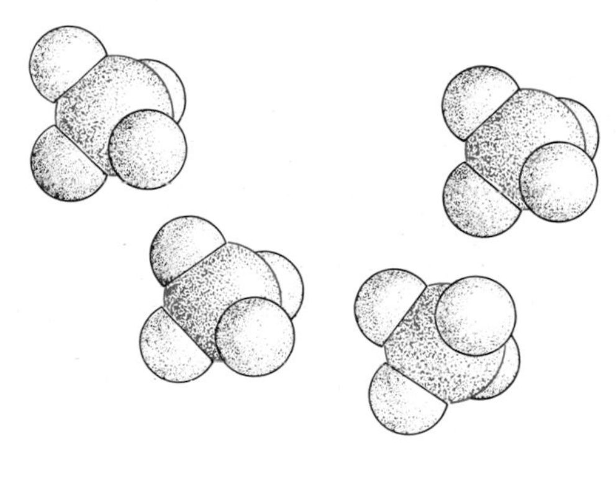

FIGURE 8.1 (a) A single layer of oppositely charged ions in the crystal lattice of sodium chloride illustrates the strong forces that hold ionic crystals together (Fig. 3.1). (b) In methane (CH_4, a gas at 25°C), the separate molecules attract each other only weakly because they are electrically neutral and the C—H bonds are not polar (Sec. 3.9).

No other element comes close to carbon in this ability to form chains. Certain silicon compounds that have six to eight silicon atoms bonded together have been prepared synthetically, but these compounds are so reactive that they burn spontaneously in air. All attempts to make longer silicon chains have failed. No element but carbon has the ability to form stable covalent chains of more than two or three atoms.

STRUCTURAL FORMULAS 8.4

A carbon atom can form four covalent bonds. To form these bonds it shares its four valence-shell electrons with four other atoms that can furnish one electron each (see Chapter 3). For example, a carbon atom can combine with four hydrogens, or with another carbon and three hydrogens, as shown below in the Lewis dot formula.

$$4\text{H}\cdot + \cdot\dot{\underset{\cdot}{\text{C}}}\cdot \rightarrow \text{H}:\overset{\text{H}}{\underset{\text{H}}{\ddot{\underset{\cdot\cdot}{\text{C}}}}}:\text{H} \quad \text{(methane)}$$

$$6\text{H}\cdot + 2\cdot\dot{\underset{\cdot}{\text{C}}}\cdot \rightarrow \text{H}:\overset{\text{H}}{\underset{\text{H}}{\ddot{\underset{\cdot\cdot}{\text{C}}}}}:\overset{\text{H}}{\underset{\text{H}}{\ddot{\underset{\cdot\cdot}{\text{C}}}}}:\text{H} \quad \text{(ethane)}$$

Why do we use more than one type of formula to describe molecular structure?

A simpler shorthand notation is the **structural formula.** This uses a dash to represent an electron-pair bond. For example, we represent methane and ethane as

```
     H            H   H
     |            |   |
  H—C—H       H—C—C—H
     |            |   |
     H            H   H
  methane        ethane
```

We write the structural formula of decane ($C_{10}H_{22}$) as

```
   H  H  H  H  H  H  H  H  H  H
   |  |  |  |  |  |  |  |  |  |
H—C—C—C—C—C—C—C—C—C—C—H
   |  |  |  |  |  |  |  |  |  |
   H  H  H  H  H  H  H  H  H  H
```

In all these compounds carbon must have four bonds to satisfy the octet rule.

We can further simplify the writing of formulas by the use of **condensed formulas.** In a condensed formula we write each carbon in the chain with its attached hydrogens, either CH_3, CH_2, or CH. Thus the condensed formulas of ethane and decane are

CH_3CH_3 (ethane)

$CH_3CH_2CH_2CH_2CH_2CH_2CH_2CH_2CH_2CH_3$ (decane)

When the chain has a branch (a hydrocarbon group attached to one of the carbon atoms of the chain), we place the branch either above or below the carbon atom to which it is attached. Some examples will illustrate chain branching (the branches are shown in color).

$$\begin{array}{c} CH_3 \\ | \\ CH_3CHCH_3 \end{array} \qquad \begin{array}{c} \;\;\;\;\;\;CH_3 \quad\;\; CH_3 \\ \;\;\;\;\;\;| \qquad\quad | \\ CH_3CHCH_2CCH_2CH_3 \\ \qquad\quad\;\; | \\ \qquad\quad\;\; CH_3 \end{array}$$

2-methylpropane   2,4,4-trimethylhexane*

The **molecular formula** of a compound simply gives its elemental composition and no indication of the arrangement of the atoms: methane is CH_4, ethane is C_2H_6, and decane is $C_{10}H_{22}$.

Problem 8.2 Complete the following skeleton structural formulas by supplying the missing hydrogen atoms.

(a) C—C—C—C—C

(b)
$$\begin{array}{c} C-C-\underset{\displaystyle C}{\underset{|}{C}}-C \end{array}$$

(c)
$$C-\overset{\displaystyle C}{\overset{|}{\underset{\displaystyle C}{\underset{|}{C}}}}-C$$

Problem 8.3 Give the molecular formula, the Lewis dot formula, and the condensed formula for each of the compounds of Problem 8.2.

TETRAHEDRAL CARBON 8.5

How are the four bonds of carbon arranged geometrically?

The geometry of carbon compounds is easy to visualize if we recall that each bond represents a shared pair of electrons, and that electron pairs repel one another because they have the same electrical charge. Therefore, the bonds in a molecule such as methane must be arranged so that they are as far apart as possible. Such an arrangement occurs when the four bonds point toward the corners of a regular tetrahedron, as shown in Figure 8.2. The carbon atom is at the center of the tetrahedron, and the hydrogen atoms are at the corners. The tetrahedral arrangement of the bonds can be visualized as lines from the center of a cube to its four diagonal corners, as shown in Figure 8.2b.

*The naming of hydrocarbons will be described in Chapter 9.

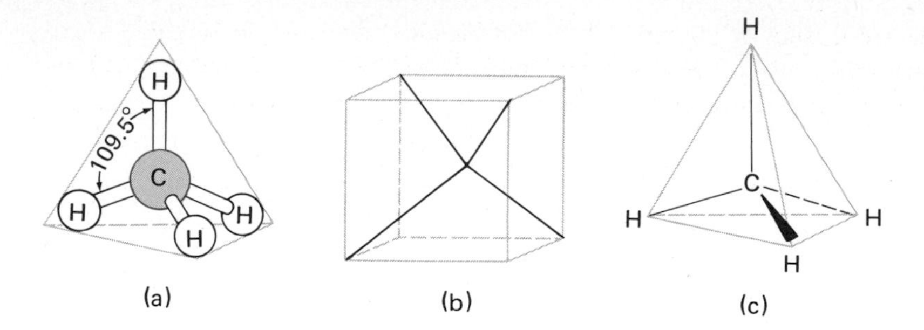

FIGURE 8.2 Tetrahedral geometry of carbon. (a) A regular tetrahedron shows the geometry of methane, CH_4. (b) Tetrahedral bonds can be visualized as lines from the center of a cube to each of the diagonal corners. (c) The line–dash–wedge formula is shown in a tetrahedron. The angle formed by any two bonds is 109.5°. This angle is often referred to as the tetrahedral angle.

One problem that confronts us in drawing chemical structures is how to represent three-dimensional objects in two dimensions. To do this, chemists have developed a system of line, dash, and wedge symbols.

H
C----H
H H

Bonds lying in the plane of the paper are shown as solid lines, bonds protruding toward the viewer are shown as wedges, and lines extending away from the viewer—out the back of the paper, in effect—are shown as dashed lines. A line–dash–wedge formula is illustrated in Figure 8.2c.

Problem 8.4 Draw the line–dash–wedge formula of each of the following compounds.

(a) CH_3Cl (b) CH_2Cl_2
(c) $CHCl_3$

8.6 ORBITAL HYBRIDIZATION

Electrons in atoms occur in orbitals that we designate *s*, *p*, *d*, and so forth. Atoms form a covalent bond by overlapping the orbitals that contain the electrons that are shared. The orbital arrangement of electrons in a free carbon atom was described in Section 2.7. This arrangement can be depicted as follows. The small arrows stand for electrons.

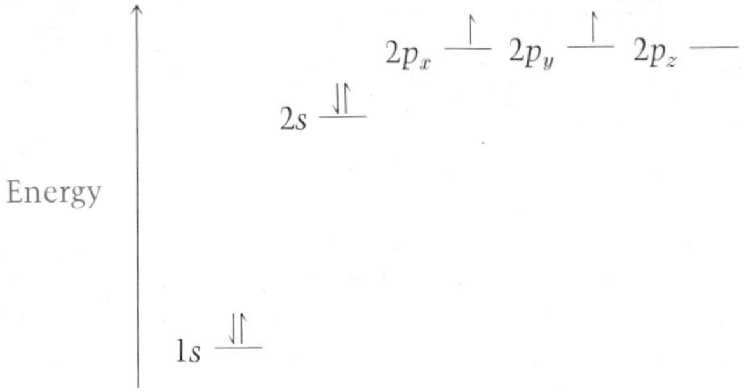

Energy

$1s$ ⇃↾ $2s$ ⇃↾ $2p_x$ ↾ $2p_y$ ↾ $2p_z$ —

Now, if carbon is going to form four covalent bonds, the four electrons in its valence shell must be in four different orbitals. One of the two $2s$ electrons must be shifted into the vacant $2p$ orbital, as shown below.

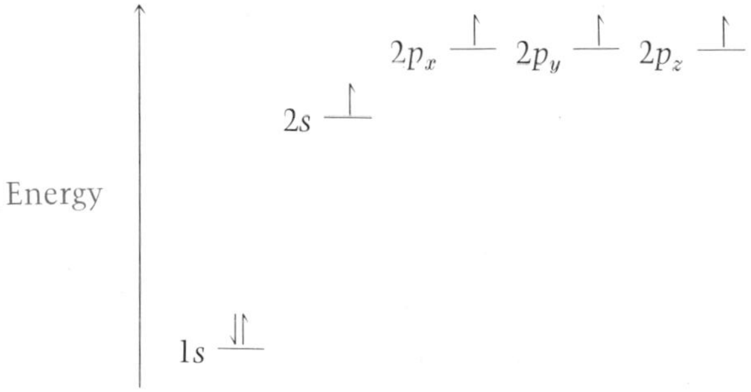

Energy

$1s$ ⇃↾ $2s$ ↾ $2p_x$ ↾ $2p_y$ ↾ $2p_z$ ↾

This description is not entirely satisfactory, however, because three of the hydrogen atoms would be bonded to carbon's p orbitals and one would be bonded to carbon's s orbital. The resulting CH_4 molecule would have three C–H bonds at right angles to one another, because the three $2p$ orbitals are directed at right angles to one another (Figure 2.5).

We can make another electron rearrangement by mixing carbon's four valence-shell orbitals—$2s$, $2p_x$, $2p_y$, and $2p_z$—to obtain four new hybrid orbitals. These hybrid orbitals are designated sp^3 orbitals, to show that they contain one part s character and three parts p character (Figure 8.3). Mixing orbitals to produce new ones is called **orbital hybridization.** Hybridization is a complicated concept, and it can only be fully described in mathematical terms. What is important for our purposes is that, if you know the orientation of bonds in a molecule, you can figure out what hybrid orbitals were used to form that molecule. (Strictly speaking, hybrid orbitals are properties of atoms, not of molecules, as we

Why is orbital hybridization a useful concept?

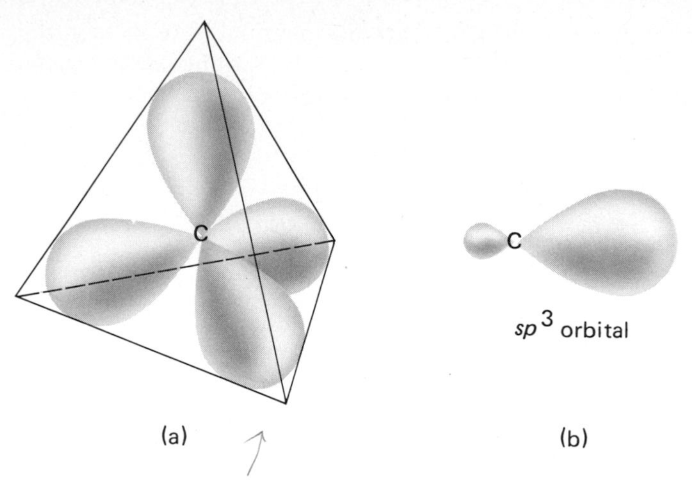

FIGURE 8.3 (a) Tetrahedral orientation of the four sp^3 orbitals (back lobes are not shown). (b) Each sp^3 orbital has a large lobe used in bonding and a small lobe directly behind it.

will see in Section 10.3.) Or, if you know the hybridization, you can figure out the orientation of the bonds in a molecule. For example, we know that methane is tetrahedral and that the carbon atom in methane starts with four valence-shell electrons. We also know that each hydrogen atom in methane starts with one electron. Plugging this information into hybridization theory, we can deduce that the carbon in methane has four sp^3 orbitals. The only way to make a tetrahedral molecule is with four sp^3 orbitals.

Hybrid orbitals permit formation of stronger bonds than pure *s* or *p* orbitals, because they stick out farther and allow a greater degree of orbital overlap when covalent bonds form. The greater the orbital overlap, the stronger the bond.

The sp^3 orbitals are not the only hybrid orbitals. For example, sp^2 hybrid orbitals are formed by the combination of a 2*s*, a $2p_x$, and a $2p_y$ orbital; *sp* hybrid orbitals are formed when a 2*s* and a $2p_x$ orbital combine. These hybrid orbitals are shown in Figure 8.4.

Because sp^2 hybrid orbitals have only *x* and *y* components (being formed from only $2p_x$ and $2p_y$ *p* orbitals), they all lie in the same plane—the *x–y* plane. Similarly, *sp* hybrid orbitals have only one component (only $2p_x$ *p* orbitals are used) and therefore lie in a straight line. There are still other hybrid orbitals, for elements with *d* orbitals, but we will be concerned only with *sp*, sp^2, and sp^3 hybrids.

Problem 8.5 Describe the geometry of the following molecules, for which the orbital hybridization of the central atom is given.

(a) CH_4 (sp^3)

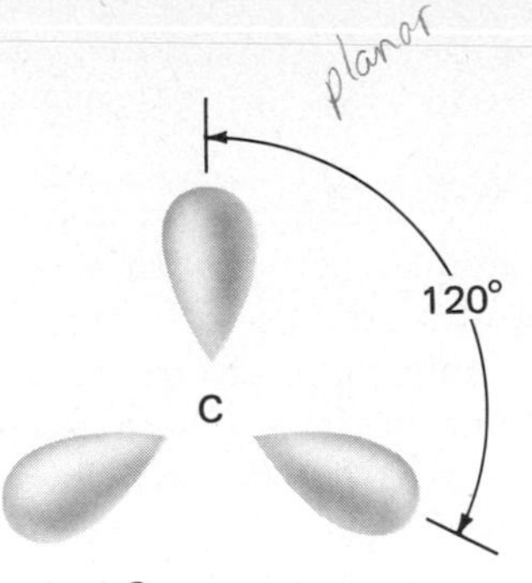

(a) sp^2 hybrid orbitals

180°
C

(b) *sp* hybrid orbitals

FIGURE 8.4 Other hybrid orbitals. (a) The sp^2 hybrid orbitals lie all in one plane, at angles of 120° to one another. (b) The *sp* hybrid orbitals lie in a straight line—that is, at 180° to one another. The individual sp^2 and *sp* hybrid orbitals have small back lobes that are not shown here (see Figure 8.3b).

(b) NH_3 (all four orbitals, including the one that contains the unshared electron pair, are sp^3-hybridized)
(c) BF_3 (sp^2)
(d) BeH_2 (*sp*)

Problem 8.6 Give the orbital hybridization of the central atom in each of the following molecules. Base your answer on the geometry shown or described.

(a)

105° H O H

(all four oxygen orbitals are tetrahedrally oriented)

(b) CF_4 (tetrahedral)
(c) BCl_3 (planar, 120° bond angles)

DOUBLE AND TRIPLE BONDS 8.7

Many molecules contain bonds in which more than one pair of electrons are shared. Ethylene, for example, has the molecular formula C_2H_4. Only one structural formula that obeys the octet rule can be drawn for C_2H_4.

$$\mathrm{H_2C::CH_2} \quad \text{or} \quad \mathrm{H_2C{=}CH_2}$$

A bond formed by the sharing of two pairs of electrons is called a **double bond.**

How do double and triple bonds affect bond angles in molecules?

The double bond uses sp^2 hybridization. We visualize the carbon–carbon double bond by first joining all the atoms together with single bonds using sp^2 hybrid orbitals of carbon and *s* orbitals of hydrogen (Fig. 8.5a). For each carbon atom, this uses only three electrons and leaves one unhybridized *p* orbital containing one electron. These two *p* orbitals can overlap with each other to form a **pi (π) bond,** which consists of an electron cloud above and below the plane of the atoms.

Figure 8.5 shows the important features of the double bond: its planar geometry and the high concentration of electrons above and below the plane of the molecule.

We have now described two ways in which orbitals can overlap to form bonds. Two orbitals that overlap in the space between the atoms form a **sigma (σ) bond.** Two orbitals that overlap in a space that is not directly between the atoms form a *pi bond.* We will examine these types of bonding further in Section 10.3

Atoms can also be joined by a bond formed from three pairs of electrons: a **triple bond.** Acetylene (C_2H_2) is an example of a molecule that has a triple bond (it is a linear molecule).

$$\mathrm{H{:}C{\vdots}{\vdots}C{:}H} \qquad \text{or} \qquad \mathrm{H{-}C{\equiv}C{-}H}$$

The triple bond is formed by the overlap of one *sp* hybrid orbital from each carbon atom to form a sigma bond and two *p* orbitals from each carbon to form two pi bonds.

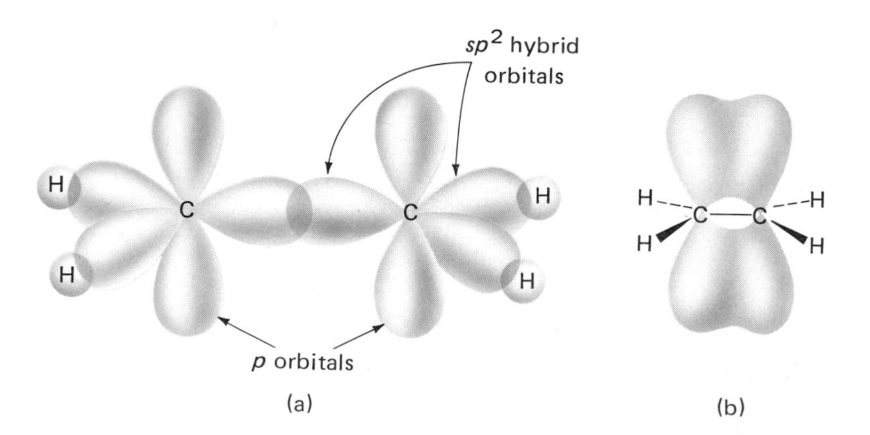

FIGURE 8.5 Geometry of the double bond showing orbital overlap. The double bond may be represented by the overlap of an sp^2 orbital of each carbon and a *p* orbital of each carbon: (a) represents the single bonds with one electron in the remaining *p* orbital of each carbon atom; (b) shows the overlap of the two *p* orbitals to form the other half of the double bond.

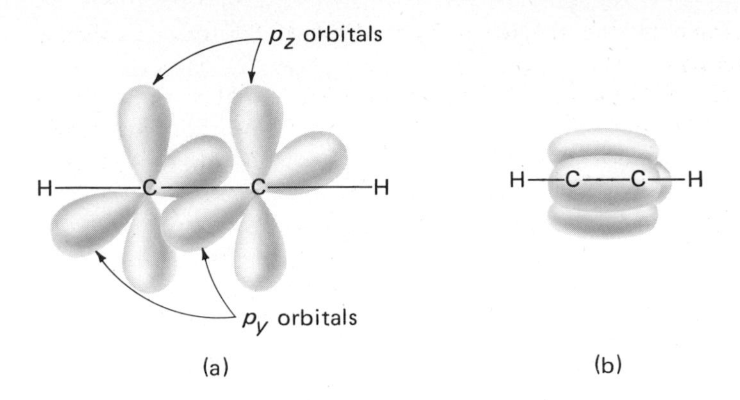

FIGURE 8.6 Geometry of the triple bond. All four atoms lie in a straight line, so the molecule is linear. (a) Each carbon atom is joined to one hydrogen atom and to the other carbon atom by *sp* hybrid orbitals. (b) The remaining two *p* orbitals (p_y and p_z) overlap to form electron clouds that surround the two carbon atoms.

Problem 8.7 Draw formulas that show the molecular geometry of the following compounds.

(a) $Cl_2C{=}CCl_2$ (b) $Cl_2C{=}CH_2$
(c) $H{-}C{\equiv}N$

ISOMERS 8.8

How do isomers differ? How are they the same?

More than one compound may have the same molecular formula. For example, both ammonium cyanate (NH_4OCN) and urea (NH_2CONH_2) have the molecular formula CH_4N_2O, but they have different structures and therefore entirely different sets of properties. Compounds that have the same molecular formula but different structures are called **isomers.**

STRUCTURAL ISOMERS 8.9

How do you decide whether two molecules are isomers?

Ammonium cyanate and urea are **structural isomers,** that is, they contain the same atoms but differ in the sequence in which their atoms are bonded. Butane and isobutane are also structural isomers. Both have the molecular formula C_4H_{10}, but their properties differ greatly, as

TABLE 8.1 Properties of Butane and Isobutane

Property	Butane	Isobutane
Density, g/mL (at −20°C)	0.622	0.604
Boiling point, °C	0	−12
Melting point, °C	−138	−159

shown in Table 8.1. The difference in properties proves that butane and isobutane are in fact different compounds.

If we try to draw all the possible structural formulas for compounds with the formula C_4H_{10}, we find only two unique ones that obey the rule that carbon has four bonds and hydrogen has one (Figure 8.7). These structures differ in the sequential arrangements of their atoms. In butane, all the carbon atoms are bonded either to one other carbon atom and three hydrogen atoms or to two of each. In isobutane, *one* of the carbon atoms is bonded to three other carbon atoms and one hydrogen atom.

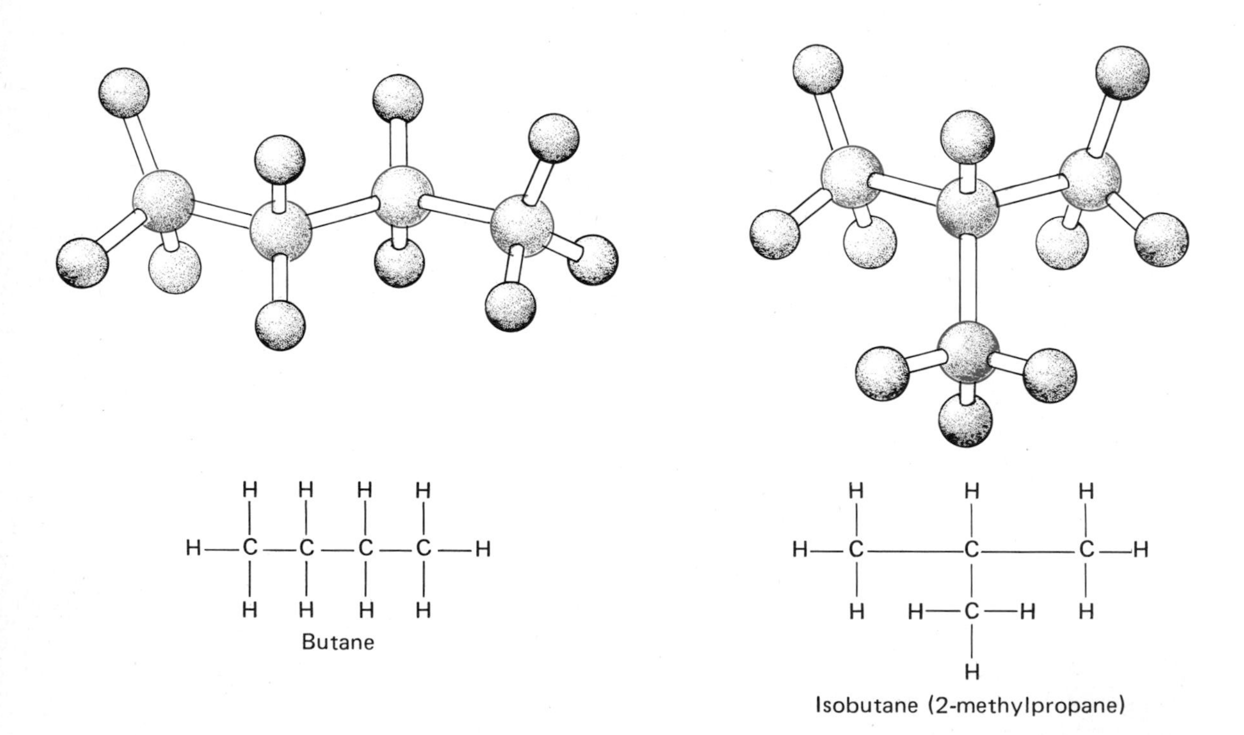

FIGURE 8.7 Butane and isobutane: formulas with models.

Problem 8.8 Identify the following pairs of compounds as isomers, the same compound, or different compounds that are not isomers.

$$\text{(a)}\quad \begin{array}{c} \text{H} \\ | \\ \text{CH}_3\text{—C—CH}_3 \\ | \\ \text{CH}_3 \end{array} \quad \text{and} \quad \begin{array}{c} \text{CH}_3 \\ | \\ \text{CH}_3\text{—C—CH}_3 \\ | \\ \text{H} \end{array}$$

$$\text{(b)}\quad \begin{array}{ccc} & \text{CH}_3 & \text{H} \\ & | & | \\ \text{CH}_3\text{—} & \text{C} & \text{—C—CH}_3 \\ & | & | \\ & \text{H} & \text{CH}_3 \end{array} \quad \text{and} \quad \begin{array}{ccc} & \text{CH}_3 & \text{CH}_3 \\ & | & | \\ \text{CH}_3\text{—} & \text{C} & \text{—C—CH}_3 \\ & | & | \\ & \text{H} & \text{H} \end{array}$$

STEREOISOMERS 8.10

What distinguishes stereoisomers from structural isomers?

There is another kind of isomerism, in which the molecules have the same sequential arrangement of atoms but differ in the orientation of their atoms in space. An example is 2-butene* ($CH_3CH{=}CHCH_3$), shown in Figure 8.8. These two isomers have their atoms bonded in the

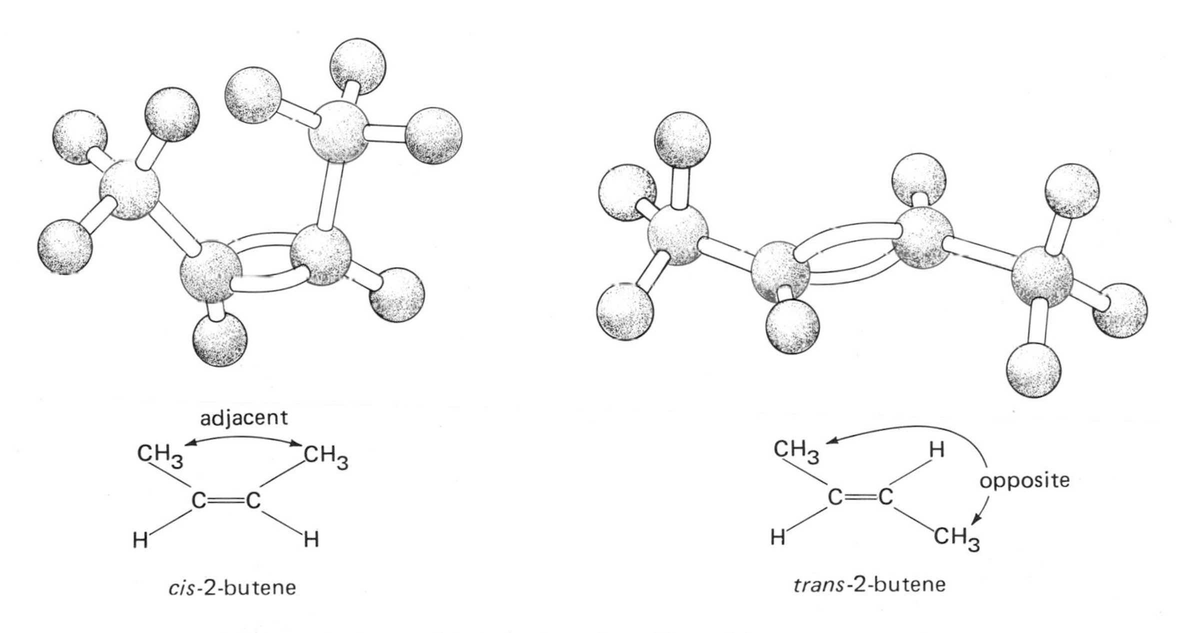

FIGURE 8.8 *Cis*-2-butene and *trans*-2-butene: formulas with models.

*We will discuss the naming of organic compounds in Chapter 9.

TABLE 8.2 Properties of *cis*-2-butene and *trans*-2-butene

	cis-2-butene	*trans*-2-butene
Density, g/mL (at −20°C)	0.6213	0.6042
Boiling point, °C	3.7	0.88
Melting point, °C	−138.9	−105.6

same sequence: $CH_3CHCHCH_3$. But in the ***cis* isomer,** the two CH_3 groups are adjacent to each other; in the ***trans* isomer,** the two CH_3 groups are on opposite sides of the molecule. Isomers like these are called *cis-trans* isomers.

Cis-trans isomers and other isomers that have the same sequential arrangement of atoms but differ in spatial orientation, are called **stereoisomers.** Stereoisomers are different compounds (see Table 8.2), but they have the same molecular formula and they are therefore isomers.

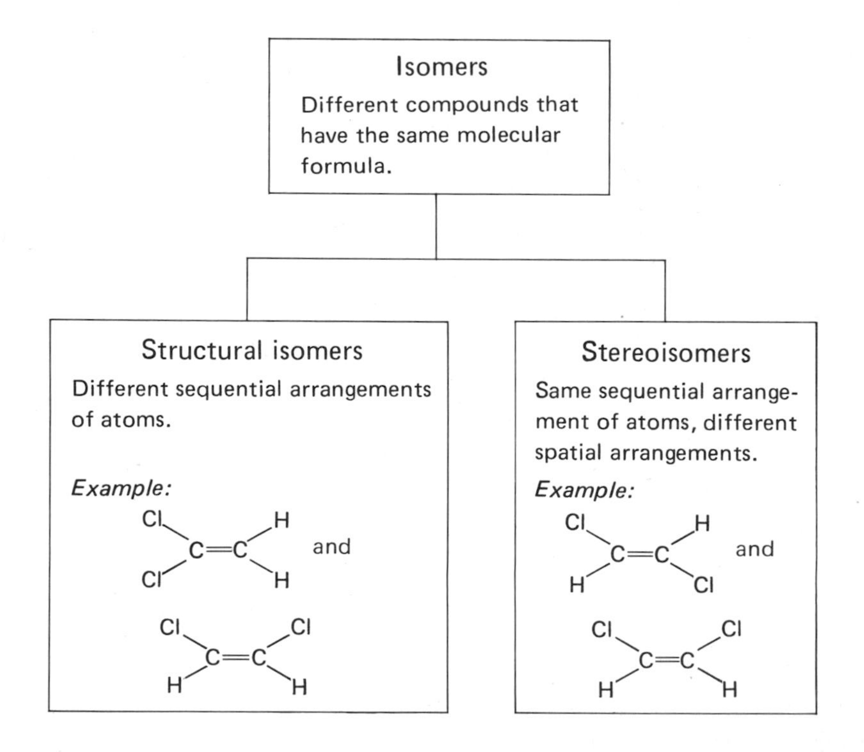

Isomers
Different compounds that have the same molecular formula.
Structural isomers
Different sequential arrangements of atoms.
Example:
Cl
Cl
C=C
H
H
and
Cl
H
C=C
Cl
H
Stereoisomers
Same sequential arrangement of atoms, different spatial arrangements.
Example:
Cl
H
C=C
H
Cl
and
Cl
H
C=C
Cl
H

FIGURE 8.9 Relationship between structural isomers and stereoisomers. We must first determine whether a pair of molecules are isomers. If they are, we then decide whether they are structural isomers or not; if they are not structural isomers, then they are stereoisomers.

Figure 8.9 shows the relationship between stereoisomers and structural isomers.

Cis and *trans* isomers like *cis*-2-butene and *trans*-2-butene do not convert into one another unless they are heated vigorously. Thus, we can conclude that the atomic groupings do not rotate about the double bond. If the atomic groupings could rotate and change their spatial position, *cis*-2-butene would readily form *trans*-2-butene, and vice versa. It seems reasonable that there should be no rotation about the double bond, because rotation would require breaking one of the bonds—the pi bond.

In general, any compound with a carbon–carbon double bond in which each of the doubly bonded carbon atoms has two *different* groups can also exist as *cis-trans* isomers. For example,

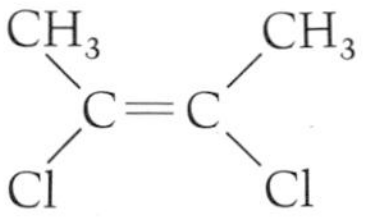

$$\begin{array}{c}CH_3 \quad\quad CH_3\\ \diagdown \;\; \diagup \\ C{=}C \\ \diagup \;\; \diagdown \\ Cl \quad\quad\quad Cl\end{array} \quad \text{and} \quad \begin{array}{c}CH_3 \quad\quad Cl\\ \diagdown \;\; \diagup \\ C{=}C \\ \diagup \;\; \diagdown \\ Cl \quad\quad\quad CH_3\end{array}$$

$(Cl \neq CH_3)$

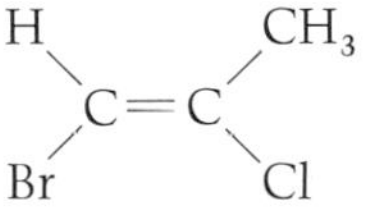

$$\begin{array}{c}H \quad\quad CH_3\\ \diagdown \;\; \diagup \\ C{=}C \\ \diagup \;\; \diagdown \\ Br \quad\quad\quad Cl\end{array} \quad \text{and} \quad \begin{array}{c}H \quad\quad Cl\\ \diagdown \;\; \diagup \\ C{=}C \\ \diagup \;\; \diagdown \\ Br \quad\quad\quad CH_3\end{array}$$

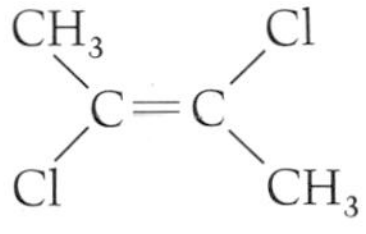

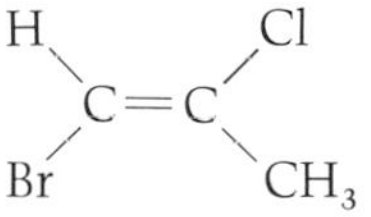

$(H \neq Br \text{ and } Cl \neq CH_3)$

Problem 8.9 Identify the following pairs of compounds as structural isomers, stereoisomers, the same compound, or not isomers.

(a) $\begin{array}{c}CH_3\\ | \\ CH_3CHCH_3\end{array}$ and $CH_3CH_2CH_2CH_3$

(b) C_4H_{10} and C_5H_{12}

(c) $\begin{array}{c}Cl \quad\quad Br\\ \diagdown \;\; \diagup \\ C{=}C \\ \diagup \;\; \diagdown \\ Cl \quad\quad\quad Br\end{array}$ and $\begin{array}{c}Cl \quad\quad Cl\\ \diagdown \;\; \diagup \\ C{=}C \\ \diagup \;\; \diagdown \\ Br \quad\quad\quad Br\end{array}$

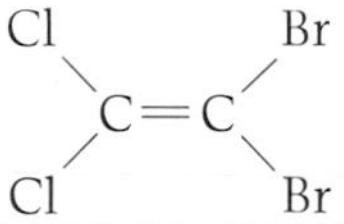

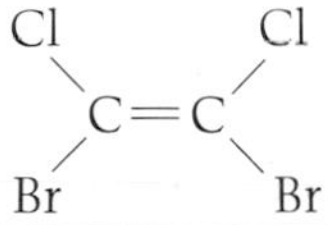

CONFORMATIONS 8.11

What distinguishes conformations from isomers?

Rotation about single bonds can and does occur. We can draw several representations for a molecule like butane. Two of them are shown in Figure 8.10. We form structure (b) by rotating one half of structure (a) about the single bond between the central pair of carbon atoms, while

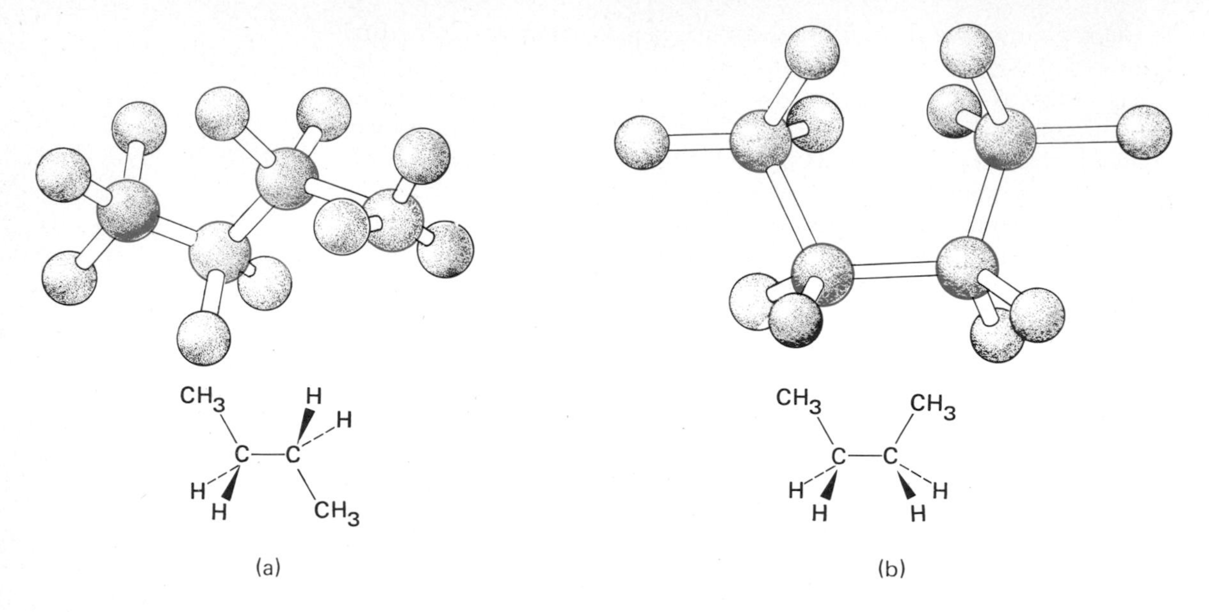

FIGURE 8.10 Butane: formulas with models.

maintaining the other half of (a) in the same position. Rotation about a single bond occurs relatively easily, because it does not affect the amount of orbital overlap. Most molecules of butane under ordinary conditions have enough energy to rotate freely, so different forms of butane spontaneously convert into one another. Any sample of butane at room temperature will contain molecules with every possible orientation about the single bond. Since the different forms cannot be isolated, they are not considered to be isomers. Instead, we call structures of this type **conformations.** Conformations are different spatial arrangements of a molecule that are caused by rotation of atomic groupings about a single bond. They are sometimes referred to as conformational isomers (a term we will not use).

Problem 8.10 Identify the following pairs of compounds as structural isomers, stereoisomers, conformations, identical molecules, or none of these.

(a)
```
 H       H            Cl      Cl
  \     /               \     /
   C=C        and        C=C
  /     \               /     \
 Cl      Cl            H       H
```

(b)
```
 H       Cl           H       H
  \     /               \     /
   C=C        and        C=C
  /     \               /     \
 Cl      H             Cl      Cl
```

```
      H       Cl              H       Cl
       \     /                 \     /
(c)     C = C         and       C = C
       /     \                 /     \
      Cl      H               H       Cl

      H       H               H       Cl
       \     /                 \     /
(d)     C = C         and       C = C
       /     \                 /     \
      Cl      Cl              Cl      Cl

      Cl      H               Cl      Cl
        \    /  H               \    /
(e)  H --C -- C        and   H --C -- C --H
        /    \                  /     \
      H       Cl              H        H

      Cl      H               Cl      Cl
        \    /  H               \    /
(f)  H --C -- C        and   H --C -- C --H
        /    \                  /     \
      Cl      H               H        H

      Cl      H               H       Cl
        \    /  H               \    /  H
(g)  H --C -- C        and   Cl--C -- C
        /    \                  /     \
      H       Cl              H        H
```

FUNCTIONAL GROUPS IN ORGANIC CHEMISTRY 8.12

Certain groupings and arrangements of atoms occur frequently in organic molecules and confer characteristic properties on the molecules. We have already encountered two of them: the carbon–carbon double and triple bonds. An atomic arrangement or grouping that occurs commonly and that behaves the same regardless of what molecule it is a part of is called a **functional group.** Functional groups may contain elements other than carbon; the most common are oxygen, nitrogen, and sulfur. Only about a dozen functional groups are important in organic chemistry. Once you have learned the properties and reactions of this small number of functional groups, you will know the properties and reactions of literally thousands of compounds.

How do functional groups aid the study of organic chemistry?

Functional groups are classified into the following categories:

1. Carbon–carbon double and triple bonds
2. Groups of atoms that contain oxygen atoms singly bonded to other atoms
3. Groups of atoms that contain sulfur atoms singly bonded to other atoms

4. Groups of atoms that contain carbon–oxygen double bonds
5. Groups of atoms that contain a nitrogen atom

Table 8.3 lists the functional groups that we will encounter in the following chapters. At this point you should familiarize yourself with these functional groups so that you will recognize them.

TABLE 8.3 Some Common Functional Groups

Name	Functional group	Examples	
	1. Groups that contain C=C and C≡C		
Alkene	$>C=C<$	$CH_2=CH_2$	(ethylene)
		$CH_2=CHCH_3$	(propene)
Alkyne	$-C\equiv C-$	$HC\equiv CH$	(acetylene)
		$HC\equiv CCH_3$	(propyne)
Arene	benzene ring with six open bonds (C_6 ring, alternating C=C)	C_6H_6 (ring of six C, each bearing H)	(benzene)
	2. Groups that contain singly bonded oxygen, —O—		
Alcohol	$-\overset{\vert}{\underset{\vert}{C}}-OH$	CH_3OH	(methanol)
		CH_3CH_2OH	(ethanol)
Phenol	ring —C₆ ring— C—OH	$CH_3-C_6H_4-OH$ (ring with H on four carbons)	(*p*-cresol)
Ether	$-\overset{\vert}{\underset{\vert}{C}}-O-\overset{\vert}{\underset{\vert}{C}}-$	$CH_3CH_2-O-CH_2CH_3$	(diethyl ether)
Peroxide	$-O-O-$	$CH_3-O-O-CH_3$	(dimethyl peroxide)
	3. Groups that contain singly bonded sulfur, —S—		
Mercaptan (also called thiol and sulfhydryl)	$-\overset{\vert}{\underset{\vert}{C}}-SH$	CH_3CH_2SH	(ethyl mercaptan)
Thiophenol	ring —C₆ ring— C—SH	$CH_3-C_6H_4-SH$ (ring with H on four carbons)	(thiocresol)

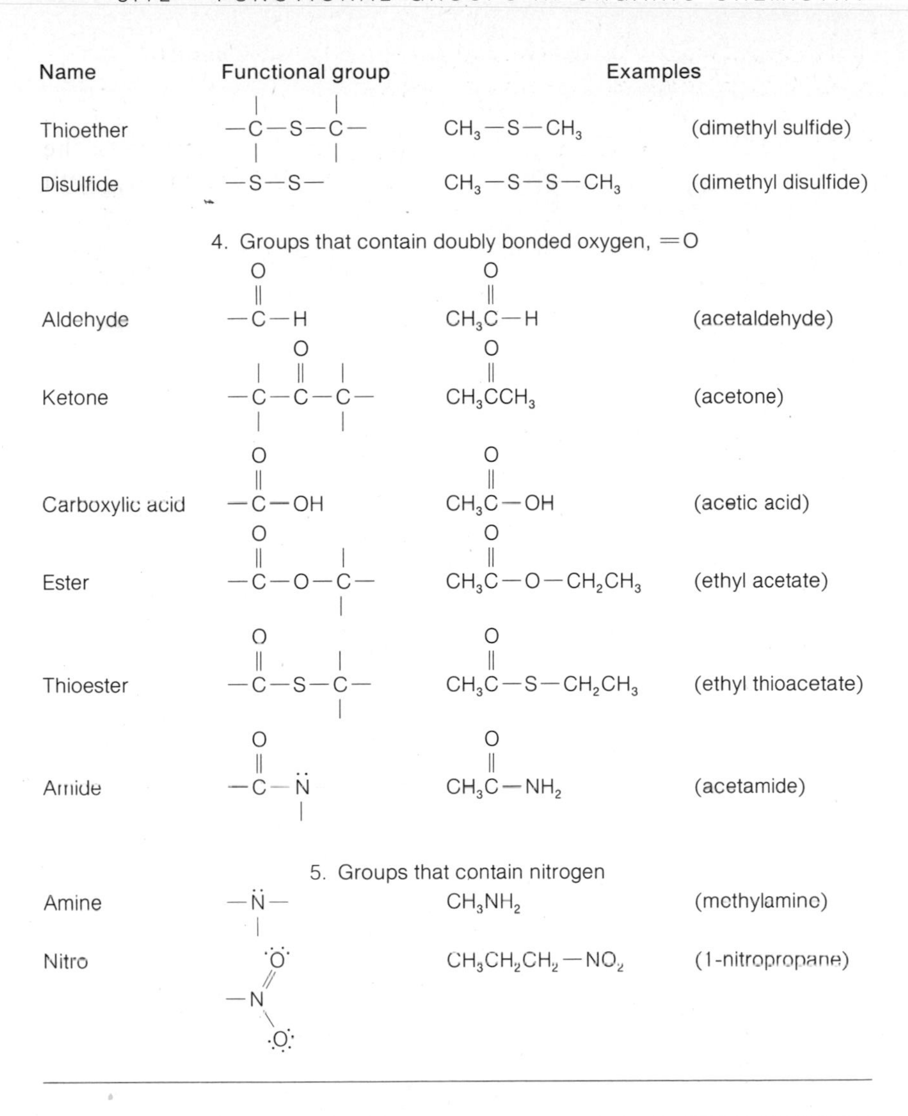

Name	Functional group	Examples	
Thioether	$-\overset{\vert}{\underset{\vert}{C}}-S-\overset{\vert}{\underset{\vert}{C}}-$	CH_3-S-CH_3	(dimethyl sulfide)
Disulfide	$-S-S-$	$CH_3-S-S-CH_3$	(dimethyl disulfide)
4. Groups that contain doubly bonded oxygen, $=O$			
Aldehyde	$-\overset{O}{\overset{\Vert}{C}}-H$	$CH_3\overset{O}{\overset{\Vert}{C}}-H$	(acetaldehyde)
Ketone	$-\overset{\vert}{\underset{\vert}{C}}-\overset{O}{\overset{\Vert}{C}}-\overset{\vert}{\underset{\vert}{C}}-$	$CH_3\overset{O}{\overset{\Vert}{C}}CH_3$	(acetone)
Carboxylic acid	$-\overset{O}{\overset{\Vert}{C}}-OH$	$CH_3\overset{O}{\overset{\Vert}{C}}-OH$	(acetic acid)
Ester	$-\overset{O}{\overset{\Vert}{C}}-O-\overset{\vert}{\underset{\vert}{C}}-$	$CH_3\overset{O}{\overset{\Vert}{C}}-O-CH_2CH_3$	(ethyl acetate)
Thioester	$-\overset{O}{\overset{\Vert}{C}}-S-\overset{\vert}{\underset{\vert}{C}}-$	$CH_3\overset{O}{\overset{\Vert}{C}}-S-CH_2CH_3$	(ethyl thioacetate)
Amide	$-\overset{O}{\overset{\Vert}{C}}-\underset{\vert}{\ddot{N}}$	$CH_3\overset{O}{\overset{\Vert}{C}}-NH_2$	(acetamide)
5. Groups that contain nitrogen			
Amine	$-\underset{\vert}{\ddot{N}}-$	CH_3NH_2	(methylamine)
Nitro	$-N(=\ddot{O}:)(-\ddot{O}:)$	$CH_3CH_2CH_2-NO_2$	(1-nitropropane)

Problem 8.11 Name the functional groups in each of the following compounds.

(benzene ring = C_6H_6 drawn with alternating single and double C–C bonds, each C bearing one H)

(a) $CH_3-O-CH_2CH_3$ (b) $CH_3CH_2-\overset{O}{\overset{\Vert}{C}}-CH_3$

(c) 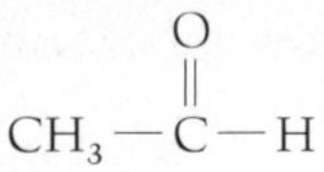 $CH_3-\overset{\overset{\displaystyle O}{\|}}{C}-H$

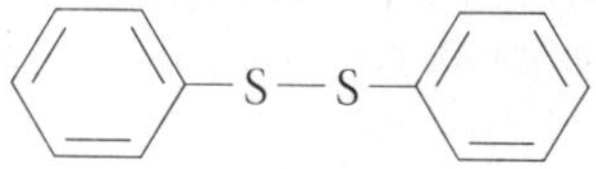

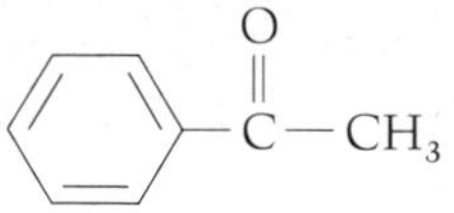

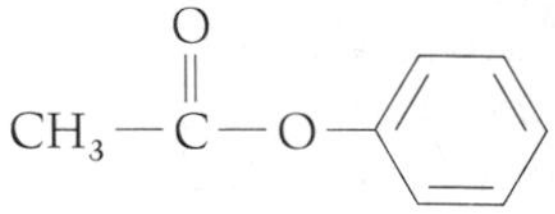

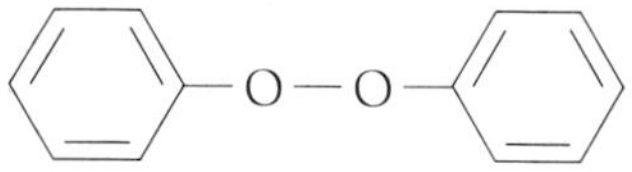

(d) $C_6H_5-S-S-C_6H_5$

(e) $C_6H_5-\overset{\overset{\displaystyle O}{\|}}{C}-CH_3$

(f) $CH_3-\overset{\overset{\displaystyle O}{\|}}{C}-O-C_6H_5$

(g) $C_6H_5-O-O-C_6H_5$

(h) $CH_3-C_6H_4-OH$

(i) 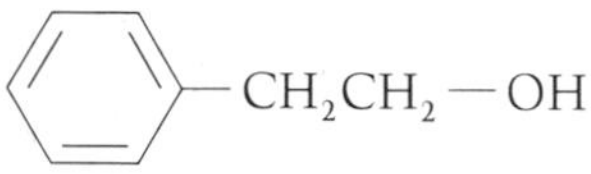 $C_6H_5-CH_2CH_2-OH$

(j) 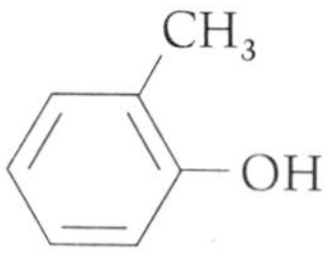 CH_3 (ortho) $-C_6H_4-OH$

(k) 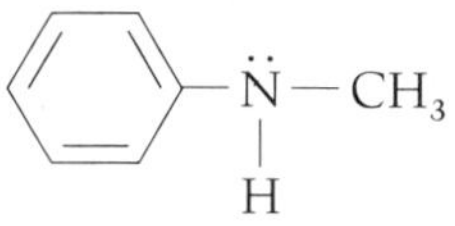 $C_6H_5-\underset{\underset{\displaystyle H}{|}}{\ddot{N}}-CH_3$

(l) 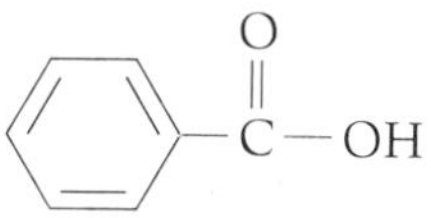 $C_6H_5-\overset{\overset{\displaystyle O}{\|}}{C}-OH$

Problem 8.12 Write the structural formula of one example for each of the following. Do not repeat any of the compounds shown in Problem 8.11.

(a) an alcohol
(b) a phenol
(c) a thiophenol
(d) an ether
(e) a peroxide
(f) a disulfide
(g) an amine
(h) a thiol
(i) an aldehyde
(j) an alkene
(k) an alkyne
(l) a ketone
(m) a carboxylic acid
(n) an ester
(o) an amide

SUMMARY

Organic compounds are compounds of carbon. They may also be defined as the **hydrocarbons** and their derivatives. The physical properties of organic compounds differ from those of inorganic salts because organic compounds are covalently bonded and therefore exhibit very weak intermolecular attractions. Inorganic salts are ionic, and their oppositely charged ions attract each other very strongly. Organic compounds are unique because carbon atoms can form long covalent chains of unlimited length.

We can represent molecular structure in several ways. **Molecular formulas** give only the atomic composition. Lewis dot formulas show a covalent bond as a pair of dots that symbolize a pair of electrons. **Structural formulas** show a covalent bond as a line joining two atoms. **Condensed formulas** of hydrocarbons simply list the carbon atoms and their attached hydrogens. **Line–dash–wedge formulas** represent the three-dimensional nature of molecules.

Orbital hybridization is the combination of atomic orbitals. Different hybrid orbitals give rise to bonds with different orientations. The types of hybrid orbitals are sp^3 (tetrahedral), sp^2 (planar, triangular), and *sp* (linear). Atoms form **double bonds** by sharing two pairs of electrons and **triple bonds** by sharing three pairs of electrons. If a pair of electrons occupy an orbital that lies between the atoms it is called a **sigma (σ) bond.** If a pair of electrons occupy an orbital that does not lie directly between the two atoms, it is called a **pi (π) bond.**

Isomers are different compounds that have the same molecular formula. Two types of isomers are **structural isomers,** which differ in their atomic sequences, and **stereoisomers,** which have the same atomic sequence but which differ in the orientation of the atoms in space. **Conformations** are different atomic arrangements that result from rotation of part of a molecule about a single bond.

The study of organic chemistry is simplified by the widespread occurrence of a few atomic groupings and arrangements called **functional groups.**

Key Terms

***Cis*-isomers** (Section 8.10)
Condensed formula (Section 8.4)
Conformation (Section 8.11)
Double bond (Section 8.7)
Functional group (Section 8.12)
Hydrocarbon (Section 8.3)
Isomer (Section 8.8)
Line–dash–wedge formula (Section 8.5)
Molecular formula (Section 8.4)
Orbital hybridization (Section 8.6)
Organic compound (Section 8.1)
Pi (π) bond (Section 8.7)
Sigma (σ) bond (Section 8.7)
Stereoisomer (Section 8.10)
Structural formula (Section 8.4)
Structural isomer (Section 8.9)
***Trans* isomer** (Section 8.10)
Triple bond (Section 8.7)

ADDITIONAL PROBLEMS

8.13 Which of the two definitions of organic compounds given in Section 8.1 gives us more useful information? Explain.

8.14 Classify each of the following compounds as organic or inorganic.

(a) KBr (b) NaSH

(c) LiCl (d) $CH_3\overset{\overset{\displaystyle O}{\|}}{C}CH_3$

(e) $CH_3\overset{\overset{\displaystyle O}{\|}}{C}{-}ONa$ (f) $HC{\equiv}CH$

8.15 How do organic compounds differ from inorganic salts in their physical properties? Explain these differences in terms of types of bonding.

8.16 On the basis of the type of bonding, decide which compound in each of the following pairs should have the higher melting point.

(a) CH_4 or NaCl

(b) $CH_3\overset{\overset{\displaystyle O}{\|}}{C}{-}NH_2$ or KCl

(c) $CH_3\overset{\overset{\displaystyle O}{\|}}{C}{-}O^-\,Na^+$ or $CH_3\overset{\overset{\displaystyle O}{\|}}{C}{-}OH$

8.17 Which compound in each pair of Problem 8.16 has (a) the higher boiling point, and (b) the greater density?

8.18 Complete the following skeleton structural formulas by supplying the missing hydrogen atoms.

(a) 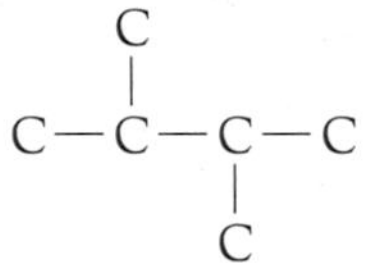 $C{-}\overset{\overset{\displaystyle C}{|}}{C}{-}\underset{\underset{\displaystyle C}{|}}{C}{-}C$ (b) $C{-}C{=}C{-}C$

(c) $C{-}C{-}C{\equiv}C$

8.19 Give the molecular formula, the Lewis dot structure, and the condensed formula of each of the compounds of Problem 8.18.

8.20 Write Lewis dot structures for the following compounds. Include all valence-shell electrons, shared and unshared.

(a) $CH_3\overset{\overset{\displaystyle O}{\|}}{C}H$ (b) CH_3NH_2

(c) CH_3OH (d) $H\overset{\overset{\displaystyle O}{\|}}{C}{-}OH$

8.21 Each of the following formulas is incorrect. Identify the error in each and draw a correct formula.

(a) $CH_3CH_2{=}CH_2CH_3$

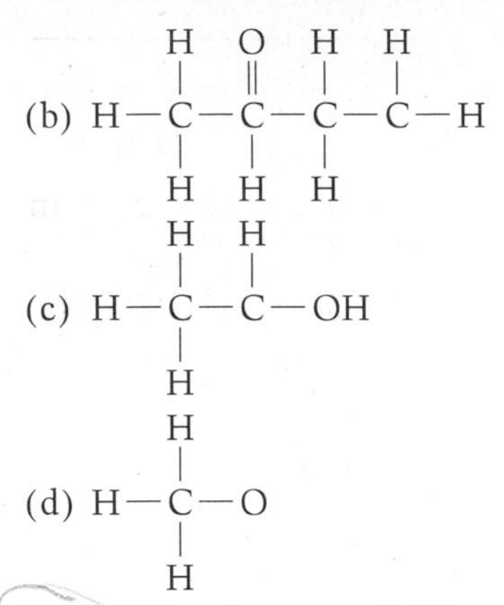

8.22 Write the structural formulas for all the isomers of C_3H_7Cl.

8.23 Draw the line–dash–wedge formulas for the following compounds.

(a) CH_3OH (b) CH_3CH_2OH

(c) $H-\overset{\overset{\displaystyle O}{\|}}{C}-H$

[*Hint:* $C{=}O$ is analogous to $C{=}C$.]

8.24 Complete the following table by supplying the missing items.

Hybridization	Geometric orientation of hybrid orbitals	An example
*sp*³		
*sp*²		
sp		

8.25 Give the orbital hybridization of the central atom in each of the following molecules.

(a) BF_3 (planar, 120° bond angles)

(b) CCl_4 (tetrahedral)

(c) $BeCl_2$ (linear, 180° bond angles)

8.26 Draw structural formulas for the following molecules; the formulas should show molecular geometry, where appropriate.

(a) $CF_2{=}CF_2$ (b) $HC{\equiv}C-C{\equiv}CH$

(c) $CH_2{=}CH-CH{=}CH_2$

8.27 Label the following pairs of compounds as structural isomers, stereoisomers, conformations, the same molecule, or different compounds that are not isomers.

(a) $CH_3\overset{\overset{\displaystyle CH_3}{|}}{C}HCH_2CH_3$ and $CH_3CH_2\overset{\overset{\displaystyle CH_3}{|}}{C}HCH_3$

(b) $CH_3CH_2CH_2CH_2CH_3$ and $CH_3CH_2CH_2\overset{\overset{\displaystyle CH_3}{|}}{C}H_2$

(c) 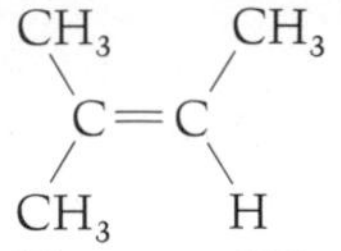
CH_3 CH_3
C=C
CH_3 H
and
CH_3 H
C=C
CH_3 CH_3

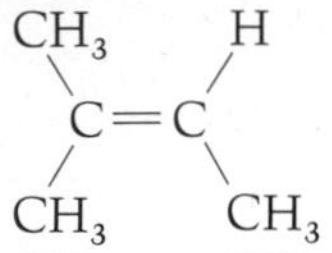

(d) 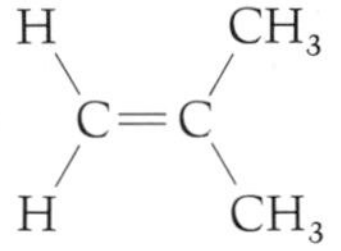
H CH_3
C=C
H CH_3
and
CH_3 CH_3
C=C
H H

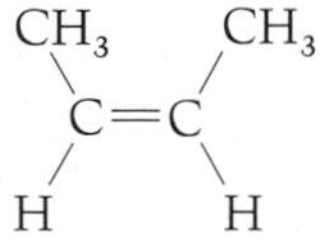

(e) 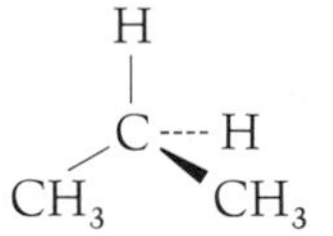
H
C---H
CH_3 CH_3
and
CH_3
C---H
CH_3 H

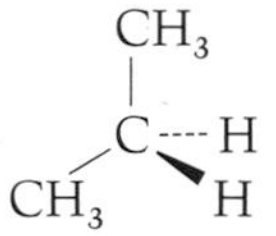

(f) 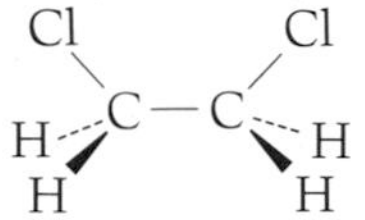
Cl Cl
H C—C H
H H
and
Cl H
H C—C Cl
H H

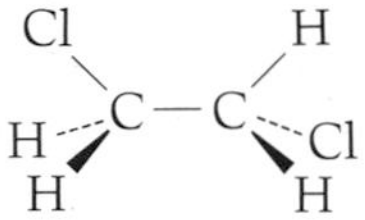

(g) 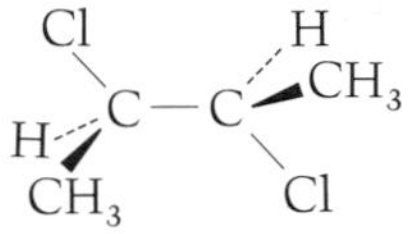
Cl H
H C—C CH_3
CH_3 Cl
and
Cl Cl
H C—C H
Cl CH_3

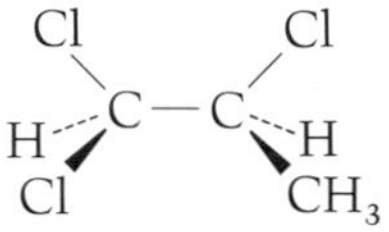

8.28 Give the structural formula for one structural isomer of each of the following compounds.

(a) CH_3CH_2OH (b) 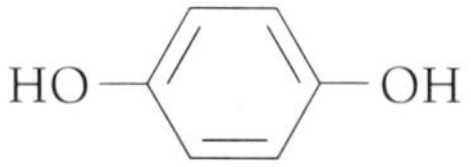 HO—⟨benzene ring⟩—OH (c) $CH_3CH_2NH_2$

(d) 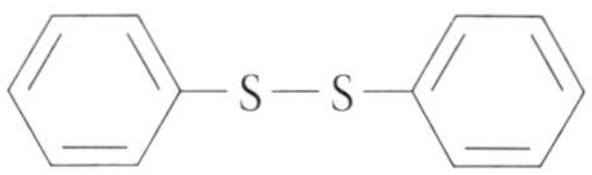 ⟨benzene ring⟩—S—S—⟨benzene ring⟩ (e)  $CH_3CH_2\overset{O}{\overset{\|}{C}}H$

8.29 Give the structural formula for one stereoisomer of each of the following compounds.

(a) 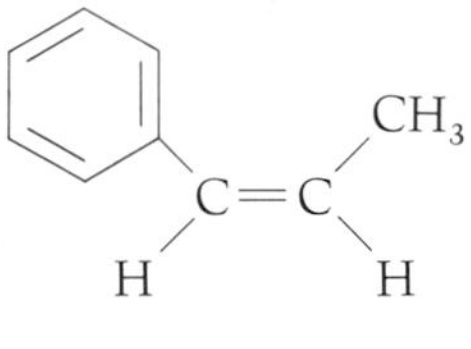
⟨benzene ring⟩ CH_3
C=C
H H

(b) 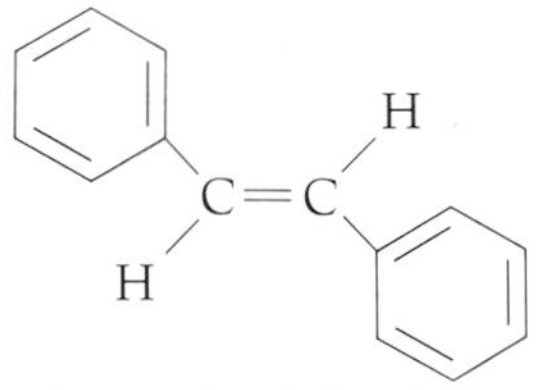
⟨benzene ring⟩ H
C=C
H ⟨benzene ring⟩

8.30 Draw another conformation for each of the following compounds.

(a) 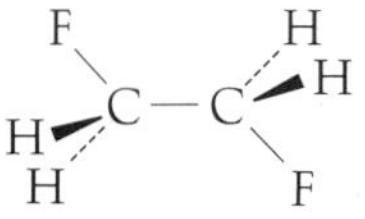
F H
H C—C H
H F

(b) 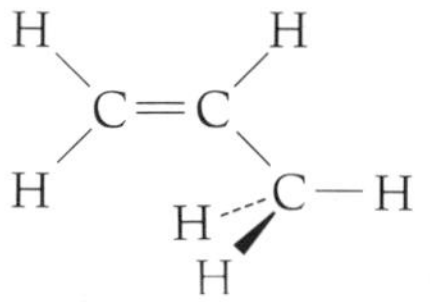
H H
C=C
H C—H
H
H

8.31 Circle and name the functional groups in each of the following compounds.

(a) 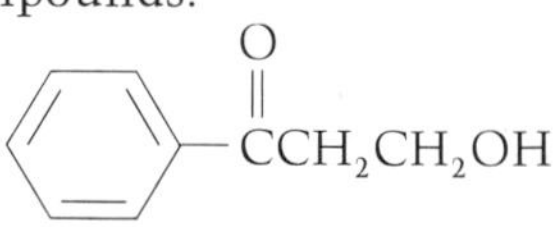 ⟨benzene ring⟩—$\overset{O}{\overset{\|}{C}}CH_2CH_2OH$ (b) 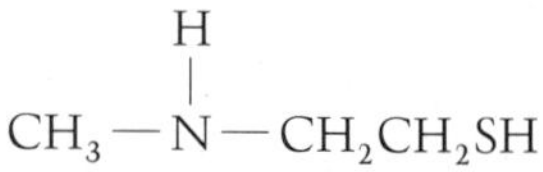 $CH_3—\overset{H}{\overset{|}{N}}—CH_2CH_2SH$

(c) 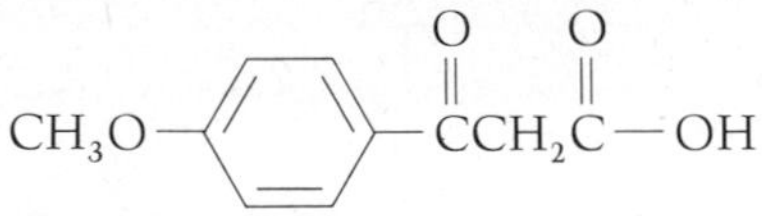 $CH_3O-C_6H_4-\overset{O}{\overset{\|}{C}}CH_2\overset{O}{\overset{\|}{C}}-OH$ (benzene ring drawn, para-substituted)   (d) $CH_3\underset{NH_2}{\underset{|}{C}}H\overset{O}{\overset{\|}{C}}-OH$

(e) $CH_3CH_2\overset{O}{\overset{\|}{C}}-NH_2$ (f) $CH_3CH_2NO_2$

(g) $CH_3CH_2CH_2-S-S-CH_3$

Problem-Solving Hints

If you have trouble solving any of the additional problems, refer to the sections listed next to the problem numbers.

8.13 8.1 **8.14** 8.1 **8.15** 8.2
8.16 8.2 **8.17** 8.2 **8.18** 8.4
8.19 2.9, 8.4 **8.20** 2.9, 8.4 **8.21** 8.4
8.22 8.4 **8.23** 8.5 **8.24** 8.6
8.25 8.6 **8.26** 8.7 **8.27** 8.8–8.11
8.28 8.9 **8.29** 8.10 **8.30** 8.11
8.31 8.12

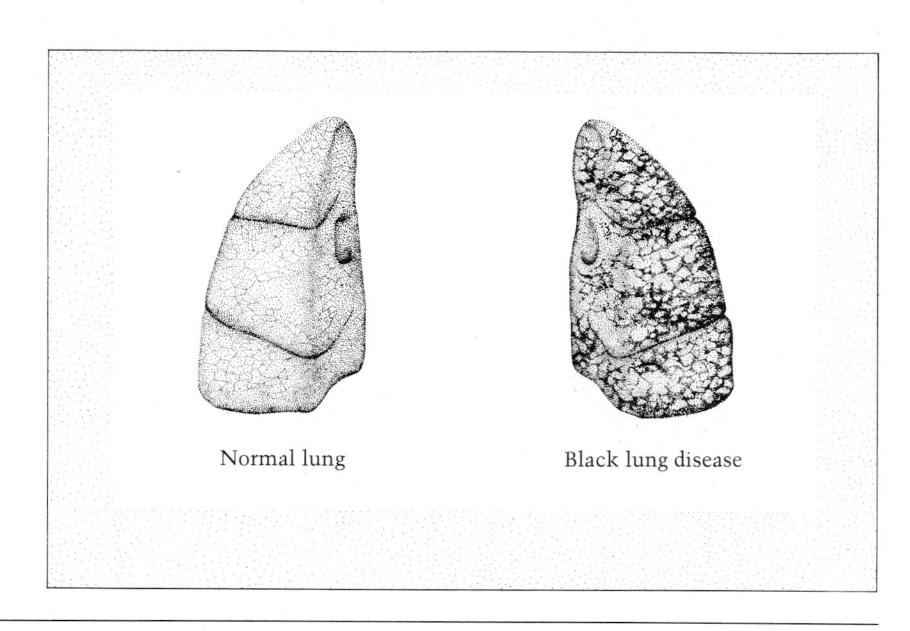

Black lung disease is a condition in which large amounts of carbon accumulate in the lungs. The disease is specific to coal miners. In England, where many people work as miners and where protective equipment is not in widespread use, black lung disease is prevalent. The primary contributing factor is coal dust, which interferes with the exchange of gases that takes place in the tissues of the lungs. A similar condition called silicosis is found quite regularly in miners who have been blasting through rock containing silicon. In these cases it is silicon dust, rather than coal dust, that accumulates in the lungs.

In the United States the diagnosis of black lung disease is often difficult to make. The initial symptoms are few. The most frequent complaint is a nagging cough that a miner will tend to accept as a simple disadvantage of the job. Eventually, additional symptoms develop. The miner may experience shortness of breath, weight loss, and weakness. At this point, when the disease is more advanced, the miner will seek help from a physician. Treatment of the affected patients is similar to that given for an upper respiratory infection. If the exposure to coal dust has been prolonged and treatment has been delayed, the patient may develop tuberculosis, making treatment even more difficult.

9

SATURATED HYDROCARBONS: ALKANES AND CYCLOALKANES

Since black lung disease is not curable, the best treatment in terms of the mining population as a whole is to direct efforts toward preventing the disease. The use of protective equipment such as dust masks will play an important role.

[Illustration: Black lung disease is characterized by the accumulation of carbon particles in lung tissue.]

SATURATED HYDROCARBONS

When we say a hydrocarbon is saturated, we don't mean that it is sopping wet. What we mean is that the carbon is saturated with other atoms: it has formed bonds to as many atoms as it possibly can. A saturated hydrocarbon can only bond to a new atom if it gives up one of the atoms it is already bonded to. Because of this requirement, a saturated hydrocarbon is unusually unreactive. It will break an old bond and form a new one only if an irresistible new partner comes along.

Because saturated hydrocarbons are so unreactive, they remain unchanged even when most functional groups are reacting. Most organic molecules have a skeletal backbone of saturated hydrocarbons that is fairly inert (in other words, it resists change). Before we can understand functional groups, we must first understand the chemistry of that backbone.

Saturated hydrocarbons come in two types: alkanes and cycloalkanes. The difference is that cycloalkanes have some of their carbon atoms bonded together to form rings. Any saturated hydrocarbon without a ring is an alkane.

9.1 ALKANES

Hydrocarbons that have only single covalent bonds and no rings of atoms are called **saturated hydrocarbons** or **alkanes.** Table 9.1 lists the simplest members of the alkane family along with some of their physical properties. Figure 9.1 shows the structures of three simple alkanes, using ball-and-stick models.

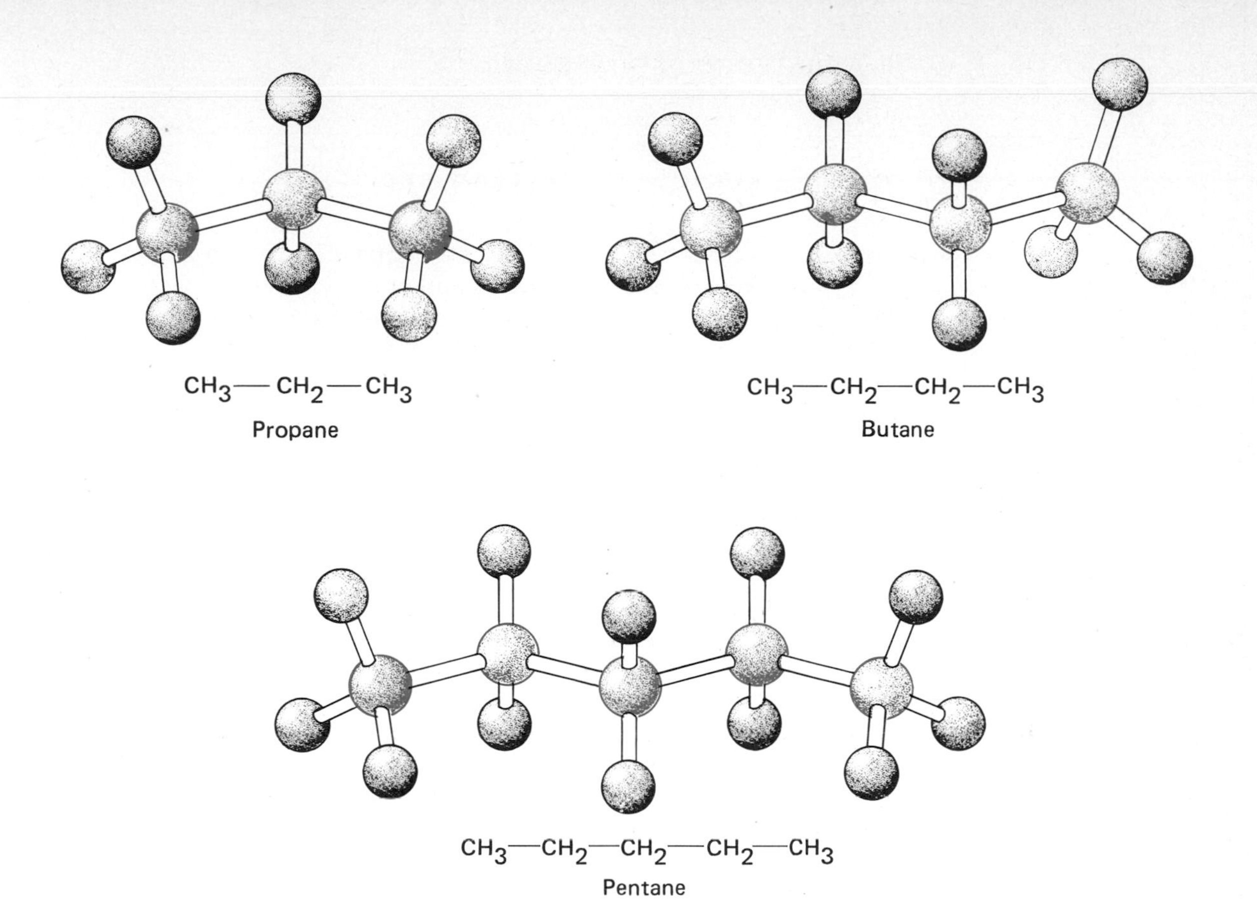

FIGURE 9.1 Propane, butane, and pentane: formulas with models.

TABLE 9.1 Some Simple Alkanes

Name	Formula	Melting point, °C	Boiling point, °C	Density, g/mL (at 20°C)
Methane	CH_4	−183	−162	0.0007
Ethane	CH_3CH_3	−172	−88.5	0.0012
Propane	$CH_3CH_2CH_3$	−187	−42	0.0018
Butane	$CH_3(CH_2)_2CH_3$	−138	0	0.0024
Pentane	$CH_3(CH_2)_3CH_3$	−130	36	0.626
Hexane	$CH_3(CH_2)_4CH_3$	−95	69	0.659
Heptane	$CH_3(CH_2)_5CH_3$	−90.5	98	0.684
Octane	$CH_3(CH_2)_6CH_3$	−57	126	0.703
Nonane	$CH_3(CH_2)_7CH_3$	−54	151	0.718
Decane	$CH_3(CH_2)_8CH_3$	−30	174	0.730
⋮				
Hexadecane	$CH_3(CH_2)_{14}CH_3$	18	280	0.775

What feature does the general name "alkane" have in common with the names of all the compounds in Table 9.1?

9.2 ALKANES: STRUCTURE

What kind of covalent bond do all alkanes have in common?

The alkanes listed in Table 9.1 are **straight-chain isomers:** that is, the carbons are strung one after another like links on a chain. The alkanes also exist as **branched-chain isomers,** in which one or more hydrocarbon groups have taken the place of hydrogen atoms (Figure 9.2).

The general formula for an alkane is C_nH_{2n+2}. We can see why this is true by looking at any unbranched alkane.

$$
\begin{array}{ccccccccc}
 & & H & & H & & H & & H & & \\
 & & | & & | & & | & & | & & \\
H & - & C & - & C & - & C & - & C & - & H \\
 & & | & & | & & | & & | & & \\
 & & H & & H & & H & & H & & \\
\end{array}
$$
$$\underbrace{\qquad\qquad\qquad}_{C_nH_{2n}}$$

Each carbon atom, except the two at the ends, has two hydrogen atoms bonded to it. Each of the end carbon atoms has three hydrogen atoms bonded to it. So the whole chain contains two hydrogen atoms for every carbon, plus two additional hydrogens—in other words, C_nH_{2n+2}.

The same general formula, C_nH_{2n+2}, also holds true for the branched-chain alkanes, as we can see if we draw all the alkane isomers that have five carbon atoms.

First, we write all the possible noncyclic skeleton structures in which five carbon atoms are bonded together with single bonds. We discover there are only three possible structures. The first is the straight-chain isomer, with five carbons in a continuous chain.

$$(1)\ C-C-C-C-C$$

Then we have a branched-chain isomer with four carbons in a continuous chain.

$$
\begin{array}{l}
(2)\ C-C-\underset{\displaystyle C}{\underset{|}{C}}-C
\end{array}
$$

Although different structural formulas can be drawn, showing the side branch in different places, as far as molecular structure is concerned, they all mean the same thing.

$$
C-C-\underset{\displaystyle C}{\underset{|}{C}}-C \quad = \quad C-\underset{\displaystyle C}{\underset{|}{C}}-C-C \quad = \quad C-C-\overset{\displaystyle C}{\overset{|}{C}}-C
$$

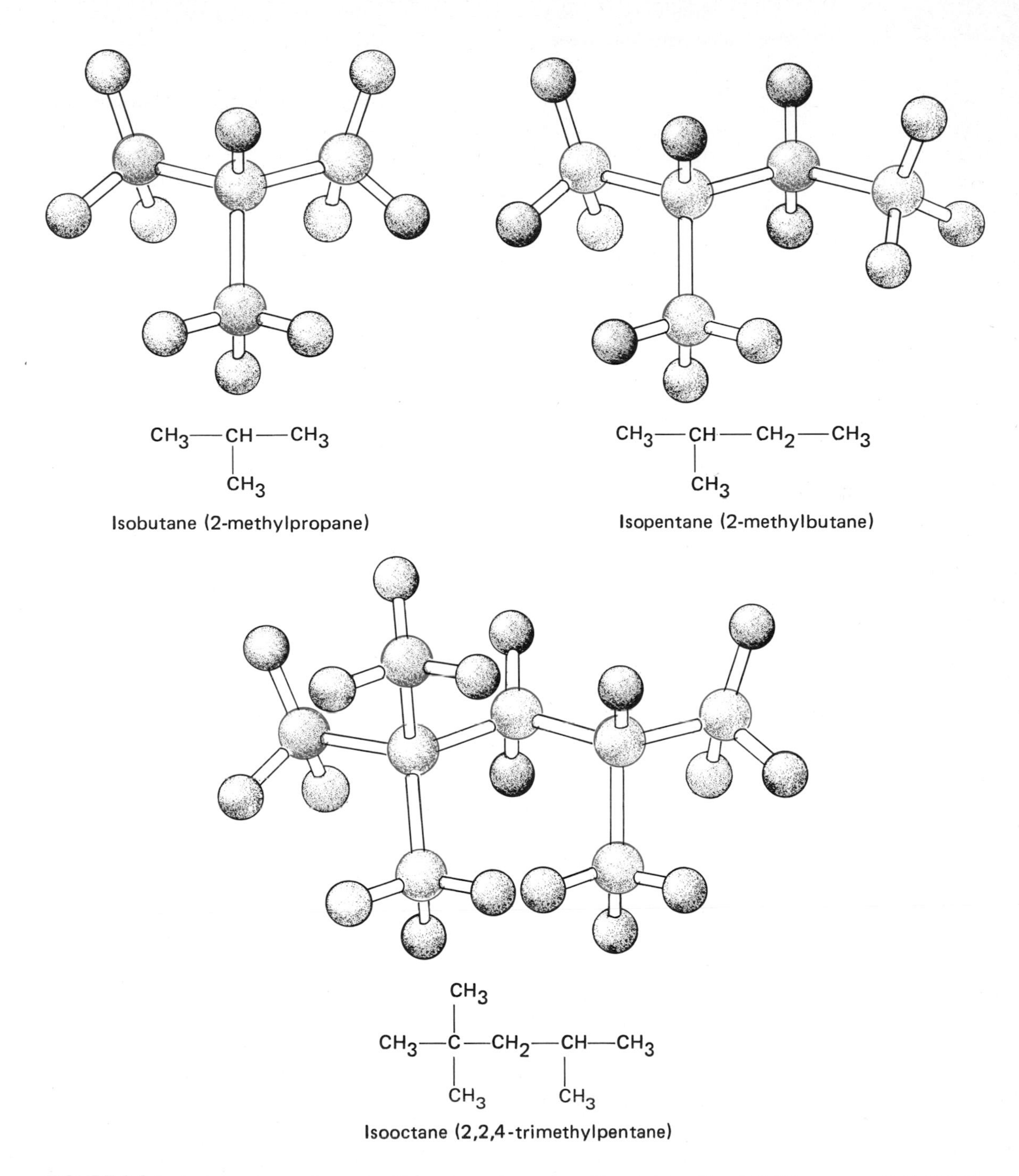

FIGURE 9.2 Isobutane, isopentane, and isooctane: formulas with models. The IUPAC names are shown in parentheses.

Finally, we have a branched-chain isomer with three carbon atoms in a continuous chain.

```
        C
        |
(3) C—C—C
        |
        C
```

This is the only way this isomer can be drawn. If you try to place the two remaining carbon atoms anywhere else, you end up with the same structure as the four-carbon chain.

```
   C
   |
   C                              C
   |                              |
C—C—C    =    C—C—C—C    =    C—C—C
                    |             |
                    C             C
```

The next step is to add hydrogen atoms to each structure to satisfy the requirement of four bonds to each carbon atom.

```
      H  H  H  H  H                  H  H  H  H
      |  |  |  |  |                  |  |  |  |
(1) H—C—C—C—C—C—H            (2) H—C—C—C—C—H
      |  |  |  |  |                  |  |  |  |
      H  H  H  H  H                  H  H  |  H
                                           |
                                         H—C—H
                                           |
                                           H

             H
             |
           H—C—H
             |
      H      |      H
      |      |      |
(3) H—C——————C——————C—H
      |      |      |
      H      |      H
             |
           H—C—H
             |
             H
```

If you count the atoms, you will see that each isomer fits the formula C_nH_{2n+2}.

Problem 9.1 Draw all the alkane isomers that have six carbon atoms, and verify that each has the formula C_6H_{14} (C_nH_{2n+2}, where $n=6$).

ALKANES: NOMENCLATURE 9.3

During the early development of organic chemistry, chemists named their newly discovered compounds rather arbitrarily. As the number of compounds increased, chemists began to realize that they needed a systematic naming scheme. Many of the early common names are still used today, like methane, ethane, propane, and butane. The system used for naming the remaining alkanes is based on the Greek or Latin numbers plus the ending *-ane* to designate that they are alk*anes*: pent*ane*, hex*ane*, hept*ane*, oct*ane*, non*ane*, and dec*ane*.

What are the advantages of the IUPAC system of naming organic compounds?

Near the beginning of the twentieth century, the International Union of Pure and Applied Chemistry (IUPAC) drafted a systematic nomenclature of organic compounds, called the IUPAC system. By using a few, very simple rules, we can give unambiguous names to a large number of compounds, so that, knowing only the name, any chemist can write the structural formula of a compound.

Rule 1: In naming an alkane, select the longest continuous carbon chain as the basis for the parent name. For example, in the structure shown below, the longest continuous chain (shown in color) has six carbon atoms; therefore the parent name is hexane.

$$\begin{array}{l} CH_3\underset{|}{C}HCH_2\underset{|}{C}HCH_2CH_3 \\ \quad\;\; CH_3 \quad\;\; CH_3 \end{array}$$

Rule 2: Name the branches, called **alkyl groups** or **alkyl substituents,** according to the number of carbon atoms in each branch. To name an alkyl group we change the alkane ending *-ane* to *-yl;* thus alkane becomes *alkyl.* A one-carbon branch (methane) becomes *methyl;* a four-carbon branch (butane) becomes *butyl;* and so forth. In the above example, both branches are methyl. Since there are two methyl groups, the compound is dimethylhexane (one word). The prefixes *di-* ("two"), *tri-* ("three"), *tetra-* ("four"), and so forth, are used to denote the number of branches of each type.

Rule 3: Number the carbon atoms in the parent chain, beginning at the end nearest to a branch. The number of the parent carbon to which each substituent is attached is indicated by a number preceding the substituent's name. In the above example, we number the chain as

$$\underset{1}{CH_3}-\underset{2}{\overset{\overset{CH_3}{|}}{CH}}-\underset{3}{CH_2}-\underset{4}{\overset{\overset{CH_3}{|}}{CH}}-\underset{5}{CH_2}-\underset{6}{CH_3}$$

2,4-dimethylhexane

The substituent methyl groups are attached to the carbon atoms numbered 2 and 4. The final IUPAC name of the compound is 2,4-dimethylhexane. Note that numbers are separated from each other by commas, and numbers are separated from words by hyphens. If different substituent groups are present, they are listed in alphabetical order (disregarding the prefixes *di-*, *tri-*, *tetra-*, and so forth).

Rule 4: Halogen substituents occur often. They are named as follows:

—F	fluoro-
—Cl	chloro-
—Br	bromo-
—I	iodo-

We can now put these rules into practice.

Example 9.1 Give the structural formula of the compound 2,3-dimethylheptane.

Solution The ending *heptane* tells us that the longest continuous chain has seven carbon atoms. We then number the carbons from either end.

$$\begin{array}{ccccccccccccc} C & - & C & - & C & - & C & - & C & - & C & - & C \\ 1 & & 2 & & 3 & & 4 & & 5 & & 6 & & 7 \end{array}$$

The prefix 2,3-dimethyl tells us that there are two methyl groups and that they are bonded to carbon atoms 2 and 3. The final structural formula with the required hydrogens is

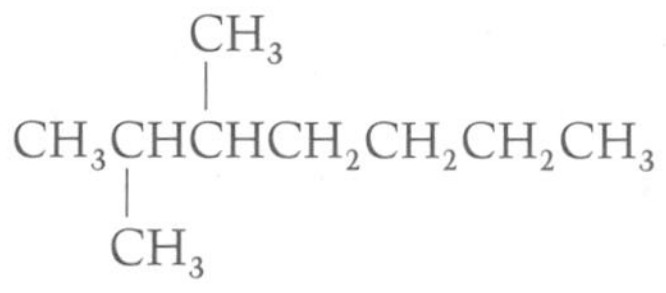

$$\begin{array}{l} \qquad\quad CH_3 \\ \qquad\quad\ | \\ CH_3CHCHCH_2CH_2CH_2CH_3 \\ \qquad | \\ \quad\ \ CH_3 \end{array}$$

Problem 9.2 Give the structural formula of the following compounds.

(a) 2,2-dimethylpentane (b) 3,3-dimethylpentane

(c) 3-ethyloctane (d) 4-propyldecane

Problem 9.3 Give the IUPAC name of each of the following compounds.

(a) 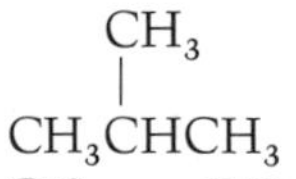 $\begin{array}{l} \quad\ CH_3 \\ \quad\ \ | \\ CH_3CHCH_3 \end{array}$

(b) 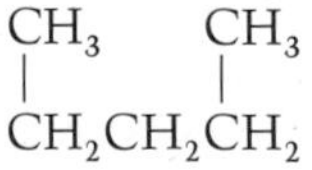 $\begin{array}{l} CH_3 \qquad CH_3 \\ \ | \qquad\qquad | \\ CH_2CH_2CH_2 \end{array}$ (be careful!)

$$\text{(c)}\ \begin{array}{l} \quad\ \ CH_3\ \ CH_3 \\ \quad\ \ \ |\qquad\ | \\ CH_3CHCHCHCH_2CH_2CH_3 \\ \qquad\ \ | \\ \qquad CH_3 \end{array}$$

$$\text{(d)}\ \begin{array}{l} \quad\ \ CH_2CH_3\quad CH_2CH_3 \\ \quad\ \ \ |\qquad\qquad\ | \\ CH_3CHCH_2CH_2CHCH_3 \end{array}$$

$$\text{(e)}\ \begin{array}{l} \qquad CH_2CH_3 \\ \qquad\ \ | \\ CH_3CH_2CCH_2CH_3 \\ \qquad\ \ | \\ \qquad CH_2CH_3 \end{array}$$

ALKANES: PROPERTIES 9.4

How would you compare the properties of the alkanes with those of water?

Table 9.1 listed some properties of the first ten unbranched alkanes. Each alkane has a lower melting and boiling point than any other organic compound of comparable molecular weight. This is because they are nonpolar and therefore exhibit very little intermolecular attraction.

Generally, the melting and boiling points of alkanes increase with increasing molecular size. The first four unbranched alkanes are gases under ordinary conditions (1 atm and 20–25°C). The next thirteen (C_5H_{12} to $C_{17}H_{36}$) are liquids, and those with eighteen or more carbon atoms are solids.

Because they are nonpolar, alkanes and other hydrocarbons are insoluble in water: "oil and water do not mix" is a common way of saying that most hydrocarbons are insoluble in water. This phenomenon is described further in Section 17.2. Alkanes and other hydrocarbons are less dense than water, so they float on its surface. These two properties of hydrocarbons are partly responsible for the devastating effects of oil spills from ships.

Problem 9.4 Of the alkanes in Table 9.1, which one most closely resembles water in its
(a) melting point (b) boiling point
(c) density

ALKANES: PREPARATION 9.5

Alkanes are found in petroleum and natural gas. The various alkanes in these sources are readily separated by fractional distillation (see Focus

What are the two general methods of preparing alkanes in the laboratory?

6). Some alkanes are not available from natural sources, however, and chemists must prepare these in the laboratory.

Hydrogenation of Alkenes

An important method for obtaining alkanes is the hydrogenation of alkenes. Alkenes are hydrocarbons with a double bond between two carbons (Section 8.12). **Hydrogenation** is simply a reaction in which one hydrogen atom is added to each carbon of the double bond. A mixture of the alkene and the gaseous hydrogen is warmed in the presence of a catalyst under a pressure of a few atmospheres. Finely divided nickel, platinum, or palladium may be used. The general reaction is as follows (R is a hydrocarbon group or hydrogen).

$$\underset{\text{an alkene}}{R_2C{=}CR_2} + H_2 \xrightarrow{\text{Ni, Pt, or Pd}} R{-}\underset{R}{\overset{H}{C}}{-}\underset{R}{\overset{H}{C}}{-}R$$

A specific example is the hydrogenation of propene.

$$CH_3CH{=}CH_2 + H_2 \xrightarrow{Ni} CH_3\underset{H}{CH}{-}\underset{H}{CH_2}$$

Wurtz Reaction

We can prepare alkanes that contain carbon chains longer than those of the reactants through the **Wurtz reaction** (named after its discoverer, Adolphe Wurtz). To carry out a Wurtz reaction, we heat a mixture of an alkyl halide (a haloalkane) and sodium. The general reaction is as follows (X is chlorine, bromine, or iodine, and R is a hydrocarbon group or hydrogen).

$$R{-}\underset{R}{\overset{R}{C}}{-}X + 2Na + X{-}\underset{R}{\overset{R}{C}}{-}R \xrightarrow{\text{heat}} R{-}\underset{R}{\overset{R}{C}}{-}\underset{R}{\overset{R}{C}}{-}R + 2NaX$$

For example, 1-bromobutane yields octane.

$$2CH_3CH_2CH_2CH_2{-}Br + 2Na \xrightarrow{\text{heat}} CH_3CH_2CH_2CH_2CH_2CH_2CH_2CH_3 + 2NaBr$$

The net effect is to link the two butyl groups together by forming a very stable ionic compound, sodium bromide. When we use two different alkyl halides, we get three alkanes.

$$CH_3{-}Cl + Cl{-}CH_2CH_3 + 2Na \xrightarrow{\text{heat}} CH_3CH_3 + CH_3CH_2CH_3 + CH_3CH_2CH_2CH_3 + 2NaCl$$

(We leave this equation unbalanced because the three alkanes are not produced in predictable amounts.) The Wurtz reaction was of great importance in the early days of organic chemistry, because Wurtz used it to show that alkanes of all lengths are possible.

In this section we have used a notation that you will encounter often in organic reactions: on or under the arrow between reactants and products we list any special solvents, catalysts, and special conditions that are required. Normally, reactants that are consumed are not written over or under the arrow. Thus, in general we write an equation as

$$\text{Reactant A} + \text{Reactant B} + \cdots \xrightarrow{\text{conditions, solvents, or catalysts}} \text{Products}$$

Problem 9.5 Complete and balance the following equations.

(a) $CH_3CH_2CH_2Br + Na \rightarrow$

(b) $CH_3CH(CH_3)Br + Na \rightarrow$ (structure: CH_3 attached to the CH carbon)

(c) $CH_3CH_2CH{=}CHCH_3 + H_2 \xrightarrow{\text{Ni}}$

(d) $\text{cyclo-}[C(CH_3){=}CH{-}CH_2{-}CH_2{-}CH_2] + H_2 \xrightarrow{\text{Ni}}$ (ring: C bearing CH_3, CH, CH_2, CH_2, CH_2)

Problem 9.6 Write equations for the preparation of butane by

(a) the Wurtz reaction and

(b) the hydrogenation of an alkene.

ALKANES: REACTIONS 9.6

Alkanes are relatively unreactive. In fact, an early name for these compounds was **paraffin,** a German word meaning "too little affinity," because alkanes have little affinity for other substances. They undergo reaction only under very vigorous conditions, and the reactions are difficult to control. We will examine two such reactions: combustion and halogenation of alkanes.

How do the products of combustion differ from those of halogenation?

Combustion

All alkanes burn in air (**combustion**) to yield carbon and oxides of carbon. Both carbon monoxide and carbon dioxide, as well as water, are produced if the supply of oxygen is limited. (The equation is unbalanced.)

$$CH_4 + O_2 \rightarrow CO + CO_2 + H_2O + \text{heat}$$

If we provide an excess of oxygen, complete combustion takes place, yielding only carbon dioxide and water (Section 2.2).

$$CH_4 + 2O_2 \xrightarrow[\text{combustion}]{\text{complete}} CO_2 + 2H_2O + \text{heat}$$

The principal use of this reaction is to produce heat energy. All alkanes give off heat during burning. We power our automobiles by burning alkanes of six to twelve carbons. In the engine, the expansion of hot gases causes pistons to move in their cylinders, thus converting the chemical energy of the gasoline into mechanical energy.

Halogenation

Chlorine and bromine react with alkanes. The conditions for halogenation are milder than for combustion. The reaction takes place at room temperature if light is present. In the dark, the reaction mixture must be heated for the reaction to take place. In the reaction, an atom of bromine or chlorine—both halogens—is substituted for a hydrogen atom of the alkane. This substitution is called a **halogenation.** The general reaction is as follows (R is a hydrocarbon group or hydrogen, and X is chlorine or bromine).

$$\begin{array}{c} R \\ | \\ R-C-H \\ | \\ R \end{array} + X-X \xrightarrow{\text{light or heat}} \begin{array}{c} R \\ | \\ R-C-X \\ | \\ R \end{array} + HX$$

An example is the chlorination of ethane.

$$CH_3CH_3 + Cl_2 \xrightarrow[\text{heat}]{\text{light or}} CH_3CH_2Cl + HCl$$

The reaction does not work so well with the other two common halogens: fluorine is so reactive that the reaction is difficult to control; iodine does not react at all. Halogenation usually gives a mixture of products,

TABLE 9.2 Some Halogenated Hydrocarbons Produced on a Large Scale

Formula	Common name	IUPAC name	Use
CH_3CH_2Cl	Ethyl chloride	Chloroethane	Synthesis of tetraethyl lead for gasoline
CH_2Cl_2	Methylene chloride	Dichloromethane	Additive in paint; varnish remover; cleaning agent
CCl_3F	Freon-11	Trichlorofluoromethane	Refrigerant
CCl_2F_2	Freon-12	Dichlorodifluoromethane	Refrigerant and propellant in aerosol sprays

because more than one substitution can occur in a given molecule. When such a mixture is produced, the process is called **polyhalogenation.** The chlorination of methane, for example, yields all four possible products (see Chapter 11).

$$CH_4 + Cl_2 \xrightarrow{\text{light}} \underset{\text{(methyl chloride)}}{CH_3Cl} + HCl$$

$$CH_3Cl + Cl_2 \xrightarrow{\text{light}} \underset{\text{(methylene chloride)}}{CH_2Cl_2} + HCl$$

$$CH_2Cl_2 + Cl_2 \xrightarrow{\text{light}} \underset{\text{(chloroform)}}{CHCl_3} + HCl$$

$$CHCl_3 + Cl_2 \xrightarrow{\text{light}} \underset{\text{(carbon tetrachloride)}}{CCl_4} + HCl$$

Halogenated hydrocarbons are very useful, but most of them are toxic. Table 9.2 lists some simple compounds of this type that are produced on a large scale, along with their uses.

In addition, the use of Freons, which are not exceptionally toxic, may be hazardous because it is suggested that they can destroy the ozone in the atmosphere. This ozone absorbs ultraviolet solar radiation and thus shields us from its harmful effects—skin cancer and an excessive warming of the earth. Freon-12 from aerosol sprays has been detected in the atmosphere, where it decomposes to form chlorine atoms.

$$\text{(1)} \qquad CCl_2F_2 \xrightarrow[\text{radiation}]{\text{ultraviolet}} \cdot CClF_2 + \underset{\text{(free atom)}}{\cdot \ddot{\underset{\cdot\cdot}{Cl}}:}$$

(2) $:\ddot{\underset{..}{Cl}}\cdot + O_3 \rightarrow \cdot\ddot{\underset{..}{O}}-\ddot{\underset{..}{Cl}}: + O_2$
(ozone)

(3) $\cdot\ddot{\underset{..}{O}}-\ddot{\underset{..}{Cl}}: + O_3 \rightarrow 2O_2 + \cdot\ddot{\underset{..}{Cl}}:$

In reaction (2) chlorine atoms destroy ozone molecules. In reaction (3) additional ozone is destroyed, and simultaneously, chlorine atoms are reformed to continue the process.

Problem 9.7 It is possible to control the chlorination of methane to obtain either CH_3Cl or CCl_4 as the major product. Suggest how we could do this. [Hint: consider varying the ratio of reactants.]

9.7 CYCLOALKANES

What is the general molecular formula of a cycloalkane?

Certain hydrocarbon molecules consist of three or more carbon atoms bonded together to form rings. These cyclic hydrocarbons with single bonds between the carbons are called **cycloalkanes** (Figure 9.3). We can simplify the drawing of the structural formulas of cycloalkanes by using **line formulas.** The lines represent C–C bonds. The meeting of two lines represents a carbon atom with enough hydrogen atoms attached to give it a total of four bonds. The structures in Figure 9.3 can be represented using line formulas as follows.

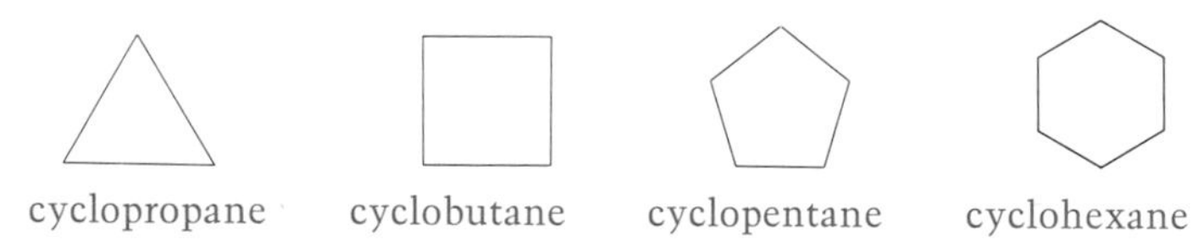

cyclopropane cyclobutane cyclopentane cyclohexane

The general formula of cycloalkanes is C_nH_{2n}.

Problem 9.8
(a) Write condensed formulas to verify that cycloheptane and cyclooctane have the formula C_nH_{2n}.
(b) Draw the line formulas of cycloheptane and cyclooctane.

9.8 CYCLOALKANES: NOMENCLATURE

The IUPAC system names the cycloalkanes by using the prefix *cyclo-*, along with the alkane name that corresponds to the number of carbons in

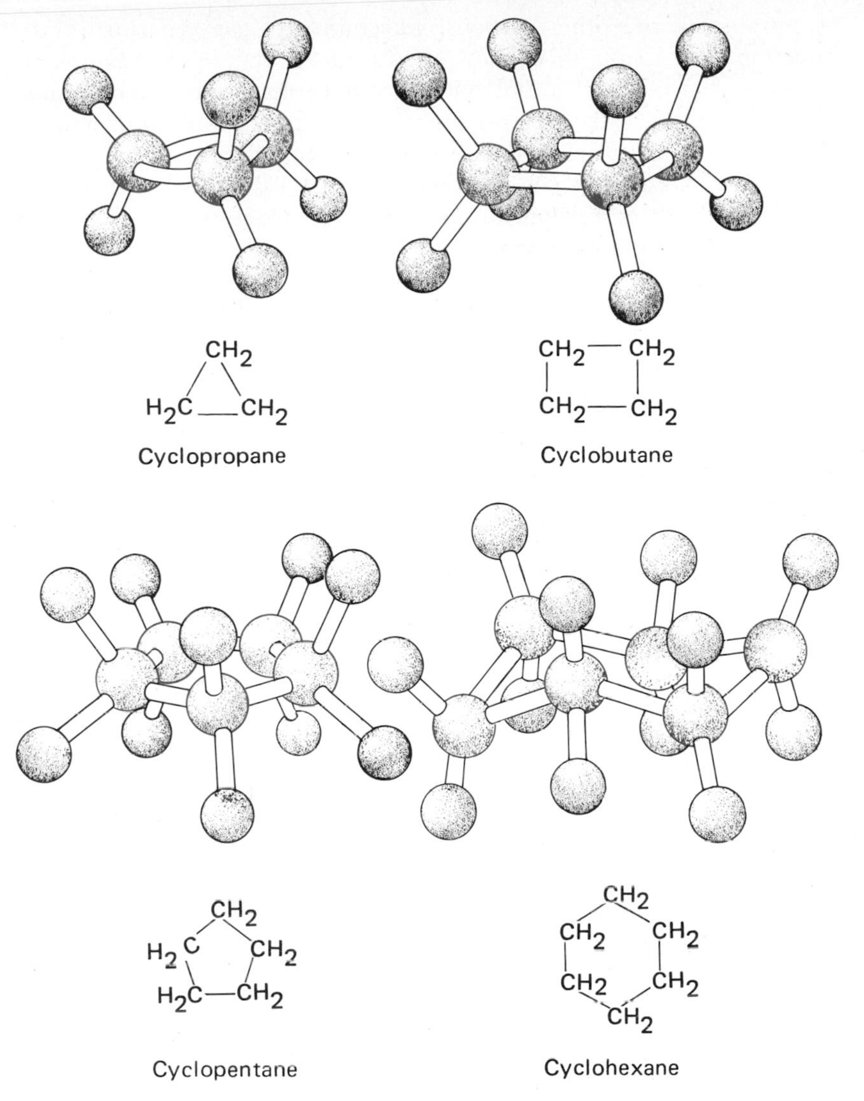

FIGURE 9.3 Cyclopropane, cyclobutane, cyclopentane, and cyclohexane: formulas with models.

the ring. Substituent groups are named in the same way as before, except that we begin to number the ring at a carbon that bears a substituent. For example,

How do we number the carbon atoms in a cycloalkane?

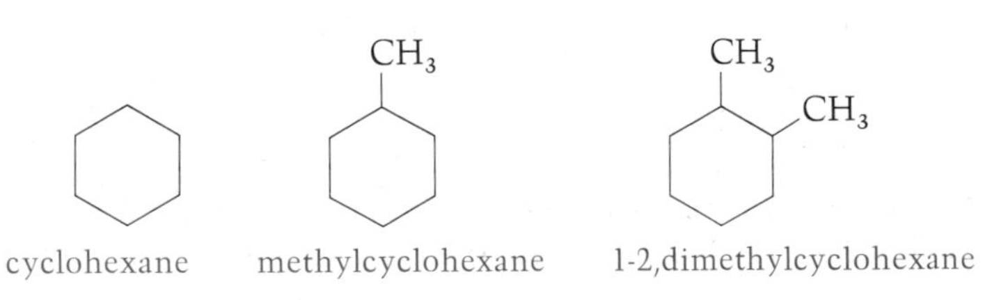

In numbering the other carbons, we count around the ring in the direction that results in the lowest sum of numbers. For example, if substituent groups are found on adjacent carbons, we don't number one of them 1 and the other 6, we number them 1 and 2. In naming methylcyclohexane we need not designate the position of the methyl substituent, because when there is only one substituent, we always consider it to be on carbon number one. We use numbers only when there is more than one substituent on the ring.

Problem 9.9 Give the IUPAC name for each of the following compounds.

(a) Cl (b) Cl Cl (c) CH_3 (d) Cl CH_3 CH_3 CH_3

Problem 9.10 Give the line formulas for the following compounds.
(a) 1,1-dibromocyclopropane (b) 1,3-dimethylcycloheptane

9.9 CYCLOALKANES: PROPERTIES

What structural feature distinguishes cyclopropane from the larger cycloalkanes?

Cycloalkanes have properties very similar to those of the open-chain alkanes. They have similar melting and boiling points, densities, and solubilities.

With the exception of cyclopropane, the cycloalkanes react exactly as the corresponding open-chain alkanes do. They undergo combustion and halogenation in the same way.

Cyclopropane, on the other hand, is more reactive than propane and it exhibits exceptional properties. In addition to the reactions characteristic of alkanes, cyclopropane undergoes reactions that result in the opening of the ring. In equation 2, X can be chlorine, bromine, or iodine. In equation 3, X can be chlorine or bromine only.

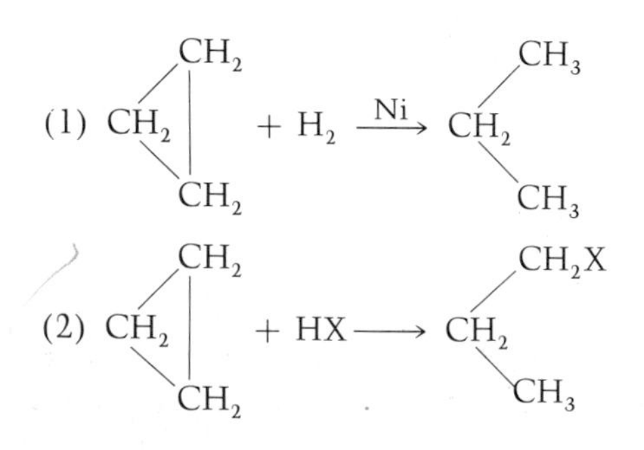

$$\text{(3)}\ \begin{matrix} & CH_2 & \\ CH_2 & \vert & \\ & CH_2 & \end{matrix} + X_2 \longrightarrow \begin{matrix} & CH_2X \\ CH_2 & \\ & CH_2X \end{matrix}$$

These reactions are possible because of the large strain in the cyclopropane molecule. The normal tetrahedral angle (Figure 8.2) is 109.5° The C—C—C angle in cyclopropane is near 60° (Figure 9.4). The distortion of the bond angles results in a high potential energy, which is released when the ring breaks. The high potential energy comes from the repulsion caused by the crowding of electron orbitals.

Problem 9.11 Complete and balance the following equations.
(a) Cyclopropane + HBr → (b) Cyclopropane + Cl_2 →

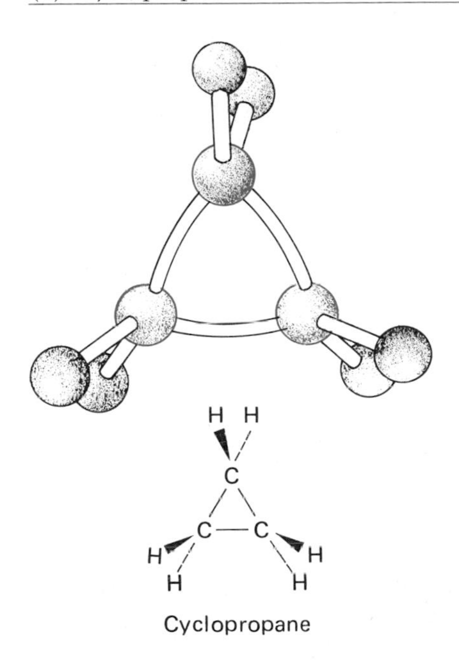

FIGURE 9.4 The carbon–carbon bonds in cyclopropane are highly strained because they are forced by the geometry of the molecule to be close to 60°. The relaxed tetrahedral carbon bond angle is 109.5° (Figure 8.2).

SUMMARY

Alkanes (general formula C_nH_{2n+2}) are noncyclic hydrocarbons that have only single bonds. They exhibit low melting and boiling points and low densities, compared with other organic compounds of similar molecular weight. Hydrocarbons are insoluble in water. Alkanes may be prepared by **hydrogenation** of alkenes and by the **Wurtz reaction.** They undergo **combustion** to produce carbon dioxide, water, and heat when burned in the presence of excess oxygen. Alkanes also undergo **halogenation** when mixed with chlorine or bromine at high temperatures in the dark or at room temperature in the light. **Cycloalkanes** (general formula C_nH_{2n}) have only single bonds, but carbon atoms are bonded together into rings. The smallest cycloalkane, cyclopropane, undergoes addition reactions. **Line formulas** provide a simplified version of the structural formulas of cycloalkanes.

Key Terms

Alkane (Section 9.1)
Alkyl group (Section 9.3)
Branched-chain isomer (Section 9.2)
Combustion (Section 9.6)
Cycloalkane (Section 9.7)
Halogenation (Section 9.6)
Hydrogenation (Section 9.5)
Line formula (Section 9.7)
Paraffin (Section 9.6)
Polyhalogenation (Section 9.6)
Saturated hydrocarbon (Section 9.1)
Straight-chain isomer (Section 9.2)
Substituent (Section 9.3)
Wurtz reaction (Section 9.5)

ADDITIONAL PROBLEMS

9.12 Give the molecular formula of the following alkanes.
(a) heptane (b) octane
(c) pentane

9.13 Give the structural formulas of the following compounds.
(a) 2-methylbutane (b) 3-methylpentane
(c) 2,2,4-trimethylpentane (d) methylcyclopropane
(e) 1,1-diethylcyclobutane (f) 1,4-dimethylcyclohexane

what?

9.14 Give the IUPAC names of the following compounds.

(a) $CH_3CH_2CH_2CHCH_3$ with a CH_3 branch on the fourth carbon

$$\begin{array}{l} CH_3CH_2CH_2\underset{\displaystyle CH_3}{\underset{|}{C}}HCH_3 \end{array}$$

(b) $CH_3CH_2CH_2C(CH_3)_2CH_3$

$$
\begin{array}{c}
\quad\quad\quad\quad\quad CH_3 \\
\quad\quad\quad\quad\quad | \\
CH_3CH_2CH_2CCH_3 \\
\quad\quad\quad\quad\quad | \\
\quad\quad\quad\quad\quad CH_3
\end{array}
$$

(c) 1,1,3-trimethylcyclopentane: cyclopentane ring bearing two CH_3 groups on one carbon and CH_3 groups (CH_3—, CH_3) on another.

(d) CH_3CH_2—cyclobutane

9.15 Explain why alkanes have lower melting and boiling points than water.
9.16 Are hydrocarbons soluble in water? Explain.
9.17 Complete and balance the following equations.

(a) $CH_3CH{=}CH_2 + H_2 \xrightarrow{Ni}$

(b) $CH_3CH_2CH_2I + Na \rightarrow$

(c) $CH_3CH_2CH_3 + O_2 \text{ (excess)} \xrightarrow{heat}$

(d) $CH_3CH_3 + Br_2 \text{ (1 mole)} \xrightarrow{light}$

9.18 Draw the line formulas of the organic reactants and organic products of Problem 9.17. (Note that line formulas can be used for noncyclic hydrocarbons.)
9.19 Write equations to show how you could prepare 2,3-dimethylbutane, starting with isopropyl bromide.
9.20 Write equations to show how you could prepare each of the following compounds, starting from cyclopropane.
(a) 1,3-dibromopropane (b) hexane
9.21 If a sample of crude petroleum oil consists of a mixture of alkanes, and if it includes all the compounds listed in Table 9.1, which compounds will be present in the fraction that boils between room temperature (25°C) and 100°C? (See Focus 6.)
9.22 Draw the condensed and line formulas of the following compounds.
(a) cyclononane (b) cyclodecane
(c) 2,3-dimethylheptane
9.23 Explain why cyclopropane is more reactive than other cycloalkanes.
9.24 Write the balanced equation for the reaction of cyclopropane with HI.
9.25 Give the structural formulas of all the products of the following Wurtz reaction.

$$
\begin{array}{l}
\quad\;\; CH_3 \\
\quad\;\;\; | \\
CH_3CHBr + BrCH_2CH_3 + Na \rightarrow
\end{array}
$$

9.26 What is the major organic product in each of the following reactions? (A large excess means that much more of the reactant is used than is called for by the balanced equation.)

(a) $CH_4 + Br_2$ (large excess) $\xrightarrow{\text{light}}$

(b) CH_4 (large excess) $+ Br_2 \xrightarrow{\text{light}}$

9.27 Section 9.5 describes two methods used for the preparation of alkanes. Name some alkanes (general types or specific compounds) that *cannot* be made by each method.

9.28 Each of the following names is incorrect. Give the correct names.

(a) 3-methylbutane
(b) 1-ethylhexane
(c) 1-chlorocyclohexane
(d) 6-methylcyclohexane

Problem-Solving Hints

If you have trouble solving any of the additional problems, refer to the sections listed next to the problem numbers.

9.12 9.2 | **9.13** 9.3 | **9.14** 9.3
9.15 9.4 | **9.16** 9.4 | **9.17** 9.5, 9.6
9.18 9.7 | **9.19** 9.5 | **9.20** 9.5, 9.9
9.21 Focus 6 | **9.22** 9.7 | **9.23** 9.9
9.24 9.9 | **9.25** 9.5 | **9.26** 9.6
9.27 9.5 | **9.28** 9.3, 9.8

FOCUS 6 Petroleum, a Potent Reservoir of Solar Energy

Petroleum is the major source of alkanes and other hydrocarbons. It consists of the remains of tiny plants and animals that lived in the sea hundreds of millions of years ago. During their lives, these organisms trapped and stored solar energy. As they died, they settled to the ocean floor, where they were mixed with mud and sand. Over the centuries, these layers packed down and formed rock. As the earth's crust buckled, enormous pressures converted these organic substances into the petroleum that lies in pools deep in the earth's crust.

Crude petroleum is obtained by tapping these subterranean pools. The crude oil is a mixture of thousands of hydrocarbons, mostly alkanes. This mixture is separated by **fractional distillation.** In this process, the crude oil is heated gradually so that the components boil off according to their boiling points (Figure 1). The most volatile components boil at the lowest temperatures. The vapor obtained at different temperatures is recondensed into liquid; in this way various fractions are recovered. A fraction is that part of the volatile mixture that boils within a specified temperature range. Table 1 shows the fractions obtained from the distillation of crude petroleum.

Some petroleum fractions are more useful than others. Chemists have developed a process for converting a less useful fraction, like kerosene, into a more useful one, like gasoline. This process, called **cracking,** consists of heating a hydrocarbon to between 250°C and 600°C in the absence of air. Under these conditions, C–C bonds break and reform in a fairly random manner. The desired fraction is removed by fractional distillation, and the remainder is returned to the cracking process.

TABLE 1 Products of the Distillation of Crude Petroleum

Name of fraction	Molecular size	Approximate boiling range, °C
Natural gas	C_1–C_4	Less than 25
Petroleum ether	C_5–C_6	20–60
Ligroin (light naphtha)	C_6–C_7	60–100
Gasoline	C_6–C_{12}	50–200
Kerosene	C_9–C_{15}	175–275
Gas oil (furnace oil, diesel oil)	C_{14}–C_{18}	Above 275
Lubricating oil	Above C_{18}	Nonvolatile liquids
Asphalt (residue)		Nonvolatile solids

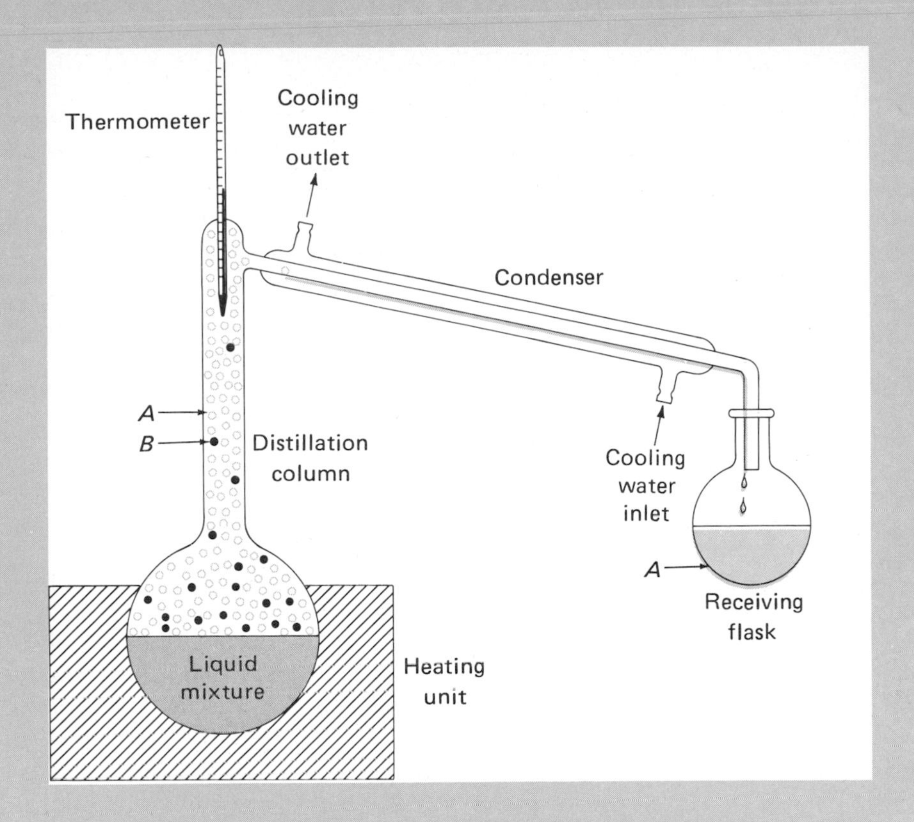

FIGURE 1 Schematic drawing of a simple apparatus for the distillation of a liquid mixture. The more volatile component, *A* (in color) vaporizes to a greater extent than the less volatile component, *B* (in black). Near the top of the column, the vapor is nearly pure *A*. As it passes through the condenser, vapor *A* condenses to liquid *A* and is collected in the receiving flask.

The most familiar use of petroleum products is as fuels—natural gas, gasoline, kerosene, and fuel oil (Table 1). Other useful compounds from petroleum, like ethylene and benzene (Chapter 10), are the fundamental raw materials for the manufacture of thousands of valuable synthetic plastics, fibers, drugs, dyes, and other consumer products. Chemicals derived from petroleum are called petrochemicals.

The decreasing supply of petroleum will undoubtedly continue to drive the price up. As the price rises, it will become less and less practical to burn this diminishing natural resource.

Key Terms

Cracking
Fractional distillation

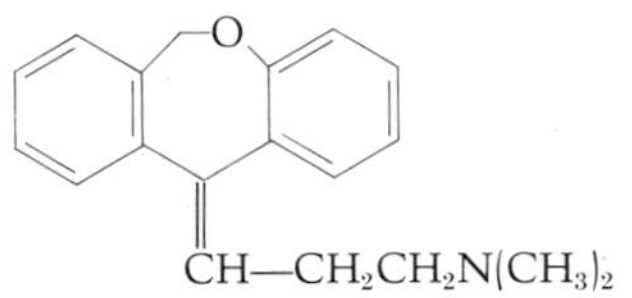

$CH—CH_2CH_2N(CH_3)_2$

Doxepin

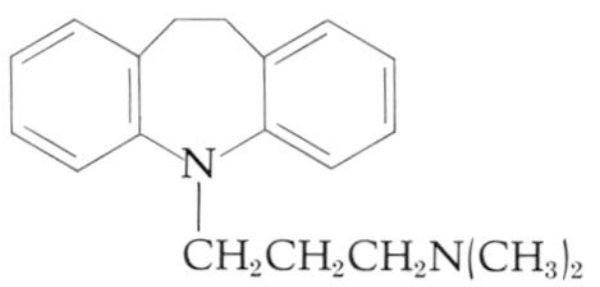

$CH_2CH_2CH_2N(CH_3)_2$

Imipramine

Depression describes the change in mood that occurs when feelings of happiness and expectation turn to feelings of sadness, loneliness, and disappointment. Many of us have been depressed at one time or another, but the mood is generally short-lived and doesn't affect us beyond the moment it is experienced. On the other hand, the psychiatric illness of depression is frequently longer lasting and more severe. Physical symptoms of depressed patients may include loss of weight, restlessness, and fatigue. In the diseased state, muscle and joint problems may also occur.

The tricyclic antidepressants are the most widely used drugs today for the treatment of depression. Among the tricyclics are amitriptyline (Elavil®), imipramine, and doxepin. Treatment with antidepressants is often used along with psychotherapy.

Structurally, the tricyclics consist of three cyclic groups (groups having a ringlike structure). Two benzene rings lie on either side of a centrally located third ring. (We will be familiar with the structure—and function—of benzene after reading this chapter.) The third cyclic ring is seven-membered and is often heterocyclic—that is, the seven members of the ring are not all alike. The only difference among the tricyclic antidepressants, in fact, is in what substituents (members) are represented in the central ring. The outer benzene rings are free of hydrocarbon substituents.

10

UNSATURATED HYDROCARBONS—ALKENES, ALKYNES, AND ARENES

Antidepressant drugs are absorbed rapidly from the gastrointestinal tract and metabolized by the liver. They have sedative effects that cause elevation of mood. These effects are directly related to the function of the neurotransmitters, which are responsible for the transmission of brain impulses throughout the body. Tricyclic antidepressants are thought to be able to increase the efficiency of these neurotransmitters.

As is true with many other drugs used therapeutically, tricyclic antidepressants can have serious toxic effects. Increased heart rate, constipation, and dizziness are among the possible side effects. Measuring the levels of the drug in the blood helps to minimize these adverse effects by providing a means of regulating drug dosage. Regulation is an important safeguard against overdoses, which can lead to heart failure, seizures, or coma.

[*Illustration: Tricyclic antidepressants are named as such because of the presence of three ring structures in their configurations.*]

UNSATURATED HYDROCARBONS

Now you know that a saturated hydrocarbon is one that is full of atoms; that is, it has only single bonds. As you might guess, then, an unsaturated hydrocarbon is one that is *not* full of atoms; that is, it has double or triple bonds. Most unsaturated hydrocarbons can easily form new bonds with new atoms, and so they are very reactive. Most of them, as you might expect, simply add new atoms to become saturated. But not all unsaturated hydrocarbons react in this way. An exception is the special group of unsaturated hydrocarbons, the arenes, which do not readily add new atoms. You may wonder why, if they are unsaturated, they do not simply form new bonds with new atoms. Isn't that what

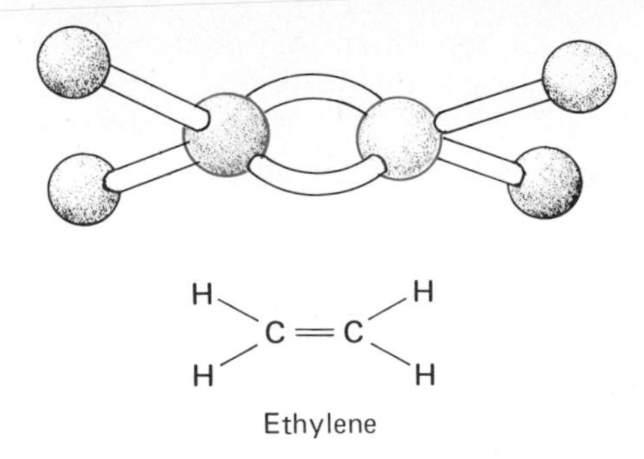

FIGURE 10.1 Ethylene (ethene): formula with model.

being unsaturated is all about? As usual, the answer is to be found in the structure of the molecule. Before we can begin to understand the aromatic hydrocarbons and how their structure gives them unique properties, we will have to learn how the other unsaturated hydrocarbons behave.

ALKENES: STRUCTURE 10.1

What is the essential structural feature of alkenes?

When two atoms share two pairs of electrons, the resulting bond is called a double bond (Section 8.7). Hydrocarbons that contain a double bond are called **alkenes.** The simplest alkene is ethene (common name, ethylene). Its structure is shown in Figure 10.1.

The presence of a double bond affects the properties of hydrocarbons. In fact, the chemistry of alkenes is the chemistry of the double bond.

ALKENES: NOMENCLATURE 10.2

The common names of a few alkenes are shown below.

$CH_2{=}CH_2$ (ethylene) $\quad CH_2{=}CHCH_3$ (propylene) $\quad \underbrace{CH_2{=}CHCH_2CH_3 \quad CH_3CH{=}CHCH_3}_{\text{butylenes}}$

$CH_2{=}C(CH_3)_2$
isobutylene

In naming alkenes the double bond has priority over alkyl substituents.

The rules for forming IUPAC names for alkenes are very similar to rules for alkanes (Section 9.3).

Rule 1: Select the longest continuous carbon chain that contains the double bond. This chain will serve as the basis for the parent name. The first part of the parent name will be the same as for an alkane with the same number of carbons in its longest chain. But the ending will be *-ene* rather than *-ane.*

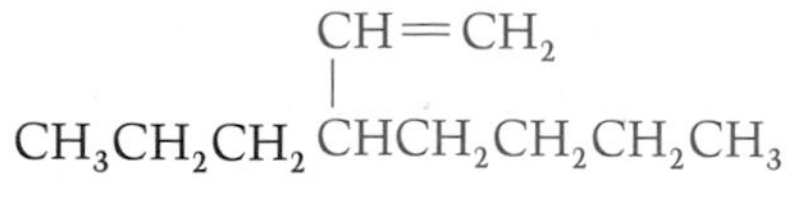

$$
\begin{array}{l}
\qquad\qquad\qquad\quad CH{=}CH_2 \\
\qquad\qquad\qquad\quad | \\
CH_3CH_2CH_2CHCH_2CH_2CH_2CH_3
\end{array}
$$

In the example above, the longest continuous chain that contains the double bond has seven carbon atoms (shown in color). The parent name is therefore heptene.

Rule 2: Name substituents exactly as for alkanes.

Rule 3: Number the parent chain from the end nearer to the double bond. The locations of substituents are designated as for alkanes—by the number of the carbon to which each substituent is attached. In addition, a number is inserted after the substituent names and before the parent name to show where the double bond is located. This number is the number of the first (lowest numbered) doubly bonded carbon. The final IUPAC name of the example above therefore is 3-propyl-1-heptene.

Rule 4: In naming cycloalkenes, number the ring so as to give the double bonded carbons the numbers 1 and 2. For example, the compound with the following structure is 3-methylcyclopentene.

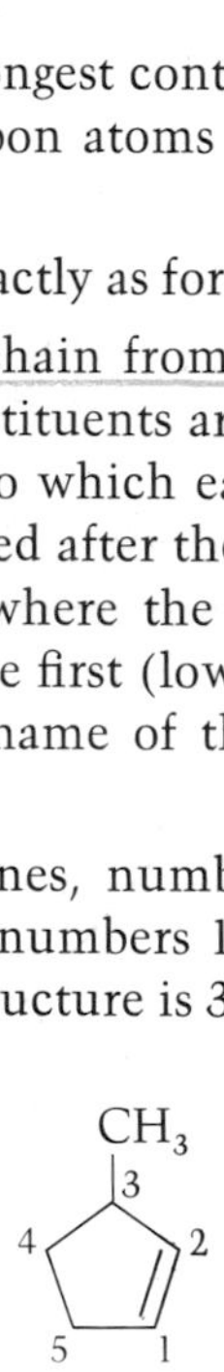

[Structure: cyclopentene ring with CH_3 at carbon 3; ring carbons numbered 1–5, double bond between 1 and 2]

We need not specify the position of the cycloalkene's double bond with a number, because it is always between carbon atoms 1 and 2.

Problem 10.1 Give the IUPAC names of the following compounds.

(a) $CH_2{=}CHCH_3$ (b) $CH_3CH{=}CHCH_3$

(c) 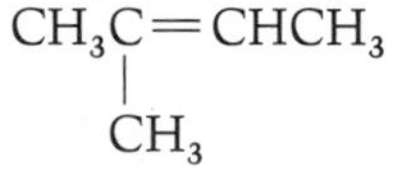 $CH_3\underset{\displaystyle CH_3}{\underset{|}{C}}{=}CHCH_3$ (d) 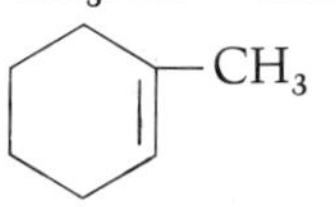 [Structure: cyclohexene ring with CH_3 on a doubly bonded carbon]

Problem 10.2 Draw the structural formulas of the following compounds.

(a) 3-methyl-2-pentene (b) 2-methyl-3-hexene

(c) 3-chlorocyclopentene

MOLECULAR ORBITALS 10.3

We can best represent the structure of the double bond by using the concept of molecular orbitals. A covalent bond between two atoms is formed by the overlap between an orbital of one atom and an orbital of the other atom. When orbitals of two atoms overlap to form a bond, the result is a **molecular orbital,** that is, an orbital that encompasses two or more atoms. You should not confuse molecular orbitals with hybrid orbitals. Molecular orbitals are formed by the combination of orbitals on *different* atoms. Hybrid orbitals are formed by the combination of orbitals *on the same atom.*

Using ethylene as an example, we can show how a double bond is constructed in molecular-orbital theory (Section 8.4). First, we join all the atoms by single bonds.

```
H       H
 \  .  . /
  C — C
 /       \
H         H
```

These single bonds are called, in molecular-orbital language, **sigma (σ) bonds.** Sigma bonds result from the overlap of orbitals on adjacent atoms along a line joining the two atomic nuclei (sigma-bond overlap is shown as a shaded area).

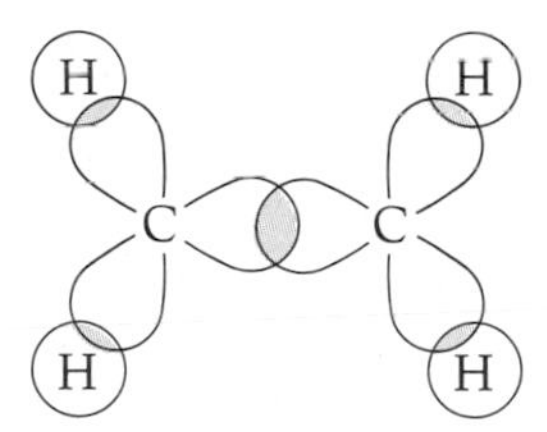

To form these sigma bonds, each hydrogen uses its 1*s* orbital, and each carbon uses three sp^2 hybrid orbitals. We know that the carbon orbitals are sp^2-hybridized, because we need three orbitals and in order to obtain them we must combine one 2*s* orbital and two 2*p* orbitals. (Note that the number of orbitals combined equals the number of hybrid orbitals formed.)

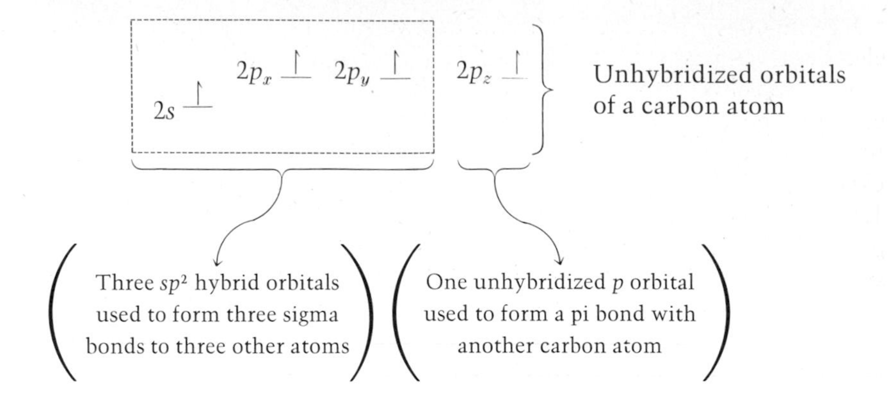

Each carbon atom now has one electron remaining in a p orbital. If p_x and p_y orbitals were used to make the sp^2 hybrid orbitals, as shown above, the remaining electron of each carbon atom must be in a p_z orbital.

The two p_z orbitals lying alongside and parallel to each other can overlap to form a **pi (π) bond,** as shown in Figure 10.2. A pi bond consists of a pi molecular orbital containing two electrons. The pi orbital has two lobes, one above and one below the plane of the molecule.

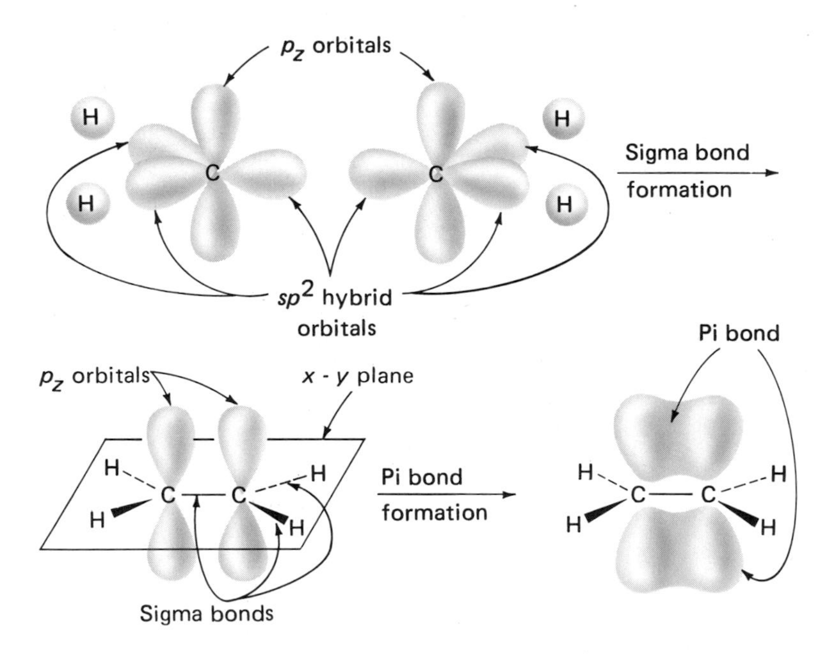

FIGURE 10.2 Construction of sigma and pi bonding of the ethylene molecule. The sigma bonds are formed when the sp^2 hybrid orbitals of each carbon overlap with an sp^2 hybrid orbital of the other carbon and the s orbitals of hydrogens. The pi bond is formed when the two remaining p_z orbitals overlap side by side to form a pi molecular orbital containing a pair of electrons. The pi molecular orbital has two lobes.

It is important to remember that all orbitals can hold a maximum of two electrons. This applies to unhybridized, hybrid, and molecular orbitals.

The molecular-orbital method of representing double bonds accurately explains the geometry of molecules. We know, for example, that sp^2 hybrid orbitals lie at 120° angles to one another. We also know, from experiments, that the H—C—H bond angle in ethylene is very close to 120°. Furthermore, because the sp^2 orbitals are constructed from p_x and p_y orbitals, they all lie in a plane (the x–y plane). This explains why all six atoms of ethylene lie in the same plane (Section 8.7).

Problem 10.3 Draw the sigma–pi molecular-orbital representation of *cis*-2-butene and *trans*-2-butene (Section 8.10).

ALKENES: PROPERTIES 10.4

The physical properties of the alkenes are very similar to those of the corresponding alkanes. They have low melting and boiling points, low densities, and are insoluble in water.

ALKENES: PREPARATION 10.5

We can prepare alkenes by several methods. We will describe two of them: dehydration of alcohols and dehydrohalogenation of alkyl halides. Both of these reactions are examples of **elimination reactions.** In an elimination reaction, two atoms or groups are lost from adjacent carbon atoms. In the following equation, *A* and *B* are eliminated to produce a double bond between the carbons to which *A* and *B* were attached (R is any hydrocarbon group or hydrogen).

What drives elimination reactions?

$$\mathrm{R{-}\underset{A}{\overset{R}{C}}{-}\underset{B}{\overset{R}{C}}{-}R} \rightarrow \mathrm{R_2C{=}CR_2} + \mathrm{A{-}B}$$

Such reactions occur more readily if the A—B molecule that is eliminated is a very stable molecule such as water. The formation of a stable product helps supply the free energy needed to drive the reaction.

If we heat alcohol in the presence of a small amount of sulfuric acid, the alcohol molecule decomposes to form an alkene and a water molecule. The reaction is called **dehydration.**

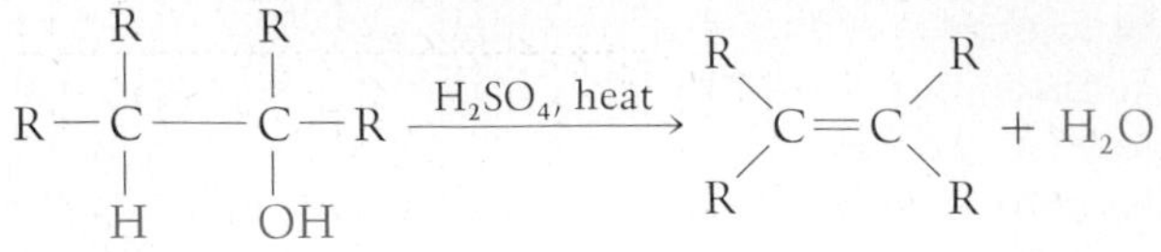

$$\begin{array}{c} \text{R} \quad \text{R} \\ | \quad\ \ | \\ \text{R}-\text{C}-\text{C}-\text{R} \\ | \quad\ \ | \\ \text{H} \quad \text{OH} \end{array} \xrightarrow{H_2SO_4,\ \text{heat}} \begin{array}{c} \text{R} \qquad \text{R} \\ \text{C}=\text{C} \\ \text{R} \qquad \text{R} \end{array} + H_2O$$

An example is the dehydration of ethyl alcohol to form ethylene.

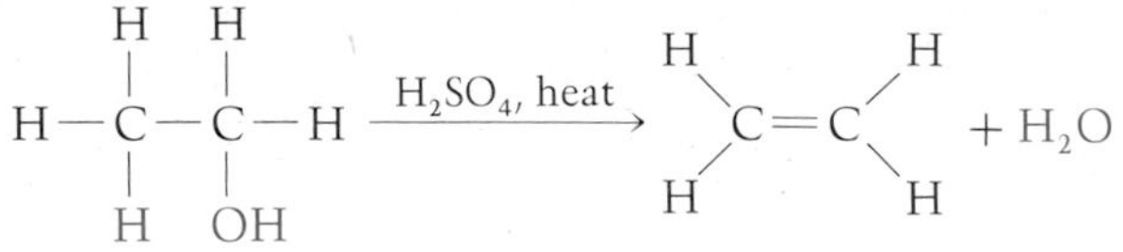

$$\begin{array}{c} \text{H} \quad \text{H} \\ | \quad\ \ | \\ \text{H}-\text{C}-\text{C}-\text{H} \\ | \quad\ \ | \\ \text{H} \quad \text{OH} \end{array} \xrightarrow{H_2SO_4,\ \text{heat}} \begin{array}{c} \text{H} \qquad \text{H} \\ \text{C}=\text{C} \\ \text{H} \qquad \text{H} \end{array} + H_2O$$

In dehydration reactions A and B are H and OH.

Another important elimination reaction is **dehydrohalogenation,** the elimination of HX, where *X* is one the halogens—chlorine, bromine, or iodine. If we heat an ethyl alcohol solution of an alkyl halide and KOH, the alkyl halide (RX) loses a molecule of HX to form an alkene.

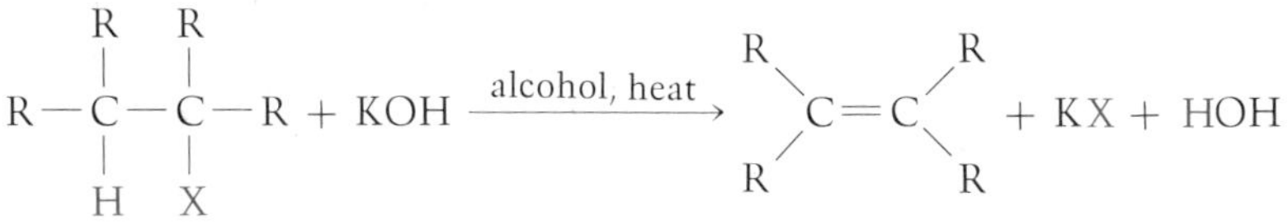

$$\begin{array}{c} \text{R} \quad \text{R} \\ | \quad\ \ | \\ \text{R}-\text{C}-\text{C}-\text{R} \\ | \quad\ \ | \\ \text{H} \quad \text{X} \end{array} + \text{KOH} \xrightarrow{\text{alcohol, heat}} \begin{array}{c} \text{R} \qquad \text{R} \\ \text{C}=\text{C} \\ \text{R} \qquad \text{R} \end{array} + \text{KX} + \text{HOH}$$

An example is the dehydrohalogenation of chloroethane.

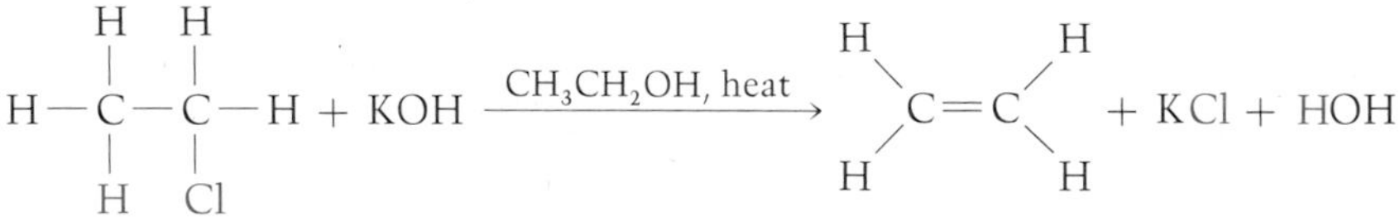

$$\begin{array}{c} \text{H} \quad \text{H} \\ | \quad\ \ | \\ \text{H}-\text{C}-\text{C}-\text{H} \\ | \quad\ \ | \\ \text{H} \quad \text{Cl} \end{array} + \text{KOH} \xrightarrow{CH_3CH_2OH,\ \text{heat}} \begin{array}{c} \text{H} \qquad \text{H} \\ \text{C}=\text{C} \\ \text{H} \qquad \text{H} \end{array} + \text{KCl} + \text{HOH}$$

In dehydrohalogenations, the formation of very stable substances like KCl and H_2O releases the large amount of free energy that drives the reaction forward.

More complicated alkyl halides may give more than one product.

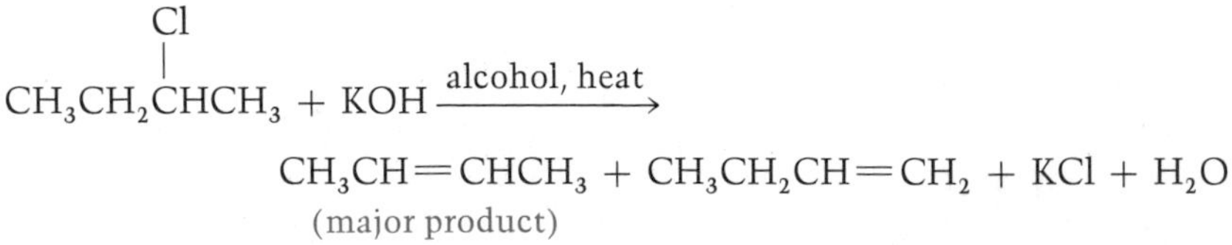

$$\begin{array}{c} \qquad\quad \text{Cl} \\ \qquad\quad | \\ CH_3CH_2CHCH_3 \end{array} + \text{KOH} \xrightarrow{\text{alcohol, heat}}$$

$$\underset{\text{(major product)}}{CH_3CH{=}CHCH_3} + CH_3CH_2CH{=}CH_2 + \text{KCl} + H_2O$$

When an alkyl halide undergoes dehydrohalogenation, the hydrogen atom is usually lost from a carbon atom adjacent to the one with the halogen. If those adjacent carbons have different numbers of hydrogens, the hydrogen is lost preferentially from the one that already has fewer hydrogens.

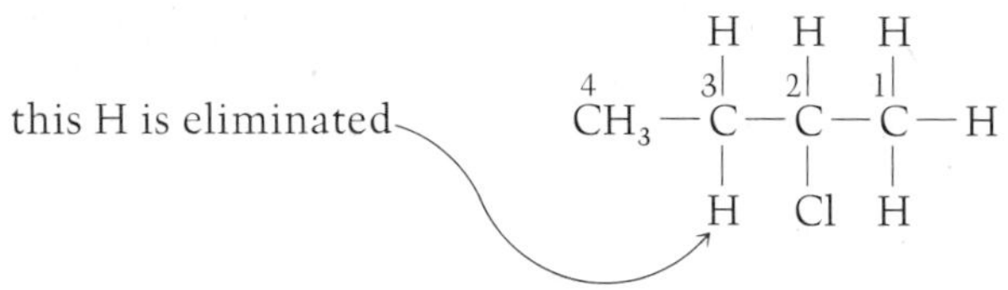

$$\underset{4}{CH_3}-\overset{H}{\underset{H}{\overset{|}{\underset{|}{C_3}}}}-\overset{H}{\underset{Cl}{\overset{|}{\underset{|}{C_2}}}}-\overset{H}{\underset{H}{\overset{|}{\underset{|}{C_1}}}}-H$$

this H is eliminated (arrow points to the lower H on carbon 3)

Carbon 3 has two hydrogen atoms, and carbon 1 has three. Therefore carbon 3 will usually be the one to lose a hydrogen, and the double bond will be formed between carbons 2 and 3.

Problem 10.4 Predict the products of the following reactions.

(a) $CH_3CH_2CH_2Cl + KOH \xrightarrow{CH_3CH_2OH,\ heat}$

(b) $CH_3CH_2CH_2OH \xrightarrow{H_2SO_4,\ heat}$

(c) $CH_3\overset{OH}{\overset{|}{C}}HCH_3 \xrightarrow{H_2SO_4,\ heat}$

(d) 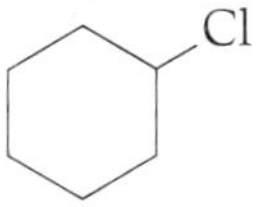 (cyclohexane ring bearing Cl) $+ KOH \xrightarrow{CH_3CH_2OH,\ heat}$

(e) $CH_3\overset{CH_3}{\overset{|}{C}}H\underset{Cl}{\underset{|}{C}}HCH_2CH_3 + KOH \xrightarrow{CH_3CH_2OH,\ heat}$

(f) 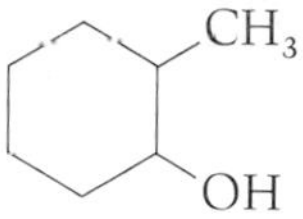 (cyclohexane ring bearing CH_3 and, on the adjacent carbon, OH) $\xrightarrow{H_2SO_4,\ heat}$

ALKENES: REACTIONS 10.6

How are addition reactions related to elimination reactions?

The high density of electrons in a double bond makes alkenes quite reactive as bases (see Chapter 7). Alkenes react with strong acids such as HX (where X is Cl, Br, or I) and H_3O^+, accepting protons from them.

In these reactions, the double bond accepts a hydrogen ion from the acid HX to form a very reactive positive ion (this ion is, of course, the conjugate acid of the alkene).

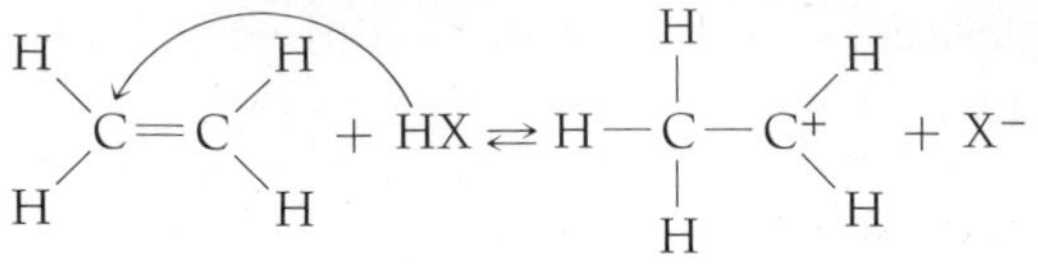

$$\mathrm{H_2C{=}CH_2 + HX \rightleftharpoons H{-}\underset{H}{\overset{H}{C}}{-}C^{+}H_2 + X^-}$$

(very reactive)

This cation is so reactive that it has a very short life. It immediately reacts with the nearest X^- anion.

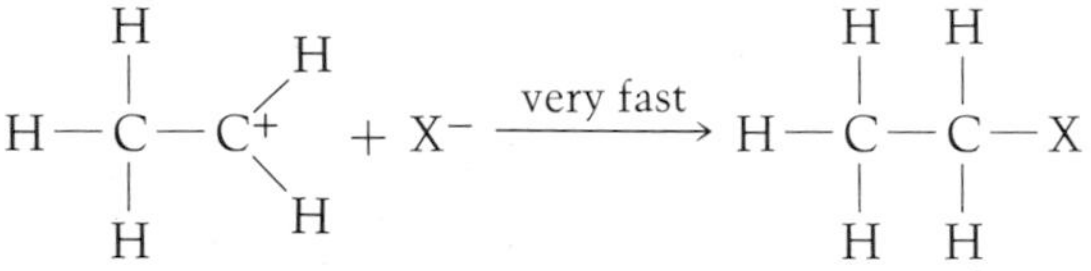

$$\mathrm{H{-}\underset{H}{\overset{H}{C}}{-}C^{+}H_2 + X^- \xrightarrow{\text{very fast}} H{-}\underset{H}{\overset{H}{C}}{-}\underset{H}{\overset{H}{C}}{-}X}$$

In these reactions, the double bond opens up to form two new single bonds; H is added to one carbon and X to the other carbon. We call the reaction an **addition reaction,** because the product contains all the atoms of the reactants.

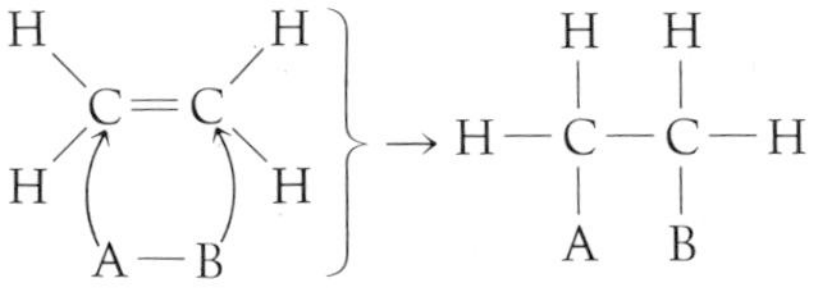

$$\left.\begin{matrix}\mathrm{H_2C{=}CH_2} \\ \mathrm{A{-}B}\end{matrix}\right\} \rightarrow \mathrm{H{-}\underset{A}{\overset{H}{C}}{-}\underset{B}{\overset{H}{C}}{-}H}$$

In summary, an addition reaction is one in which an alkene reacts with a molecule A—B so that the alkene's double bond is broken and A and B are added to the alkene. The general reaction is as follows (*R* is any hydrocarbon group or hydrogen).

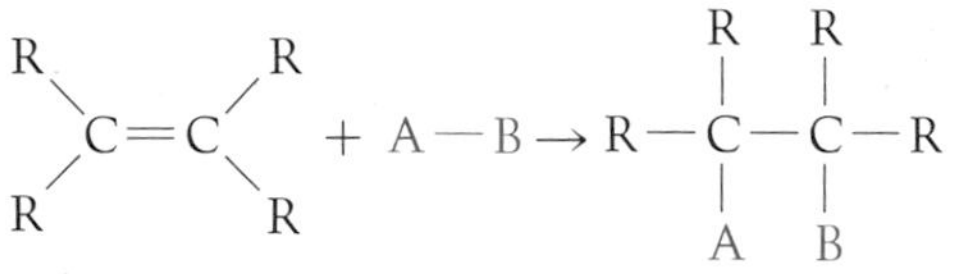

$$\mathrm{R_2C{=}CR_2 + A{-}B \rightarrow R{-}\underset{A}{\overset{R}{C}}{-}\underset{B}{\overset{R}{C}}{-}R}$$

An addition reaction is the reverse of an elimination reaction (Section 10.5).

It is because alkenes have this ability to add molecules that we say they are unsaturated. As we mentioned in the Introduction, unsaturated compounds are those whose molecules have one or more double or triple bonds.

Here are some examples of addition reactions.

Hydrohalogenation

The addition of a hydrogen halide (HCl, HBr, or HI) is called **hydrohalogenation;** it is the reverse of dehydrohalogenation.

$$CH_2{=}CH_2 + HCl \rightarrow \underset{\displaystyle H}{\underset{|}{CH_2}}{-}\underset{\displaystyle Cl}{\underset{|}{CH_2}}$$

Addition reactions of alkenes are usually fast and occur at room temperature.

Alkenes larger than ethylene undergo addition reactions in a very specific way. For example, when propene is hydrohalogenated, only one product, 2-chloropropane, is formed. 1-chloropropane is not produced.

$$CH_3CH{=}CH_2 + HCl \rightarrow CH_3\overset{\displaystyle Cl}{\overset{|}{C}}HCH_3 \quad \text{(only product)}$$

This phenomenon was first observed by the Russian chemist Vladimir Markovnikov. **Markovnikov's rule** states that *when HX is added to an alkene, the H becomes attached to the doubly bonded carbon atom that already has more hydrogen atoms.* The HX can be HCl, HBr, HI, or H_2O. In our example above, the H of HCl is added to the end carbon, which was doubly bonded and already had two H's attached.

$$CH_3CH{=}CH_2$$

X adds here (C has one hydrogen atom) — H adds here (C has two hydrogen atoms)

Hydration

The addition of H_2O to an alkene is called **hydration.** It requires an acidic catalyst (usually H_2SO_4), because water is not acidic enough to react with the alkene.

$$CH_2{=}CH_2 + H{-}OH \xrightarrow{H_2SO_4} H{-}CH_2{-}CH_2{-}OH$$

The stronger acid, H_2SO_4, reacts with the double bond to form the cation, which then reacts with water to yield the product.

$$H_2C{=}CH_2 + H_2SO_4 \rightleftharpoons H{-}\overset{H}{\underset{H}{C}}{-}C^{+}H_2 + HSO_4^-$$

$$CH_3{-}CH_2^+ + H_2O \rightleftharpoons CH_3{-}CH_2{-}O^{+}H_2$$

$$CH_3{-}CH_2{-}O^{+}H_2 + H_2O \rightleftharpoons CH_3{-}CH_2{-}OH + H_3O^+$$

The hydration of alkenes obeys Markovnikov's rule. For example, 1-butene gives only the product shown.

$$CH_3CH_2CH{=}CH_2 + H_2O \xrightarrow{H_2SO_4} CH_3CH_2\underset{OH}{\underset{|}{C}}HCH_3$$

Hydrogenation

In **hydrogenation,** hydrogen gas is added to alkenes in the presence of a finely powdered nickel catalyst to give alkanes (Section 9.5).

$$CH_2{=}CH_2 + H{-}H \xrightarrow{Ni} H{-}CH_2{-}CH_2{-}H$$

Halogenation

In **halogenation,** the halogens chlorine and bromine are added to alkenes, usually in a carbon tetrachloride solution, to give dichloroalkanes and dibromoalkanes, respectively.

$$CH_2{=}CH_2 + Cl_2 \xrightarrow{CCl_4} Cl{-}CH_2{-}CH_2{-}Cl$$

$$CH_2{=}CH_2 + Br_2 \xrightarrow{CCl_4} Br{-}CH_2{-}CH_2{-}Br$$

If you want to find out whether a hydrocarbon is unsaturated, a good visual test is to add bromine. Bromine has a dark reddish-brown color, even when dissolved in CCl_4. But when bromine reacts with an alkene a colorless product, dibromoalkane, is formed. We test for an alkene (which is colorless to start with) by adding a few drops of bromine in carbon tetrachloride solution (written Br_2/CCl_4) to the sample. If the sample contains a double bond, the color of the bromine will disappear immediately (Figure 10.3).

Another useful test to see whether a sample is an alkene is to add aqueous potassium permanganate ($KMnO_4$). At room temperature, an alkene will then be oxidized to form a glycol (a dihydroxy compound).

$$3CH_2{=}CH_2 + 2KMnO_4 + 4H_2O \rightarrow 3\overset{\displaystyle OH}{\overset{|}{C}H_2}-\overset{\displaystyle OH}{\overset{|}{C}H_2} + \underline{2MnO_2} + 2KOH$$

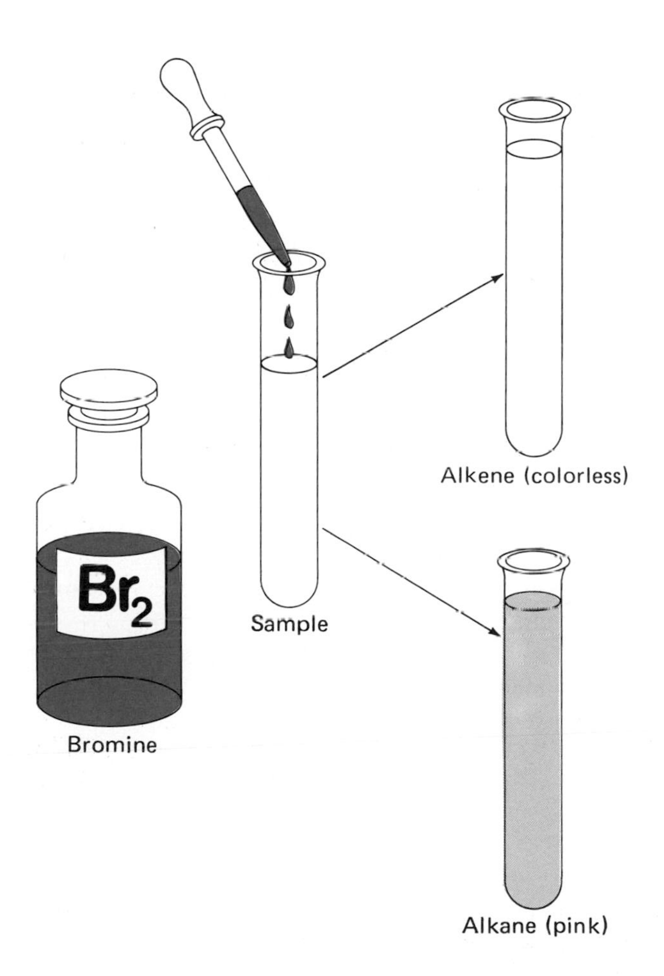

FIGURE 10.3 To test whether a hydrocarbon is unsaturated (has a double or triple carbon–carbon bond), we add a few drops of bromine solution to the sample. The bromine solution is red. If the red color disappears, the sample is unsaturated.

Here again, the test is visual. The $KMnO_4$ solution has a deep purple color, but the product MnO_2 is a dark brown to black precipitate. The presence of a double bond is signaled by the rapid loss of the purple color in the solution.

Polymerization

One of the most important commercial reactions of alkenes is **polymerization.** In polymerization, many alkene molecules are added to each other to form long hydrocarbon chains called polymers. These chains can contain thousands of carbons. Polymers have unusual properties, such as high elasticity and tensile strength; they are also easily molded and transparent or translucent. Polyethylene is one of the most common polymers; it is used in a wide variety of household and commercial products. Other functional groups, besides the carbon–carbon double bond of the alkenes, can react to form polymers. Some important biological polymers are discussed in Chapters 17 and 18.

Polymerization may be initiated by a number of different reagents. Acids are commonly used as initiators. The reaction is simply an addition reaction. If HCl is used, the reaction takes the following form.

$$H^+ \; CH_2{=}CH_2 \; CH_2{=}CH_2 \; CH_2{=}CH_2 \rightarrow H{-}CH_2CH_2CH_2CH_2CH_2\overset{+}{C}H_2$$

After many such steps, the positive ion combines with Cl^- to form $H\text{—}(CH_2{-}CH_2)_n Cl$, where n is a large number, often in the thousands. The curved arrows show the movement of a pair of electrons during the reaction. The chain may be represented by the formula $\text{—}(CH_2CH_2)_n\text{—}$, omitting the remnants of the initiator at the ends of the chain.

Here are some more alkene polymers, together with the reactants (called **monomers**) that produce them.

Polypropylene (used in making containers, cups, etc.)

$$nCH_2{=}CH(CH_3) \rightarrow \text{—}(CH_2{-}CH(CH_3))_n\text{—}$$

propylene polypropylene

Polystyrene (used in making moldable plastics and foams like styrofoam)

$$nCH_2{=}CH(C_6H_5) \rightarrow \text{—}(CH_2{-}CH(C_6H_5))_n\text{—}$$

styrene polystyrene

Poly(vinyl chloride) (used in making phonograph records)

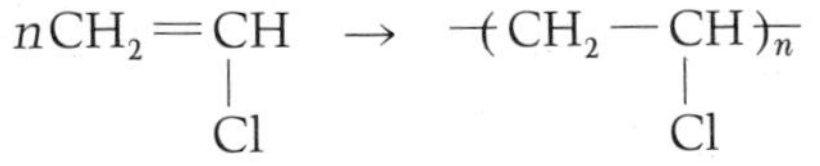

$$n\mathrm{CH_2{=}CH(Cl)} \rightarrow \mathrm{\text{+}CH_2{-}CH(Cl)\text{+}}_n$$

vinyl chloride — poly(vinyl chloride) (PVC)

Polyacrylonitrile (used in making fabrics such as Orlon)

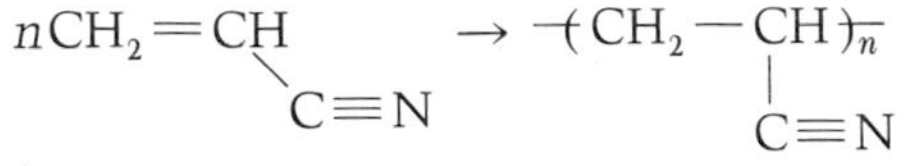

$$n\mathrm{CH_2{=}CH{-}C{\equiv}N} \rightarrow \mathrm{\text{+}CH_2{-}CH(C{\equiv}N)\text{+}}_n$$

acrylonitrile — polyacrylonitrile

Poly(methyl methacrylate) (clear hard plastics, such as Plexiglas and Lucite)

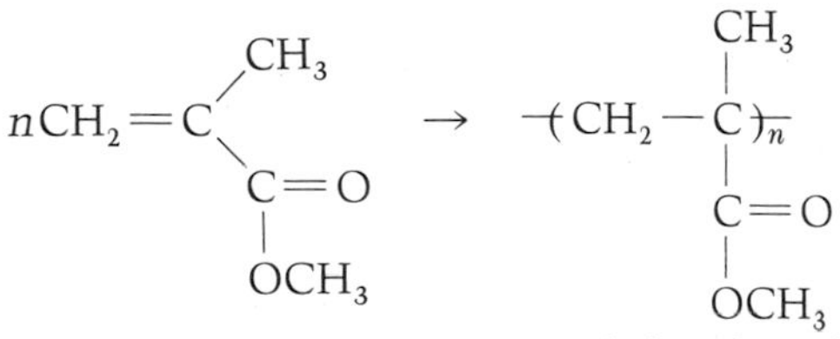

$$n\mathrm{CH_2{=}C(CH_3)(C{=}O)OCH_3} \rightarrow \mathrm{\text{+}CH_2{-}C(CH_3)(C{=}O)OCH_3\text{+}}_n$$

methyl methacrylate — poly(methyl methacrylate)

Problem 10.5 Predict the products of the reaction of 2-butene with each of the following reagents.

(a) $KMnO_4/H_2O$ (b) Br_2/CCl_4
(c) H_2O/H_2SO_4 (d) HCl

Problem 10.6 Repeat Problem 10.5, using cyclopentene in place of 2-butene.

Problem 10.7 Repeat Problem 10.5, using 1-butene in place of 2-butene.

Problem 10.8 What simple chemical tests would allow you to distinguish between the members of the following pairs of compounds? Write equations and tell what you would do and observe.

(a) cyclohexane and 1-hexene (b) ethylene and polyethylene

ALKENES: INTERACTION WITH LIGHT 10.7

An interesting property of compounds that have several double bonds alternating with single bonds (called **conjugated double bonds**) is their interaction with light. Long conjugated systems of double bonds impart color to substances; examples are the red color of tomatoes and the yellow color of carrots.

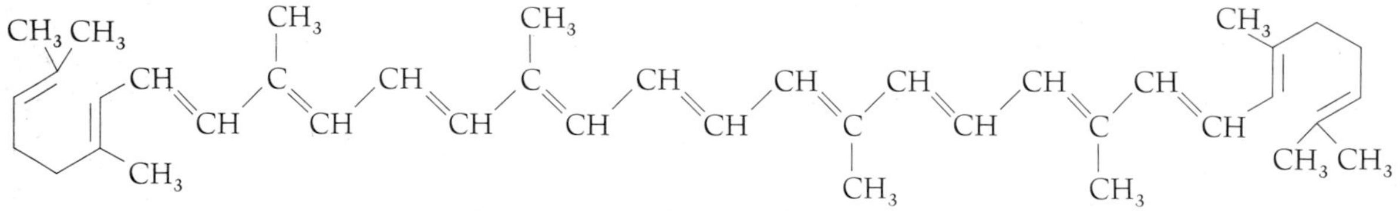

lycopene, the red pigment in tomatoes

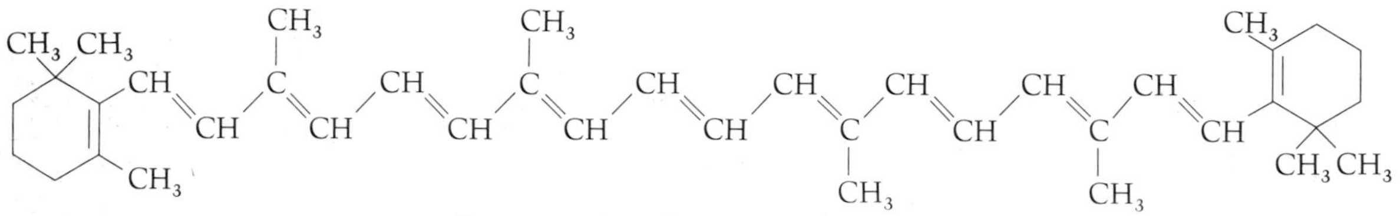

β-carotene, the yellow pigment in carrots

When such conjugated systems absorb light energy, their electrons are excited to higher energies. This light absorption occurs in the visible region of the spectrum, so we perceive these molecules as colored.

Conjugated double bonds have this property because the adjacent pi molecular orbitals can overlap with each other.

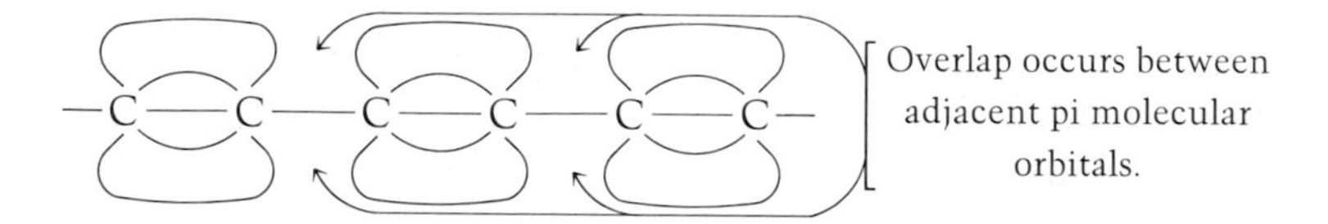
—C—C—C—C—C—C—
Overlap occurs between adjacent pi molecular orbitals.

Because of such overlap, pi electrons are able to move all along the conjugated chain. Such "loose" electrons interact with light at wavelengths that are visible to our eyes, so we are able to see color in compounds that contain long conjugated systems of double bonds.

Problem 10.9 What is the difference between the structural formulas of lycopene and β-carotene?

Mammals have enzymes that split β-carotene—the yellow pigment in carrots—in half, to give two molecules of retinal. The retinal, in turn, is reduced (by picking up hydrogens) to vitamin A_1.

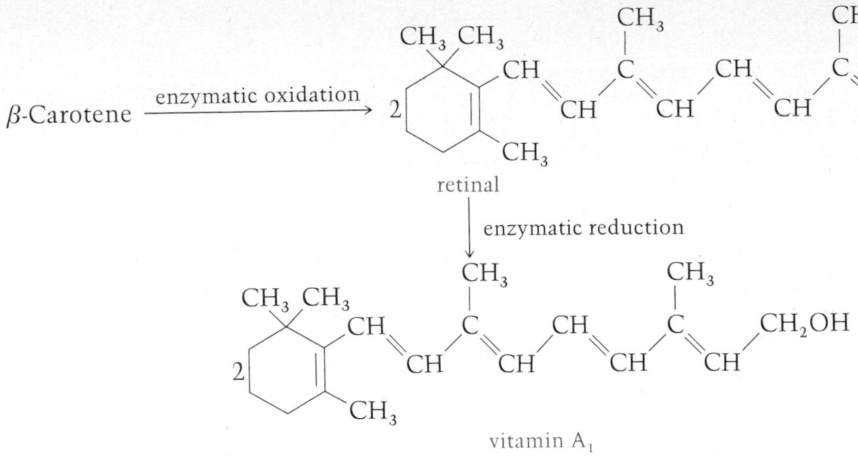

β-Carotene —enzymatic oxidation→ 2 retinal

CH_3 CH_3 CH CH CH_3 C CH CH CH CH_3 C CH H C O CH CH_3

retinal

↓ enzymatic reduction

CH_3 CH_3 CH CH_3 C CH CH CH CH_3 C CH CH_2OH CH CH_3 2

vitamin A_1

Problem 10.10 Draw the structure of β-carotene and show which bond breaks when β-carotene is converted to retinal.

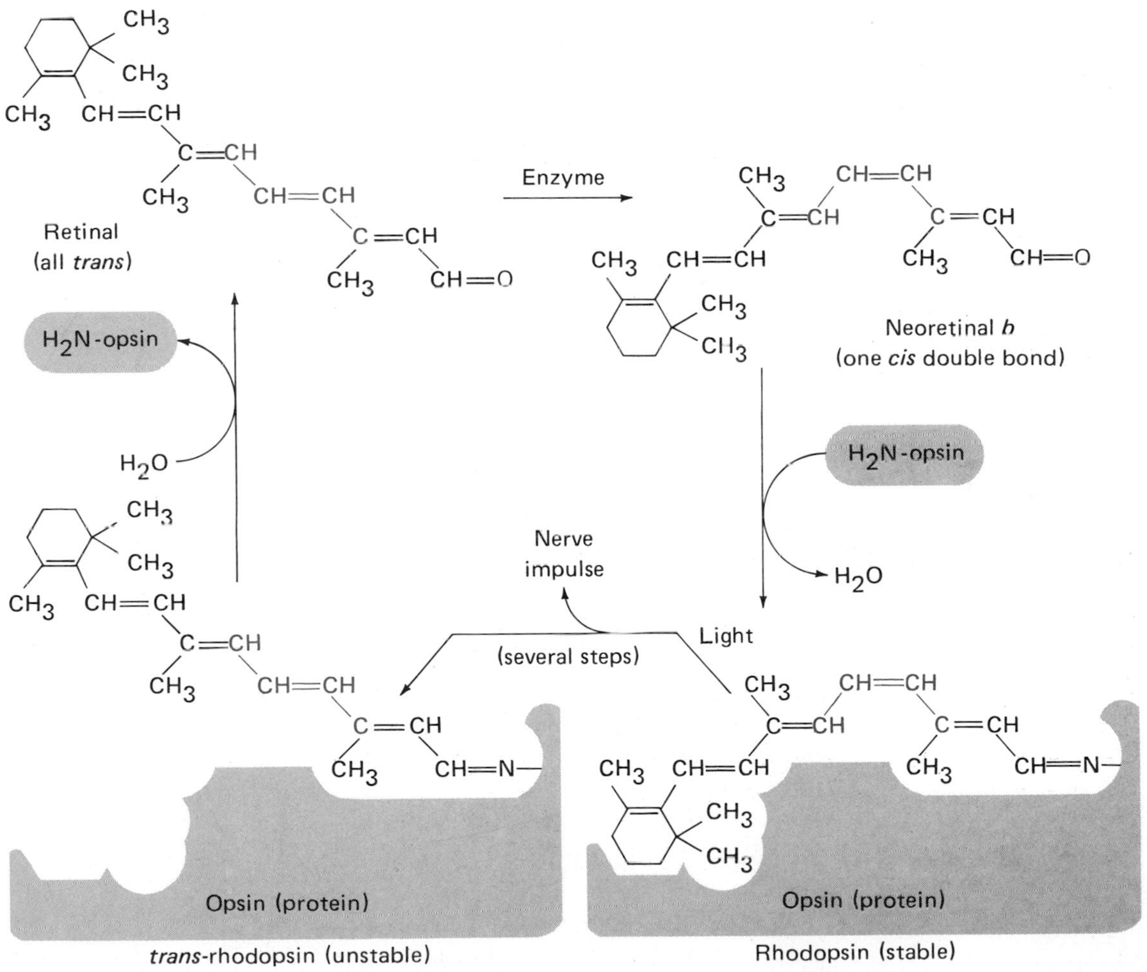

CH_3 CH_3 CH_3 CH=CH C=CH CH_3 CH=CH C=CH CH_3 CH=O

Retinal (all *trans*)

Enzyme →

CH_3 CH=CH C=CH C=CH CH_3 CH=CH CH_3 CH=O CH_3 CH_3

Neoretinal *b* (one *cis* double bond)

H_2N-opsin

H_2O

H_2N-opsin

H_2O

Nerve impulse

Light

(several steps)

CH_3 CH_3 CH_3 CH=CH C=CH CH_3 CH=CH C=CH CH_3 CH=N—

Opsin (protein)

trans-rhodopsin (unstable)

CH_3 CH=CH C=CH C=CH CH_3 CH=CH CH_3 CH=N— CH_3 CH_3

Opsin (protein)

Rhodopsin (stable)

FIGURE 10.4 Summary of the reactions involved in the chemistry of vision. The double bond that undergoes *cis–trans* isomerization is shown in color.

Retinal plays an important role in vision. First, retinal's double bonds are all *trans.* If one *trans* double bond is converted to the *cis* configuration, retinal is changed to neoretinal b. Neoretinal b has a unique shape because of its *cis* double bond (Figure 10.4). This shape permits the neoretinal b molecule to interact with opsin, a protein, to give rhodopsin. When light strikes the retina of the eye it causes the *cis* double bond of neoretinal b to revert back to the *trans* configuration to give *trans*-rhodopsin, which is unstable. This isomerization of neoretinal b to *trans*-rhodopsin releases energy in the form of a nerve impulse from the optic nerve to the brain. Because *trans*-rhodopsin is unstable, it readily breaks down to opsin and retinal, and a new cycle begins (Figure 10.4).

10.8 ALKYNES: STRUCTURE

Why are the atoms in H—C≡C—H arranged in a straight line?

Hydrocarbons that have a triple bond are called **alkynes.** As we saw in Section 8.7, all the atoms involved in a triple bond lie in a straight line. Acetylene (C_2H_2), the simplest alkyne, has the structure shown in Figure 10.5.

Using molecular-orbital theory, we first connect the four atoms of acetylene with sigma bonds (see Section 10.3).

sigma bonds

H—Ċ—Ċ—H

Two sigma bonds are needed for each carbon. These sigma bonds require two *sp* hybrid orbitals (formed by combining a 2*s* orbital and a 2*p* orbital of each carbon).

Problem 10.11 What kind of carbon hybrid orbitals are needed for the sigma bonds of acetylene? [Hint: see Section 8.7.]

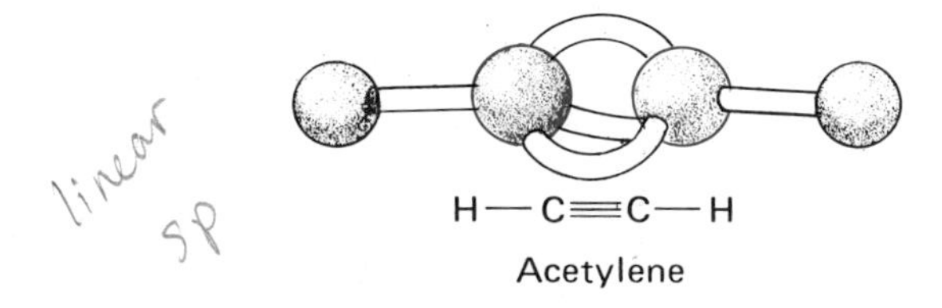

FIGURE 10.5 Acetylene (ethyne): formula with model.

Each carbon atom is left with two electrons in two orbitals. If we assume that p_x orbitals were used to form the sigma bonds, then p_y and p_z orbitals remain. These p orbitals overlap to form two pi molecular orbitals, as shown.

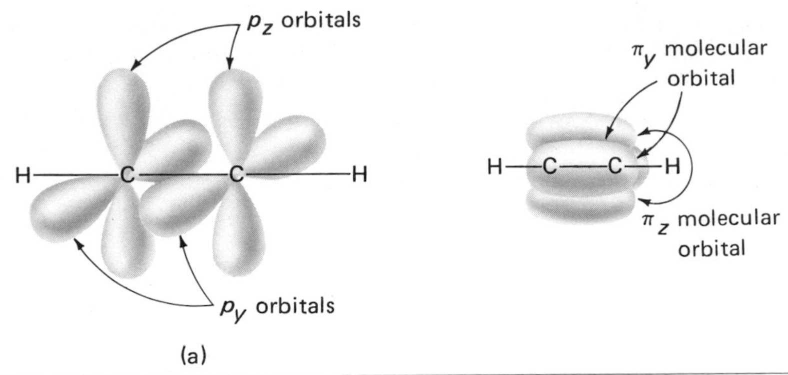

Problem 10.12 It is difficult to prepare cyclohexyne, which has the structure [cyclohexyne]. Attempts to do so usually result in the formation of 1,3-cyclohexadiene.* Explain. [Hint: Consider the bond angles.]

ALKYNES: NOMENCLATURE 10.9

The IUPAC system names alkynes exactly as it does alkenes except that, for the parent name, the ending *-ene* becomes *-yne*. Some simple alkynes are shown below.

What is the only difference between the naming of alkenes and alkynes?

$HC{\equiv}CH$	Ethyne
$HC{\equiv}CCH_3$	Propyne
$HC{\equiv}CCH_2CH_3$	1-Butyne
$CH_3C{\equiv}CCH_3$	2-Butyne
$HC{\equiv}C{-}C{\equiv}CH$	1,3-Butadiyne†

The simplest member, ethyne, is nearly always called by its common name, acetylene.

*The ending *-diene* is used to represent a hydrocarbon with two double bonds.

†The ending *-diyne* is used to represent a hydrocarbon with two triple bonds.

Problem 10.13 Name the following compounds.

(a) $CH_3C{\equiv}CCH(CH_3)CH_3$ (b) $C_6H_{11}{-}C{\equiv}CCH_2CH_2C(CH_3)_2CH_2CH_3$ (cyclohexyl group)

(c) $CH_3C{\equiv}CCH_2C{\equiv}CCH_3$

10.10 ALKYNES: PROPERTIES

The physical properties of the alkynes are nearly the same as those of the corresponding alkenes and alkanes; they are insoluble in water, less dense than water, and have relatively low melting and boiling points.

10.11 ALKYNES: REACTIONS

What general reaction do alkynes undergo?

Alkynes resemble alkenes in their addition reactions. A typical addition reaction is halogenation (with chlorine and bromine only).

$$HC{\equiv}CH + Br_2 \rightarrow (H)(Br)C{=}C(Br)(H)$$

$$(H)(Br)C{=}C(Br)(H) + Br_2 \rightarrow HCBr_2{-}CHBr_2$$

These reactions can usually be carried out separately.

Another addition reaction is hydrogenation.

$$HC{\equiv}CH + H_2 \xrightarrow{Ni} CH_2{=}CH_2$$

$$CH_2{=}CH_2 + H_2 \xrightarrow{Ni} CH_3CH_3$$

Problem 10.14 Write equations for the reactions of 2-butyne with (a) Cl_2 and (b) H_2/Ni.

AROMATIC HYDROCARBONS: STRUCTURE 10.12

What is the molecular structure of an aromatic system?

The **arenes** are an interesting group of compounds. They are highly unsaturated, and yet they do not undergo the addition reactions typical of most unsaturated compounds. The simplest example is benzene (C_6H_6), which is a planar, six-membered ring with equal-length C—C bonds. The benzene ring can be represented in four different ways.

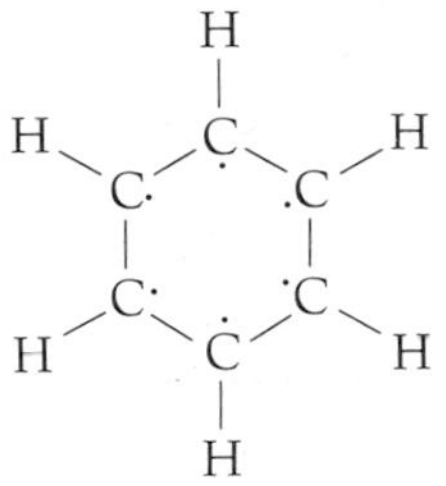

I 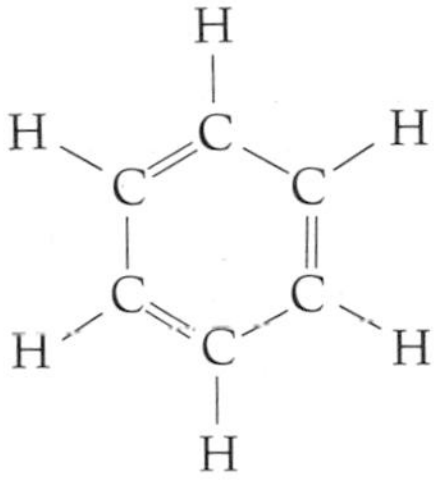 II 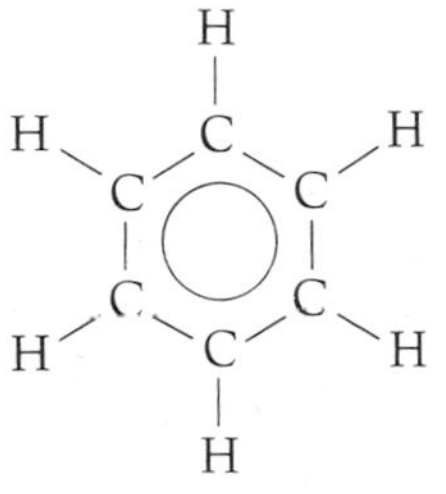 III  IV

Structure I (not a valid representation of benzene) shows the electron each carbon has left over after it has formed bonds with the adjoining carbons and with a hydrogen. If we assume that these six extra electrons form three double bonds, as in structure II, we run into a problem: why doesn't benzene react like a molecule with double bonds? Since benzene doesn't act like an unsaturated compound, and since all the C—C bonds in benzene are the same length, a better representation is structure III. Structure III can be shortened to the line formula IV, in which hydrogen atoms are not shown. Structures III and IV imply that the six electrons involved in the pi bonding between carbons are able to move around all the carbon atoms of the ring. Such an arrangement of six electrons in a ring makes the benzene molecule unusually unreactive, compared with other unsaturated systems. If the molecule underwent an addition reaction, some of these electrons would be used in bond formation and the integrity of the system would be lost. As a result, addition reactions do not normally occur, and the structure of benzene (as shown in III or IV) is preserved in most reactions.

Because the benzene ring was first found in compounds such as methyl salicylate (oil of wintergreen), vanillin, and other substances with strong aromas, early chemists called these ring compounds "aromatic." The name **aromatic hydrocarbon** has persisted to the present time, but now it refers to a cyclic system of alternating single and double bonds (usually a six-membered ring) that does not undergo the addition reactions expected of an unsaturated compound. In other words, an aromatic compound is any compound that contains a benzene ring or similar structure. When an aromatic group occurs as a substituent, we call it an **aryl group.** Aryl groups in aromatic compounds correspond to alkyl groups in aliphatic compounds. An **aliphatic hydrocarbon** is any hydrocarbon that is not aromatic. This includes the alkanes, alkenes, alkynes, and their cyclic derivatives.

10.13 MOLECULAR-ORBITAL REPRESENTATION OF BENZENE

The structure of benzene can be represented more accurately by using molecular orbitals. In structure I, above, the bonds shown will all be sigma bonds. The sigma bonds use up three of the four electrons each carbon has available for bonding. The remaining electron of each carbon is in a p_z orbital. These six p_z orbitals combine to form molecular orbitals, some of which are very stable. One of them is shown as two doughnut-shaped lobes, one above the carbon ring and one below it. Other molecular orbitals exist, but we won't examine them.

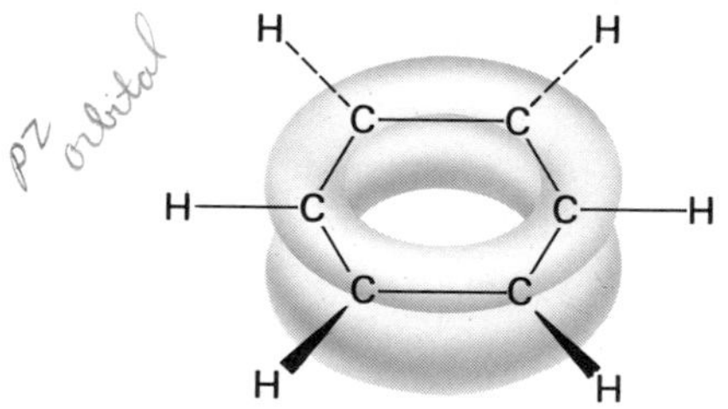

If benzene underwent addition reactions, its stable molecular structure would be destroyed. The pi electrons that were left would not be free to move around the entire ring. They would be forced to occupy different molecular orbitals of higher energy, resulting in a less stable molecule. For example, the addition of a molecule of chlorine to benzene would give a nonaromatic compound.

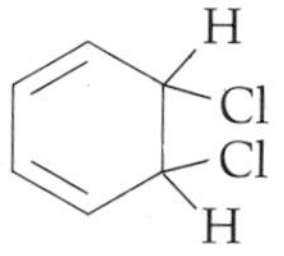

This compound is not aromatic, because it does not have alternating single and double bonds (conjugated double bonds) all the way around the ring.

Problem 10.15 Which of the following compounds are aromatic? Which are conjugated? Explain your choices.

(a) 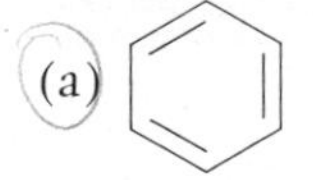 (b) 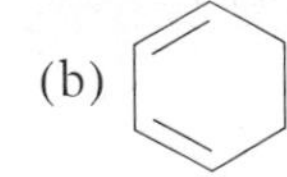 (c) 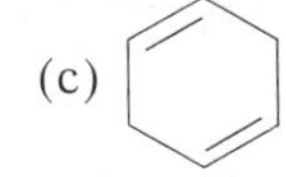 (d) 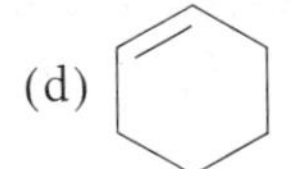 (e) 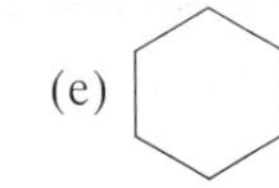

(f) $CH_2{=}CH{-}CH{=}CH{-}CH{=}CH_2$

RESONANCE 10.14

We can represent the structure of benzene in yet another way. If, for a given molecule, we can write two or more reasonable structural formulas that differ only in the placement of electrons, then neither formula represents the actual molecule. The actual molecule in such a case is represented as a **resonance** hybrid of all the structural formulas that we can write.

Do the structures that contribute to the resonance hybrid actually exist?

We can write two equivalent structures, IIa and IIb, for benzene.

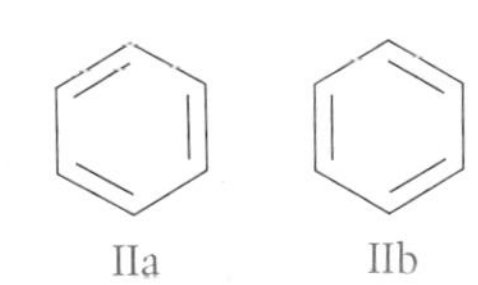

Neither diagram alone represents the true structure of benzene, because benzene has six equivalent carbon–carbon bonds. The hybrid of the two, however, gives a good representation. (Note that the hybrid predicts that each bond will be intermediate between a single and a double bond.) To imagine the hybrid structure, we must superimpose IIa on IIb in our minds. The resulting hybrid has six equal carbon–carbon bonds (shown as structures III and IV in Section 10.12). This hybrid structure is called a resonance hybrid. A resonance hybrid does not change back and forth between the two structures; it has the same structure all the time.

One important aspect of resonance is its stabilizing influence. Any molecule or ion that requires resonance for its description (that is, any molecule or ion that can be represented by two or more acceptable structural formulas) is more stable than we would otherwise expect any of the contributing structures to be. The resonance hybrid structure of benzene is more stable (less reactive) than the molecule would be if it had either structure IIa or IIb alone.

Problem 10.16 Naphthalene is a resonance hybrid of three structures. One of them is

Draw the structural formulas of the other two.

10.15 BENZENE DERIVATIVES: NOMENCLATURE

What similarities are there between the names of benzene derivatives and the names of cycloalkanes?

In naming benzene derivatives, we use the name *benzene* as the parent name. We name and number substituents as we did for cycloalkanes.

Cl

chlorobenzene

Cl 1 2 3 Cl 4 5 6

1,3-dichlorobenzene

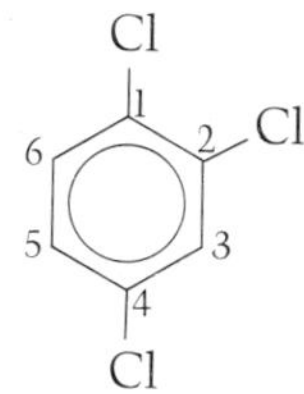

1,2,4-trichlorobenzene

When the benzene ring contains only two substituents, chemists often use the terms *ortho, meta,* and *para* to designate where the substituents are located. *Ortho* indicates that the substituents are on carbons 1 and 2, *meta* that they are on 1 and 3, and *para* on 1 and 4. These prefixes are abbreviated *o*-, *m*-, and *p*-, and are always italicized.

NO_2 NO_2

o-dinitrobenzene

NO_2 Cl

m-chloronitrobenzene (or *m*-nitrochlorobenzene)

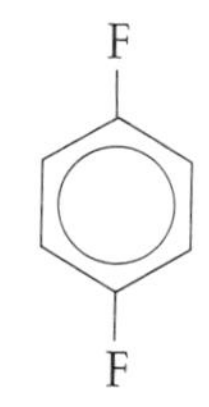

p-difluorobenzene

Certain benzene derivatives have common names that are not based on the IUPAC system. The most important of these are:

CH_3	OH	NH_2	CO_2H	OCH_3
toluene	phenol	aniline	benzoic acid	anisole

Derivatives of these special compounds are named by assigning the number 1 to the ring carbon that bears the special substituent. Some examples of toluene derivatives will illustrate this.

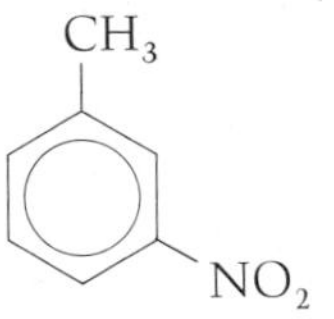

CH_3 / NO_2 — 3-nitrotoluene (or *m*-nitrotoluene)

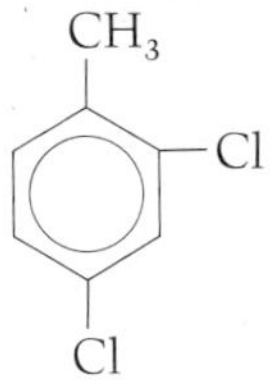

CH_3 / Cl / Cl — 2,4-dichlorotoluene

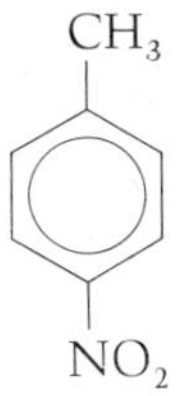

CH_3 / NO_2 — 4-nitrotoluene (or *p*-nitrotoluene)

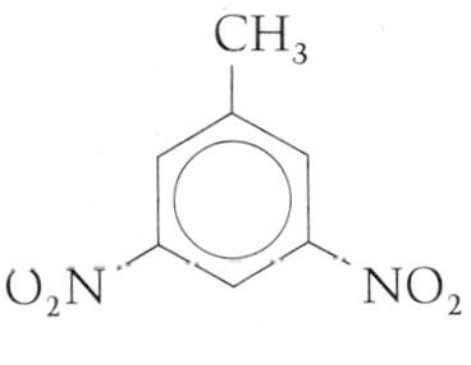

CH_3 / O_2N / NO_2 — 3,5-dinitrotoluene

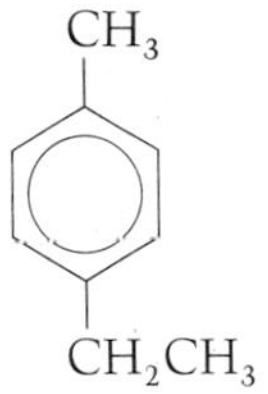

CH_3 / CH_2CH_3 — 4-ethyltoluene (or *p*-ethyltoluene)

Problem 10.17 Name the following compounds.

(a) 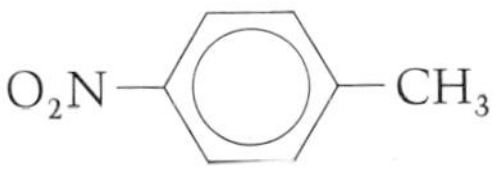 O_2N—⟨benzene⟩—CH_3 (b) 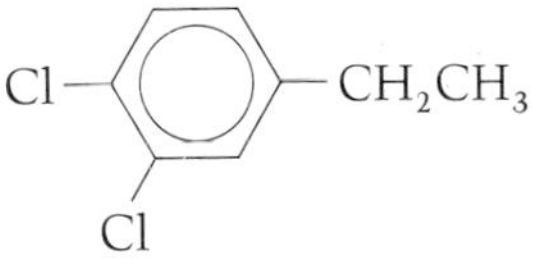 Cl—⟨benzene⟩—CH_2CH_3, Cl

(c) 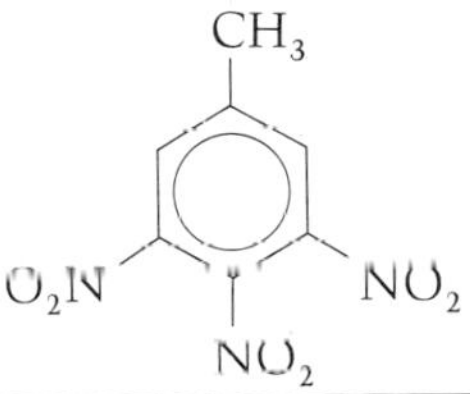 CH_3, O_2N, NO_2, NO_2 (d) 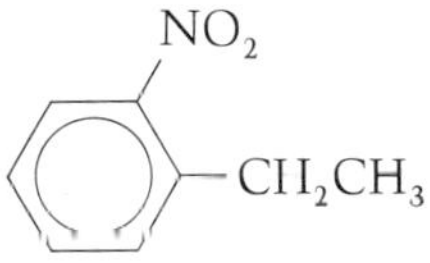 NO_2, —CH_2CH_3

BENZENE: REACTIONS 10.16

Benzene undergoes substitution reactions in which a hydrogen atom is replaced by a different group. Among these are halogenation, sulfonation, and nitration.

Why does benzene undergo substitution and not addition?

Halogenation

Benzene reacts with chlorine or bromine when a small amount of iron or iron salts are present. Light has no effect on this reaction.

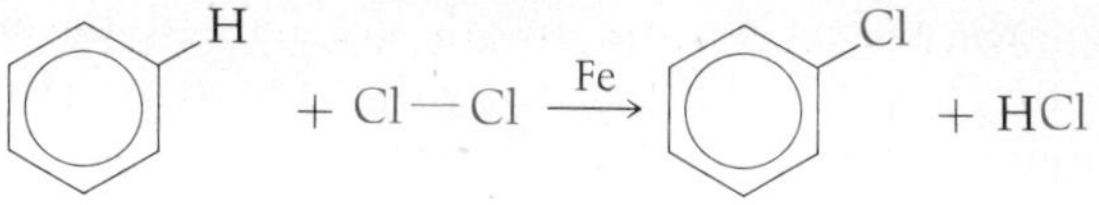

$$C_6H_5\text{—}H + Cl\text{—}Cl \xrightarrow{Fe} C_6H_5\text{—}Cl + HCl$$

chlorobenzene

Sulfonation

When benzene is heated with fuming sulfuric acid (a solution of SO_3 dissolved in concentrated H_2SO_4), benzenesulfonic acid is formed. The process is called **sulfonation.**

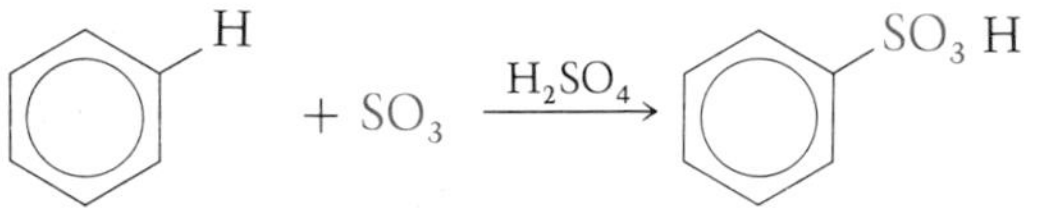

$$C_6H_5\text{—}H + SO_3 \xrightarrow{H_2SO_4} C_6H_5\text{—}SO_3H$$

benzenesulfonic acid

Although this appears to be an addition reaction, it is not, because the sulfur atom replaces the hydrogen on the ring carbon.

Nitration

Sulfuric acid also promotes the reaction of benzene with concentrated nitric acid (**nitration**) to give nitrobenzene.

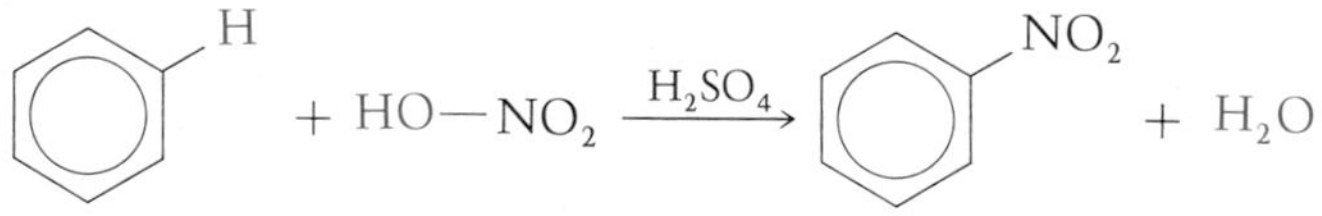

$$C_6H_5\text{—}H + HO\text{—}NO_2 \xrightarrow{H_2SO_4} C_6H_5\text{—}NO_2 + H_2O$$

nitrobenzene

As we mentioned before, benzene does not undergo addition reactions; if it did, the aromatic ring system would be destroyed. Addition of Cl_2, for example, would result in a nonaromatic ion that would be very reactive.

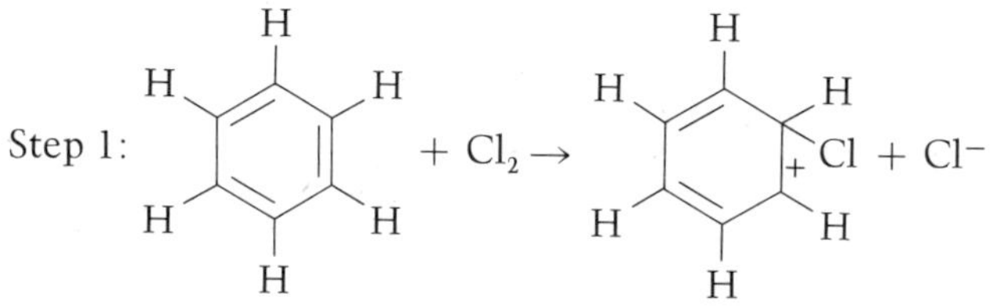

Step 1: $C_6H_6 + Cl_2 \rightarrow C_6H_6Cl^+ + Cl^-$

very reactive
ionic intermediate

This reactive ion would react very rapidly in one of two ways.

Step 2: (arenium ion, H H H H on ring; H, Cl, + at ring carbon) + Cl^- — arrows (a) and (b)

(a) addition → nonaromatic

(b) substitution → aromatic + HCl

Substitution of chlorine for hydrogen (step 2b) is energetically more favored than addition, because it gives a product with an intact aromatic ring.

Problem 10.18 Write equations to show the preparation of each of the following compounds, starting with benzene and any other necessary reagents.

(a) (benzene ring)–Br (b) (benzene ring)–SO_3H (c) (benzene ring)–NO_2

AROMATIC HYDROCARBONS WITH MORE THAN ONE RING 10.17

Other aromatic compounds are formed of several six-membered conjugated rings. These are called **polynuclear aromatic hydrocarbons.** The following examples are written in two equivalent ways.

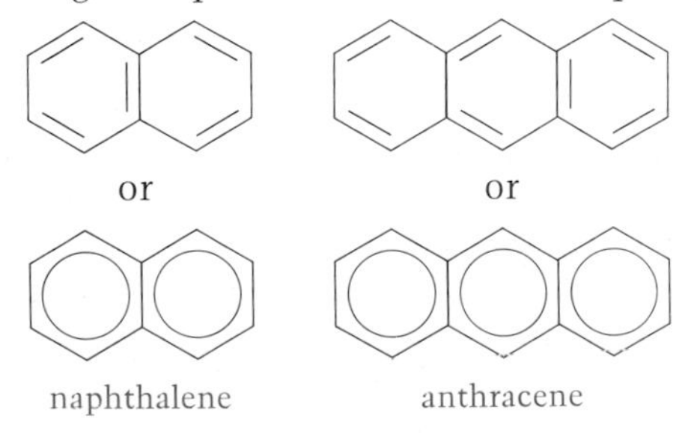

or

or

phenanthrene

pyrene

or

1,2-benzopyrene

Most of these compounds occur in coal (see Section 10.18). Many of them are of extreme interest because they and their derivatives can cause cancer.

Problem 10.19 Draw one other resonance structure of the following compounds.

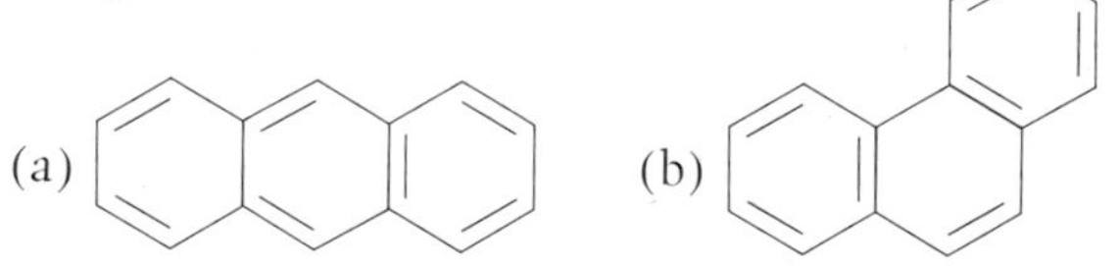

10.18 COAL

Why is graphite a good lubricant?

Coal is the most abundant source of aromatic hydrocarbons on earth. It was formed from freshwater plants that lived several hundred million years ago. Once they had died and were covered over with water, bacterial action converted these plants into peat. Later, when the oxygen had been depleted, time and intense pressure converted these peat beds into coal, an impure form of graphite.

Graphite consists of layers of six-membered carbon rings (Figure 10.6). The molecular layers are not bonded to each other, so they attract each other only weakly. Because of this weak attraction, they are able to slide

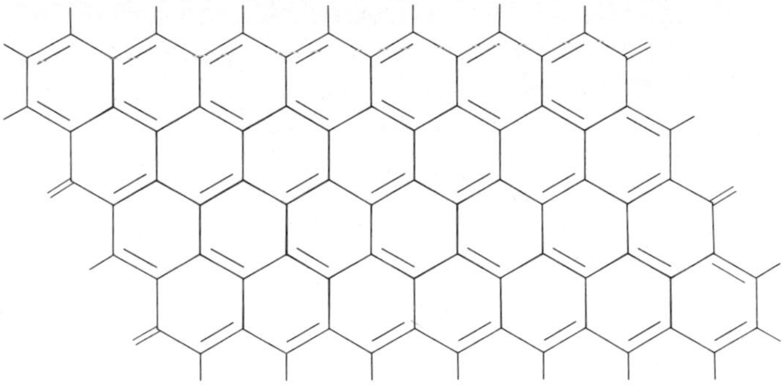

FIGURE 10.6 Partial structure of graphite molecular layer.

past one another. It is this quality that makes graphite such a good lubricant. Each molecular layer has the aromatic structure shown in Figure 10.6; this structure extends indefinitely in two dimensions.

If we heat coal to 1,000–1,300°C in the absence of oxygen, these large graphite molecules decompose into smaller, volatile molecules and a nonvolatile residue. The volatile molecules condense as *coal tar.* The nonvolatile residue is *coke,* which is used as a smokeless fuel. The process of heating a substance or a mixture in the absence of oxygen is called **destructive distillation,** because large molecules are broken down into smaller ones. Distillation of coal tar yields many polynuclear aromatic hydrocarbons (Section 10.17), along with benzene, toluene, and other useful organic compounds including hydrocarbons that may be useful as gasoline substitutes.

SUMMARY

Alkenes are hydrocarbons that have a carbon–carbon double bond. In molecular-orbital theory, the double bond may be represented as a **sigma** (σ) **bond** and a **pi** (π) **bond.** Alkenes are prepared by **elimination reactions,** such as **dehydration** (loss of H_2O) and **dehydrohalogenation** (loss of HCl, HBr, or HI). The most important reactions of alkenes are **addition reactions,** such as **hydration** (addition of H_2O), **hydrohalogenation** (addition of HCl, HBr, or HI), **hydrogenation** (addition of H_2), halogenation (addition of Cl_2 or Br_2), oxidation and **polymerization. Unsaturated hydrocarbons** are compounds that have one or more double or triple bonds. Alkenes add HX (HCl, HBr, HI, or H_2O) in accordance with **Markovnikov's rule:** the H of HX is added to the carbon that has more hydrogens already bonded to it. **Alkynes** are hydrocarbons that have a carbon–carbon triple bond. Alkynes undergo addition reactions like those of alkenes, except that an alkyne can add two molecules of the reagent. **Aromatic hydrocarbons** (**arenes**) are cyclic, conjugated ring systems. They are highly unsaturated compounds that are surprisingly unreactive. **Aliphatic hydrocarbons** are all hydrocarbons that are not aromatic. Aromatic compounds are unreactive because of their **resonance.** They undergo substitution reactions—**halogenation, sulfonation,** and **nitration**—rather than addition reactions. Molecules that have multiple, six-membered, conjugated ring systems are called **polynuclear aromatic hydrocarbons.** Coal is a mixture of large polynuclear aromatic compounds in which the molecules extend indefinitely in two dimensions.

Key Terms

Addition reaction (Section 10.6)
Aliphatic hydrocarbon (Section 10.12)
Alkene (Section 10.1)
Alkyne (Section 10.8)
Arene (Section 10.12)
Aromatic hydrocarbon (Section 10.12)
Aryl group (Section 10.12)
Conjugated double bond (Section 10.7)
Dehydration (Section 10.5)
Dehydrohalogenation (Section 10.5)
Destructive distillation (Section 10.18)
Elimination reaction (Section 10.5)
Hydration (Section 10.6)
Hydrogenation (Section 10.6)
Hydrohalogenation (Section 10.6)
Markovnikov's rule (Section 10.6)
Molecular orbital (Section 10.3)
Nitration (Section 10.16)
Pi (π) bond (Section 10.3)
Polymerization (Section 10.6)
Polynuclear aromatic hydrocarbon (Section 10.17)
Resonance (Section 10.14)
Sigma (σ) bond (Section 10.3)
Sulfonation (Section 10.16)
Unsaturated hydrocarbon (Introduction and Section 10.6)

ADDITIONAL PROBLEMS

10.20 Draw structural formulas and give IUPAC names for all the alkene isomers of C_6H_{12}.

10.21 Give the structural formulas of the following compounds.

(a) 2-methyl-2-butene
(b) 3-methyl-1-butyne
(c) *p*-dinitrobenzene
(d) *o*-nitrotoluene
(e) *m*-fluorotoluene
(f) 2-fluoro-4,6-dinitrotoluene
(g) 1,3-dimethylcyclohexene
(h) cyclobutene
(i) 1,4-dichlorocyclohexane
(j) 2-pentyne
(k) 1-chloro-1-butene

10.22 Give the IUPAC names for the following compounds.

(a) $CH_3CH{=}CHCH(CH_3){-}CH(CH_2CH_3)CH_2CH_2CH_2CH_3$

(b) $CH_3CH_2C{\equiv}CCH_2CH_3$
(c) $CH_3CH_2C(CH_3){=}CH_2$

(d) 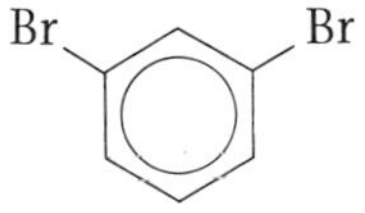 Br, Br (benzene ring)

(e) 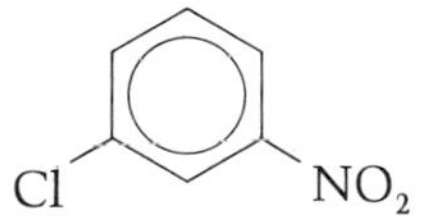 Cl, NO_2 (benzene ring)

(f) 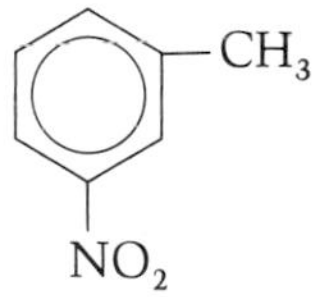 CH_3, NO_2 (benzene ring)

(g) 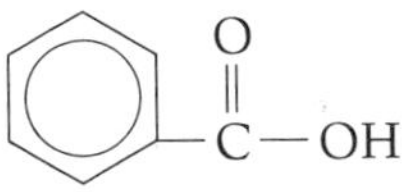 benzene ring $-\overset{O}{\overset{\|}{C}}-OH$

(h) 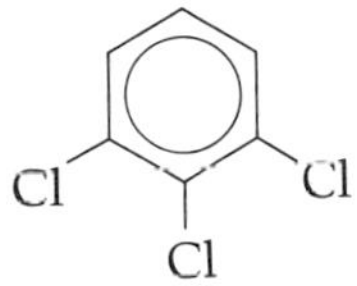 Cl, Cl, Cl (benzene ring)

(i) $HC{\equiv}CCH_2CH_2CH_3$

10.23 Complete the following reactions.

(a) $CH_3C{\equiv}CCH_3 + Cl_2$ (2 moles) $\rightarrow$

(b) $CH_3C{\equiv}CCH_3 + H_2$ (1 mole) $\xrightarrow{Ni}$

(c) Product of (b) $+ H_2$ (1 mole) $\xrightarrow{Ni}$

(d) $CH_3CH_2CH_2CH_2Br + KOH \xrightarrow{CH_3CH_2OH,\ heat}$

(e) $CH_3C(CH_3)(OH)CH_3 \xrightarrow{H_2SO_4\ +\ heat}$

(f) $CH_3CH{=}CH_2 + Br_2 \xrightarrow{CCl_4}$

(g) 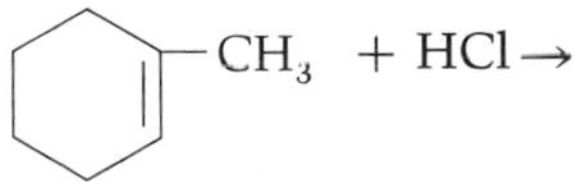 1-methylcyclohexene ($-CH_3$) $+ HCl \rightarrow$

(h) $+ KMnO_4 + H_2O \rightarrow$

(i) $CH_2{=}CHCH_3 \xrightarrow{\text{acid (polymerization)}}$

(j) $-CH_3 + H_2 \xrightarrow{Ni}$

(k) $-CH_3 + H_2O \xrightarrow{H_2SO_4}$

***10.24** The resonance hybrid of naphthalene consists of three structures (Problem 10.16). Assuming that the three structures contribute equally to the resonance hybrid, do you expect each bond to be the average of a single and a double bond, as is the case with benzene? Explain.

10.25 Give at least one resonance structure for 1,2,7,8-dibenzopyrene by completing the following structure. (See Focus 7.)

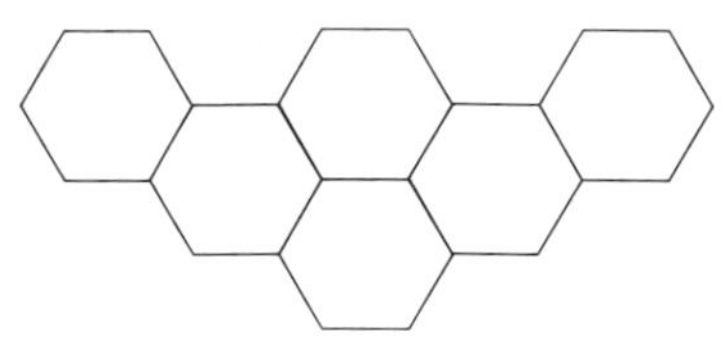

10.26 Would you expect a C≡C triple bond to be longer or shorter than a C=C double bond? Explain.

10.27 Write equations to show how you could prepare each of the following compounds, starting with benzene. You can use any other chemicals that are necessary.

(a) nitrobenzene (b) benzenesulfonic acid

(c) bromobenzene

10.28 Show the double bonds in the aromatic molecules whose carbon skeletons are shown below.

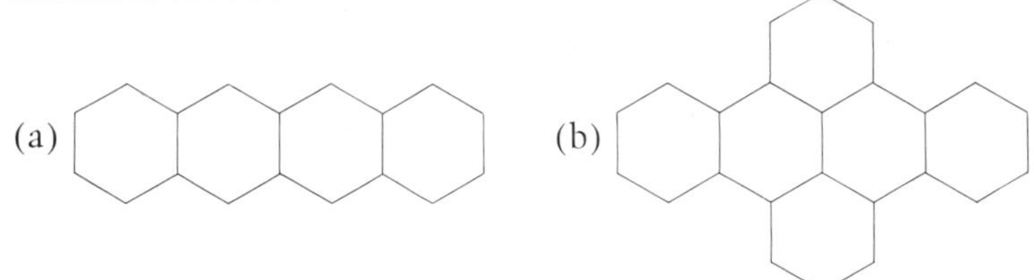

10.29 Describe the major chemical constituent of coal.

10.30 How do the molecular structures of graphite and diamond differ? (See Section 4.13 for the structure of diamond.)

10.31 Explain, in terms of electron repulsions, why the bond system —C≡ is linear.

10.32 Explain, in terms of the *sp* hybrid orbitals that are used, why the —C≡ bond system is linear.

10.33 Which of the following molecules would you expect to be more stable? Explain in terms of bond angles. (See Problem 10.12.)

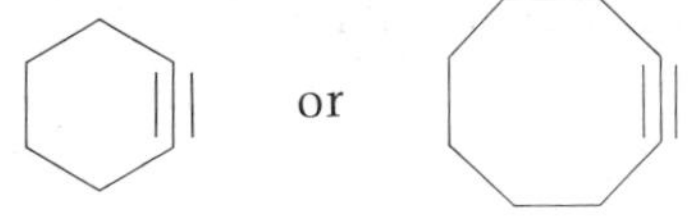

10.34 Which of the following molecules are aromatic?

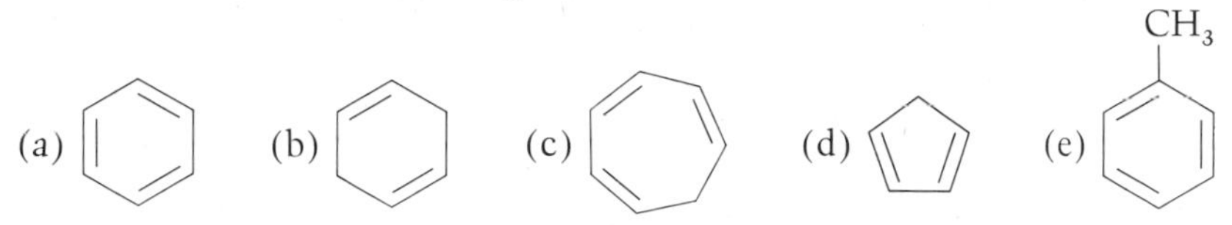

10.35 Explain, in terms of resonance theory, why all the carbon–carbon bonds in benzene are equal.

10.36 Explain why benzene does not readily undergo addition reactions.

10.37 Draw the sigma–pi molecular-orbital structure of the following compounds.

(a) $CH_3CH{=}CH_2$ (b) chloroethene

*__10.38__ Cycloalkanes and alkenes have the same general molecular formula. Are there any examples of cycloalkanes that behave like unsaturated compounds? Explain with an example.

10.39 Explain how elimination and addition reactions are related. Include specific examples of reactions in your explanation.

10.40 Describe simple chemical tests that would allow you to distinguish between the compounds in each of the following pairs. Write equations and tell what you would do and observe.

(a) cyclopentane and 1-pentene (b) pentane and 2-pentene

Problem-Solving Hints

If you have trouble solving any of the additional problems, refer to the sections listed next to the problem numbers.

10.20 9.7
10.21 9.8, 10.2, 10.15
10.22 10.2, 10.9, 10.15
10.23 10.5, 10.6, 10.11
10.24 10.14, 10.17
10.25 10.17
10.26 10.8
10.27 10.16
10.28 10.17
10.29 10.18
10.30 4.13, 10.18
10.31 10.8
10.32 10.8
10.33 10.8
10.34 10.13
10.35 10.14
10.36 10.13
10.37 10.3
10.38 9.9
10.39 10.5, 10.6
10.40 10.6

FOCUS 7 Aromatic Hydrocarbons and Cancer

At the turn of the century, coal-tar distillation was an important process that yielded many commercially useful compounds, most notably benzene. By 1915, it became evident that workers in these plants were developing skin cancers at an alarming rate. Chemists immediately began to study the various components of coal-tar distillates in order to determine the cause of these cancers. To determine which specific compounds were carcinogenic, they applied the compounds to the skin of laboratory animals. The first carcinogenic hydrocarbon to be isolated from coal tar was 1,2,5,6-dibenzanthracene. Soon thereafter, 1,2-benzopyrene, an even more potent carcinogen, was discovered in coal tar.

One of the most potent, carcinogens known is methylcholanthrene. It is not found in coal tar, but it can be formed by heating cholic acid, a component of bile (which is one of our digestive fluids).

1,2,5,6-dibenzanthracene

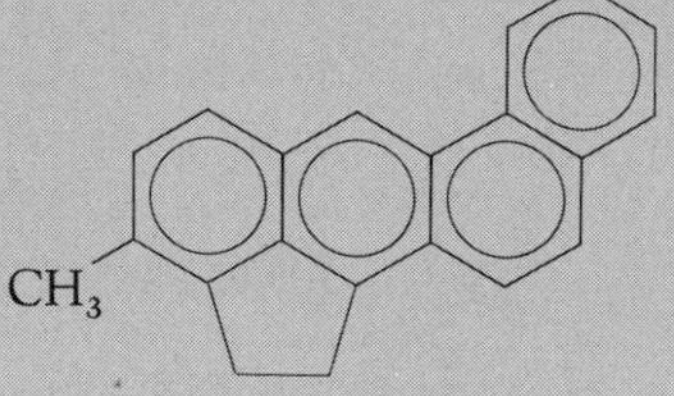

methylcholanthrene

1,2,7,8-dibenzopyrene

1,2-benzopyrene

1,2-benzopyrene and 1,2,7,8-dibenzopyrene, also a carcinogen, have been isolated from the tar of cigarette smoke. Carcinogenic polynuclear aromatic hydrocarbons have also been isolated from automobile exhaust fumes.

Carcinogens and mutagens will be discussed later in Focus 15.

FIGURE 1 Automobile exhaust fumes contain aromatic hydrocarbons. (Hugh Rogers from Monkmeyer Press Photo Service)

FOCUS 8 Natural Rubber

Another organic compound of great importance to all of us is natural rubber. Rubber is a polymer of the hydrocarbon isoprene (C_5H_8). The polymer exists as two isomers, the all-*cis* and the all-*trans*. The all-*cis* isomer is the elastic rubber produced by the tree *Hevea brasiliensis*, which, as its name suggests, was originally found in Brazil. The all-*trans* isomer is a hard, brittle solid rubber, called gutta-percha, used in making combs and other rigid objects. It comes from various Malayan trees.

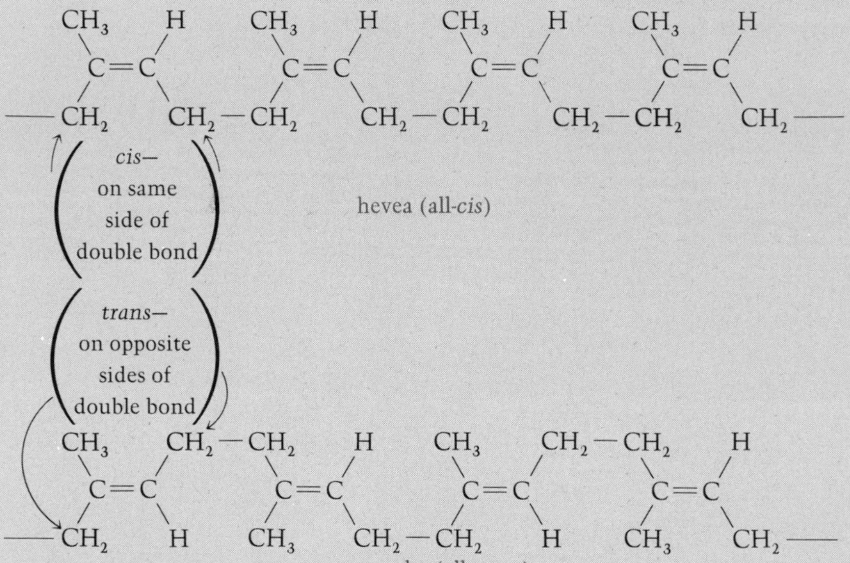

```
     CH3     H       CH3     H       CH3     H       CH3     H
       \    /          \    /          \    /          \    /
        C=C             C=C             C=C             C=C
       /    \          /    \          /    \          /    \
 -- CH2      CH2 -- CH2      CH2 -- CH2      CH2 -- CH2      CH2 --
```

cis—
on same
side of
double bond

hevea (all-*cis*)

trans—
on opposite
sides of
double bond

```
     CH3     CH2 -- CH2      H       CH3     CH2 -- CH2      H
       \    /          \    /          \    /          \    /
        C=C             C=C             C=C             C=C
       /    \          /    \          /    \          /    \
 -- CH2      H       CH3     CH2 -- CH2      H       CH3     CH2 --
```

gutta-percha (all-*trans*)

In the all-*cis* isomer, the CH_3 groups interfere with each other, so the molecule is not actually a straight chain. Instead, it forms large coils that tangle. Such tangled molecules impart the elastic properties to *Hevea* rubber. In the all-*trans* isomer, the molecules tend to remain linear, as shown above, and they align themselves into a fairly regular crystal network. For this reason, gutta-percha is hard and brittle.

FIGURE 1 A worker is shown harvesting rubber on a Malayan plantation. (Forbert from Monkmeyer Press Photo Service)

Paraquat is a water-soluble organic compound that is known to be an effective herbicide. Although this simple compound is not a halogenated hydrocarbon, it exists as a cation that is associated with two chloride ions. Paraquat is extremely toxic to human beings and has no antidote. Since paraquat is not readily metabolized in the body, most of it is excreted in the urine absolutely unchanged. This indicates that the toxicity of the compound is due to the characteristics of paraquat itself rather than to the effects of any by-products of its breakdown.

Lung, liver, and kidney damage show up in all cases of paraquat poisoning. Initially, the victim experiences irritation of mucous membranes, mouth, and stomach. Then gastrointestinal upset sets in. In fatal cases, death occurs as a result of respiratory failure that is caused by the damage to the lungs.

In 1975 the United States and Mexican governments experimented with paraquat to find out if the herbicide could be used advantageously to destroy Mexican marijuana and opium fields. Paraquat has a delayed effect of approximately three days on marijuana plants. As a result, no difference can be seen between a treated and an untreated plant for this period of time.

11

HYDROCARBON DERIVATIVES

Therefore, if a plant is harvested very soon after it has been sprayed with paraquat, it could be quite toxic and no one would have any reason to suspect this development. The consequences could be unfortunate since ingestion of paraquat in the lungs by smoking marijuana has the potential of producing irreversible lung damage.

[Illustration: Dusting the marijuana plants with the herbicide paraquat.]

We've described all the different kinds of compounds you can make using only carbon and hydrogen: the single-bonded alkanes, the double-bonded alkenes, the triple-bonded alkynes, as well as the aromatic hydrocarbons with their unusual bonding. We have also described some of the reactions these hydrocarbons can undergo. Now, we will examine in more detail the products of those reactions—the molecules you get when a hydrocarbon reacts with another substance. And we will look at the properties such molecules have.

These hydrocarbon derivatives are probably more important to us than the hydrocarbons themselves. They have provided us with anesthetics, insecticides, medicines, bombs, and, inadvertently, health problems. When you hear about people getting sick because of the dumping of industrial wastes in our environment, chances are the wastes included hydrocarbon derivatives. Let's look at a few important typical hydrocarbon derivatives, how they are made, how they are used, and how they affect us.

11.1 CHLORINATED METHANES

What are the general uses and properties of the chlorinated methanes?

Chlorinated hydrocarbons are compounds that may be prepared by the reaction of chlorine with hydrocarbons (Section 9.6). We have seen that methane reacts with chlorine in the presence of light to form a mixture of compounds. (We have not balanced the following equation, because the ratio of products depends on the conditions under which the reaction is carried out.)

$$CH_4 + Cl_2 \xrightarrow{\text{light}} CH_3Cl + CH_2Cl_2 + CHCl_3 + CCl_4 + HCl$$

The mixture can be separated into the individual components by fractional distillation (Focus 6). Table 11.1 lists these four chlorinated methanes, together with some of their physical properties.

TABLE 11.1 Chlorinated Methane Derivatives

Name*	Formula	Density, g/mL (at 20°C)	Melting point, °C	Boiling point, °C
Methyl chloride (chloromethane)	CH_3Cl	0.916	−97	−24
Methylene chloride (dichloromethane)	CH_2Cl_2	1.33	−95	39.8
Chloroform (trichloromethane)	$CHCl_3$	1.48	−63.5	61
Carbon tetrachloride (tetrachloromethane)	CCl_4	1.59	−23	76.7

*The common name is given first, with the IUPAC name in parentheses.

Problem 11.1 Draw the structural formulas and give the IUPAC names of all the possible products of the chlorination of ethane (C_2H_6).

Methyl chloride (CH_3Cl) vaporizes very rapidly at room temperature, absorbing the heat of vaporization from the area around it (see Section 4.8). In the process, the surrounding area may be cooled to temperatures below 0°C. For this reason methyl chloride is useful as a refrigerant in special applications. It is also a local anesthetic, causing numbness when sprayed on the skin. Like most chlorinated hydrocarbons, it can injure the liver, kidneys, and the central nervous system. High concentrations have a narcotic effect: they dull the senses, reduce pain, and produce sleep.

Methylene chloride (CH_2Cl_2) is a liquid at room temperature and is a good solvent for hydrocarbon oils. It is commonly used as a cleaning fluid and for degreasing. It causes anesthesia when inhaled. In high concentration, it is a narcotic.

Chloroform ($CHCl_3$) is probably the best known of the chlorinated methanes. It is an inhalation anesthetic, but it is no longer used in humans because it causes hypotension (low blood pressure), respiratory and myocardial depression, and even death. Chloroform is also a good solvent for nonpolar organic compounds and serves as a cleansing agent.

Carbon tetrachloride (CCl_4) is used as a solvent. Until recently it was used as a cleaning fluid and fire extinguisher. Because of its low electrical conductivity, it is especially useful in electrical fires and as a drying agent for wet automobile spark plugs. Carbon tetrachloride has found some use as an insecticide and as a cheap raw material for the production of synthetic organic chemicals. It is also effective against hookworms and tapeworms in humans and animals. However, carbon tetrachloride is very damaging to the liver and kidneys, and continued inhalation can be fatal. Contact with the skin causes irritation by dissolving the fat in the skin cells.

The chlorinated methanes are good solvents for oils and greases and are therefore good cleaning agents. However, they affect the central nervous system and can cause liver and kidney damage. Most of them are narcotic.

11.2 DDT

What characteristics of DDT make it dangerous to our environment?

Probably the best-known chlorinated hydrocarbon is **DDT.** Its IUPAC name is 1,1,1-trichloro-2,2-bis(*p*-chlorophenyl)ethane.

Cl
Cl—C—Cl
Cl—⌬—C—⌬—Cl
H

DDT

DDT was first used in the 1940s. At first it seemed to be the final solution to the pest problem, but by the early 1970s its dangers had become well known. In 1972 the Environmental Protection Agency banned the use of DDT.

How does DDT work on insects, and why does it pose such a serious threat? It affects the nervous system of insects. Soon after absorbing or ingesting DDT, the insect develops problems in coordination that eventually lead to convulsions and death. At first, it appeared that DDT was only slightly toxic to humans but highly toxic to the insects that caused the most severe health problems to humans (for example, mosquitoes, which spread malaria, or the tsetse fly, which carries sleeping sickness).

Two characteristics of DDT led to its becoming an ecological threat: its insolubility in water and its persistence in the environment. Although DDT is very insoluble in water, it is highly soluble in organic substances, especially fats. Because it is not metabolized, DDT, when it gets into the body, finds its way into those organs that are rich in fatty tissues, such as the liver, kidneys, brain, and gonads. DDT also concentrates in the milk of mammals, because of the high fat content of the milk.

DDT decomposes very slowly. It does not decompose by the natural processes of photochemical decomposition (decomposition through the

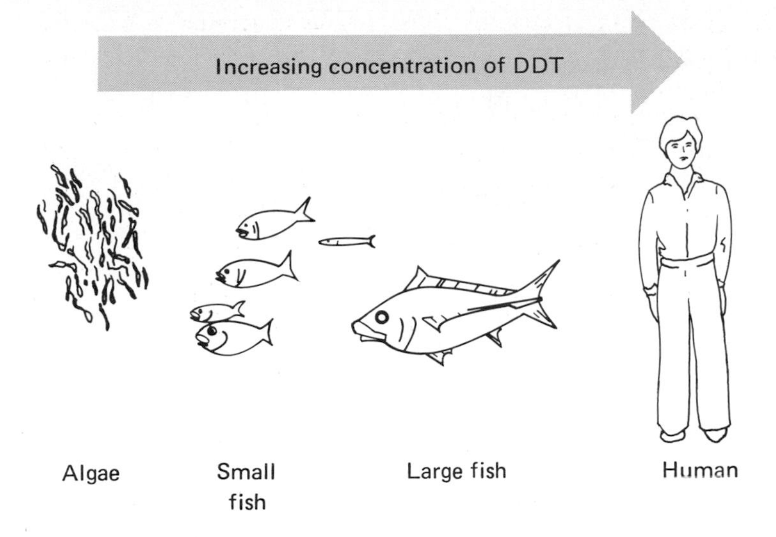

FIGURE 11.1 Concentration of DDT along the food chain.

influence of sunlight), oxidation (reaction with atmospheric oxygen), or biodegradation (breakdown by living organisms). At first, the persistence of DDT seemed to be an advantage; however, it was soon found to be a problem, because DDT tends to concentrate along the food chain (Figure 11.1). Whereas a small dose may be harmless to an animal near the bottom of the food chain, the carnivores further along the chain may ingest lethal doses. For example, DDT in grass eaten by a cow becomes concentrated in the milk. The milk thus contains a higher percentage of DDT than was in the grass. This DDT is transmitted to the calf, whose organs further accumulate DDT. Finally, when the calf has grown and the beef goes to market, to be consumed by us, its DDT content is much higher than that of the milk or of the grass.

DDT has affected certain insect-eating birds in an interesting way. The populations of these birds began to decline in areas where DDT was used; DDT interfered with the metabolism of calcium and thus caused their eggshells to be thinner than normal. This resulted in the loss of a greater than normal percentage of eggs and reduced the number of birds. With fewer birds to eat the insects, DDT actually led in some cases to an increase in the number of insects.

Another problem is that insects have become increasingly resistant to DDT. More and more DDT must be applied to kill the same number of insects. Resistance develops through natural selection: initially most insects are killed, but a few that have the ability to resist the effects of DDT survive, reproduce, and pass on this resistance to their offspring. Later generations thus contain a higher percentage of insects that can resist the insecticide.

11.3 OTHER CHLORINATED HYDROCARBON INSECTICIDES

Many other chlorinated hydrocarbons have similar uses as insecticides. They also have similar side effects (Table 11.2). All of the compounds shown in Table 11.2 affect the central nervous system of humans, and prolonged exposure results in liver and kidney damage.

TABLE 11.2 Important Chlorinated Hydrocarbons

Name	Formula	Properties	Uses
Lindane		Solid, melting point 112.5°C, slight musty odor, insoluble in water	Used against skin parasites on animals
Heptachlor		Solid, melting point 95°C, insoluble in water	Used to control cotton-boll weevil
Chlordane		Liquid, insoluble in water	Used against fleas, lice, ticks, and mange
Aldrin*		Solid, melting point 104°C, insoluble in water	Used as general insecticide

*The line formula represents the structure

TABLE 11.2 (Continued)

Name	Formula	Properties	Uses
Dieldrin	[structure: epoxide ring fused to bicyclic system with CCl_2 bridge and four Cl]	Solid, melting point 177°C, practically insoluble in water	Used against lice, and keds in sheep
Polychlorinated biphenyl (PCB)	[structure: biphenyl with ten Cl substituents]	Insoluble in water; very persistent and tends to accumulate in fatty tissue	Used in paints, plastics, insulators, and as a plasticizer (a plasticizer alters the properties of a plastic to make it more pliable)
Poly(vinyl chloride) (PVC)	$\text{-}(CH_2CH(Cl))_n\text{-}$	Very persistent; incineration produces HCl gas, a serious irritant	Used to make garbage bags, piping, and phonograph records

POLYVINYL CHLORIDE: A LINK TO CANCER 11.4

One final important point about chlorinated hydrocarbons is that, more and more, they are being linked to cancer. Vinyl chloride ($CH_2{=}CHCl$) has been used since the 1930s to produce a plastic called **poly(vinyl chloride)** (PVC).

$$nCH_2{=}\underset{\displaystyle Cl}{\underset{|}{CH}} \rightarrow \sim\!\!(CH_2{-}\underset{\displaystyle Cl}{\underset{|}{CH}})_n\!\!\sim$$

vinyl chloride   poly(vinyl choride) (PVC)

At present, over 10 million tons of PVC are produced annually. PVC is used in containers, phonograph records, garden hoses, and plastic wraps. Until recently, it was considered harmless. But in the 1960s, workers in PVC plants began to show some unusual and severe symptoms: bone atrophy of fingers and toes, a decrease in the number of blood platelets, poor functioning of the lungs and spleen, varicose veins in the throat, and cirrhosis of the liver. By the end of 1974, more than thirty workers in PVC plants around the world had died of a rare liver cancer caused by vinyl chloride. Tests with laboratory animals have shown that concentrations of vinyl chloride as low as 50 parts per million are carcinogenic.

11.5 FLUOROCARBONS

How do fluorocarbons differ from chlorinated hydrocarbons?

Elemental fluorine is the most reactive of all the elements. It is so reactive that it decomposes glass. When fluorine was first isolated, its storage posed a serious problem. How do you store a chemical that reacts with all the common containers, such as glass and plastics? Chemists found the answer in the **fluorocarbons**—compounds that contain only carbon and fluorine.

CF_4 — perfluoromethane

$F_3C{-}CF_3$ (F—C—C—F, each carbon bearing two further F atoms) — perfluoroethane

$F_2C{=}CF_2$ — 1,1,2,2-fluoroethene (or perfluoroethylene)

We call these compounds fluorocarbons because their structures resemble those of hydrocarbons, except that fluorine atoms have replaced all the hydrogen atoms. To name a particular fluorocarbon, we simply attach the prefix *perfluoro-* to the name of the corresponding hydrocarbon. The prefix *per* indicates that *all* the hydrogen atoms have been replaced by fluorine atoms.

Fluorocarbons are inert to fluorine and, in fact, to nearly all chemical reagents. Chemists at the Du Pont Company developed plastics made from fluorocarbons, which are useful because of their inertness. The best-known perfluorinated plastic is **Teflon,** a polymer of perfluoroethene, as shown below ($n = 1{,}000$).

$$nCF_2{=}CF_2 \rightarrow \text{+}CF_2{-}CF_2\text{+}_n$$

Teflon

Teflon does not react with fluorine. The carbon-carbon bonds are not attacked by fluorine, and there are no hydrogen atoms to be replaced by fluorine. Its chemical inertness gives Teflon unusual properties. It does not adhere chemically to any substance, so it exhibits very low friction (Teflon feels oily, although it is not). This low friction makes Teflon suitable for gaskets and bearings for machinery. Teflon is also used to coat pots and pans. Eggs can be fried without grease in a Teflon-coated frying pan, because the eggs do not stick to the Teflon. It is also used in heart valves, linings for joint sockets, and other artificial body parts. Because Teflon is inert, the body does not usually reject it.

Freons are compounds that contain carbon, fluorine, and chlorine. They are useful primarily as refrigerants and aerosol propellants. The common names of Freons follow the formula Freon-*xy*, where x is the

TABLE 11.3 Some Common Freons

Name	Formula	Properties	Uses	Toxicity
Freon-14 (tetrafluoromethane)	CF_4	Colorless, odorless gas; chemically very inert	Used as low-temperature refrigerant	Narcotic in high concentrations
Freon-12 (dichlorodifluoromethane)	CCl_2F_2	Colorless, practically odorless, nonirritating, chemically inert gas	Used as refrigerant and aerosol propellant	Little, if any, toxicity
Freon-11 (trichlorofluoromethane)	CCl_3F	Boiling point 23.7°C; nonflammable	Used as refrigerant and aerosol propellant	May be narcotic at high temperatures

number of carbon atoms and y the number of fluorine atoms in the molecule (Table 11.3). Section 9.6 discussed the possible risks in the use of Freons.

Problem 11.2 Compare the toxic properties of fluorinated hydrocarbons with those of chlorinated hydrocarbons.
Problem 11.3 Compare the physical properties of fluorinated hydrocarbons with those of chlorinated hydrocarbons.

NITRO COMPOUNDS: EXPLOSIVES 11.6

Why is nitroglycerin explosive?

Combustion involves the reaction of a compound with oxygen—rapid oxidation, usually with the evolution of heat and light. A familiar example is the combustion of methane.

$$CH_4 + 2O_2 \rightarrow CO_2 + 2H_2O + \text{energy}$$

When combustion occurs very rapidly, an explosion may result. Compounds that give off large amounts of energy and gas on combustion tend to explode if they are thoroughly mixed with oxygen and ignited. If a mixture of methane and oxygen is ignited in a closed container, the reaction liberates so much energy so suddenly that the gas mixture quickly heats up and expands. The reaction proceeds even more rapidly as the temperature rises (Section 6.3), and a thermal explosion results.

Compounds that contain the nitro group (NO_2) are often explosive.

$$-N\begin{matrix} \nearrow\!\!\!\!= O \\ \searrow O \end{matrix}$$

During combustion a lot of energy escapes in the form of heat. The oxygen in the nitro group serves as a source of oxygen for combustion. As a result, combustion is extremely rapid because the necessary oxygen is present where it is needed.

One of the most powerful chemical high explosives is **glyceryl trinitrate,** which is commonly known by the informal name *nitroglycerin.* It is the nitrate ester (Chapter 15) of glycerol, $CH_2OHCHOHCH_2OH$.

$$4\begin{pmatrix} CH_2-O-NO_2 \\ | \\ CH-O-NO_2 \\ | \\ CH_2-O-NO_2 \end{pmatrix} \rightarrow 12CO_2 + 10H_2O + 6N_2 + O_2 + \text{energy}$$

glyceryl trinitrate

Glyceryl trinitrate, an oily liquid, can explode in the absence of atmospheric oxygen because its own molecules contain enough oxygen atoms to convert its hydrocarbon part into the combustion products CO_2 and H_2O. The reaction moves through a sample of glyceryl trinitrate at a rate of about 6,000 m/s. A few grams of glyceryl trinitrate will react completely in less than one-millionth of a second. Pure glyceryl trinitrate is so sensitive to shock that it is very dangerous to handle.

The explosive power of glyceryl trinitrate arises from the extremely rapid conversion of a small volume of liquid into several thousand times as large a volume of hot gases. (Note that all of the products of the reaction are gases.)

Alfred Nobel, who endowed the Nobel Prizes, discovered a way to moderate the sensitivity of glyceryl trinitrate without decreasing its explosive power. By impregnating an inert, porous solid (diatomaceous earth) with glyceryl trinitrate, he arrived at a shock-resistant, solid high explosive that could be handled safely. He called this product **dynamite.**

Cellulose trinitrate, also called "nitrocellulose," is similar to glyceryl trinitrate, but it is less sensitive. It is a plastic made by treating cellulose (in the form of cotton fibers) with nitric acid. The product is actually the ester cellulose nitrate. "Nitrocellulose" is a misnomer.

$$\underset{\text{cellulose}}{(C_6H_{10}O_5)_n} + 2nHNO_3 \rightarrow \underset{\text{cellulose nitrate}}{(C_6H_8N_2O_9)_n} + 2nH_2O$$

Cellulose nitrate is a fibrous solid that resembles cotton. It is very flammable, and it is used in solution with alcohol or acetone. When the solution is spread on a surface and the solvent allowed to evaporate, it forms a plastic film. This has been used to coat light-sensitive surfaces in electronics, microscopy, and photography.

One of the best-known high explosives is 2,4,6-trinitrotoluene, better known as **TNT.** TNT has a high chemical potential energy, which it releases on combustion.

$$4\ \text{(2,4,6-}(NO_2)_3C_6H_2CH_3\text{)} + 21O_2 \rightarrow 28CO_2 + 6N_2 + 10H_2O + \text{energy}$$

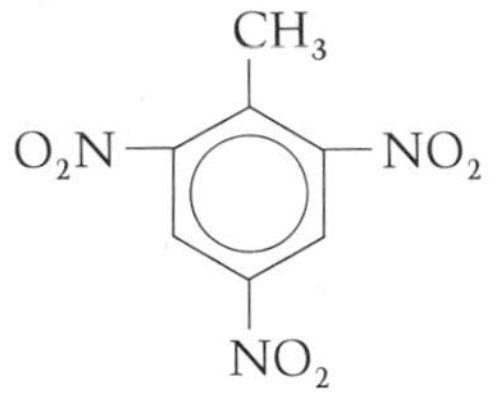

TNT is a true nitro compound, because it has the nitro group ($-NO_2$), rather than the nitrate group ($-ONO_2$).

NITROGEN COMPOUNDS AS VASODILATORS 11.7

Another interesting use of glyceryl trinitrate is as a coronary **vasodilator,** a substance that dilates blood vessels. By relaxing the smooth muscles of the blood vessels glyceryl trinitrate lowers resistance to blood flow and thereby lowers blood pressure. Glyceryl trinitrate is used to relieve angina pectoris. Angina pectoris consists of sharp chest pains of short duration. It is caused by poor circulation (resulting from degeneration of the heart muscle). By dilating the blood vessels, glyceryl trinitrate aids blood flow and usually stops the attack.

Glyceryl trinitrate is made nonexplosive by diluting it with alcohol or other solvent.

Isoamyl nitrite,

$$CH_3\underset{\displaystyle CH_3}{\overset{}{C}}HCH_2CH_2{-}O{-}N{=}O$$

is also a vasodilator. Isoamyl nitrite is not a nitro compound, because the hydrocarbon group is bonded to oxygen rather than to nitrogen; it is the nitrite ester of isoamyl alcohol.

Problem 11.4 Write the structural formula of isoamyl alcohol.

SUMMARY

Chlorinated hydrocarbons are compounds containing chlorine atoms in place of one or more of the hydrogen atoms. They may be produced by the reaction of hydrocarbons with chlorine. Chlorinated hydrocarbons are, in general, soluble in oils, greases, and fatty substances and insoluble in water. They are often narcotic, and they can cause liver and kidney damage. Many chlorinated hydrocarbons are useful as insecticides, but they are dangerous to the environment because they persist in animal fatty tissues, because they decompose very slowly in the environment, and because insects tend to become resistant to them reducing their effectiveness. **Fluorocarbons** are compounds of carbon and fluorine. Fluorocarbons are inert to most chemical reagents, a property that makes them useful in many applications, such as bearings, gaskets, cooking utensils, and artificial body parts. The **Freons** are compounds of carbon, fluorine, and chlorine. They are used as refrigerants and aerosol propellants. Fluorocarbons and Freons are much less toxic than chlorinated hydrocarbons. Nitro compounds ($R-NO_2$, where R is a hydrocarbon group) and nitrate esters ($R-ONO_2$) are useful as explosives and as **vasodilators.**

Key Terms

Chlorinated hydrocarbons (Section 11.1)
DDT (Section 11.2)
Dynamite (Section 11.6)
Fluorocarbons (Section 11.5)
Freons (Section 11.5)
Glyceryl trinitrate (Section 11.6)
Polyvinyl chloride (Section 11.4)
Teflon (Section 11.5)
TNT (Section 11.6)

ADDITIONAL PROBLEMS

11.5 Compare the physical properties of the chlorinated methanes with those of alkanes of comparable molecular weights (Table 9.1). Explain why the chlorinated compounds are so much more dense than the hydrocarbons.
11.6 Summarize the toxic properties of the chlorinated hydrocarbons.
11.7 List briefly some of the natural processes by which substances decompose in the environment.
11.8 Explain how elemental fluorine can be stored, in spite of its being highly reactive to glass and other ordinary container materials.
11.9 Compare the toxic properties of chlorinated hydrocarbons with those of fluorinated hydrocarbons.
11.10 Compare the physical properties of chlorine-containing compounds with those of fluorine-containing compounds.
11.11 Describe the chemical factors that cause a compound to be a high explosive.
11.12 Why are nitro compounds generally explosive?

Problem-Solving Hints

If you have trouble solving any of the additional problems, refer to the sections listed next to the problem numbers.

11.5 11.1 | **11.6** 11.3 | **11.7** 11.2
11.8 11.5 | **11.9** 11.3, 11.5 | **11.10** 11.3, 11.5
11.11 11.6 | **11.12** 11.6

It seems logical to assume that two substances that closely resemble each other structurally would share a similarity in function. As we will learn in this chapter, methanol (methyl alcohol) and ethanol (ethyl alcohol) are two very similar chemical compounds. However, when ingested, there is quite a striking difference in the immediate physiological consequences of each. Ethanol can intoxicate you, while methanol can cause blindness or death in fairly short order. Obviously, structure in this case translates into life or death in terms of function.

Methanol and ethanol differ chemically by one carbon and two hydrogens—a very slight structural difference, but one that has caused, and will undoubtedly continue to cause, numerous deaths to unsuspecting victims. (Chronic alcoholics, for example, frequently will drink whatever is available, and methanol is not an exception.) Methanol is the simpler of the two compounds; it is referred to as wood alcohol because it is produced in the destructive distillation of wood as charcoal is formed. Methanol is oxidized by the human enzyme alcohol dehydrogenase to produce formaldehyde and formic acid, both of which have a toxic effect on the body.

Ethanol, which is present in beer, wine, and other alcoholic beverages, is oxidized by the same enzyme to produce two less toxic compounds, acetic acid and acetaldehyde. Because the same enzyme is involved in both oxidations, ethanol and methanol will, in effect, compete for the alcohol dehydrogenase if both are present in the body at the same time. Therefore,

12

ALCOHOLS, PHENOLS, ETHERS, AND THEIR SULFUR ANALOGUES

the oxidation of methanol to toxic products might be less destructive if ethanol, too, is present. The presence of ethanol will allow methanol to be excreted unchanged by the body. For this reason, ethanol is frequently used in the treatment of methanol poisoning. Although ethanol can have the same deadly effect as methanol, a much larger quantity and higher concentration are necessary to produce a fatal result.

[Illustration: Methanol and ethanol have surprisingly similar structures but lethal differences in function.]

ALCOHOLS, PHENOLS, ETHERS

The alcohols, phenols, and ethers all have one feature in common: they all contain an oxygen that is linked to carbon by a single bond. Sulfur is an analogue of oxygen (that is, they are similar) because they are in the same chemical family (take a glance at the periodic table). Since sulfur and oxygen have a lot in common, the compounds they form have a lot in common too.

The alcohols are the best known of these compounds. In daily life, we often say "alcohol" no matter which alcohol we mean. However, there are several different alcohols, and they all have quite dramatically different effects. We have just seen how the simple difference in structure between methanol and ethanol accounts for methanol being a deadly poison and ethanol being less toxic. Here is direct proof that molecular structure deserves special emphasis, which we will give it in this chapter.

ALCOHOLS 12.1

Alcohols have a **hydroxyl group** (—OH) bonded to an aliphatic group (Section 10.12). Table 12.1 lists some common alcohols and their properties. We classify alcohols into three groups, according to the nature of the carbon that is bonded to the hydroxyl group (R may be aliphatic or aromatic).

$$\underset{\text{primary alcohols}}{R-\overset{\displaystyle H}{\underset{\displaystyle OH}{\overset{|}{\underset{|}{C}}}}-H} \qquad \underset{\text{secondary alcohols}}{R-\overset{\displaystyle H}{\underset{\displaystyle OH}{\overset{|}{\underset{|}{C}}}}-R} \qquad \underset{\text{tertiary alcohols}}{R-\overset{\displaystyle R}{\underset{\displaystyle OH}{\overset{|}{\underset{|}{C}}}}-R}$$

How do primary, secondary, and tertiary alcohols differ?

If the OH is attached to a carbon that is bonded to only one other carbon atom, it is a **primary alcohol.** If the OH is attached to a carbon that is bonded to two other carbon atoms, it is a **secondary alcohol.** If the OH is attached to a carbon that is bonded to three other carbon atoms, it is a **tertiary alcohol.** We abbreviate these terms by using a degree symbol, 1° for primary, 2° for secondary, and 3° for tertiary.

TABLE 12.1 Properties of Some Simple Alcohols

Name	Formula	Boiling point, °C	Density, g/mL (at 20°C)	Solubility,* g/100 mL In water	In ether
Methanol	CH_3OH	65	0.79	∞	∞
Ethanol	CH_3CH_2OH	78	0.79	∞	∞
1-Propanol	$CH_3CH_2CH_2OH$	97	0.80	∞	∞
1-Butanol	$CH_3CH_2CH_2CH_2OH$	118	0.81	7.9	∞
1-Pentanol	$CH_3CH_2CH_2CH_2CH_2OH$	138	0.81	2.7	∞
1-Hexanol	$CH_3(CH_2)_4CH_2OH$	157	0.82	0.59	∞
1-Heptanol	$CH_3(CH_2)_5CH_2OH$	176	0.82	0.09	∞
1-Octanol	$CH_3(CH_2)_6CH_2OH$	195	0.82	Insoluble	∞
1-Nonanol	$CH_3(CH_2)_7CH_2OH$	213	0.83	Insoluble	∞
1-Decanol	$CH_3(CH_2)_8CH_2OH$	231	0.83	Insoluble	∞

*∞ means "infinitely soluble"; that is, the alcohol is soluble in water or ether in all proportions at room temperature.

Problem 12.1 Classify each of the following alcohols as primary, secondary, or tertiary.

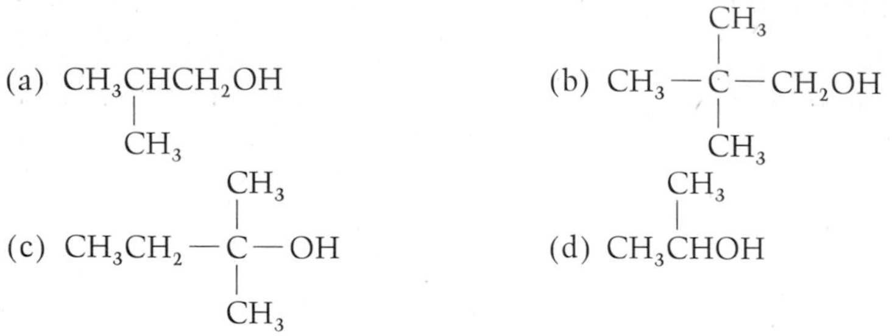

(a) $CH_3CH(CH_3)CH_2OH$

(b) $CH_3-C(CH_3)_2-CH_2OH$

(c) $CH_3CH_2-C(CH_3)_2-OH$

(d) $CH_3CH(CH_3)OH$

12.2 ALCOHOLS: NOMENCLATURE

Most simple alcohols are known by their common names—methyl alcohol, ethyl alcohol, propyl alcohol, and isopropyl alcohol (an isopropyl group is a three-carbon group in which the substituent is bonded to the second carbon atom).

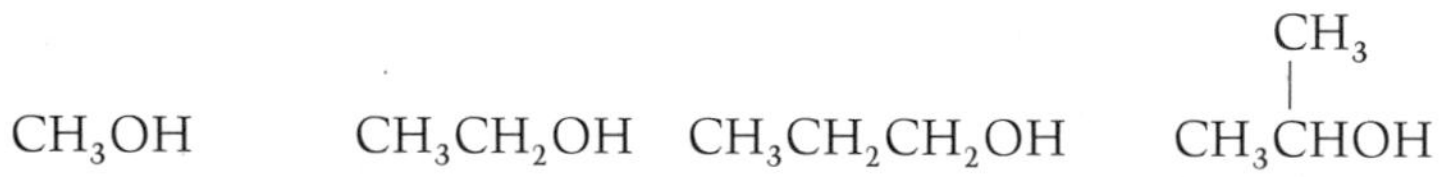

CH_3OH	CH_3CH_2OH	$CH_3CH_2CH_2OH$	$CH_3CH(CH_3)OH$
methyl alcohol	ethyl alcohol	propyl alcohol	isopropyl alcohol

What modification to the IUPAC rules is needed with alcohols?

The following IUPAC rules for nomenclature apply to alcohols.

Rule 1: Select the longest continuous carbon chain that includes the carbon that bears the OH group. Give that parent chain the name of the alkane, but replace the final *-e* with *-ol.* For example, CH_3CH_2OH is ethanol, $CH_3CH_2CH_2OH$ is propanol.

Rule 2: Name the branches as before (Section 9.3).

Rule 3: Number the carbons of the parent chain starting at the end nearer to the hydroxyl group. Designate the position of the hydroxyl group by a number immediately preceding the parent name. Designate the position of each branch by a number preceding the branch name. For example,

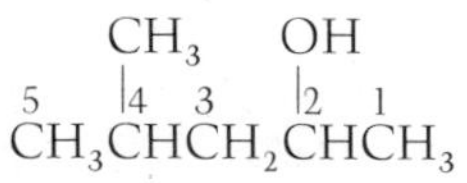

$$\overset{5}{C}H_3\overset{4}{C}H(CH_3)\overset{3}{C}H_2\overset{2}{C}H(OH)\overset{1}{C}H_3$$

is 4-methyl-2-pentanol.

In cyclic compounds, the carbon to which the hydroxyl group is attached is assigned the number 1. We do not need to include this number in the name of the alcohol. For example,

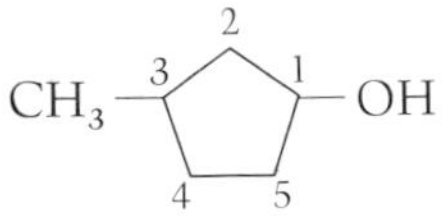

$$CH_3\text{—(cyclopentane ring, C-3)}\ldots\text{(C-1)—}OH$$

(ring carbons numbered 1, 2, 3, 4, 5)

is 3-methylcyclopentanol.

Rule 4: If an alcohol contains more than one hydroxyl group, change the ending from *-ol* to *-diol* (if two groups), *-triol* (three groups), *-tetraol* (four groups), and so forth. Insert a number before the parent name to indicate the location of each OH group. For example,

$$\overset{4}{C}H_3\overset{3}{C}H(OH)\overset{2}{C}H_2\overset{1}{C}H_2OH$$

is 1,3-butanediol. (Note that in these cases the final *-e* of the alkane name is kept.)

Problem 12.2 Draw the structural formulas of the following compounds.

(a) 2,2-dimethyl-1-butanol (b) isopropyl alcohol
(c) cyclopentanol (d) 3,3-dimethylcyclohexanol

Problem 12.3 Give the IUPAC names of the following compounds.

(a) 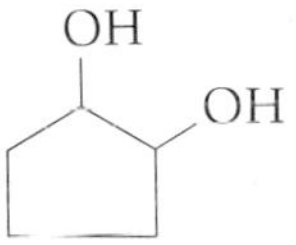 cyclohexane ring with OH on two adjacent carbons

(b) $CH_3CH_2C(CH_3)(CH_2CH_3)CH_2CH_2OH$

(c) $CH_3(CH_2)_8CH_2OH$ (d) $HOCH_2(CH_2)_6CH_2OH$

ALCOHOLS: PROPERTIES 12.3

How does hydrogen bonding explain solubility and boiling points?

Alcohols are similar to water in many ways. For example, the lower alcohols are soluble in water, and all the alcohols have boiling points higher than those of hydrocarbons of comparable molecular weight. (CH_3CH_2OH has a molecular weight of 46 and boils at 78°C; $CH_3CH_2CH_3$ has a molecular weight of 44 and boils at −42°C.)

These properties are the result of hydrogen bonding (Section 5.4). Alcohols can form hydrogen bonds because they have an OH group.

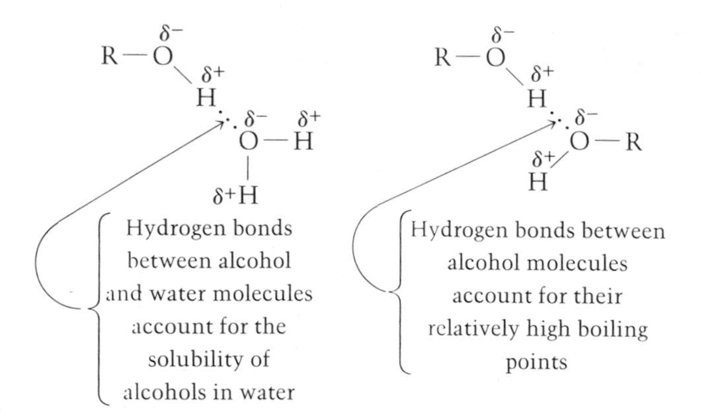
Hydrogen bonds between alcohol and water molecules account for the solubility of alcohols in water
Hydrogen bonds between alcohol molecules account for their relatively high boiling points

Hydrocarbons are insoluble in water and they have low boiling points because they cannot form hydrogen bonds. With alcohols, the longer the carbon chain, the less soluble the alcohol is in water (Table 12.1). This happens because the long-chain alcohols resemble hydrocarbons. The lower alcohols have fewer carbon atoms and therefore resemble water. This rule of solubility is summarized by the phrase, "Like dissolves like."

$CH_3—O—H$

$H—O—H$

Methanol is very much like water

$H—O—H$

$CH_3CH_2CH_2CH_2CH_2CH_2CH_2—O—H$

$CH_3CH_2CH_2CH_2CH_2CH_2CH_3$

1-Heptanol resembles heptane more than it does water

Diethyl ether ($CH_3CH_2—O—CH_2CH_3$) resembles both water and hydrocarbons in its properties (see Section 12.12). For this reason, alcohols are generally soluble in ether.

Problem 12.4 Which should be more soluble in water, butanol or octanol? Explain.

ALCOHOLS: REACTIONS 12.4

What two chemical bonds are included in the hydroxyl group of alcohols?

Alcohols react chemically in three important ways: reactions of the O—H bond; oxidation; and reactions of the C—O bond.

Reactions of the O—H bond

How acidic are alcohols?

Alcohols react with the very reactive alkali metals to produce **alkoxides** and hydrogen gas. In this reaction, alcohols behave like water. The general reaction is as follows (where R is an alkyl group)

$$2R\text{—}OH + 2Na \rightarrow \underset{\text{sodium alkoxide}}{2R\text{—}ONa} + H_2$$

A specific example is

$$\underset{\text{methanol}}{2CH_3OH} + 2Na \rightarrow \underset{\text{sodium methoxide}}{2CH_3ONa} + H_2$$

Alcohols are not acidic enough to react with inorganic hydroxides. The conjugate base of an alcohol, the alkoxide ion, is a very strong base.

$$R\text{—}OH + OH^- \not\rightleftarrows \underset{\text{alkoxide ion}}{R\text{—}O^-} + H_2O$$

The equilibrium lies so far to the left that for all practical purposes we may say that no reaction occurs.

Problem 12.5 Label the conjugate acid–base pairs in the following equation.

$$CH_3OH + OH^- \rightleftarrows CH_3O^- + H_2O$$

By reacting alcohols with carboxylic acids in the presence of small amounts of a strong acid (usually H_2SO_4) we prepare esters (see Chapter 15). We call this reaction **esterification.**

$$R\text{—}OH + HO\text{—}\overset{\overset{\displaystyle O}{\|}}{C}\text{—}R' \xrightarrow{H_2SO_4} \underset{\text{an ester}}{R\text{—}O\text{—}\overset{\overset{\displaystyle O}{\|}}{C}\text{—}R'} + H_2O$$

Experiments using the oxygen isotope ^{18}O prove that the C—O bond of the alcohol is not broken during this reaction.

$$CH_3CH_2-{}^{18}OH + HO-\overset{\overset{\large O}{\|}}{C}-CH_3 \xrightarrow{H_2SO_4} CH_3CH_2-{}^{18}O-\overset{\overset{\large O}{\|}}{C}-CH_3 + H_2O$$

The ^{18}O remains entirely in the ester molecule; none of it ends up in the water. Thus the water molecules must be getting their oxygen atoms from the carboxylic acid. We can summarize this reaction as follows.

$$R-O-H + H-O-\overset{\overset{\large O}{\|}}{C}-R' \xrightarrow{H^+} R-O-\overset{\overset{\large O}{\|}}{C}-R' + H-O\diagdown H$$

Oxidation

Are all types of alcohols easily oxidized?

We can oxidize primary and secondary alcohols with oxidizing agents such as alkaline potassium permanganate ($KMnO_4$) or chromic acid (H_2CrO_4). Tertiary alcohols are not oxidized under these conditions. Primary alcohols yield aldehydes (see Chapter 14). If the aldehydes are not removed from the reaction mixture, they are oxidized further to yield carboxylic acids.

$$(1)\quad \underset{\text{ethanol (1°)}}{3CH_3CH_2OH} + 2H_2CrO_4 \rightarrow \underset{\text{acetaldehyde}}{3CH_3\overset{\overset{\large O}{\|}}{C}H} + Cr_2O_3 + 5H_2O$$

$$(2)\quad 3CH_3\overset{\overset{\large O}{\|}}{C}H + 2H_2CrO_4 \rightarrow \underset{\text{acetic acid}}{3CH_3\overset{\overset{\large O}{\|}}{C}-OH} + Cr_2O_3 + 2H_2O$$

Most aldehydes have a lower boiling point than either the corresponding alcohol or carboxylic acid. Aldehydes, therefore, can be removed by distillation as they are formed.

Secondary alcohols yield ketones on oxidation (see Chapter 14). Ketones are stable to oxidation and do not react further.

$$\underset{\text{isopropyl alcohol (2°)}}{3CH_3\overset{\overset{\large OH}{|}}{C}HCH_3} + 2H_2CrO_4 \rightarrow \underset{\text{acetone}}{3CH_3\overset{\overset{\large O}{\|}}{C}CH_3} + Cr_2O_3 + 5H_2O$$

Reactions of the C—O bond

What reaction can be used to weaken the C—O bond of an alcohol?

The C—O bond in alcohols is very strong and hard to break. However, a strong acid such as H_2SO_4 can break the bond and dissociate the alcohol by the following reactions.

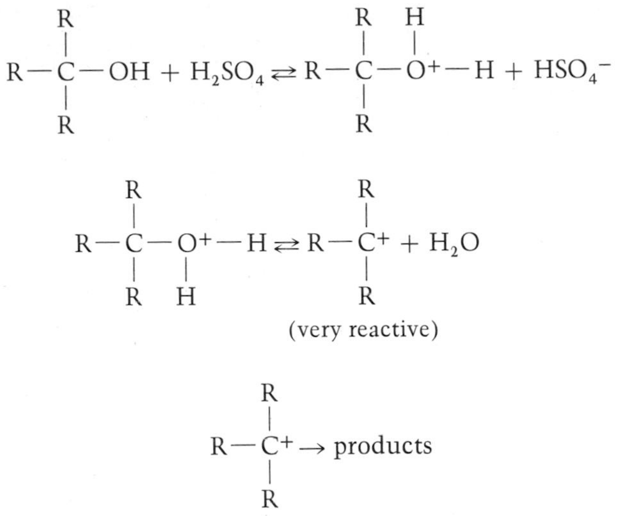

$$R_3C{-}OH + H_2SO_4 \rightleftharpoons R_3C{-}O^{+}H{-}H + HSO_4^-$$

$$R_3C{-}O^{+}H{-}H \rightleftharpoons R_3C^{+} + H_2O$$

(very reactive)

$$R_3C^{+} \longrightarrow \text{products}$$

In the presence of a strong acid, the alcohol acts as a base by accepting a hydrogen ion. The ion

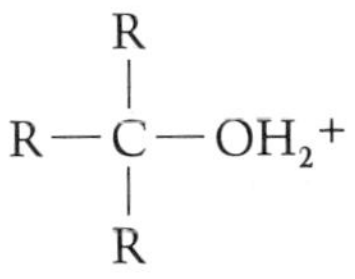

$$R_3C{-}OH_2^{+}$$

can dissociate readily, because in doing so, it releases the very stable water molecule.

We can use this behavior of alcohols toward strong acids to bring about several reactions. For example, if we react isopropyl alcohol with a small amount of sulfuric acid, we produce propene through an elimination reaction (Section 10.5).

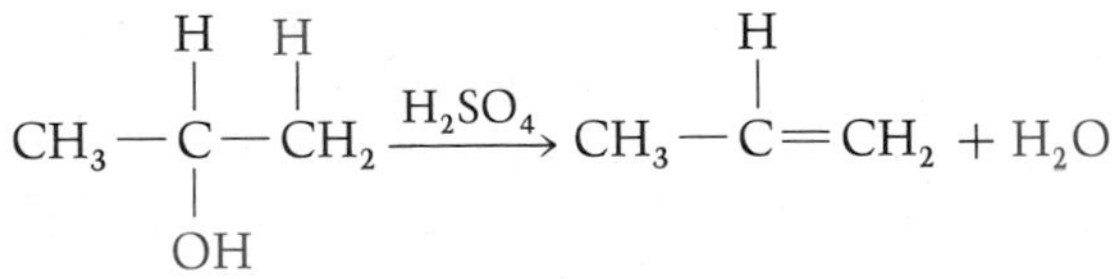

$$CH_3{-}CH(OH){-}CH_2H \xrightarrow{H_2SO_4} CH_3{-}CH{=}CH_2 + H_2O$$

In this reaction, the very reactive ion $CH_3\overset{+}{C}HCH_3$ loses its charge and becomes a stable molecule by donating an H^+ from the adjacent carbon to the base H_2O.

$$\underset{\text{base}}{H_2O} + \underset{\text{acid}}{CH_3-\overset{+}{C}H-\underset{\substack{|\\H}}{CH_2}} \rightarrow \underset{\text{base}}{CH_3-CH=CH_2} + \underset{\text{acid}}{H_3O^+}$$

If we use at least 1 mole of HCl or HBr per mole of alcohol, we can replace the OH group by chlorine or bromine in a substitution reaction.

$$CH_3-\underset{\substack{|\\OH}}{\overset{\substack{H\\|}}{C}}-CH_3 + HCl \rightarrow CH_3-\underset{\substack{|\\Cl}}{\overset{\substack{H\\|}}{C}}-CH_3 + H-OH$$

Here, the $CH_3\overset{+}{C}HCH_3$ ion reacts with the Cl^- that is present in large quantities in the reaction mixture.

$$Cl^- + CH_3-\overset{+}{C}H-CH_3 \rightarrow CH_3-\underset{\substack{|\\Cl}}{CH}-CH_3$$

We can summarize the path of the reactions of alcohols with strong acids as follows.

$$R-\underset{\substack{|\\H}}{\overset{\substack{R\\|}}{C}}-\underset{\substack{|\\R}}{\overset{\substack{R\\|}}{C}}-OH + HX \rightleftarrows R-\underset{\substack{|\\H}}{\overset{\substack{R\\|}}{C}}-\underset{\substack{|\\R}}{\overset{\substack{R\\|}}{C}}-\overset{+}{O}H_2 + X^-$$

$$R-\underset{\substack{|\\H}}{\overset{\substack{R\\|}}{C}}-\underset{\substack{|\\R}}{\overset{\substack{R\\|}}{C}}-\overset{+}{O}H_2 \rightleftarrows R-\underset{\substack{|\\H}}{\overset{\substack{R\\|}}{C}}-\underset{\substack{|\\R}}{\overset{\substack{R\\|}}{C^+}} + H_2O$$

where R is a hydrocarbon group and X = Cl, Br, or I.

$$R-\underset{\substack{|\\H}}{\overset{\substack{R\\|}}{C}}-\underset{\substack{|\\R}}{\overset{\substack{R\\|}}{C^+}} \xrightarrow{-H^+} \begin{matrix} R & & R \\ & C=C & \\ R & & R \end{matrix} \quad \text{(elimination)}$$

$$R-\underset{\substack{|\\H}}{\overset{\substack{R\\|}}{C}}-\underset{\substack{|\\R}}{\overset{\substack{R\\|}}{C^+}} \xrightarrow{+X^-} R-\underset{\substack{|\\H}}{\overset{\substack{R\\|}}{C}}-\underset{\substack{|\\R}}{\overset{\substack{R\\|}}{C}}-X \quad \text{(substitution)}$$

Problem 12.6 Supply the products of the following reactions. You need not balance the equations. If no reaction occurs, write "no reaction."

(a) 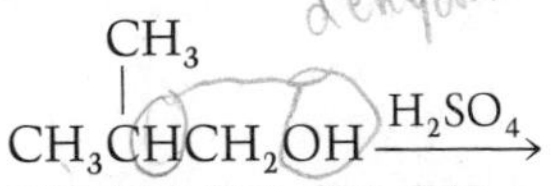 $CH_3CH(CH_3)CH_2OH \xrightarrow{H_2SO_4}$

(b) $CH_3CH_2CH_2CH_2OH + HBr \rightarrow$

(c) 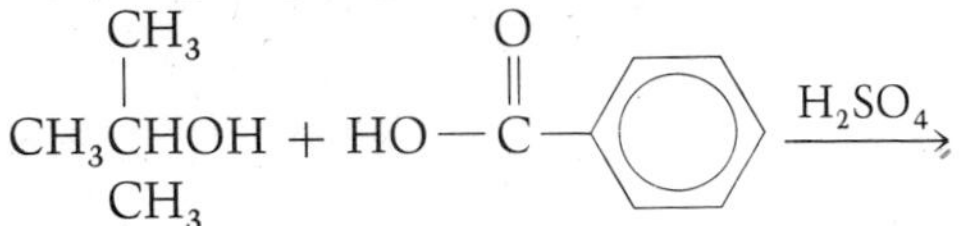 $CH_3CH(CH_3)OH + HO{-}C(=O){-}C_6H_5 \xrightarrow{H_2SO_4}$

(d) 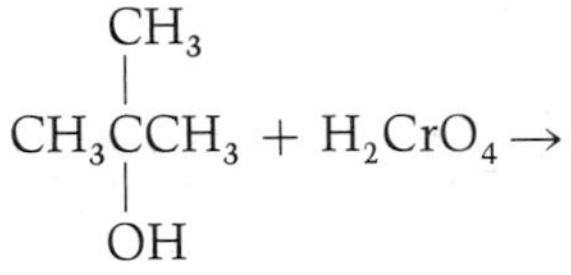 $CH_3C(CH_3)(OH)CH_3 + H_2CrO_4 \rightarrow$

(e) 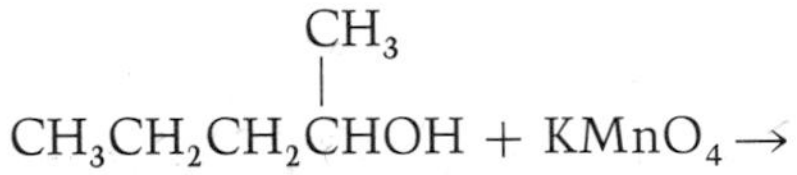 $CH_3CH_2CH_2CH(CH_3)OH + KMnO_4 \rightarrow$

(f) 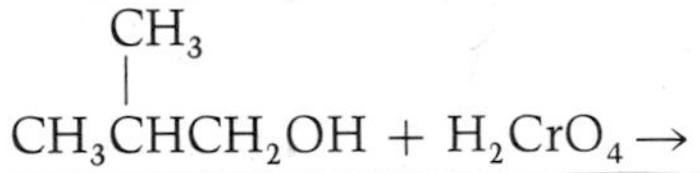 $CH_3CH(CH_3)CH_2OH + H_2CrO_4 \rightarrow$

PHENOLS 12.5

What is the difference between a phenol and an alcohol?

Phenols are compounds in which a hydroxyl group is attached directly to an aromatic ring.

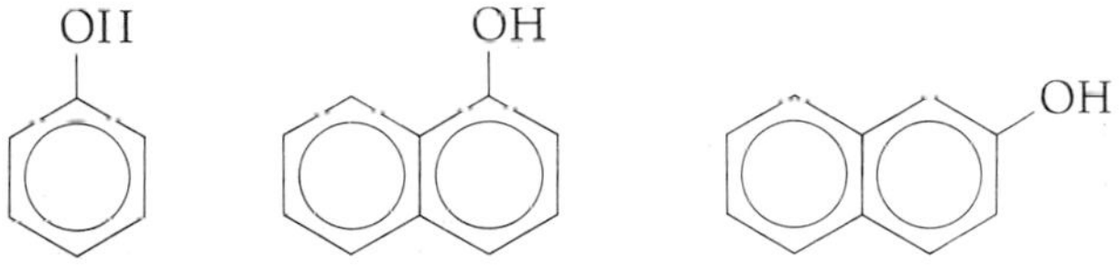

phenol 1-naphthol 2-naphthol

Phenols behave quite differently from alcohols, as we will see in the sections that follow.

PHENOLS: NOMENCLATURE 12.6

Phenols are named as derivatives of the parent compound (which itself is a derivative of benzene). An early name for phenol was *carbolic acid,* a name still used today in hospitals. We discussed the naming of phenols in Section 10.15; it would be well to review that section briefly before continuing. Some examples are

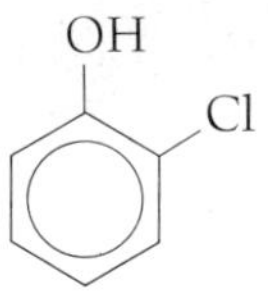

2-chlorophenol 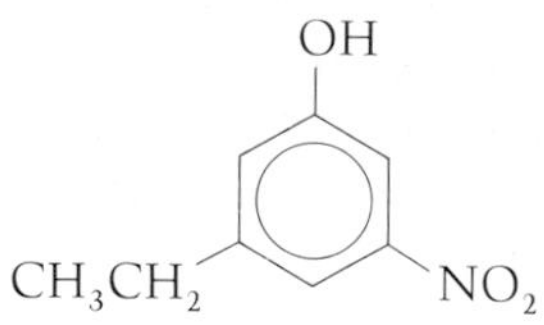 3-nitro-5-ethylphenol

Certain phenols have special names, for example, the methylphenols are called cresols.

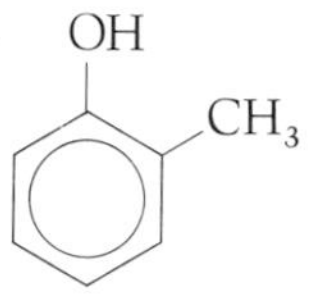

o-cresol 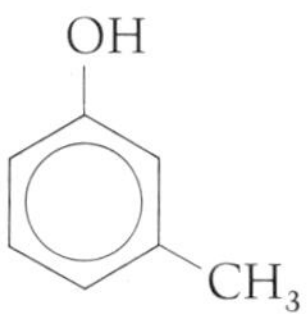 *m*-cresol 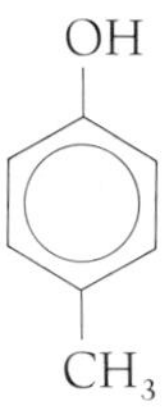 *p*-cresol

Problem 12.7 Give an acceptable name for each of the following compounds.

(a) 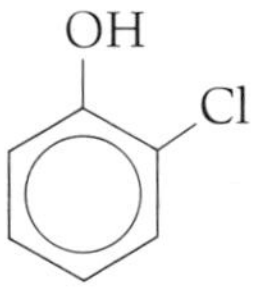 (b) 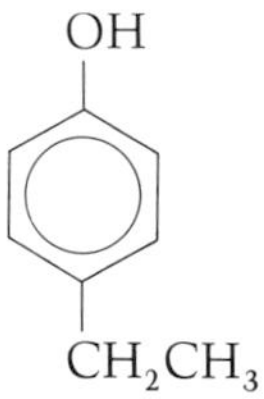 (c)

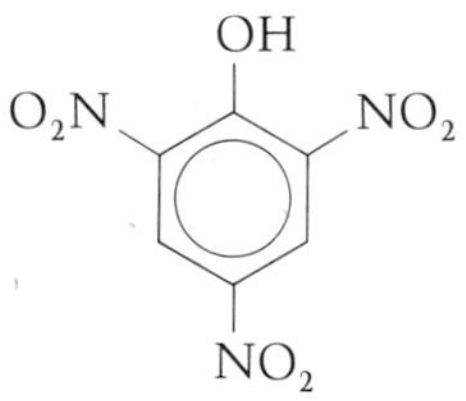

(d)

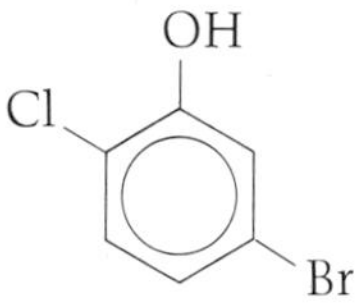

12.7 PHENOLS: PROPERTIES

Phenols are weakly acidic solids. Phenol itself is irritating to the skin and toxic if ingested. It is used in solutions and ointments as a general disinfectant.

PHENOLS: REACTIONS 12.8

What are the two reactive sites of phenols?

The most striking chemical property of phenols is their acidity. In contrast to the alcohols, which are neutral in water solution, phenols are weak acids (Section 7.4). This means that whereas alcohols do not give up H^+ ions to water, phenols do, to a slight extent, to produce the **phenoxide ion.**

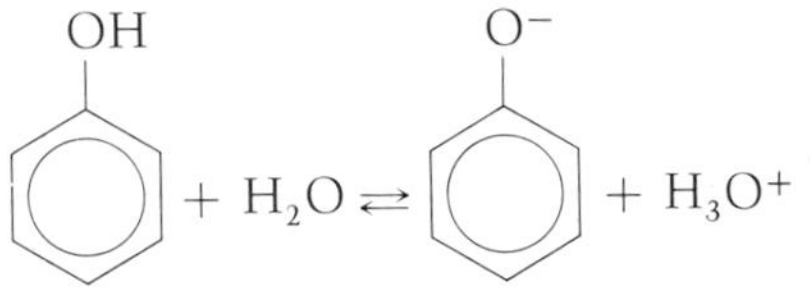

$$C_6H_5OH + H_2O \rightleftharpoons C_6H_5O^- + H_3O^+$$

phenoxide ion

The phenols are acidic because their conjugate bases are stabilized by resonance. This is illustrated by the resonance structures below (Section 10.14). Recall that, when a molecule or ion can be represented by several resonance structures, that molecule or ion is more stable than any of the individual structures would suggest. Thus a phenol can give off an H^+ ion without becoming too unstable.

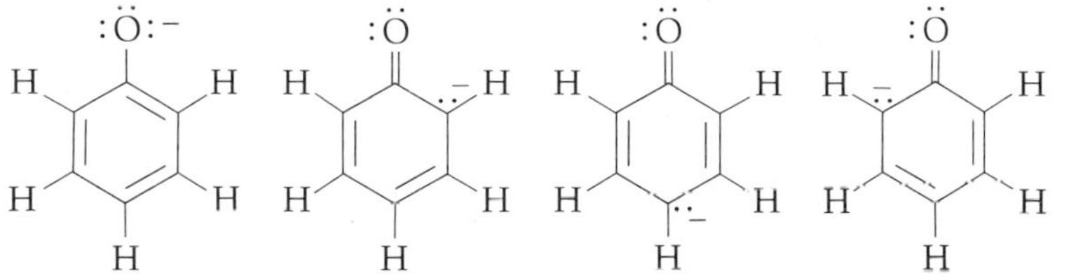

Alcohols are weaker acids because alkoxide ions—the ions formed when an alcohol gives off a proton—are not stabilized by resonance. That is, we can draw only one structure for the alkoxide ion.

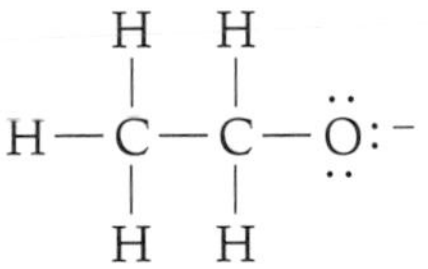

$$CH_3CH_2O^-$$

Consequently, the negative charge of an alkoxide ion is localized on the oxygen, whereas in a phenoxide ion the negative charge is delocalized (spread out) over the ring. The localization of negative charge solely on the oxygen atom makes the alkoxide ion less stable than the phenoxide ion.

Phenols react completely with alkali metal hydroxides to form water and phenoxide ions.

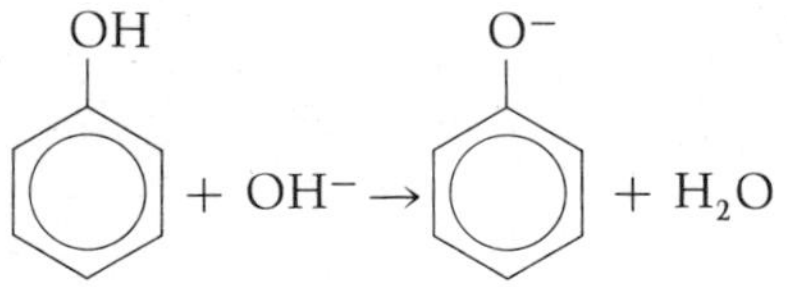

$$C_6H_5OH + OH^- \rightarrow C_6H_5O^- + H_2O$$

Phenols also react with alkali metals to yield phenoxide ions and hydrogen.

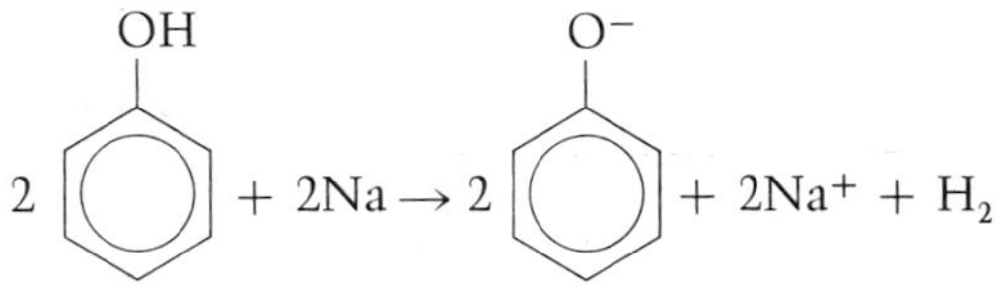

$$2\,C_6H_5OH + 2Na \rightarrow 2\,C_6H_5O^- + 2Na^+ + H_2$$

Ring Substitution in Phenols

Phenol undergoes substitution reactions typical of the benzene ring (Section 10.16). The presence of the OH group supplies electron density to the ring. The higher density of electrons renders the phenol ring much more reactive than benzene, which has a much lower electron density because it does not have resonance structures corresponding to III, IV, and V.

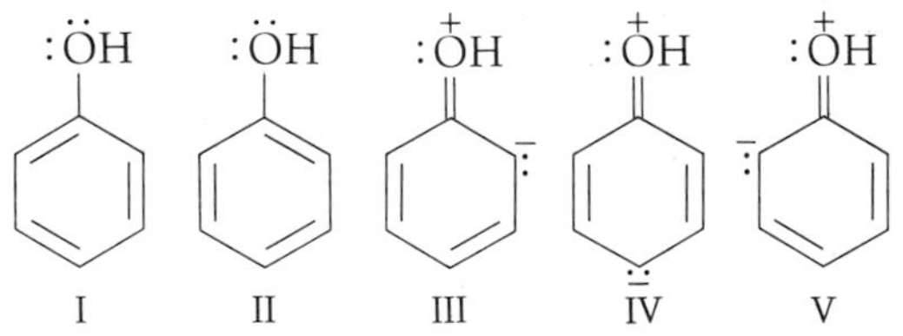

I II III IV V

For example, a phenol reacts more readily with bromine than benzene does. Bromination of benzene at room temperature with a catalyst is a slow reaction that gives bromobenzene. At room temperature and with no catalyst, phenol reacts very rapidly to give 2,4,6-tribromophenol. It is, in fact, difficult to obtain the monobrominated phenol because the reaction is so vigorous.

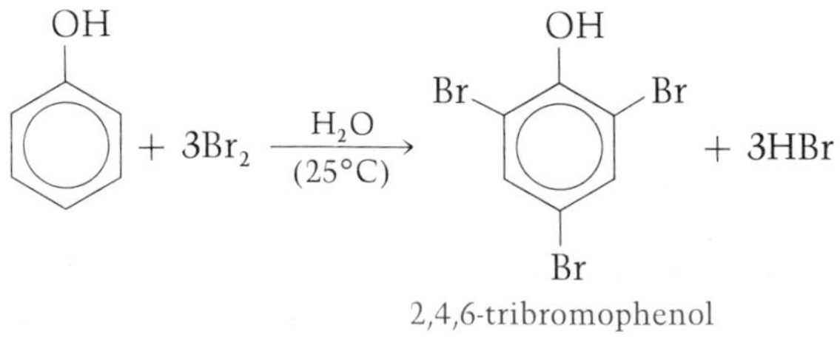

$$C_6H_5OH + 3Br_2 \xrightarrow[(25°C)]{H_2O} C_6H_2Br_3OH + 3HBr$$

2,4,6-tribromophenol

Nitration and sulfonation also occur more readily with phenol than with benzene. When an $—NO_2$ or $—SO_3H$ group is substituted in the phenol ring, however, the ring is deactivated and further reaction is much slower. For this reason, nitration and sulfonation are easier to stop at the monosubstitution stage than bromination is. (The percentages in the equations below show the percentage of each product formed.)

Under mild conditions (0°C):

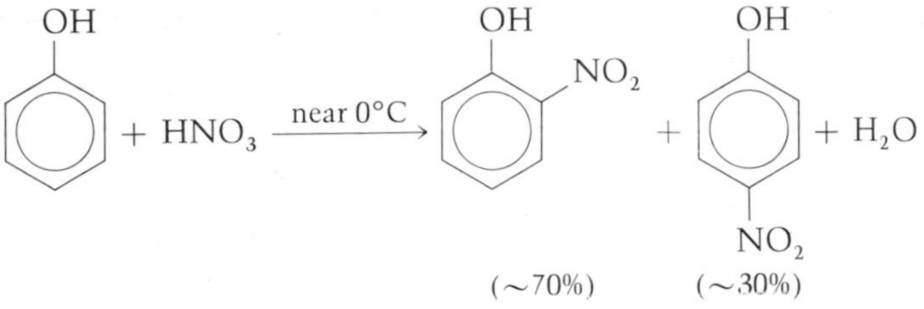

$$C_6H_5OH + HNO_3 \xrightarrow{\text{near } 0°C} o\text{-}NO_2C_6H_4OH + p\text{-}NO_2C_6H_4OH + H_2O$$

(~70%) (~30%)

Under more vigorous conditions:

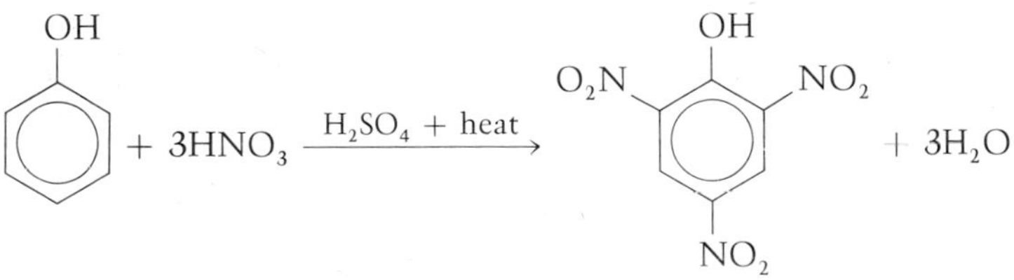

$$C_6H_5OH + 3HNO_3 \xrightarrow{H_2SO_4 + \text{heat}} 2,4,6\text{-}(NO_2)_3C_6H_2OH + 3H_2O$$

2,4,6-trinitrophenol (picric acid)

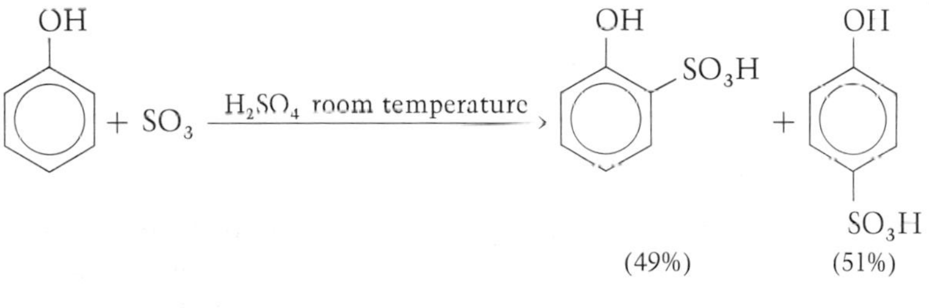

$$C_6H_5OH + SO_3 \xrightarrow{H_2SO_4 \text{ room temperature}} o\text{-}SO_3HC_6H_4OH + p\text{-}SO_3HC_6H_4OH$$

(49%) (51%)

Problem 12.8 Name the products of the mononitration reaction above.

Problem 12.9 Predict the products of the reaction of phenol with each of the following reagents.

(a) Na
(b) NaOH (aqueous)
(c) H_2O
(d) Br_2 at room temperature
(e) HNO_3 at 0°C
(f) HNO_3/H_2SO_4 plus heat
(g) SO_3/H_2SO_4 at room temperature

It is interesting to note that the $—OH$ group always directs entering groups to the *ortho* and *para* positions; little or no *meta* isomer is formed (see Section 10.15). We call a group that has this effect an *ortho–para director.* This means that when such a group is attached to an aromatic ring, if a second group enters the ring, it will become bonded to a position ortho or para to the one that is already there.

12.9 IMPORTANT ALCOHOLS AND PHENOLS

What physical property do all these alcohols share?

Methanol finds its greatest use as a solvent. Since at one time methanol was made exclusively by the destructive distillation of wood, it is often called "wood alcohol." Most of the methanol used today is manufactured by the catalytic hydrogenation of carbon monoxide.

$$CO + 2H_2 \xrightarrow[\text{ZnO or } CrO_3 \text{ catalyst}]{300\text{–}400°C;\ 200\text{–}300 \text{ atm};} CH_3OH$$

Methanol is highly toxic. Drinking 30 to 100 mL of it may cause blindness or even death.

Ethanol ("grain alcohol"), the alcohol found in alcoholic beverages, is manufactured by fermentation of sugar solutions, such as fruit juices. Fermentation is usually carried out by adding yeast to the sugar solution. The enzymes in yeast catalyze a series of reactions that convert sugar molecules to ethanol.

$$\underset{\text{glucose}}{C_6H_{12}O_6} \xrightarrow{\text{yeast}} 2CH_3CH_2OH + 2CO_2$$

Fermentation creates solutions in which the alcohol content is usually less than 15%. Brandy, whiskey, and other strong beverages are made by distillation of the dilute solutions produced by fermentation. The alcoholic content of distilled beverages is usually given as the **proof.** The proof is twice the volume percentage of alcohol. Thus, an 86-proof whiskey is 43% ethanol by volume.

Most commercial ethanol is manufactured by the reaction of ethylene with water in the presence of an acid as catalyst.

$$CH_2{=}CH_2 + H_2O \xrightarrow{H^+} CH_3CH_2OH$$

Besides its use in alcoholic beverages, ethanol is widely used as a solvent and an antiseptic. It is toxic, and, when ingested, it depresses the central nervous system. It can cause intoxication, nausea, vomiting, flushing, mental excitement or depression, drowsiness, impaired perception and coordination, and even death.

Isopropyl alcohol is used primarily as a solvent and in antifreeze products. Medically, it is used as a rubbing alcohol. It is more toxic than either methanol or ethanol.

Isopropyl alcohol is prepared by the reaction of propene with water.

$$\underset{\text{propene}}{CH_3CH{=}CH_2} + H_2O \xrightarrow{H^+} \underset{\text{2-propanol}}{CH_3\overset{\overset{CH_3}{|}}{C}HOH}$$

Ethylene glycol has two hydroxyl groups.

$$HO-CH_2-CH_2-OH$$

Ethylene glycol is manufactured by allowing ethylene to react with atmospheric oxygen to form the cyclic ether, ethylene oxide; the ethylene oxide then reacts with water to form ethylene glycol.

$$2\ CH_2{=}CH_2 + O_2 \xrightarrow{\text{catalyst}} 2\ \underset{\text{ethylene oxide}}{CH_2-CH_2\ (\text{bridged by } O)}$$

$$CH_2-CH_2\ (\text{bridged by } O) + H_2O \rightarrow \underset{\text{ethylene glycol}}{\underset{|\qquad\ |}{CH_2-CH_2}\atop OH\quad OH}$$

Because of its two hydroxyl groups, ethylene glycol is very soluble in water. Its most important use is as antifreeze for automobiles. It is also widely used as a solvent and as a humectant (a substance used in small quantities that aids in retaining moisture, for example, in cakes and tobacco). If it is ingested, ethylene glycol attacks the central nervous system, damages the kidneys, and can result in death.

Glycerol, also called "glycerin," is a trihydroxy alcohol. It is a by-product in the manufacture of soaps from fat. **Fats** consist of esters of glycerol and various carboxylic acids (called "fatty acids"). When a fat is heated with strong alkali hydroxide, the fat hydrolyzes to produce glycerol and the salt of the fatty acid, which is a soap.

$$\underset{\text{a fat}}{\begin{matrix} CH_2-O-\overset{\overset{O}{\|}}{C}-R \\ | \\ CH-O-\overset{\overset{O}{\|}}{C}-R \\ | \\ CH_2-O-\overset{\overset{O}{\|}}{C}-R \end{matrix}} + 3NaOH \longrightarrow \underset{|\qquad\quad |\qquad\quad |}{CH_2-CH-CH_2}\atop{OH\qquad OH\qquad OH} + \underset{\text{a soap}}{3R\overset{\overset{O}{\|}}{C}-O^-Na^+}$$

Glycerol is a nontoxic, sweet-tasting, viscous liquid. It is used as a solvent, humectant, sweetener, as a liquid medium for medicines, and in the manufacture of glyceryl trinitrate (Section 11.7). It is also used in skin creams and lotions. Because of its ability to absorb moisture from the air, it helps to keep the skin soft and moist.

What use do phenols share?

Hexylresorcinol is an important commercial phenol that is used as an antiseptic. It is a strong skin irritant.

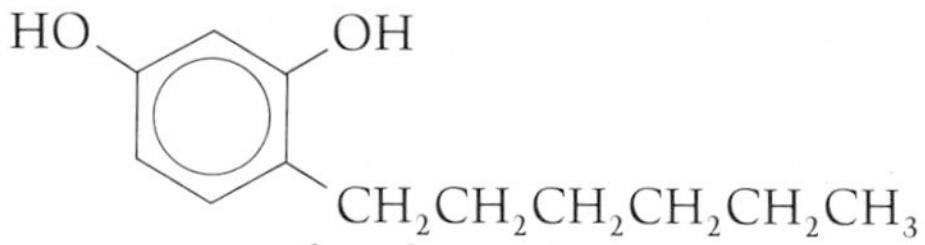

HO, OH, $CH_2CH_2CH_2CH_2CH_2CH_3$

hexyl resorcinol
(4-hexyl-1,3-dihydroxybenzene)

Hexachlorophene was used chiefly in germicidal soaps, deodorants, and toothpastes. But it proved to be so toxic to humans that its sale has been rigidly controlled.

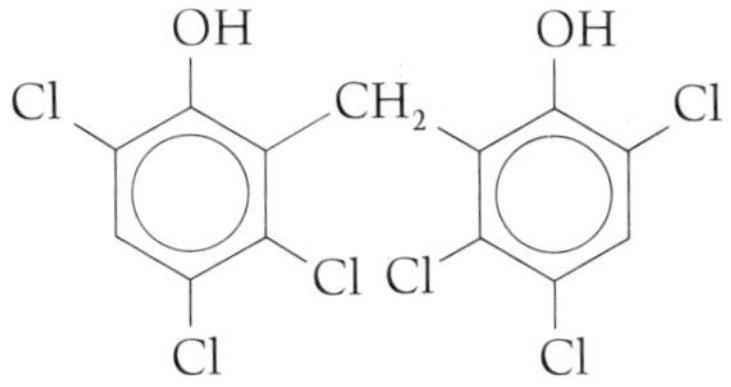

OH, OH, Cl, CH_2, Cl, Cl, Cl, Cl, Cl

hexachlorophene

12.10 ETHERS

We may think of an **ether** as a molecule in which two alkyl or aryl groups or one of each are bonded to an oxygen atom.

$$R{-}O{-}R \qquad R{-}O{-}Ar \qquad Ar{-}O{-}Ar \qquad (R = \text{alkyl};\ Ar = \text{aryl})$$

12.11 ETHERS: NOMENCLATURE

How does the IUPAC naming of ethers differ from that of alcohols?

We form the common name of an ether by giving the names of the two hydrocarbon groups, followed by the word "ether." Some examples are

$$CH_3{-}O{-}CH_2CH_3 \qquad CH_3CH_2{-}O{-}CH_2CH_3$$

methyl ethyl ether — diethyl ether

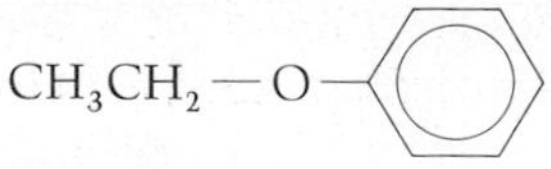

$CH_3CH_2-O-C_6H_5$ ethyl phenyl ether

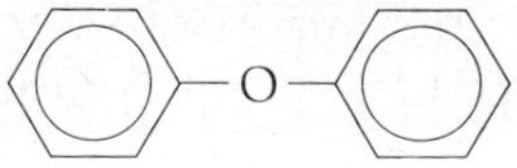

$C_6H_5-O-C_6H_5$ diphenyl ether

Under the IUPAC system of nomenclature, the $-OR'$ group of ROR' ethers is treated as a substituent on the hydrocarbon $R-H$, where R is larger than R′. In this system, we name the $-OR'$ group by replacing the ending *-ane* of the alkane name of R′H by the ending *-oxy*. The general name of $R'O-$ is **alkoxy.** Some simple $-OR'$ groups are $-OCH_3$ (methoxy), $-OCH_2CH_3$ (ethoxy), $-OCH_2CH_2CH_3$ (propoxy), and $-O-C_6H_5$ (phenoxy). The IUPAC names for the above ethers are as follows (alkoxy groups are shown in color).

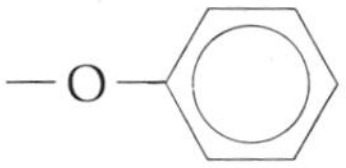

$CH_3-O-CH_2CH_3$ methoxyethane

$CH_3CH_2-O-CH_2CH_3$ ethoxyethane

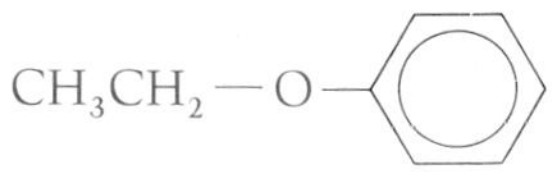

$CH_3CH_2-O-C_6H_5$ ethoxybenzene

$C_6H_5-O-C_6H_5$ phenoxybenzene

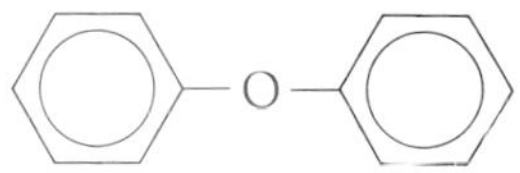

If a molecule has two ether groups, we follow the above system for each one individually.

$CH_3-O-\overset{1}{C}H_2\overset{2}{C}H_2-O-CH_3$ 1,2-dimethoxyethane

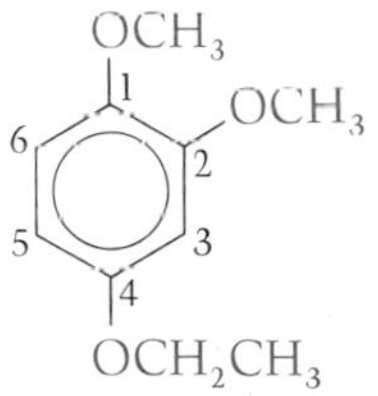

Benzene ring numbered 1–6 with OCH_3 at C-1, OCH_3 at C-2, and OCH_2CH_3 at C-4: 4-ethoxy-1,2-dimethoxybenzene

Problem 12.10 Give the common and the IUPAC names of the following compounds.

(a) $CH_3CH_2CH_2-O-CH_2CH_3$

(b) 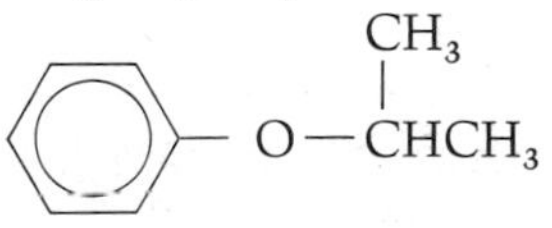 $C_6H_5-O-CH(CH_3)CH_3$

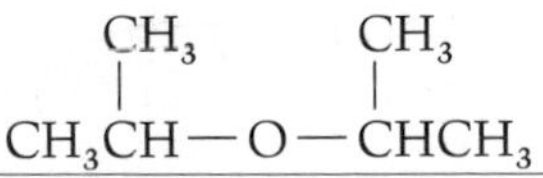

(c) $CH_3CH(CH_3)-O-CH(CH_3)CH_3$

12.12 ETHERS: PROPERTIES

How do ethers and alcohols differ in their hydrogen bonding?

Because ethers have no O—H groups, their molecules cannot hydrogen-bond to each other. For this reason, ethers have low boiling points. Their boiling points resemble those of hydrocarbons of similar molecular weight (which also cannot form hydrogen bonds with each other). For example, diethyl ether, whose molecular weight is 74, boils at 35°C; pentane, whose molecular weight is 72, boils at 36°C.

Although ether molecules cannot form hydrogen bonds with one another, they can form hydrogen bonds with water molecules, because the water molecules supply the necessary O—H groups.

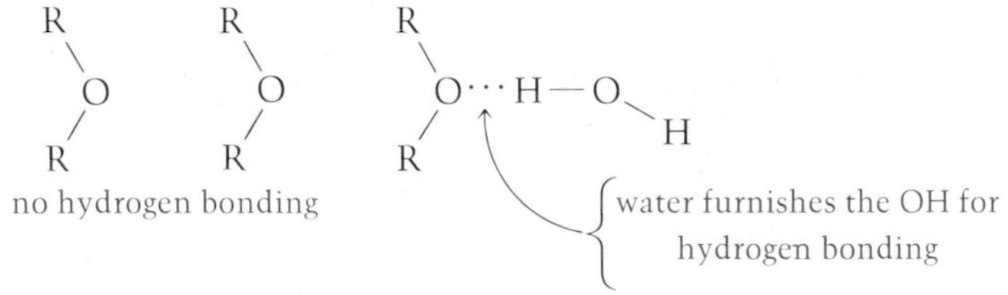

R₂O   R₂O   (no hydrogen bonding)

$R_2O \cdots H-O-H$ (water furnishes the OH for hydrogen bonding)

As a result of this ether–water hydrogen bonding, ethers and alcohols of the same molecular weight have similar solubilities in water. For example, diethyl ether ($C_4H_{10}O$), has a solubility of 7.5 g/100 mL of water; butanol ($C_4H_{10}O$) has a solubility of 7.9 g/100 mL of water. These properties are summarized in Table 12.2.

TABLE 12.2 Comparison of Properties of Hydrocarbons, Ethers, and Alcohols of Similar Molecular Weights

Name	Formula	Molecular weight	Boiling point, °C	Solubility, g/100 mL water (at 25° C)
Propane	$CH_3CH_2CH_3$	44	−42	Insoluble
Dimethyl ether	CH_3OCH_3	46	−24	3,700 mL per 100 mL H_2O
Ethanol	CH_3CH_2OH	46	78	∞
Pentane	$CH_3(CH_2)_3CH_3$	72	36	0.036
Diethyl ether	$CH_3CH_2OCH_2CH_3$	74	35	7.5
Methyl propyl ether	$CH_3OCH_2CH_2CH_3$	74	39	3.1
Butanol	$CH_3CH_2CH_2CH_2OH$	74	118	7.9

Problem 12.11 Predict which compound in each of the following pairs is more soluble in water and which has the higher boiling point.
(a) heptane and dipropyl ether (b) heptane and 1-hexanol
(c) 1-hexanol and dipropyl ether

ETHERS: REACTIONS 12.13

What reaction occurs with all ethers?

Ethers are fairly unreactive compounds. They generally react only with acids. Ethers react as bases with strong acids, just as water and alcohols do. For example, ethers dissolve in cold, concentrated sulfuric acid. They do so because they react to form ions that are soluble in H_2SO_4.

$$R-\ddot{\underset{..}{O}}-R + H_2SO_4 \rightleftharpoons \underbrace{R-\overset{+}{\underset{|\atop H}{\ddot{O}}}-R + HSO_4^-}_{\text{soluble}}$$

More specifically, here is how diethyl ether reacts with sulfuric acid:

$$CH_3CH_2-\ddot{\underset{..}{O}}-CH_2CH_3 + H_2SO_4 \rightarrow \underbrace{CH_3CH_2-\overset{H\atop |}{\underset{+}{\underset{..}{O}}}-CH_2CH_3 + HSO_4^-}_{\text{soluble}}$$

The reaction is reversible, and the ether can be recovered by adding water to the mixture.

We can cleave ethers by heating them with concentrated inorganic acids such as HCl, HBr, or HI.

$$R-O-R' + 2HCl \rightarrow RCl + R'Cl + H_2O$$

Dipropyl ether, for example, gives propyl chloride.

$$CH_3CH_2CH_2-O-CH_2CH_2CH_3 + 2HCl \rightarrow 2CH_3CH_2CH_2Cl + H_2O$$

In this process, both C—O bonds are broken, and an alkyl halide results. Cleavage of aromatic ethers does not yield aryl halides:

$$C_6H_5-OCH_2CH_3 + HBr \rightarrow C_6H_5-OH + BrCH_2CH_3$$

Ethers (except epoxides, Section 12.14) do not react with bases.

Aromatic ethers undergo ring-substitution reactions similar to those of phenols, to give mixtures of *ortho* and *para* isomers.

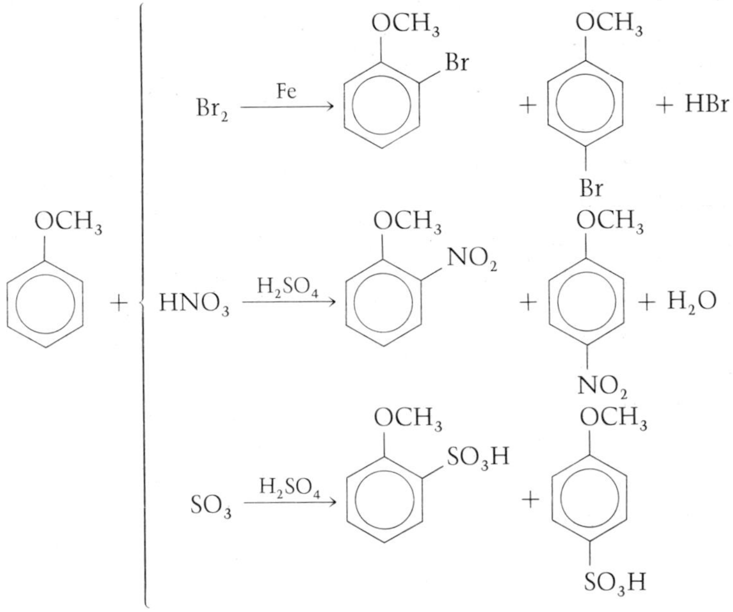

$$C_6H_5OCH_3 + Br_2 \xrightarrow{Fe} o\text{-}BrC_6H_4OCH_3 + p\text{-}BrC_6H_4OCH_3 + HBr$$

$$C_6H_5OCH_3 + HNO_3 \xrightarrow{H_2SO_4} o\text{-}O_2NC_6H_4OCH_3 + p\text{-}O_2NC_6H_4OCH_3 + H_2O$$

$$C_6H_5OCH_3 + SO_3 \xrightarrow{H_2SO_4} o\text{-}HO_3SC_6H_4OCH_3 + p\text{-}HO_3SC_6H_4OCH_3$$

Aromatic ethers do not undergo substitution reactions as readily as phenols do. It is easy to stop bromination of aromatic ethers at the monobromination stage, when only one bromine has been substituted.

Problem 12.12 Complete the following equations. You need not balance them.

(a) $CH_3CH_2OCH_3 + HI \xrightarrow{\text{heat}}$

(b) $CH_3CH_2O\text{—}C_6H_5 + Cl_2 \xrightarrow{Fe}$

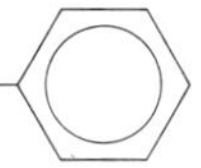

12.14 CYCLIC ETHERS

In **cyclic ethers,** the two carbon atoms to which the oxygen is bonded are in the same hydrocarbon chain. Cyclic ethers in which the oxygen is part of a three-membered ring are called **epoxides.**

How are epoxides different from other ethers?

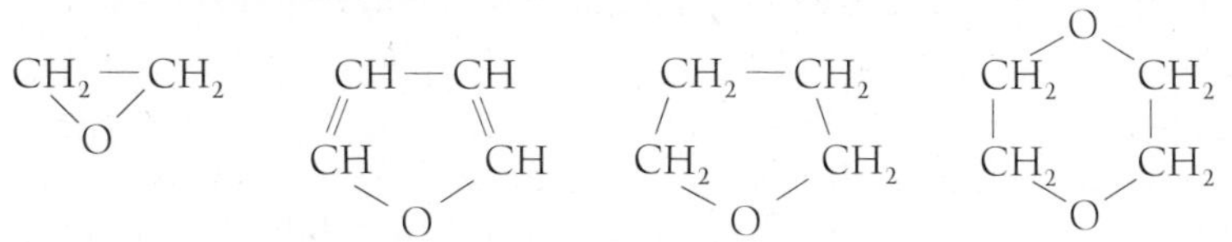

$$\underset{\text{ethylene oxide}}{\overset{CH_2-CH_2}{\underset{O}{\diagdown\ \diagup}}} \qquad \underset{\text{furan}}{\begin{matrix} CH=CH \\ CH \quad CH \\ O \end{matrix}} \qquad \underset{\text{tetrahydrofuran}}{\begin{matrix} CH_2-CH_2 \\ CH_2 \quad CH_2 \\ O \end{matrix}} \qquad \underset{\text{dioxane}}{\begin{matrix} O \\ CH_2 \quad CH_2 \\ CH_2 \quad CH_2 \\ O \end{matrix}}$$

Except for epoxides, cyclic ethers behave chemically exactly like the noncyclic (or acyclic) ethers. That is, they react with acids, and most of them do not react with bases.

Because they have a highly strained three-membered ring (Section 9.9), epoxides are more reactive than other ethers. Most reactions of epoxides involve opening the ring. In contrast to other ethers, epoxides are reactive toward bases. The following are some examples of the reactions of an epoxide, ethylene oxide.

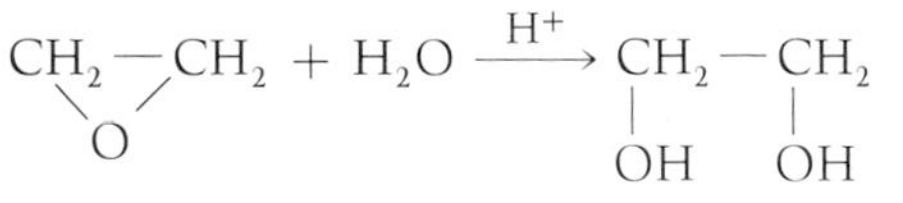

$$\underset{O}{CH_2-CH_2} + H_2O \xrightarrow{H^+} \underset{\text{ethylene glycol}}{\underset{OH \quad\ OH}{CH_2-CH_2}}$$

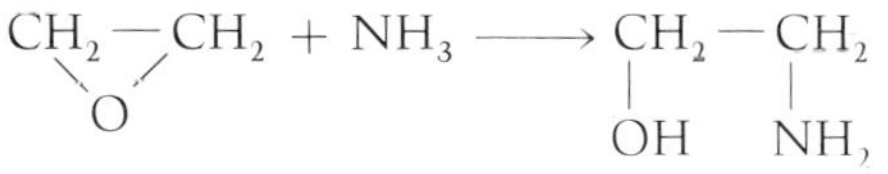

$$\underset{O}{CH_2-CH_2} + NH_3 \longrightarrow \underset{\text{ethanolamine}}{\underset{OH \quad\ NH_2}{CH_2-CH_2}}$$

Problem 12.13 Supply the products of the following reactions.

(a) 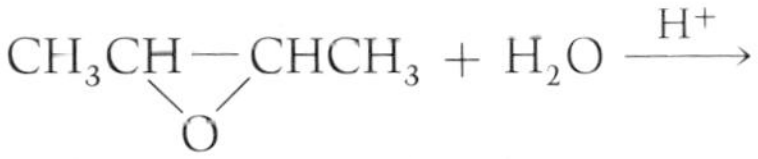 $\underset{O}{CH_3CH-CHCH_3} + H_2O \xrightarrow{H^+}$

(b) 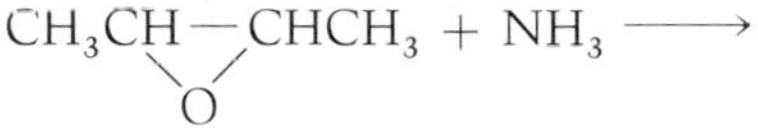 $\underset{O}{CH_3CH-CHCH_3} + NH_3 \longrightarrow$

IMPORTANT ETHERS 12.15

Ethers are important commercial products, used primarily as solvents. Many ethers occur naturally in plants.

Diethyl ether (also called ethyl ether, or simply ether) is produced on a large scale by the dehydration of ethanol in the presence of an acidic catalyst.

$$CH_3CH_2{-}O{-}H + H{-}O{-}CH_2CH_3 \xrightarrow{H^+} CH_3CH_2{-}O{-}CH_2CH_3 + H_2O$$

Diethyl ether was once widely used as a general anesthetic, because it is safe even in long surgical procedures; its only side effects are irritation of the mucous membranes, plus nausea and vomiting after the operation. But because of these side effects and the additional danger of explosion, diethyl ether is no longer used as an anesthetic. It is still a useful solvent for waxes and fats. One big advantage it has as a solvent is its low boiling point (35°C)—this allows it to be easily removed by distillation. However, because ether vaporizes so easily and is very flammable, it poses a serious fire hazard.

Ethylene oxide is a water soluble, colorless gas that liquefies at 12°C. It is manufactured by the reaction of ethylene with atmospheric oxygen in the presence of a catalyst.

$$2CH_2{=}CH_2 + O_2 \xrightarrow{Ag} 2\underset{\diagdown\ O\ \diagup}{CH_2{-}CH_2}$$

Ethylene oxide is used to fumigate textiles and grains. It also finds wide use as a raw material in the synthesis of other organic compounds.

Dioxane is a colorless liquid that boils at 101°C. It is soluble in water and organic solvents, and it is very useful as a solvent for cellulose acetate and other cellulose derivatives. Dioxane is prepared by the acid-catalyzed dehydration of ethylene glycol.

$$\begin{matrix} CH_2{-}O{-}H \\ | \\ CH_2{-}O{-}H \end{matrix} + \begin{matrix} H{-}O{-}CH_2 \\ | \\ H{-}O{-}CH_2 \end{matrix} \xrightarrow{H^+,\ heat} \begin{matrix} CH_2{-}O{-}CH_2 \\ | \quad\quad | \\ CH_2{-}O{-}CH_2 \end{matrix} + 2H_2O$$

Dioxane can cause depression of the central nervous system and kidney and liver damage.

Tetrahydrofuran is a valuable solvent for polymers, especially polyvinyl chloride. It is manufactured by the reaction of furan with hydrogen.

$$\underset{\text{furan}}{\begin{matrix} CH{-}CH \\ /\!/ \quad \backslash\!\backslash \\ CH \quad\ CH \\ \diagdown\ O\ \diagup \end{matrix}} + 2H_2 \xrightarrow{catalyst} \underset{\text{tetrahydrofuran}}{\begin{matrix} CH_2{-}CH_2 \\ / \quad\quad \backslash \\ CH_2 \quad\ CH_2 \\ \diagdown\ O\ \diagup \end{matrix}}$$

Furan itself is obtained from plant products like pine-wood rosin, oat hulls, corncobs, and other cereal products.

Tetrahydrofuran is used as a solvent in histological techniques. It irritates the skin, eyes, and mucous membranes, and it can cause kidney and liver damage.

Many biologically important molecules contain the ether functional group. Two examples are tetrahydrocannabinol (the active ingredient of marijuana) and morphine (the major component of opium). The heavy line in the morphine structure represents bonds that protrude from the page.

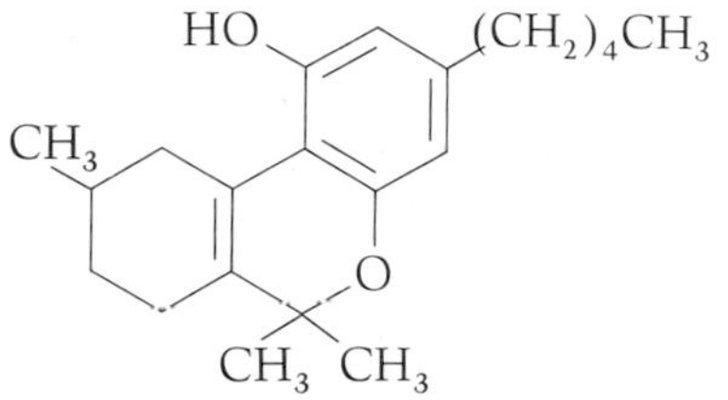

tetrahydrocannabinol

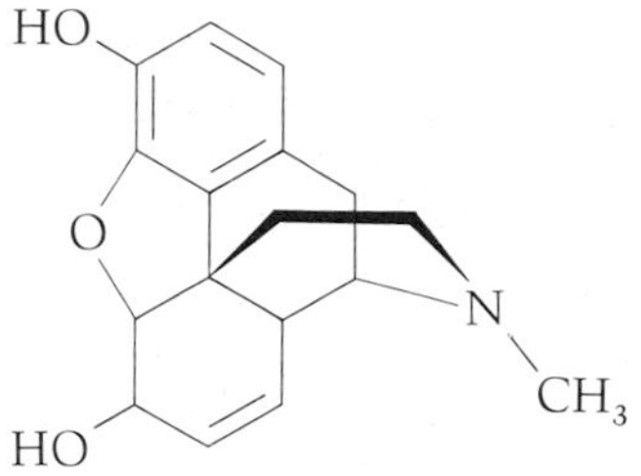

morphine

Problem 12.14 Identify all the functional groups in tetrahydrocannabinol and morphine (see Table 8.3).

SULFUR ANALOGUES OF ALCOHOLS, PHENOLS, AND ETHERS 12.16

How are the elements sulfur and oxygen similar?

Because sulfur lies just beneath oxygen in the periodic table, we might expect these two elements to be similar. In fact, many oxygen compounds have their sulfur counterparts, in which oxygen has simply been replaced by sulfur. However, there are interesting differences between oxygen and sulfur compounds.

SULFUR ANALOGUES: NOMENCLATURE 12.17

The sulfur analogues of alcohols are called **thiols;** they are also commonly called **mercaptans. Thioethers,** called **sulfides,** and **thiophenols** are the sulfur counterparts of ethers and phenols.

CH_3CH_2SH — ethanethiol, ethyl mercaptan (a thiol)

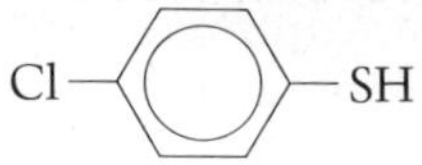

$CH_3SCH_2CH_3$ — methyl ethyl sulfide (a thioether)

Cl—C_6H_4—SH — *p*-chlorothiophenol (a thiophenol)

The —SH group is called a **sulfhydryl** group. We name thiols in the IUPAC system by simply adding the ending *-thiol* to the name of the alkane to which the sulfhydryl group is attached. We name phenols by replacing *phenol* in the parent name with *thiophenol.* We name thioethers by naming the two hydrocarbon groups, as with ethers, except that we use the word **sulfide** in place of *ether.* Some examples follow.

CH_3SH — methanethiol

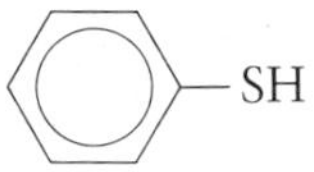

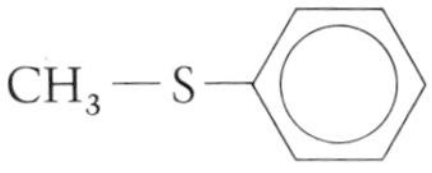

C_6H_5—SH — thiophenol

CH_3—S—C_6H_5 — methylphenyl sulfide

$\underset{3}{CH_3}\underset{2}{CH(CH_3)}\underset{1}{CH_2}SH$ — 2-methyl-1-propanethiol

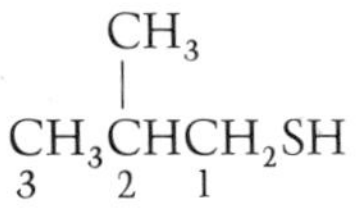

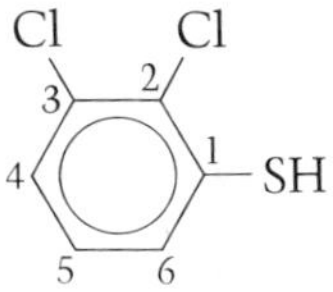

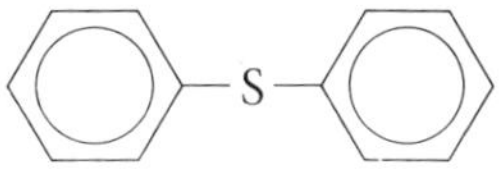

$C_6H_3Cl_2$—SH (Cl at positions 2 and 3; SH at 1) — 2,3-dichlorothiophenol

C_6H_5—S—C_6H_5 — diphenyl sulfide

Problem 12.15 Give structural formulas for the following compounds.
(a) dipropyl sulfide
(b) 2,4-dinitrothiophenol
(c) cyclohexanethiol

Problem 12.16 Give the IUPAC and, where possible, the common name of each of the following compounds.

(a) CH_3CH_2SH

(b) 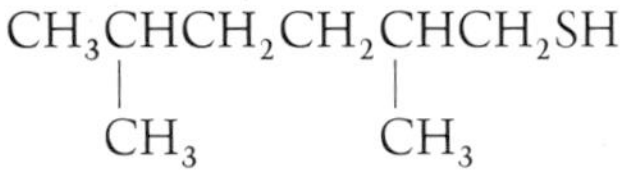 $CH_3CH(CH_3)CH_2CH_2CH(CH_3)CH_2SH$

(c) 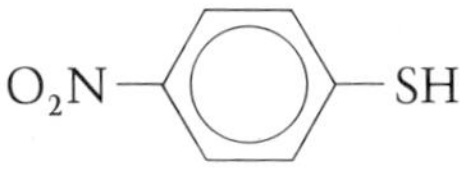 O_2N—C_6H_4—SH

(d) 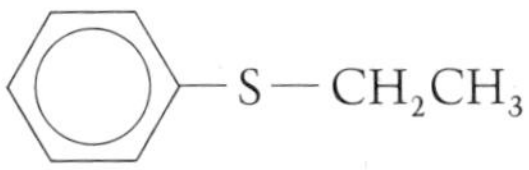 C_6H_5—S—CH_2CH_3

(e) 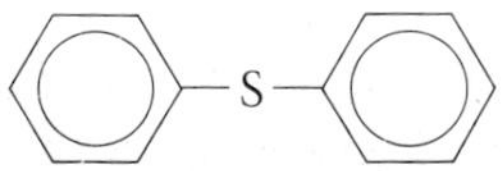 C_6H_5—S—C_6H_5

SULFUR ANALOGUES: PROPERTIES 12.18

How do oxygen compounds differ from their sulfur analogues?

Sulfur does not form strong hydrogen bonds. Because of this, sulfur compounds have different physical properties from their oxygen counterparts. Thiols and thiophenols, for example, have lower melting and boiling points than similar alcohols and phenols, respectively (Table 12.3). Sulfides boil at higher temperatures than ethers because they have higher molecular weights. Recall that the atomic weights of oxygen and sulfur are 16 and 32, respectively.

Problem 12.17 On the basis of molecular weight alone, what ether would you expect to have the same boiling point as methyl ethyl sulfide? Methyl ethyl sulfide has a boiling point of 66°C.

TABLE 12.3 Boiling Points of Sulfur Analogues of Oxygen Compounds

Oxygen compound	Boiling point, °C	Sulfur analogue	Boiling point, °C
Alcohols		*Thiols*	
CH_3OH	65	CH_3SH	7
CH_3CH_2OH	78	CH_3CH_2SH	37
$CH_3CH_2CH_2CH_2OH$	118	$CH_3CH_2CH_2CH_2SH$	98
Ethers		*Thioethers (Sulfides)*	
CH_3-O-CH_3	−24	CH_3-S-CH_3	38
$CH_3CH_2-O-CH_2CH_3$	35	$CH_3CH_2-S-CH_2CH_3$	92
Phenols		*Thiophenols*	
C_6H_5-OH	182	C_6H_5-SH	170
$CH_3-C_6H_4-OH$	203	$CH_3-C_6H_4-SH$	195

SULFUR COMPOUNDS: REACTIONS 12.19

Analogous oxygen and sulfur compounds react differently because of differences in the strengths of C—O and C—S bonds and O—H and

Name two reactions in which alcohols and thiols differ.

S—H bonds. Thiols are stronger acids than alcohols, because the S—H bond is weaker than the O—H bond. For this same reason, oxidation of thiols can lead to the formation of **disulfides** (R—S—S—R).

$$2CH_3CH_2SH + H_2O_2 \rightarrow CH_3CH_2S{-}SCH_2CH_3 + 2H_2O$$

ethanethiol — diethyl disulfide

Alcohols do not undergo this reaction, because the O—H bond is too strong. Oxidation of alcohols occurs instead at the C—H bond, to give a carbonyl group (Section 14.1).

A simple piece of evidence for the acidity of thiols is that NaOH converts a thiol to its sodium salt.

$$CH_3CH_2SH + OH^- \rightarrow CH_3CH_2S^- + H_2O$$

OH^-: stronger base; $CH_3CH_2S^-$: weaker base

Recall that alcohols are not acidic enough to react with NaOH (Section 12.4). Similarly, thiophenols are stronger acids than phenols.

Problem 12.18 Predict whether the following equilibrium reaction will proceed to the right to any great extent.

$$CH_3CH_2ONa + CH_3CH_2SH \rightleftharpoons CH_3CH_2OH + CH_3CH_2SNa$$

[Hint: Which is the stronger acid, CH_3CH_2SH or CH_3CH_2OH? The stronger acid and base will react more than the weaker acid and base.]

12.20 SULFUR COMPOUNDS IN LIVING SYSTEMS

Thiols and disulfides are important in living systems. For example, let's consider the oxidation–reduction reaction between lipoic acid and dihydrolipoic acid. This reaction can be made to proceed in either direction by the use of the proper enzyme. (An enzyme, as you may know, is a protein that catalyzes a specific reaction.)

$$\text{(cyclic S–S)}{-}(CH_2)_4{-}\overset{O}{\overset{\|}{C}}{-}OH \underset{\text{oxidation}}{\overset{\text{reduction}}{\rightleftharpoons}} HS{-}CH_2CH_2{-}CH(SH){-}(CH_2)_4{-}\overset{O}{\overset{\|}{C}}{-}OH$$

lipoic acid — dihydrolipoic acid

Lipoic acid is an important substance (which will be covered in Focus 14) in biological oxidations.

The amino acids cysteine and cystine undergo a similar interconversion. This reaction can also be made to go in either direction when the proper enzyme is present.

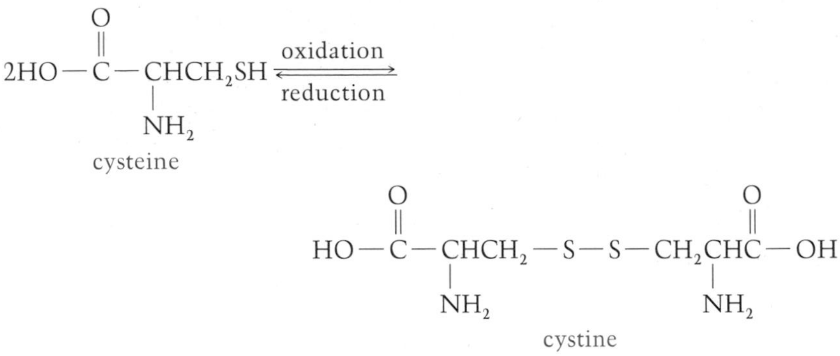

$$2HO-\overset{\overset{\displaystyle O}{\|}}{C}-\underset{\underset{\displaystyle NH_2}{|}}{C}HCH_2SH \underset{\text{reduction}}{\overset{\text{oxidation}}{\rightleftharpoons}}$$

cysteine

$$HO-\overset{\overset{\displaystyle O}{\|}}{C}-\underset{\underset{\displaystyle NH_2}{|}}{C}HCH_2-S-S-CH_2\underset{\underset{\displaystyle NH_2}{|}}{C}H\overset{\overset{\displaystyle O}{\|}}{C}-OH$$

cystine

The disulfide linkage is one of the means by which protein molecules acquire their individual shapes (Chapter 17).

Organic sulfur compounds are known for their offensive odors. Some familiar substances that contain sulfur compounds are rotten eggs, the fluid that skunks emit as a defense, and certain cheeses.

SUMMARY

Alcohols have a hydroxyl group ($-OH$) bonded to an aliphatic carbon atom. Alcohols are classified as **primary (1°), secondary (2°),** or **tertiary (3°).** Because of hydrogen bonding, alcohols have higher boiling points and are more soluble in water than the corresponding hydrocarbons. Alcohols are not appreciably acidic; they undergo **esterification** and they react with reactive metals. Primary alcohols can be oxidized to aldehydes; secondary alcohols can be oxidized to ketones. Tertiary alcohols are not readily oxidized. In strong acid solutions, alcohols undergo dehydration to form alkenes and substitution to form alkyl halides.

Phenols have a hydroxyl group bonded to an aromatic ring. They are solids with low melting points and are used as disinfectants. Phenols are acidic, compared with alcohols. They neutralize NaOH solutions and react with alkali metals. The acidity of phenols is explained by resonance. Phenols undergo substitution reactions typical of the benzene ring—nitration, sulfonation, and bromination—except that phenols are more reactive than benzene.

Ethers are compounds in which an oxygen atom is bonded to two hydrocarbon groups, either alkyl or aryl. Ether molecules cannot hydrogen-bond to each other, so they have lower boiling points than corresponding alcohols. They can hydrogen-bond to water molecules, so ethers have water solubilities comparable to alcohols of similar molecular weight. Ethers undergo cleavage reactions when heated with concentrated acids. Aromatic ethers undergo substitution reactions in the benzene ring. **Epoxides** are three-membered cyclic ethers. They are the only ethers that react with both bases and acids.

The sulfur analogues of oxygen compounds are called **thiols, thiophenols,** and **thioethers** (also called **sulfides**). Thiols and thiophenols are more acidic than their oxygen analogues. Thiols undergo oxidation to form **disulfides** (R—S—S—R).

Key Terms

Alcohol (Section 12.1)
Alkoxide (Section 12.4)
Alkoxy (Section 12.11)
Cyclic ether (Section 12.14)
Disulfide (Section 12.19)
Epoxide (Section 12.14)
Esterification (Section 12.4)
Ether (Section 12.10)
Fat (Section 12.9)
Hydroxyl group (Section 12.1)
Mercaptan (Section 12.17)
Phenol (Section 12.5)
Phenoxide ion (Section 12.8)
Primary (1°) alcohol (Section 12.1)
Proof (Section 12.9)
Secondary (2°) alcohol (Section 12.1)
Sulfhydryl group (Section 12.17)
Sulfide (Section 12.17)
Tertiary (3°) alcohol (Section 12.1)
Thioether (Section 12.17)
Thiol (Section 12.17)
Thiophenol (Section 12.17)

ADDITIONAL PROBLEMS

12.19 Draw the structural formulas of the following compounds.

(a) 3-methyl-3-pentanol
(b) 2,4,6-trimethylphenol
(c) diisopropyl ether
(d) dicyclohexyl ether
(e) 2,4-dimethyl-2-hexanethiol
(f) 3,4,5-trinitrothiophenol
(g) 1,4-dimethoxybenzene
(h) dipentyl sulfide
(i) 2,4-pentanediol
(j) 3-ethoxy-1-propanethiol
(k) 3,3-dimethyl-1-pentanol
(l) 2-methyl-2-pentanol

12.20 Identify the primary, secondary, and tertiary alcohols in Problem 12.19.

12.21 Explain why alcohols with higher molecular weights are less soluble in water than those with lower molecular weights. (See Table 12.1.)

12.22 Find the alkane in Table 9.1 whose molecular weight is nearest to each of the following alcohols (Table 12.1) and construct a table to compare their boiling points and densities. Explain the differences.

(a) 1-butanol
(b) 1-pentanol
(c) 1-hexanol
(d) 1-nonanol

12.23 Explain why alcohols have higher boiling points than ethers of comparable molecular weights.

12.24 Complete the following equations.

(a) Primary alcohol $\xrightarrow{\text{oxidation}}$

(b) Secondary alcohol $\xrightarrow{\text{oxidation}}$

(c) Tertiary alcohol $\xrightarrow{\text{oxidation}}$

(d) $C_6H_5-CH_2CH_2CH_2-OH + HBr \rightarrow$

(e) $CH_3CH(CH_3)OH + H_2CrO_4 \rightarrow$

(f) $C_6H_5-CH(OH)CH_3 \xrightarrow{H_2SO_4}$

(g) $C_6H_5-C(=O)-OH + CH_3OH \xrightarrow{H_2SO_4}$

(h) $C_6H_5-OH + NaOH \rightarrow$

(i) 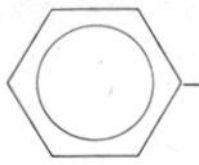 $C_6H_5{-}OH + Na \rightarrow$

(j) 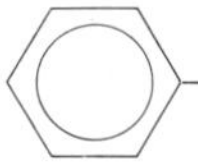 $C_6H_5{-}OH + HNO_3 \xrightarrow{H_2SO_4,\ heat}$

(k) 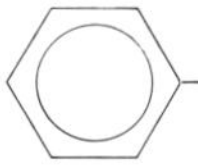 $C_6H_5{-}OH + SO_3 \xrightarrow{H_2SO_4\ \text{(at room temperature)}}$

(l) 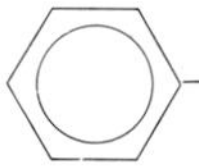 $C_6H_5{-}OH + Br_2 \xrightarrow{25°C}$

(m) $CH_3CH_2{-}O{-}CH_3 + HBr \xrightarrow{boil}$

(n) 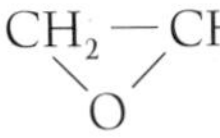 $\underset{\diagdown O \diagup}{CH_2{-}CH_2} + H_2O \xrightarrow{H^+}$

(o) 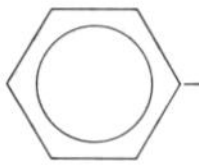 $C_6H_5{-}CH_2SH + H_2O_2 \rightarrow$

12.25 Write equations for two methods of manufacturing ethanol.
12.26 How is ethylene glycol manufactured? Give equations.
12.27 How is glycerol manufactured? What other product is obtained in this process? Give unbalanced equations. (See Section 12.9.)
12.28 Ethers and alcohols of comparable molecular weights have similar water solubilities but different boiling points. Explain.
12.29 What is the principal reaction of ethers? Write an equation for one example.
***12.30** In Section 12.13 the reaction of diethyl ether with H_2SO_4, shown below, was said to be reversible by the addition of water to the mixture.

$$CH_3CH_2{-}\ddot{\underset{\cdot\cdot}{O}}{-}CH_2CH_3 + H_2SO_4 \rightarrow CH_3CH_2{-}\overset{H}{\underset{\ddot{+}}{\overset{|}{O}}}{-}CH_2CH_3 + HSO_4^-$$

(a) Write the equation for the reaction with water of

$$CH_3CH_2{-}\overset{H}{\underset{\ddot{+}}{\overset{|}{O}}}{-}CH_2CH_3$$

(b) Label the acid–base pairs in the equation.

(c) Assuming that the reaction goes predominantly to completion, that is, to form diethyl ether, which of the following ions is the stronger acid?

$$H_3O^+ \quad \text{or} \quad CH_3CH_2-\overset{\overset{\displaystyle H}{|}}{\underset{\overset{\cdot\cdot}{+}}{O}}-CH_2CH_3$$

12.31 Why are epoxides more reactive than other ethers?

12.32 What structural features account for the following differences between sulfur and oxygen compounds?

(a) Alcohols have higher boiling points than the corresponding thiols.
(b) Ethers have lower boiling points than the corresponding sulfides.
(c) Alcohols are less acidic than thiols.

***12.33** Suggest a reason why sulfur compounds exhibit weaker hydrogen bonding than the corresponding oxygen compounds.

Problem-Solving Hints

If you have trouble solving any of the additional problems, refer to the sections listed next to the problem numbers.

12.19 12.2, 12.6, 12.11 12.17
12.20 12.1
12.21 12.3
12.22 12.3
12.23 12.3, 12.12
12.24 12.4, 12.8, 12.13, 12.14, 12.19
12.25 12.9
12.26 12.9
12.27 12.9
12.28 12.12
12.29 12.13
12.30 12.13
12.31 12.14
12.32 12.18
12.33 12.18, 12.19

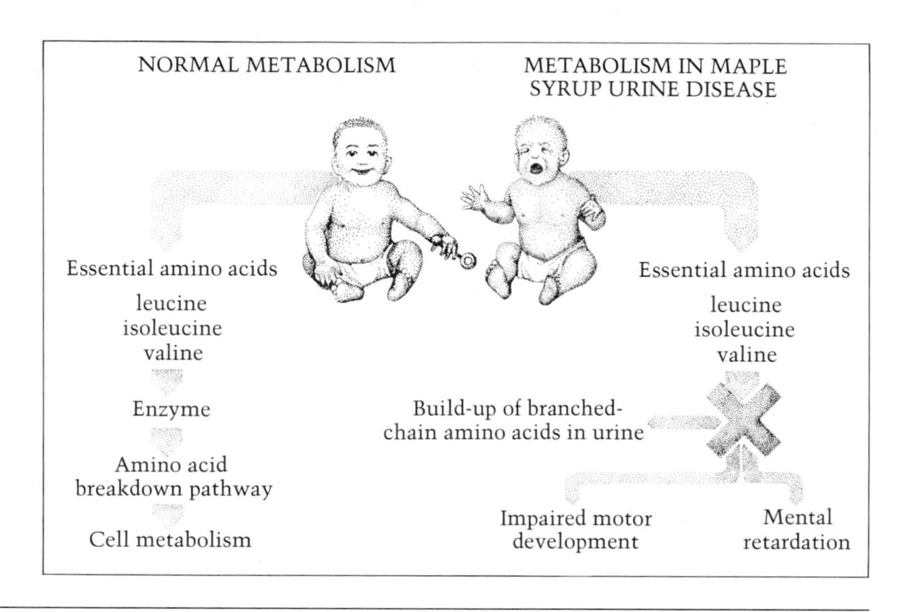

The following case study will add a new dimension to our study of the amine group in this chapter.

An apparently normal infant is born to a family with a history of past neurological disease and mental retardation. The baby feeds poorly, often vomits, and appears drowsy. At about the fifth day of life, it is noted that the infant's urine has the odor of burned sugar. Suspecting the possibility of a genetic defect, the physician orders tests to check the baby's blood and urine for amino acids. The laboratory tests show high levels of branched-chain amino acids (leucine, isoleucine, and valine). The diagnosis of maple syrup urine disease is established.

Maple syrup urine disease is a rare, inherited metabolic disease that was first described in 1954 when it was recognized as having affected four children in one family. Neurological findings for these children (and for other subsequent cases) were similar, and in all cases there was the characteristic urine odor that gave the disease its name. In its classic form the disease occurs with a frequency of approximately 1 in 250,000 live births. Afflicted children lack the enzyme that normally would break down branched-chain amino acids by oxidation. As a result, these amino acids accumulate in the blood, urine, and spinal fluid, where their levels can be easily measured. The accumulating amino acids seem to impair the synthesis of certain substances that are vital for normal brain function. However, the exact relationship between the patient's symptoms and the chemical findings remains unclear.

13

AMINES

The disease is usually fatal unless treated early. Treatment requires extensive control of amino acids in the diet, which is difficult to follow and often is not successful. Untreated children who survive may have seriously retarded mental and motor development.

[Illustration. The contrast between the carbohydrate metabolism of the normal infant versus that of the child with maple syrup urine disease.]

AMINES

Carbon, hydrogen, oxygen, and nitrogen are essential to life as we know it. It should be obvious by now why carbon, hydrogen, and oxygen are so important. However, nitrogen also deserves some of the limelight. The most important nitrogen compounds found in living organisms (as well as dead ones) are the amines. The amines contain the amine functional group, which we mentioned briefly in Chapter 8 and will describe in more detail in just a moment. Such substances as proteins, caffeine, cocaine, DNA, RNA, plus many other biologically important compounds contain the amine group.

The amine group has two characteristics that make it vital to living organisms. First, the nitrogen atom is moderately electronegative and has an unshared pair of electrons, so it forms hydrogen bonds. Second, amines are more basic than water, phenols, alcohols, or ethers and thus can react with acids.

These striking chemical properties of amines will lead us to an understanding of the role of amines in life processes.

13.1 AMINES: STRUCTURE

Amines are derivatives of ammonia in which one or more hydrogen atoms have been replaced by alkyl or aryl groups.

What distinguishing feature do all amines share?

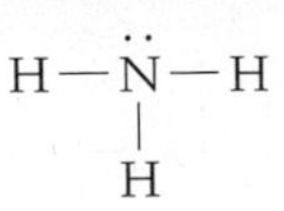

ammonia

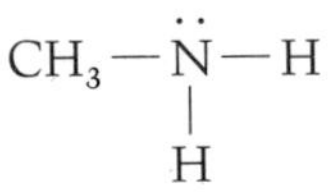

methylamine

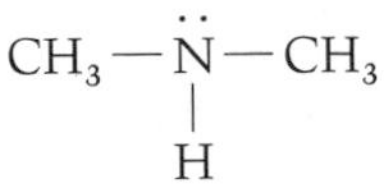

dimethylamine

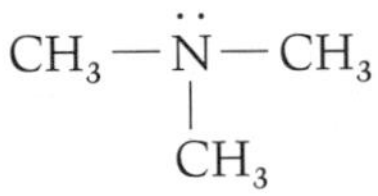

trimethylamine

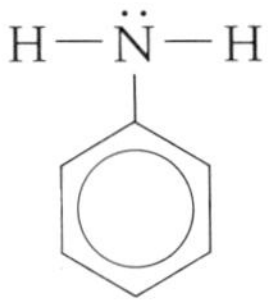

aniline

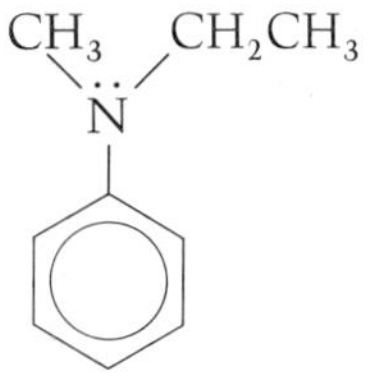

N-methyl-N-ethylaniline

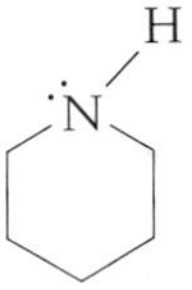

piperidine

AMINES: CLASSIFICATION 13.2

Do we classify amines the way we classify alcohols?

We classify amines according to the number of hydrocarbon groups attached to the nitrogen atom. **Primary (1°) amines** have one hydrocarbon group attached to the nitrogen atom ($R-NH_2$). **Secondary (2°) amines** have two hydrocarbon groups attached to the nitrogen atom ($R-NHR'$). **Tertiary (3°) amines** have three hydrocarbon groups attached to the nitrogen atom

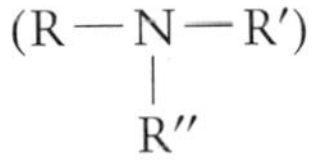

Problem 13.1 Classify as 1°, 2°, or 3° the amines whose structural formulas are shown in Section 13.1.

Problem 13.2 Classify the following amines as 1°, 2°, or 3°.

(a) 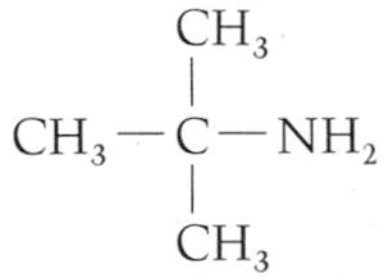

(b) 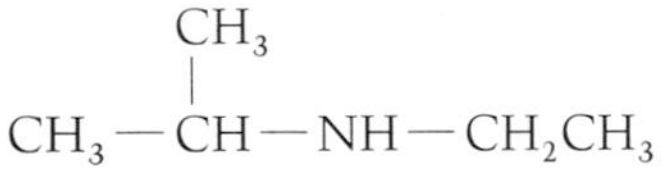

(c) 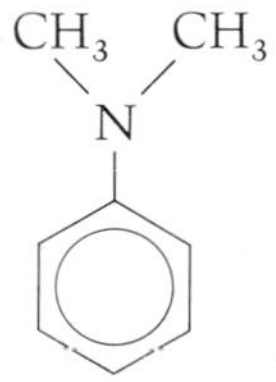

Problem 13.3 How does the classification of amines as 1°, 2°, or 3° differ from the classification of alcohols as 1°, 2°, or 3°. (See Section 12.1.)

13.3 AMINES: NOMENCLATURE

Amines are usually called by their common names. The common names of simple amines consist of the names of the hydrocarbon groups, followed by the word *amine* (Section 13.1). Some examples will illustrate this.

CH_3NH_2	$CH_3CH_2NH_2$	$CH_3CH_2CH_2NH_2$	$CH_3CH_2CH_2CH_2NH_2$
methylamine	ethylamine	propylamine	butylamine

$CH_3—NH(CH_3)$	$CH_3—N(CH_3)—CH_3$	$CH_3—N(CH_2CH_3)—CH_2CH_2CH_3$
dimethylamine	trimethylamine	methylethylpropylamine

The IUPAC system is less often used and we will not include it here.

Some amines have special names, for example, aniline and piperidine, shown in Section 13.1.

Just as we use numbers to locate the positions of substituents in a carbon chain, we use the letter *N* to show that a substituent is attached to nitrogen when there could be some question about where the substituent is attached. For example, the name methylaniline is ambiguous because the methyl group could be attached either to the nitrogen or to the benzene ring. But when an *N* is added, to form *N*-methylaniline, we know that the methyl group is bonded to nitrogen. (Do not confuse this with *n*-, which denotes *normal* or unbranched.) Some other examples are shown below.

$C_6H_5—N(H)(CH_2CH_3)$	$C_6H_5—N(CH_3)(CH_3)$	$C_6H_5—N(CH_3CH_2)(CH_2CH_3)$
N-ethylaniline	N,N-dimethylaniline	N,N-diethylaniline

Problem 13.4 Give the structural formulas of the following amines.
(a) tripropylamine
(b) methylpropylbutylamine
(c) *p*-fluoroaniline
(d) N-butylaniline
(e) *p*-ethylaniline
(f) N-methyl-N-ethylaniline

Problem 13.5 Name each of the following amines.
(a) $CH_3CH_2CH_2—NHCH_2CH_3$
(b) $CH_3CH_2CH_2—NH_2$
(c) $C_6H_5—NH—C_6H_5$
(d) $O_2N—C_6H_4—NH_2$

AMINES: PHYSICAL PROPERTIES 13.4

Table 13.1 lists several amines, with their boiling points and water solubilities. In both of these properties, amines, especially primary and secondary amines, resemble alcohols more than they resemble hydrocarbons (given similar molecular weights). We can explain this behavior on the basis of hydrogen bonding.

What kind of molecular attraction accounts for the high boiling points and water solubilities of amines?

Problem 13.6 Examine the data of Tables 10.1, 12.1, and 13.1, to see whether amines resemble alcohols more than they resemble hydrocarbons in their boiling points and water solubilities. Be sure to compare an amine, an alcohol, and an alkane that have nearly the same molecular weight, for example, methylamine (molecular weight 31), methanol (molecular weight, 32), and ethane (molecular weight, 30).

Problem 13.7 Label each amine in Table 13.1 as 1°, 2°, or 3°.

TABLE 13.1 Physical Properties of Amines

Name	Formula	Boiling point, °C	Solubility in water
Methylamine	$CH_3—NH_2$	−6.5	1,154 mL of gas/ 100 mL of liquid at 12°C
Dimethylamine	$CH_3—N(H)—CH_3$	7.4	Very soluble
Trimethylamine	$CH_3—N(CH_3)—CH_3$	3.5	Very soluble
Ethylamine	$CH_3CH_2—NH_2$	16.6	Very soluble
Diethylamine	$CH_3CH_2—N(H)—CH_2CH_3$	55.5	Very soluble
Triethylamine	$CH_3CH_2—N(CH_2CH_3)—CH_2CH_3$	89.5	Slightly soluble
Piperidine	(cyclohexane ring with N)N—H	106	Very soluble
Morpholine	O(ring)N—H	129	Very soluble
Aniline	(benzene ring)—NH_2	184	Slightly soluble

Boiling Point

Primary and secondary amines have relatively high boiling points because of hydrogen bonding (Figure 13.1). Tertiary amines have no hydrogen atoms bonded to the nitrogen atom, so their molecules cannot form hydrogen bonds to each other. Tertiary amines therefore have lower boiling points than primary and secondary amines. For example, trimethylamine (molecular weight 59) has a lower boiling point than dimethylamine (molecular weight 45). Even though trimethylamine cannot form hydrogen bonds, its boiling point is higher than that of pentane, because amine molecules are polar and therefore attract one another more strongly than do hydrocarbons.

Nitrogen is not as electronegative as oxygen; therefore, hydrogen bonds in amines are not as strong as hydrogen bonds in alcohols. For this reason, the boiling points of amines are somewhat lower than those of comparable alcohols.

Solubility

Amines are more soluble in water than hydrocarbons of similar molecular weight. Amines are soluble in water because amine molecules form hydrogen bonds with water molecules (Figure 13.2). Even tertiary amine molecules can form hydrogen bonds to water molecules, although they cannot hydrogen-bond to one another.

Problem 13.8

(a) Draw the structural formulas of butylamine and diethylamine.

(b) Dipropylamine boils at 111°C; triethylamine boils at 89.5°C. Explain

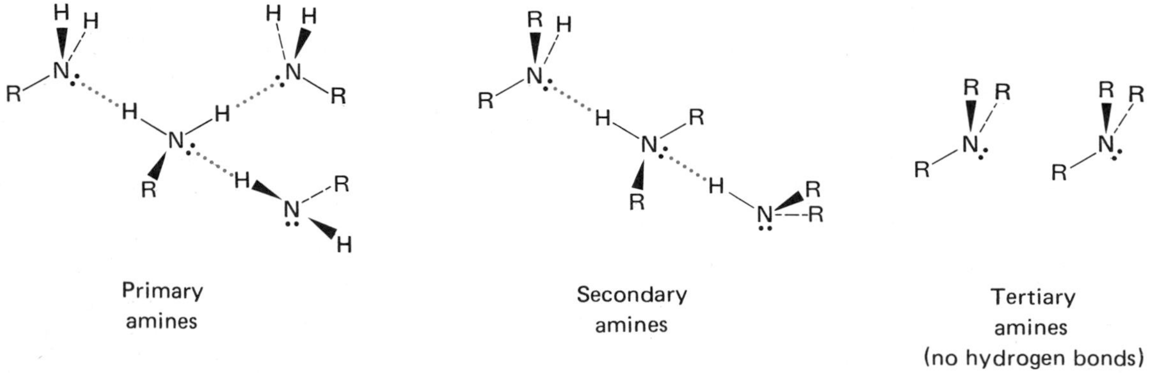

FIGURE 13.1 (a) Primary and (b) secondary amines have high boiling points because of hydrogen bonding. (c) Tertiary amines have no hydrogen atoms bonded to nitrogen, and they do not form hydrogen bonds. Hydrogen bonds are shown as dotted lines.

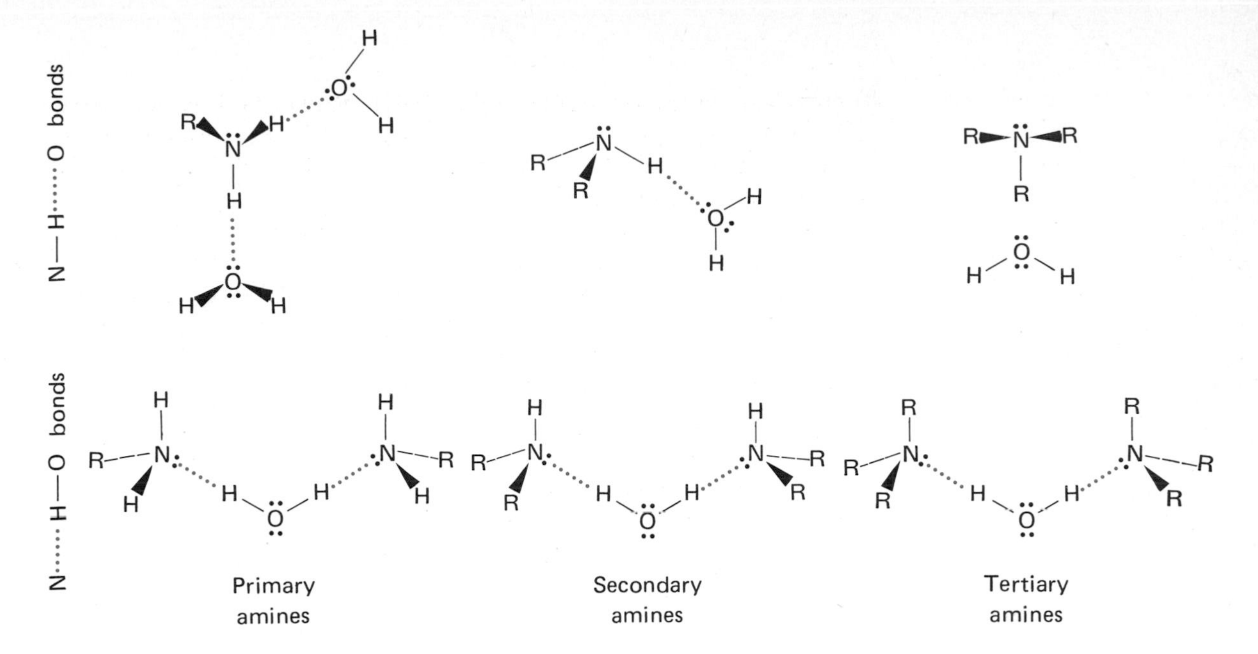

FIGURE 13.2 Hydrogen bonds between amines and water. Primary and secondary amines can form both N—H···O and N···H—O hydrogen bonds. Secondary amines form fewer N—H···O bonds because they have only one N—H bond, whereas primary amines have two N—H bonds per molecule. Tertiary amines cannot form N—H···O bonds because they have no N—H bonds. Tertiary amines form only N···H—O bonds. Hydrogen bonds are shown as dotted lines.

why these compounds have such different boiling points when they are isomers and therefore have the same molecular weight.

Problem 13.9 Predict which compound in each of the following pairs will be more soluble in water.

(a) propylamine and tripropylamine
(b) aniline and diphenylamine

Problem 13.10 Predict which compound in each of the pairs of Problem 13.9 will have the higher boiling point.

Problem 13.11 Explain the difference in boiling points of piperidine and morpholine (Table 13.1).

AMINES AS BASES 13.5

What structural feature of amines accounts for their basicity?

Nearly all the chemical properties of amines result from the unshared pair of electrons on the nitrogen atom. The most striking characteristic of amines is their basicity. Like ammonia, amines react with acids in water solution to form conjugate acids.

$$H-\underset{\displaystyle H}{\overset{\cdot\cdot}{N}}-H + H_3O^+ \rightleftharpoons H-\underset{\displaystyle H}{\overset{\displaystyle H}{N^+}}-H + H_2O$$

base (ammonia) | acid | conjugate acid (ammonium ion) | conjugate base

$$CH_3-\underset{\displaystyle H}{\overset{\cdot\cdot}{N}}-H + H_3O^+ \rightleftharpoons CH_3-\underset{\displaystyle H}{\overset{\displaystyle H}{N^+}}-H + H_2O$$

methylamine | methylammonium ion

Problem 13.12 The anions are not shown in the preceding examples. What anion is present with the ammonium ions in these examples? Where does it come from?

Amines, like ammonia, also react with water, because water is a weak acid.

$$H-\underset{\displaystyle H}{\overset{\cdot\cdot}{N}}-H + H_2O \rightleftharpoons H-\underset{\displaystyle H}{\overset{\displaystyle N}{N^+}}-H + OH^-$$

ammonia | ammonium ion

$$CH_3-\underset{\displaystyle H}{\overset{\cdot\cdot}{N}}-H + H_2O \rightleftharpoons CH_3-\underset{\displaystyle H}{\overset{\displaystyle H}{N^+}}-H + OH^-$$

methylamine | methylammonium ion

These equations show why water solutions of amines are basic: because they contain excess OH^- ions. Removal of an H^+ ion from a water molecule leaves an OH^- ion.

Problem 13.13 Complete the equations for the following acid–base equilibria. Label the conjugate acid–base pairs for each equation.

(a) dimethylamine + $H_3O^+ \rightleftharpoons$

(b) trimethylamine + $H_3O^+ \rightleftharpoons$

(c) dimethylamine + $H_2O \rightleftharpoons$

(d) trimethylamine + $H_2O \rightleftharpoons$

We name the conjugate acid of an amine by changing the *-amine* ending to **-ammonium ion.** For example:

Amine	*Conjugate acid*
$CH_3-\ddot{N}H-H$ (structure: $CH_3-\ddot{N}(H)-H$)	$CH_3-N^+H_2-H$ (structure: $CH_3-N^+(H)(H)-H$)
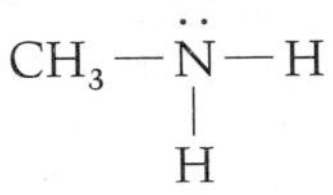 methylamine	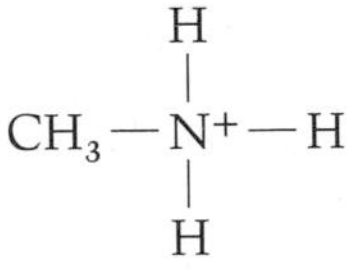 methylammonium ion
$CH_3-\ddot{N}(H)-CH_2CH_3$	$CH_3-N^+(H)(H)-CH_2CH_3$
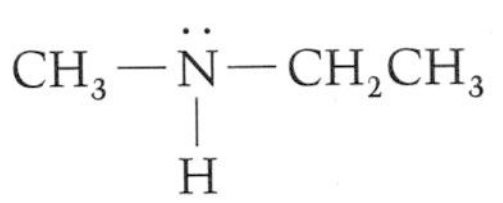 methylethylamine	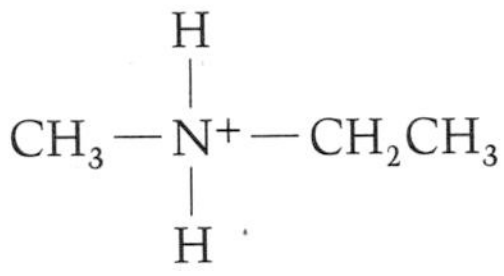 methylethylammonium ion

13.6 AMINE SALTS 13.6

The solid product of the reaction of an amine with an acid is an **amine salt.** We name these amine salts by combining the name of the ammonium ion with the name of the anion.

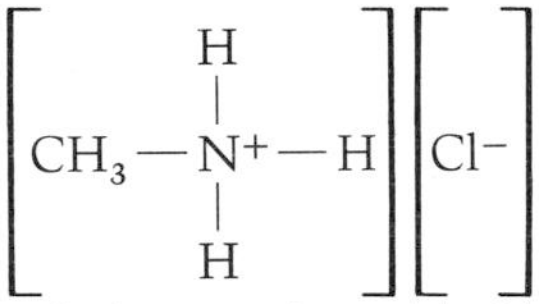

Do amine salts behave like inorganic salts or like organic compounds?

$$\left[CH_3-N^+(H)(H)-H\right]\left[Cl^-\right]$$

methylammonium chloride

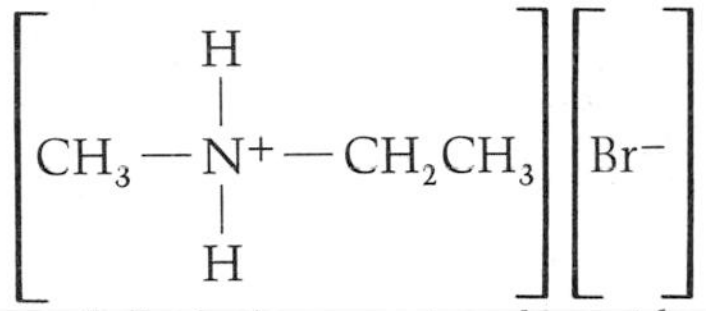

$$\left[CH_3-N^+(H)(H)-CH_2CH_3\right]\left[Br^-\right]$$

methylethylammonium bromide

We can make these salts by the direct reaction of a gaseous acid with an amine under anhydrous conditions.

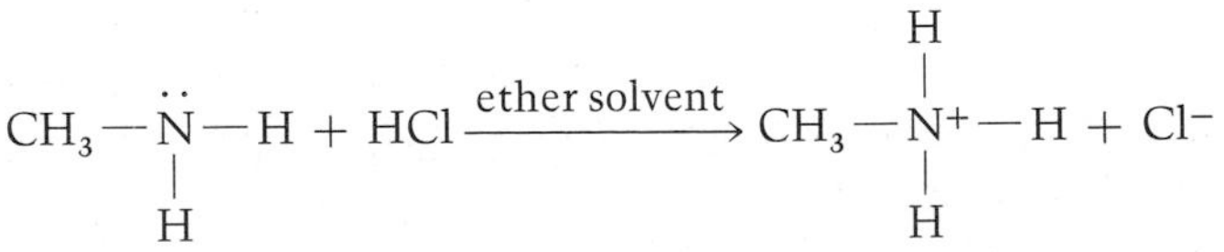

$$CH_3-\ddot{N}(H)-H + HCl \xrightarrow{\text{ether solvent}} CH_3-N^+(H)(H)-H + Cl^-$$

Amine salts are crystalline solids and they are generally more soluble in water than the amines from which they were prepared. Many amines used as drugs are administered in the form of a salt, because the salt is more soluble in the body fluids, which are, of course, mostly water. Morphine sulfate and codeine phosphate are examples of amine salts that are used as drugs.

Problem 13.14 Write the structural formula and give the name of the salt that is produced in each of the following reactions involving gaseous acids.

(a) triethylamine + HCl $\xrightarrow[\text{solvent}]{\text{ether}}$

(b) butylamine + HI $\xrightarrow[\text{solvent}]{\text{ether}}$

13.7 ALKYLATION OF AMINES

What is a displacement reaction?

Amines react with alkyl halides to produce ammonium ions that have an additional alkyl group attached to the nitrogen. This reaction is called **alkylation,** because an alkyl group becomes bonded to the nitrogen atom.

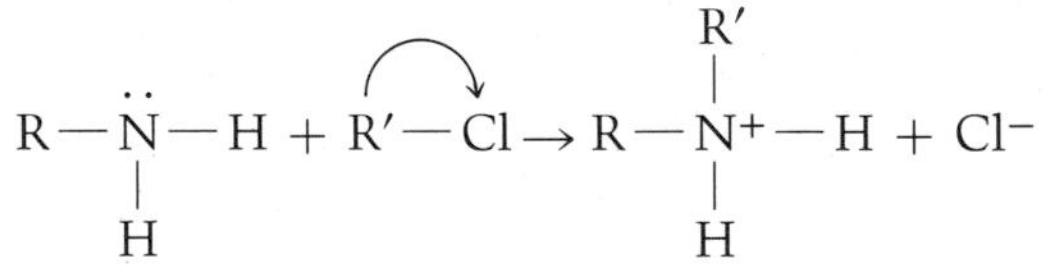

$$\underset{1^\circ\text{ amine}}{\mathrm{R{-}\ddot{N}(H){-}H}} + \mathrm{R'{-}Cl} \rightarrow \mathrm{R{-}N^{+}(R')(H){-}H} + \mathrm{Cl^-}$$

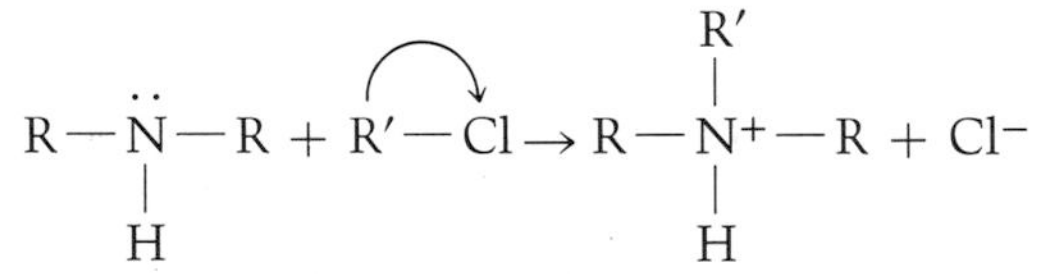

$$\underset{2^\circ\text{ amine}}{\mathrm{R{-}\ddot{N}(H){-}R}} + \mathrm{R'{-}Cl} \rightarrow \mathrm{R{-}N^{+}(R')(H){-}R} + \mathrm{Cl^-}$$

In this alkylation reaction, the chlorine atom and the pair of electrons that bonded it to R′ are displaced by the amine. Thus, a chloride ion is one of the products of the reaction. The reaction is called a **displacement reaction** because the amine displaces the chlorine.

13.8 QUATERNARY AMMONIUM SALTS

When we carry out an alkylation reaction with a tertiary amine, we get a **quaternary ammonium ion**—an ammonium ion that has four alkyl groups bonded to the nitrogen atom.

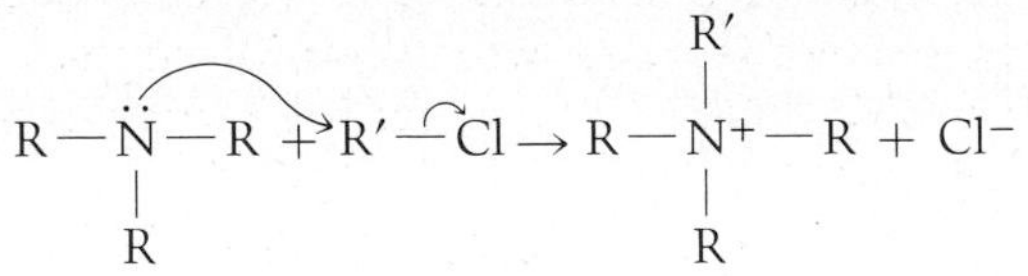

$$R-\ddot{N}(R)-R + R'-Cl \rightarrow R-N^{+}(R')(R)-R + Cl^-$$

How do quaternary ammonium ions differ from other ammonium ions?

Problem 13.15 Write equations to show the preparation of the following compounds.
(a) trimethylammonium bromide from dimethylamine and methyl bromide
(b) butylammonium iodide from ammonia and butyl iodide
(c) tetraethylammonium chloride from triethylamine and ethyl chloride

Problem 13.16 Complete the equation for the following alkylation reaction.

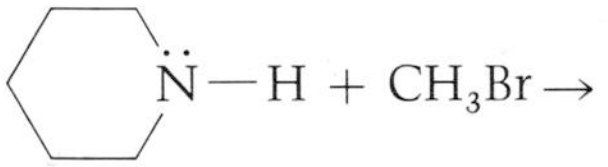

$$\text{(piperidine ring)}\ddot{N}-H + CH_3Br \rightarrow$$

piperidine

Quaternary ammonium salts are different from other ammonium salts because they are not in equilibrium with the amine. For example, in water solution the butylammonium ion (which is not quaternary) is in equilibrium with butylamine.

$$CH_3CH_2CH_2CH_2-N^{+}H_2-H + Br^- + H_2O \rightleftarrows CH_3CH_2CH_2CH_2-\ddot{N}H-H + H_3O^+ + Br^-$$

Quaternary ammonium salts cannot undergo such a reaction, because the quaternary ammonium ion has no H^+ ion to donate. In other words, a quaternary ammonium ion has no acid properties at all.

Quaternary ammonium hydroxides are strong bases, as strong as the inorganic hydroxides NaOH and KOH.

HETEROCYCLIC AMINES 13.9

Heterocyclic compounds are cyclic compounds that contain one or more atoms other than carbon in the ring. We already encountered a number of heterocyclic compounds in Chapter 12.

What is the meaning of the prefix hetero-?

CH_2—CH_2 O
ethylene oxide

O
furan

O
tetrahydrofuran

O O
dioxane

Piperidine and morpholine, mentioned previously in this chapter, are heterocyclic amines.

N H
piperidine

O N H
morpholine

Heterocyclic amines occur widely in biological systems. Some of the more important ones are shown below.

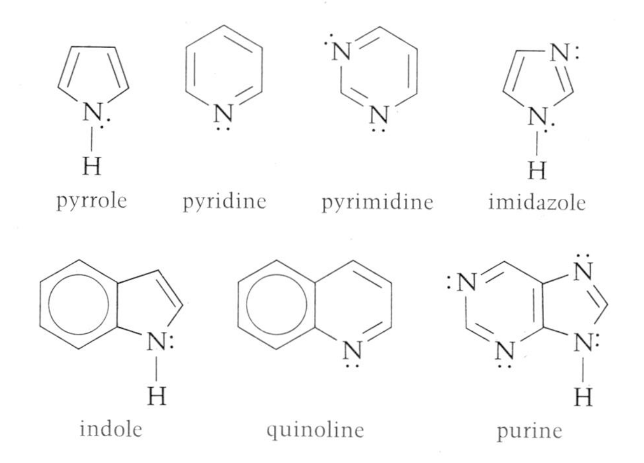

Pyrimidine and purine derivatives occur in DNA (deoxyribonucleic acid, see Section 18.2).

Problem 13.17 Describe the structure of purine, in terms of the structures of pyrimidine and imidazole.
Problem 13.18 Which nitrogen atom in imidazole resembles the nitrogen in pyrrole, and which resembles the nitrogen in pyridine?
Problem 13.19 Label the two nitrogen atoms of imidazole as 1°, 2°, or 3°. (Treat the double bond as if there were two separate carbon atoms bonded to nitrogen.)

ALKALOIDS 13.10

Many compounds that occur in the leaves, bark, roots, and fruit of plants react with acids to produce water-soluble ammonium ion derivatives. Because of this alkali-like behavior, these compounds are called **alkaloids.**

What functional group do all alkaloids have in common?

Alkaloids have very pronounced but varied physiological effects. Some raise blood pressure, others lower it. Some stimulate the central nervous system, others act as tranquilizers. Many of the alkaloids relieve pain. Nearly all alkaloids are toxic in high enough doses.

The names of alkaloids all end in *-ine* because they are all amines. There follow descriptions of some of the more important alkaloids.

Nicotine occurs in the tobacco plant. In small doses, nicotine is a stimulant. At higher doses it can cause extreme nausea, vomiting, convulsions, and even death. Nicotine is used as an agricultural insecticide. Oxidation of nicotine yields *nicotinic acid,* also called niacin. Nicotinic acid (which is not an alkaloid) is one of the vitamins of the B complex; a deficiency of it causes pellagra. Like the other B vitamins, nicotinic acid is found in various foods (for instance, red meat, liver, yeast, and tomato juice).

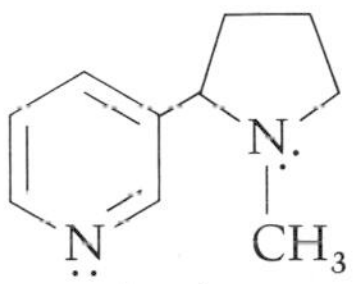

nicotine

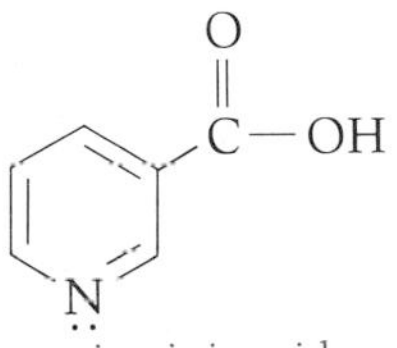

nicotinic acid

Atropine, which is obtained from the belladonna plant ("deadly nightshade"), is highly toxic. It is used in a weak solution to dilate the pupil of the eye during ophthalmic examinations.

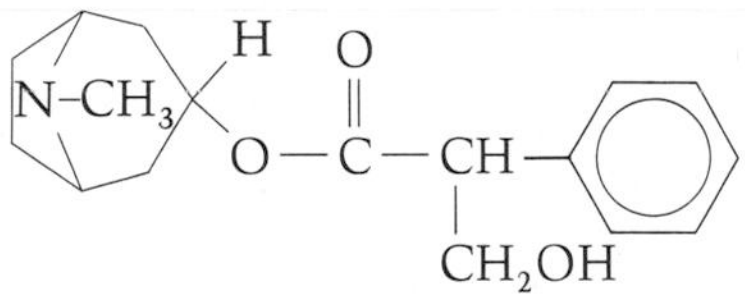

atropine

Strychnine comes from the seeds of the nux vomica tree. It is used as a rat poison. It is highly toxic to man too. Since it stimulates the central nervous system, in small doses it is used medically to counteract poisoning by central nervous system depressants.

Morphine is a derivative of opium, which is prepared from the opium poppy. It has been used for many years as a pain reliever (analgesic).

It is usually taken in the form of the salt, morphine sulfate. Morphine was named after the Roman god Morpheus, god of dreams. Unfortunately, morphine is addictive and it can cause respiration to become fatally slow.

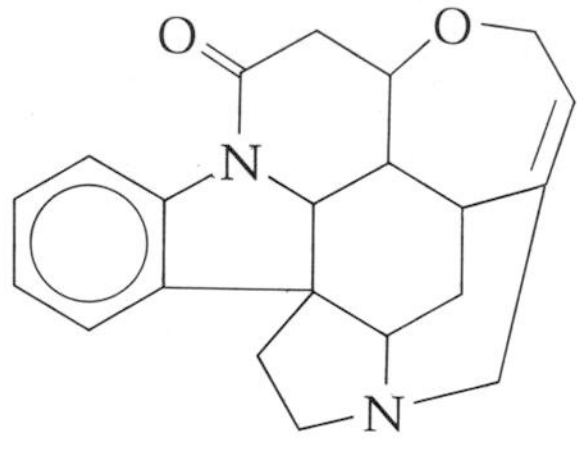

strychnine

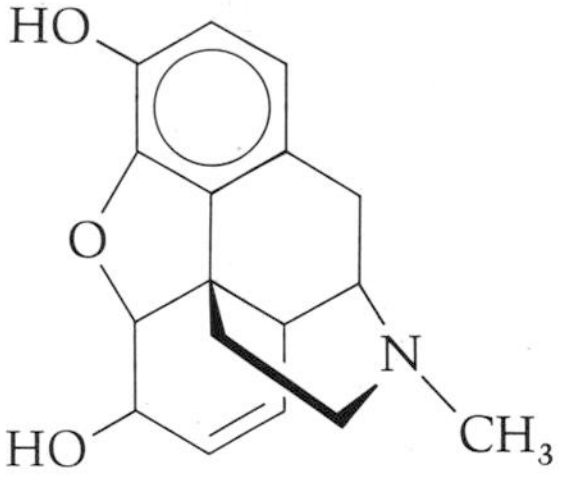

morphine

Codeine has the same source (the opium poppy) and the same structure as morphine, except that the phenolic OH is replaced by a methoxy group. Codeine is also a narcotic and can cause nausea, vomiting, and dizziness. It is used as an analgesic and antitussive (cough suppressant).

Problem 13.20 Draw the structural formula of codeine. [Hint: see the previous paragraph.]

Reserpine can be extracted from the Indian snake root, *Rauwolfia serpentina,* which has been used for centuries in native medicines. Reserpine is a tranquilizer and is used to lower blood pressure.

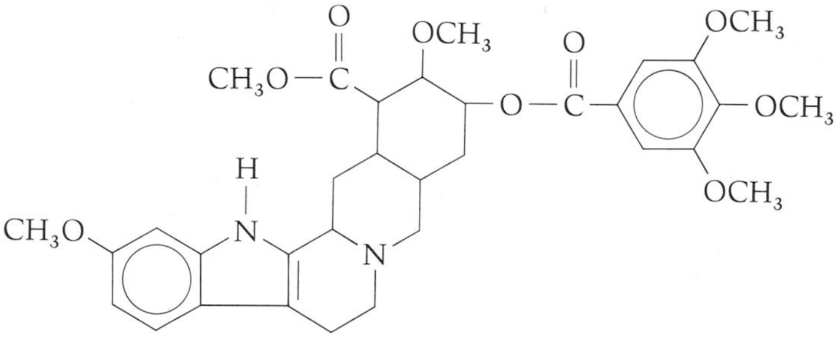

reserpine

Ergotamine is one of the alkaloids present in ergot. Ergot is a fungus that grows on certain rye grains. Ergotamine is an analgesic vasoconstrictor—a pain-relieving drug that causes constriction of the blood vessels. It is used to treat vascular headaches (migraine). Ergotamine sometimes causes nausea and vomiting.

Problem 13.21 Locate the amine groups in each of the alkaloids whose structures have been shown in this section. State whether each is 1°, 2°, or 3°. (See Problem 13.19.)

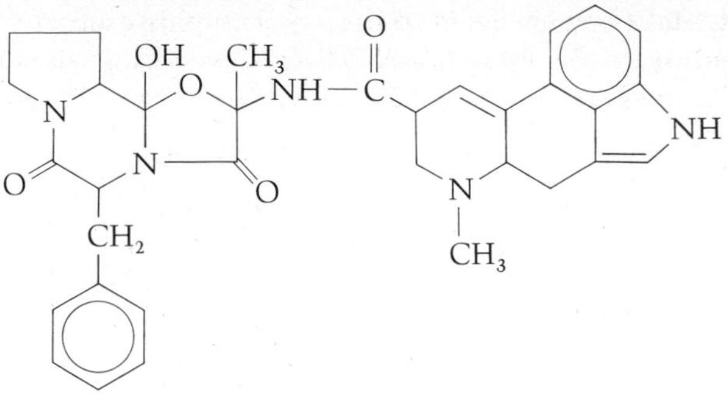

ergotamine

AMINO ACIDS 13.11

Probably the most important compounds that contain the amine group are the **amino acids.** They are the building blocks of proteins (Chapter 17). Each amino acid has both an amino group and a carboxyl group. The general structure of naturally occurring amino acids is

What structural features do all naturally occurring amino acids have?

$$H_2N-\underset{}{\overset{R}{\overset{|}{C}}}H-\overset{O}{\overset{\|}{C}}-OH$$

The R groups in amino acids vary widely. In addition to H, alkyl, or aryl groups, R may also contain other amino or carboxyl groups. Some simple amino acids are:

$$H_2N-CH_2\overset{O}{\overset{\|}{C}}-OH$$

glycine

$$H_2N-\overset{CH_3}{\overset{|}{C}}H-\overset{O}{\overset{\|}{C}}-OH$$

alanine

$$H_2N-\overset{H_2NCH_2CH_2CH_2CH_2}{\overset{|}{C}}H-\overset{O}{\overset{\|}{C}}-OH$$

lysine

Because amino acids contain both amino and carboxyl groups, they react with both acids and bases.

$$H_2NCHRCO_2H + HCl \rightarrow H_3\overset{+}{N}CHRCO_2H + Cl^-$$

$$H_2NCHRCO_2H + NaOH \rightarrow H_2NCHRCO_2^- + Na^+ + H_2O$$

Problem 13.22 Amino acids contain both a basic group ($-NH_2$) and an acidic group

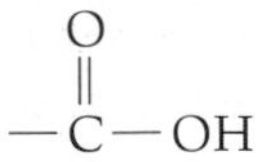

$$-\overset{\overset{\displaystyle O}{\|}}{C}-OH$$

In water solution, amino acids exist in the salt form (in which these two groups have neutralized each other). Give the structural formula of the salt form of

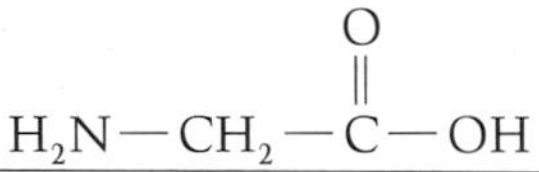

$$H_2N-CH_2-\overset{\overset{\displaystyle O}{\|}}{C}-OH$$

13.12 N-NITROSAMINES

Inorganic nitrites, such as sodium nitrite ($NaNO_2$), are used as preservatives and coloring agents for cooked meats and cold cuts. A solution of sodium nitrite in water imparts a reddish color to meat that would otherwise turn gray with prolonged exposure to air. The typical red color of hot dogs is due mainly to sodium nitrite treatment.

The use of nitrite salts in foods can be dangerous. In the presence of an acid, nitrites react with certain amines to produce carcinogenic compounds called **N-nitrosamines.**

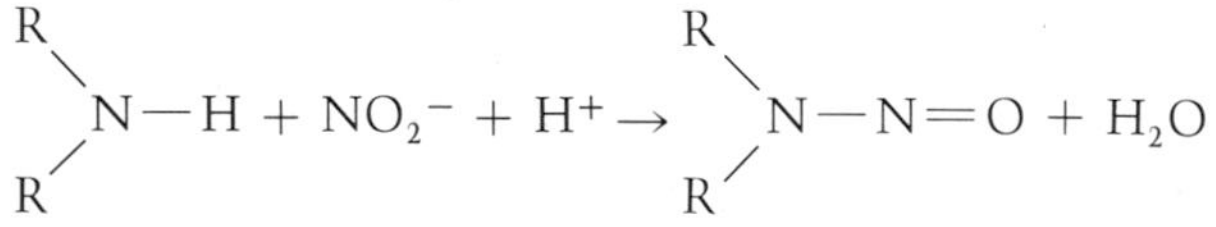

$$R_2N-H + NO_2^- + H^+ \longrightarrow \underset{\text{N-nitrosamine}}{R_2N-N{=}O} + H_2O$$

The human stomach is highly acidic. It also contains amines produced in the digestion of proteins, so conditions are ideal for the synthesis of N-nitrosamines. Although no one has yet shown a carcinogenic effect of N-nitrosamines in humans, laboratory animals are known to develop cancers upon exposure to nitrosamines. Thus, it is possible that these compounds are the cause of many cancers in humans.

SUMMARY

Amines are derivatives of ammonia in which one or more hydrogen atoms have been replaced by alkyl or aryl groups. We classify amines as

primary (1°), secondary (2°), and **tertiary (3°).** Because amines can form hydrogen bonds, they have higher boiling points and higher water solubilities than hydrocarbons of comparable molecular weight.

The most distinguishing characteristic of amines is their basicity. They react with acids to produce **amine salts.** Amines undergo **alkylation** reactions when they are treated with alkyl halides; the product is an ammonium ion that has one more alkyl group than the original amine. An ammonium ion that has four hydrocarbon groups attached to nitrogen is called a **quaternary ammonium ion.** Quaternary ammonium hydroxides are strong bases. **Heterocyclic compounds** are cyclic compounds that have an atom other than carbon in the ring. **Alkaloids** are basic amine compounds found in plants. They have pronounced but varied physiological effects. **Amino acids** have both an amino group and a carboxyl group. Inorganic nitrites are used as preservatives and coloring agents for meats. Inorganic nitrites are dangerous because in the stomach they may produce **N-nitrosamines,** which are known to be carcinogens.

Key Terms

Alkaloid (Section 13.10)
Alkylation (Section 13.7)
Amine (Section 13.1)
Amine salt (Section 13.6)
Amino acid (Section 13.11)
Ammonium ion (Section 13.5)
Displacement reaction (Section 13.7)
Heterocyclic compound (Section 13.9)
N-nitrosamines (Section 13.12)
Primary (1°) amine (Section 13.2)
Quaternary ammonium ion (Section 13.8)
Secondary (2°) amine (Section 13.2)
Tertiary (3°) amine (Section 13.2)

ADDITIONAL PROBLEMS

13.23 Draw the structure of each of the following amines.
(a) diethylamine (b) dipropylamine
(c) ethylisopropylamine (d) triphenylamine
(e) aniline (f) *p*-nitroaniline

13.24 Classify each of the amines in Problem 13.23 as 1°, 2°, or 3°.

13.25 Consider the following boiling points. Explain why the two dimethylhydrazines have lower boiling points than hydrazine even though they have higher molecular weights.

Compound	*Formula*	*Boiling point, °C*
Hydrazine	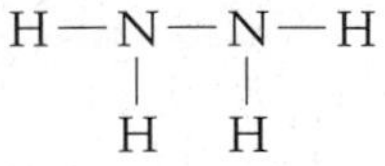 $H{-}N(H){-}N(H){-}H$	113
1,2-Dimethylhydrazine	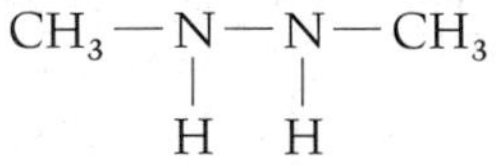 $CH_3{-}N(H){-}N(H){-}CH_3$	81
1,1-Dimethylhydrazine	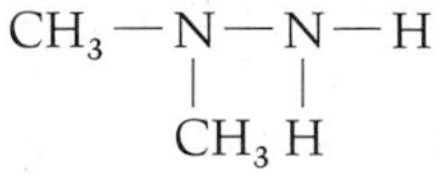 $CH_3{-}N(CH_3){-}N(H){-}H$	64

13.26 Complete each of the following reactions.

(a) $CH_3CH_2NH_2 + H_3O^+ \rightleftarrows$

(b) $CH_3CH_2NH_2 + HCl(gaseous) \longrightarrow$

(c) 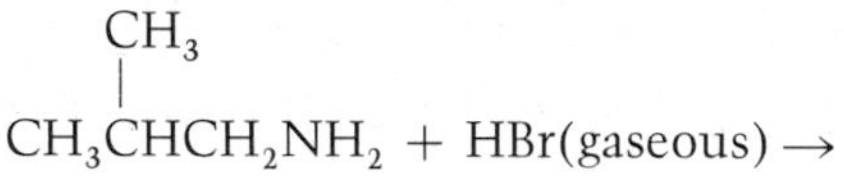 $CH_3CH(CH_3)CH_2NH_2 + HBr(gaseous) \longrightarrow$

(d) 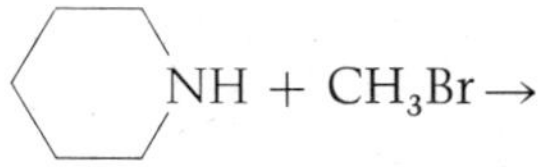 (piperidine ring) NH $+ CH_3Br \longrightarrow$

(e) Product of reaction (d) + NaOH(aqueous) $\longrightarrow$

(f) 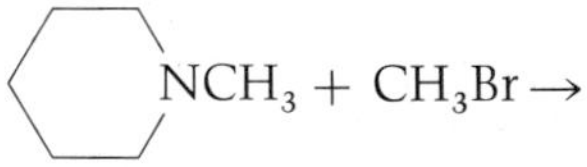 (piperidine ring) $NCH_3 + CH_3Br \longrightarrow$

(g) 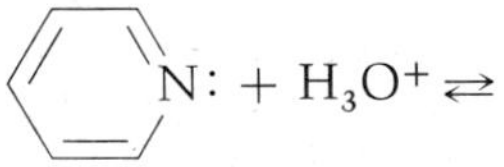 (pyridine ring) $N\colon + H_3O^+ \rightleftarrows$

(h) Morpholine + HCl(aqueous) $\longrightarrow$

***13.27** Tell whether, at equilibrium, reactants or products are favored in each of the following reactions.

(a) 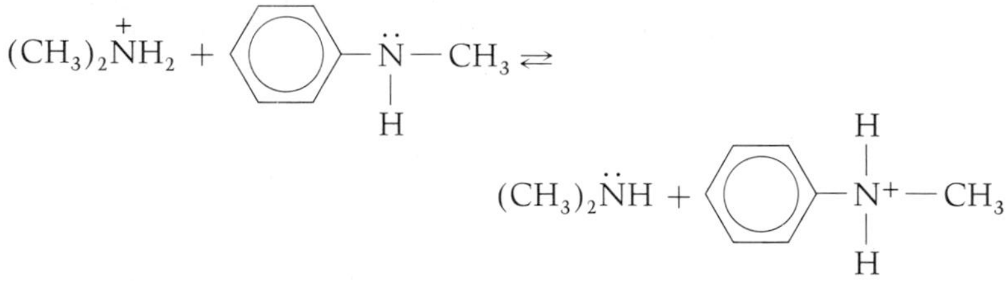 $(CH_3)_2\overset{+}{N}H_2 + C_6H_5{-}\ddot{N}H{-}CH_3 \rightleftarrows (CH_3)_2\ddot{N}H + C_6H_5{-}\overset{+}{N}H_2{-}CH_3$

(Dimethylamine is a stronger base than N-methylaniline.)

(b) 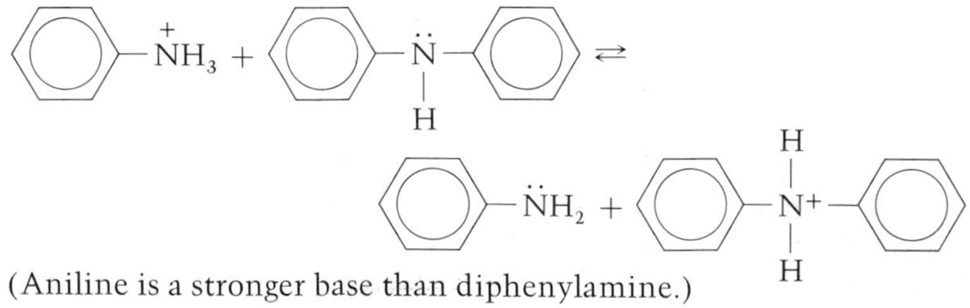 $C_6H_5{-}\overset{+}{N}H_3 + C_6H_5{-}\ddot{N}H{-}C_6H_5 \rightleftarrows C_6H_5{-}\ddot{N}H_2 + C_6H_5{-}\overset{+}{N}H_2{-}C_6H_5$

(Aniline is a stronger base than diphenylamine.)

13.28 Write the structural formula and give the name of the salt that is produced in each of the following reactions.

(a) Pentylamine + HI(g) $\xrightarrow[\text{solvent}]{\text{ether}}$

(b) Tributylamine + HI(g) $\xrightarrow[\text{solvent}]{\text{ether}}$

(c)  (cyclohexane ring)N—H + HI(g) $\xrightarrow[\text{solvent}]{\text{ether}}$

13.29 Explain how quaternary ammonium salts differ from other ammonium salts.

13.30 Compare the base strengths of trimethylammonium hydroxide and tetramethylammonium hydroxide.

13.31 What are alkaloids? Describe them. What properties do they have in common?

13.32 Identify and name all the functional groups in morphine. Classify the nitrogen atom as 1°, 2°, or 3°.

13.33 Identify and name all the functional groups in reserpine. Classify the nitrogen atoms as 1°, 2°, or 3°.

*__13.34__ When water solutions of N-methylaniline and methylammonium chloride are mixed, the following equilibrium is established.

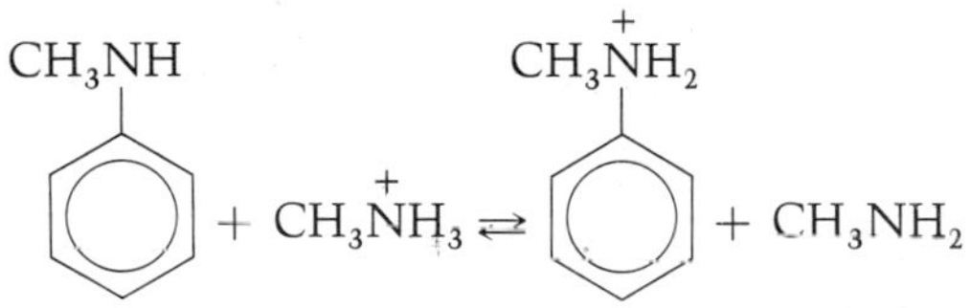

$$C_6H_5NHCH_3 + CH_3\overset{+}{N}H_3 \rightleftarrows C_6H_5\overset{+}{N}H_2CH_3 + CH_3NH_2$$

(a) Label the two acids and the two bases in the equilibrium.

(b) Analysis of the equilibrium mixture shows that the concentrations of reactants are greater than the concentrations of products. Of the two acids, which is stronger?

(c) Of the two bases, which is stronger?

Problem-Solving Hints

If you have trouble solving any of the additional problems, refer to the sections listed next to the problem numbers.

13.23 13.1, 13.3
13.24 13.2
13.25 13.4
13.26 13.5, 13.6
13.27 13.5
13.28 13.6
13.29 13.8
13.30 13.8
13.31 13.10
13.32 8.12, 13.2
13.33 8.12, 13.2
13.34 13.5

Acetylcholine, an Important Quaternary Ammonium Ion

FOCUS 9

A quaternary ammonium compound of great biological importance is acetylcholine, which is one of the compounds involved in the transmission of nerve impulses from one nerve cell to another. Every action that we make, no matter how slight, involves the transmission of a complex series of nerve impulses. Acetylcholine is released when a nerve cell is stimulated. The acetylcholine released at one nerve cell stimulates an adjacent nerve cell, which in turn releases acetylcholine to the next nerve cell. The impulse is thus carried along to the brain. After the nerve impulse has been transmitted, the acetylcholine must be destroyed, otherwise the nerve impulses would continue indefinitely. The enzyme cholinesterase is produced, when needed, to destroy acetylcholine. Cholinesterase acts by catalyzing the hydrolysis of the ester group of acetylcholine.

$$\underset{\text{acetylcholine}}{CH_3\overset{\overset{O}{\|}}{C}-O-CH_2CH_2-\underset{\underset{CH_3}{|}}{\overset{\overset{CH_3}{|}}{N^+}}-CH_3} \xrightarrow{\text{cholinesterase}}$$

$$CH_3\overset{\overset{O}{\|}}{C}-O^- + \underset{\text{choline}}{HO-CH_2CH_2-\underset{\underset{CH_3}{|}}{\overset{\overset{CH_3}{|}}{N^+}}-CH_3}$$

If the process is disrupted—either by inhibiting cholinesterase or by blocking the release of acetylcholine—paralysis and death may result. Nerve gases operate by inhibiting cholinesterase.

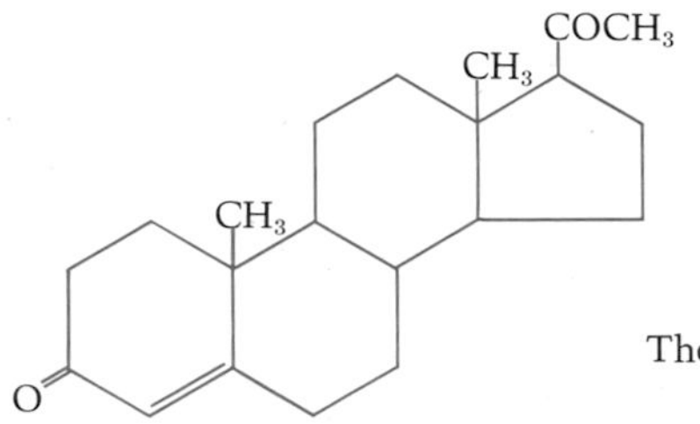

The compound (hormone) progesterone

Progesterone is a human hormone that belongs to a group of compounds known as steroids (these will be discussed in Chapter 20). However, progesterone also has a carbonyl group and is an example of a ketone. Although a build-up of ketone compounds in a diabetic patient can lead to brain or kidney damage, there are ketones in the human body that perform necessary and useful functions. Progesterone is one such ketone.

In the first three months of pregnancy, progesterone is produced by the ovaries and is essential to maintaining the pregnancy in its early stages. Later in the pregnancy the placenta undertakes the production of this crucial hormone. The chief functions of progesterone are to stop ovulation during pregnancy and to provide a stable environment in the uterus for the growth of an embryo. In addition, progesterone causes the body to retain sodium chloride (salt) and water—which can be an annoying side effect.

The human uterus has muscular walls, which produce rhythmic contractions throughout the uterus. Progesterone acts as a sedative on these muscular movements, thus helping to create conditions that are favorable for a growing fetus. However, when there is a shortage of the hormone (a progesterone deficiency) in the mother, uterine activity can increase, leading to dislodging of the embryo and eventual termination of the pregnancy. In fact, the deficiency may be responsible for a pattern of repeated miscarriages.

14

CARBONYL COMPOUNDS I: ALDEHYDES AND KETONES

Progesterone can be synthesized commercially from such common sources as soybeans and cholesterol. When there is a threat of miscarriage, progesterone may be given to the mother until the symptoms subside. Thus, in cases of progesterone deficiency, the timely administration of chemically synthesized progesterone may result in a successful pregnancy.

[*Illustration: Progesterone is an essential hormone for the maintenance of a full-term pregnancy.*]

CARBONYL COMPOUNDS I: ALDEHYDES AND KETONES

Aldehydes and ketones, which are not aromatic in chemical terms (except for the few that do have a benzene ring), still provide us with some of life's more memorable smells. Garbage, vanilla, formaldehyde, and apricots all gain their odors from aldehydes and ketones.

What aldehydes and ketones have in common, besides strong smells, is that each has a carbonyl group. A carbonyl group is just an oxygen and a carbon with a double bond between them. This group gives the aldehydes and ketones their distinctive odors and properties. The aldehydes and ketones are not the only organic compounds that contain the carbonyl group. In the next chapter we will go on to consider the carboxylic acids and related compounds that also feature the carbonyl group. Even then, we will not be through with the carbonyl group. Throughout the rest of this book, whenever we discuss a compound that is crucial to life, we will almost always discover that it has a carbonyl group.

14.1 CARBONYL GROUP STRUCTURE

The carbonyl group has a carbon–oxygen double bond.

Why is the carbonyl group more reactive than the carbon–carbon double bond?

Because oxygen is much more electronegative than carbon, the C=O double bond is highly polar—the oxygen atom has a rather large concentration of negative charge, and the carbon atom has an equal concentration of positive charge. In addition, we can write two resonance structures for the carbonyl group.

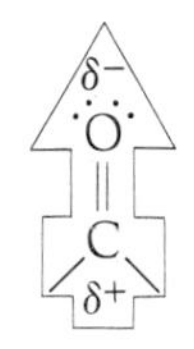

resonance structures

This resonance makes the carbonyl group even more polar. We will see that the polar nature of the carbonyl group accounts for most of the physical properties (Section 14.11) and chemical properties (Sections 14.7 and 14.10) of carbonyl compounds.

Problem 14.1 What factors contribute to the polarity of the carbonyl group? Explain each.

CARBONYL COMPOUNDS 14.2

The compounds that contain the carbonyl group are aldehydes, ketones, carboxylic acids, and carboxylic acid derivatives (acyl halides, carboxylic anhydrides, esters, and amides). A carboxylic acid actually contains a combination of a carbonyl group and a hydroxyl group; this combination is called a **carboxyl group.** We will discuss carboxylic acids and their derivatives in Chapter 15. In the structures below, for the acyl halide X is chlorine, bromine or iodine, and for the amide R′ is hydrogen, an alkyl group or an aryl group.

What structural feature do aldehydes and ketones have in common?

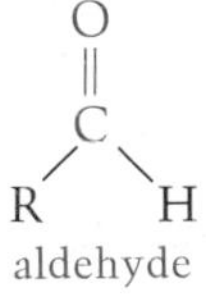

aldehyde

ketone

carboxylic acid

Carboxylic acid derivatives

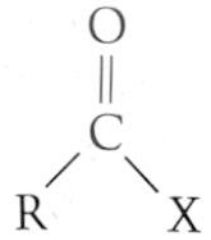

$$\mathrm{R{-}C({=}O){-}X}$$
acyl halide

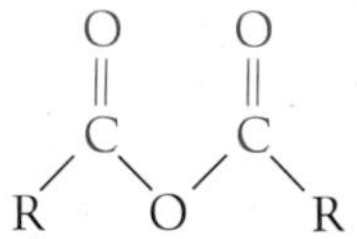

$$\mathrm{R{-}C({=}O){-}O{-}C({=}O){-}R}$$
carboxylic anhydride

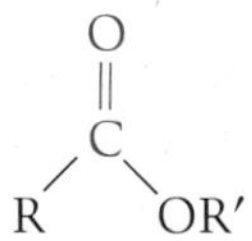

$$\mathrm{R{-}C({=}O){-}OR'}$$
ester

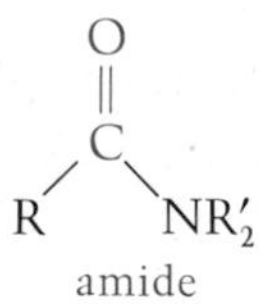

$$\mathrm{R{-}C({=}O){-}NR'_2}$$
amide

Problem 14.2 Identify each of the following compounds as an aldehyde, ketone, carboxylic acid, ester, carboxylic anhydride, amide, or acyl halide.

(a) 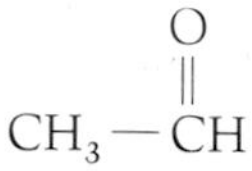 $CH_3{-}\overset{O}{\overset{\|}{C}}H$ (b) $CH_3{-}\overset{O}{\overset{\|}{C}}{-}OH$ (c) $C_6H_5{-}\overset{O}{\overset{\|}{C}}{-}C_6H_5$

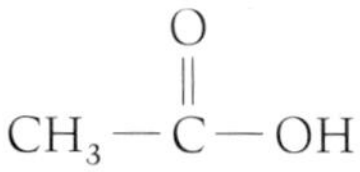

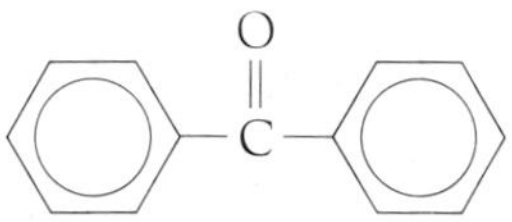

(d) 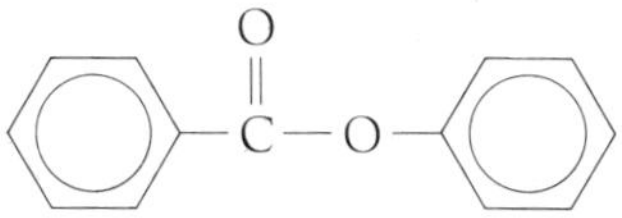 $C_6H_5{-}\overset{O}{\overset{\|}{C}}{-}O{-}C_6H_5$ (e) 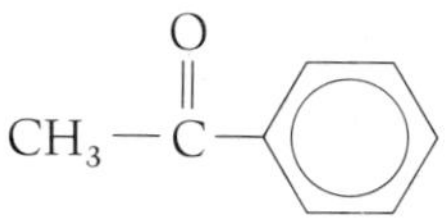 $CH_3{-}\overset{O}{\overset{\|}{C}}{-}C_6H_5$

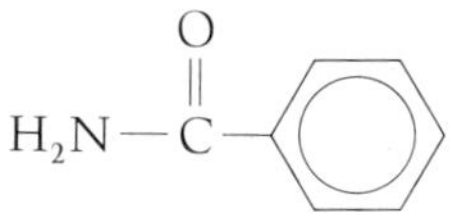

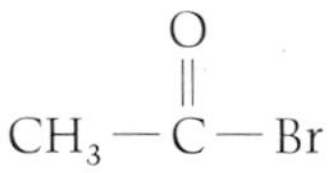

(f) 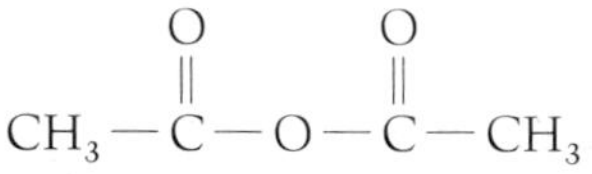 $CH_3{-}\overset{O}{\overset{\|}{C}}{-}O{-}\overset{O}{\overset{\|}{C}}{-}CH_3$ (g) $H_2N{-}\overset{O}{\overset{\|}{C}}{-}C_6H_5$

(h) $CH_3{-}\overset{O}{\overset{\|}{C}}{-}Br$ (i) 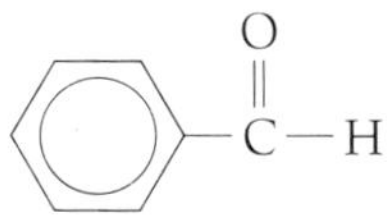 $C_6H_5{-}\overset{O}{\overset{\|}{C}}{-}H$

14.3 ALDEHYDES: STRUCTURE

In **aldehydes,** at least one H is bonded to the carbon of the carbonyl group. The —CHO group is thus called the aldehyde group. The aldehyde group is usually bonded to a hydrocarbon group, either alkyl or aryl. The one exception is formaldehyde, the simplest aldehyde, in which the aldehyde group is bonded to a hydrogen atom. Some examples of aldehydes are shown below (the aldehyde group is in color).

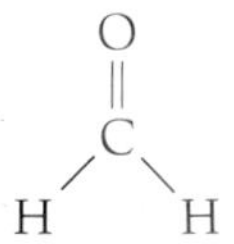

$$\mathrm{H{-}C({=}O){-}H}$$
formaldehyde

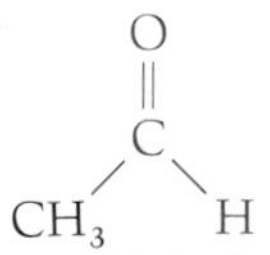

$$\mathrm{CH_3{-}C({=}O){-}H}$$
acetaldehyde

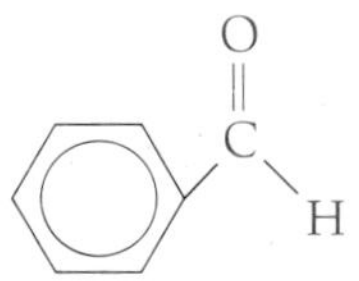

$$\mathrm{C_6H_5{-}C({=}O){-}H}$$
benzaldehyde

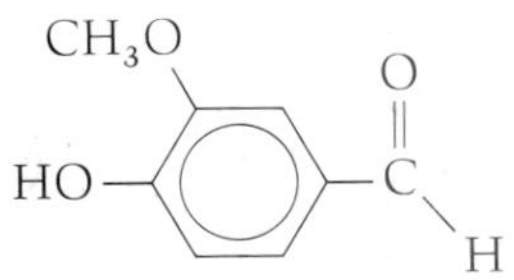

$$\mathrm{HO{-}C_6H_3(OCH_3){-}C({=}O){-}H}$$
vanillin

ALDEHYDES: NOMENCLATURE 14.4

The names used above—formaldehyde, acetaldehyde, benzaldehyde, and vanillin—are common names. Benzaldehyde is an IUPAC name as well. We will see later (Chapter 15) that these names are similar to the common names of the related carboxylic acids.

Why don't we indicate the position of the aldehyde group with a number?

We form the IUPAC names of the aldehydes by using the ending *-al* in place of the final *-e* of the corresponding hydrocarbon.

$H-\overset{\overset{O}{\|}}{C}-H$ methanal

$CH_3\overset{\overset{O}{\|}}{C}-H$ ethanal

$CH_3CH_2\overset{\overset{O}{\|}}{C}-H$ propanal

$CH_3CH_2CH_2\overset{\overset{O}{\|}}{C}-H$ butanal

$\overset{4}{C}H_3\overset{3}{C}H(CH_3)\overset{2}{C}H_2\overset{1}{\overset{\overset{O}{\|}}{C}}-H$ 3-methylbutanal

$CH_2=CH\overset{\overset{O}{\|}}{C}-H$ propenal

$\overset{5}{C}H_2=\overset{4}{C}H\overset{3}{C}H_2\overset{2}{C}H_2\overset{1}{\overset{\overset{O}{\|}}{C}}-H$ 4-pentenal

$O_2N-C_6H_4-\overset{\overset{O}{\|}}{C}-H$ *m*-nitrobenzaldehyde

Note that, in naming aldehydes, the carbonyl carbon atom is always numbered 1.

Aromatic aldehydes are named as derivatives of benzaldehyde. Most chemists use common names for complex aldehydes like vanillin (Section 14.3).

Problem 14.3 Give the structural formulas for the following compounds.
(a) 3-hydroxybutanal
(b) 2,3-dimethylpentanal
(c) *p*-nitrobenzaldehyde

Problem 14.4 Give the IUPAC name for each of the following compounds.

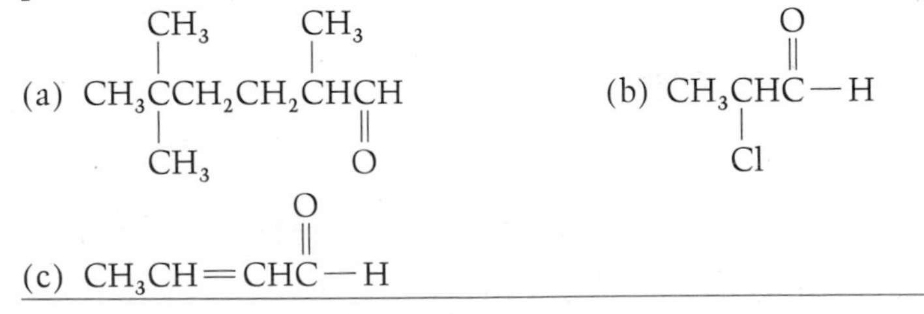

14.5 ALDEHYDES: REACTIONS

What bonds are involved in the two general reactions of aldehydes?

Aldehydes undergo two types of reactions: oxidation and addition. In one, the $-\overset{\overset{O}{\|}}{C}-H$ group is oxidized to a carboxyl group, $-\overset{\overset{O}{\|}}{C}-OH$.

$$R-\overset{\overset{O}{\|}}{C}-H \xrightarrow{\text{(oxidation)}} R-\overset{\overset{O}{\|}}{C}-OH$$

In the second type of reaction, the reagent adds to the C=O bond of the carbonyl group.

$$R-\overset{\overset{O}{\|}}{C}-H + AB \xrightarrow{\text{(addition)}} R-\underset{\underset{B}{|}}{\overset{\overset{OA}{|}}{C}}-H$$

These reactions are described in more detail in the next two sections.

14.6 ALDEHYDES: OXIDATION

What reagents are most useful for oxidizing aldehydes?

Aldehydes are among the most easily oxidized of organic compounds. Even atmospheric oxygen can oxidize most aldehydes to the corresponding carboxylic acid.

$$2R-\overset{\overset{O}{\|}}{C}-H + O_2 \rightarrow 2R-\overset{\overset{O}{\|}}{C}-OH$$

For this reason, aldehydes must be stored away from atmospheric oxygen.

Most oxidizing agents oxidize aldehydes to carboxylic acids. Two of the most common oxidizing agents are potassium permanganate and chromic acid.

$$3R-\overset{\overset{O}{\|}}{C}-H + 2MnO_4^- + OH^- \rightarrow 3R-\overset{\overset{O}{\|}}{C}-O^- + 2MnO_2 + 2H_2O$$

$$3R-\overset{\overset{O}{\|}}{C}-H + 2H_2CrO_4 \rightarrow 3R-\overset{\overset{O}{\|}}{C}-OH + Cr_2O_3 + 2H_2O$$

Because aldehydes are so easily oxidized, we can usually oxidize them without affecting other functional groups that may be present in the molecule. Several mild oxidizing agents are available that specifically oxidize aldehydes—Tollen's reagent, Fehling's reagent, and Benedict's reagent. Each of these oxidizing agents gives a visual change during the reaction (see following paragraphs), so we can also use them as chemical tests for the presence of aldehydes. These tests are also positive for other easily oxidizable groups, notably ketones with a hydroxyl group on the adjacent carbon atom.

Tollen's Test

Tollen's reagent is a water solution of silver nitrate and ammonia. This solution contains the complex ion $Ag(NH_3)_2^+$, which is a mild oxidizing agent. **Tollen's test** consists of adding Tollen's reagent to an alcohol solution of the aldehyde. The reaction produces silver metal, which coats the walls of the test tube. A silver mirror is formed inside the tube if the glass surface is clean. The equation for this reaction is

$$\underset{\text{aldehyde}}{RCHO} + 2H_2O + \underset{\text{Tollen's reagent}}{2Ag(NH_3)_2^+} \rightarrow \underset{\substack{\text{carboxylic}\\ \text{acid salt}}}{R\overset{\overset{\displaystyle O}{\|}}{C}-O^-} + \underset{\substack{\text{silver}\\ \text{mirror}}}{2Ag} + 4NH_4^+ + OH^-$$

Mirrors are manufactured commercially by using this reaction.

Fehling's Test

Fehling's test is also an oxidation–reduction reaction. The aldehyde is oxidized by copper (II) hydroxide ($Cu(OH)_2$), a mild oxidizing agent.

$$\underset{\text{aldehyde}}{RCHO} + \underset{\substack{\text{Fehling's}\\ \text{solution}}}{2Cu(OH)_2} + OH^- \rightarrow R\overset{\overset{\displaystyle O}{\|}}{C}-O^- + \underset{\substack{\text{reddish-brown}\\ \text{precipitate}}}{Cu_2O} + 3H_2O$$

However, copper (II) hydroxide is insoluble in water. To keep the Cu^{2+} ions in solution, sodium tartrate is added to the copper (II) hydroxide–water mixture. The tartrate ion forms a complex with copper (II) ion and keeps it dissolved in the aqueous OH^- solution.

Water solutions that contain Cu^{2+} ions and Cu^{2+} complex ions are blue. Cu_2O is a reddish-brown precipitate. The test consists of adding the Fehling's solution to the aldehyde and seeing whether the blue color disappears from the solution and a reddish-brown precipitate appears.

Fehling's solution decomposes on standing, so it must be prepared fresh each time it is used.

Benedict's Test

Benedict's test uses the same oxidizing agent and reaction as Fehling's test. The difference is that, instead of sodium tartrate, sodium citrate is used to maintain the $Cu(OH)_2$ in solution. The result is visually the same as for Fehling's test, but Benedict's test is generally preferred, because the reagent is more stable and can be stored for long periods of time.

Glucose contains an aldehyde group, so Fehling's and Benedict's reagents are used clinically to test for glucose in urine. The presence of glucose in urine is an indication of several conditions, including diabetes. Benedict's reagent can detect as little as 0.25% glucose in urine.

Problem 14.5 Which of the following compounds would give a positive test with either Tollen's, Fehling's, or Benedict's solution?

(a) 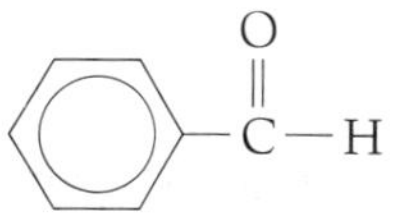 $C_6H_5-\overset{\overset{\large O}{\|}}{C}-H$

(b) $H-\overset{\overset{\large O}{\|}}{C}-H$

(c) 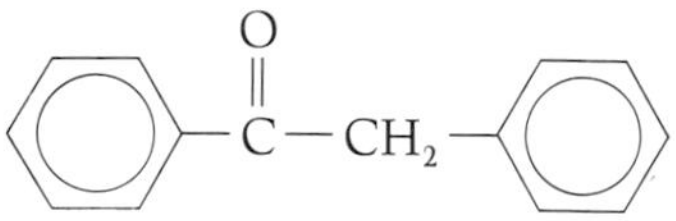 $C_6H_5-\overset{\overset{\large O}{\|}}{C}-CH_2-C_6H_5$

(d) $CH_3\overset{\overset{\large O}{\|}}{C}-OH$

(e) $\underset{\large OH}{CH_2}\underset{\large OH}{CH}\underset{\large OH}{CH}\underset{\large OH}{CH}\underset{\large OH}{CH}\overset{\overset{\large O}{\|}}{C}-H$ (glucose)

Problem 14.6 Write the equations for the reactions of glucose in Fehling's and Benedict's tests.

Problem 14.7 Tell how the two test reagents—Benedict's and Fehling's—differ. Explain the purpose of the additives sodium tartrate and sodium citrate. Which of the two tests is more convenient? Why?

ALDEHYDES: ADDITION REACTIONS 14.7

The carbon–oxygen double bond of the carbonyl group undergoes addition reactions even more readily than a carbon–carbon double bond does. The reason for this great reactivity is the polar nature of the carbonyl group (Section 14.1). When a polar reagent is added to a carbonyl group, the negative part of the reagent becomes bonded to carbon, and the positive part becomes bonded to oxygen. In general terms, we can show the reaction as

What happens to the C=O bond in addition reactions of aldehydes?

$$\overset{\delta+}{A}-\overset{\delta-}{B} + R-\underset{\delta+}{\overset{\overset{\delta-}{O}}{\overset{\|}{C}}}-R' \rightarrow R-\underset{B}{\overset{O-A}{\overset{|}{\underset{|}{C}}}}-R'$$

polar reagent

Some typical polar reagents are hydrogen cyanide, alcohols, and Grignard reagents (see below). A few examples will illustrate this reaction.

Hydrogen Cyanide

Aldehydes add hydrogen cyanide (HCN) to produce **cyanohydrins.** Cyanohydrins have a hydroxyl group and a cyano group ($-C{\equiv}N:$) bonded to the same carbon atom.

$$R-\overset{\overset{O}{\|}}{C}-H + H-CN \rightarrow R-\underset{CN}{\overset{OH}{\overset{|}{\underset{|}{C}}}}-H$$

cyanohydrin

Chemists use this reaction to extend the carbon chains of aldehydes by one unit.

Grignard Reaction

Alkyl and aryl halides react with magnesium metal in dry ether (ether that is free of moisture) solution to produce **Grignard reagents** (named after their discoverer, the French chemist Victor Grignard). Alkyl or aryl chlorides, bromides, or iodides can be used.

$$\mathrm{R{-}Cl + Mg \xrightarrow{ether} RMgCl}$$
an alkyl Grignard reagent

$$\mathrm{Ar{-}Br + Mg \xrightarrow{ether} ArMgBr}$$
an aryl Grignard reagent

The covalent bond between the magnesium and carbon atoms is highly polar. Carbon is much more electronegative than magnesium, so the carbon atom has a partial negative charge.

$$\overset{\delta-}{-\underset{|}{\overset{|}{\mathrm{C}}}}-\overset{\delta+}{\mathrm{MgBr}}$$

Grignard reagents are very reactive. They must be kept dry because they react with water to produce a hydrocarbon.

$$\mathrm{RMgBr + H_2O \rightarrow R{-}H + MgBrOH}$$

Grignard reagents react with aldehydes in ether solution to produce alcohols. The reaction occurs in two steps. If we react formaldehyde with a Grignard reagent, we obtain a primary alcohol. All other aldehydes produce secondary alcohols.

Step 1: $$\mathrm{RMgCl} + \underset{\text{formaldehyde}}{\mathrm{H{-}\overset{\overset{\displaystyle O}{\|}}{C}{-}H}} \xrightarrow{\text{ether}} \mathrm{R{-}CH_2{-}OMgCl}$$

Step 2: $$\mathrm{R{-}CH_2{-}OMgCl + H_2O \rightarrow R{-}CH_2{-}OH + MgClOH}$$

Step 1: $$\mathrm{RMgCl + R'{-}\overset{\overset{\displaystyle O}{\|}}{C}{-}H \xrightarrow{ether} R'{-}\underset{\underset{\displaystyle R}{|}}{\overset{\overset{\displaystyle OMgCl}{|}}{CH}}}$$

Step 2: $$\mathrm{R'{-}\underset{\underset{\displaystyle R}{|}}{\overset{\overset{\displaystyle OMgCl}{|}}{CH}} + H_2O \rightarrow R'{-}\overset{\overset{\displaystyle OH}{|}}{CH}{-}R + MgClOH}$$

This is a very useful reaction in organic synthesis, because we can use it to form a carbon–carbon bond and thus to construct large molecules from smaller ones.

Acetals

When we mix an aldehyde and an alcohol, they react to form a compound called a **hemiacetal.** The hemiacetal is the product of the addition of an alcohol molecule to the carbonyl double bond.

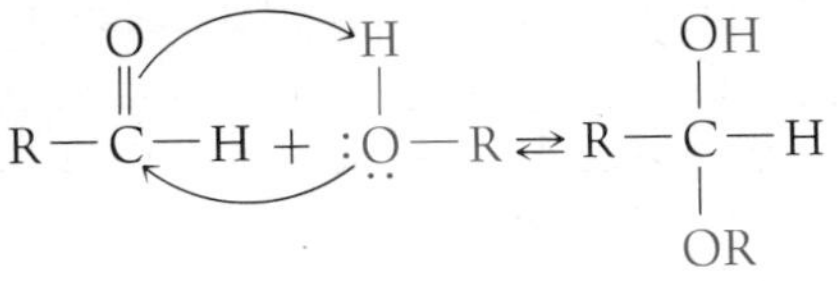

$$R-\overset{\overset{O}{\|}}{C}-H + :\ddot{O}(H)-R \rightleftharpoons R-\underset{OR}{\overset{OH}{\overset{|}{\underset{|}{C}}}}-H$$

an aldehyde an alcohol a hemiacetal

The reaction is reversible, and most hemiacetals are too unstable to be isolated. Cyclic hemiacetals (formed from a molecule that has *both* an aldehyde *and* an alcohol group), however, are generally stable, especially if they have five- or six-membered rings.

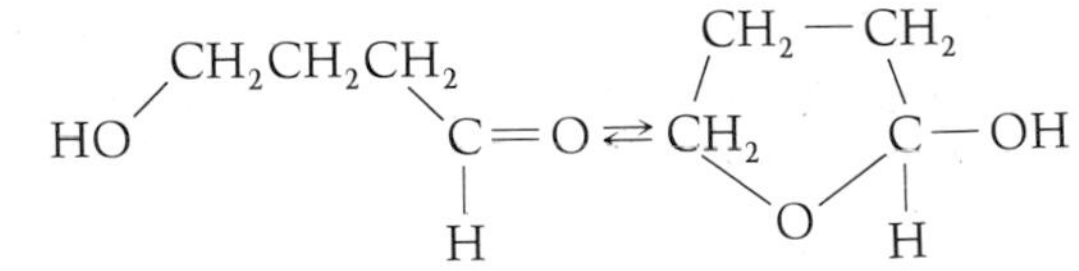

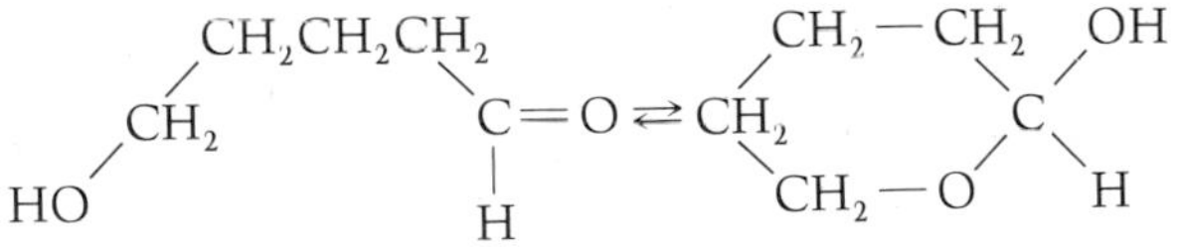

$$HO-CH_2CH_2CH_2-\underset{H}{\underset{|}{C}}=O \rightleftharpoons \text{(five-membered ring: } CH_2-CH_2,\ CH_2,\ O,\ \underset{H}{C}-OH)$$

$$HO-CH_2-CH_2CH_2CH_2-\underset{H}{\underset{|}{C}}=O \rightleftharpoons \text{(six-membered ring: } CH_2-CH_2,\ CH_2,\ CH_2-O,\ C(OH)(H))$$

Most simple sugars, such as glucose, exist primarily in the cyclic hemiacetal form, in equilibrium with the aldehyde form.

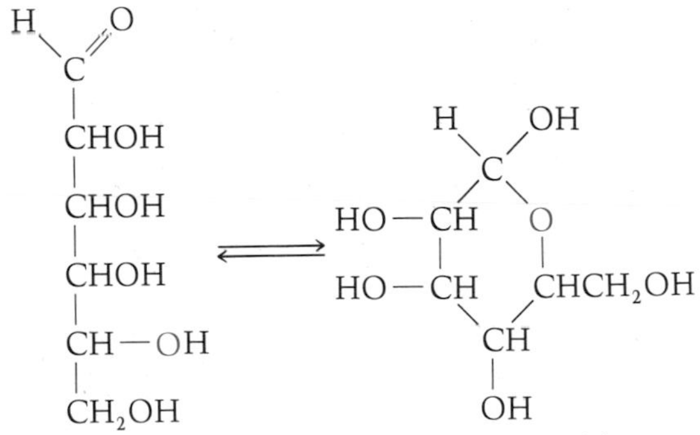

$$\begin{array}{c} H\ \ O \\ \diagdown\!\!\!/\!\!/ \\ C \\ | \\ CHOH \\ | \\ CHOH \\ | \\ CHOH \\ | \\ CH-OH \\ | \\ CH_2OH \end{array} \rightleftharpoons \begin{array}{c} H\ \ \ OH \\ C \\ HO-CH\ \ \ O \\ |\ \ \ \ \ \ | \\ HO-CH\ \ \ CHCH_2OH \\ CH \\ | \\ OH \end{array}$$

a simple sugar (small amount at equilibrium) cyclic hemiacetal form (large amount at equilibrium)

Because sugars exist as hemiacetals, the study of hemiacetals is important to the study of the chemicals of life.

If we acidify an aldehyde–alcohol solution by mixing in some dry, gaseous HCl, a second reaction occurs; this time, the hemiacetal reacts with another molecule of alcohol to produce an **acetal.**

$$\text{R}-\underset{\displaystyle \text{OCH}_2\text{CH}_3}{\overset{\displaystyle \text{OH}}{\underset{|}{\overset{|}{\text{C}}}}}-\text{H} + \text{HOCH}_2\text{CH}_3 \xrightleftharpoons{\text{HCl}} \text{R}-\underset{\displaystyle \text{OCH}_2\text{CH}_3}{\overset{\displaystyle \text{OCH}_2\text{CH}_3}{\underset{|}{\overset{|}{\text{C}}}}}-\text{H} + \text{H}_2\text{O}$$

an acetal

The equilibrium exists only if acid is present. If we neutralize the acid with a base, we suspend the equilibrium.* The reaction slows down in both directions. As long as the reaction medium is basic, the acetal does not equilibrate with hemiacetal or aldehyde.

Acetals are a special class of ethers, but because two alkoxy groups (—OR) are bonded to the same carbon atom, acetals and simple ethers behave differently. In an acetal, each alkoxy group increases the reactivity of the other, so acetals are more reactive than ordinary ethers.

Sugars also form acetals. We shall study the formation of sugar acetals, as well as other reactions of sugars, in Chapter 16.

Problem 14.8 Give the major organic products of each of the following reactions.

(a) $\text{C}_6\text{H}_5-\overset{\displaystyle \text{O}}{\overset{\|}{\text{C}}}-\text{H} + \text{HCN} \longrightarrow$

(b) $\text{C}_6\text{H}_5-\text{Cl} + \text{Mg} \xrightarrow{\text{ether}}$

(c) Product of (b) $+ \text{CH}_3\overset{\displaystyle \text{O}}{\overset{\|}{\text{C}}}-\text{H} \xrightarrow{\text{ether}}$

(d) Product of (c) $+ \text{H}_2\text{O} \longrightarrow$

(e) Product of (b) $+ \text{H}-\overset{\displaystyle \text{O}}{\overset{\|}{\text{C}}}-\text{H} \xrightarrow{\text{ether}}$

(f) Product of (e) $+ \text{H}_2\text{O} \longrightarrow$

(g) $\text{CH}_3\text{CH}_2\overset{\displaystyle \text{O}}{\overset{\|}{\text{C}}}-\text{H} + \text{CH}_3\text{OH} \rightleftarrows$

---

*Acid catalyzes both the forward and the reverse reaction by providing a pathway with a low free energy of activation. Removal of the acid forces both reactions to occur along a path of very high free energy of activation. The practical result is that both reactions are so slow in basic solutions that reactants and products do not attain a state of equilibrium with each other.

(h) Product of (g) + $CH_3OH \xrightleftharpoons{HCl(g)}$

(i) $HOCH_2CH_2CH_2\overset{\overset{O}{\|}}{C}-H \rightleftharpoons$

KETONES: STRUCTURE 14.8

How do ketones differ structurally from aldehydes?

A **ketone** has two hydrocarbon groups bonded to the carbonyl carbon atom. The hydrocarbon groups can be alkyl or aryl. Some common examples are shown below.

$CH_3-\overset{\overset{O}{\|}}{C}-CH_3$ acetone

$CH_3-\overset{\overset{O}{\|}}{C}-CH_2CH_3$ methyl ethyl ketone

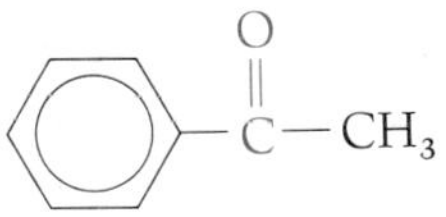

$C_6H_5-\overset{\overset{O}{\|}}{C}-CH_3$ acetophenone (or methyl phenyl ketone)

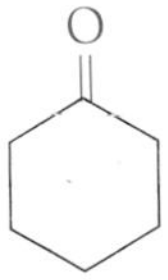

cyclohexanone

Problem 14.9 Which of the following compounds are *not* ketones?

(a) 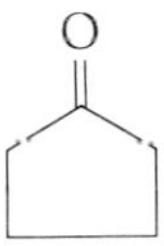 cyclopentanone (b) 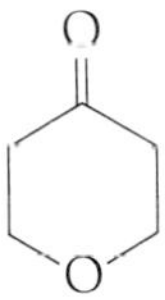 tetrahydro-4H-pyran-4-one (ring O) (c) $CH_3\overset{\overset{O}{\|}}{C}-OCH_3$ (d) 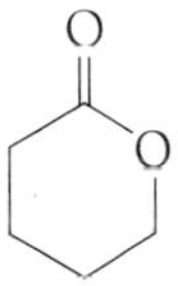 six-membered ring lactone (C=O adjacent to ring O)

(e) $CH_3\overset{\overset{O}{\|}}{C}CH_2\overset{\overset{O}{\|}}{C}CH_3$

KETONES: NOMENCLATURE 14.9

The common name of a ketone consists simply of the names of the two hydrocarbon groups attached to the carbonyl group, followed by the word "ketone" (see the examples in Section 14.8).

Is it necessary to specify the position of the carbonyl group in naming ketones?

We form the IUPAC name by replacing the ending *-e* of the comparable alkane with *-one.* We number the chain beginning at the end nearer to the carbonyl group. The following examples illustrate IUPAC names for ketones.

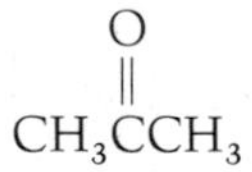

$CH_3C(=O)CH_3$ — propanone

$CH_3C(=O)CH_2CH_3$ — butanone

$\overset{1}{C}H_3\overset{2}{C}(=O)\overset{3}{C}H_2\overset{4}{C}H(CH_3)\overset{5}{C}H_3$ — 4-methyl-2-pentanone

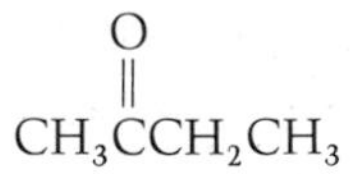

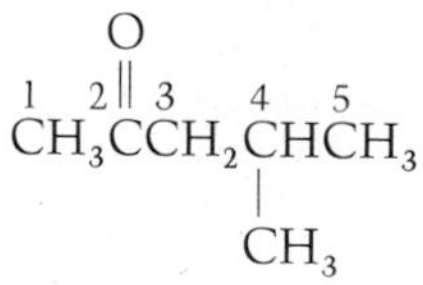

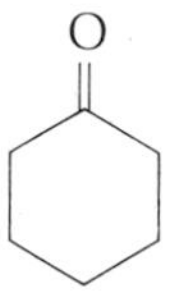

cyclohexanone

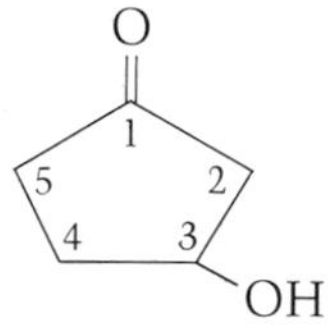

3-hydroxycyclopentanone (ring numbered 1–5, C=O at 1, OH at 3)

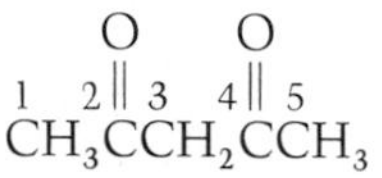

$\overset{1}{C}H_3\overset{2}{C}(=O)\overset{3}{C}H_2\overset{4}{C}(=O)\overset{5}{C}H_3$ — 2,4-pentanedione

Problem 14.10 Write the structural formulas of the following ketones.

(a) diethyl ketone
(b) phenyl ethyl ketone
(c) methyl isopropyl ketone
(d) 3,4-dimethylcyclopentanone
(e) 3-hexanone
(f) 1,4-cyclohexanedione

Problem 14.11 Give the IUPAC and, where possible, the common name of each of the following compounds.

(a) 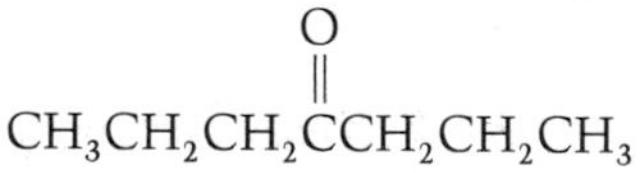 $CH_3CH_2CH_2C(=O)CH_2CH_2CH_3$

(b) 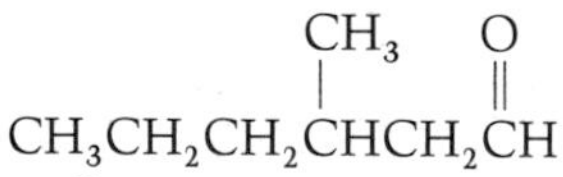 $CH_3CH_2CH_2CH(CH_3)CH_2C(=O)H$

(c) 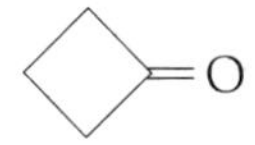 cyclobutane=O

(d) 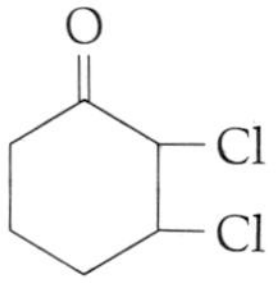 cyclohexanone with Cl, Cl substituents

(e) 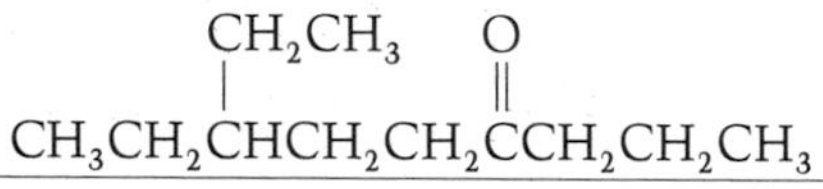 $CH_3CH_2CH(CH_2CH_3)CH_2CH_2C(=O)CH_2CH_2CH_3$

14.10 KETONES: REACTIONS

What reaction of aldehydes fails with ketones?

Ketones do not have the carbonyl C—H bond that is so easily oxidized in aldehydes. Oxidation of a C—C bond is much more difficult, so ketones do not readily undergo oxidation. However, they decompose to give complex mixtures of products when they are heated with either potassium permanganate or chromic acid.

$$\mathrm{R{-}\overset{\overset{\displaystyle O}{\|}}{C}{-}R} \begin{cases} \xrightarrow{\text{(a) ordinary oxidizing agents at room temperature}} \text{no reaction} \\ \xrightarrow{\text{(b) } KMnO_4 \text{ or } H_2CrO_4 \text{ + heat}} \text{decomposition products} \end{cases}$$

Ketones undergo the same addition reactions as aldehydes. We will only summarize them here (see Section 14.7 for more details).

Addition of HCN

$$\mathrm{R{-}\overset{\overset{\displaystyle O}{\|}}{C}{-}R' + HCN \rightarrow R{-}\underset{\underset{\displaystyle CN}{|}}{\overset{\overset{\displaystyle OH}{|}}{C}}{-}R'}$$

a cyanohydrin

Grignard reaction

$$\mathrm{R{-}\overset{\overset{\displaystyle O}{\|}}{C}{-}R' + R''MgBr \xrightarrow{ether} R{-}\underset{\underset{\displaystyle R''}{|}}{\overset{\overset{\displaystyle OMgBr}{|}}{C}}{-}R' \xrightarrow{H_2O} MgBrOH + R{-}\underset{\underset{\displaystyle R''}{|}}{\overset{\overset{\displaystyle OH}{|}}{C}}{-}R'}$$

The Grignard reaction with ketones always gives tertiary alcohols.

A convenient way to write a series of chemical transformations, such as the Grignard reaction above, is to use curved-arrow notation. The curved arrows show where chemicals were added or by-products formed at each step.

$$\text{Reactant} \xrightarrow{\text{chemical added} \searrow\!\nearrow \text{by-product formed}} \text{major product}$$

Using this notation, we may write the Grignard reaction as follows.

$$\mathrm{R{-}\overset{\overset{\displaystyle O}{\|}}{C}{-}R' \xrightarrow{R''MgBr\ (ether)} R{-}\underset{\underset{\displaystyle R''}{|}}{\overset{\overset{\displaystyle OMgBr}{|}}{C}}{-}R' \xrightarrow{H_2O \searrow\!\nearrow MgBrOH} R{-}\underset{\underset{\displaystyle R''}{|}}{\overset{\overset{\displaystyle OH}{|}}{C}}{-}R'}$$

Equations written in this form need not be balanced.

Problem 14.12 Express the following equations by using curved-arrow notation.

(a) $$CH_3\overset{\overset{OH}{|}}{\underset{\underset{H}{|}}{C}}CH_3 + H_2CrO_4 \rightarrow CH_3\overset{\overset{O}{\|}}{C}CH_3 + Cr_2O_3 + H_2O$$

(b) $$CH_3\overset{\overset{O}{\|}}{C}CH_3 + HCN \rightarrow CH_3\overset{\overset{OH}{|}}{\underset{\underset{CN}{|}}{C}}CH_3$$

The products of the reaction of a ketone with an alcohol are called **hemiketals** and **ketals.**

$$R-\overset{\overset{O}{\|}}{C}-R' + CH_3OH \rightleftharpoons R-\overset{\overset{OH}{|}}{\underset{\underset{OCH_3}{|}}{C}}-R'$$

a hemiketal

$$R-\overset{\overset{OH}{|}}{\underset{\underset{OCH_3}{|}}{C}}-R' + CH_3OH \overset{H^+}{\rightleftharpoons} R-\overset{\overset{OCH_3}{|}}{\underset{\underset{OCH_3}{|}}{C}}-R' + H_2O$$

a ketal

Using curved-arrow notation, the above reactions are written

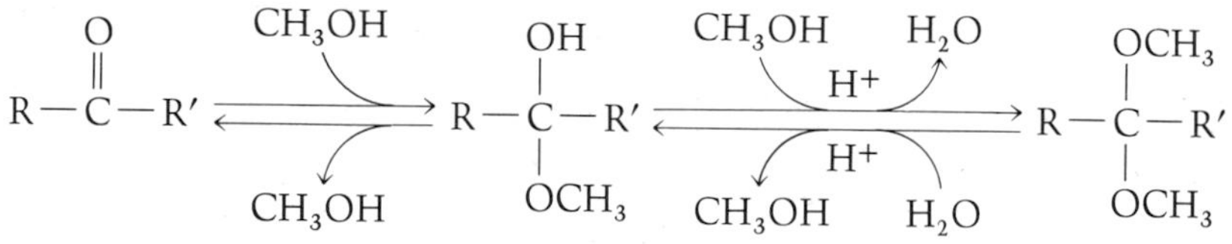

$$R-\overset{\overset{O}{\|}}{C}-R' \underset{CH_3OH}{\overset{CH_3OH}{\rightleftharpoons}} R-\overset{\overset{OH}{|}}{\underset{\underset{OCH_3}{|}}{C}}-R' \underset{CH_3OH \quad H^+ \quad H_2O}{\overset{CH_3OH \quad H^+ \quad H_2O}{\rightleftharpoons}} R-\overset{\overset{OCH_3}{|}}{\underset{\underset{OCH_3}{|}}{C}}-R'$$

Problem 14.13 Give the major organic product of each of the following reactions.

(a) 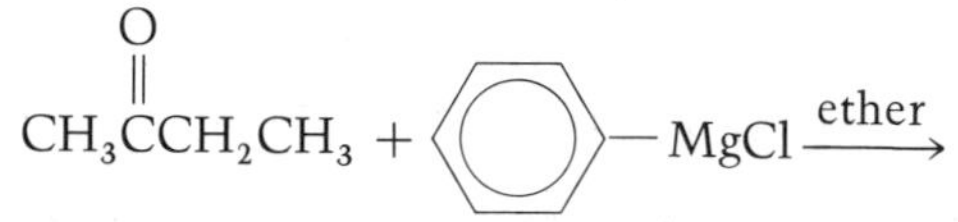 $$CH_3\overset{\overset{O}{\|}}{C}CH_2CH_3 + C_6H_5-MgCl \xrightarrow{\text{ether}}$$

(b) Product of (a) + $H_2O \rightarrow$

(c) 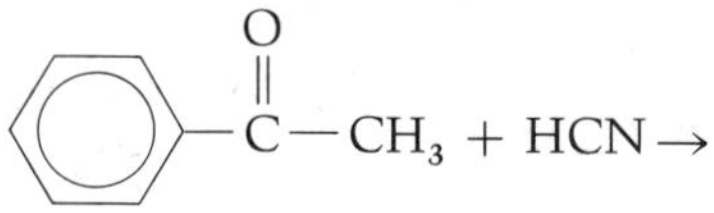 $$C_6H_5-\overset{\overset{O}{\|}}{C}-CH_3 + HCN \rightarrow$$

(d) $C_6H_5{-}CH_2\overset{O}{\overset{\|}{C}}CH_3 + CH_3OH \rightleftarrows$

(e) Product of (d) + CH_3OH + HCl(gaseous) $\rightleftarrows$

(f) Express reactions (a) and (b) in the curved-arrow notation.
(g) Express reactions (d) and (e) in the curved-arrow notation.

ALDEHYDES AND KETONES: PHYSICAL PROPERTIES 14.11

Neither aldehydes nor ketones have an O—H group that can engage in hydrogen bonding. This absence of hydrogen bonding in aldehydes and ketones explains why these compounds, when pure, have lower boiling points than alcohols of comparable molecular weight (Table 14.1). However, because of the highly polar carbonyl group, aldehydes and ketones have higher boiling points than alkanes of comparable molecular weight.

Do aldehydes and ketones more closely resemble ethers or alcohols in their physical properties?

TABLE 14.1 Physical Properties of Some Aldehydes, Ketones, and Other Hydrocarbons

Name	Formula	Molecular weight	Boiling point, °C	Solubility, g/100 mL H_2O, 20°C
Propanal	$CH_3CH_2\overset{O}{\overset{\|}{C}}{-}H$	58	48.8	20
Acetone	$CH_3\overset{O}{\overset{\|}{C}}CH_3$	58	56.5	∞
1-Propanol	$CH_3CH_2CH_2OH$	60	97.2	∞
Butane	$CH_3CH_2CH_2CH_3$	58	−0.6	Insoluble (~0.04 at 17°C)
Butanal	$CH_3CH_2CH_2\overset{O}{\overset{\|}{C}}{-}H$	72	75.6	3.7
Butanone	$CH_3CH_2\overset{O}{\overset{\|}{C}}CH$	72	79.6	35
1-Butanol	$CH_3CH_2CH_2CH_2OH$	74	117.7	7.9
2-Butanol	$CH_3CH_2\underset{OH}{\underset{\|}{C}}HCH_3$	74	99.5	12.5
Pentane	$CH_3CH_2CH_2CH_2CH_3$	72	36.2	Insoluble (0.036 at 16°)

The highly polar carbonyl bond also allows aldehydes and ketones to form strong hydrogen bonds with water molecules.

For this reason, aldehydes and ketones are more soluble in water than hydrocarbons of comparable molecular weight (Table 14.1). Aldehydes and ketones resemble alcohols in their water solubility.

Problem 14.14 On the basis of hydrogen bonding or polarity, explain the differences and similarities in boiling points and water solubilities of each of the following pairs of compounds.

	Compound	*Boiling point, °C*	*Solubility in water*
(a)	$CH_3C(=O)-H$	21	Soluble in all proportions
	CH_3CH_2OH	78.5	Soluble in all proportions
(b)	$CH_3CHOHCH_3$	82.4	Soluble in all proportions
	$CH_3C(=O)CH_3$	56.2	Soluble in all proportions
(c)	$H-C(=O)-H$	−21	Soluble
	CH_3CH_3	−88.6	Insoluble

14.12 SOME IMPORTANT ALDEHYDES AND KETONES

Formaldehyde is the simplest aldehyde. It is a pungent gas that readily forms polymers on standing. The polymers decompose on heating to give formaldehyde gas again.

$$n\,H_2C{=}O \rightarrow \sim\!\sim CH_2-O\!\left(CH_2-O\right)_n CH_2-O\sim\!\sim$$

Formaldehyde is used mainly in the manufacture of plastics, notably Bakelite, one of the earliest commercial plastic materials. It also finds wide use as a germicide, fungicide, and embalming agent. In these applications a 37 to 40% solution of formaldehyde in water is used. This solution, called *Formalin*, also contains a little methanol. Water solutions of formaldehyde polymerize on standing unless they contain methanol. The methanol prevents polymerization by reacting with the formaldehyde to produce the hemiacetal, and probably some acetal.

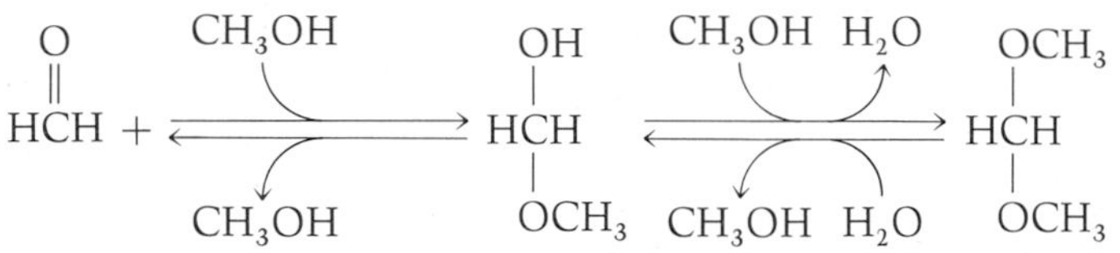

$$\mathrm{HCH{=}O} + CH_3OH \underset{CH_3OH}{\overset{CH_3OH}{\rightleftharpoons}} \mathrm{HCH(OH)(OCH_3)} \underset{CH_3OH\ \ H_2O}{\overset{CH_3OH\ \ H_2O}{\rightleftharpoons}} \mathrm{HCH(OCH_3)(OCH_3)}$$

Benzaldehyde occurs in the kernels of bitter almonds. Synthetic benzaldehyde is used as artificial oil of almond. It is also used widely in the manufacture of dyes, perfumes, flavors, and solvents. In high doses, it is a narcotic.

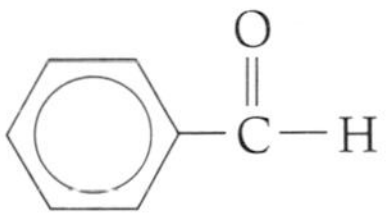

$$C_6H_5-\overset{O}{\overset{\|}{C}}-H$$

benzaldehyde

Acetaldehyde is a commercially important synthetic intermediate. It finds wide use in the manufacture of other industrial chemicals, perfumes, flavors, plastics, drugs, and fibers. It is a narcotic in large doses. Symptoms of chronic exposure to acetaldehyde resemble the symptoms of chronic alcoholism. The reason for this similarity becomes apparent if we compare the structures of these two compounds. We see that acetaldehyde is just oxidized ethanol.

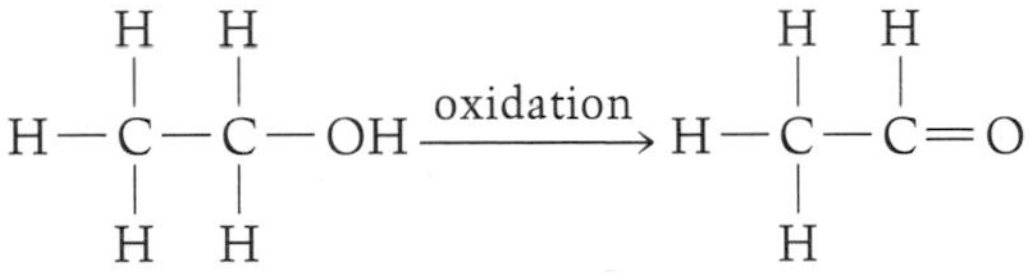

$$\underset{\text{ethanol}}{H-\overset{H}{\underset{H}{C}}-\overset{H}{\underset{H}{C}}-OH} \xrightarrow{\text{oxidation}} \underset{\text{acetaldehyde}}{H-\overset{H}{\underset{H}{C}}-\overset{H}{C}=O}$$

Acetaldehyde is known to be an intermediate in the metabolism of ethanol in the body.

Acetone is one of the most widely used solvents. It is also an important synthetic intermediate for the manufacture of such diverse products as explosives, rayon, photographic films, and paint and varnish removers.

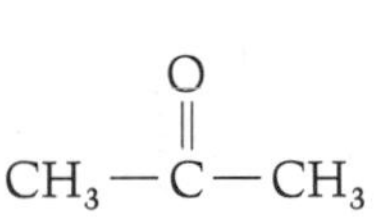

$$CH_3-\overset{O}{\overset{\|}{C}}-CH_3$$

acetone

SUMMARY

The properties and reactions of **aldehydes** and **ketones** result from the polarity of the carbonyl group. In aldehydes, the **carbonyl group** can be oxidized by a variety of reagents. Ketones are not readily oxidized. Both aldehydes and ketones undergo the following addition reactions: addition of HCN to form **cyanohydrins,** addition of **Grignard reagents** to form alcohols, and addition of alcohols, first to form **hemiacetals** and **hemiketals,** then to form **acetals** and **ketals,** respectively.

Key Terms

Acetal (Section 14.7)
Aldehyde (Section 14.3)
Benedict's test (Section 14.6)
Carboxyl group (Section 14.1)
Cyanohydrin (Section 14.7)
Fehling's test (Section 14.6)
Grignard reagent (Section 14.7)
Hemiacetal (Section 14.7)
Hemiketal (Section 14.10)
Ketal (Section 14.10)
Ketone (Section 14.8)
Tollen's test (Section 14.6)

ADDITIONAL PROBLEMS

14.15 Identify each of the following compounds as an aldehyde, ketone, carboxylic acid, ester, or acid anhydride.

(a) 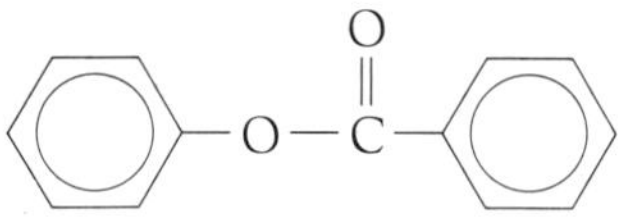 $C_6H_5-O-C(=O)-C_6H_5$ (b) 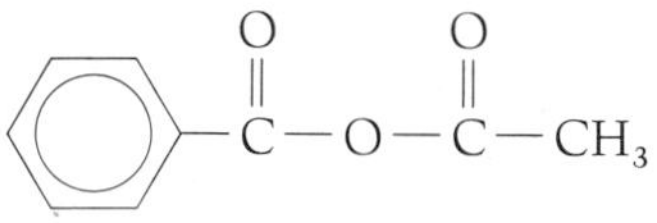 $C_6H_5-C(=O)-O-C(=O)-CH_3$

(c) 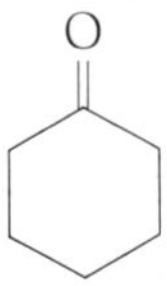 (d) 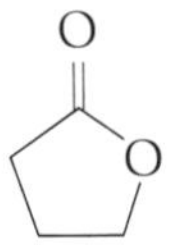 (e) 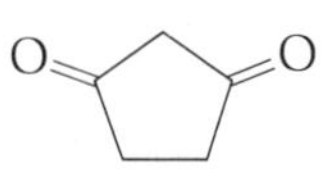 (f)

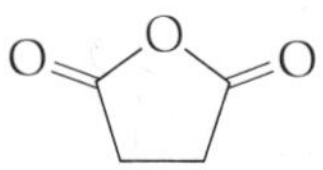

(g) 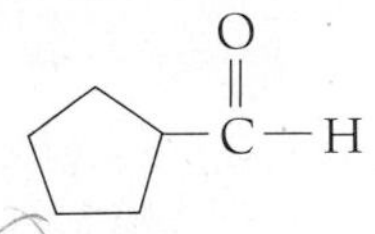 cyclopentyl—$\overset{O}{\overset{\|}{C}}-H$ (h) phenyl—$CH_2\overset{O}{\overset{\|}{C}}-H$

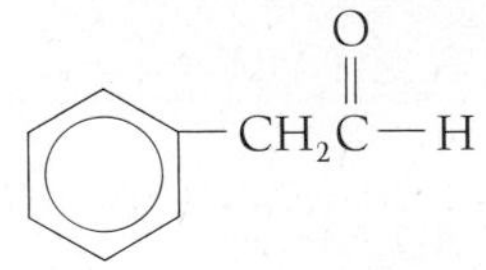

14.16 Give the structural formulas of the following compounds.
(a) propanal (b) propanone
(c) diphenyl ketone (d) *p*-chlorobenzaldehyde
(e) 3-chlorobutanal (f) 3-bromo-2-butanone
(g) 4-bromocyclohexanone

14.17 Give the IUPAC names of the following compounds.

(a) $CH_3\underset{\underset{CH_3}{|}}{C}HCH_2\overset{O}{\overset{\|}{C}}-H$ (b) $H-\overset{O}{\overset{\|}{C}}-H$

(c) $CH_3\overset{O}{\overset{\|}{C}}-H$ (d) $CH_2=CH\overset{O}{\overset{\|}{C}}CH_2CH_3$

(e) $ClCH_2CH_2\overset{O}{\overset{\|}{C}}CH_2CH_2Cl$

14.18 Describe what you *do* and *see* in
(a) Tollen's test (b) Fehling's test
(c) Benedict's test.
Write equations to show the chemical reactions that occur in each test.

14.19 Give the major organic product (if any) of each of the following reactions.

(a) $CH_3\overset{O}{\overset{\|}{C}}-H + Ag(NH_3)_2^+ \xrightarrow{H_2O}$

(b) $CH_3\overset{O}{\overset{\|}{C}}CH_3 + \text{Cu(II)tartrate complex} \rightarrow$

(c) $H-\overset{O}{\overset{\|}{C}}-H + \text{Cu(II)citrate complex} \rightarrow$

(d) $CH_3\overset{O}{\overset{\|}{C}}CH_3 + HCN \rightarrow$

(e) $CH_3CH_2Br + Mg \xrightarrow{\text{ether}}$

(f) Product of (e) + $CH_3\overset{O}{\overset{\|}{C}}H \xrightarrow{\text{ether}}$

(g) Product of (f) + $H_2O \rightarrow$

(h) 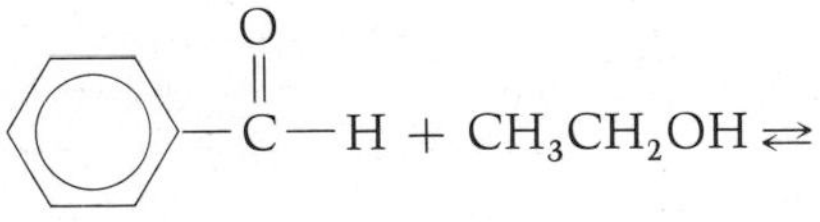 phenyl—$\overset{O}{\overset{\|}{C}}-H + CH_3CH_2OH \rightleftarrows$

(i) Product of (h) + $CH_3CH_2OH \xrightleftharpoons{HCl(g)}$

(j) $CH_3\overset{O}{\overset{\|}{C}}CH_3 + Ag(NH_3)_2^+ \xrightarrow{H_2O}$

(k) 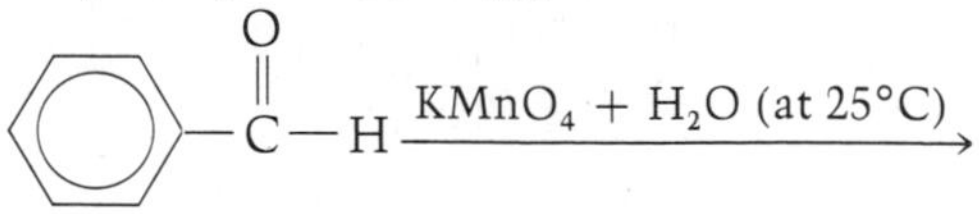 $C_6H_5-\overset{O}{\overset{\|}{C}}-H \xrightarrow{KMnO_4 + H_2O \text{ (at 25°C)}}$

(l) $CH_3CH_2\overset{O}{\overset{\|}{C}}-H \xrightarrow{H_2CrO_4 + H_2O \text{ (at 25°C)}}$

(m) $CH_3\overset{O}{\overset{\|}{C}}CH_3 \xrightarrow{KMnO_4 + H_2O \text{ (at 25°C)}}$

(n) 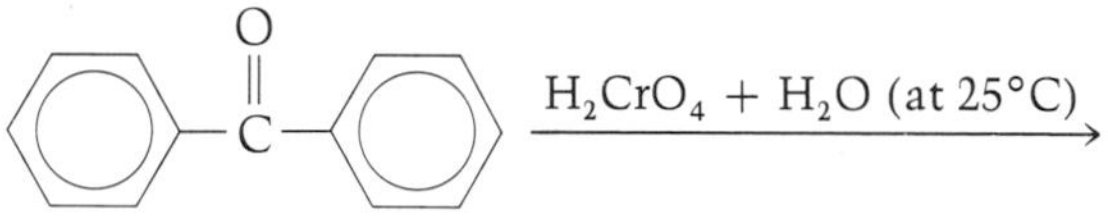 $C_6H_5-\overset{O}{\overset{\|}{C}}-C_6H_5 \xrightarrow{H_2CrO_4 + H_2O \text{ (at 25°C)}}$

14.20 Express the following reaction sequences from Problem 14.19 in the curved-arrow notation.

(a) Problem 14.19 (e), (f), and (g).

(b) Problem 14.19 (h) and (i).

14.21 The boiling point of 1-propanol is 97°C; the boiling point of propanal is 49°C. Explain this large difference.

14.22 Would you expect the water solubilities of propanal and propanol to be as different as their boiling points? Actually, both are very soluble in water. Explain.

14.23 Tell which compound in each of the following pairs would have the higher boiling point, and which would be more soluble in water.

(a) $H-\overset{O}{\overset{\|}{C}}-H$ and CH_3OH

(b) $H-\overset{O}{\overset{\|}{C}}-H$ and $CH_2=CH_2$

(c) $H-\overset{O}{\overset{\|}{C}}-H$ and $H-\underset{OCH_3}{\underset{|}{\overset{OH}{\overset{|}{C}}}}-H$ (the hemiacetal of $H-\overset{O}{\overset{\|}{C}}-H$)

Problem-Solving Hints

If you have trouble solving any of the additional problems, refer to the sections listed next to the problem numbers.

14.15 14.2
14.16 14.4, 14.9
14.17 14.4, 14.9
14.18 14.6
14.19 14.6, 14.7, 14.10
14.20 14.10
14.21 14.11
14.22 14.11
14.23 14.11

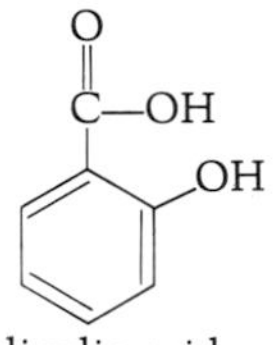

Salicylic acid

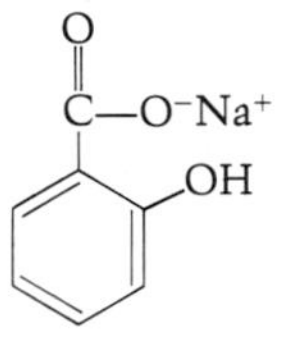

Sodium salicylate

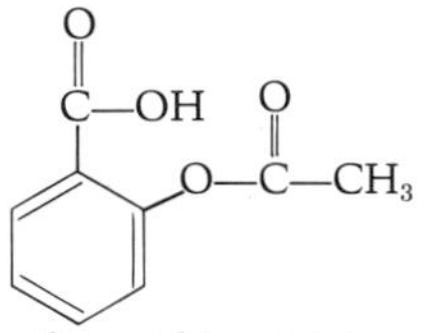

Acetylsalicylic acid (aspirin)

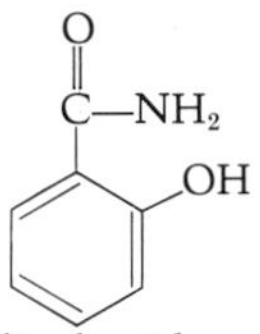

Salicylamide

Salicylic acid, acetylsalicylic acid, sodium salicylate, and salicylamide are all members of the group of compounds called salicylates that will enter in our discussion of carboxylic acids and their derivatives in this chapter. Although the structures of the compounds in the group are quite similar, simple chemical differences result in considerable variation in the action of each compound. (Remember when we made the same observation in our comparison of methanol and ethanol in Chapter 12.)

Salicylic acid is a carboxylic acid that is soluble in water, and its IUPAC chemical name is o-hydroxybenzoic acid. The primary use of salicylic acid is in lotions and ointments for local treatment of dandruff, psoriasis, and parasitic skin diseases. Salicylic acid is extremely irritating to the more sensitive internal body tissues and therefore is only used externally. Sodium salicylate and salicylamide are derivatives of salicylic acid that are synthesized for safe internal use.

Acetylsalicylic acid is the salicylate ester that is commonly referred to as aspirin. It, too, is somewhat water soluble and therefore is rapidly absorbed in the body. Aspirin affects the central nervous system by lowering body temperature, alleviates pain from certain sources, and has an anti-inflammatory effect. However, it can be lethal and is responsible for thousands of accidental poisonings each year.

15

CARBONYL COMPOUNDS II: CARBOXYLIC ACIDS AND THEIR DERIVATIVES

In the strictest structural terms, salicylamide (o-hydroxybenzamide) would not qualify as a salicylate since it is missing an extra oxygen atom that is characteristic of the salicylates. Nonetheless, its effects and its similarity in structure do justify including it in a discussion of the salicylates. Salicylamide has properties that resemble those of aspirin, and yet it is strikingly more potent with regard to depression of the central nervous system.

As the most commonly prescribed drugs, salicylates are frequently combined with other drugs to increase the effectiveness of pain relief. In this form, they are used to treat inflammatory illnesses and such conditions as rheumatoid arthritis. Possible side effects of the salicylates are reduced by the use of sodium salicylate, a buffer which reduces any liver-damaging reactions that normally might occur. In Chapter 20 we will discuss how aspirin and other salicylates are believed to act.

[Illustration: The various therapeutic derivatives of salicylic acid are quite similar in structure but differ in function in the body.]

CARBONYL COMPOUNDS II: CARBOXYLIC ACIDS

Now we move on to some more compounds that contain the carbonyl group—the carboxylic acids, esters, and amides.

Many of the everyday acids we discussed in Chapter 7 are really carboxylic acids. Acetic acid, the main component of vinegar, is probably the most familiar carboxylic acid. As you may know, wine turns to acetic acid if it is exposed to air. The acetic acid is formed by the oxidation of ethanol, the alcohol found in wine. In fact, the major reaction for forming a carboxylic acid is simply to oxidize a 1° alcohol.

If a carboxylic acid reacts with an alcohol, we get an ester. Although the carboxylic acids are notoriously foul smelling (besides the distinctive reek of vinegar, the even more unpleasant smell of rancid butter is due to a carboxylic acid), the esters are noted for their pleasant odors. Bananas, pineapples, oranges, rum, and jasmine all get their fragrances from esters. The structural difference between a carboxylic acid and an ester is that where you had an H in the acid you have an R group in the ester. Just swapping an H for an R can transform a vile stench into a sweet scent.

THE CARBOXYL GROUP 15.1

We have examined organic compounds that have the hydroxyl group (alcohols and phenols) and compounds that have the carbonyl group (aldehydes and ketones). Now we will look at compounds that have the **carboxyl group**—a functional group that is a combination of a carbonyl group and a hydroxyl group.

How are carboxylic acids related to aldehydes?

$$-\overset{\overset{\Large O}{\|}}{C}-OH$$

But the carboxyl group does not exhibit just the composite properties of these two simpler groups. It has unique properties of its own.

CARBOXYLIC ACIDS: STRUCTURE 15.2

Carboxylic acids have the carboxyl group attached to an alkyl or an aryl group (or to hydrogen in the case of formic acid, the simplest carboxylic acid).

$$H-\overset{\overset{\Large O}{\|}}{C}-OH \qquad CH_3\overset{\overset{\Large O}{\|}}{C}-OH \qquad CH_3CH_2\overset{\overset{\Large O}{\|}}{C}-OH \qquad C_6H_5-\overset{\overset{\Large O}{\|}}{C}-OH$$

formic acid, acetic acid, propionic acid, benzoic acid

We will often find the carboxyl group written as $-COOH$ or as $-CO_2H$.

CARBOXYLIC ACIDS: NOMENCLATURE 15.3

The names we used above--formic acid, acetic acid, and so forth—are common names. The common names of the carboxylic acids are related to the common names of the corresponding aldehydes (Table 15.1).

What is the general IUPAC name for a carboxylic acid?

We form the IUPAC names of carboxylic acids by replacing the *-e* of the corresponding alkane with *-oic acid* (Table 15.1). In the IUPAC

Table 15.1 Naming of Carboxylic Acids

Carboxylic acid (common name)	Corresponding aldehyde (common name)	IUPAC name of carboxylic acid
$H-C(=O)-OH$ (Formic acid)	$H-C(=O)-H$ (Formaldehyde)	Methanoic acid
$CH_3C(=O)-OH$ (Acetic acid)	$CH_3C(=O)-H$ (Acetaldehyde)	Ethanoic acid
$CH_3CH_2C(=O)-OH$ (Propionic acid)	$CH_3CH_2C(=O)-H$ (Propionaldehyde)	Propanoic acid
$CH_3CH_2CH_2C(=O)-OH$ (Butyric acid)	$CH_3CH_2CH_2C(=O)-H$ (Butyraldehyde)	Butanoic acid
$C_6H_5-C(=O)-OH$ (Benzoic acid)	$C_6H_5-C(=O)-H$ (Benzaldehyde)	Benzoic acid

system, the general name for carboxylic acids is alkanoic acids. Many aromatic acids, such as benzoic acid, have the same IUPAC and common name. The common names of carboxylic acids are widely used, and the student should memorize those in Table 15.1.

Problem 15.1 Give the structural formulas of the following acids.

(a) 2-methylpropanoic acid (b) acetic acid
(c) *p*-nitrobenzoic acid (d) formic acid
(e) 3-hydroxybutanoic acid

Problem 15.2 Give an acceptable name for each of the following compounds.

(a) $Cl-C_6H_4-C(=O)-OH$ (b) $Cl-CH_2-C(=O)-OH$

(c) $CH_3CH_2CH_2CH_2-C(=O)-OH$

CARBOXYLIC ACIDS: PHYSICAL PROPERTIES 15.4

The carboxyl group has both an O—H group and a carbonyl oxygen atom, and each can engage in hydrogen bonding (Figure 15.1). As we might expect from this fact, carboxylic acids have higher boiling points and are more soluble in water than hydrocarbons, aldehydes, and even alcohols of comparable molecular weight. Table 15.2 compares the properties of some carboxylic acids with those of other organic compounds of comparable molecular weight.

How do the boiling points and solubilities of carboxylic acids compare with those of other compounds that we have studied?

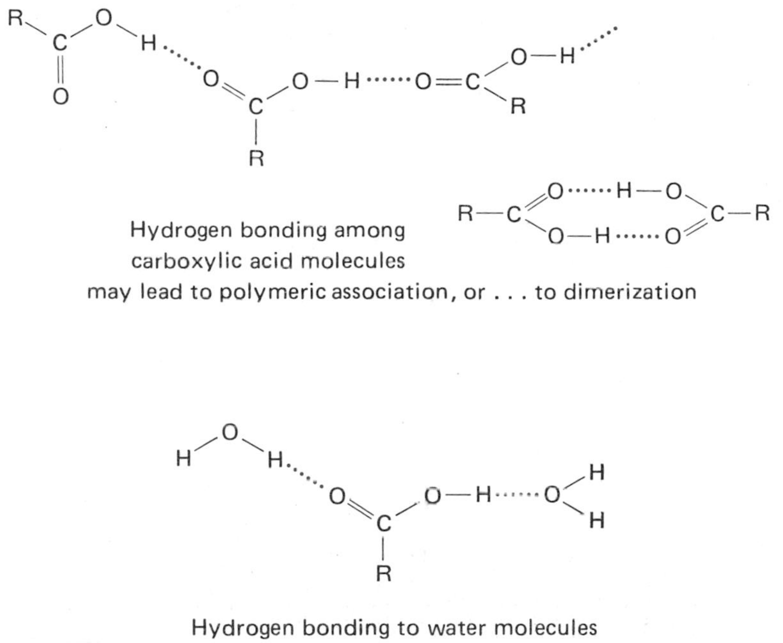

Hydrogen bonding among carboxylic acid molecules may lead to polymeric association, or . . . to dimerization

Hydrogen bonding to water molecules

FIGURE 15.1 Hydrogen bonding accounts for the high boiling points of carboxylic acids and for their high solubility in water.

Problem 15.3 Which compound in each of the following pairs has the higher boiling point, and which is more soluble in water?

(a) $CH_3CH_2CO_2H$ and CH_3CH_2CHO
(b) $CH_3CH_2CO_2H$ and $CH_3CH_2CH_2CHO$
(c) CH_3CHO and CH_3CH_2OH
(d) 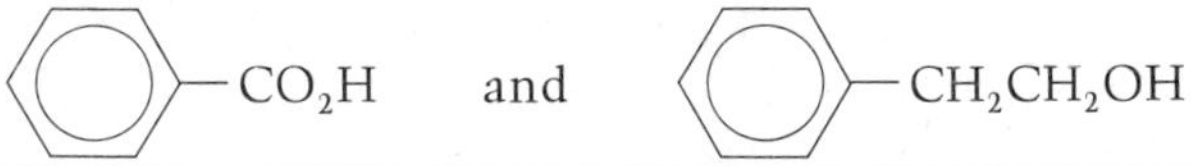 C_6H_5—CO_2H and C_6H_5—CH_2CH_2OH

Table 15.2 Comparison of Physical Properties of Carboxylic Acids with those of Other Compounds

Compound	Formula	Molecular weight	Boiling point, °C	Solubility, g/100 mL H_2O
Formic acid	HCO_2H	46	101	∞
Ethanol	CH_3CH_2OH	46	78	∞
Acetaldehyde	CH_3CHO	44	21	∞
Propane	$CH_3CH_2CH_3$	44	−42	Insoluble
Acetic acid	CH_3CO_2H	60	118	∞
Propanol	$CH_3CH_2CH_2OH$	60	97	∞
Propionaldehyde	CH_3CH_2CHO	58	49	20
Butane	$CH_3CH_2CH_2CH_3$	58	−0.6	Insoluble

15.5 CARBOXYLIC ACIDS: REACTIONS WITH BASES

How can we distinguish chemically between carboxylic acids and phenols?

Of the compounds that have the hydroxyl group, carboxylic acids are the strongest acids. Alcohols do not react appreciably with ordinary strong bases like NaOH; phenols react essentially completely with solutions of strong bases, but not with solutions of weaker bases like $NaHCO_3$ (sodium bicarbonate). Carboxylic acids react completely with both the strongly basic OH^- ion and the more weakly basic HCO_3^- ion to form the **carboxylate ion**

$$R-\overset{\overset{O}{\|}}{C}-O^-$$

These differences in their reactions with bases serve to distinguish alcohols, phenols, and carboxylic acids. We can summarize the reactions as follows.

Alcohols

$$R-OH + \underset{\text{strong base}}{OH^-} \rightarrow \text{no reaction}$$

$$R-OH + \underset{\text{weak base}}{HCO_3^-} \rightarrow \text{no reaction}$$

Phenols

$$\mathrm{Ar{-}OH + OH^- \rightarrow Ar{-}O^- + H_2O}$$

strong base; phenoxide ion

$$\mathrm{Ar{-}OH + HCO_3^- \rightarrow}\ \text{no reaction}$$

weak base

Carboxylic acids

$$\mathrm{R{-}\overset{\overset{\displaystyle O}{\|}}{C}{-}OH + OH^- \rightarrow R{-}\overset{\overset{\displaystyle O}{\|}}{C}{-}O^- + H_2O}$$

strong base; carboxylate ion

$$\mathrm{R{-}\overset{\overset{\displaystyle O}{\|}}{C}{-}OH + HCO_3^- \rightarrow R{-}\overset{\overset{\displaystyle O}{\|}}{C}{-}O^- + H_2CO_3}$$

weak base; carboxylate ion

Carbonic acid (H_2CO_3), once formed, undergoes the equilibrium reaction

$$\mathrm{H_2CO_3 \rightleftharpoons CO_2 + H_2O}$$

to form gaseous carbon dioxide, which escapes as bubbles. Because phenols do not react with HCO_3^- to give off CO_2, we can use aqueous sodium bicarbonate solution to distinguish between phenols and carboxylic acids. First we test our compound with litmus paper to see if it is an acid. If it is, we then add some $NaHCO_3$ solution to the sample. If bubbles form, we know that the sample is a carboxylic acid and not a phenol.

CARBOXYLIC ACIDS: ACIDITY 15.6

Water solutions of carboxylic acids are only weakly acidic.

Why are carboxylic acids stronger acids than phenols?

$$\mathrm{R{-}\overset{\overset{\displaystyle O}{\|}}{C}{-}OH + H_2O \rightleftharpoons R{-}\overset{\overset{\displaystyle O}{\|}}{C}{-}O^- + H_3O^+}$$

acid; base; base; acid

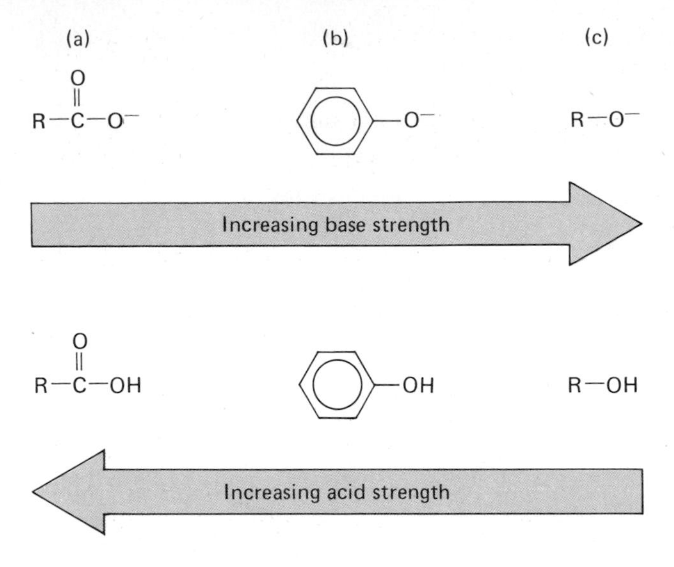

FIGURE 15.2 Correlation of acid strengths of carboxylic acids, phenols, and alcohols with the ability of the conjugate bases to accommodate a negative charge. (a) The negative charge is dispersed effectively. (b) The negative charge is dispersed less effectively than in RCO_2^-. (c) Almost no dispersal of negative charge.

In dilute water solutions of acetic acid, only about 0.4% of the acetic acid molecules exist as the acetate ion ($CH_3CO_2^-$) and the hydronium ion (H_3O^+). The pH of a 0.1 *M* acetic acid solution is about 2.8. The pH of a 0.1 *M* solution of a strong acid like HCl is 1.0.

Although carboxylic acids are weak acids, they are stronger acids than phenols (Section 12.8), because the carboxyl group can accommodate the negative charge better than the benzene ring can. The anions formed from alcohols are even less able to accommodate a negative charge, so alcohols are even weaker acids than phenols (Figure 15.2). The ability of the carboxylate ion (RCO_2^-) to stabilize the negative charge can be explained by resonance. Two resonance forms of the ion are

$$R-C\begin{matrix} \nearrow\!\!\!=\ddot{O}: \\ \searrow \ddot{O}:^- \end{matrix} \quad \text{and} \quad R-C\begin{matrix} \nearrow \ddot{O}:^- \\ \searrow\!\!\!= \ddot{O}: \end{matrix}$$

These two resonance structures are equivalent. The hybrid structure

$$R-C\begin{matrix} \overset{\delta-}{O} \\ \overset{\delta-}{O} \end{matrix}$$

represents a high degree of electron delocalization, and this delocalization stabilizes the ion (Section 10.14).

Problem 15.4 Complete each of the following equations by supplying the equilibrium products. Give products even if they are not formed in appreciable amounts.

no reaction

(a) $CH_3CH_2OH + H_2O \rightleftarrows$ (as acid) (as base)

(b) $C_6H_5-OH + H_2O \rightleftarrows$

(c) $CH_3CO_2H + H_2O \rightleftarrows$

(d) $CH_3CH_2OH + OH^- \rightleftarrows$

(e) $C_6H_5-OH + OH^- \rightleftarrows$

(f) $CH_3CO_2H + OH^- \rightleftarrows$

(g) $CH_3CH_2OH + HCO_3^- \rightleftarrows$ (as acid) (as base)

(h) $C_6H_5-OH + HCO_3^- \rightleftarrows$

(i) $CH_3CO_2H + HCO_3^- \rightleftarrows$

CARBOXYLIC ACID DERIVATIVES: ESTERS AND AMIDES 15.7

What is an acyl group? Are esters and amides acyl derivatives?

Carboxylic acids form a number of derivatives in which the OH group is replaced by another group. Some of these derivatives are the **acyl halides** (recall that the **acyl group** is RCO), **carboxylic anhydrides, esters,** and **amides** (see Section 14.2). Because the esters and amides are so important in biological processes, we will look at them in some detail.

Esters

An ester can be formed by heating a carboxylic acid with an alcohol in the presence of a small amount of strong acid. This reaction, called esterification, is reversible, but we may drive it to completion by removing the water as it forms or by using a large excess of alcohol or RCO_2H.

$$R-\overset{O}{\overset{\|}{C}}-OH + HO-R' \underset{\text{heat}}{\overset{H^+}{\rightleftarrows}} R-\overset{O}{\overset{\|}{C}}-O-R' + H_2O$$

An ester can also be prepared by reacting an alcohol with either an acyl halide or a carboxylic anhydride. The reaction of an acyl halide with an alcohol goes like this.

$$\underset{\text{acyl chloride}}{R-\overset{O}{\overset{\|}{C}}-Cl} + \underset{\text{alcohol}}{HO-R'} + NaOH \rightarrow \underset{\text{ester}}{R-\overset{O}{\overset{\|}{C}}-O-R'} + NaCl + H_2O$$

The base is used to neutralize the HCl as it forms. And here is how a carboxylic anhydride reacts with an alcohol:

$$\underset{\text{carboxylic anhydride}}{R-\overset{O}{\overset{\|}{C}}-O-\overset{O}{\overset{\|}{C}}-R} + \underset{\text{alcohol}}{HO-R'} \rightarrow \underset{\text{ester}}{R-\overset{O}{\overset{\|}{C}}-O-R'} + \underset{\text{carboxylic acid}}{R-\overset{O}{\overset{\|}{C}}-OH}$$

Amides

Amides can be prepared by the reaction of ammonia, a primary amine, or a secondary amine with any of the three other carboxylic-acid derivatives: an acyl halide, a carboxylic anhydride, or an ester. (In the following reactions R′ may be a hydrogen, an alkyl group, or an aryl group.)

$$\underset{\text{acyl chloride}}{R-\overset{O}{\overset{\|}{C}}-Cl} + \underset{\text{amine}}{2H-NR'_2} \rightarrow \underset{\text{amide}}{R-\overset{O}{\overset{\|}{C}}-NR'_2} + \underset{\text{ammonium salt}}{R'_2NH_2{}^+ + Cl^-}$$

$$\underset{\text{carboxylic anhydride}}{R-\overset{O}{\overset{\|}{C}}-O-\overset{O}{\overset{\|}{C}}-R} + \underset{\text{amine}}{2H-NR'_2} \rightarrow$$

$$\underset{\text{amide}}{R-\overset{O}{\overset{\|}{C}}-NR'_2} + \underset{\text{ammonium carboxylate}}{{}^-O-\overset{O}{\overset{\|}{C}}-R + H_2\overset{+}{N}R'_2}$$

$$\underset{\text{ester}}{R-\overset{O}{\overset{\|}{C}}-O-R''} + \underset{\text{amine}}{H-NR'_2} \rightarrow \underset{\text{amide}}{R-\overset{O}{\overset{\|}{C}}-NR'_2} + \underset{\text{alcohol}}{R''-OH}$$

In each of the preceding reactions, the electron pair of the amine nitrogen attacks the carbonyl carbon atom, which has a partial positive charge. We may show the overall reactions diagrammatically as follows.

Acyl chloride

$$\left.\begin{matrix} \overset{\delta-}{\mathrm{O}} \\ \| \\ \mathrm{R-C \wr Cl} \\ \uparrow\delta+ \\ \mathrm{R_2'\ddot{N} \wr H} \end{matrix}\right\} \longrightarrow \mathrm{R-\overset{\overset{\displaystyle O}{\|}}{C}(NR_2')} + \mathrm{HCl} \xrightarrow{\mathrm{R_2'NH}} \mathrm{R_2'NH_2^+ + Cl^-}$$

Carboxylic anhydride

$$\left.\begin{matrix} \overset{\delta-}{\mathrm{O}} \quad\quad \mathrm{O} \\ \| \quad\quad\quad \| \\ \mathrm{R-C \wr O-C-R} \\ \uparrow\delta+ \\ \mathrm{R_2'\ddot{N} \wr H} \end{matrix}\right\} \longrightarrow \mathrm{R-\overset{\overset{\displaystyle O}{\|}}{C}(NR_2')} + \mathrm{HO-\overset{\overset{\displaystyle O}{\|}}{C}-R}$$

(the acid reacts further with $R_2'NH$ to form the salt)

Ester

$$\left.\begin{matrix} \overset{\delta-}{\mathrm{O}} \\ \| \\ \mathrm{R-C \wr O-R''} \\ \uparrow\delta+ \\ \mathrm{R_2'\ddot{N} \wr H} \end{matrix}\right\} \longrightarrow \mathrm{R-\overset{\overset{\displaystyle O}{\|}}{C}(NR_2')} + \mathrm{HO-R''}$$

In each case, the amine must have at least one hydrogen atom; in other words, tertiary amines do not form amides.

If we react a carboxylic acid with ammonia or an amine, we obtain a salt, because ammonia is a base.

$$\underset{\text{carboxylic acid}}{\mathrm{R-\overset{\overset{\displaystyle O}{\|}}{C}-OH}} + \mathrm{NH_3} \longrightarrow \underset{\text{an ammonium salt}}{\mathrm{R-\overset{\overset{\displaystyle O}{\|}}{C}-O^-} + \mathrm{NH_4^+}}$$

We can prepare the amide by heating the dry ammonium salt.

$$\underset{\text{ammonium carboxylate}}{\mathrm{R-\overset{\overset{\displaystyle O}{\|}}{C}-ONH_4}} \xrightarrow{\text{heat}} \underset{\text{amide}}{\mathrm{R-\overset{\overset{\displaystyle O}{\|}}{C}-NH_2}} + \mathrm{H_2O}$$

Nylon, a polyamide, is manufactured by reacting a dicarboxylic acid with a diamine (Section 15.17).

Amides are the least reactive of all the carboxylic acid derivatives.

Problem 15.5 Give the names of the following carboxylic-acid derivatives (see Section 14.2).

(a) $CH_3-\overset{O}{\overset{\|}{C}}-O-\overset{O}{\overset{\|}{C}}-CH_3$ (b) $CH_3-\overset{O}{\overset{\|}{C}}-NH_2$

(c) $CH_3-\overset{O}{\overset{\|}{C}}-Cl$ (d) $CH_3-\overset{O}{\overset{\|}{C}}-OCH_2CH_3$

Problem 15.6 Supply the products of the following reactions.

(a) 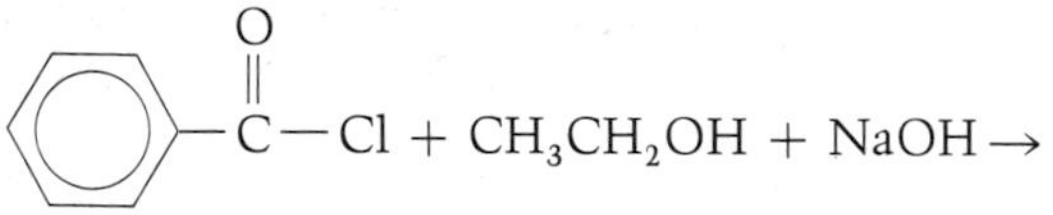 $C_6H_5-\overset{O}{\overset{\|}{C}}-Cl + CH_3CH_2OH + NaOH \rightarrow$

(b) $CH_3CH_2\overset{O}{\overset{\|}{C}}-Cl + CH_3NH_2 \text{ (excess)} \rightarrow$

(c) $CH_3-\overset{O}{\overset{\|}{C}}-O-\overset{O}{\overset{\|}{C}}-CH_3 + CH_3OH \rightarrow$

(d) $CH_3-\overset{O}{\overset{\|}{C}}-O-\overset{O}{\overset{\|}{C}}-CH_3 + CH_3NH_2 \text{ (excess)} \rightarrow$

(e) 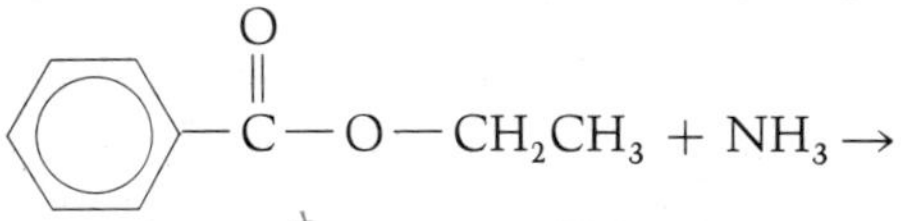 $C_6H_5-\overset{O}{\overset{\|}{C}}-O-CH_2CH_3 + NH_3 \rightarrow$

(f) $CH_3\overset{O}{\overset{\|}{C}}-OH + CH_3CH_2NH_2 \rightarrow$

(g) Product of (f) $\xrightarrow{\text{heat, dry}}$

15.8 ESTERS: NOMENCLATURE

Ester names have two parts. Which comes first?

We name esters by giving first the name of the hydrocarbon part (derived from the alcohol), then the name of the acyl part (derived from the acid). We name the acyl part by taking the name of the corresponding carboxylic acid and replacing the ending *-ic acid* with *-ate.*

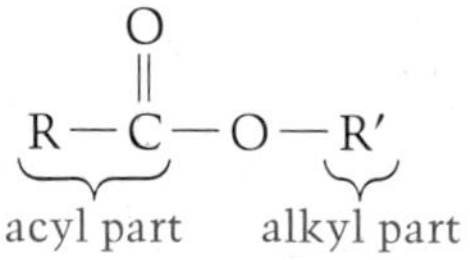

$$\underbrace{R-\overset{O}{\overset{\|}{C}}}_{\text{acyl part}}-O-\underbrace{R'}_{\text{alkyl part}}$$

The following example illustrates the naming of esters.

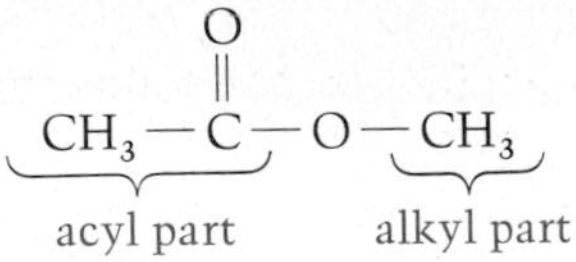

$$\underbrace{CH_3-\overset{\overset{\displaystyle O}{\|}}{C}}_{\text{acyl part}}-O-\underbrace{CH_3}_{\text{alkyl part}}$$

The alkyl part is methyl, the acyl part is acet*ate* (from acet*ic acid*). The name is *methyl acetate* (two words). In the IUPAC system the name would be methyl ethanoate, since the IUPAC name of the acid is ethanoic acid. Note that the rules are the same for forming common and IUPAC names of esters. Here are the names of some other esters (see Table 15.1 for names of carboxylic acids).

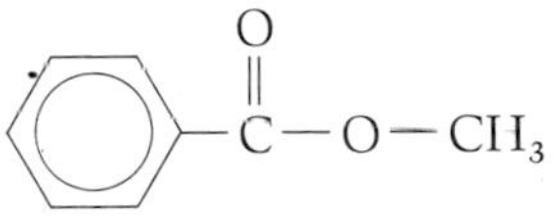

$$C_6H_5-\overset{\overset{\displaystyle O}{\|}}{C}-O-CH_3 \qquad CH_3-\overset{\overset{\displaystyle O}{\|}}{C}-O-C_6H_5$$

methyl benzoate 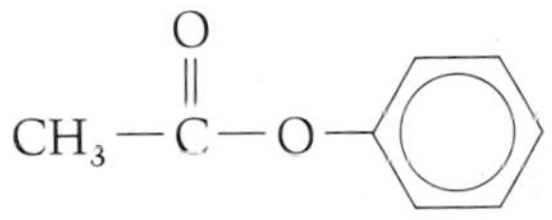 — phenyl acetate or phenyl ethanoate

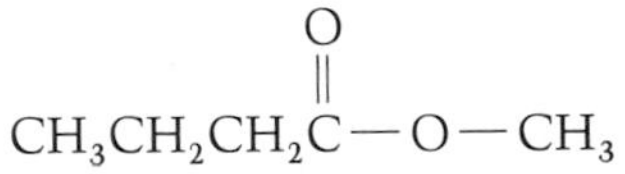

$$CH_3CH_2CH_2\overset{\overset{\displaystyle O}{\|}}{C}-O-CH_3 \qquad CH_3CH_2\overset{\overset{\displaystyle O}{\|}}{C}-O-\overset{\overset{\displaystyle CH_3}{|}}{C}HCH_3$$

methyl butyrate or methyl butanoate — isopropyl propionate or isopropyl propanoate

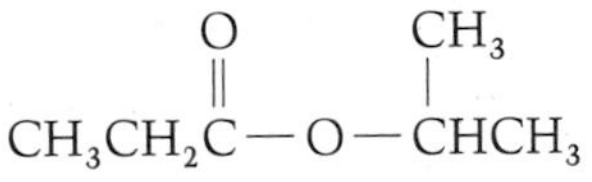

Problem 15.7 Give the structural formulas of the following esters.

(a) butyl formate (b) phenyl butyrate

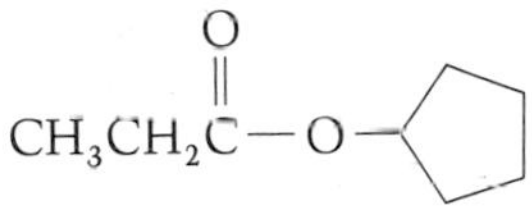

(c) cyclohexyl acetate

Problem 15.8 Give an acceptable name for each of the following esters.

(a) 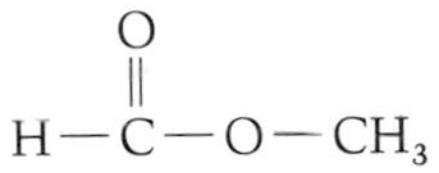 $H-\overset{\overset{\displaystyle O}{\|}}{C}-O-CH_3$

(b) $CH_3CH_2\overset{\overset{\displaystyle O}{\|}}{C}-O-C_5H_9$ (cyclopentyl)

(c) $CH_3\overset{\overset{\displaystyle O}{\|}}{C}-O-CH_2CH_2CH_2CH_2CH_3$

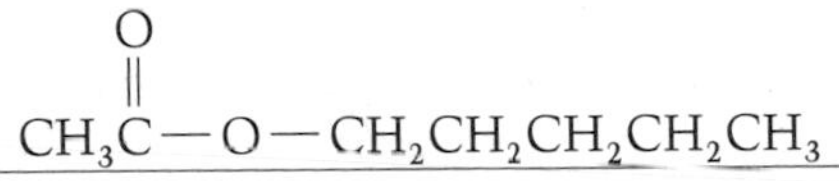

ESTERS: PROPERTIES 15.9

In physical properties, do esters more closely resemble carboxylic acids or ketones?

Esters generally are pleasant-smelling compounds. Many of them have odors that resemble the odors of fruits. They are often used as artificial flavors and in the manufacture of perfumes (Table 15.3).

Esters have no O—H group, so they cannot form intermolecular hydrogen bonds. However, because they have two oxygen atoms, they

TABLE 15.3 Properties of Some Esters

Name	Formula	Boiling point, °C	Odor
Isopentyl acetate	$CH_3-\overset{\overset{O}{\|}}{C}-O-CH_2CH_2\overset{\overset{CH_3}{\mid}}{C}HCH_3$	143	Banana
Isopentyl 3-methylbutanoate	$CH_3\overset{\overset{CH_3}{\mid}}{C}HCH_2\overset{\overset{O}{\|}}{C}-O-CH_2CH_2\overset{\overset{CH_3}{\mid}}{C}HCH_3$	194	Apple
Butyl butanoate	$CH_3CH_2CH_2\overset{\overset{O}{\|}}{C}-O-CH_2CH_2CH_2CH_3$	166	Pineapple
2-Methyl-1-propyl propanoate	$CH_3CH_2\overset{\overset{O}{\|}}{C}-O-CH_2\overset{\overset{CH_3}{\mid}}{C}HCH_3$	137	Rum
Methyl salicylate	$HO-C_6H_4-\underset{\underset{O}{\|}}{C}-OCH_3$ (ortho)	223	Oil of wintergreen
Glyceryl trimyristate	$CH_2-O-\overset{\overset{O}{\|}}{C}-(CH_2)_{12}CH_3$ $\mid$ $CH-O-\overset{\overset{O}{\|}}{C}-(CH_2)_{12}CH_3$ $\mid$ $CH_2-O-\overset{\overset{O}{\|}}{C}-(CH_2)_{12}CH_3$	Solid; melting point 56.5°C	Nutmeg

can accept hydrogen bonds from water molecules. For this reason, esters have boiling points and water solubilities that resemble those of aldehydes and ketones of comparable molecular weight.

15.10 ESTERS: HYDROLYSIS

What is the reverse of an acid-catalyzed hydrolysis reaction?

An ester will react with water, in the presence of either an acid or a base catalyst, to give a carboxylic acid and an alcohol. In acid solution, the reaction is reversible.

$$R-\overset{\overset{O}{\|}}{C}-O-R' + H_2O \xrightleftharpoons{\text{acid catalyst}} R-\overset{\overset{O}{\|}}{C}-OH + R'OH$$

In basic solution, the product is a carboxylate ion and the reaction is not reversible.

$$R-\overset{O}{\overset{\|}{C}}-O-R' + OH^- \rightarrow R-\overset{O}{\overset{\|}{C}}-O^- + R'OH$$

The acid-catalyzed reaction is simply the reverse of what happens when an ester is formed from the reaction of an acid and an alcohol. It is called a hydrolysis reaction, because a bond is broken by reaction with water. During hydrolysis, the ester bond that breaks is the C—O single bond.

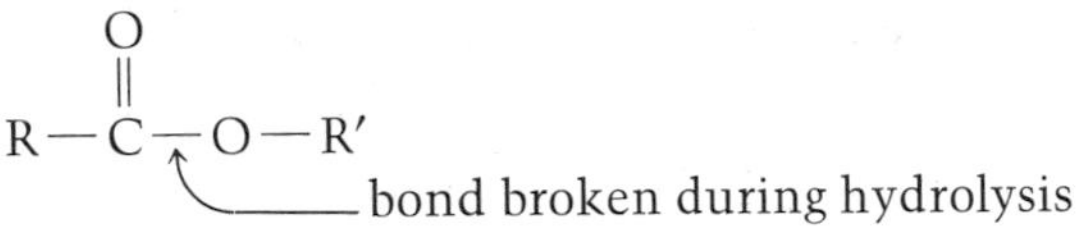

$$R-\overset{O}{\overset{\|}{C}}\underset{\uparrow}{-}O-R'$$

bond broken during hydrolysis

Problem 15.9 Give the products of the hydrolysis of each of the following esters in acid solution.

(a) $CH_3-\overset{O}{\overset{\|}{C}}-O-CH_2CH_3$ (b) 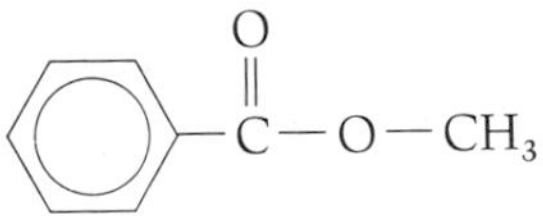 $C_6H_5-\overset{O}{\overset{\|}{C}}-O-CH_3$

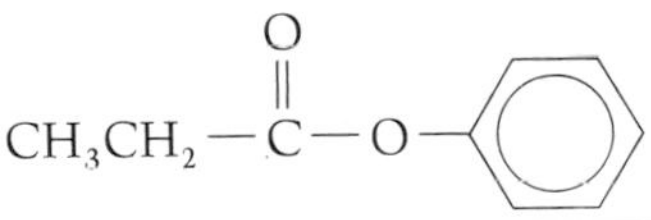

(c) $CH_3CH_2-\overset{O}{\overset{\|}{C}}-O-C_6H_5$

SAPONIFICATION 15.11

When we hydrolyze an ester in the presence of NaOH or KOH, the carboxylic acid salt is produced instead of the free acid. We call this process **saponification,** from the Latin *sapo,* "soap." *To saponify* means to make into soap. Soaps can be made by boiling fats with alkali. Fats are esters of the trihydroxy alcohol glycerin and long-chain acids. Because many carboxylic acids are obtained by hydrolysis of fats, carboxylic acids are in general called fatty acids. Soaps are the sodium or potassium salts of fatty acids.

Is saponification the same as hydrolysis?

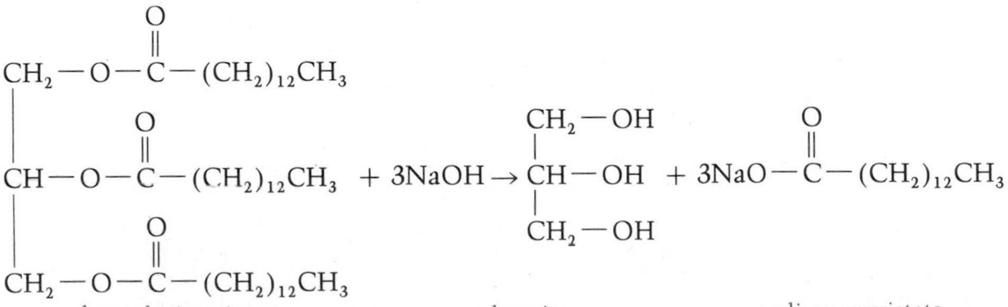

$$\begin{array}{l} CH_2-O-\overset{O}{\overset{\|}{C}}-(CH_2)_{12}CH_3 \\ | \\ CH-O-\overset{O}{\overset{\|}{C}}-(CH_2)_{12}CH_3 \\ | \\ CH_2-O-\overset{O}{\overset{\|}{C}}-(CH_2)_{12}CH_3 \end{array} + 3NaOH \rightarrow \begin{array}{l} CH_2-OH \\ | \\ CH-OH \\ | \\ CH_2-OH \end{array} + 3NaO-\overset{O}{\overset{\|}{C}}-(CH_2)_{12}CH_3$$

glyceryl trimyristate (a fat) — glycerin — sodium myristate (a soap)

15.12 SOME IMPORTANT ESTERS

Besides being used in flavors and fragrances, esters also provide us with those popular plastics and fibers called **polyesters.** Among the well-known polyesters are Mylar, Dacron, and Plexiglas. These polyesters are produced by the esterification reactions that we examined in Section 15.7.

Mylar is a very tough polyester used to make films and magnetic tapes. The acid (acyl) and alcohol (alkyl) parts are shown in color.

$$-\overset{\displaystyle O}{\overset{\|}{C}}-C_6H_4-\overset{\displaystyle O}{\overset{\|}{C}}-O-CH_2CH_2-O-\overset{\displaystyle O}{\overset{\|}{C}}-C_6H_4-\overset{\displaystyle O}{\overset{\|}{C}}-O-CH_2CH_2-O-$$

acyl part — alkyl part — acyl part — alkyl part

Here, both the original acid and alcohol had *two* functional groups, so each could react at both ends to form an endless chain. If they did not have two groups each, a polyester could not be formed.

$$HO-\overset{\displaystyle O}{\overset{\|}{C}}-C_6H_4-\overset{\displaystyle O}{\overset{\|}{C}}-OH + HO-CH_2CH_2-OH \rightarrow$$

terephthalic acid — ethylene glycol

$$HO-\overset{\displaystyle O}{\overset{\|}{C}}-C_6H_4-\overset{\displaystyle O}{\overset{\|}{C}}-O-CH_2CH_2-OH + H_2O$$

$$\xrightarrow{\text{and so on}} \text{polyester}$$

Dacron is a polyester chemically identical to Mylar, but it is manufactured in the form of a filament rather than a sheet.

Plexiglas is a different kind of polymer. Its ester groups are not part of the main polymer chain. Instead, they are substituents on that chain.

$$CH_2=C(CH_3)-C(=O)OCH_3 + CH_2=C(CH_3)-C(=O)OCH_3 + \cdots \rightarrow \sim CH_2-\underset{C(=O)OCH_3}{\overset{CH_3}{C}}-CH_2-\underset{C(=O)OCH_3}{\overset{CH_3}{C}}-CH_2-\underset{C(=O)OCH_3}{\overset{CH_3}{C}}\sim$$

methyl methacrylate

poly(methyl methacrylate)
(hundreds of repeating units)

One of the oldest and most widely used synthetic drugs is acetylsalicylic acid, an ester more commonly known as *aspirin*. Aspirin is a mild analgesic (pain reliever) and it reduces fever.

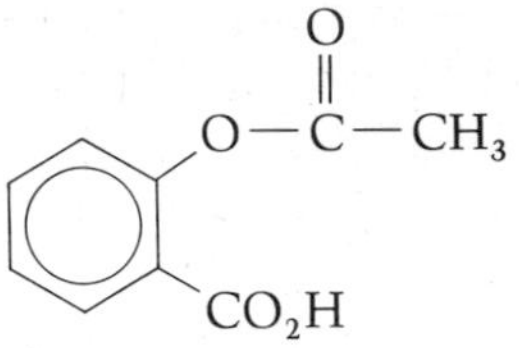

$$\text{C}_6\text{H}_4(\text{O}-\overset{\overset{\text{O}}{\|}}{\text{C}}-\text{CH}_3)(\text{CO}_2\text{H})$$

aspirin

Problem 15.10 What are the products of the hydrolysis of aspirin? Give their structural formulas.

ESTERS OF PHOSPHORIC ACID 15.13

The subject of this chapter is carboxylic acids and their derivatives. Phosphoric acid is not a carboxylic acid, but its derivatives are so important in the chemistry of life processes that we will examine some of them here.

Phosphoric acid (*ortho*-phosphoric acid) has the structural formula

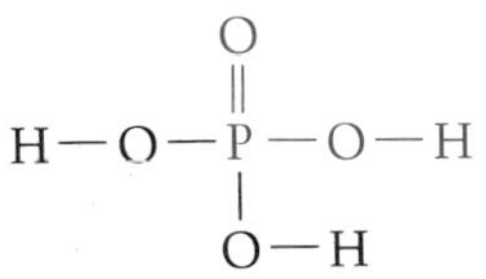

$$\text{H}-\text{O}-\overset{\overset{\text{O}}{\|}}{\underset{\underset{\text{O}-\text{H}}{|}}{\text{P}}}-\text{O}-\text{H}$$

o-phosphoric acid

There are other forms of phosphoric acid that differ from o-phosphoric acid in the degree of hydration. We will not be concerned with these. Phosphoric acid includes a group, shown in color, that resembles the carboxyl group, except that the carbon has been replaced by a phosphorus atom.

Phosphates (phosphoric acid esters) occur extensively in the body in the form of monoalkyl esters and their anhydrides.

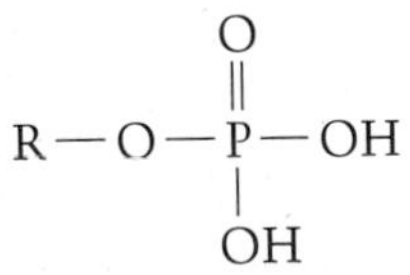

$$\text{R}-\text{O}-\overset{\overset{\text{O}}{\|}}{\underset{\underset{\text{OH}}{|}}{\text{P}}}-\text{OH}$$

a monoalkyl phosphate ester

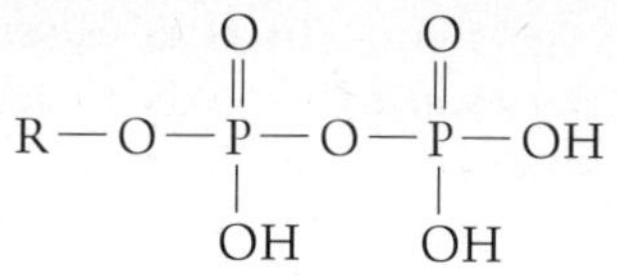

$$\mathrm{R{-}O{-}\overset{\overset{\displaystyle O}{\|}}{\underset{\underset{\displaystyle OH}{|}}{P}}{-}O{-}\overset{\overset{\displaystyle O}{\|}}{\underset{\underset{\displaystyle OH}{|}}{P}}{-}OH}$$

a monoalkyl diphosphate ester*

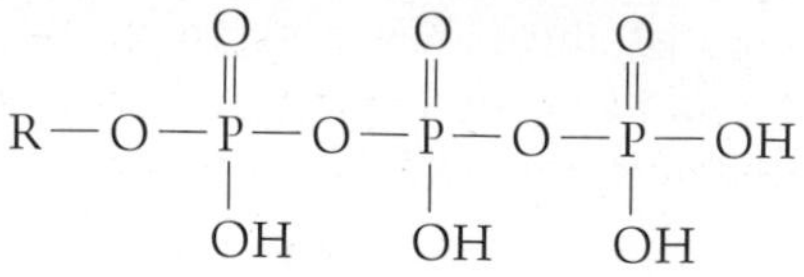

$$\mathrm{R{-}O{-}\overset{\overset{\displaystyle O}{\|}}{\underset{\underset{\displaystyle OH}{|}}{P}}{-}O{-}\overset{\overset{\displaystyle O}{\|}}{\underset{\underset{\displaystyle OH}{|}}{P}}{-}O{-}\overset{\overset{\displaystyle O}{\|}}{\underset{\underset{\displaystyle OH}{|}}{P}}{-}OH}$$

a monoalkyl triphosphate ester

These esters are produced in the body by enzyme-catalyzed reactions. The body stores energy temporarily in the form of diphosphate and triphosphate esters. These compounds are very energy-rich (Section 22.7). When they break down to the monophosphate or diphosphate forms, they release this stored energy.

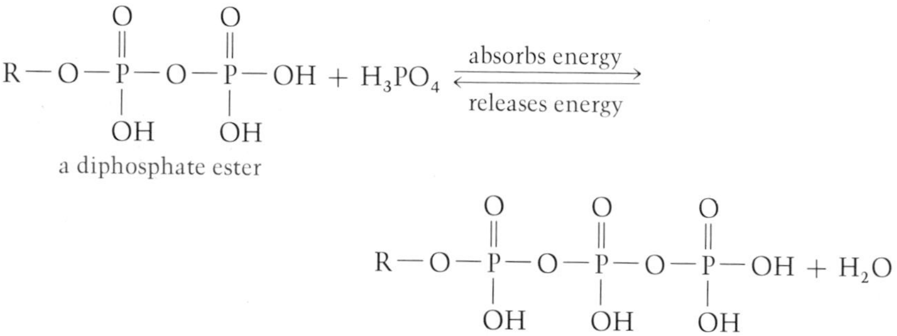

$$\underset{\text{a diphosphate ester}}{\mathrm{R{-}O{-}\overset{\overset{\displaystyle O}{\|}}{\underset{\underset{\displaystyle OH}{|}}{P}}{-}O{-}\overset{\overset{\displaystyle O}{\|}}{\underset{\underset{\displaystyle OH}{|}}{P}}{-}OH}} + H_3PO_4 \underset{\text{releases energy}}{\overset{\text{absorbs energy}}{\rightleftarrows}} \underset{\text{a triphosphate ester}}{\mathrm{R{-}O{-}\overset{\overset{\displaystyle O}{\|}}{\underset{\underset{\displaystyle OH}{|}}{P}}{-}O{-}\overset{\overset{\displaystyle O}{\|}}{\underset{\underset{\displaystyle OH}{|}}{P}}{-}O{-}\overset{\overset{\displaystyle O}{\|}}{\underset{\underset{\displaystyle OH}{|}}{P}}{-}OH}} + H_2O$$

Problem 15.11 Write an equation similar to the one above for the conversion of a monophosphate ester into a diphosphate ester.

15.14 AMIDES: NOMENCLATURE

Amides have the general formula

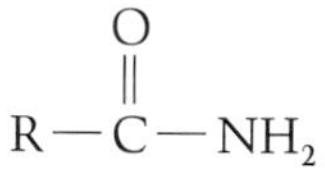

$$\mathrm{R{-}\overset{\overset{\displaystyle O}{\|}}{C}{-}NH_2}$$

*Diphosphates are also called pyrophosphates or phosphate anhydrides. In structure they are analogous to the carboxylic anhydrides (Section 15.7).

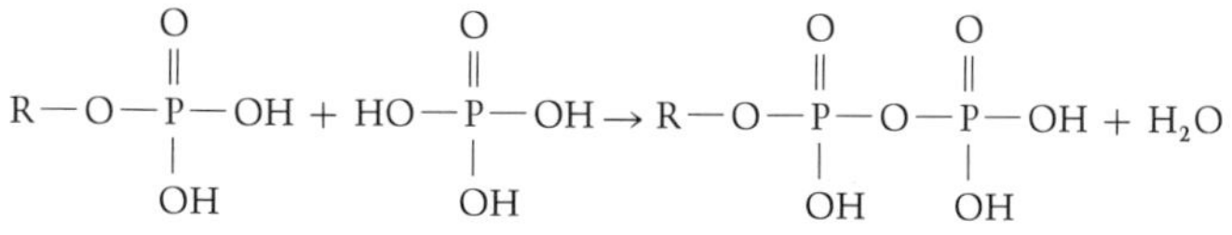

$$\mathrm{R{-}O{-}\overset{\overset{\displaystyle O}{\|}}{\underset{\underset{\displaystyle OH}{|}}{P}}{-}OH + HO{-}\overset{\overset{\displaystyle O}{\|}}{\underset{\underset{\displaystyle OH}{|}}{P}}{-}OH \rightarrow} \underset{\text{a phosphate anhydride}}{\mathrm{R{-}O{-}\overset{\overset{\displaystyle O}{\|}}{\underset{\underset{\displaystyle OH}{|}}{P}}{-}O{-}\overset{\overset{\displaystyle O}{\|}}{\underset{\underset{\displaystyle OH}{|}}{P}}{-}OH}} + H_2O$$

We name amides by replacing the ending *-oic acid* (of the IUPAC name, or *-ic acid* of the common name) of the corresponding carboxylic acid with the ending *-amide*. We may show the structure of amides as

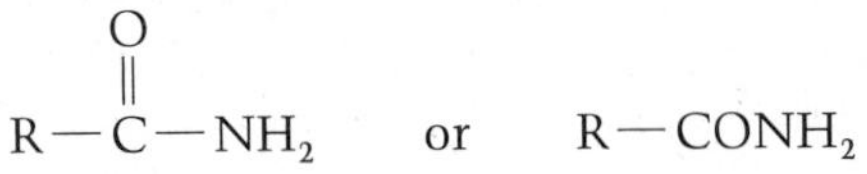

$$R{-}\overset{\overset{\displaystyle O}{\|}}{C}{-}NH_2 \quad \text{or} \quad R{-}CONH_2$$

Some examples will illustrate this system of nomenclature.

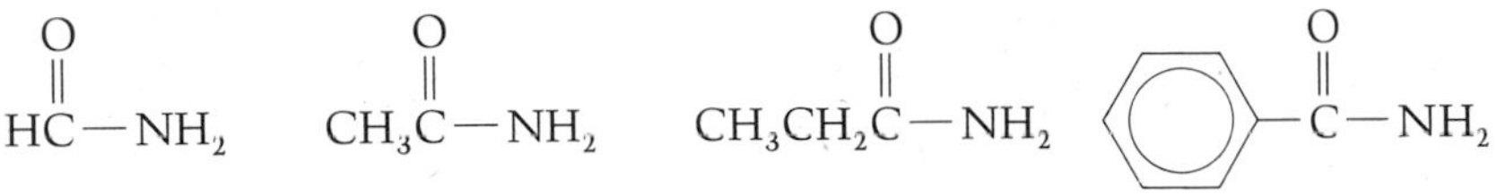

$$H\overset{\overset{\displaystyle O}{\|}}{C}{-}NH_2 \qquad CH_3\overset{\overset{\displaystyle O}{\|}}{C}{-}NH_2 \qquad CH_3CH_2\overset{\overset{\displaystyle O}{\|}}{C}{-}NH_2 \qquad C_6H_5{-}\overset{\overset{\displaystyle O}{\|}}{C}{-}NH_2$$

formamide acetamide propionamide benzamide

Problem 15.12 Give the structural formulas of the following amides.
(a) butyramide (b) *p*-chlorobenzamide

Problem 15.13 Name the following compounds.

(a) 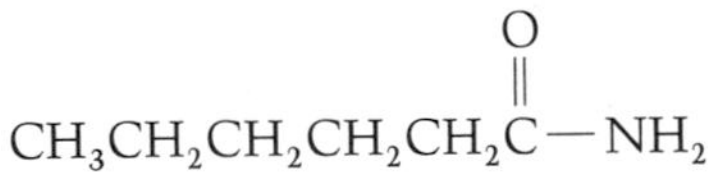 $CH_3CH_2CH_2CH_2CH_2\overset{\overset{\displaystyle O}{\|}}{C}{-}NH_2$ (b) 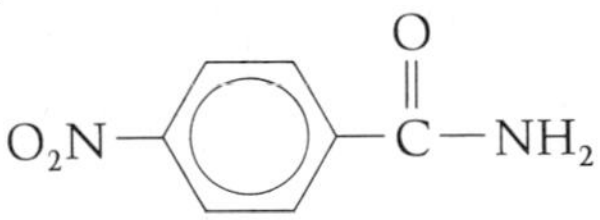 $O_2N{-}C_6H_4{-}\overset{\overset{\displaystyle O}{\|}}{C}{-}NH_2$

AMIDES: PROPERTIES 15.15

How do the physical properties of amides compare with those of carboxylic acids?

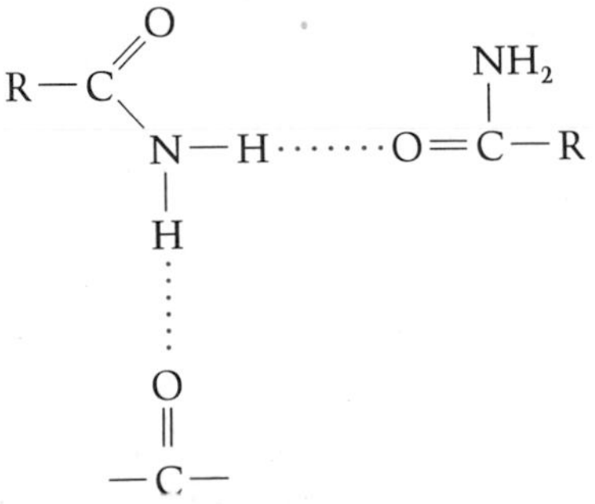

Amides can engage in hydrogen bonding through both the N—H group and the carbonyl oxygen atom. They therefore have high boiling points.

R—C(=O)—N—H······O=C(NH₂)—R, with N—H······O=C— (hydrogen bonding to other amide molecules)

hydrogen bonding to other amide molecules

R—C(=O······H—O—H)—N(—H······OH₂)—H······OH₂

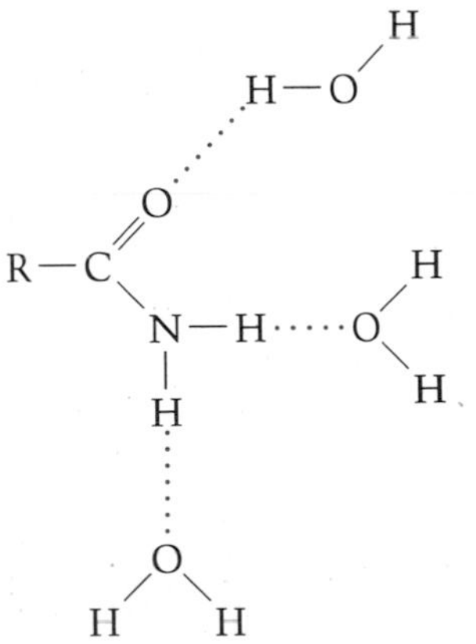

hydrogen bonding to water

TABLE 15.4 Properties of Some Amides

Name	Formula	Melting point, °C	Boiling point, °C	Solubility, g/100 mL H_2O (at 20°C)
Formamide	$HCONH_2$	2.6	211	∞
Acetamide	CH_3CONH_2	81	222	98
Propionamide	$CH_3CH_2CONH_2$	79	213	Soluble
Benzamide	$C_6H_5-CONH_2$	130	290	1.35

They are also highly soluble in water, unless their hydrocarbon part (R—) is large. Hydrogen bonding is probably more extensive in amides than in carboxylic acids. One important piece of evidence for this strong hydrogen bonding is that most amides are solids at room temperature (Table 15.4).

Problem 15.14 Draw structural formulas to demonstrate why hydrogen bonding accounts for the high water solubility of acetamide.

Problem 15.15 Propose an explanation based on hydrogen bonding to explain why amides have higher melting and boiling points than the corresponding carboxylic acids.

15.16 AMIDES: CHEMICAL BEHAVIOR

Why aren't amides basic?

Although amides have an $-\ddot{N}H_2$ group, they are not basic. We can explain this lack of basicity by noting that the amides also contain a carbonyl group. The carbonyl group draws electrons away from the nitrogen, so that they are not available for bonding with a hydrogen ion. This dispersion of electrons is shown by resonance structures.

$$R-\overset{\overset{:\ddot{O}}{\|}}{C}-\underset{\underset{H}{|}}{\ddot{N}}-H \quad R-\overset{\overset{:\ddot{O}:^-}{|}}{C}=\underset{\underset{H}{|}}{\overset{+}{N}}-H \quad R-\overset{\overset{O\,\delta-}{\|}}{C}\text{=}\overset{\delta+}{N}H_2$$

resonance structures

electrons are dispersed over colored area

Amides can be hydrolyzed in much the same way as esters. This hydrolysis occurs with either an acidic or basic catalyst.

Acid-catalyzed hydrolysis

$$R-\overset{\overset{\displaystyle O}{\|}}{C}-NH_2 + H_3O^+ \rightarrow R-\overset{\overset{\displaystyle O}{\|}}{C}-OH + NH_4^+$$

Base-catalyzed hydrolysis

$$R-\overset{\overset{\displaystyle O}{\|}}{C}-NH_2 + OH^- \rightarrow R-\overset{\overset{\displaystyle O}{\|}}{C}-O^- + NH_3$$

Proteins are large molecules that contain many amide groups. The bond that joins the carbonyl carbon atom to the amide nitrogen atom in protein molecules is called the **peptide bond.** During hydrolysis, the peptide bond breaks.

$$R-\overset{\overset{\displaystyle O}{\|}}{C}-\underset{\underset{\displaystyle H}{|}}{N}-H$$

peptide bond (the C—N bond)

This reaction is very important in protein chemistry. For example, during digestion of proteins in our stomachs, the peptide bond is broken (Section 23.4). In living systems, however, hydrolysis of the peptide bond is catalyzed by enzymes, not by acids or bases.

Problem 15.16 Explain why acid- and base-catalyzed hydrolysis of amides give different products (see the above equations).

Problem 15.17 Complete the following equations.

(a) $CH_3\overset{\overset{\displaystyle O}{\|}}{C}-NH_2 + H_2O \xrightarrow{OH^-}$ (b) $C_6H_5-\overset{\overset{\displaystyle O}{\|}}{C}-NH_2 + H_2O \xrightarrow{H^+}$

SOME IMPORTANT AMIDES 15.17

Polyamides are polymers synthesized by reacting a dicarboxylic acid with a diamine. Nylon-66 is the polyamide formed when adipic acid reacts with hexamethylenediamine. The name Nylon-66 refers to the starting acid and amine, each of which has six carbon atoms. The nylon polymer has about fifty repeating amide units. "Nylon" is a general term for fiber-forming polyamides.

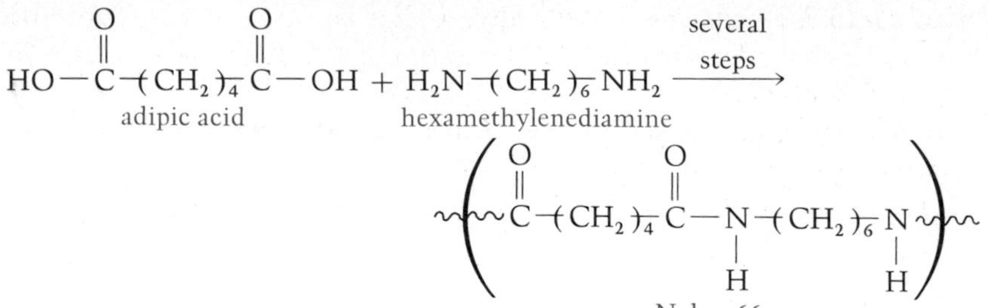

$$\underset{\text{adipic acid}}{HO-\overset{\overset{\displaystyle O}{\|}}{C}-(CH_2)_4-\overset{\overset{\displaystyle O}{\|}}{C}-OH} + \underset{\text{hexamethylenediamine}}{H_2N-(CH_2)_6-NH_2} \xrightarrow[\text{steps}]{\text{several}}$$

$$\underset{\text{Nylon-66}}{\sim\left(\sim\overset{\overset{\displaystyle O}{\|}}{C}-(CH_2)_4-\overset{\overset{\displaystyle O}{\|}}{C}-\underset{\underset{\displaystyle H}{|}}{N}-(CH_2)_6-\underset{\underset{\displaystyle H}{|}}{N}\sim\right)\sim}$$

As we will see in Chapter 17, nylons are structurally similar to protein. But a nylon is a regular polyamide containing the same repeating unit, and proteins are complex polyamides with many different repeating units.

The strength of nylon as a textile arises from the strong hydrogen bonds that join one molecular chain to another.

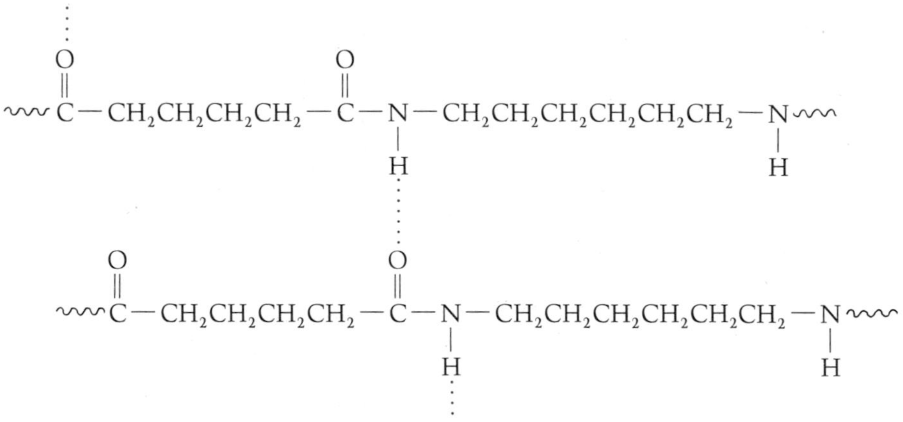

$$\sim\overset{\overset{\displaystyle \vdots}{\overset{\displaystyle O}{\|}}}{C}-CH_2CH_2CH_2CH_2-\overset{\overset{\displaystyle O}{\|}}{C}-\underset{\underset{\displaystyle H}{|}}{N}-CH_2CH_2CH_2CH_2CH_2CH_2-\underset{\underset{\displaystyle H}{|}}{N}\sim$$

$$\vdots$$

$$\sim\overset{\overset{\displaystyle O}{\|}}{C}-CH_2CH_2CH_2CH_2-\overset{\overset{\displaystyle O}{\|}}{C}-\underset{\underset{\displaystyle \vdots}{\underset{\displaystyle H}{|}}}{N}-CH_2CH_2CH_2CH_2CH_2CH_2-\underset{\underset{\displaystyle H}{|}}{N}\sim$$

Urea occurs in the urine of animals, including humans. It is a unique amide that has two NH_2 groups attached to the same carbonyl group.

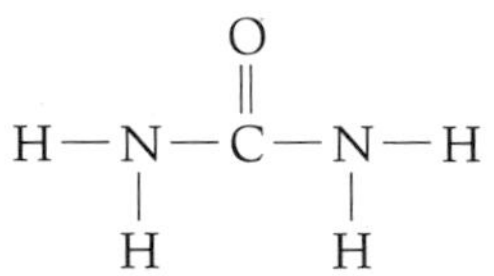

$$\underset{\text{urea}}{H-\underset{\underset{\displaystyle H}{|}}{N}-\overset{\overset{\displaystyle O}{\|}}{C}-\underset{\underset{\displaystyle H}{|}}{N}-H}$$

You will recall (Section 8.1) that urea was the first organic compound synthesized from inorganic materials. It is a product of the metabolism of protein. Urea is used as a fertilizer because of its high nitrogen content. It is also used in large quantities as a raw material in the manufacture of plastics. Medically, it has been used as a diuretic (an agent that increases the flow of urine) and to reduce intracranial and intraocular pressure.

Barbituric acid is formed when urea reacts with diethylmalonate, the diethyl ester of malonic acid.

```
           H                        O             H       O
           |                        ||             \     //
           N—H   CH3CH2—O—C                         N—C
          /                  \                     /     \
     O=C            +          CH2 → O=C                  CH2 + 2CH3CH2OH
          \                  /                     \     /
           N—H   CH3CH2—O—C                         N—C
           |                        ||             /     \\
           H                        O             H       O
          urea          diethyl malonate         barbituric acid
```

$$\text{urea} + \text{diethyl malonate} \rightarrow \text{barbituric acid} + 2CH_3CH_2OH$$

Barbituric acid itself has no sedative properties, but many of its derivatives, the barbiturates, do.

```
       H     O                                   H     O
        \   //                                    \   //
         N—C     CH2CH3                            N—C     CH2CH3
        /    \  /                                 /    \  /
    O=C        C                              O=C        C
        \    /  \                                 \    /  \
         N—C     CH2CH3                            N—C     C6H5 (phenyl)
        /   \\                                    /   \\
       H     O                                   H     O
```

5,5-diethylbarbituric acid
(barbital)

5-ethyl-5-phenylbarbituric acid
(phenobarbital)

```
       H     O                                   H     O
        \   //                                    \   //
         N—C     CH2CH=CH2                         N—C     CH2CH3
        /    \  /                                 /    \  /
    O=C        C                              O=C        C
        \    /  \                                 \    /  \
         N—C     CHCH2CH2CH3                       N—C     CHCH2CH2CH3
        /   \\   |                                /   \\   |
       H     O   CH3                             H     O   CH3
```

secobarbital
(Seconal)

pentobarbital
(Nembutal)

Barbiturates are sedatives that depress the central nervous system. They induce sleep by causing deep breathing and slowing of the heartbeat. At higher doses, they cause intoxication; at sufficiently high doses, they cause death.

Barbiturates are addictive and have been widely abused. They are commonly called "downers." Taken together with alcohol, they are extremely dangerous and often fatal.

Phenacetin is an analgesic, like aspirin. The combination pain relievers sold over the counter often contain phenacetin and aspirin.

$$CH_3CH_2O-C_6H_4-\underset{\underset{H}{|}}{N}-\overset{\overset{O}{\|}}{C}-CH_3$$

phenacetin

SUMMARY

Carboxylic acids contain the **carboxyl group,** $-CO_2H$. They are called alkanoic acids in the IUPAC system. Carboxylic acids engage in extensive hydrogen bonding and therefore exhibit relatively high boiling points and water solubilities. Carboxylic acids are weak acids, but they are stronger acids than phenols. **Esters** are **acyl** derivatives of alcohols. They are prepared by acid-catalyzed esterification (reaction of a carboxylic acid with an alcohol) or by reaction of an alcohol with an acyl halide or a **carboxylic anhydride. Amides** are acyl derivatives of ammonia, primary amines, or secondary amines. Amides can be prepared by heating a mixture of carboxylic acid and either amine or ammonia, or by reacting the amine or ammonia with an acyl halide, a carboxylic anhydride, or an ester. Ester molecules cannot hydrogen-bond to each other, so they have lower boiling points and water solubilities than carboxylic acids. Amide molecules, however, can form hydrogen bonds with each other, and they are soluble in water. Hydrolysis of esters can be carried out with either an acidic or basic catalyst. Hydrolysis of an ester with an aqueous base is called **saponification. Polyesters** are polymers in which the repeating units are connected by ester groups. Phosphate esters are esters of phosphoric acid. Amides are neutral substances. They can be hydrolyzed by either an acid or a base. **Polyamides** are polymers in which the repeating units are connected by amide groups. The amide linkage in proteins is called a **peptide bond.**

Key Terms

Acyl group (Section 15.7)
Acyl halide (Section 15.7)
Amide (Introduction and Section 15.7)
Carboxyl group (Section 15.1)
Carboxylate ion (Section 15.5)

Carboxylic acid (Introduction and Section 15.2)
Carboxylic anhydride (Section 15.7)
Ester (Introduction and Section 15.7)
Peptide bond (Section 15.16)
Polyamide (Section 15.17)
Polyester (Section 15.12)
Saponification (Section 15.11)

ADDITIONAL PROBLEMS

15.18 Give the structural formula, the common name, and the IUPAC name of the carboxylic acid that is produced by the oxidation of each of the following aldehydes.
(a) formaldehyde
(b) acetaldehyde
(c) propionaldehyde
(d) butyraldehyde
(e) benzaldehyde

15.19 Give the structural formula and an acceptable name for the ester produced by the reaction of ethanol with each carboxylic acid in Problem 15.18.

15.20 Give the structural formula and an acceptable name for the unsubstituted amide of each of the acids in Problem 15.18.

15.21 Explain why carboxylic acids have higher boiling points than esters of similar molecular weight.

15.22 Explain why amides have higher boiling points and water solubilities than esters of similar molecular weight.

15.23 Of the two polymers mylar (a polyester) and nylon (a polyamide), which would you expect to have the greater strength? Base your explanation on hydrogen bonding.

15.24 Give structural formulas for the following compounds.
(a) phenylacetic acid
(b) formyl chloride
(c) benzamide
(d) butanoic acid

15.25 Name the following compounds.

(a) 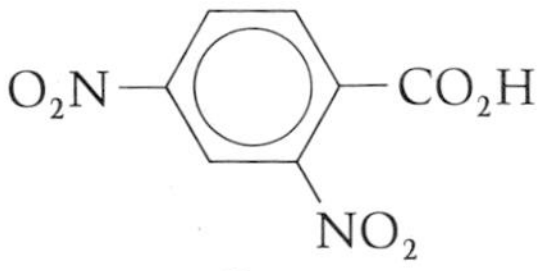 $O_2N-C_6H_3(NO_2)-CO_2H$ (benzene ring bearing O_2N and CO_2H para to each other, with NO_2 ortho to CO_2H)

(b) $CH_3-CCl_2-CH_2CO_2H$

(c) $Cl-CH_2C(=O)-OH$

(d) 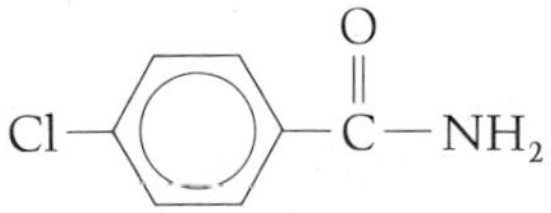 $Cl-C_6H_4-C(=O)-NH_2$

(e) $H-C(=O)-NH_2$

(f) $H-C(=O)-O-C(=O)-H$

15.26 Write the equation for a chemical reaction that would allow you to distinguish between the two compounds in each of the following pairs.

(a) acetic acid and benzene
(b) acetic acid and phenol
(c) acetic acid and pentanol
(d) phenol and cyclohexanol

15.27 Give the major organic product for each of the following reactions.

(a) 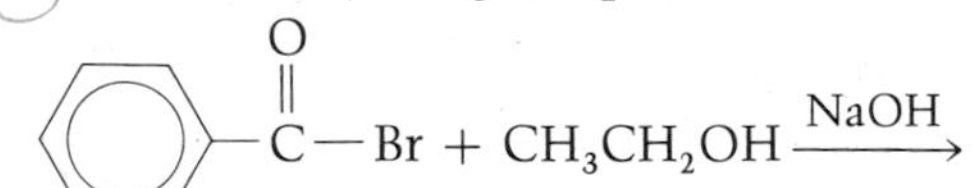 $C_6H_5-\overset{O}{\overset{\|}{C}}-Br + CH_3CH_2OH \xrightarrow{NaOH}$

(b) $CH_3CH_2CO_2H + CH_3CH_2CH_2OH \xrightarrow{H^+}$

(c) $CH_3\overset{O}{\overset{\|}{C}}-OH + CH_3NH_2\ (excess) \rightarrow$

(d) Product of (c) $\xrightarrow{\text{heat, dry}}$

(e) $CH_3\overset{O}{\overset{\|}{C}}-O-CH_2CH_3 + CH_3NH_2 \rightarrow$

(f) $Cl-\overset{O}{\overset{\|}{C}}-(CH_2)_4-\overset{O}{\overset{\|}{C}}-Cl + H_2N-(CH_2)_6-NH_2 \xrightarrow{\text{base}}$

(g) $CH_3CH_2\overset{O}{\overset{\|}{C}}-N(CH_3)_2 + H_2O \xrightarrow{OH^-}$

(h) $CH_3CH_2\overset{O}{\overset{\|}{C}}-N(CH_3)_2 + H_2O \xrightarrow{H_3O^+}$

(i) $CH_3\overset{O}{\overset{\|}{C}}-O-CH_3 + H_2O \xrightarrow{H_3O^+}$

(j) $CH_3\overset{O}{\overset{\|}{C}}-O-CH_3 + H_2O \xrightarrow{OH^-}$

(k) $CH_3\overset{O}{\overset{\|}{C}}-O-CH_3 + NaOH \rightarrow$

(l) Plexiglas + $H_2O \xrightarrow{OH^-}$

(m) Phenacetin + $H_2O \xrightarrow{OH^-}$

(n) Phenacetin + $H_2O \xrightarrow{H_3O^+}$

15.28 Benzoic acid is only slightly soluble in water. If we add NaOH solution to a mixture of benzoic acid and water, the benzoic acid dissolves. If we add HCl solution to the resulting basic solution, a precipitate forms. Describe, using equations, what happens in each step, and explain why it happens.

15.29 An unknown compound has the molecular formula $C_7H_6O_2$. It is an aromatic compound that dissolves in aqueous NaOH solution but not in aqueous $NaHCO_3$ solution. The unknown compound is known to be one of the following.

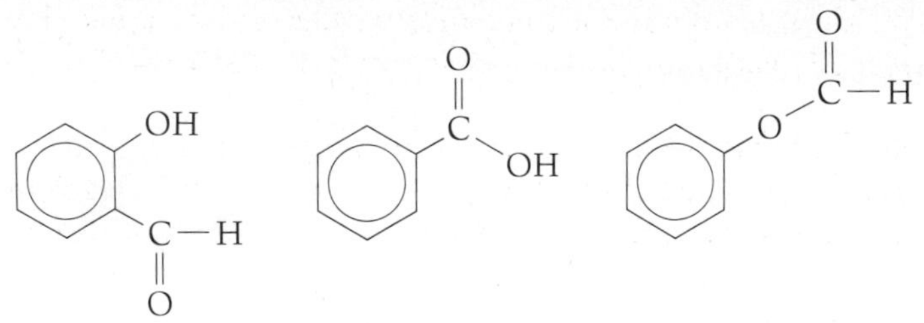

I II III

Which of these structures corresponds with the observed reactions of the unknown compound?

***15.30** An unknown compound **X** has the molecular formula $C_9H_9O_2Cl$. It is aromatic and is insoluble in water, aqueous NaOH solution, and aqueous $NaHCO_3$ solution. Compound **X** reacts with excess ammonia to yield compound **Y,** whose molecular formula is C_8H_8ONCl. Compound **X,** is known to have one of the following structures.

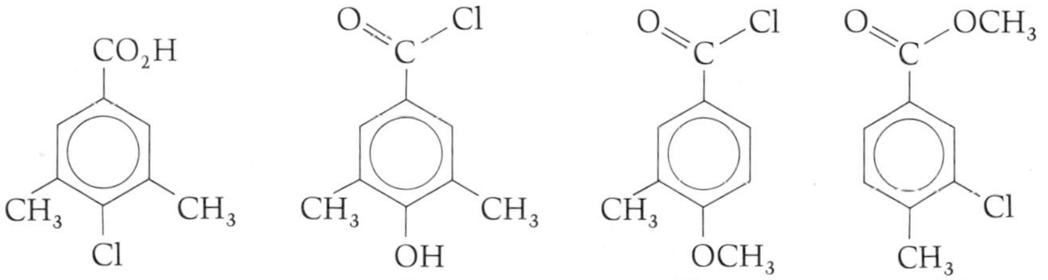

I II III IV

Which of these is the probable structure of compound **X**? What is the structural formula of compound **Y**? Write the equation for the reaction of compound **X** to form compound **Y.**

Problem-Solving Hints

If you have trouble solving any of the additional problems, refer to the sections listed next to the problem numbers.

15.18 15.3
15.19 15.8
15.20 15.14
15.21 15.9
15.22 15.15
15.23 15.12, 15.17
15.24 15.3, 15.7, 15.14
15.25 15.3, 15.14
15.26 15.5
15.27 15.5–15.7, 15.9–15.11, 15.16
15.28 15.5
15.29 15.5
15.30 15.5, 15.7

FOCUS 10 Organic Phosphate Insecticides

Halogenated hydrocarbon insecticides pose a serious threat to the environment (Chapter 11). One problem is their persistence. Another is that organisms become resistant to them. A highly toxic group of compounds that have partially replaced the halogenated hydrocarbons are certain phosphate esters. Because they are esters, they hydrolyze readily and therefore do not persist for long. They pose a special hazard, however, in that they are extremely toxic to humans. Organic phosphoric ester insecticides have caused many accidental deaths among humans. Because of their similarity to phosphates needed by the body, they are readily absorbed and interfere with metabolic processes. Most of them act by inhibiting the activity of cholinesterase (Focus 9), which acts in the transmission of nerve impulses. Thus phosphate ester insecticides attack the nervous system and can cause paralysis and death.

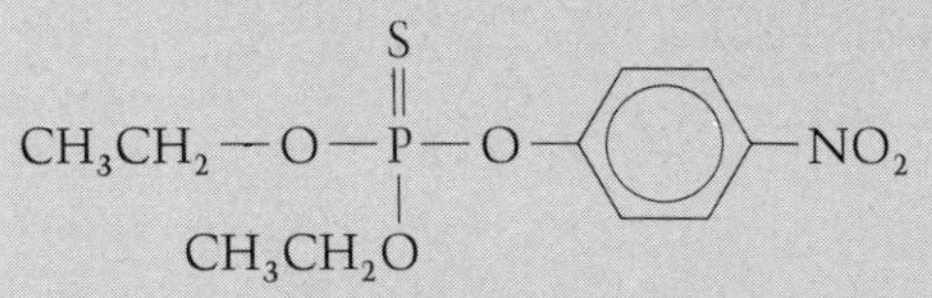

$$CH_3CH_2-O-\underset{\underset{\displaystyle CH_3CH_2O}{|}}{\overset{\overset{\displaystyle S}{\|}}{P}}-O-C_6H_4-NO_2$$

Parathion (This is a thiophosphate—it has a sulfur atom in place of an oxygen.)

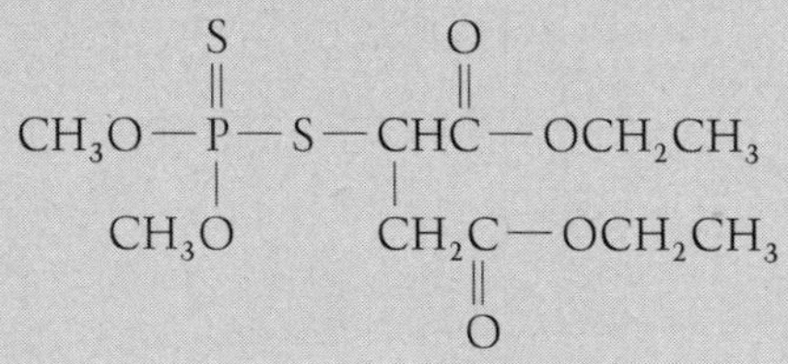

$$CH_3O-\underset{\underset{\displaystyle CH_3O}{|}}{\overset{\overset{\displaystyle S}{\|}}{P}}-S-\underset{\underset{\displaystyle CH_2\overset{}{\underset{\underset{\displaystyle O}{\|}}{C}}-OCH_2CH_3}{|}}{CH}\overset{\overset{\displaystyle O}{\|}}{C}-OCH_2CH_3$$

Malathion (a thiophosphate)

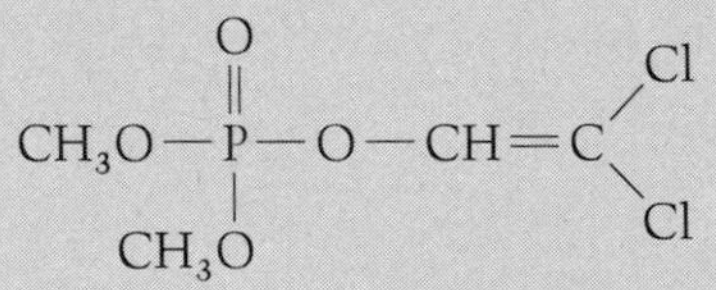

$$CH_3O-\underset{\underset{\displaystyle CH_3O}{|}}{\overset{\overset{\displaystyle O}{\|}}{P}}-O-CH=CCl_2$$

DDVP (Dichlorovox; used in Shell "No Pest Strip")

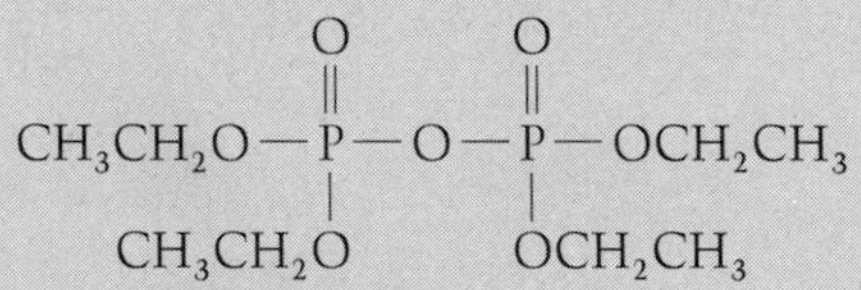

$$CH_3CH_2O-\underset{\underset{\displaystyle CH_3CH_2O}{|}}{\overset{\overset{\displaystyle O}{\|}}{P}}-O-\underset{\underset{\displaystyle OCH_2CH_3}{|}}{\overset{\overset{\displaystyle O}{\|}}{P}}-OCH_2CH_3$$

TEPP (*T*etra*e*thyl*p*yro*p*hosphate)

FIGURE 1 Crop dusting with certain insecticides can pose a serious health threat. (Paul Conklin)

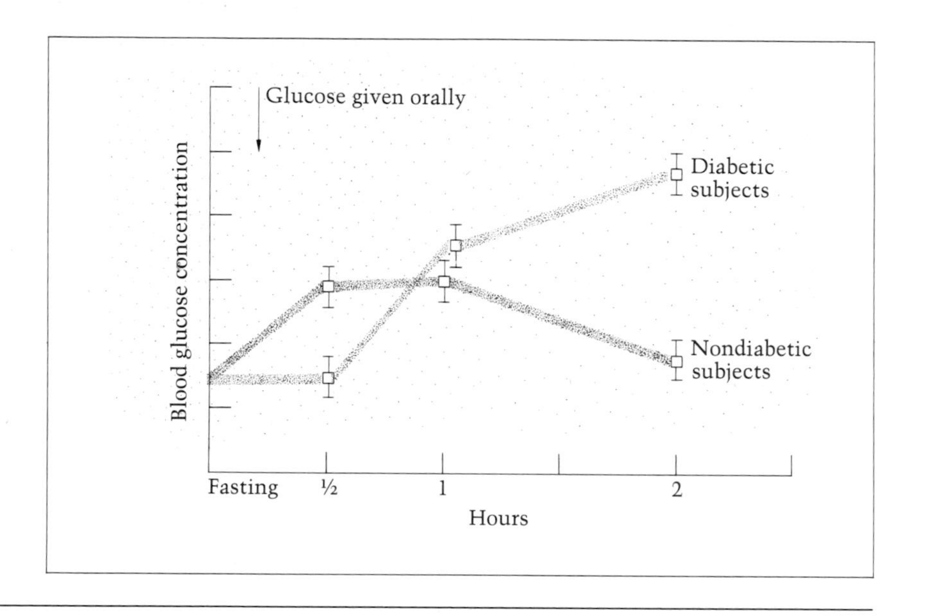

The glucose tolerance test is a measurement of a patient's tolerance of glucose, a simple carbohydrate that we will introduce in this chapter. Early cases of diabetes mellitus can be detected by this test, which shows the patient's ability to metabolize large amounts of glucose in a short period of time. The liver is the primary organ involved in the regulation of the glucose level in the blood. It takes up glucose when excess insulin is present and allows glucose to enter the blood stream when insulin is at a low level or absent altogether. Insulin is necessary to help in the transportation of glucose into the body cells. Without glucose, other sources of energy must be exhausted to support the life of the cells.

During the glucose tolerance test a specified dose of glucose is given to the patient, and blood samples are taken at several hourly intervals. The blood sugar level is measured as each specimen is drawn; then a diagnosis is made from the results as a group. (The normal blood glucose level is between 70 mg and 125 mg per 100 mL of plasma, with variations that depend on the particular test method used.) In the normal person there is a large increase in the level of blood sugar at the start of the test, when the glucose is administered. Then the glucose level returns to normal values within two to three hours. The diabetic patient also shows an increased blood sugar when the glucose is given, but this increase may be more gradual than in the normal person. Furthermore, the blood sugar in the diabetic patient usually does not return to a normal level by the end of the test.

16

CARBOHYDRATES

A patient's inability to tolerate glucose effectively is most likely due to a low level of active insulin. Treatment involves administering insulin to the patient on a routine basis.

[*Illustration: There is a marked difference in the metabolism of glucose between diabetics and nondiabetics.*]

CARBOHYDRATES

We hear carbohydrates mentioned in so many different contexts that it can be difficult to get a grip on exactly what they are. The problem, in part, is that there are many different carbohydrates, which vary widely in size and function.

In simplest terms a carbohydrate is a compound of carbon, hydrogen, and oxygen. Most carbohydrates (but not all of them, as we will see shortly) have the formula $C_n(H_2O)_m$. All sugars are carbohydrates. Starch is a carbohydrate. Cellulose is a carbohydrate. The smallest carbohydrate is a sugar that has only three carbons. The largest is cellulose, which can come in chains of over 90,000 carbons and has a molecular weight of over 2,000,000. But big as cellulose is, its structure is simple—it is just a long chain of glucose molecules strung together in a certain way. Starch is another long chain of glucose molecules connected in yet a different way. A starch molecule is impressive in size, but not as big as a cellulose molecule. Over half of the organic carbon on earth is found in cellulose. Trees are almost all cellulose. When you burn wood, you are obtaining energy (in the form of heat) from cellulose. When you eat bread, you are obtaining energy from starch as well as from sugars, which are also carbohydrates.

Carbohydrates are, in fact, the most important source of energy in the human diet. The body stores energy that is not used, either in the form of glycogen (another carbohydrate) or as body fat. Unfortunately, the combination of too many carbohydrates and too little exercise favors excess energy storage, which results in weight gain. For this reason, people looking at this cause-and-effect relationship tend to think, perhaps hastily, of carbohydrates as an evil. Yet we cannot survive without them.

In this chapter we will discuss only the simpler carbohydrates. Later, in Chapter 18, we will describe the more complex ones.

CARBOHYDRATES: CLASSIFICATION 16.1

What chemical reaction is used to classify carbohydrates?

The **carbohydrates** got their name because their general formula, $C_n(H_2O)_m$, suggested that they were hydrates of carbon. Actually, carbohydrates are not hydrates of carbon. They are covalent compounds.

We can classify carbohydrates according to whether they can be hydrolyzed into simpler carbohydrate units when heated in aqueous acid, and how many units are produced. **Disaccharide** molecules react in acidic solutions to produce two simple units that do not break down further. These simple, nonhydrolyzable carbohydrates are called **monosaccharides. Trisaccharide** molecules hydrolyze to form three monosaccharide molecules. **Polysaccharide** molecules hydrolyze to form hundreds and even thousands of monosaccharide molecules.

We can illustrate these terms with a couple of common examples.

1 mole sucrose ($C_{12}H_{22}O_{11}$) $\xrightarrow{\text{hydrolysis}}$
(a disaccharide)

1 mole glucose ($C_6H_{12}O_6$) + 1 mole fructose ($C_6H_{12}O_6$)
(monosaccharides)

1 mole starch [$(C_6H_{10}O_5)_n$] $\xrightarrow{\text{hydrolysis}}$ glucose (more than 100 moles)
(a polysaccharide)

One example of carbohydrate hydrolysis in the body is the digestion of starches. When starch is digested (Section 23.2), it breaks down into glucose, which is released from the digestive tract and absorbed into the blood stream. Starches and sugars such as sucrose, maltose, and lactose are the major carbohydrate foodstuffs.

Problem 16.1 Hydrolysis of 1 mole of an unknown carbohydrate yields 2 moles of glucose and 1 mole of fructose. Classify the unknown carbohydrate as a mono-, di-, tri-, or polysaccharide.

CARBOHYDRATES: PROPERTIES 16.2

What functional groups do monosaccharides have?

The monosaccharides are all very soluble in water, because of their molecular structure. Monosaccharides are molecules that have one aldehyde or keto group and one hydroxyl group on each of the remaining carbon atoms.

$$\begin{array}{ll} \text{aldehyde } H\diagdown C{=}O & CH_2OH \\ \quad | & | \\ \quad CHOH & C{=}O \text{ ketone} \\ \quad | & | \\ \quad CHOH & CHOH \\ \quad | & | \\ \quad CHOH & CHOH \\ \quad | & | \\ \quad CHOH & CHOH \\ \quad | & | \\ \quad CH_2OH & CH_2OH \\ \quad \text{glucose} & \text{fructose} \end{array}$$

The large number of hydroxyl groups permit extensive hydrogen bonding with water molecules, which makes the monosaccharides so soluble.

Hydrogen bonding causes the monosaccharides to form tight clusters. This, in turn, accounts for their solid physical state and crystalline structure. The smaller polysaccharides, called **oligosaccharides** (up to ten monosaccharide units), are also water-soluble crystalline solids, although solubility decreases with size. The monosaccharides and smaller oligosaccharides are called sugars. Most monosaccharides and smaller oligosaccharides have a sweet taste.

The larger oligosaccharides and the polysaccharides, such as starch and cellulose, are insoluble in water and do not taste sweet. Starch forms a suspension in water, but it does not actually dissolve.

Problem 16.2 An unknown carbohydrate is crystalline, somewhat soluble in water, and slightly sweet to the taste. How would you classify this carbohydrate?

16.3 MONOSACCHARIDES

What are the three parts of a general monosaccharide name?

Monosaccharides are classified into families according to the number of carbon atoms and whether there is an aldehyde or keto group. All sugars have names ending in *-ose.* If a monosaccharide contains an aldehyde group, it belongs to the **aldose** family; if it contains a keto group, it belongs to the **ketose** family. We designate the number of carbon atoms by inserting *tri*-(3 carbons), tetr-(4), pent-(5), hex-(6), and so forth, before the *-ose.* An aldehyde monosaccharide that has six carbon atoms therefore belongs to the *aldohexose* family. Here are the names of some other families of monosaccharides.

Number of carbon atoms	Aldose family	Ketose family
3	Aldotriose	Ketotriose
4	Aldotetrose	Ketotetrose
5	Aldopentose	Ketopentose
6	Aldohexose	Ketohexose

Problem 16.3 Classify the following monosaccharides by family.

(a)
H–C=O
|
CHOH
|
CHOH
|
CH_2OH

(b)
CH_2OH
|
C=O
|
CHOH
|
CHOH
|
CHOH
|
CH_2OH

(c)
H–C=O
|
CHOH
|
CHOH
|
CHOH
|
CHOH
|
CH_2OH

(d)
CH_2OH
|
C=O
|
CHOH
|
CHOH
|
CH_2OH

MIRROR-IMAGE MOLECULES 16.4

Some objects look exactly like their mirror images—a perfect sphere, for example. Other objects differ from their mirror images. For example, the mirror image of your right hand does not look like your right hand; it looks like your left hand (Figure 16.1). Objects that differ from their mirror images—hands, feet, ears, screws, propellers, and scissors, for example—are described as **chiral objects** (from the Greek word *cheir*, "hand"). A chiral object is *not superposable* on its mirror image (Figure 16.2). "Superposable" means that one object can be placed upon the other so that every part of one coincides with the corresponding part of the other (Figure 16.3). We cannot actually superpose two objects, because they would have to occupy the same space. Superposition is an imaginary exercise.

How do we determine whether a molecule is chiral?

Problem 16.4 Which of the following objects are chiral?
(a) spoon
(b) fork
(c) corkscrew
(d) the twist of a wire coathanger

Many chiral molecules exist. The most common chiral molecules are those in which a carbon atom is bonded to four different groups or

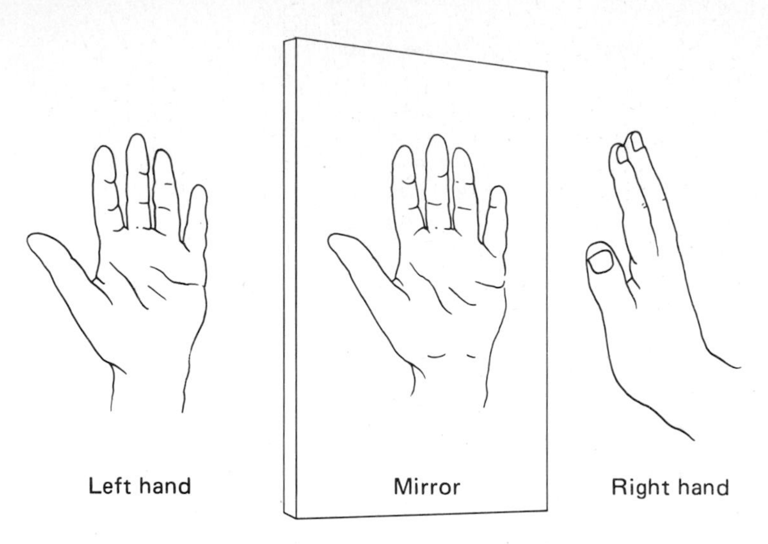

FIGURE 16.1 The mirror image of your right hand (a chiral object) looks like your left hand.

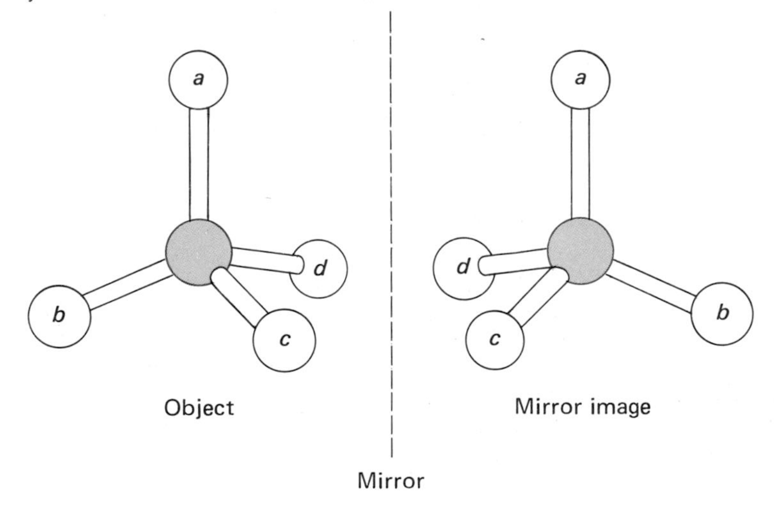

FIGURE 16.2 A chiral object and its mirror image are not superposable.

atoms. Such a carbon is called a chiral or **asymmetric carbon atom.**

Problem 16.5 Which of the following molecules are chiral?

(a) CH_4 (b) CH_3Cl

(c) CH_2ClBr (d) CHClBrF

Each chiral object can exist in two forms, a right-handed form and a left-handed one. For example, we can write two different line–dash–wedge formulas for 2-butanol.

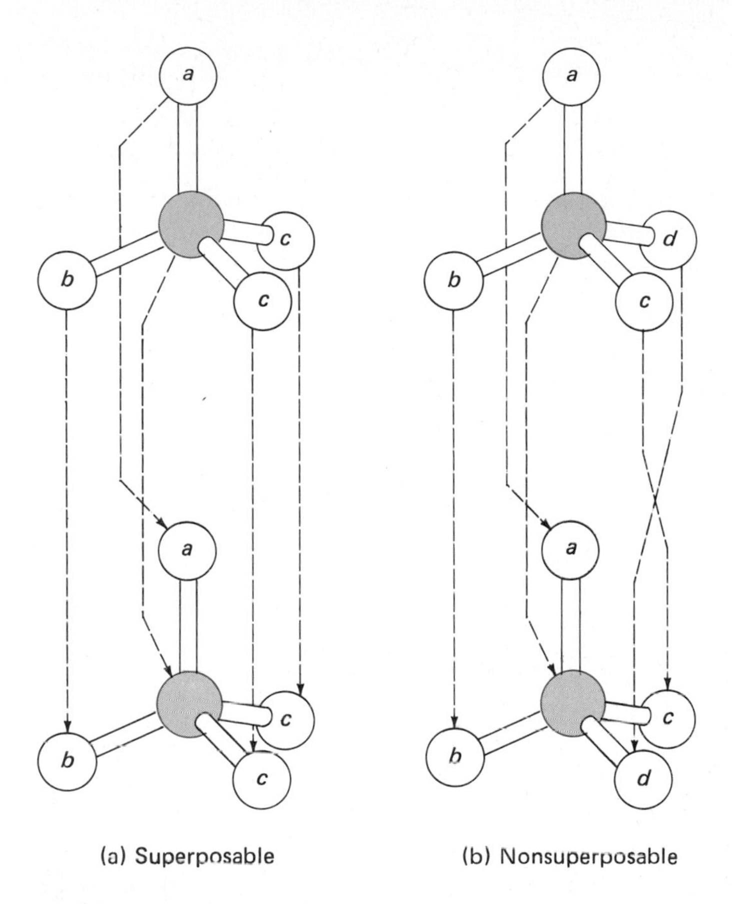

FIGURE 16.3 (a) The two molecular models are mirror images of one another and superposable. Therefore they are *achiral* (not chiral). (b) The molecular models (also mirror images of one another) are chiral because they are not superposable.

$$\begin{array}{c} CH_3 \\ | \\ CH_3CH_2\!-\!C\!\cdots\!H \\ \quad\quad\;\; \backslash OH \end{array} \qquad \begin{array}{c} CH_3 \\ | \\ H\!\cdots\!C\!-\!CH_2CH_3 \\ HO/ \quad\quad\quad \end{array}$$

These two forms are called **enantiomers,** or mirror-image isomers.

Problem 16.6 Identify the four different groups attached to the chiral carbon in 2-butanol.

Problem 16.7 Identify the chiral carbon atoms in the following molecules.

(a) $CH_3CHBrCH_2CH_3$ (b) $CH_3CHOH\overset{\overset{\displaystyle O}{\|}}{C}H$

(c) C_6H_5—$\underset{\displaystyle CHCO_2H}{\overset{\displaystyle CH_3}{|}}$

Enantiomers have the same melting point, boiling point, density, and color. But they differ in one very important respect: the way they interact with **plane-polarized light.**

16.5 PLANE-POLARIZED LIGHT

A light beam consists of electromagnetic vibrations. Ordinary light vibrates in all planes (Figure 16.4). It is possible to filter ordinary light so that the light waves that emerge from the filter are vibrating in only one plane. Such a filter is found in Polaroid sunglasses (Figure 16.4). A beam of light that vibrates in only one plane is said to be plane-polarized.

16.6 OPTICAL ACTIVITY

How do enantiomers compare in their optical activities?

When plane-polarized light passes through a sample of a pure enantiomer of a chiral substance, the plane of polarization rotates so that the plane of the emerging light is at an angle to the plane of the entering light. This ability to rotate the plane of plane-polarized light is called **optical activity** (Figure 16.5). The instrument used to measure the angle of rotation of plane-polarized light is called a **polarimeter.**

The rotations caused by the two enantiomers of a chiral substance are exactly equal in magnitude, but go in opposite directions. The enantiomer that rotates the plane of polarization of light to the right (clockwise) is designated (+); the one that rotates the plane to the left (counterclockwise) is designated (−). Thus the two enantiomers of 2-butanol are named (+)-2-butanol and (−)-2-butanol. We should note that we cannot tell which of the structural formulas in Section 16.4 corresponds to the (+) enantiomer and which to the (−) enantiomer. Such absolute configurations (that is, complete three-dimensional structures) are known only in a few cases.

Problem 16.8 A pure sample of (+)-2-butanol rotates the plane of polarization +13.52°. What rotation should you observe for a pure sample of (−)-2-butanol?

Problem 16.9 Two pure isomers have the following optical rotations: sample A, +29.2°; sample B, −56.5°. Can these isomers be enantiomers? Explain.

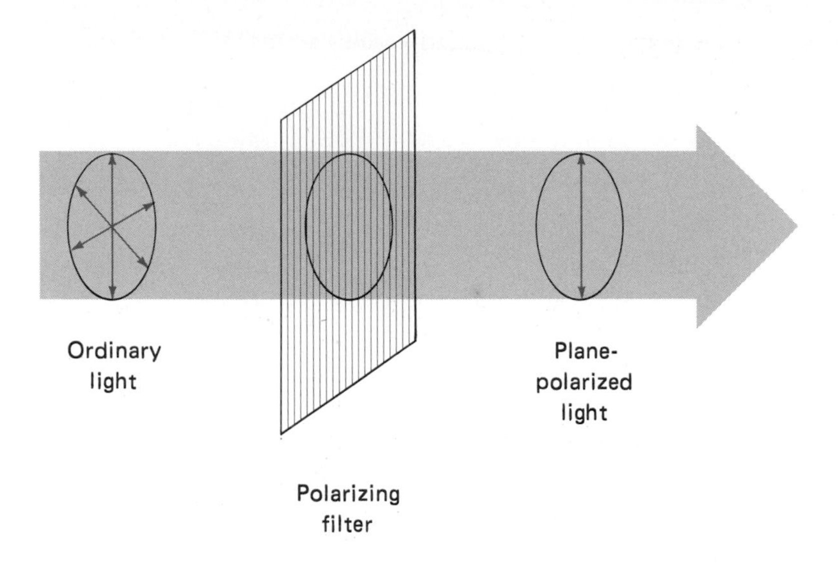

FIGURE 16.4 Ordinary light consists of light waves that vibrate in all possible planes. When it passes through a polarizing filter, it emerges as plane-polarized light; that is, its light waves vibrate in only one plane.

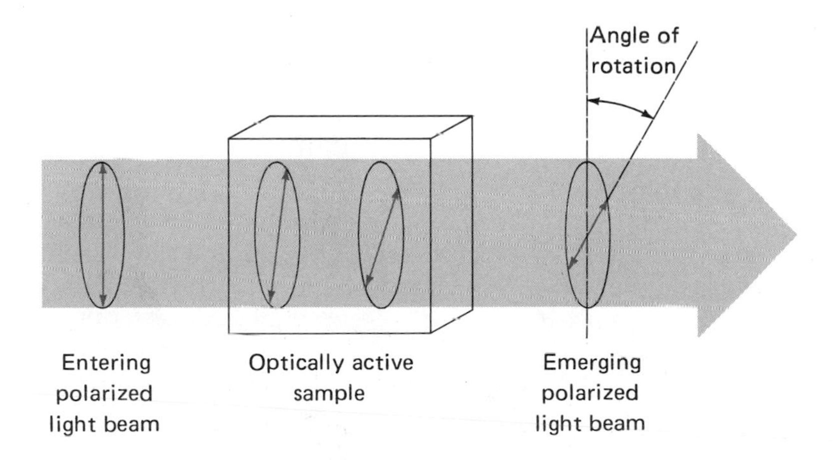

FIGURE16.5 The plane of a plane-polarized light beam is rotated gradually as it passes through an optically active sample.

OPTICAL ACTIVITY IN MONOSACCHARIDES 16.7

How do we determine which atoms in a molecule are chiral?

We have only to draw the structural formula of a simple monosaccharide to see that nearly all monosaccharides have chiral carbon atoms.

Glucose, for example, has four chiral carbon atoms; they are marked with an asterisk (*).

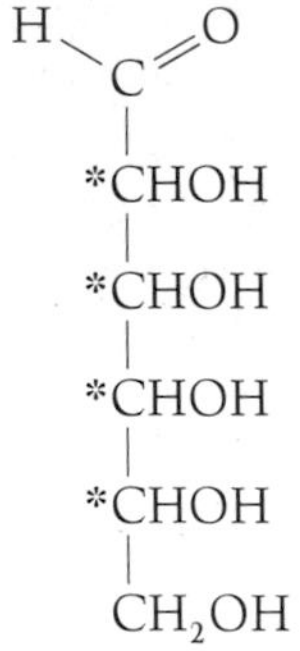

H—C=O
*CHOH
*CHOH
*CHOH
*CHOH
CH_2OH
glucose

Problem 16.10
(a) What is the simplest chiral aldose?
(b) What is the simplest chiral ketose?

16.8 FISCHER PROJECTION FORMULAS

What operation is not *allowed when working with Fischer projection formulas?*

Drawing structural formulas of molecules that have several chiral atoms can be a problem. We solve the problem by the use of projection formulas. A projection formula is the two-dimensional projection of a three-dimensional object. One of the two enantiomers of the simplest aldose, glyceraldehyde, is shown below in three-dimensional formulas and two-dimensional projections.

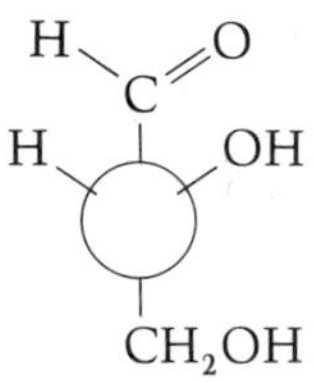

H—C=O; H, OH; CH_2OH
ball-and-stick type formula
I

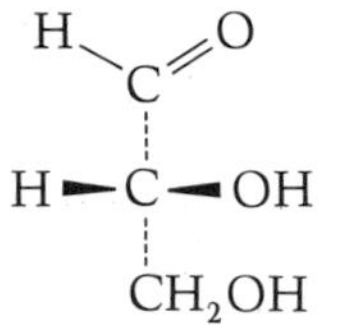

H—C=O; H—C—OH; CH_2OH
three-dimensional formula
II

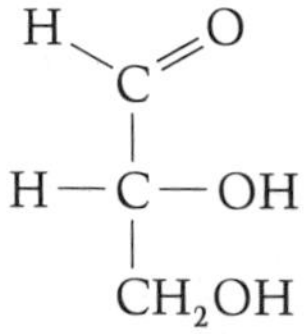

H—C=O; H—C—OH; CH_2OH
two-dimensional projection of three-dimensional formula
III

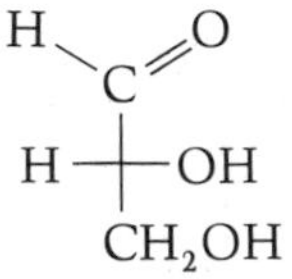

H—C=O; H—+—OH; CH_2OH
Fischer projection formula
IV

Emil Fischer was a German chemist who did much of the pioneering research on the structure and chemistry of sugars around 1900. He proposed formula IV above, and this way of representing molecules has come to be known as a **Fischer projection formula.** In a Fischer formula, the vertical bond lines are understood to extend back behind the plane of the page, and the horizontal bond lines to extend toward the reader. We must not rotate a Fischer projection, because, if we do, we change its meaning. For example, rotating the Fischer formula for glyceraldehyde yields a different enantiomer.

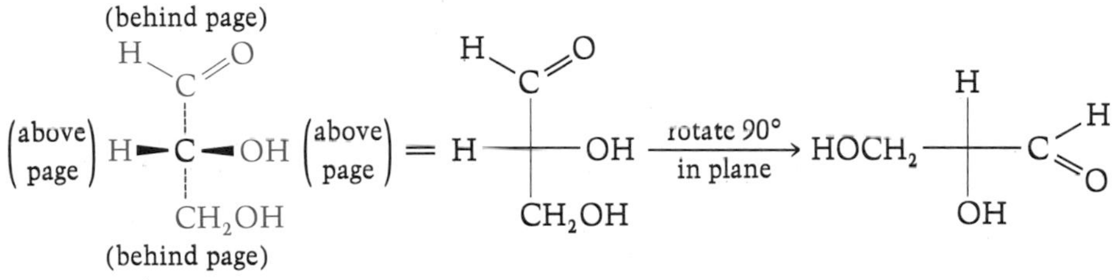

(behind page)
H–C=O
(above page) H–C–OH (above page) = H–C(–OH) with CHO above and CH_2OH below $\xrightarrow[\text{in plane}]{\text{rotate } 90°}$ $HOCH_2$–C(–CHO) with H above and OH below
CH_2OH
(behind page)
(*a*)

which means

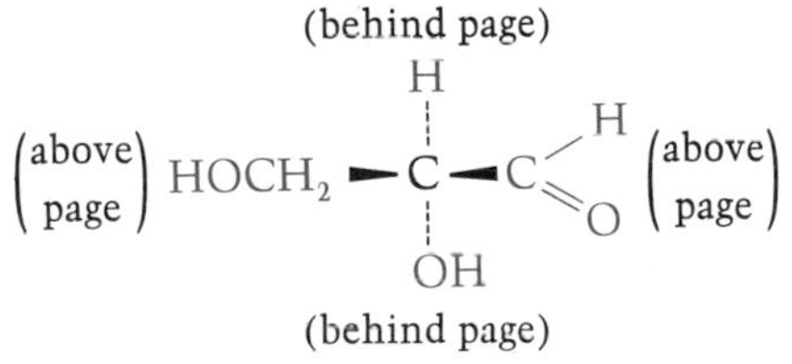

(behind page)
H
(above page) $HOCH_2$–C–C(H)=O (above page)
OH
(behind page)
(*b*)

We have apparently changed *a* into its enantiomer *b* simply by rotating the formula 90°—clearly an impossibility!

Problem 16.11 Does the meaning of the Fischer projection formula change if we rotate the formula 180°? Demonstrate by carrying out transformations like those just shown for glyceraldehyde.

Fischer formulas of the naturally occurring enantiomers of glucose and fructose are

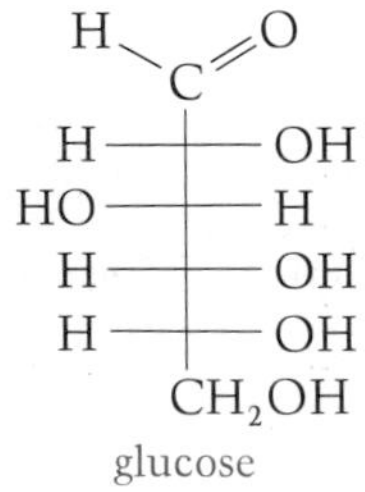

glucose			fructose		
	H–C=O			CH_2OH	
H		OH		=O	
HO		H	HO		H
H		OH	H		OH
H		OH	H		OH
	CH_2OH			CH_2OH	

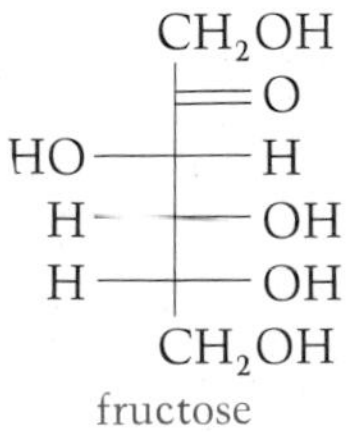

Problem 16.12 Draw the Fischer projection formula of the enantiomer of naturally occurring glucose. [Hint: Imagine that the mirror is placed vertically beside the Fischer formula of glucose and draw the mirror image.]

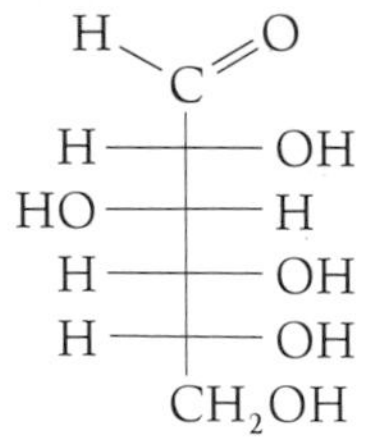

H–C=O
H — OH
HO — H
H — OH
H — OH
CH_2OH

mirror

(Draw enantiomer here.)

16.9 D- AND L-GLUCOSE

What determines whether a monosaccharide is D- or L-?

Fischer first proposed the convention that the (+)-glyceraldehyde enantiomer be called D-glyceraldehyde and that the (−)-glyceraldehyde enantiomer be called L-glyceraldehyde. We now know that these compounds have the absolute configurations shown below. By convention, we name all monosaccharides D- or L-, depending on the orientation of the hydroxyl group on the highest-numbered chiral carbon (the one farthest away from the carbonyl group). If that carbon has the hydroxyl group on the right side (with the molecules drawn as shown below), as the chiral carbon of D-glyceraldehyde does, the monosaccharide is called a D-sugar. If the chiral carbon has the opposite configuration, we have an L-sugar. The D- and L- designations depend only on the configuration of the next-to-last carbon; none of the others matter. The hydroxyl group that determines the designation appears in color in the formulas below.

D-sugars

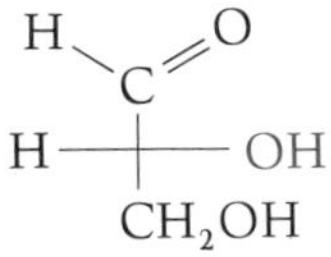

H–C=O
H — OH
CH_2OH
D-glyceraldehyde

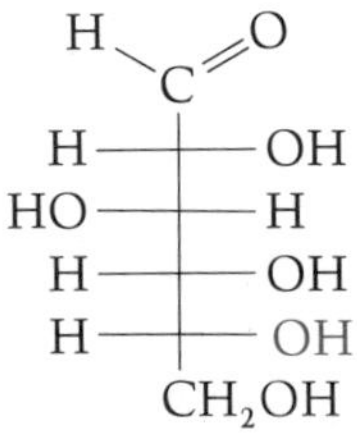

H–C=O
H — OH
HO — H
H — OH
H — OH
CH_2OH
D-glucose

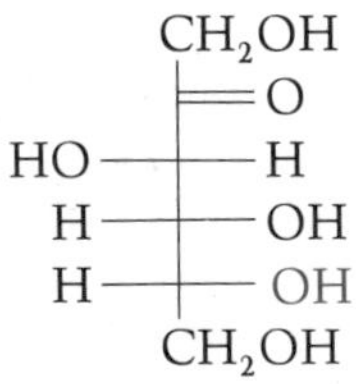

CH_2OH
=O
HO — H
H — OH
H — OH
CH_2OH
D-fructose

L-*sugars*

```
   H   O                   H   O                  CH2OH
    \ //                    \ //                    |
     C                       C                      ==O
     |                 OH ---+--- H           H ----+---- OH
HO --+-- H              H ---+--- OH         HO ----+---- H
     |                 HO ---+--- H          HO ----+---- H
   CH2OH               HO ---+--- H                 |
                             |                    CH2OH
L-glyceraldehyde           CH2OH                 L-fructose
                         L-glucose
```

Problem 16.13 The Fischer projection formula of D-galactose is

```
      H   O
       \ //
        C
        |
  H ----+---- OH
 HO ----+---- H
 HO ----+---- H
  H ----+---- OH
        |
      CH2OH
```

Draw the Fischer projection formula of L-galactose.

Except for rare cases we find only D-sugars in nature. We will therefore confine our discussion to them.

MUTAROTATION OF GLUCOSE 16.10

What happens to the ring during mutarotation?

The structure of D-glucose is complicated by the fact that in water solution it exists in three forms in equilibrium with each other. The chain form that we have used before exists only to a slight extent in solution. Instead, the most common forms of D-glucose in solution are two cyclic hemiacetal forms (Section 14.7).

```
  HO   H                        H   O                  H    OH
    \ /                          \ //                   \  /
     C ---------.                 C                      C ---------.
     |          |                 |                      |          |
 H --+-- OH     |            H ---+--- OH            H --+-- OH     |
HO --+-- H      O  <==>     HO ---+--- H    <==>    HO --+-- H      O
 H --+-- OH     |            H ---+--- OH            H --+-- OH     |
 H --+----------'            H ---+--- OH            H --+----------'
     |                            |                      |
   CH2OH                        CH2OH                  CH2OH
β-D-glucose                   D-glucose              α-D-glucose
```

Notice that this reaction converts the carbonyl carbon into a new chiral center. These two hemiacetals have different optical rotations, and they convert into one another through a process called **mutarotation** (see Problem 16.26).

The use of Fischer formulas for the cyclic forms implies unrealistic bond lengths. The cyclic monosaccharides can be better represented by using **Haworth formulas,** designed by W. N. Haworth.

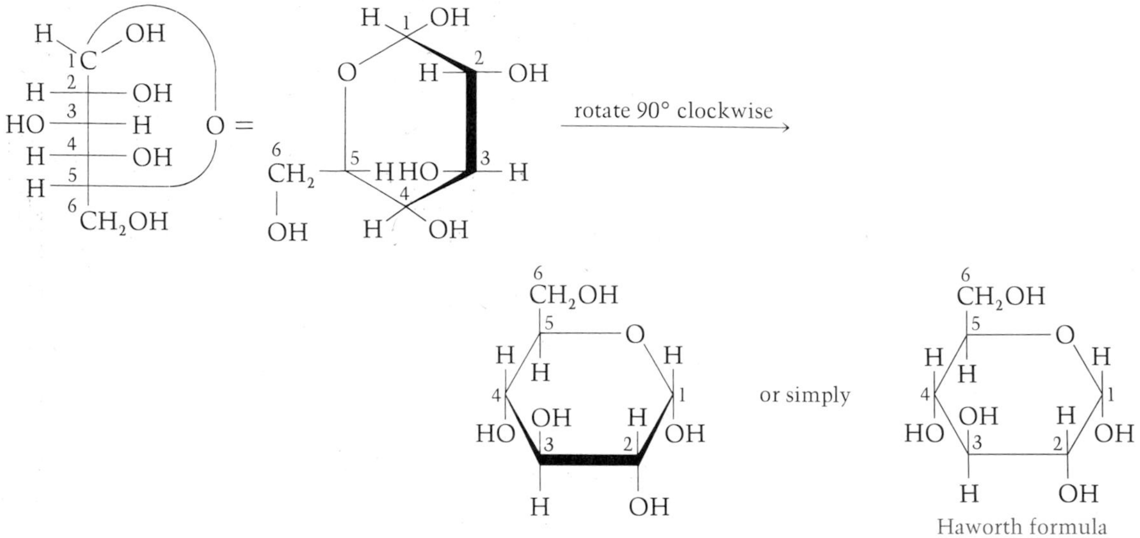

rotate 90° clockwise

or simply

Haworth formula

Mutarotation results from the reversible reaction between the 5-hydroxyl group and the carbonyl carbon to form α and β hemiacetals. Both hemiacetals are very stable six-membered rings. The hemiacetal group is outlined in the following structure.

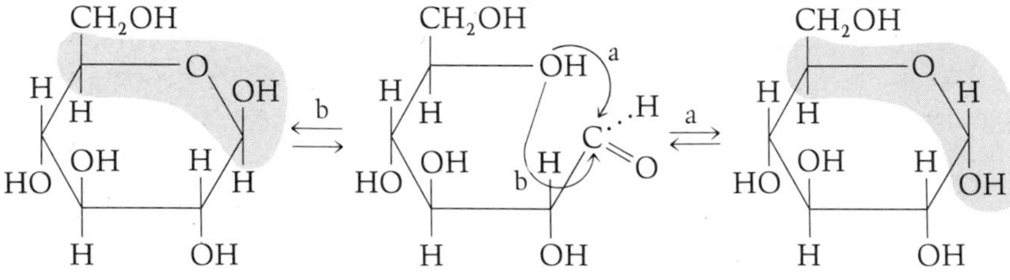

β-D-glucose   open-chain form α-D-glucose

The α form of a D-aldohexose is the one in which the hemiacetal hydroxyl group is on the side of the ring *trans* to carbon 6 (that is, on the opposite side from carbon 6). In the β form, the hydroxyl group is *cis* to carbon 6.

The α and β forms are known as **anomers,** and carbon 1 is known as the **anomeric carbon.** It still retains some of the high reactivity that you would expect from a carbon in a carbonyl group, as compared with a hydroxyl group. Most pentoses and hexoses exist in the hemiacetal forms. A six-membered ring containing one oxygen atom is called a **pyranose.** Thus, to be more precise, we may name the two glucose anomers α-D-glucopyranose and β-D-glucopyranose. A five-membered ring containing one oxygen atom is called a **furanose.**

Problem 16.14 Draw the Haworth formulas of the α and β anomers of D-galactose (see Problem 16.13).

GLYCOSIDES 16.11

Through what functional group are monosaccharides joined to form disaccharides and larger carbohydrates?

Monosaccharides in the hemiacetal form may react further with an alcohol to form an acetal, if the reaction mixture is acidic (Section 14.7).

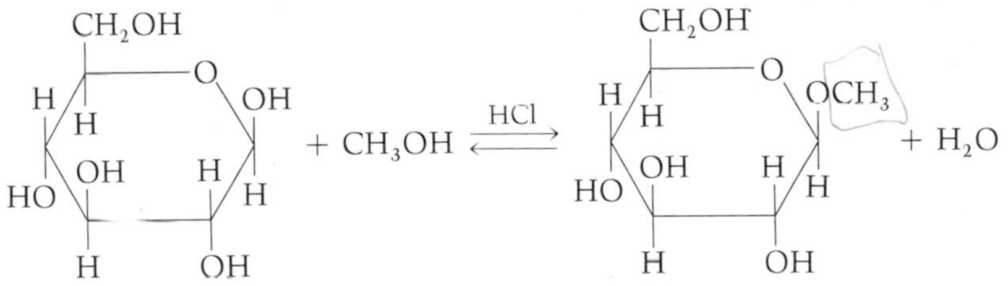

CH_2OH ... $+ CH_3OH \xrightleftharpoons{HCl}$ ... OCH_3 ... $+ H_2O$

Such acetals of carbohydrates are called **glycosides,** which is a general name. Specific glycoside names are *glucoside* from glucose, *mannoside* from mannose, *galactoside* from galactose, and so forth.

The OH group that reacts with the hemiacetal may be part of another monosaccharide molecule. In this way, monosaccharide units can unite to form disaccharides, trisaccharides, and higher polysaccharides. In the example below, the anomeric carbons of the two monosaccharide units are designated by an asterisk (*). In this case the OH group that reacts is the 4-hydroxyl group of ring B.

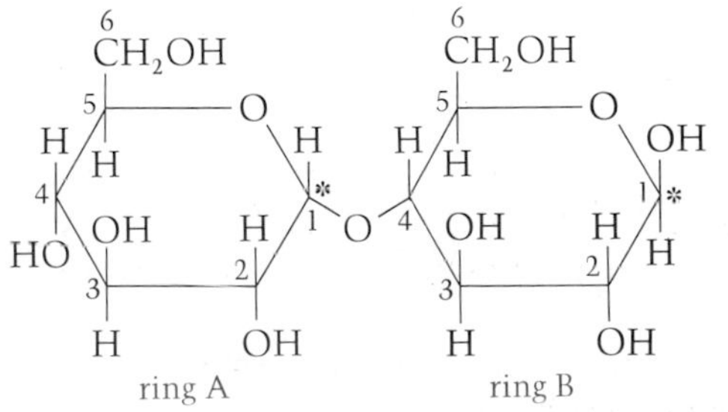

ring A ring B

Problem 16.15

(a) The formula shown above is for the α-glycoside. Draw the Haworth formula of the β anomer.

(b) Draw the α- and β-glycosides of the acetal formed using the hydroxyl group on carbon 6 of ring B.

SUMMARY

Carbohydrates have the general formula $C_n(H_2O)_m$. They are classified as **monosaccharides, disaccharides, trisaccharides,** and so forth, depending on the number of simple monosaccharide units obtained by hydrolysis. A monosaccharide cannot be hydrolyzed to simpler units. Monosaccharides are polyhydroxy aldehydes (**aldoses**) or polyhydroxy ketones (**ketoses**). Monosaccharides generally taste sweet and are water-soluble.

Molecules that have one **chiral** carbon atom (a carbon atom that is bonded to four different atoms or groups) exist as **enantiomers.** Enantiomers are related to each other as an object is related to its non-superposable mirror image. They are also called mirror-image isomers. Chiral molecules rotate the plane of polarization of **plane-polarized light.** This phenomenon is called **optical activity.**

Fischer projection formulas are used to represent open-chain carbohydrate structures. **Haworth formulas** are used to represent cyclic hemiacetal and **glycoside** structures. D- and L- designations are used to identify the configuration of the chiral carbon farthest from the carbonyl group. **Mutarotation** is the process in which a sample containing only the α or β form is converted spontaneously into an equilibrium mixture of α and β forms. Monosaccharides exist in cyclic form as five-membered rings (**furanoses**) and as six-membered rings (**pyranoses**). The acetal forms of carbohydrates are called glycosides.

Key Terms

Anomer (Section 16.10)
Aldose (Section 16.3)
Anomeric carbon (Section 16.10)
Asymmetric carbon atom (Section 16.4)
Carbohydrate (Section 16.1)
Chiral object (Section 16.4)
Disaccharide (Section 16.1)
Enantiomer (Section 16.4)

Fischer projection formula (Section 16.8)
Furanose (Section 16.10)
Glycoside (Section 16.11)
Haworth formula (Section 16.10)
Ketose (Section 16.3)
Monosaccharide (Section 16.1)
Mutarotation (Section 16.10)
Oligosaccharide (Section 16.2)
Optical activity (Section 16.6)
Plane-polarized light (Section 16.5)
Polarimeter (Section 16.6)
Polysaccharide (Section 16.1)
Pyranose (Section 16.10)
Trisaccharide (Section 16.1)

ADDITIONAL PROBLEMS

16.16 Write the molecular formulas of the following carbohydrates in the form $C_n(H_2O)_m$.

(a) D-glucose

(b) CHO
|
CHOH
|
CH_2OH

(c)
CH_2OH
=O
HO—|—H
H—|—OH
H—|—OH
CH_2OH

(d) $C_{12}H_{22}O_{11}$

16.17 Classify the following carbohydrates as monosaccharides, disaccharides, and so forth.

(a) fructose
(b) D-glyceraldehyde
(c) sucrose
(d) cellulose
(e) starch

16.18 Hydrolysis of 1 mole of an unknown carbohydrate yields 3 moles of D-glucose and 2 moles of D-galactose. Classify the unknown carbohydrate.

16.19 A certain carbohydrate is very sweet and very soluble in water. What can you say about the structure of this carbohydrate?

16.20 Draw the Fischer projection formulas of the following carbohydrates.

(a) a D-aldotriose
(b) an L-ketopentose
(c) a D-aldotetrose
(d) a D-ketohexose

16.21 Draw the Haworth formulas of the following carbohydrates.

(a) a β-pyranose made from a D-aldohexose other than D-glucose
(b) the α anomer of (a)

16.22 Which of the following molecules are chiral?

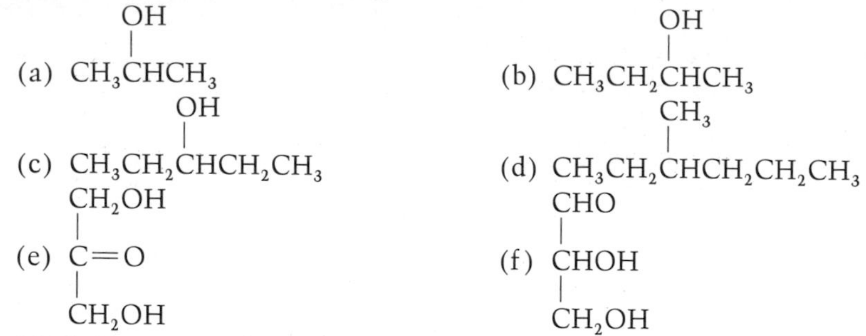

16.23 Two pure substances, *A* and *B*, have the following optical rotations under identical conditions: A, +82°; B, −82°. Are *A* and *B* necessarily enantiomers? Explain.

16.24 Which carbon atom determines whether a sugar is D or L? Explain.

16.25 Compare these two monosaccharides, *A* and *B*.

(a) Are *A* and *B* isomers?
(b) Are they stereoisomers?
(c) Are they enantiomers?
(d) Which one is a D-sugar?

A:

	CHO	
H	—	OH
HO	—	H
H	—	OH
H	—	OH
	CH_2OH	

B:

	CHO	
H	—	OH
HO	—	H
H	—	OH
HO	—	H
	CH_2OH	

***16.26** Two different forms of D-glucose exist. When we dissolve ordinary D-glucose (the α form) in water, the solution initially has an optical rotation of +112°. As time passes, the rotation gradually changes, until it reaches a final value of +53°. A second form of D-glucose (the β form) can be obtained by evaporating a water solution of D-glucose at 98°C. When we dissolve this second form of D-glucose in water, it has an initial optical rotation of +19°; as time passes, this rotation changes gradually until it reaches a final value of +53°. This process is called mutarotation.

(a) Explain mutarotation in terms of the α and β hemiacetal structures.
(b) Which form do you think is present in greater amount at equilibrium? Explain.

***16.27** Draw the Haworth formula of a carbohydrate whose molecules hydrolyze to two molecules of glucose, one of which is connected by an α-glycoside linkage to the 6-hydroxyl group of the other.

16.28 Sucrose has the structural formula

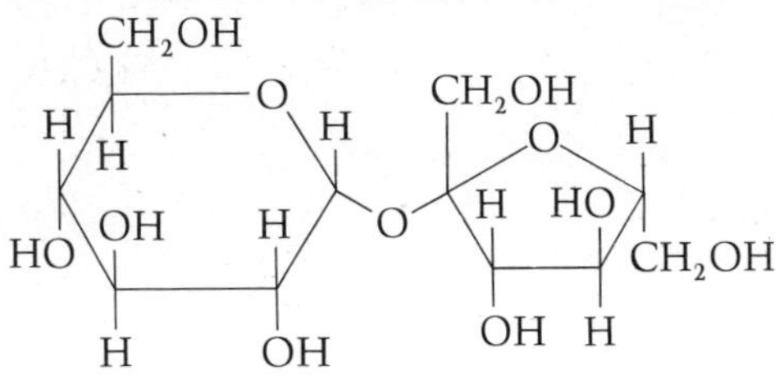

CH_2OH O   H CH_2OH O   H
H   H   OH   H   O   H   HO CH_2OH
HO   H   OH   OH   H

A   *B*

(a) Circle the anomeric carbons in the two monosaccharide units.

(b) Ring *A* is a pyranose ring. What is ring *B*?

Problem-Solving Hints

If you have trouble solving any of the additional problems, refer to the sections listed next to the problem numbers.

16.16 16.1	**16.17** 16.1	**16.18** 16.1
16.19 16.2	**16.20** 16.8	**16.21** 16.10
16.22 16.4	**16.23** 16.6	**16.24** 16.9
16.25 16.8, 16.9	**16.26** 16.10	**16.27** 16.10, 16.11
16.28 16.10, 16.11		

FOCUS 11 Food Additives

In discussing food additives we will consider four questions:

1. What are food additives?
2. Why do manufacturers add them to foods?
3. Are they necessary?
4. What steps are taken to ensure that they are safe?

What Are Food Additives?

In the late 1950s, the U.S. Congress adopted a definition of a food additive; it states in simple terms that a food additive is a substance, not generally considered to be a food, that is intentionally added to foods to improve their characteristics. For example, sugar added to canned fruits is not a food additive, whereas saccharin is.

Why Do Manufacturers Add Additives to Foods?

Food additives are used for a variety of reasons. (a) To keep foods that might deteriorate in storage unspoiled for long periods of time, manufacturers use antioxidants and preservatives. (b) To make foods more attractive, manufacturers use coloring and flavoring agents, emulsifiers, stabilizers, thickeners, clarifiers, and bleaching agents. (c) To aid in food processing, manufacturers use acids, alkalis, buffers, and other chemical agents.

Are Food Additives Necessary?

In many parts of the world, people eat fruits and vegetables only when they are in season. In developed countries, we can consume meats, fruits, and vegetables not only in every season, but also from distant lands. We enjoy this luxury to a great extent because we are able to preserve foods. The capacity to preserve foods also allows us to support a much greater population than would otherwise be possible.

How Is Safety of Food Additives Monitored?

The most hotly debated question about food additives is, "Are they safe?" We hear it asked daily, in one form or another. The Pure Food and Drug

FIGURE 1 Many of the foods we eat have additives in them. (FDA photo)

Act of 1906 required that our food supply be safe. That law requires that food additives satisfy two conditions: they must be safe for human consumption, and they must serve a useful purpose. In 1938 an amendment to the law set up the Food and Drug Administration to enforce the regulations. The 1938 law was weak, because the Food and Drug Administration was required to prove that the additive was harmful or useless before it could take action to discontinue its use. Manufacturers could use an additive until FDA agents could prove that the product was harmful. The law was changed in 1960 and now requires the manufacturer to prove to the FDA that a new product is safe *before* it is placed on the market. There have been further refinements of the law in subsequent years. One of the most notable is the Delaney Amendment, which prohibits the use of any food additive that causes cancer in any animal, even at very high dosages.

Older Methods of Treating Foods

Foods were preserved and modified by natural means long before synthetic chemicals came into use. For thousands of years people have preserved meats by salting and by smoking. The use of salt inhibits the growth of organisms in the meat. Smoking adds preservatives to the meat.

Sugar is used in high concentrations to make jams and jellies. Most bacteria cannot grow in solutions with high concentrations of sugar, and spoilage is thus prevented. Sugar is also used to make condensed milk. Here again the high sugar concentration inhibits the growth of organisms. Vinegar, which is a water solution of acetic acid, prevents spoilage, since most bacteria cannot survive in acidic conditions.

Flavoring agents have been known for centuries. The search for spices launched the expeditions that led to the discovery and colonization of America. In Europe in the fifteenth century spices were not merely decorations for food; they were necessary to mask the taste of spoiled foods and make them edible. Saffron, for example, has been used for centuries, especially in the Mediterranean basin, to color and flavor soups and rice dishes.

Acids and alkalis are used to intensify the flavors of sherbets, cheeses, and other foods. Fruit acids (mainly citric acid from lemon and orange juices) have long been used as acidifying agents. Sodium and calcium carbonates have been used as alkalis to control the acidity of wines and to preserve vegetables, such as olives and peas.

Problem 1 In what other ways can foods be preserved without using synthetic food additives?

Preservatives

Preservatives, such as sugar and vinegar, are chemicals that inhibit the growth of microorganisms. Cooking normally kills any that are present in foods. However, to prevent the reintroduction of organisms, manu-

TABLE 1 Some Commonly Used Preservatives

Name	Formula	Uses
Calcium propionate	$(CH_3CH_2CO_2)_2Ca$	Added to bread to increase calcium content and to avoid gas formation by yeasts
Sodium propionate	$CH_3CH_2CO_2Na$	Used in cakes instead of calcium propionate
Lactic acid	$CH_3CH(OH)CO_2H$	Used in bread to inhibit "ropiness," a bacterial contamination
Sorbic acid Sodium sorbate	$CH_3CH{=}CHCH{=}CHCO_2H$ $CH_3CH{=}CHCH{=}CHCO_2Na$	Inhibits mold and yeast in bread, cheese, syrups, jams, mayonnaise, fruit juices, pie fillings, and pickle products
Benzoic acid Sodium benzoate	$C_6H_5{-}CO_2H$ $C_6H_5{-}CO_2Na$	Used in highly acid food products—carbonated beverages, orange drinks, apple cider, fruit cocktail, margarine, maraschino cherries
Calcium disodium EDTA*	$(\bar{O}_2CCH_2)_2NCH_2CH_2N(CH_2C\bar{O}_2)_2$ $Ca^{2+}, 2Na^+$	Combines with trace metals such as iron, copper, and cobalt, and renders them chemically inactive; used in red wine vinegar, imitation mayonnaise, and other products
Sulfur dioxide Sodium sulfite	SO_2 Na_2SO_3	Used in lemon juice, instant mashed potatoes, cookies

*EDTA = *E*thylene*d*iamine*t*etraacetic *a*cid.

facturers add preservatives. Table 1 lists some of the more important preservatives and their uses.

Food Coloring

Tests have shown repeatedly that people reject foods that do not have their customary color. You would probably appreciate this problem if you were confronted with a blue or green steak. Manufacturers add coloring agents to many foods—soft drinks, puddings, candy, sausages, canned meats, cereals, cheeses, butter, ice cream, and many others. Coloring agents are often used because manufacturing processes have altered the natural color of a food.

Coloring agents must not only impart the desired color to a product, they must also remain stable under conditions of manufacture, storage, and cooking. Table 2 lists some food coloring agents, many of which are known by their color and a number.

TABLE 2 Some Food Coloring Agents

Color	Name	Formula
Red no. 2 (discontinued)	Amaranth	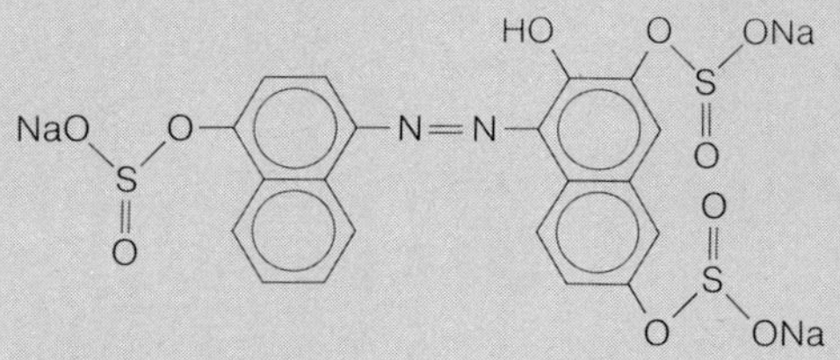
Red no. 3	Erythrosine	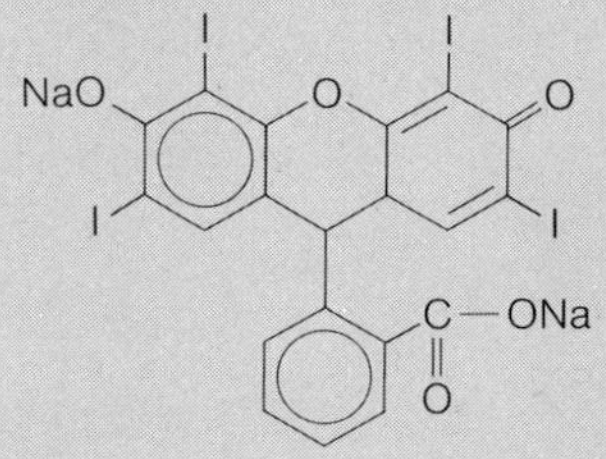
Yellow no. 5	Tartrazine	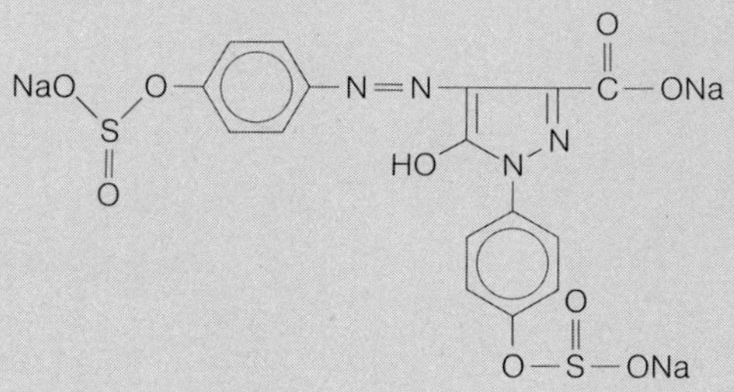
Yellow no. 6	Sunset yellow	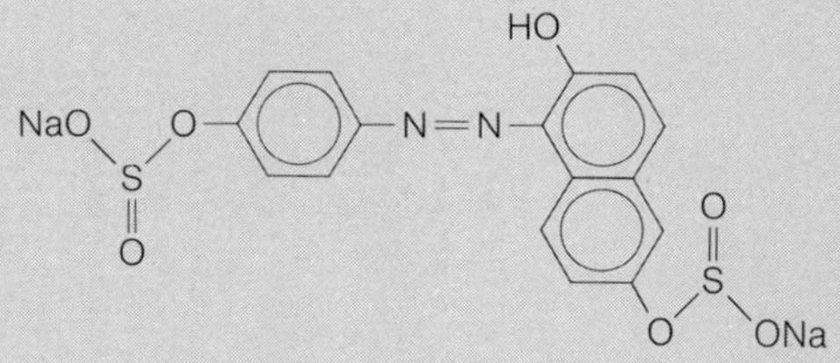

TABLE 2 Some Food Coloring Agents (Continued)

Color	Name	Formula
Green no. 3	Fast green	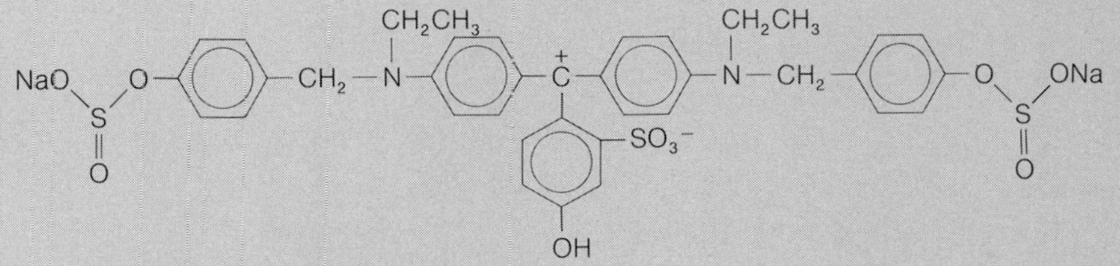 NaO, S, O, O, CH_2, N, CH_2CH_3, $\overset{+}{C}$, SO_3^-, OH, CH_2CH_3, N, CH_2, O, S, O, ONa
Violet no. 1	Benzylviclet	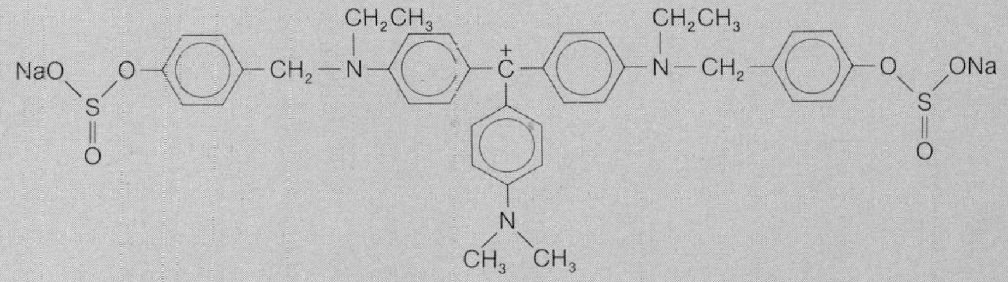 NaO, S, O, O, CH_2, N, CH_2CH_3, $\overset{+}{C}$, N, CH_3, CH_3, CH_2CH_3, N, CH_2, O, S, O, ONa
Blue no. 2	Indigo carmine	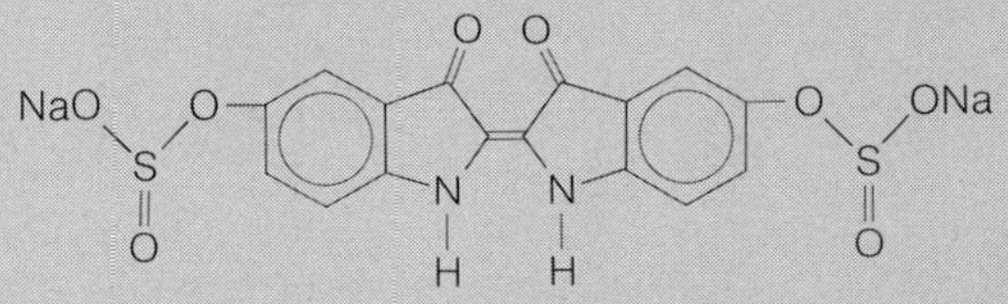 NaO, S, O, O, O, N, H, O, N, H, O, S, O, ONa

Highly colored compounds often possess aromatic rings and highly conjugated systems of double and single bonds. Compounds with those structural features are often carcinogenic (see Focus 7). For this reason, food coloring agents are carefully screened by the Food and Drug Administration. In 1977, the widely used Red dye No. 2 was removed from the market because it was found to produce cancers in laboratory animals.

Flavoring Agents

The largest class of food additives is the flavoring agents. About 1,400 natural and synthetic flavors are known. The use of flavors has grown in the last few decades, as new foods have been devised and new processing techniques developed. Some flavoring substances are listed in Table 3.

Flavor Enhancers

Flavor enhancers are substances that improve the natural flavors of foods. The best-known flavor enhancer is monosodium glutamate (MSG), which is often used in Chinese cooking and can cause a condition known as "Chinese restaurant syndrome". The symptoms of this syndrome are headaches and a feeling of weakness after eating food with MSG in it. Maltol is another substance used to enhance the flavor of fruits, desserts, and soft drinks that are high in carbohydrates.

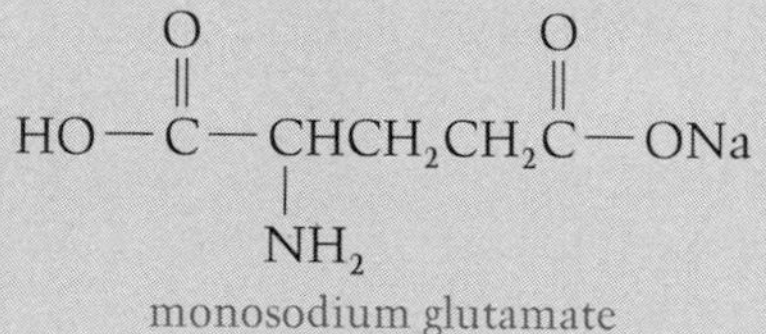

$$HO-\overset{\overset{O}{\|}}{C}-\underset{\underset{NH_2}{|}}{C}HCH_2CH_2\overset{\overset{O}{\|}}{C}-ONa$$

monosodium glutamate

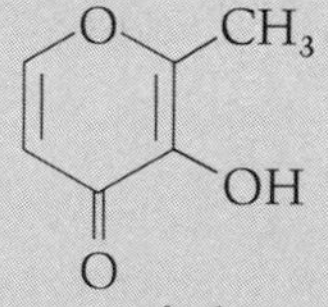

maltol (ring structure: O, CH_3, OH, O)

maltol

Emulsifiers

Oil and water do not form a solution, and mixtures of the two tend to separate into two layers. Not too long ago peanut butter always had an oily layer on top, which had to be mixed with the semisolid peanut matter before eating. Today nearly all commercial peanut butter products are emulsified; they are smooth, easy to spread, and do not separate even after standing on the shelf for long periods. Milk is another mixture of oil (milk fat) and water that is usually kept from separating by use of

TABLE 3 Some Flavoring Compounds

Name	Flavor	Formula
Cinnamaldehyde	Cinnamon	$C_6H_5{-}CH{=}CH\overset{O}{\overset{\|}{C}}H$
Vanillin	Vanilla	$\overset{O}{\overset{\|}{HC}}{-}C_6H_3(OCH_3)(OH)$
Citral	Lemon	$(CH_3)_2C{=}CH{-}CH_2{-}CH_2{-}C(CH_3){=}CH{-}\overset{O}{\overset{\|}{C}}H$
Furfuryl mercaptan	Coffee	$C_4H_3O{-}CH_2SH$
Capsaicin	Red pepper	$HO(CH_3O)C_6H_3{-}CH_2O\overset{O}{\overset{\|}{C}}CH_2CH_2CH_2CH_2CH{=}CHCHCH_3$ with CH_3 on the last CH
α-Ionone	Strawberry and raspberry	cyclohexene ring with CH_3,CH_3; $CH{=}CH\overset{O}{\overset{\|}{C}}CH_3$; CH_3
Propyl disulfide	Onion	$CH_3CH_2CH_2{-}S{-}S{-}CH_2CH_2CH_3$
Anethole	Anise and licorice	$CH_3O{-}C_6H_4{-}CH{=}CHCH_3$
Piperidine	Pepper	$(CH_2)_5NH$ (ring with N–H)
Eugenol	Clove	$HO(CH_3O)C_6H_3{-}CH_2CH{=}CH_2$

TABLE 3 Some Flavoring Compounds (Continued)

Name	Flavor	Formula
Ethyl *trans-2-cis*-4-decadienoate	Pear	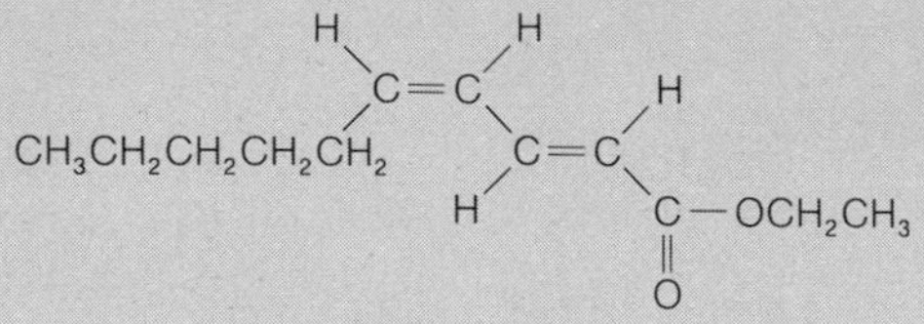 $CH_3CH_2CH_2CH_2CH_2(H)C{=}C(H)-(H)C{=}C(H)-\underset{\displaystyle O}{\overset{\Vert}{C}}-OCH_2CH_3$
Allyl isothiocyanate	Mustard and horseradish	$CH_2{=}CHCH_2-N{=}C{=}S$

an emulsifier. A mixture of two mutually insoluble liquids that do not separate, such as oil and water, is called an *emulsion*. An *emulsifier* is a substance that causes two such liquids to remain in an emulsion. One end of the emulsifier molecule has an affinity for water, and the other end has an affinity for oils (hydrocarbons). The emulsifier therefore bonds the two insoluble liquids together into tiny globules. Many food products are emulsions—mayonnaise, margarine, chocolate products, ice cream, custards, cake icings, fillings, and toppings. The most widely used emulsifiers are lecithin and the synthetic monoglycerides and diglycerides. In the formulas for these emulsifiers shown below, R stands for hydrocarbon groups of fifteen to nineteen carbon atoms.

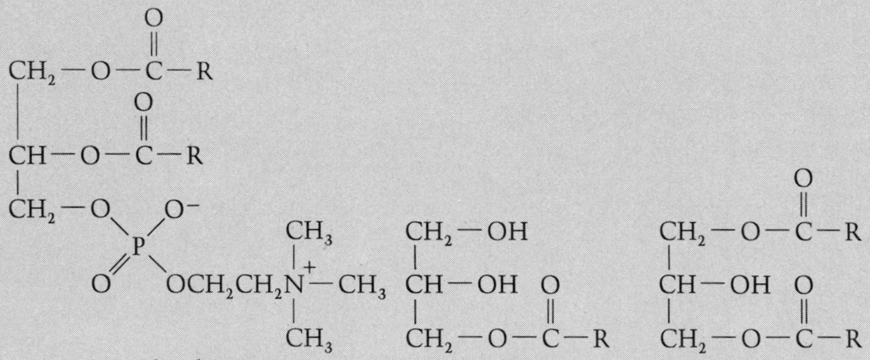

$$\begin{array}{l}
CH_2-O-\overset{\displaystyle O}{\overset{\Vert}{C}}-R \\
\vert \\
CH-O-\overset{\displaystyle O}{\overset{\Vert}{C}}-R \\
\vert \\
CH_2-O-\underset{\displaystyle \Vert \atop \displaystyle O}{\overset{\displaystyle O^-}{\overset{\vert}{P}}}-OCH_2CH_2\overset{+}{N}(CH_3)_3
\end{array}$$

lecithin

$$\begin{array}{l}
CH_2-OH \\
\vert \\
CH-OH \\
\vert \\
CH_2-O-\overset{\displaystyle O}{\overset{\Vert}{C}}-R
\end{array}$$

monoglyceride

$$\begin{array}{l}
CH_2-O-\overset{\displaystyle O}{\overset{\Vert}{C}}-R \\
\vert \\
CH-OH \\
\vert \\
CH_2-O-\overset{\displaystyle O}{\overset{\Vert}{C}}-R
\end{array}$$

diglyceride

Antioxidants

Some fatty foods become rancid when exposed to oxygen in the air because they undergo oxidation. In the process, carbon–carbon double bonds are oxidized to aldehydes and ketones, which have strong odors. Other foods, like butter, become rancid when their esters hydrolyze.

TABLE 4 Some Common Antioxidants

Name	Formula
Ascorbic acid	O, O, OH, $CHCH_2OH$, HO, OH
Butylated hydroxyanisole (BHA)	OH, $C(CH_3)_3$, OCH_3 and OH, $C(CH_3)_3$, OCH_3
Butylated hydroxytoluene (BHT)	OH, $(CH_3)_3C$, $C(CH_3)_3$, CH_3
Propyl gallate	O, $OCH_2CH_2CH_3$, C, HO, OH, OH

In rancid butter, butyric acid is produced. Among the foods that oxidize readily are margarine, cooking oils, bacon, potato chips, salted nuts, cereals, precooked meats, and fish.

Antioxidants are compounds that react with atmospheric oxygen to form harmless, nonoxidizing products. Cooks have for centuries used citrus juices to control the oxidation of sliced fruits, such as peaches and apples. The effective antioxidant in citrus juices is ascorbic acid (vitamin C). Table 4 lists some antioxidants that are used commercially in foods.

Nitrites

Prepared meat products such as bacon and sausages often turn a brown or gray color on exposure to air. The color change occurs because of their high fat content. The C=C double bonds in fats oxidize when exposed to oxygen in the air. If these meats are treated with sodium nitrite ($NaNO_2$) before they are packaged, they retain their red color

(Section 13.12). The nitrite ions produce a red color by reacting with the myoglobin (chemically similar to hemoglobin) in the meat to form nitrosomyoglobin, which is red. The nitrite ion attaches to the iron atom of myoglobin in much the same way that oxygen binds to iron. The nitrite ion also prevents the growth of *Clostridium botulinum,* the organism that causes botulism poisoning.

Nitrite ions are potentially dangerous, though, because in the body they react with secondary amines (Section 13.2) to produce *nitrosamines.* First, the high acidity in the stomach converts nitrite ions to nitrous acid.

$$\underset{\text{nitrite ion}}{NO_2^-} + H_3O^+ \rightarrow \underset{\text{nitrous acid}}{H{-}O{-}N{=}O} + H_2O$$

Nitrous acid then reacts with secondary amines to produce nitrosamines.

$$H{-}O{-}N{=}O + R_2N{-}H \rightarrow \underset{\text{a nitrosamine}}{R_2N{-}N{=}O} + H_2O$$

Nitrosamines are known to be carcinogenic, so the danger exists that nitrites in foods could lead to the formation of cancers. Some people argue that the quantities of nitrites used in foods are too small to pose a health hazard. On the other hand, nitrosamines have been detected in meats that were treated with nitrites.

SUMMARY

Although some food additives may appear frivolous and unnecessary, others are essential. We must use preservatives if we are to continue to feed our growing population. As long as we continue to use them, however, food additives will continue to pose problems of safety.

The Food and Agriculture Organization of the World Health Organization of the United Nations has proposed the following recommendations concerning the use of food additives:

1. A food additive should be technologically effective.
2. It should be safe to use.
3. It should not be used in any quantity greater than is necessary to achieve its stated effect.

4. It should never be used to mislead the consumer as to the nature and quality of food.
5. The use of nonnutrient food additives should be kept to the minimum.

Problem 2 List some common food additives and their functions.

Problem 3 Make a list of the canned or packaged foods you have, and alongside each food list all the food additives shown on the label. Classify each additive according to the types described in this focus.

Problem 4 Write the equation for the reaction of nitrous acid with diethylamine.

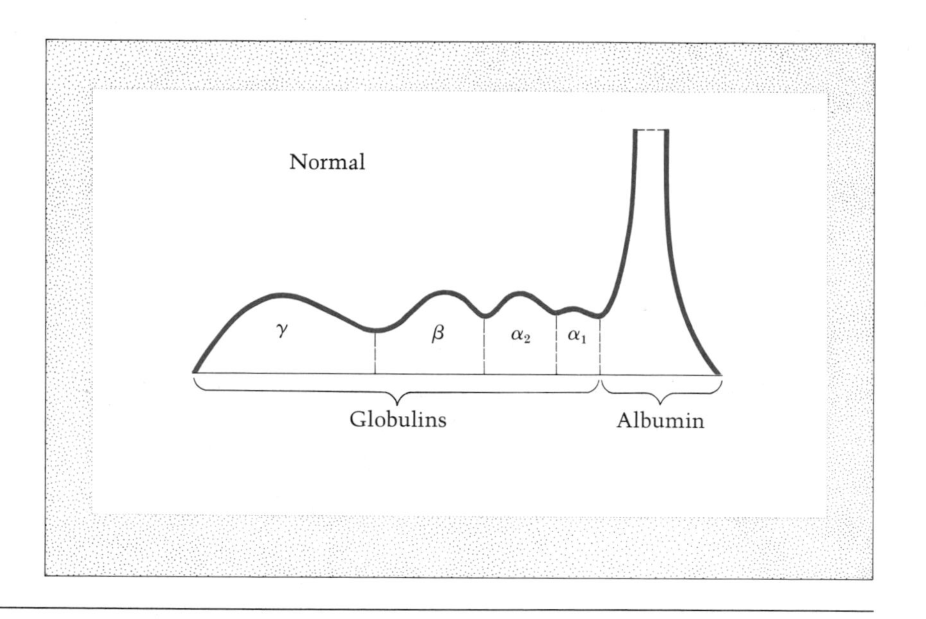

In the early 1800s, one protein was thought to be responsible for all biochemical systems in plants and animals. Since that time well over a hundred such proteins have been discovered in the human body. Interestingly enough, we have come to understand the functions of less than half of them. Some of these proteins are called serum proteins because they are present in the blood serum. Serum proteins are responsible for many biochemical processes and can aid in the control of disease.

There are five major serum protein groups: alpha 1-globulin, alpha 2-globulin, beta-globulin, gamma-globulin, and albumin. These groups can now be separated by a laboratory method called electrophoresis. The basis of this technique involves separation of molecules according to their different electrical charges. At the same time that each of the protein electrophoresis components are measured, the total protein level in the serum must also be measured in order to calculate the relative amount of each protein present.

Various medical diagnoses are possible from protein electrophoresis studies. For example, albumin levels help to detect liver and kidney diseases, alpha-globulins can detect trauma, and beta-globulins can signal lipid (fat) disorders. The gamma-globulin component represents perhaps the best diagnostic aid to the physician because it contains most of the antibodies. Increases in gamma-globulin levels can indicate chronic infections and various malignancies, whereas decreases in these levels can point to leukemia.

17

INTRODUCTION TO BIOLOGICAL MACROMOLECULES; PROTEINS

Taken individually, the readings of serum protein levels are not always a valid basis for a diagnosis. However, when they are combined and considered along with additional results from related tests, the results are highly informative.

[Illustration: The protein electrophoresis pattern produces graphic levels of globulins and albumin that can be used in the diagnosis of possible disease.]

INTRODUCTION TO BIOLOGICAL MACROMOLECULES

A **macromolecule** is simply a very big molecule. There are three classes of biological macromolecules: the proteins, which we will learn about in this chapter, and the nucleic acids and the polysaccharides, which we will consider in the next two chapters. One of the remarkable characteristics of living organisms is how many macromolecules they utilize and what different and important roles these macromolecules play.

A macromolecule can contain as many as 200 million individual atoms. In contrast, ordinary organic molecules contain no more than a few hundred atoms. One advantage of a big molecule is that it can assume different shapes and do different things. In general, a small molecule can have only one shape or can vary its shape only slightly. But a macromolecule such as a protein can change shape dramatically—just as a mile of rope can be arranged in more different ways than a foot of rope.

The functioning of all biological macromolecules depends on their shapes, or three-dimensional structures. If you change the three-dimensional structure of a protein, for example, you may destroy its ability to function. You see simple evidence of this every time you cook an egg.

In uncooked egg white, the proteins are curled into globular shapes. When the egg white is cooked, these proteins uncurl. This uncurling is reflected in the changing color of the egg white—the previously clear substance indeed becomes white. When the egg white proteins change shape, they lose their ability to function. We say that they have become denatured, since, in effect, they have lost their nature.

Usually, a denatured protein has permanently lost its functional capability—it cannot return to what it was before. But many proteins do change shape only temporarily, resuming their original structure in time and under the right conditions. On a humid day, your hair may curl because the dampness causes the proteins in the hair to change shape. When the humidity drops, your hair straightens again.

Many proteins and other macromolecules change shape in the course of performing their biological roles. The biological macromolecules provide us with the clearest example and the most sensitive expression of the relationship between molecular structure and chemical and physical properties.

INTRODUCTION TO MACROMOLECULES 17.1

Proteins, nucleic acids, and **polysaccharides** are polymers (Section 10.6). These polymers are formed by condensation reactions between many similar building-block molecules. A condensation reaction occurs when a covalent bond is formed and a water molecule is lost simultaneously. For example, the building blocks of proteins are **amino acids,** compounds containing both a carboxyl and an amino group (Section 13.11). Two amino acids can condense in the following reaction between the amino and carboxyl groups:

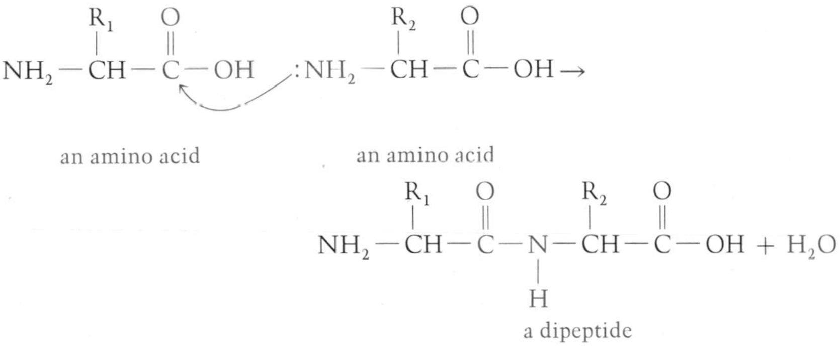

$$\underset{\text{an amino acid}}{NH_2-\overset{\overset{R_1}{|}}{CH}-\overset{\overset{O}{\|}}{C}-OH} \quad \underset{\text{an amino acid}}{:NH_2-\overset{\overset{R_2}{|}}{CH}-\overset{\overset{O}{\|}}{C}-OH} \rightarrow$$

$$\underset{\text{a dipeptide}}{NH_2-\overset{\overset{R_1}{|}}{CH}-\overset{\overset{O}{\|}}{C}-\underset{\underset{H}{|}}{N}-\overset{\overset{R_2}{|}}{CH}-\overset{\overset{O}{\|}}{C}-OH} + H_2O$$

The product is a **peptide.** A peptide consists of two or more amino acids linked by peptide bonds (Section 15.17). More specifically, this particular product is a *di*peptide, because it consists of *two* amino acids linked by a peptide bond. The carboxyl group of the amino acid on the right side of the dipeptide is free to condense with the amino group of a third amino acid, to form a *tri*peptide. This tripeptide can condense with a fourth amino acid, and so on. In this way, very long polymers of amino acids—**polypeptides**—can be formed:

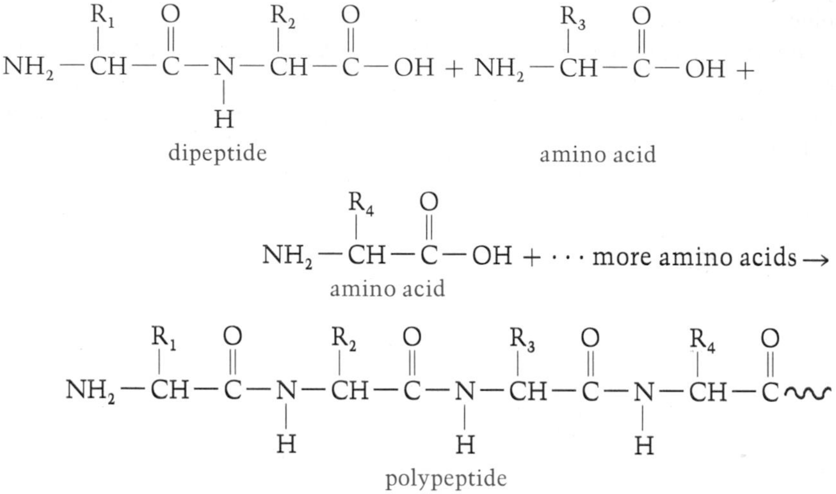

$$\underbrace{NH_2-\overset{R_1}{\overset{|}{CH}}-\overset{O}{\overset{\|}{C}}-\underset{\underset{H}{|}}{N}-\overset{R_2}{\overset{|}{CH}}-\overset{O}{\overset{\|}{C}}-OH}_{\text{dipeptide}} + \underbrace{NH_2-\overset{R_3}{\overset{|}{CH}}-\overset{O}{\overset{\|}{C}}-OH}_{\text{amino acid}} +$$

$$\underbrace{NH_2-\overset{R_4}{\overset{|}{CH}}-\overset{O}{\overset{\|}{C}}-OH}_{\text{amino acid}} + \cdots \text{ more amino acids} \longrightarrow$$

$$\underbrace{NH_2-\overset{R_1}{\overset{|}{CH}}-\overset{O}{\overset{\|}{C}}-\underset{\underset{H}{|}}{N}-\overset{R_2}{\overset{|}{CH}}-\overset{O}{\overset{\|}{C}}-\underset{\underset{H}{|}}{N}-\overset{R_3}{\overset{|}{CH}}-\overset{O}{\overset{\|}{C}}-\underset{\underset{H}{|}}{N}-\overset{R_4}{\overset{|}{CH}}-\overset{O}{\overset{\|}{C}}\sim}_{\text{polypeptide}}$$

Similarly, nucleic acids are polymers made from nucleotide building blocks, and polysaccharides are polymers of simple sugars (monosaccharides). We will discuss these building blocks and their reactions in Chapters 18 and 19. Meanwhile, we will introduce some concepts necessary to an understanding of the three-dimensional structures of the macromolecules.

Problem 17.1 Write the reaction of three amino acids to form a tripeptide. (Use structural formulas.)

17.2 MACROMOLECULES AND NONCOVALENT BONDS

What noncovalent bonds are important in the three-dimensional structure of macromolecules?

The bonds that link the building blocks of a polymer are covalent bonds. However, the functional structures of macromolecules depend just as much on noncovalent bonds as on covalent ones. By **functional structure,** we mean the unique three-dimensional structure of the macromolecule that is necessary for it to perform its function: to catalyze a reaction if it is an enzyme, to bind a hormone molecule if it is a hormone receptor, to transmit genetic information accurately if it is a chromosome, and so on. Therefore, we need to consider the nature of such noncovalent interactions in some detail.

There are three types of noncovalent bonds: ionic bonds, hydrogen bonds, and the so-called hydrophobic interactions. You should already be familiar with at least two of these bonds from earlier discussions in this book. Ionic bonds (Section 3.4) are those found in salts, in which positively charged ions and negatively charged ions are held together by electrostatic attraction.

Hydrogen bonds (Section 5.4) are the noncovalent bonds that are so important in the structure of water, for example. The formation of a hydrogen bond requires an electronegative atom such as oxygen or nitrogen. This atom must be covalently bonded to hydrogen to give —OH or $—NH_2$, and another oxygen or nitrogen must be nearby. The result is that the hydrogen is shared by the two electronegative atoms. For example,

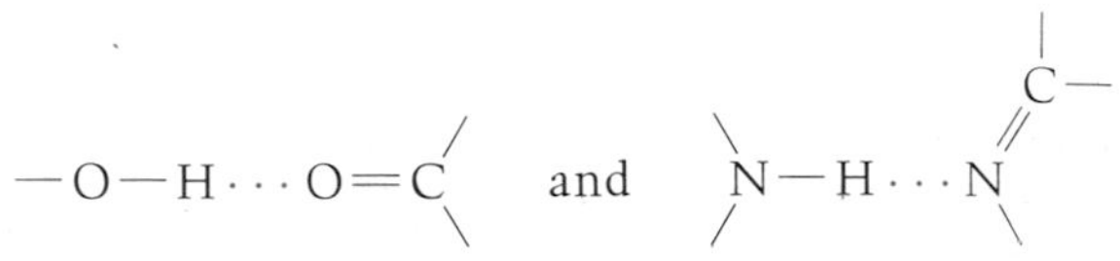

$$—O—H \cdots O=C\langle \quad \text{and} \quad \rangle N—H \cdots N(=C—)\langle$$

Individual hydrogen bonds are very weak—about one-tenth as strong as covalent bonds and considerably weaker than ionic bonds. Nonetheless, they exert a powerful influence on the structure of macromolecules because there are so many of them.

The third type of noncovalent bond is the **hydrophobic interaction.** Hydrophobic—literally, "water-hating"—substances like oils or hydrocarbons seem to prefer to interact with one another rather than with water. This phenomenon is the direct result of the great attraction of water molecules for one another. Hydrogen bonding takes place between water molecules so readily that even liquid water contains a good deal of icelike structure. At any given moment, most water molecules are interacting with up to four other water molecules via hydrogen bonds, as the following structure shows:

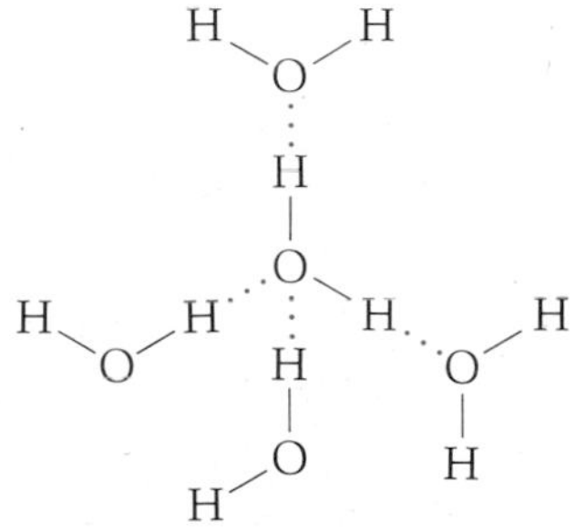

```
        H     H
         \   /
           O
           :
           H
           |
           O
        .·   \
H     H'   :   H.    H
 \   /     H    ·O  /
   O       |     |
           O     H
         /
        H
```

In the liquid state these bonds are constantly forming, breaking, and forming again with new water molecules. But solid ice has a rigid structure because the maximally hydrogen-bonded molecules are not free to move.

We are all familiar with the idea that oil and water do not mix (see Figure 17.1). For example, if you add a nonpolar liquid hydrocarbon such as hexane to some water in a flask, the hydrocarbon will not dissolve in the water (Section 9.4). Instead, the two substances will continue to seek separate phases, even if you have vigorously shaken them together. We can explain this observation as follows: The nonpolar hexane molecules have little attraction for one another. On the other hand, the water molecules have a lot of attraction for one another because they can hydrogen-bond. Thus water molecules tend to stick together as much as they can. As a result, the water structure shuts out the hexane molecules, which are forced to take up a layer of their own. Consequently, minimum contact between water and hydrocarbon molecules takes place. In other words, nonpolar molecules or nonpolar parts of large organic molecules tend to be largely excluded from contact with an aqueous solution. The other side of this coin is that such hydrophobic substances tend to be found in close association with one another simply because there are no alternatives. Although in its literal sense of water-hating, "hydrophobic" is not really an accurate description of this phenomenon, it is the term we generally use.

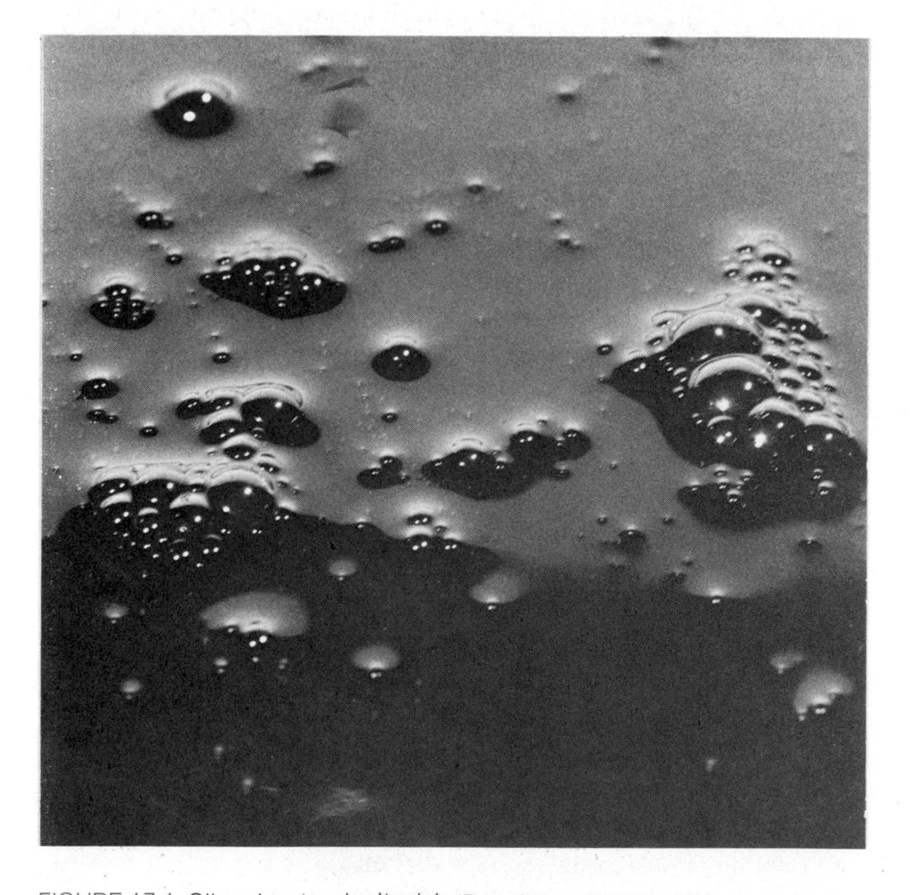

FIGURE 17.1 Oil and water don't mix! (Forsyth from Monkmeyer)

INTRODUCTION TO PROTEINS 17.3

What are some characteristics of proteins?

Proteins play many roles in the structure and function of living things. Every cell contains numerous **enzymes,** that is, proteins that catalyze the many different chemical reactions that are necessary for life. Other proteins serve as structural material. Bone, for example, consists of a strong protein matrix into which calcium and magnesium salts are embedded for hardness. The blood contains a variety of proteins that act as transport vehicles for oxygen, nutrients, hormones, and other materials. The antibodies that confer immunity to diseases (immunoglobulins) are also protein in nature. Some hormones are proteins.

Here are some examples of proteins whose names may be familiar to you: pepsin (a digestive enzyme of the stomach), collagen (a structural protein of cartilage and bone), gelatin (the soluble protein obtained by boiling soup bones, for example, is a solubilized collagen), and insulin (a hormone necessary for normal metabolism of foods).

Proteins as a class are very large molecules. The smallest proteins have molecular weights of perhaps 5,000, and the largest have molecular weights much in excess of a million. All proteins are polymers of amino acids and can be hydrolyzed by strong acid to yield a mixture of their constituent amino acids. **Simple proteins** contain nothing other than amino acids. **Conjugated proteins** contain one or more nonprotein components such as metal ions, sugar or lipid groups, or other, more complex organic structures. Table 17.1 lists some examples of simple and conjugated proteins.

Problem 17.2 List and describe the functions that proteins serve in living systems.

TABLE 17.1 Classification of Proteins by Composition

Protein class	Examples	Nonprotein components
Simple protein	Serum albumin,	None
	collagen	None
Conjugated protein		
Metalloprotein	Alcohol dehydrogenase	Zn^{2+} ions
	Ferritin	Fe^{3+} ions
Lipoprotein	Serum β-lipoproteins	Fat and other lipids
Glycoprotein	Immunoglobulins	Carbohydrate
Nucleoprotein	Chromosomes, viruses	Nucleic acids
Chromoprotein	Hemoglobin, cytochromes	Heme group (an iron-containing tetrapyrrole)

17.4 AMINO ACIDS: BUILDING BLOCKS OF PROTEINS

Twenty amino acids are commonly found in proteins. All are alpha amino acids, which have the following general structure:

$$\underset{\text{amino group}}{H_2N}-\overset{R}{\underset{H}{\overset{|}{\underset{|}{C_{\alpha}}}}}-\underset{\text{carboxyl group}}{C(=O)OH}$$

The amino group is bonded to the carbon adjacent to (α to) the carboxyl group. The R group may be aliphatic or aromatic, or it may be a hydrogen atom. It may also contain one of several functional groups: hydroxyl, carboxyl, amino, sulfhydryl, and so on. The structures of these twenty amino acids are illustrated in Table 17.2. All have at least one amino group and one carboxyl group. All the naturally occurring amino acids in proteins are L-enantiomers (Section 16.9), except for glycine, which has no asymmetric carbon and thus no enantiomers.

$$H_2N-\overset{C(=O)OH}{\underset{CH_3}{\overset{|}{\underset{|}{C}}}}-H \qquad HO-\overset{C(=O)H}{\underset{CH_2OH}{\overset{|}{\underset{|}{C}}}}-H$$

L-alanine L-glyceraldehyde

TABLE 17.2 The Fundamental Amino Acids

Name	Structure at pH 7*	Abbreviation
Hydrophobic R group†		
Alanine	$CH_3-C(\overset{+}{N}H_3)(H)-CO_2^-$	Ala
Valine	$(CH_3)_2CH-C(\overset{+}{N}H_3)(H)-CO_2^-$	Val

*At pH 7 (that is, at near neutral pH, as in the human body), all acidic groups (CO_2H) will have lost a hydrogen ion and all basic groups (NH_2) will have gained a hydrogen ion.
†Importance of hydrophobic and hydrophilic R groups is discussed in Section 17.6. Note that the R groups are shown in color.

TABLE 17.2 The Fundamental Amino Acids (Continued)

Name	Structure at pH 7*	Abbreviation
Leucine	$(CH_3)_2CH-CH_2-\overset{\overset{+}{N}H_3}{C}H-CO_2^-$	Leu
Isoleucine	$CH_3-CH_2-CH(CH_3)-\overset{\overset{+}{N}H_3}{C}H-CO_2^-$	Ile
Proline	$CH_2-CH_2-CH_2-\overset{+}{N}H_2-CH-CO_2^-$ (five-membered ring: CH_2-CH_2, CH_2, $CH-CO_2^-$, N^+ with two H)	Pro
Phenylalanine	$C_6H_5-CH_2-\overset{\overset{+}{N}H_3}{C}H-CO_2^-$	Phe
Tryptophan	indole ring (N–H) $-CH_2-\overset{\overset{+}{N}H_3}{C}H-CO_2^-$	Trp
Methionine	$CH_3-S-CH_2CH_2-\overset{\overset{+}{N}H_3}{C}H-CO_2^-$	Met
Hydrophilic R group		
Glycine	$H-\overset{\overset{+}{N}H_3}{C}H-CO_2^-$	Gly
Serine	$HO-CH_2-\overset{\overset{+}{N}H_3}{C}H-CO_2^-$	Ser
Threonine	$CH_3-CH(HO)-\overset{\overset{+}{N}H_3}{C}H-CO_2^-$	Thr

TABLE 17.2 The Fundamental Amino Acids (Continued)

Name	Structure at pH 7*	Abbreviation
Cysteine	$HS-CH_2-\underset{H}{\overset{\overset{+}{N}H_3}{C}}-CO_2^-$	Cys
Tyrosine	$HO-C_6H_4-CH_2-\underset{H}{\overset{\overset{+}{N}H_3}{C}}-CO_2^-$	Tyr
Asparagine	$H_2N\overset{O}{\overset{\Vert}{C}}CH_2-\underset{H}{\overset{\overset{+}{N}H_3}{C}}-CO_2^-$	Asn
Glutamine	$H_2N\overset{O}{\overset{\Vert}{C}}CH_2CH_2-\underset{H}{\overset{\overset{+}{N}H_3}{C}}-CO_2^-$	Gln
Hydrophilic and acidic R group		
Aspartic acid	$^-O_2CCH_2-\underset{H}{\overset{\overset{+}{N}H_3}{C}}-CO_2^-$	Asp
Glutamic acid	$^-O_2CCH_2CH_2-\underset{H}{\overset{\overset{+}{N}H_3}{C}}-CO_2^-$	Glu
Hydrophilic and basic R group		
Lysine	$H_3\overset{+}{N}CH_2CH_2CH_2CH_2-\underset{H}{\overset{\overset{+}{N}H_3}{C}}-CO_2^-$	Lys
Arginine	$H_2N-\overset{\overset{+}{N}H_2}{\overset{\Vert}{C}}-NHCH_2CH_2CH_2-\underset{H}{\overset{\overset{+}{N}H_3}{C}}-CO_2^-$	Arg
Histidine	(imidazolium ring: $H\overset{+}{N}$, NH)$-CH_2-\underset{H}{\overset{\overset{+}{N}H_3}{C}}-CO_2^-$	His

Problem 17.3 Draw Fischer projection formulas (Section 16.8) of the following amino acids (in the L-configuration if enantiomers exist):

(a) alanine (b) serine

(c) glycine (d) aspartic acid

Problem 17.4 One of the fundamental amino acids actually does *not* possess a free amino group. Find it in Table 17.2. What is its name? How does its alpha nitrogen differ from a free amino group?

Problem 17.5 Find and draw the structures of the amino acids in Table 17.2 that

(a) have aromatic character (b) are alcohols

(c) contain sulfur.

ACID–BASE CHEMISTRY OF THE AMINO ACIDS 17.5

Can the same amino acid exist in +, −, and uncharged forms?

Each amino acid has both an acidic functional group (the carboxyl group $-CO_2H$, discussed in Section 15.6), and a basic functional group (the amino group $-NH_2$, covered in Section 13.5). Several of the amino acids have an additional acidic or basic functional group.

The typical amino acid (containing one amino and one carboxyl group) exists as a zwitterion near neutral pH, such as in the cell or bloodstream.

$$^{+}H_3N-\underset{R}{CH}-C(=O)OH \;\underset{H_3O^+}{\overset{H_2O}{\longleftrightarrow}}\; ^{+}H_3N-\underset{R}{CH}-C(=O)O^- \;\underset{H_3O^+}{\overset{H_2O}{\longleftrightarrow}}\; H_2N-\underset{R}{CH}-C(=O)O^-$$

cation, low pH | zwitterion, pH ~ 6–8 | anion, high pH

A **zwitterion** has both negative and positive charges (*zwitter* is German for "hybrid"). The carboxyl group is a fairly strong acid, so it will be completely ionized or dissociated near neutral pH. The free amino group is also a rather strong base, meaning that it has a strong attraction for H^+ ions. Thus, the pH will have to be somewhat greater than 8 before many $-NH_3^+$ groups will lose H^+ ions, and these groups will be positively charged at neutral pH.

Example 17.1 Write the structures of alanine, glycine, and phenylalanine as they would exist at a pH of 7.

Solution

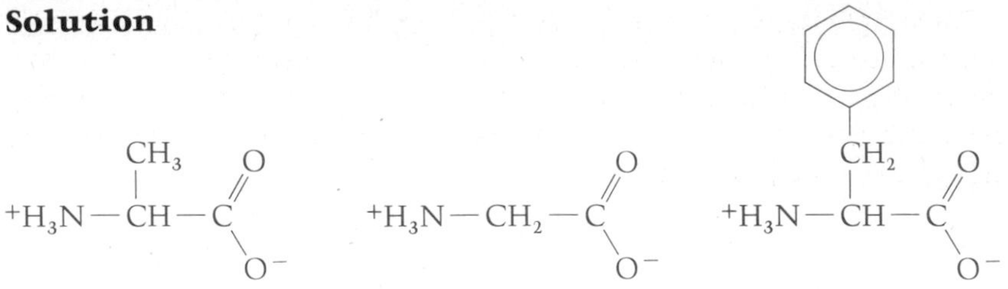

$$^{+}H_3N-\underset{|}{\overset{CH_3}{C}}H-C(=O)O^{-}$$

alanine

$$^{+}H_3N-CH_2-C(=O)O^{-}$$

glycine

$$^{+}H_3N-\overset{C_6H_5-CH_2}{CH}-C(=O)O^{-}$$

phenylalanine

The pK_a (Section 7.9) of an H^+ ion-donating group such as $-CO_2H$ or $-NH_3^+$ gives a measure of the tendency of that group to give up the H^+ ion, that is, a measure of the acid strength of that group. (Recall that $pK_a = -\log K_a$.) A stronger acid will have a lower pK_a value than a weaker acid. Typical pK_a values for the carboxyl groups of monoamino, monocarboxylic acids (those with one amino group and one carboxyl group) are between 2 and 2.5. Typical pK_a values for the corresponding $-NH_3^+$ groups are between 9 and 10.

An important aspect of the acid–base chemistry of amino acids (and of proteins, which are mostly or entirely made up of amino acids) is that it determines the net charge the molecules bear. A monoamino, monocarboxylic amino acid molecule has no *net* charge when in the zwitterion form, since the positive and negative charges cancel out. The pH at which the molecules of an amino acid in solution bear no net charge is called the **isoelectric pH.** The actual value of the isoelectric pH depends on the exact values of the pK_as of the particular amino acid. If the amino acid is in solution at a pH lower than that of the isoelectric pH, some of the molecules will be in the following form, in which they have a charge of $+1$:

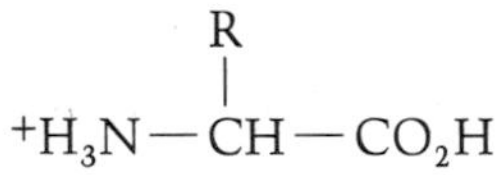

$$^{+}H_3N-\overset{R}{\overset{|}{C}}H-CO_2H$$

On the other hand, at a pH greater than the isoelectric pH, some of the amino acid molecules will be in the form

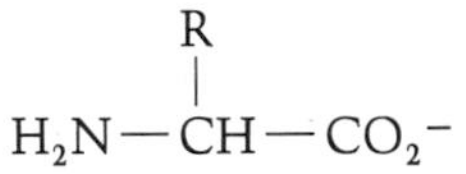

$$H_2N-\overset{R}{\overset{|}{C}}H-CO_2^-$$

in which the molecules in solution have an average negative charge.

We can summarize by saying that when the pH is greater than the isoelectric pH, the net charge on the amino acid molecules is negative, whereas when the pH is lower than the isoelectric pH, the net charge is positive. This relationship holds for any compound that has both negative and positive charges. The different net charges provide the basis for many of the methods used for separating and identifying amino acids and proteins.

For simplicity, we have thus far neglected the acid–base properties of the amino acids with ionizable R groups. This does not mean they are insignificant. In fact, the ionizable R groups of acidic and basic amino acids are very important, because they are often directly involved in carrying out the functions of proteins.

POLAR AND NONPOLAR AMINO ACIDS 17.6

When amino acids are linked into a polymer, the carboxyl and alpha amino groups are bound up in peptide linkages and no longer capable of ionizing. The properties of amino acids in the protein then depend mainly on the nature of their R groups: whether they are *hydrophilic* and polar or *hydrophobic* and nonpolar. The hydrophilic (water-loving) amino acid R groups are those that are charged, or polar, and can hydrogen-bond with water. The hydrophobic (water-hating) R groups are those that have hydrocarbon character and cannot hydrogen-bond with water. The arrangement and number of hydrophobic and hydrophilic amino acids in a polypeptide have a very important influence on the structure of the molecule, as we will see in Section 17.11. Table 17.3 shows which amino acids are polar and which are nonpolar.

How can an amino acid with $-CO_2^-$ and $-NH_3^+$ groups be called hydrophobic?

TABLE 17.3 Classification of Amino Acids on the Basis of Polarity

Nonpolar (hydrophobic)	Polar (hydrophilic)	Polar group
Alanine	Aspartic acid	(CO_2H)
Valine	Glutamic acid	(CO_2H)
Leucine	Serine	$(-OH)$
Isoleucine	Threonine	$(-OH)$
Phenylalanine	Tyrosine	$(-OH)$
Glycine	Lysine	$(-NH_2)$
Methionine	Arginine	$(-NH-\overset{NH}{\overset{\Vert}{C}}NH_2)$
Proline	Histidine	$(>NH)$
Tryptophan	Cysteine	$(-SH)$
	Asparagine	$(-\overset{O}{\overset{\Vert}{C}}-NH_2)$
	Glutamine	$(-\overset{O}{\overset{\Vert}{C}}-NH_2)$

17.7 FIBROUS AND GLOBULAR PROTEINS

There are two broad classes of proteins: the fibrous proteins and the globular proteins. This classification is based on their chemical and physical properties. The **fibrous proteins**—such as silk fibroin, collagen of animal connective tissues, and the alpha keratins of wool and hair—are highly insoluble in water. As the name implies, they take the form of fibers, which are relatively strong mechanically. Fibrous proteins generally act as structural or supporting materials.

We will concentrate mainly on the **globular** class of proteins, since they are much more numerous and of more general importance. All enzymes, plasma proteins, and membrane proteins are globular. The word *globular* in this context means just what you might imagine: the molecule has a more or less globelike, compact shape. Some globular protein molecules are nearly spherical, others are more football- or egg-shaped, still others rather flattened and disclike.

17.8 LEVELS OF STRUCTURE IN PROTEINS

Since we introduced proteins as long polymers of amino acids, you might have first imagined them as long, flexible molecules. The idea of globular shape does not seem to match up with this notion, and rightly so. Those long, flexible chains of covalently linked amino acids must fold up very compactly to acquire a globular form.

The covalent structure—that is, the sequence of amino acids in a polypeptide—is called its primary structure. Superimposed on the primary structure are the secondary and tertiary structures, which depend on noncovalent interactions and lead to a folding of the polymer into a unique globular form. Let us consider these levels of structure one by one.

17.9 AMINO ACID SEQUENCE: PRIMARY STRUCTURE

What kind of bonding accounts for the primary structure of a protein?

The specific sequence of amino acids in a polypeptide is known as the **primary structure** of the polypeptide. We can illustrate our meaning quite simply with the structure of a small peptide:

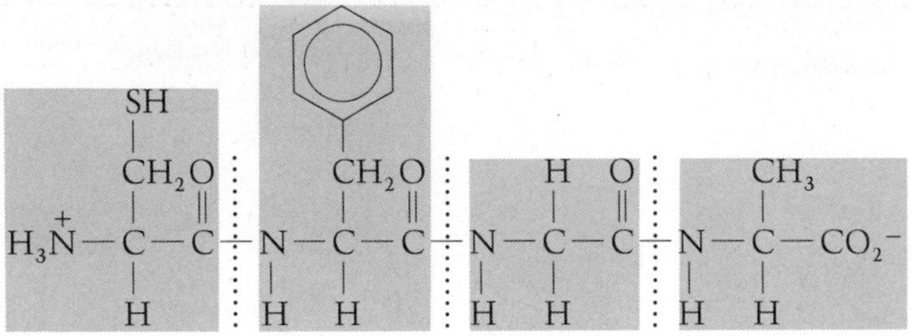

$$\overset{+}{H_3N}-\underset{H}{\overset{\overset{SH}{|}}{\underset{|}{\overset{CH_2}{|}}{C}}}-\overset{O}{\overset{\|}{C}}-\underset{H}{\underset{|}{N}}-\underset{H}{\underset{|}{\overset{\overset{C_6H_5}{|}}{\overset{CH_2}{|}}{C}}}}-\overset{O}{\overset{\|}{C}}-\underset{H}{\underset{|}{N}}-\underset{H}{\underset{|}{\overset{H}{\overset{|}{C}}}}-\overset{O}{\overset{\|}{C}}-\underset{H}{\underset{|}{N}}-\underset{H}{\underset{|}{\overset{CH_3}{\overset{|}{C}}}}-CO_2^-$$

or

cysteine–phenylalanine–glycine–alanine

or

H_3N^+–Cys–Phe–Gly–Ala–CO_2^-

N terminus (amino end of peptide) *C* terminus (carboxyl end of peptide)

As you can see, there are several ways of writing such an amino acid sequence. Whichever way is used, it is conventional to place the *N*-terminal amino acid at the left and the *C*-terminal amino acid at the right. Other peptides can be formed from the same four amino acids placed in different sequences. Those peptides would not be the same as this one, and they would not have the same properties. A typical protein is a much larger molecule, with from perhaps 40 up to thousands of amino acids. The amino acids in each protein are arranged in a unique sequence, just as we specified a unique sequence of Cys–Phe–Gly–Ala for our peptide.

Clever chemical methods have been developed for sequencing proteins, that is, for determining their unique amino acid sequences. As a result, the primary structures of many large and small proteins are now known. Those of proinsulin and lysozyme (an egg white protein) are shown in Figure 17.2. Note that in both structures, pairs of cysteines are linked covalently. These linkages, called **disulfide linkages,** are formed in the following manner:

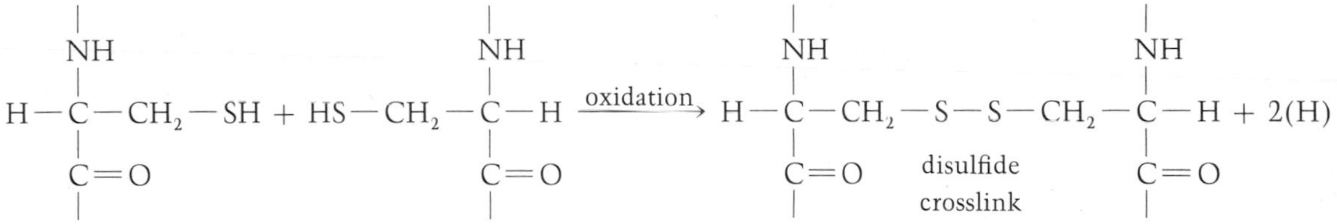

$$H-\underset{\underset{|}{C=O}}{\underset{|}{\overset{\overset{|}{NH}}{\overset{|}{C}}}}-CH_2-SH + HS-CH_2-\underset{\underset{|}{C=O}}{\underset{|}{\overset{\overset{|}{NH}}{\overset{|}{C}}}}-H \xrightarrow{\text{oxidation}} H-\underset{\underset{|}{C=O}}{\underset{|}{\overset{\overset{|}{NH}}{\overset{|}{C}}}}-CH_2-S-S-CH_2-\underset{\underset{|}{C=O}}{\underset{|}{\overset{\overset{|}{NH}}{\overset{|}{C}}}}-H + 2(H)$$

cysteine residue cysteine residue disulfide crosslink cystine

The primary structure completely dictates the higher levels of structure. The final unique structure is necessary for a given protein to function; thus, a single change in the primary structure can have profound effects. A good example is found in the inherited disease *sickle cell anemia.* This serious and often fatal disease is caused by the

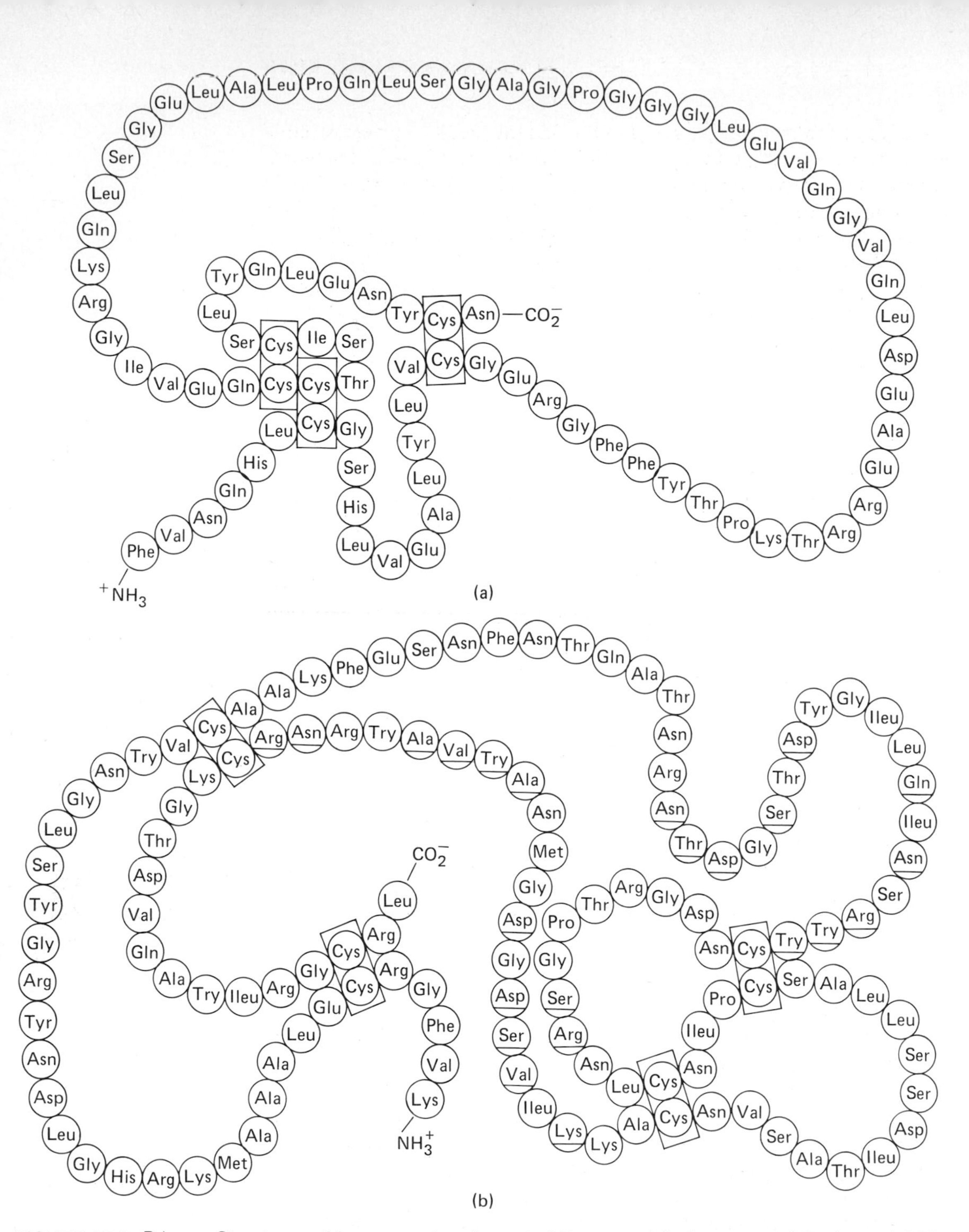

FIGURE 17.2 Primary Structures of human proinsulin and of the egg white lysozyme of the hen. (a) The primary structure of human proinsulin. The formula is arranged to show the position of the disulfide linkages, and is not an indication of three dimensional shape. (b) Primary structure of hen egg white lysozyme. (Part a from *Biochemical Concepts* by Robert W. McGillvery, Copyright © 1975 by W. B. Saunders Company. Reprinted by permission of Holt, Rinehart and Winston. Part b from Blake et al., *Nature* 206, 757, 1965.)

presence of abnormal hemoglobin in the patient's red blood cells. Hemoglobin is the protein that is responsible for oxygen transport in the bloodstream. One hemoglobin molecule contains four polypeptide chains: two alpha chains and two beta chains. (The *alpha* and *beta* prefixes simply differentiate between the two types of chains.) Sickle cell hemoglobin (Hb S) differs from normal hemoglobin (Hb A) in that it has a valine in place of the normal glutamic acid at a particular location in the beta chains. This single amino acid substitution causes significant changes in the physical properties of the hemoglobin. As a result, the actual shape of the red blood cells is altered, which restricts their passage through the blood vessels and produces the circulatory problems characteristic of the disease. (In Chapter 27 we will discuss this disease further.)

Problem 17.6 A tetrapeptide contains the amino acids alanine, tryptophan, glycine, and aspartic acid. Using the three-letter abbreviations for amino acids from Table 17.2, show all the possible primary structures this tetrapeptide could have.

Problem 17.7 Choose one of your suggested sequences from Problem 17.6 and draw the complete molecule, showing the peptide linkages and the charges that would be present at a pH near 7.

HELIX AND SHEET STRUCTURES: SECONDARY STRUCTURE 17.10

Secondary structures derive from the formation of hydrogen bonds between peptide groups, that is, between the carboxyl oxygen of one peptide bond and the amino hydrogen of another.

$$>C{=}O \cdots\cdots H{-}N<$$

There are two types of secondary structures: the **alpha helix** (α helix) and the **beta sheet** (β sheet). An alpha helix is a particular type of coiled polypeptide structure, which is by far the most common kind of polypeptide coil. As you can see from the helical structure pictured in Figure 17.3a, every peptide group in the alpha helix is hydrogen bonded to another peptide group located four amino acids farther along the chain. The helix has 3.6 amino acids per complete turn.

Whereas hydrogen bonding in the helical structure takes place between peptide groups that are only four amino acids apart, the hydrogen bonding in the beta-sheet structure takes place between groups somewhat farther apart, and sometimes between groups in entirely different peptide chains. Here, too, all the peptide groups are hydrogen bonded. In the

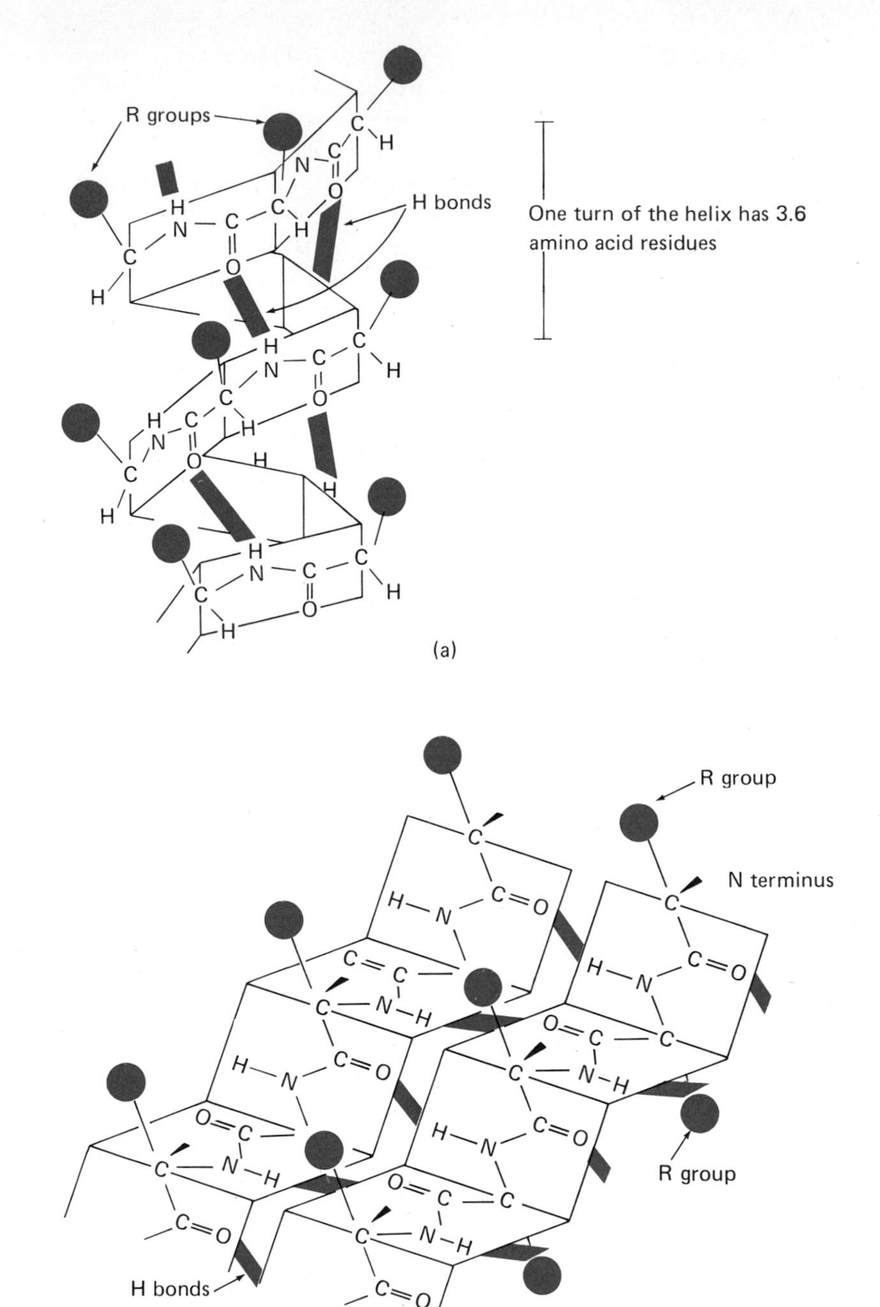

FIGURE 17.3 Hydrogen-bond stabilized polypeptide structures: secondary structures. (a) The α helix. (b) The β sheet. (From *Organic Chemistry of Biological Compounds* by Robert Baker, © 1971. Reprinted by permission of Prentice-Hall Inc., Englewood Cliffs, N.J.)

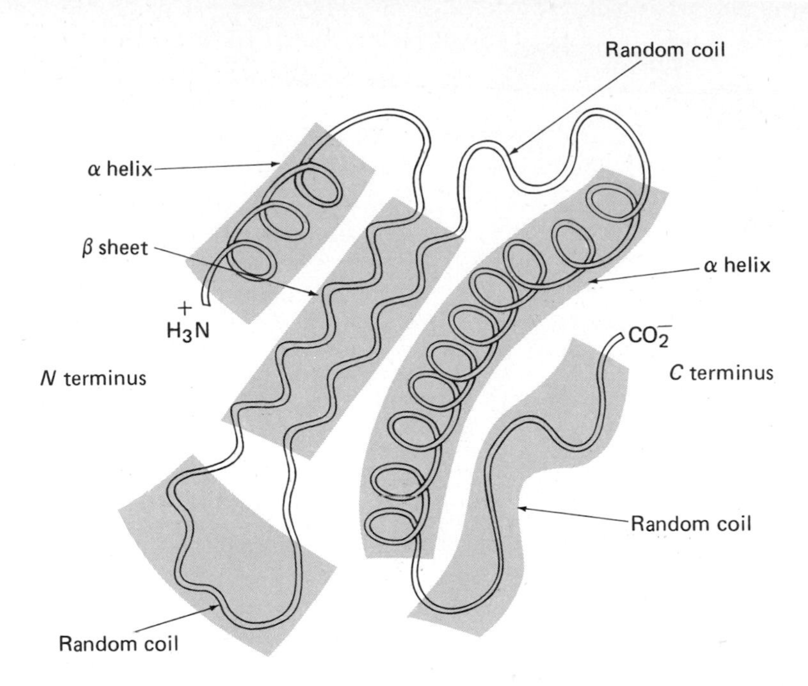

FIGURE 17.4 Secondary structures in globular protein.

beta sheet, the polypeptide backbone is maximally stretched out or extended rather than coiled. Figure 17.3b is a diagram representing beta sheet structure.

A globular protein will typically have some regions of alpha-helical structure, some regions of beta-sheet structure, and some regions in which neither form of secondary structure exists. These regions are said to be in **random coil** form. Figure 17.4 is a schematic diagram of secondary structure in a globular protein.

FOLDING OF POLYPEPTIDES: TERTIARY STRUCTURE 17.11

The tertiary level of structure is the final level that brings an individual polypeptide into its unique globular form. Hydrophobic interaction and, to a lesser extent, other noncovalent interactions are responsible for **tertiary structure.** Look back at Table 17.2, which shows the structures of the amino acid R groups. Notice that about half the R groups are hydrocarbon and are thus hydrophobic. The others are either charged at physiological pH (a pH of about 7), such as glutamic acid and lysine, or are polar but neutral, such as serine (with its hydroxyl group) or

What kinds of interactions account for tertiary structures?

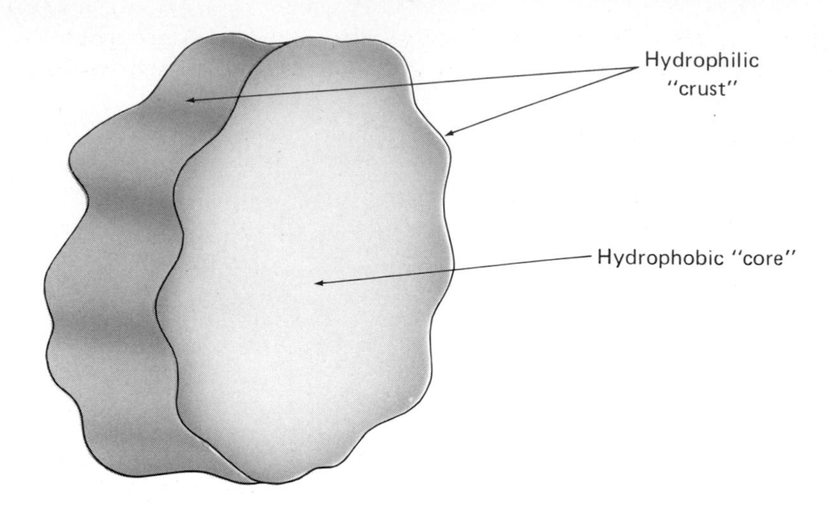

FIGURE 17.5 Imaginary cutaway view of globular protein molecule.

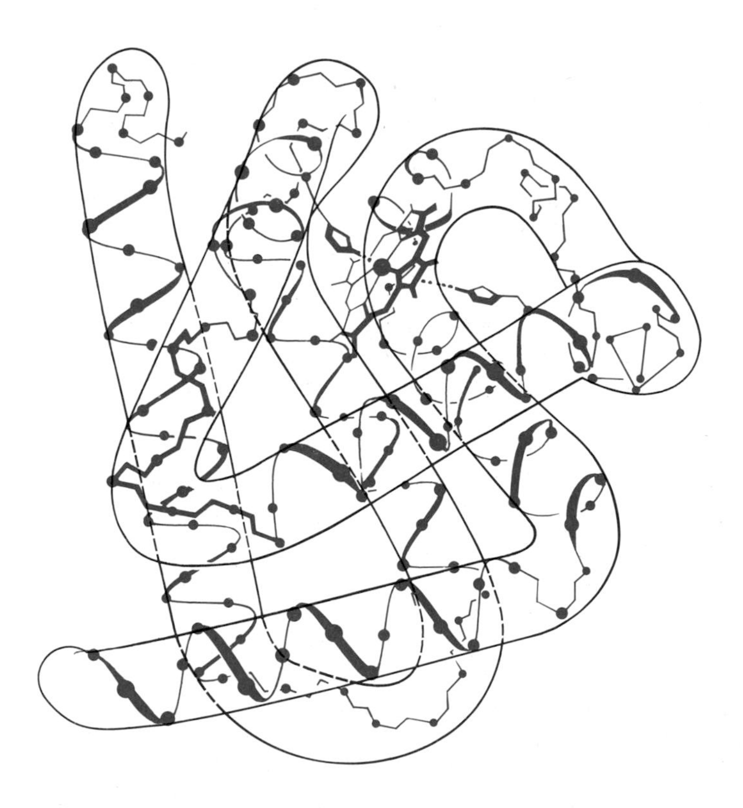

FIGURE 17.6 Three-dimensional structure of the protein myoglobin. Note the regions of α-helical and random coil structures. This structure was deduced from x-ray diffraction data. (From R. E. Dickerson in H. Neurath (ed.), *The Proteins,* Vol. 2, c. 1964, Academic Press)

cysteine (with its sulfhydryl group). These charged or polar R groups can form hydrogen bonds or ionic bonds with one another or with water molecules.

If you could put a completely unfolded polypeptide into water or a dilute salt solution, it would spontaneously fold up in a very predictable way. The hydrophobic portions of the molecule would tend to become buried in the interior of the folded structure, and the polar and charged groups would shift to the surface, becoming hydrogen bonded to a layer of water molecules. The driving force for this folding process is the tremendous attraction water molecules have for each other (as well as for other polar groups). This attraction tends to squeeze the hydrophobic regions of the chain together as a nonpolar core and bury them inside a crust consisting of the hydrophilic portions of the chain (Figure 17.5). Do not forget that within the globular tertiary structure there are likely to be distinct regions of secondary structure (both alpha helixes and beta sheets), as shown in Figures 17.4 and 17.6.

There is nothing random about the three-dimensional folded structure of a protein. It is a highly specific structure dictated by the primary structure, that is, the amino acid sequence of the polypeptide. This unique three-dimensional structure is called the **native conformation** of the particular protein (see Figure 17.6). If it is lost, the protein is said to be **denatured.** A denatured protein is no longer capable of functioning in its normal way, as a catalyst or whatever. We'll discuss denaturation further in Section 17.15.

PROTEIN QUATERNARY STRUCTURE 17.12

In Section 17.9 we pointed out that a molecule of the protein hemoglobin contains four polypeptide chains. A protein like hemoglobin that contains more than one polypeptide in its functional form is said to possess **quaternary structure.** Each of the polypeptides involved has its own primary, secondary, and tertiary structure. The polypeptides are held together mainly by noncovalent interactions. Figure 17.7 illustrates the structure of the hemoglobin molecule—a good example of quaternary structure.

The polypeptide chains (or subunits) of a protein with quaternary structure must be arranged in a certain way in order for the protein to function properly. If the association of the subunits is altered (by weakening their binding to one another or changing their orientation), the functional activity of the protein is also modified. In Section 21.15 we will see that this is an important feature of the regulation of enzyme functions and thus of the body's functions.

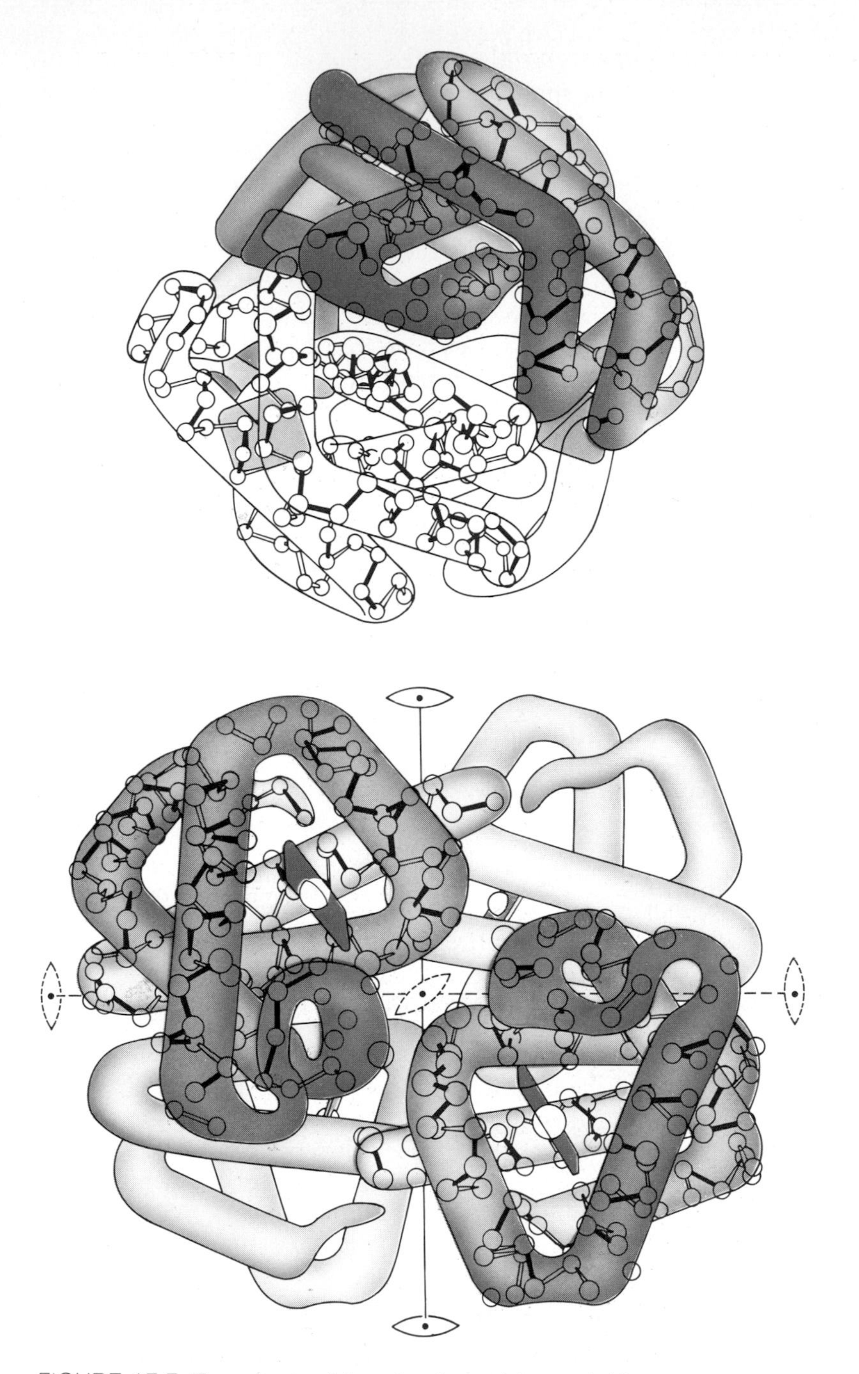

FIGURE 17.7 Two views of the structure of hemoglobin as revealed by x-ray diffraction analysis. The four polypeptides α_1, α_2, β_1, β_2, each with its own heme group, are organized to give a characteristic functional structure (quaternary structure).

PROPERTIES OF TYPICAL GLOBULAR PROTEINS 17.13

Are globular proteins likely to be water soluble?

Protein molecules with globular structures tend to be water soluble because of their polar surfaces. At neutral pH they usually bear a net charge, which can be either positive or negative. If the protein contains more basic amino acid residues, the charge will be positive. If it has more acidic amino acid residues, the charge will be negative. The charge also contributes to the water solubility of the protein. Since individual molecules of the same type bear the same charge, they tend to repel each other. This repulsion prevents them from clumping together and precipitating out of solution. As shown in Figure 17.8, a protein will be least soluble when the pH of the solution is such that the net charge on the molecule is zero. Each protein has a characteristic isoelectric pH, defined as the pH when the net charge on the molecule is zero. A protein in solution at its isoelectric pH will not migrate in an electric field. Electrophoresis is a valuable technique that is used to purify and identify proteins by subjecting them to an electric field. Since different proteins have different charges, they behave differently in an electric field and may often be separated from one another.

In cellulose acetate electrophoresis, proteins are separated from a small sample on the surface of a strip of cellulose acetate (a paperlike material).

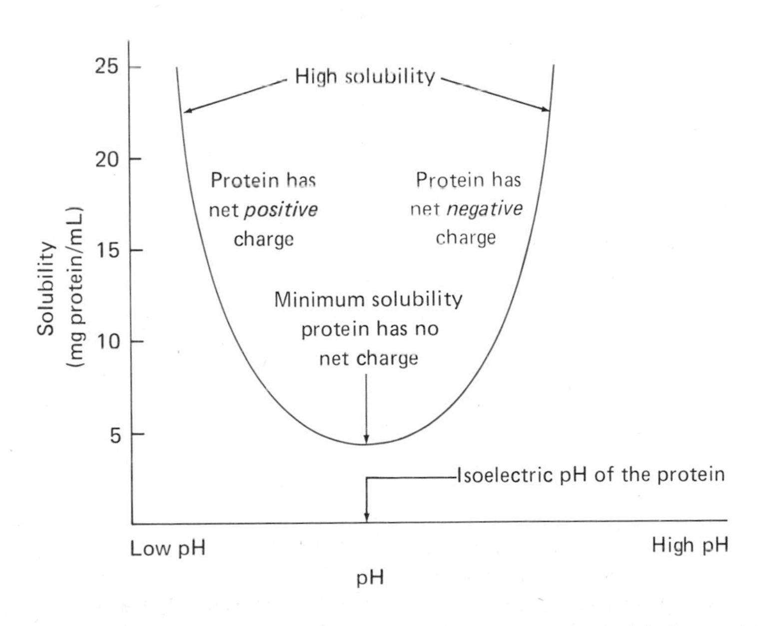

FIGURE 17.8 Effect of pH on the solubility of a typical globular protein.

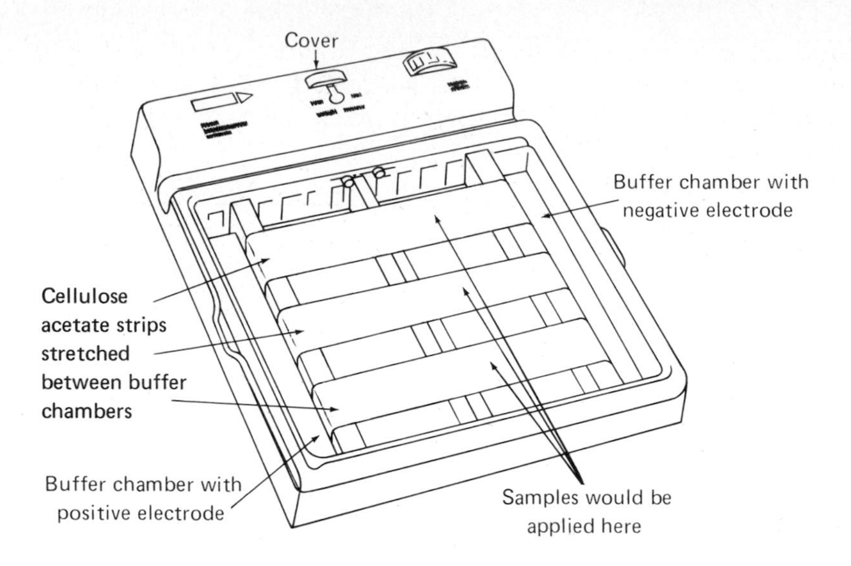

FIGURE 17.9 Equipment for cellulose acetate electrophoresis.

In hospital laboratories this technique is used on samples of blood serum to detect particular proteins that are present in abnormally low or high amounts in certain diseases (Figure 17.9).

The size and shape of globular proteins depend on their molecular weights and their specific folded structures. A variety of experimental techniques can be used to determine these properties. Differences in size and shape as well as solubility and acid–base chemistry are the bases for the many procedures available for separating and purifying proteins from complex mixtures. Electrophoresis is one of the most important of those techniques.

17.14 PROPERTIES OF FIBROUS PROTEINS

Let us return to the subject of fibrous proteins. These proteins might almost be said to stop at the secondary level of structure. Here we find examples of pure alpha-helix or beta-sheet structure. Alpha keratin of wool is a pure alpha helical protein; many helical polypeptides are bundled together and fastened with covalent crosslinks (disulfide linkages between cysteines) to yield the visible wool or hair strands. Because of their coiled structure, alpha keratins are elastic (stretchy). Silk fibroin is the primary example of pure beta structure. A silk fiber consists of numerous parallel, fully extended fibroin polypeptide chains that are hydrogen bonded to one another in one dimension, and stacked on top of one another in a second dimension. Since the chains are already fully extended, little stretch is possible, and silk is not elastic. A third important fibrous protein is collagen, the chief protein of mammalian connective tissue such as bone and tendon. This protein has neither an alpha helical nor a beta-sheet structure, but rather a unique ropelike structure (Figure 17.10). It is not elastic, as the keratins are, but instead very tough and strong. In case you are wondering, there are other fibrous proteins in connective tissue that are elastic, allowing some kinds of connective tissue to stretch very nicely.

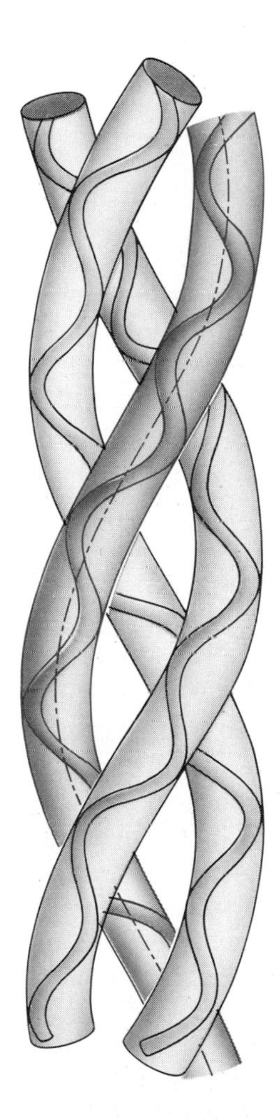

FIGURE 17.10 The supercoiled structure of collagen, where three coiled polypeptides coil one about another in a triple coil, or helix. (From *The Structure and Action of Proteins* by R. E. Dickerson and I. Geis, Benjamin/Cummings, Menlo Park, California, Publishers. © 1969 by Dickerson and Geis.)

17.15 DENATURATION OF PROTEINS

In Section 17.11, denaturation was defined as a loss of the native folded structure of a protein. Loss of this structure is accompanied by loss of ability to function, since function so closely depends on the three-dimensional structure of macromolecules.

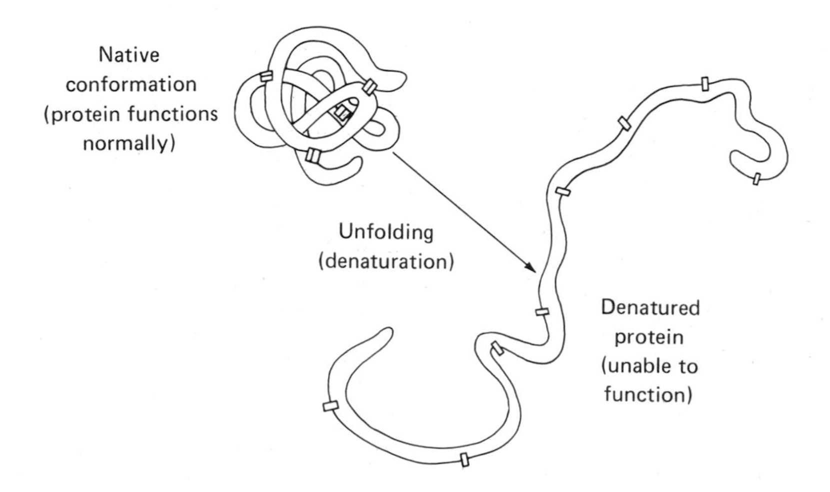

In denaturation, proteins with quaternary structure (that is, proteins that consist of two or more polypeptide subunits) first break apart into separate subunits, which then unfold. Since a protein's three-dimensional folded structure is held together mainly by noncovalent bonds, agents or treatments that break noncovalent bonds will denature proteins. What are some of these denaturing agents? Heat is a powerful one. When you cook an egg or any other protein-rich food, the change in appearance, texture, and other properties of the food is due to denaturation. When surgical instruments or other medical supplies are sterilized at high temperatures, bacterial proteins are denatured. The organisms die because their enzymes are no longer functional. Strong acid or base denatures proteins. If you burn your hands with hydrochloric acid or lye (or heat, for that matter), skin proteins are denatured and skin cells die because they cannot function normally. Organic solvents (nonpolar solvents) denature proteins because the normally buried nonpolar portions of the polypeptide chains spontaneously unfold to undergo hydrophobic interaction with the solvent. Strong detergents are also good denaturing agents. Thus, disinfectants, which often contain both a detergent and some organic solvent like pine oil, kill many bacteria by denaturing their proteins.

Problem 17.8 People in some parts of the world prepare fish for eating by pickling it rather than cooking it. Explain why pickling in vinegar is similar to cooking the fish.

SUMMARY

Proteins, nucleic acids, and **polysaccharides** are biological polymers formed by the condensation of building-block molecules. The three-dimensional **functional structures** of **macromolecules** are the result of noncovalent bonds: ionic bonds, hydrogen bonds, and **hydrophobic interactions.** Proteins play many roles in living things, as **enzymes** (catalysts), transport vehicles, hormones, and structural materials. The building blocks of proteins are **amino acids.** The amino acids in a peptide are connected by peptide bonds. Long polymers of amino acids connected by peptide bonds are called **polypeptides.** Twenty fundamental amino acids are found in proteins. An amino acid is both an acid and a base, by virtue of its carboxyl and amino groups. Near neutral pH, the amino acid exists as a **zwitterion** in which the carboxyl group has a negative charge and the amino group has a positive charge. An amino acid can be classified as hydrophobic or hydrophilic on the basis of the nonpolar or polar nature of its R group. The R group is the main determinant of the properties of an amino acid residue in a protein. Proteins may be classified as **fibrous** or **globular.** They may also be classified as **simple** (proteins that contain only amino acids) and **conjugated** (proteins that contain other non-amino acid structures as well). Protein structure can be considered at four levels: **primary, secondary, tertiary,** and **quaternary.** Secondary, tertiary, and quaternary structures are mainly determined by noncovalent bonds. The two types of secondary structures are the **alpha helix** and the **beta-sheet.** In globular proteins there may be regions of alpha-helical structure and beta-sheet structure as well as regions in which neither type of secondary structure exists—the latter are regions of **random coil** structure. The tertiary structure is the unique folding of the polypeptide chain into its **native conformation.** Proteins that contain more than one polypeptide chain also possess quaternary structure. A covalent bond important in the three-dimensional structure of many proteins is the **disulfide linkage** between cysteine residues. Proteins are **denatured** when noncovalent bonds are destroyed by heat, acidity, or other conditions. A denatured protein is no longer capable of performing its normal functions.

Key Terms

Alpha helix (Section 17.10)
Amino acid (Section 17.1)
Beta sheet (Section 17.10)
Conjugated protein (Section 17.3)
Denaturation (Section 17.11, 17.15)
Disulfide linkage (Section 17.9)
Enzyme (Section 17.3)
Fibrous protein (Section 17.7)
Functional structure (Section 17.2)
Globular protein (Section 17.7)
Hydrophobic interaction (Section 17.2)
Isoelectric pH (Section 17.5, 17.13)
Macromolecule (Introduction)
Native conformation (Section 17.11)
Nucleic acid (Section 17.1)
Peptide (Section 17.1)
Polypeptide (Section 17.1)
Polysaccharide (Section 17.1)
Primary structure (Section 17.9)
Protein (Section 17.1)
Quaternary structure (Section 17.12)
Random coil (Section 17.10)
Secondary structure (Section 17.10)
Simple protein (Section 17.3)
Tertiary structure (Section 17.11)
Zwitterion (Section 17.5)

ADDITIONAL PROBLEMS

17.9 Draw the structure of the dipeptide Ile–Lys. Show the charges present on the molecule at a pH of 7. What is the *net* charge on the molecule?

17.10 Draw the structure of the tetrapeptide Trp–Phe–Tyr–Gly. Would this compound be as water soluble as the dipeptide Asp–Glu? Why would there be a difference in water solubilities?

17.11 Write the structure of each of the following amino acids as they would exist at a pH of 1.0. Indicate the net charge on each molecule.

(a) arginine
(b) proline
(c) histidine
(d) isoleucine
(e) glutamic acid

***17.12** Paper electrophoresis, in which moistened filter paper is used as the solid support for migration of molecules in an electric field, is often used to separate mixtures of amino acids. Could you separate aspartic acid and glutamic acid from one another with paper electrophoresis at a pH of 12? Could you separate aspartic acid from alanine under these conditions?

17.13 Sketch and label a diagram of a hypothetical globular protein, indicating where nonpolar and polar R groups, aspects of secondary structure, and some disulfide linkages might be found.

17.14 What is responsible for secondary structure in polypeptides? How do alpha helical and beta-sheet structure differ?

17.15 Proline does not take part in the formation of secondary structures. Draw a dipeptide containing proline and any other amino acid. Do you see why proline cannot participate directly in forming an alpha helix or a beta-sheet structure?

17.16 Which of the following is more water soluble?

(a) tryptophan or serine
(b) a globular protein or a fibrous protein
(c) a globular protein at its isoelectric pH or the same protein at a higher pH

17.17 Why does a wool sweater shrink and become matted when washed in very hot water? Explain.

17.18 Describe the alpha helix.

17.19 How are the subunits of a protein with quaternary structure usually held together? Suggest agents that would cause them to dissociate from one another.

17.20 Name the building blocks of

(a) proteins (b) polypeptides.

17.21 Why is the amino acid sequence of a polypeptide or protein so important?

17.22 What is the difference between amino acid sequence and amino acid composition?

17.23 Trypsin is a digestive enzyme that catalyzes the hydrolysis of peptide linkages in which the carbonyl group is part of a lysine or an arginine residue. What peptides would be produced by the action of trypsin on the following peptide:

Gly–Lys–Pro–Val–Lys–Asp–Pro–Arg–Glu–Gly

Problem-Solving Hints

If you have trouble solving any of the additional problems, refer to the sections listed next to the problem numbers.

17.9 17.5
17.10 17.5, 17.6
17.11 17.5
17.12 17.5
17.13 17.10, 17.11
17.14 17.10
17.15 17.4, 17.10
17.16 17.13, 17.14
17.17 17.15
17.18 17.10
17.19 17.12
17.20 17.4
17.21 17.9
17.22 17.9
17.23 17.1, 17.4

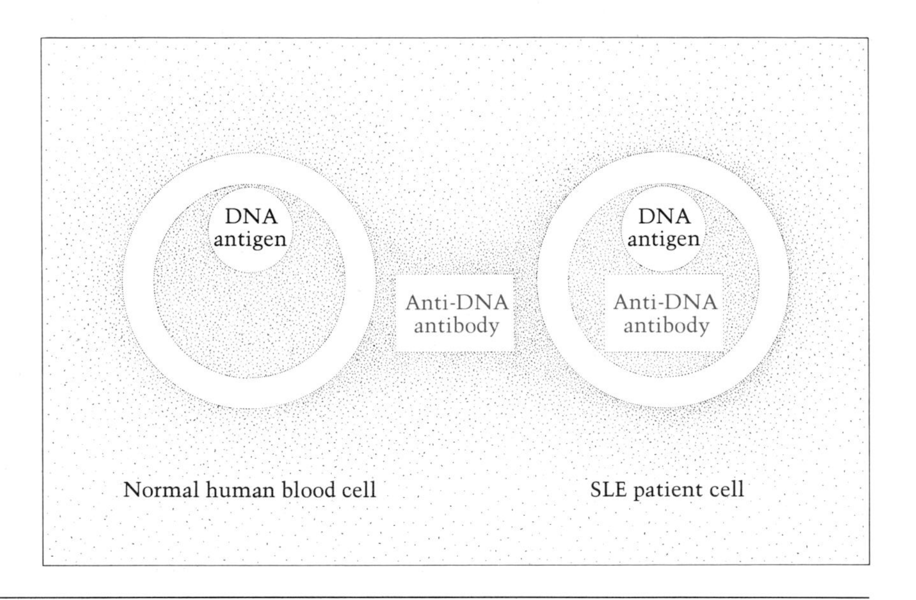

Systemic lupus erythematosus (SLE) is a disease that is characterized by the activity of a special antibody in the patient's blood. Most antibodies, as we know them, are present in the blood to ward off disease and form the basis of an immune system. The particular antibody that is active in SLE destroys DNA (deoxyribonucleic acid), a necessary hereditary constituent of all human cells. The antibody is called anti-DNA.

Anti-DNA is present in everyone's blood. In healthy people, anti-DNA is unable to permeate the cell. However, in patients with SLE, the cells break down and allow anti-DNA to enter. Once inside the cell, the anti-DNA can attach to the DNA molecule and render it inactive.

Although the origin of this antibody is unclear at this time, it is found to be active in most patients with SLE. Since knowledge of the antibody's position inside or outside of the cell is used as a diagnostic aid, it is important to be able to detect anti-DNA activity by laboratory techniques.

Several methods are available, but, to date, immunofluorescence is the most popular approach. With the immunofluorescence method, a fluorescing reagent is added to the patient's blood (in a test tube), where it will literally light up the antibody. The fluorescing reagent used is one that reacts specifically with this antibody. So when the blood cells are then examined in ultraviolet light under the microscope, the illuminated antibody can be clearly seen when it is positioned on the DNA.

18

NUCLEIC ACIDS

[Illustration: Systemic lupus erythematosus allows anti-DNA to permeate the cell wall and destroy the DNA molecule.]

NUCLEIC ACIDS

Nucleic acids are macromolecules. The name suggests that they are found in the nuclei of cells. Indeed, they were first isolated from nuclei, but we now know that certain nucleic acids are more often found in the cytoplasm of cells. All the nucleic acids play important roles in the storage and transmission of genetic information. Nucleic acids are of two fundamental types: **deoxyribonucleic acid (DNA)** and **ribonucleic acid (RNA).** DNA is the nucleic acid found in the chromosomes of all cells; each chromosome in a cell contains a different DNA molecule. The DNA of a cell is the genetic material; that is, it contains the information needed to specify the complete genetic makeup of that organism. Basically, that information is simply a set of directions for the manufacture of the hundreds of different proteins from which the cells of the organism are constructed, as well as those proteins that catalyze the necessary reactions for the oxidation of foods and other vital cell functions. Individual differences in these genetically transmitted directions are responsible for the fact that you may have blue eyes and your best friend has brown, that you may be shorter or taller than your sister, and so forth.

Ribonucleic acids are of three types: messenger RNA, transfer RNA, and ribosomal RNA. All forms of RNA are involved in the process by which the genetic information of DNA is used for directing the manufacture of proteins. In this chapter we will study the structures of the nucleic acids. We will consider their functions later, in Chapter 26.

18.1 NUCLEOTIDES: THE BUILDING BLOCKS OF NUCLEIC ACIDS

Nucleic acids are polymers of nucleotides. A **nucleotide** consists of three different molecules covalently linked to form a unit: (1) an organic base, (2) a five-carbon sugar (a pentose), and (3) a phosphoric acid group. Nucleic acids have molecular weights ranging from about 25,000 up to many millions.

phosphoric acid group

base

pentose

a nucleotide

THE ORGANIC BASES 18.2

The five organic bases commonly found in nucleic acids are shown in Figure 18.1. Two of these—adenine and guanine—are **purine** bases. The other three—cytosine, uracil, and thymine—are **pyrimidine** bases. DNAs generally have only adenine, guanine, cytosine, and thymine nucleotides. Uracil nucleotides are not found in DNA. On the other hand, most RNAs have little thymine; uracil nucleotides are present instead.

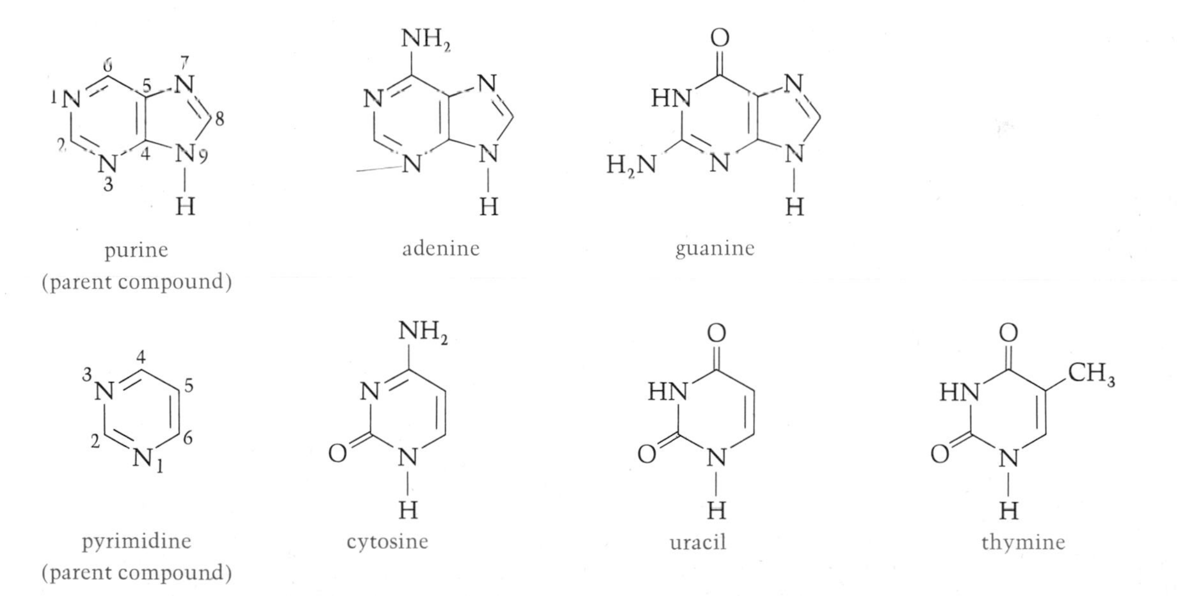

FIGURE 18.1 The organic bases of DNA and RNA.

Problem 18.1 How does each of the purine bases, adenine and guanine, differ from purine?
Problem 18.2 How does each of the pyrimidine bases differ from pyrimidine?

18.3 THE PENTOSES

The two closely related pentoses found in nucleotides are shown in Figure 18.2. The deoxyribonucleic acids (DNAs) contain, as the name suggests, 2-deoxy-D-ribose. The ribonucleic acids (RNAs) contain D-ribose. The only difference between these two sugars is the absence of the oxygen atom at the 2 position of the ring in 2-deoxy-D-ribose. Note that we did not specify the position of the hydroxyl group on carbon 1. The position of this group distinguishes the two isomers, alpha and beta (Section 16.10). In nucleotides the beta isomers are present. In a nucleotide the base is linked to the sugar by an *N*-glycosidic bond. If the base is a pyrimidine, the sugar is always linked to the nitrogen at the 1 position of the base. If it is a purine, the nitrogen at the 9 position of the base is bonded to the sugar. The resulting base–sugar compound is called a **nucleoside.**

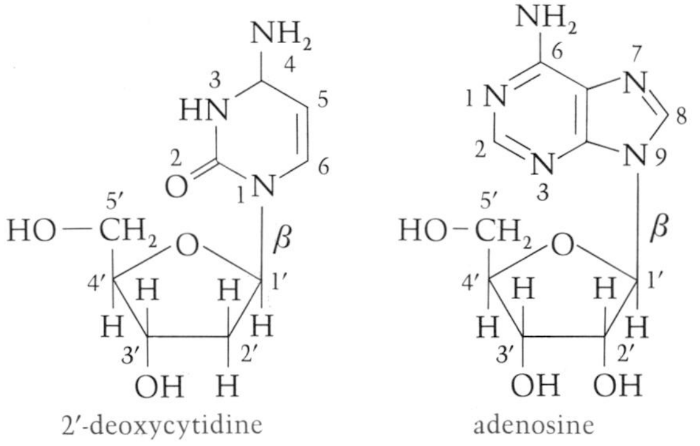

2′-deoxycytidine

adenosine

The names of the ribose-containing nucleosides are adenosine, guanosine, thymidine, cytidine, and uridine. For the corresponding deoxyribose nucleosides, the names are 2′-deoxyadenosine, 2′-deoxyguanosine, and so forth. The prime following the 2 indicates that the numbering refers to the sugar ring rather than to the organic base.

Problem 18.3 Draw the complete structural formulas of the following nucleosides:
(a) 2′-deoxythymidine (b) guanosine
(c) 2′-deoxyadenosine (d) uridine.

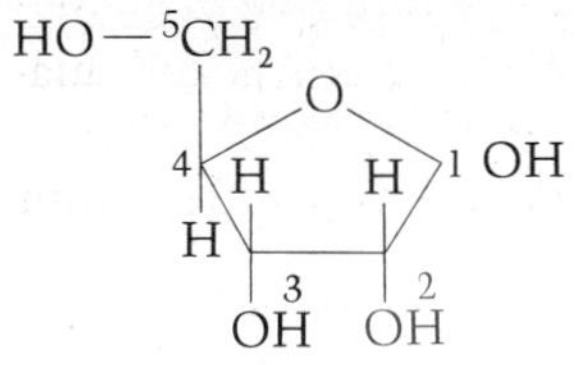

D-ribose
(found in RNA)

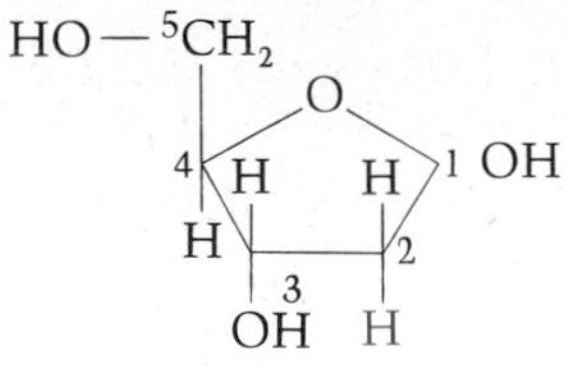

2-deoxy-D-ribose
(found in DNA)

FIGURE 18.2 The pentoses of DNA and RNA.

THE PHOSPHORIC ACID–ESTER LINKAGE 18.4

What is a nucleotide?

The phosphate group is bonded to the hydroxyl group on the 5′ carbon of the pentose to give a *phosphate ester* (Section 15.13).

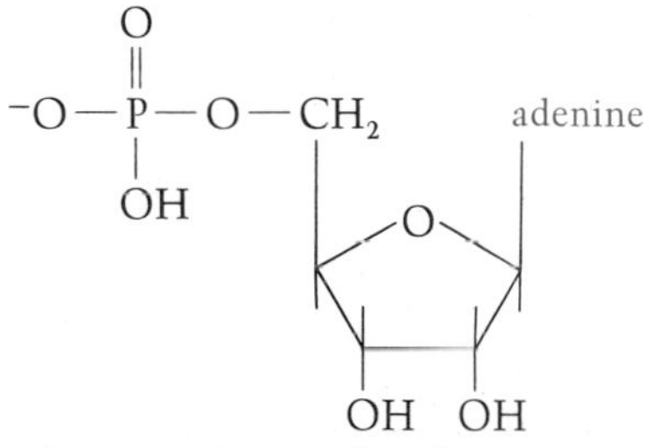

adenosine-5′-monophosphate (AMP)

Since the phosphoric acid group is a fairly strong acid, it exists in the negatively charged ionized form at physiological pH. As mentioned in Section 18.1, a base–sugar–phosphate compound is a nucleotide. Nucleotides are usually named as derivatives of the corresponding nucleosides, for example, adenosine-5′-monophosphate or 2′-deoxycytidine-5′-monophosphate. The names are frequently abbreviated: adenosine-5′-monophosphate as AMP, guanosine-5′-monophosphate as GMP, and so forth. Table 18.1 summarizes the naming of the bases, nucleosides, and nucleotides.

Problem 18.4 Draw complete structures of the following nucleotides:

(a) GMP (b) CMP
(c) dTMP (d) dAMP

TABLE 18.1 Naming of Bases, Nucleosides, and Nucleotides

	Nucleoside containing:		5′-Nucleotide containing:	
Organic base	**Ribose**	**2′-deoxyribose**	**Ribose (abbreviation)**	**2′-deoxyribose (abbreviation)**
Adenine	Adenosine	Deoxyadenosine	Adenosine-5′-monophosphate (AMP)	Deoxyadenosine-5′-monophosphate (dAMP)
Guanine	Guanosine	Deoxyguanosine	Guanosine-5′-monophosphate (GMP)	Deoxyguanosine-5′-monophosphate (dGMP)
Cytosine	Cytidine	Deoxycytidine	Cytidine-5′-monophosphate (CMP)	Deoxycytidine-5′-monophosphate (dCMP)
Uracil	Uridine	(Deoxyuridine)*	Uridine-5′-monophosphate (UMP)	Deoxyuridine-5′-monophosphate* (dUMP)
Thymine	(Ribothymidine)*	Thymidine	Thymidine-5′-monophosphate* (TMP)	Thymidine-5′-monophosphate (dTMP)

*Uncommon forms

18.5 POLYNUCLEOTIDES

What is the name of the bond formed by the condensation of two nucleotides?

The covalent backbone of a nucleic acid is formed via condensation reactions occurring between the 5′-phosphate group of one nucleotide and the 3′-hydroxyl group of another. Recall that in a condensation reaction a water molecule is lost as the new covalent bond is formed. The resulting linkage between the two pentoses via the phosphate is known as a **phosphodiester linkage.** Through the consecutive formation of many such linkages, a **polynucleotide,** or nucleic acid, arises (Figure 18.3). Thus, we have arrived at a definition of "nucleic acid primary structure." Just as the primary structure of a protein is the specific sequence of amino acids in the polypeptide chain, the primary structure of a nucleic acid is the specific sequence of nucleotides in the polynucleotide chain. In both cases, the sequence is a unique characteristic of a particular kind of molecule. If we have a different sequence, we have a different protein or a different nucleic acid.

Problem 18.5 The nucleotide content of a trinucleotide is AMP, CMP, UMP. What are all the possible primary structures (sequences) such a trinucleotide might have?

Problem 18.6 Choose one of the sequences in Problem 18.5 and draw the complete structure of the molecule.

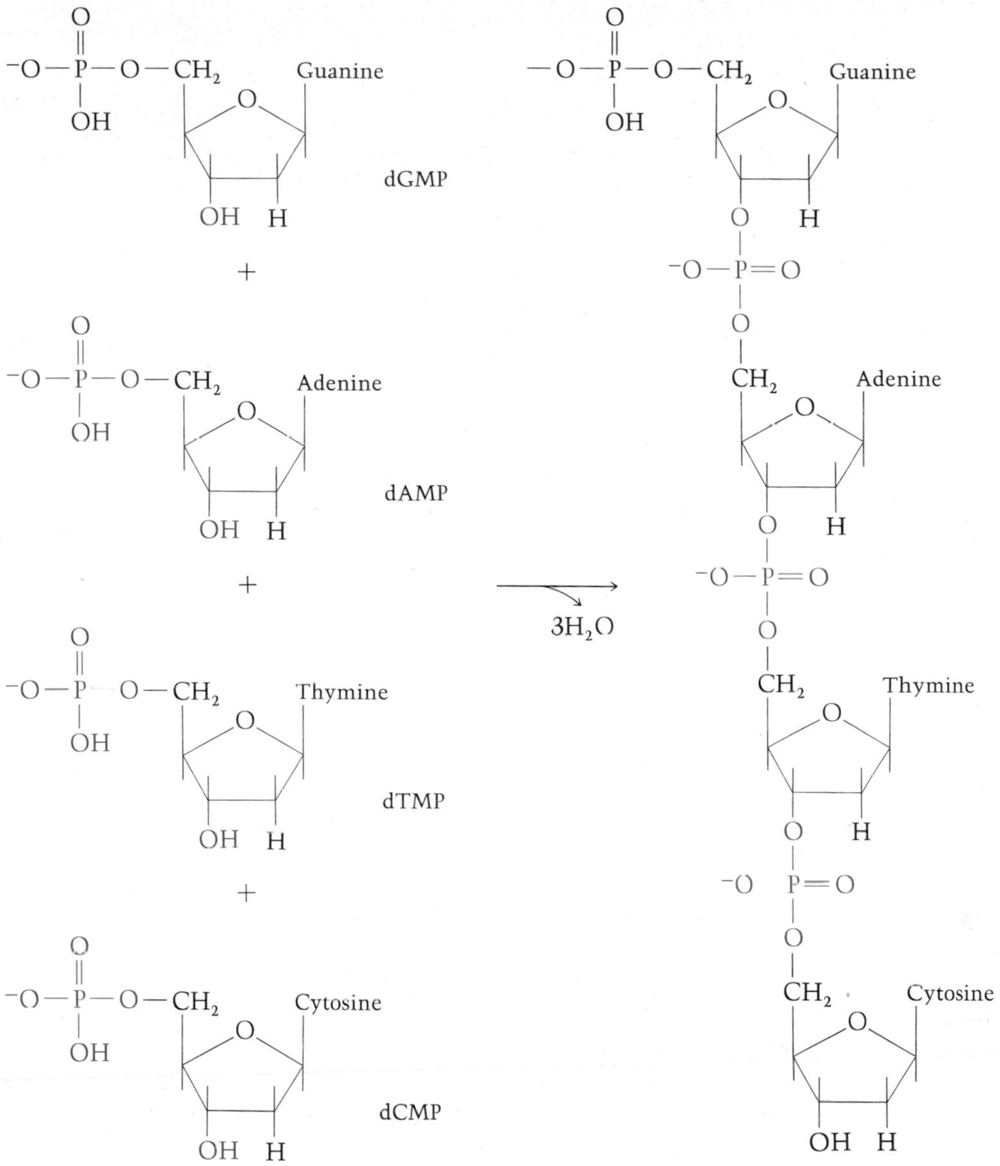

dGMP + dAMP + dTMP + dCMP $\xrightarrow{}$ $3H_2O$ + tetranucleotide

(Structures: each mononucleotide $^-O-P(=O)(OH)-O-CH_2$–deoxyribose (3′-OH, 2′-H) bearing Guanine, Adenine, Thymine, Cytosine respectively; product: Guanine–Adenine–Thymine–Cytosine chain linked by $O-P(=O)(O^-)-O$ phosphodiester bridges between 3′-O and 5′-CH_2, terminal 3′-OH H.)

FIGURE 18.3 Formation of a tetranucleotide by condensation of four mononucleotides. We should note that in the cell polynucleotides are not formed directly from nucleoside monophosphates as shown here. In Chapter 26 we will study this in more detail.

18.6 HIGHER LEVELS OF STRUCTURE IN POLYNUCLEOTIDES

What factors stabilize base pairing?

Higher levels of structure—that is, three-dimensional structures—of polynucleotides or nucleic acids arise because of something called **base pairing.** Base pairing is the pairing up of particular bases (in nucleotide chains) by means of hydrogen bonding. The structures of the bases are such that only certain ones can pair up: adenine with thymine, adenine with uracil, and guanine with cytosine.

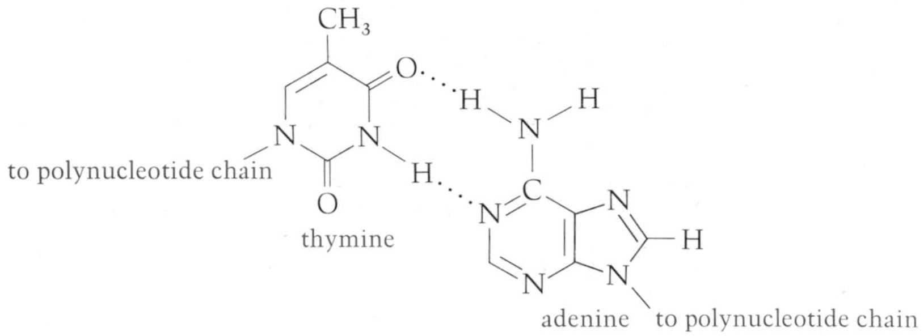

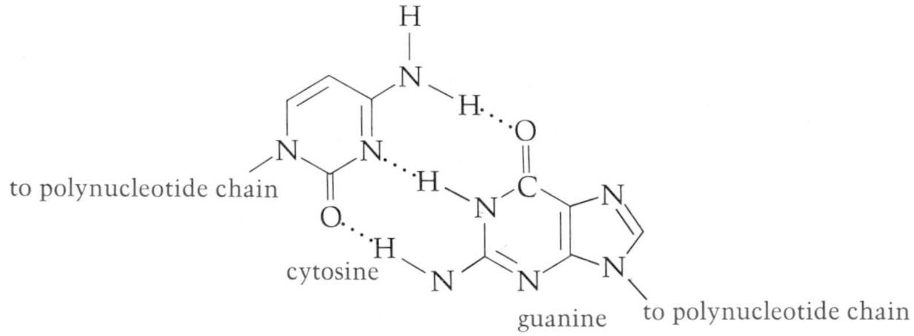

Each of these pairs of hydrogen-bonded bases has about the same dimensions. Say we have a long polynucleotide chain in which two segments contain sequences of nucleotides with bases capable of pairing. We can readily imagine formation of a base-paired region in which the chain is folded back upon itself to give a particular three-dimensional structure.

In fact, the two chains coil about one another in the base-paired region, but for simplicity we have omitted the coiling from the diagram. Sequences capable of undergoing base pairing in this manner are called **complementary sequences.** Base-paired structures can be maintained indefinitely if they are more stable than the unpaired structures. We know that numerous hydrogen bonds acting together can be a strong stabilizing force. Water competes with the bases for hydrogen-bonding positions, but there is still another stabilizing influence: hydrophobic interaction.

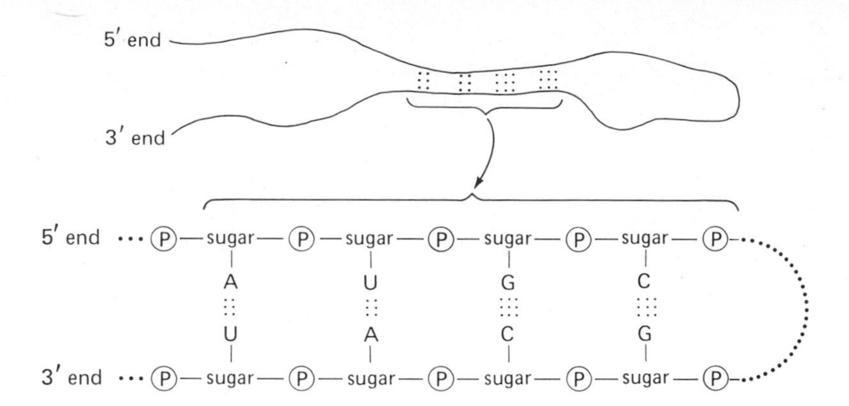

The nucleic acid bases have much in common with aromatic hydrocarbons, even though they have polar functional groups. If two parts of a polynucleotide chain coil about one another, not only do the complementary bases have the opportunity for pairing by hydrogen bonding, but also adjacent bases come into positions where they are more or less stacked on top of one another (Figure 18.4). The nonpolar rings of the bases become isolated from the water molecules in the surroundings. The water molecules can no longer compete for hydrogen bond formation, and the complementary bases become more strongly associated. In addition, the polar sugar–phosphate backbones of the polynucleotide chains are on the surface of this coil, where they can hydrogen-bond with water and with polar solutes. Thus, the hydrogen bonding between bases is reinforced by the hydrophobic base stacking, which creates a stable secondary structure in the base-paired region of the polynucleotide.

The base-pairing arrangement between segments of the same chain is a common feature of the structures of various forms of RNA (Figure 18.5). In contrast, a DNA molecule consists of two separate polynucleotide chains that are complementary along their entire lengths. These two chains undergo uninterrupted base pairing to form a complete **double helix** (Figure 18.6). An important feature of the double-helical DNA structure is the fact that the strands are oriented in opposite directions; that is, the arrangement of the strands is antiparallel.

First strand: 5′ end ———————————— 3′ end
Second strand: 3′ end ———————————— 5′ end

Note that the same thing was true in the RNA structures shown above and in Figure 18.5, in which the same chain was folded back on itself to undergo base pairing in one region.

This extremely specific base pairing provides the basis for the accurate transmission of genetic information (which is contained in the sequence of nucleotides in DNA) from one generation to the next. In Chapter 26

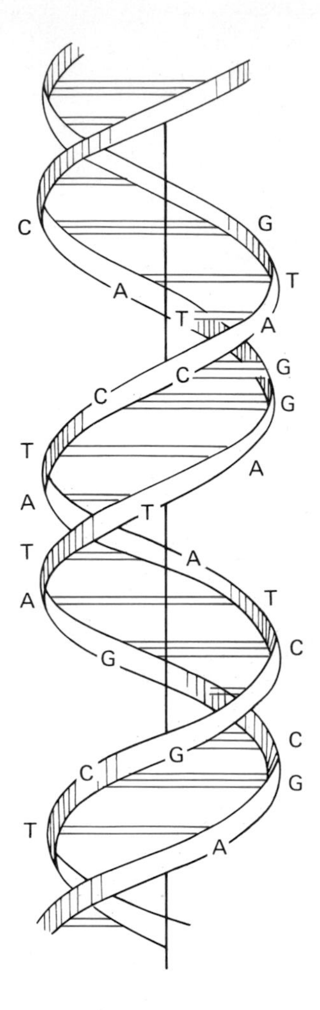

FIGURE 18.4 A schematic representation of the double helix of DNA (Watson-Crick model). The two ribbons represent the phosphate-sugar chains, and the horizontal rods represent the bonding between the pairs of bases. The vertical line indicates the long axis of the molecule. (From *Principles of Biochemistry,* White, Handler, Smith, and Lehman, 6th ed., 1978. Used with permission of McGraw Hill Book Company.)

we will return to the subject of genetic information, its storage, and its transmission. The functional structures of DNA and RNA will also be described in more detail there.

Problem 18.7 Give the complementary sequence of the following ribonucleotide sequence:

5′ end: ACGAUGAGUCG: 3′ end

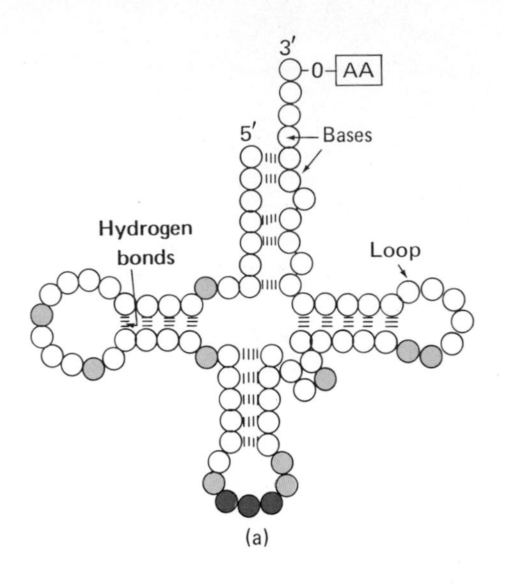

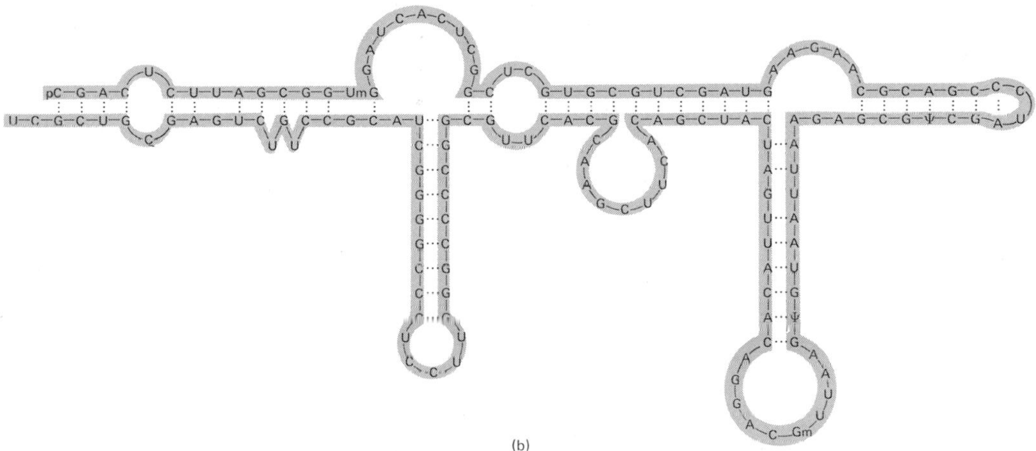

FIGURE 18.5 Examples of secondary structures in (a) a transfer RNA (*t*RNA) and (b) a ribosomal RNA (*r*RNA). (Part a from E. Borek, *Trends in Biochemical Sciences* 2, 4 (1977). Part b from Nazar, Sitz, and Brusch (1975), *Journal of Biological Chemistry* 250: 8591. Copyright The American Society of Biological Chemists, Inc.)

(Note: By convention, nucleic acid sequences are written as follows:

$$5' \rightarrow 3'$$
$$\text{A G C U}$$

The complement of this short sequence is

$$\text{U C G A}$$
$$3' \leftarrow 5'$$

FIGURE 18.6 A space-filling model of a portion of a DNA double helix. (Courtesy of Ealing Corporation, Natick, MA.)

which is read by convention as AGCU, that is, with the 5′ end given first.)

Problem 18.8 In a sample of DNA taken from yeast, it was found that 31% of the bases in the DNA were adenine and 19% were guanine. What would you expect the percentages of thymine and cytosine to be?

SUMMARY

Nucleic acids are polymers of **nucleotides** and are of two fundamental types: **DNA (deoxyribonucleic acid)** and **RNA (ribonucleic acid).** A nucleotide consists of a **purine** base (adenine or guanine) or a **pyrimidine** base (cytosine, uracil, or thymine) linked by a glycosidic bond to a pentose (ribose or 2-deoxyribose), which is bonded in turn to a phosphoric acid group. The nucleotides in a **polynucleotide** are connected by **phosphodiester linkages.** Higher levels of structure in polynucleotides result from **base pairing** between specific **pyrimidine** and **purine** bases. In RNA a single chain may fold back on itself to undergo pairing in one or more regions, but a DNA molecule consists of two base-paired **complementary** polynucleotide chains completely coiled into a **double helix.**

Key Terms

Base pairing (Section 18.6)
Complementary sequence (Section 18.6)
Deoxyribonucleic acid (DNA) (Introduction)
Double helix (Section 18.6)
Nucleic acid (Introduction)
Nucleoside (Section 18.3)
Nucleotide (Section 18.1)
Phosphodiester linkage (Section 18.5)
Polynucleotide (Section 18.5)
Pyrimidine (Section 18.2)
Purine (Section 18.2)
Ribonucleic acid (RNA) (Introduction)

ADDITIONAL PROBLEMS

18.9 What is the bond in the polynucleotide backbone that is analogous to the bond in the polypeptide backbone? The R groups of amino acid residues in a polypeptide are often called side chains. What are the analogous side-chain structures of polynucleotides?

18.10 Draw the complete structure of the trinucleotide dAdCdT. Draw the structure of its complementary trinucleotide. (Make the complement a *ribo-* compound rather than a *deoxyribo-* compound).

18.11 Draw the complete structure of the base pair A–U with hydrogen bonds in place.

18.12 Draw the structure of uracil. Substitute a fluorine atom for the hydrogen atom at the 5 position. The compound you have drawn is 5-fluorouracil, a potent anticancer drug.

18.13 Compare the expected water solubility of adenine, adenosine, and adenosine-5′-monophosphate. Explain your answer.

18.14 Name the nucleosides and mononucleotides that contain this organic base:

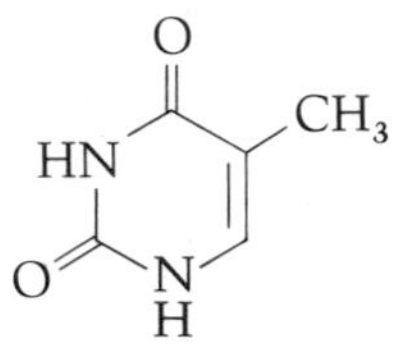

18.15 Give the complementary sequence of the following deoxyribonucleotide sequence:

CATGAUTCGTTAGCA

18.16 Write an equation, using complete structural formulas, for the formation of a dinucleotide from CMP and dAMP.

18.17 Determine the approximate molecular weight of a purine ribonucleotide by adding the atomic weights of all the atoms present. Do the same for a pyrimidine ribonucleotide. If a particular nucleic acid is about 50% purine and 50% pyrimidine nucleotides and has a molecular weight of 25,000, about how many nucleotides does it contain?

18.18 What is the net charge on a nucleoside near neutral pH? What is the net charge on a nucleotide near neutral pH? Could electrophoresis be used to separate nucleosides and nucleotides from a mixture?

18.19 5-Fluorouracil is effective as an antitumor agent because it is incorporated into the DNA of the tumor cells in place of one of the bases that is normally present, which prevents the proper functioning of the DNA. Which of the bases normally found in DNA would you expect 5-fluorouracil to replace?

Problem-Solving Hints

If you have trouble solving any of the additional problems, refer to the sections listed next to the problem numbers.

18.9 18.4
18.10 18.5
18.11 18.6
18.12 18.2
18.13 18.2, 18.3, 18.4
18.14 18.3, 18.4
18.15 18.6
18.16 18.5
18.17 18.5
18.18 18.4, 18.5
18.19 18.2

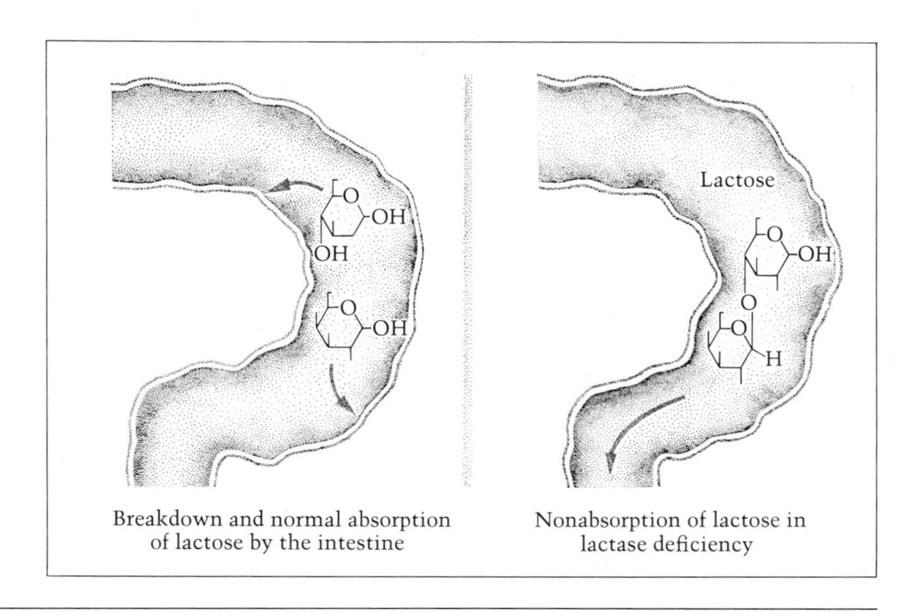

Breakdown and normal absorption of lactose by the intestine

Nonabsorption of lactose in lactase deficiency

In this chapter, we will be looking at the structure and function of the carbohydrates that are known as polysaccharides. Under normal conditions, some of these carbohydrates are broken down in the body so that they can be absorbed. There are also three disaccharides present in the normal diet: maltose, composed of two monosaccharide (single-sugar) units of glucose; sucrose, consisting of glucose and fructose; and lactose, consisting of glucose and galactose. Under normal conditions, these too are broken down in the body in a manner similar to the breakdown of polysaccharides. Specific enzymes help in the breaking down of disaccharides into their respective components, and absorption occurs primarily in the intestine.

Lactose, the disaccharide that is composed of one unit of galactose and one unit of glucose, is found mainly in milk and milk products. The enzyme that catalyzes the breakdown of lactose into these two simpler units is called lactase. Deficiency of this enzyme has been reported among Jewish and black adults, although it is quite uncommon in children.

When there is a decrease in the activity of a particular enzyme that is necessary to break down its corresponding disaccharide, the sugar involved will not be absorbed by the intestine. In fact, the sugar just stays in its original form in the walls of the intestine, where it causes retention of large amounts of water, which leads to diarrhea. Patients with a lactase deficiency will often have abdominal cramps and diarrhea after consuming lactose.

19

POLYSACCHARIDES

The diagnosis of lactase deficiency is made by having the patient ingest lactose and then measuring the level of the enzyme lactase in the intestinal tissue. A low level of enzyme is a clear indication that a deficiency exists. Once the diagnosis is certain, the patient can be given appropriate amounts of commercially prepared lactase. Treatment may also require that the diet be restricted to foods low in disaccharides.

[*Illustration: Lactase deficiency does not allow normal absorption of lactose in the intestine, and the sugar is readily passed out of the body.*]

POLYSACCHARIDES

We talked about the monosaccharides, the smaller carbohydrates, in Chapter 16. The polysaccharides are long chains of monosaccharides—in other words, they are just lots of little sugars hooked together. Starch and cellulose are well-known examples of polysaccharides. Starch serves mainly as an energy-storage material, and cellulose serves as a structural material. This should give you an idea of the wide variety of the roles that the polysaccharides play. Some form strong fibers that lend strength to the cell walls of plants and bacteria. Others form jellylike materials that serve as protective padding in animal tissues and in plants such as seaweed.

19.1 INTRODUCTION TO POLYSACCHARIDES

Polysaccharides are polymers of sugar (monosaccharide) molecules. They are large carbohydrates, with molecular weights up to several million. Most polysaccharides perform either a structural role (as supporting material for cell walls of plant and bacterial cells, or in animal connective tissues) or an energy-storage role. **Starch,** for example, is an energy-storage polysaccharide found in the cells of vegetables such as potatoes and grains.

19.2 BUILDING BLOCKS OF POLYSACCHARIDES

The majority of naturally occurring polysaccharides contain only a single kind of building block, the six-carbon sugar D-glucose. The structure of α-D-glucose is shown here in a Haworth formula (Section 16.10).

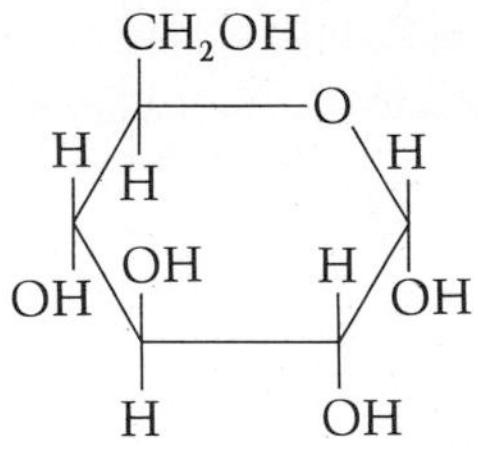

α-D-glucose

What is the most common building block in polysaccharides?

Problem 19.1 Draw the Haworth structural formula of β-D-glucose.
Problem 19.2 Draw the straight-chain Fischer projection formula of D-glucose.

HOW THE SUGARS ARE LINKED 19.3

When D-glucose or other monosaccharide units are linked to form a polymer, a condensation reaction occurs between the anomeric carbon (Section 16.10) of one unit (the carbonyl carbon in the straight-chain molecule) and a hydroxyl group of the second unit (such as that at carbon 4). In this reaction an acetal is formed from a hemiacetal and an alcohol (Section 14.7).

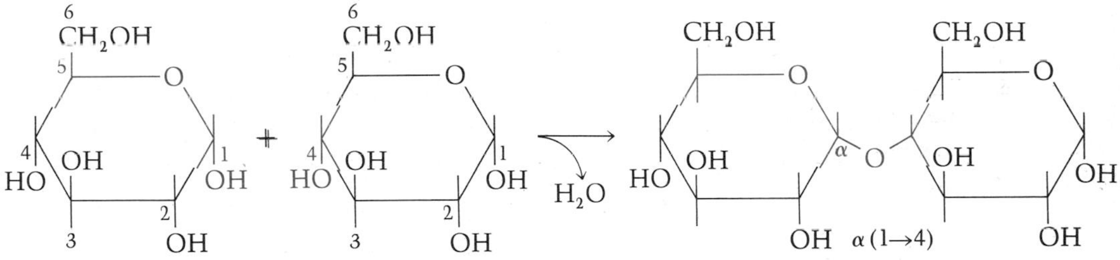

α-D-glucose
(hemiacetal)

α-D-glucose
(alcohol)

maltose
a disaccharide
(acetal)

In the example shown, the anomeric carbon of one sugar, in the α-configuration, is linked to the oxygen attached to the 4 carbon of the second sugar. The resulting bond is a glycosidic bond (Section 16.11). This is the linkage connecting glucose molecules in amylose, which is one form of plant starch. In amylose molecules, thousands of glucose molecules have condensed in this way to form a polymer.

Glycogen, or "animal starch," is similar to amylose except that it is a branched rather than a straight-chain polymer. The occasional branches arise from $\alpha(1 \rightarrow 6)$ linkages. A small segment of a glycogen molecule is shown here:

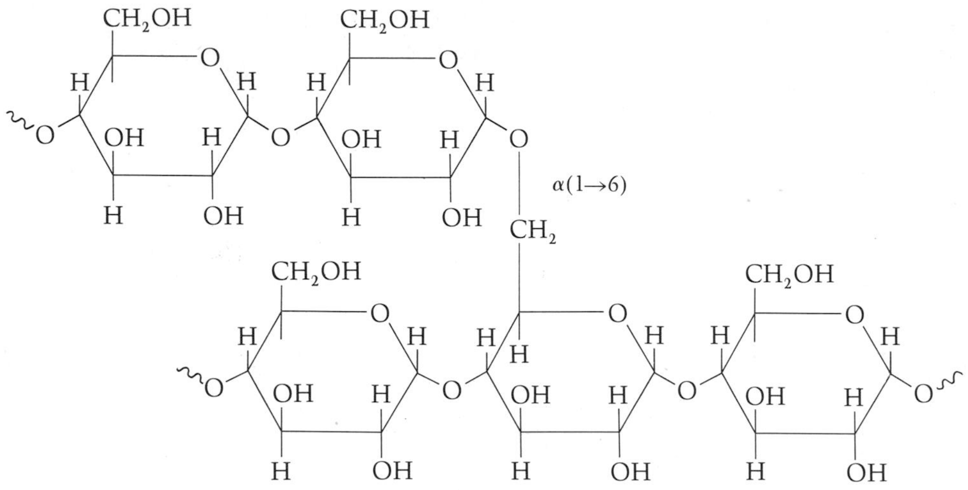

glycogen

As well as amylose, plant starch contains amylopectin, which is branched like glycogen. But whereas branching occurs every 8 to 10 glucose residues in glycogen, it occurs somewhat less frequently in amylopectin.

Cellulose is the fibrous structural polysaccharide of cotton and plant cell walls. It contains β-D-glucose, and its anomeric carbon undergoes beta-glycosidic linkage to the oxygen on the 4 carbon of the adjacent β-D-glucose.

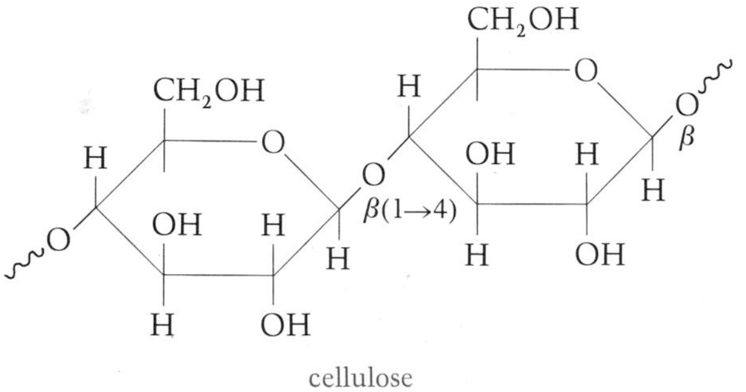

cellulose

It is interesting that although most animals, including humans, can digest starch by means of digestive enzymes called **amylases,** they cannot digest cellulose in this way. Cellulose, in fact, cannot be digested at all. Amylases are unable to catalyze hydrolysis of beta-glycosidic linkages. Ruminants (cows, sheep) can digest cellulose only because of the bacteria that live in their digestive tracts. These bacteria possess cellulases, enzymes that catalyze hydrolysis of the cellulose beta linkages.

Cellulose is the most abundant polysaccharide on earth. Some day we will have the technology to convert it to food sugars, probably by using a cellulase produced in quantity from large bacterial cultures (See Focus 19, Genetic Engineering). As a result, we should be able to increase the world's food supplies substantially.

SOME IMPORTANT POLYSACCHARIDES 19.4

Another important polysaccharide is **chitin,** the chief structural material of an insect's exoskeleton. The structure of chitin is very similar to that of cellulose. The only difference is that the monosaccharide building block unit is N-acetyl-β-D-glucosamine, a derivative of β-D-glucose. A glucosamine is a glucose in which an amino group has replaced one of the hydroxyl groups, usually the one at the 2 carbon. The prefix *N-acetyl* indicates that an acetyl group has been added to the amino group.

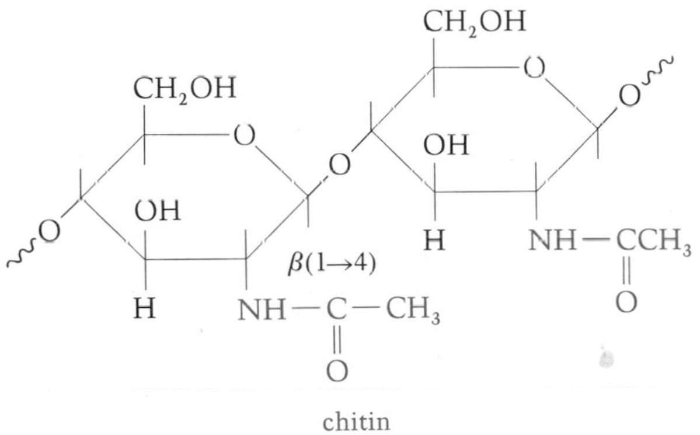

CH_2OH O OH H NH—C—CH_3 O β(1→4) CH_2OH O OH H NH—CCH_3 O

chitin

Hyaluronic acid is an important polysaccharide of animal connective tissues. It is one of a group of related substances that usually occur in combination with specific proteins. These complexes, called **mucoproteins,** are jellylike, slippery, or sticky substances. Some of them function as lubricants, such as in the synovial fluid of the joints. Others act as cement or glue between cells in a tissue, forming the cell coat.

In hyaluronic acid the repeating unit is a disaccharide of β-D-glucuronic acid and N-acetyl-β-D-glucosamine, which are linked by a

$\beta(1 \rightarrow 3)$ bond. Each of these disaccharide units is linked to the next by a $\beta(1 \rightarrow 4)$ bond.

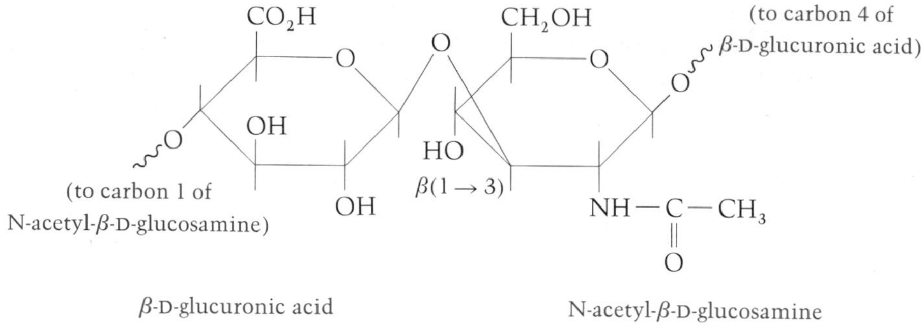

CO_2H CH_2OH (to carbon 4 of β-D-glucuronic acid)
O O O O
OH HO
(to carbon 1 of N-acetyl-β-D-glucosamine) OH $\beta(1 \rightarrow 3)$ $NH-C(=O)-CH_3$

β-D-glucuronic acid N-acetyl-β-D-glucosamine

hyaluronic acid

Another group of polysaccharides are the **chondroitins,** which contain *N*-acetyl-β-D-galactosamine instead of the similar glucose derivative found in hyaluronic acid. Certain of these have sulfate groups ($-O-SO_3$) substituted at particular hydroxyl positions. The chondroitins are major structural materials in bone, cartilage, cell coats, and other connective tissue structures. Heparin is a related polysaccharide that prevents blood clotting; it is found in the lung and in artery walls.

Problem 19.3 Draw the disaccharide repeating unit of a chondroitin molecule. The Haworth structural formula of β-D-galactose is as follows:

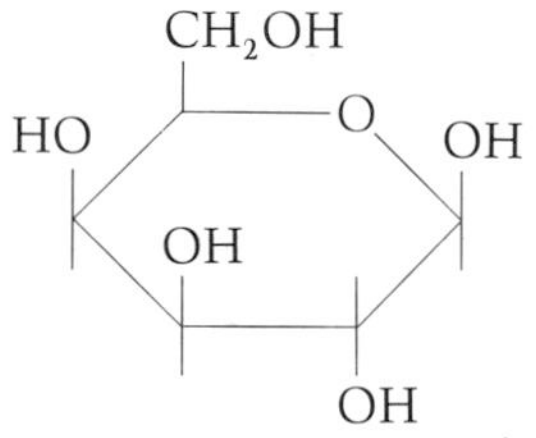

CH_2OH
HO O OH
OH
OH

19.5 HIGHER LEVELS OF POLYSACCHARIDE STRUCTURE

Polysaccharides have primary structures—that is, sequences of building blocks—that are very simple in comparison with those of proteins and nucleic acids. We should mention, however, that polysaccharides do have secondary and sometimes higher levels of structure based on noncovalent interactions, just as the proteins and nucleic acids do. Here, too, three-dimensional structure dictates the properties of the molecule and therefore the functions for which a particular polysaccharide is suited.

Do polysaccharides have three-dimensional "functional structures"?

We will look briefly at two major types of secondary structure: extended and helical. Cellulose and chitin have the extended, ribbonlike structure (Figure 19.1). In this form, the polysaccharide chains are rather flat, and the glucose rings are arranged in such a way that hydrogen bonding can easily take place between hydroxyl groups of chains that are lying side by side. The presence of many such hydrogen bonds between separate polysaccharide chains gives rise to a strong fibrous structure.

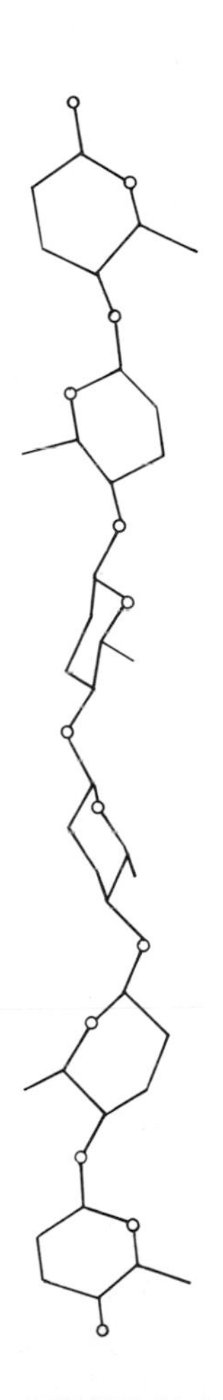

FIGURE 19.1 The extended conformation of cellulose. In a cellulose fiber, many of these molecules are lined up and hydrogen-bonded together. (From Rees and Scott, *Journal of the Chemical Society* (13) 469, 1971.)

In amylose or hyaluronic acid, however, the chains form regions of double helical structure by folding back on themselves and coiling together by means of hydrogen bonding. Figure 19.2 is a schematic representation of the helical structure of amylose. In these polymers the coiled structure is more stable than the extended structure would be. The networks of helical structure can trap many water molecules, so that such polysaccharides form gels. You are probably familiar with the gel form of starch and the rather jellylike consistency of some animal connective tissues. These properties are the direct result of the coiled structure of such polysaccharides that favors the absorption of water.

Problem 19.4 What type of bonding accounts for the secondary structures of polysaccharides?

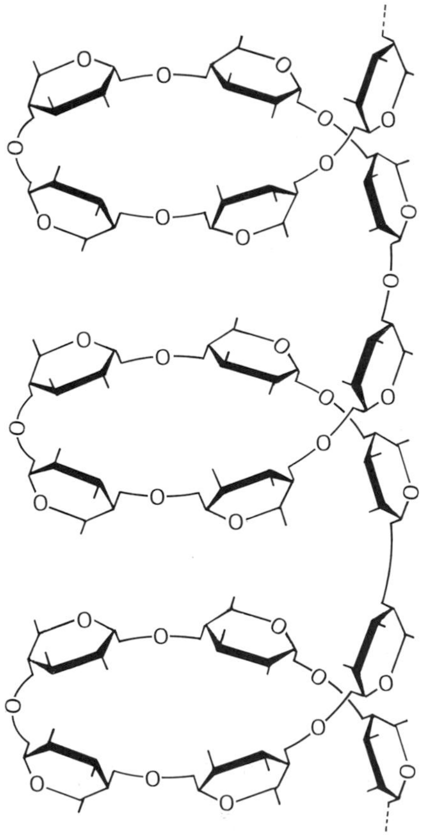

FIGURE 19.2 The helical conformation of amylose. Hydrogen bonds (not shown) hold the coils in place. (From Lehninger, A. L., *Biochemistry,* 2nd ed., Worth Publishers, New York, 1975, p. 264.)

SUMMARY

Polysaccharides are high-molecular-weight polymers made up of monosaccharide building blocks linked by covalent glycosidic bonds. Polysaccharides may serve as structural components (as **cellulose, chitin,** and **hyaluronic acid** do) or as energy-storage molecules (**glycogen** or **starch**). Polysaccharides are sometimes found associated with proteins—these complexes are called **mucoproteins.** They are found in connective tissues. Polysaccharides have simpler primary structures than protein and nucleic acids, but they also have secondary structures, for which hydrogen bonding is mainly responsible.

Key Terms

Amylopectin (Section 19.3)
Amylose (Section 19.3)
Cellulose (Section 19.3)
Chitin (Section 19.4)
Chondroitins (Section 19.4)
Glycogen (Section 19.3)
Hyaluronic acid (Section 19.4)
Mucoprotein (Section 19.4)
Polysaccharide (Section 19.1)
Starch (Section 19.1)

ADDITIONAL PROBLEMS

19.5 Glucose is by no means the only biologically important six-carbon monosaccharide. The following structure is D-galactose (shown in the Fischer representation):

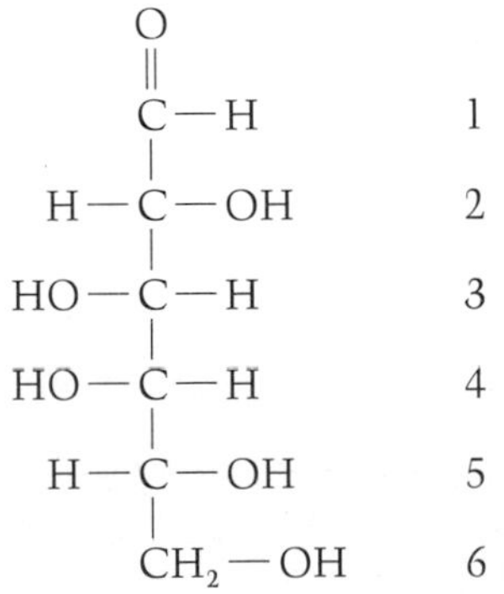

```
          O
          ‖
          C—H          1
          |
      H—C—OH           2
          |
     HO—C—H            3
          |
     HO—C—H            4
          |
      H—C—OH           5
          |
          CH2—OH       6
```

Write the reaction for the formation of the two hemiacetal anomers of D-galactose.

19.6 The disaccharide made up of glucose and galactose is called lactose. As you might guess, lactose is "milk sugar." In lactose, glucose and galactose are linked $\beta(1\rightarrow4)$, that is, with the 1 carbon of galactose linked to the 4 carbon of glucose. Draw the structure of lactose in the Haworth form.

19.7 Draw a segment of an amylopectin molecule, showing at least six glucose residues and including one $\alpha(1\rightarrow6)$ branch point.

19.8 Draw a segment of a hyaluronic acid molecule, including at least three disaccharide repeating units.

19.9 Write an equation for the hydrolysis of the disaccharide maltose to its monosaccharide units. Use structural formulas.

19.10 Write an equation for the complete hydrolysis of a molecule of cellulose containing 100 β-D-glucose units.

19.11 Do you imagine that a cellulase could catalyze hydrolysis of amylose?

19.12 Most proteins and some polysaccharides contain sulfur. In what form is sulfur present in proteins? In what form is sulfur present in the polysaccharides called chondroitins?

19.13 Polysaccharides are, generally speaking, very hydrophilic. Cotton (cellulose) fabrics absorb water very easily. Starch absorbs a lot of water, too. What kind of bond is responsible for the attraction between polysaccharides and water? What polysaccharide functional groups are involved?

***19.14** When cellulose is treated with acetic anhydride, a product is formed which can be made into a number of useful materials, such as photographic film and textile yarns. What type of functional group is formed in this reaction?

Problem-Solving Hints

If you have trouble solving any of the additional problems, refer to the sections listed next to the problem numbers.

19.5	19.2	**19.6**	19.3	**19.7**	19.3
19.8	19.4	**19.9**	19.3	**19.10**	19.3
19.11	19.3	**19.12**	19.4, 17.4	**19.13**	19.5

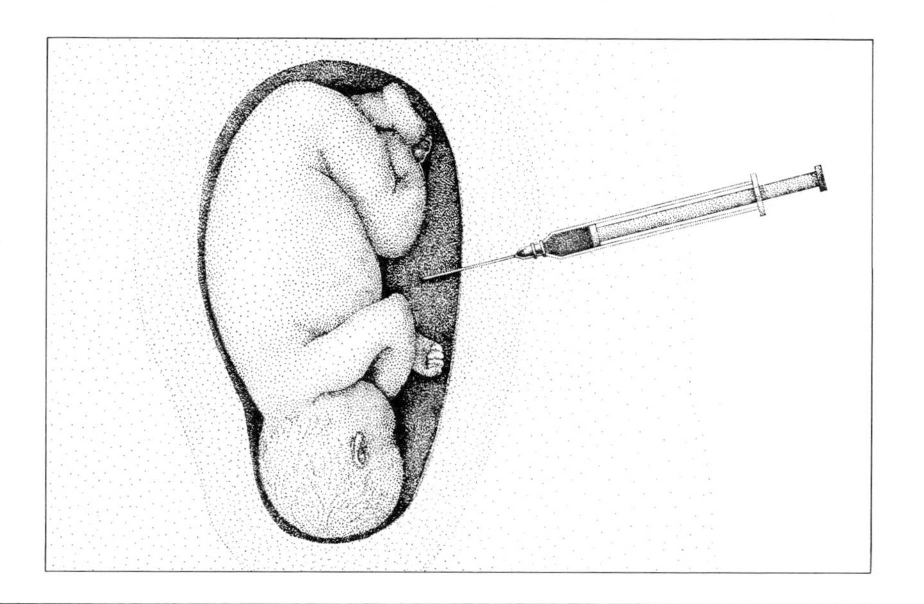

Amniotic fluid surrounds the fetus during its development in the uterus. The fluid serves as a protective cushion for the fetus and allows the fetus to change position in the mother. Amniocentesis is a clinical procedure in which amniotic fluid is removed from the amniotic cavity of a pregnant woman at any time prior to the delivery of the baby. The fluid, which contains many biochemical compounds, can then be analyzed chemically and biologically to assess a wide variety of genetic problems.

There are three major reasons to perform an amniocentesis: (1) to find out how mature the fetus is—such as in the case of an expected cesarian-section delivery; (2) to determine fetal problems—for example, in the case of Rh blood incompatibility; and (3) to detect genetic disease—such as mongolism, cystic fibrosis, and disorders of lipid metabolism.

Recently the phospholipids, a group of necessary lipids that we will discuss in this chapter, have been identified among the biochemical compounds present in the amniotic fluid. Two phospholipids, lecithin and sphingomyelin, are present in the aminiotic fluid in low concentrations until the last few weeks of fetal development. At this time, lecithin is produced in much larger quantities than sphingomyelin.

We now know that lecithin contributes to making the fetal lungs capable of respiration at the time of birth. Without an increase in the level of lecithin, the fetus has a chance of developing respiratory distress syndrome (RDS), in which the fetal lungs cannot function adequately. If a cesarian

20

LIPIDS

section is considered, the levels of lecithin and sphingomyelin (expressed as the L/S ratio) are measured regularly. If possible, the physician will delay the cesarian-section procedure until the L/S ratio indicates that there is little chance the baby will develop respiratory problems. With the ability to monitor such important compounds as the phospholipids, many more babies are now born with normal lung capacities.

[*Illustration: Amniocentesis involves the withdrawal of amniotic fluid from the area surrounding a fetus for analysis of possible defects.*]

Lipids are organic compounds that are quite soluble in organic solvents—such as ether, methanol, and chloroform—but not very soluble in water. Many diverse compounds fall into this class, including fatty acids, fats and oils, sterols, the fat soluble vitamins, prostaglandins, and phospholipids. Lipids play a variety of roles in biological systems. For example, phospholipids and cholesterol are important components of biological membranes, fats and oils are energy-storage materials, and various sterol derivatives serve as hormones.

20.1 FATTY ACIDS

Why do naturally-occurring unsaturated fatty acids have melting points much lower than those of saturated fatty acids?

A fatty acid (Section 15.11) is a compound represented by the general formula $R—CO_2H$, where the R stands for an unbranched alkyl hydrocarbon chain from 1 to 25 carbons in length. Such alkyl chains may be fully saturated (Section 9.1) or unsaturated. A saturated alkyl chain contains only single-bonded carbon atoms, whereas an unsaturated alkyl chain contains one or more double-bonded carbon atoms. An unsaturated fatty acid's alkyl chain may contain from one to four double bonds, depending on its length.

Table 20.1 shows the structures of the most common naturally occurring fatty acids. You will notice that these fatty acids have an even number of carbons. Those with 2 to 4 carbons are called short-chain fatty acids; those with 6 to 10 carbons, medium-chain fatty acids; and those with 12 or more carbons, long-chain fatty acids. In naturally occurring unsaturated fatty acids, the configuration of the double bond or bonds is always *cis* rather than *trans* (Section 8.10).

Unsaturated fatty acids with *cis* double bonds have melting points about 50°C lower than those of saturated fatty acids or *trans* unsaturated

TABLE 20.1 Some Naturally Occurring Fatty Acids

Structure	Common name	m.p., °C
Saturated fatty acids		
$CH_3(CH_2)_{10}CO_2H$	Lauric acid	44.2
$CH_3(CH_2)_{12}CO_2H$	Myristic acid	53.9
$CH_3(CH_2)_{14}CO_2H$	Palmitic acid	63.1
$CH_3(CH_2)_{16}CO_2H$	Stearic acid	69.6
$CH_3(CH_2)_{18}CO_2H$	Arachidic acid	76.5
$CH_3(CH_2)_{22}CO_2H$	Lignoceric acid	86.0
Unsaturated fatty acids		
$CH_3(CH_2)_5CH{=}CH(CH_2)_7CO_2H$	Palmitoleic acid	−0.5
$CH_3(CH_2)_7CH{=}CH(CH_2)_7CO_2H$	Oleic acid	13.4
$CH_3(CH_2)_4CH{=}CHCH_2CH{=}CH(CH_2)_7CO_2H$	Linoleic acid	−5
$CH_3CH_2CH{=}CHCH_2CH{=}CHCH_2CH{=}CH(CH_2)_7CO_2H$	Linolenic acid	−11
$CH_3(CH_2)_4(CH{=}CHCH_2)_3CH{=}CH(CH_2)_3CO_2H$	Arachidonic acid	−49.5

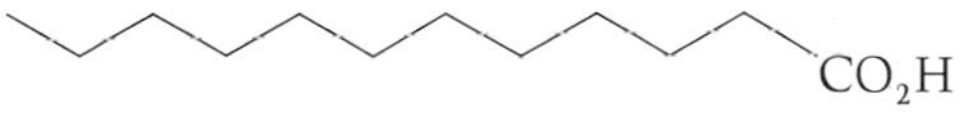

CO_2H

saturated fatty acid

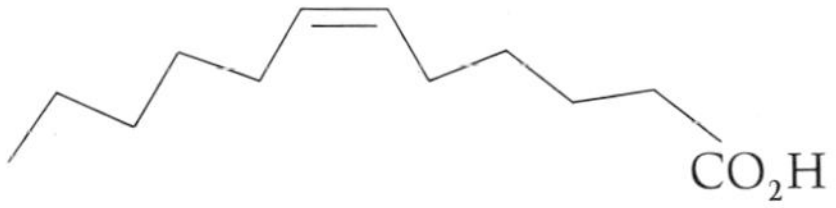

CO_2H

cis monounsaturated fatty acid

CO_2H

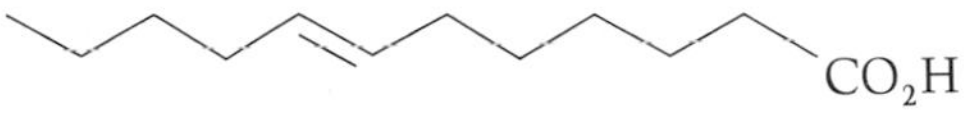

trans monounsaturated fatty acid (not a common fatty acid)

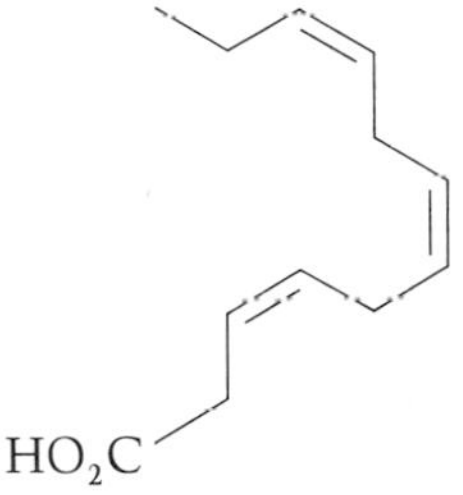

HO_2C

all *cis* polyunsaturated fatty acid

fatty acids of the same chain length. Just by looking at the structural formulas you can probably see that the molecules of saturated and *trans* unsaturated fatty acids can pack together in a regular fashion, and thus solidify, rather easily. On the other hand, the kinked *cis* unsaturated fatty acid chains are going to find it harder to pack together into an organized structure. Thus the temperature must be considerably lower to solidify the *cis* unsaturated fatty acids.

The normal positions of the double bonds are quite well defined, as should be evident from Table 20.1. In **polyunsaturated fatty acids,** the double bonds (two or more) are not conjugated—that is, they never occur on alternate carbon atoms.

Compounds containing unsaturated fatty acids play important roles in the structure of membranes in all cells. They are also essential raw material for the synthesis of prostaglandins, hormone-like lipids that have recently been discovered (Section 20.8). Linoleic acid, the 18-carbon fatty acid with two *cis* double bonds, has been classified as an **essential fatty acid.** If it is missing from the diet, pathological changes occur; some of these changes affect the skin, the body fluids, and reproductive functions. The changes disappear when linoleic acid is added back into the diet. Apparently, the body cannot synthesize this fatty acid and it must therefore be obtained from foods.

Problem 20.1 Draw and name one short-chain, one medium-chain, and one long-chain saturated fatty acid. If necessary, refer to Section 15.3 for naming hints.

20.2 FATS AND OILS

What is the difference between a fat and an oil?

Because fatty acids have a carboxyl group, they can form esters (esterify) by reacting with alcohols. When three fatty acid molecules esterify with the trihydroxy compound, glycerol, a fat or an oil is produced.

$$\begin{array}{l} CH_2-OH \\ | \\ CH-OH \\ | \\ CH_2-OH \end{array} + \left\{ \begin{array}{l} R-CO_2H \\ \\ R'-CO_2H \\ \\ R''-CO_2H \end{array} \right. \rightarrow \begin{array}{r} \\ \\ \overset{\displaystyle O}{\overset{\|}{R'-C-O}}- \\ \\ \\ \end{array} \begin{array}{l} CH_2O-\overset{\displaystyle O}{\overset{\|}{C}}-R \\ | \\ CH \\ | \\ CH_2O-\overset{\displaystyle O}{\overset{\|}{C}}-R'' \end{array} + 3H_2O$$

glycerol — fatty acids — triacylglycerol

The product—whether it be a fat or an oil—is called a **triacylglycerol** or triglyceride. (*Triglyceride* is the common name.)

Many different triacylglycerols exist. Their nature depends on the particular fatty acids present in their structures. Oils, which are liquid at room temperature, contain more polyunsaturated fatty acids than do fats. On the other hand, fats, which are solid at ordinary temperatures,

TABLE 20.2 Component Fatty Acids in the Depot Fats of Different Animals (values are as percent weight)

Animal	Saturated			Unsaturated			
	C_{14}	C_{16}	C_{18}	C_{16}	Oleic acid	C_{18} (diene)	C_{20-22}
Chicken	1	25	4	7	43	18	1
Rat	7	24	5	6	49	5	1
Horse	5	26	5	7	34	5	2
Pig	1	28	12	3	48	6	2
Human	6	25	6	7	45	8	2
Cat	4	29	17	4	41	2	—

contain more saturated and monounsaturated fatty acids. The same reasoning we used in explaining the different melting points of saturated and *cis* unsaturated fatty acids in Section 20.1 applies here. Table 20.2 gives the fatty acid compositions of several animal fats. Typically, in most triacylglycerols from plant or animal sources, two of the three fatty acids will be unsaturated. Plant-derived oils tend to have a greater content of polyunsaturated fatty acids, that is, fatty acids that have more than one double bond.

A high intake of saturated fats has been correlated with an increased tendency to heart disease. Thus, most doctors recommend reducing the intake of saturated fats and including more polyunsaturated fats in the diet. See Focus 12 for more about dietary lipids and health.

Problem 20.2 Write an equation for the formation of a triacylglycerol containing linolenic acid and palmitic acid in a 2:1 ratio.

Problem 20.3 Proteins, nucleic acids, and polysaccharides are all optically active. Is it possible for a triacylglycerol to be optically active? Explain.

PHOSPHOGLYCERIDES 20.3

A diacylglycerol is simply glycerol esterified with fatty acids at carbons 1 and 2. If a phosphate group is esterified at the hydroxyl on carbon 3 of a diacylglycerol, the resulting compound is a **phosphatidic acid.**

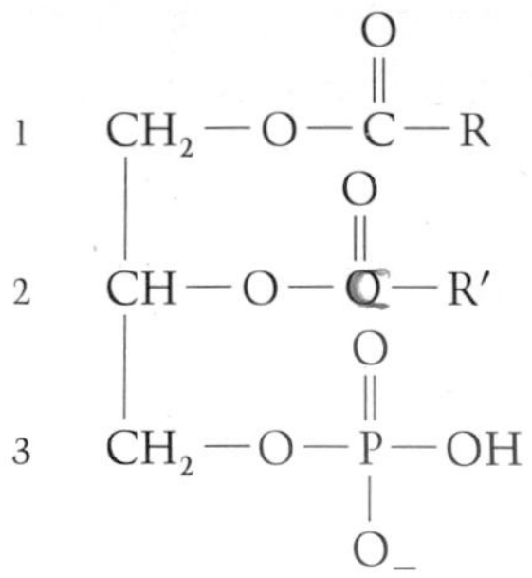

$$
\begin{array}{lllll}
 & & & \text{O} & \\
 & & & \| & \\
1 & \text{CH}_2\text{—O—} & & \text{C—R} & \\
 & | & & \text{O} & \\
 & | & & \| & \\
2 & \text{CH—O—} & & \text{C—R}' & \\
 & | & & \text{O} & \\
 & | & & \| & \\
3 & \text{CH}_2\text{—O—} & & \text{P—OH} & \\
 & & & | & \\
 & & & \text{O}^- &
\end{array}
$$

a phosphatidic acid

The phosphatidic acids are the parent compounds of the phosphate-containing lipids called phosphoglycerides (Table 20.3). A **phosphoglyceride** is formed when an alcohol is esterified to the phosphate group of a phosphatidic acid. Phosphoglycerides are prevalent in cell membranes. Phosphatidyl choline, also known as lecithin, for example, is an important membrane phosphoglyceride.

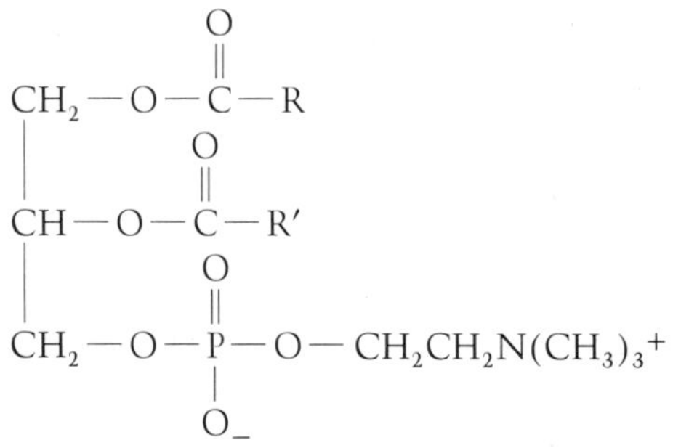

$$
\begin{array}{llll}
 & & \text{O} & \\
 & & \| & \\
\text{CH}_2\text{—O—} & & \text{C—R} & \\
| & & \text{O} & \\
| & & \| & \\
\text{CH—O—} & & \text{C—R}' & \\
| & & \text{O} & \\
| & & \| & \\
\text{CH}_2\text{—O—} & & \text{P—O—CH}_2\text{CH}_2\text{N(CH}_3)_3{}^+ & \\
 & & | & \\
 & & \text{O}^- &
\end{array}
$$

phosphatidyl choline (lecithin)

These phosphoglycerides are both hydrophilic (polar) and hydrophobic (nonpolar), just as the free fatty acids themselves are. Previously, we saw that this is true of proteins and nucleic acids, but the lipids are much smaller molecules, so it is more remarkable that they have both properties. Clearly, the fatty acid side chains have the hydrophobic, nonpolar character. It is the free carboxyl group (of a fatty acid) or the phosphate ester group (of a phosphoglyceride) that has a highly polar character. The hydrophilic–hydrophobic structures of these molecules give them important properties, as we will see in the next section.

Problem 20.4 Draw complete structures for two different phosphoglycerides, including complete unsaturated fatty acyl chains of 16 or 18 carbons.

TABLE 20.3 Common Phosphoglycerides

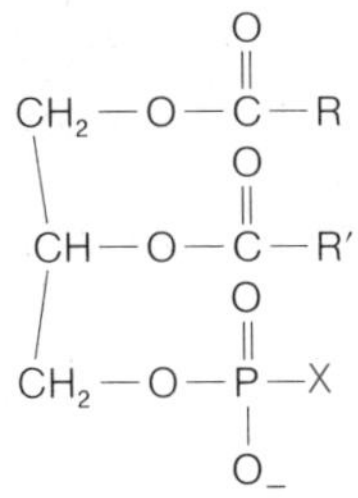

$$
\begin{array}{l}
CH_2-O-\overset{\overset{\displaystyle O}{\|}}{C}-R \\
\;|\\
CH-O-\overset{\overset{\displaystyle O}{\|}}{C}-R' \\
\;|\\
CH_2-O-\overset{\overset{\displaystyle O}{\|}}{\underset{\underset{\displaystyle O^-}{|}}{P}}-X
\end{array}
$$

a phosphoglyceride,
where X may come from any one of these alcohols

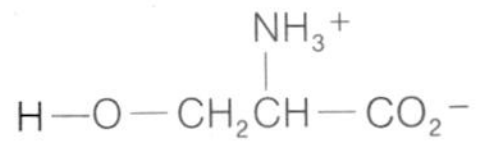

$H-O-CH_2\overset{\overset{\displaystyle NH_3^+}{|}}{C}H-CO_2^-$

serine
(phosphatidyl serine)

$H-O-CH_2CH_2NH_3^+$

ethanolamine
(phosphatidyl ethanolamine)

$H-O-CH_2CH_2N(CH_3)_3^+$

choline
(phosphatidyl choline)

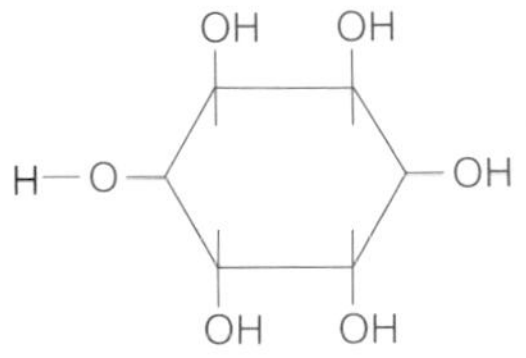

H—O—(cyclohexane ring bearing OH, OH, OH, OH, OH)

inositol
(phosphatidyl inositol)

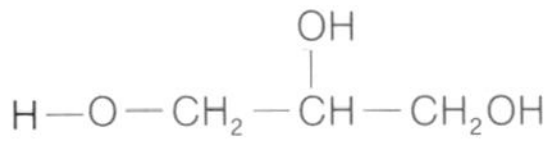

$H-O-CH_2-\overset{\overset{\displaystyle OH}{|}}{C}H-CH_2OH$

glycerol
(phosphatidyl glycerol)

LIPID STRUCTURES 20.4

When fatty acids or phosphoglycerides are mixed into water under suitable conditions, the molecules spontaneously stick together to form **micelles, monolayers,** or **bilayers.** Figure 20.1 shows what these terms mean. The driving force that causes the molecules to come together into one of these forms is the same as the force that leads a polypeptide to fold up in aqueous solution. That is, the hydrophobic portions of the molecules (the hydrocarbon chains) become arranged so as to minimize their contact with water molecules. At the same time, the polar groups (carboxyl or phosphate ester groups) arrange

Is hydrophobic interaction important in the formation of lipid structures?

$CH_2-NH_3^+$
CH_2
O
$^-O-P=O$
O
$CH_2-CH-CH_2$

Polar head group

Nonpolar hydrocarbon tails

Polar head group

Nonpolar hydrocarbon tails

A phosphoglyceride
(phosphatidyl ethanolamine)

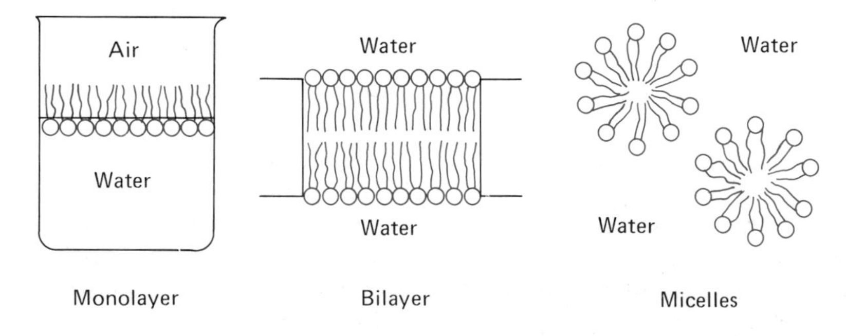

FIGURE 20.1 Cross-sectional diagrams of monolayer, bilayer, and micelle structures formed by phosphoglycerides in contact with water.

themselves so as to maximize their hydrogen bonding with the surrounding water. Biological membranes are examples of lipid structure. They are basically lipid bilayers in which numerous protein molecules are embedded. We will consider membrane structure in more detail in Chapter 21.

Problem 20.5 Define the following in your own words: micelle, bilayer, monolayer.

SOME OTHER IMPORTANT LIPIDS 20.5

How are phosphoglycerides and sphingomyelins structurally related?

Sphingolipids are structurally related to phosphoglycerides. No glycerol is present in sphingolipids, but sphingosine, a long-chain amino alcohol, plays a role very similar to that of glycerol. That is, it provides a backbone to which a fatty acid hydrocarbon chain and another group are attached.

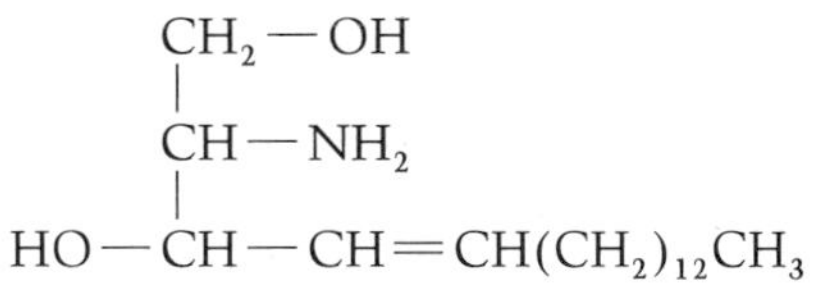

$$\begin{array}{l} \qquad\quad CH_2-OH \\ \qquad\quad | \\ \qquad\quad CH-NH_2 \\ \qquad\quad | \\ HO-CH-CH=CH(CH_2)_{12}CH_3 \end{array}$$

sphingosine

In sphingolipids, a fatty acid forms a peptide bond with the amino group:

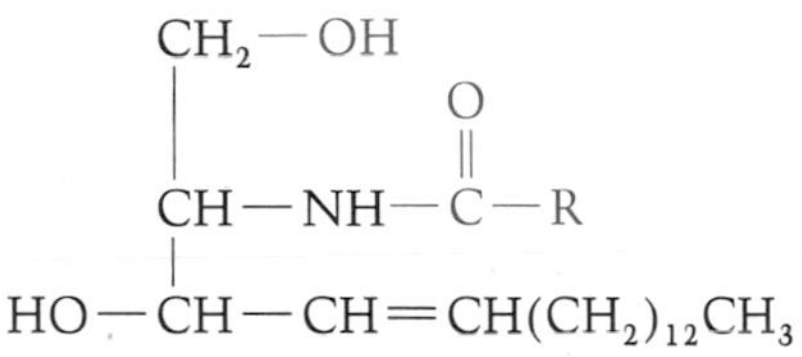

$$\begin{array}{l} \qquad\quad CH_2-OH \\ \qquad\quad | \qquad\qquad\quad O \\ \qquad\quad | \qquad\qquad\quad \| \\ \qquad\quad CH-NH-C-R \\ \qquad\quad | \\ HO-CH-CH=CH(CH_2)_{12}CH_3 \end{array}$$

The hydroxyl group (in color) is bonded to some polar group—perhaps a phosphate group linked as in a phosphoglyceride to a nitrogen-containing alcohol like ethanolamine or choline. The polar group may also be a sugar molecule or a short chain of sugars. The sugar-containing sphingolipids are called **cerebrosides** (from *cerebrum*, brain) because brain cell membranes are rich in them. The sphingolipids in which phosphate is esterified to a nitrogen-containing alcohol are called **sphingomyelins.**

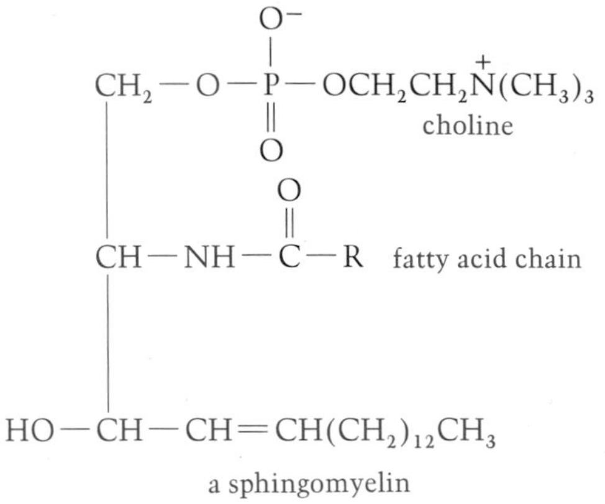

$$\begin{array}{l} \quad\quad\quad\quad\quad O^- \\ \quad\quad\quad\quad\quad | \quad\quad\quad\quad\quad + \\ CH_2-O-P-OCH_2CH_2N(CH_3)_3 \\ | \quad\quad\quad\quad \| \quad\quad\quad \text{choline} \\ | \quad\quad\quad\quad O \\ | \quad\quad\quad\quad\quad O \\ | \quad\quad\quad\quad\quad \| \\ CH-NH-C-R \quad \text{fatty acid chain} \\ | \\ HO-CH-CH{=}CH(CH_2)_{12}CH_3 \end{array}$$

a sphingomyelin

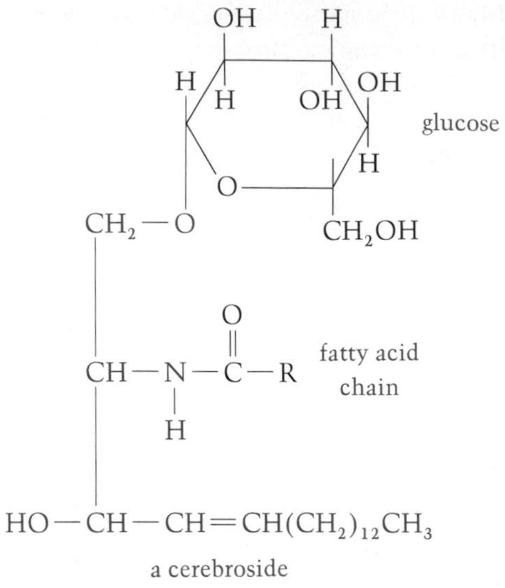

$$\begin{array}{l} CH_2-O-\text{glucose} \\ | \quad\quad\quad\quad\quad O \\ | \quad\quad\quad\quad\quad \| \\ CH-N-C-R \quad \text{fatty acid chain} \\ | \quad\quad | \\ | \quad\quad H \\ HO-CH-CH{=}CH(CH_2)_{12}CH_3 \end{array}$$

glucose

a cerebroside

Sphingolipids can work with phosphoglycerides to form bilayers and micelles, and they are important components of biological membranes, especially in nerve tissues. Although we do not yet understand why, sphingolipids must give nerve cell membranes some particularly favorable properties. The sphingomyelins and the phosphoglycerides are both called **phospholipids,** since both contain phosphate.

Problem 20.6 Identify each of the following in the structural formula of a sphingomyelin shown above:
(a) ester linkage
(b) phosphate group
(c) amide

Problem 20.7 Draw a complete sphingomyelin structure, with ethanolamine in ester linkage with the phosphate group, and with linolenic acid in amide linkage to the nitrogen in sphingosine.

20.6 STEROIDS

What are some of the functions of steroids in the body?

Steroids are derivatives of the following multi-ring skeletal structure:

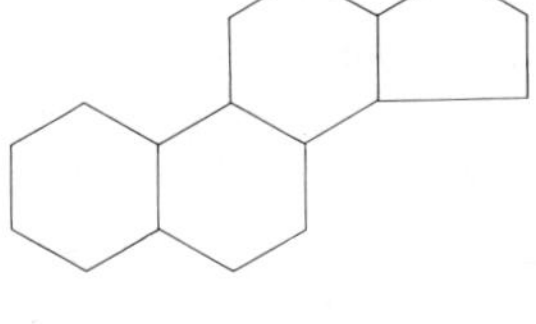

Many different steroids occur naturally. Cholesterol is the major steroid in animal tissues, and all other animal steroids are derived from it.

CH_3 CH_2 CH_2 CH_2 CH CH_3 CH_3

HO

cholesterol

Problem 20.8 Locate the steroid skeletal structure in the structural formula of cholesterol.

Problem 20.9 Some of the major components of the deposits that clog the arteries in atherosclerosis are esters of cholesterol and saturated fatty acids. Draw the structure of the ester formed from cholesterol and palmitic acid.

Table 20.4 gives the structures and names of a number of important steroids. Steroids have a variety of functions in the body. The bile acids are made in the liver from cholesterol and used to make bile salts. Bile salts are secreted into the intestine, where they are very important in the digestion of fats. They act like detergents to break up large globules of fatty material into small micelles.

The adrenal gland makes cortisol and aldosterone, two more important steroids. Cortisol promotes the manufacture of carbohydrate in the liver, and aldosterone acts on the kidney to reduce the loss of sodium ions into the urine.

The male and female sex hormones are steroids as well. The major male sex hormone is testosterone, which is necessary for the development and maintenance of the male reproductive system and secondary sex characteristics (like the beard). The major female sex hormones are estradiol and progesterone. Estradiol is necessary for the development and maintenance of the female reproductive organs and secondary sex characteristics. Progesterone is responsible for preparing the uterus for pregnancy following ovulation (the release of an ovum, or egg, from one of the ovaries).

Estrogen and progestin are general terms that describe hormones that act like estradiol and progesterone, respectively. You may know that the common oral contraceptive formulas contain both types of hormones. They fool the body into thinking that pregnancy has already occurred, thus preventing ovulation. The hormones of the oral contraceptive tablets are usually synthetic estrogens and progestins that have been developed to decrease undesirable side effects like water retention and weight gain.

TABLE 20.4 Some Important Steroids

Name	Formula
Two bile acids	
Cholic acid	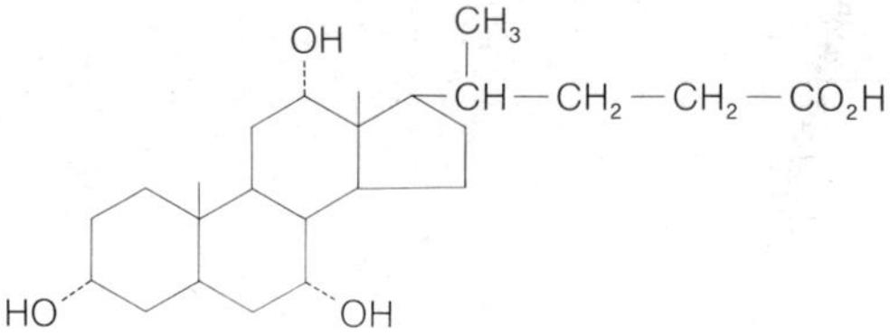
Deoxycholic acid	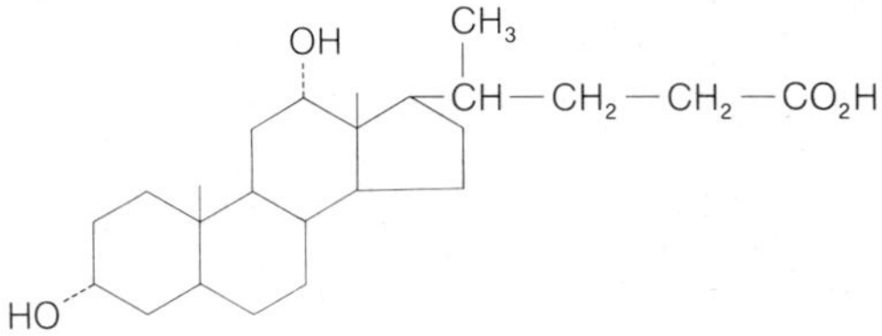
Two adrenal steroids	
Cortisol	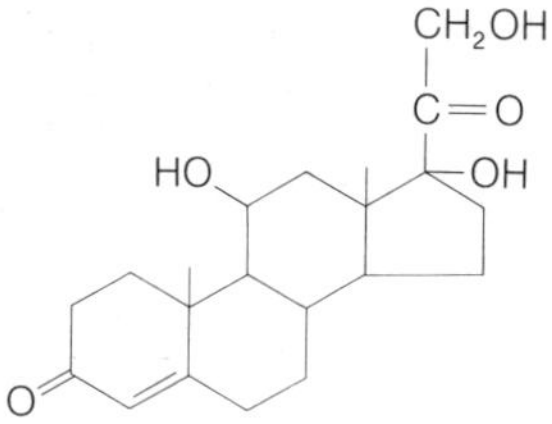
Aldosterone	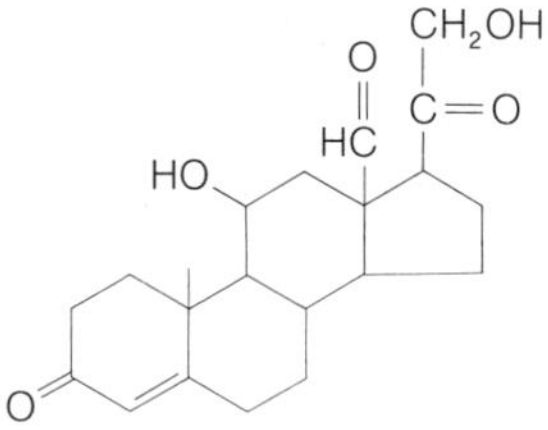
Two estrogens (female sex hormones)	
Estrone	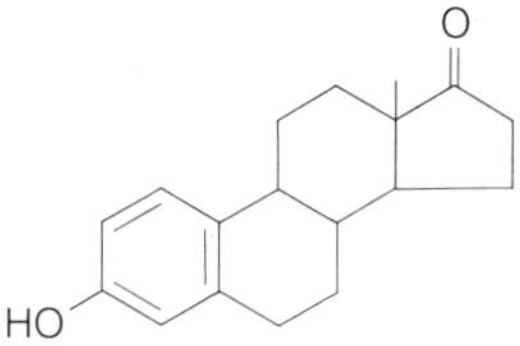
β-Estradiol	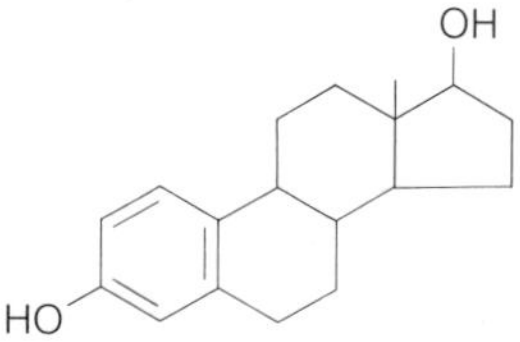

TABLE 20.4 Some Important Steroids (Continued)

Name	Formula
An androgen (male sex hormone) Testosterone	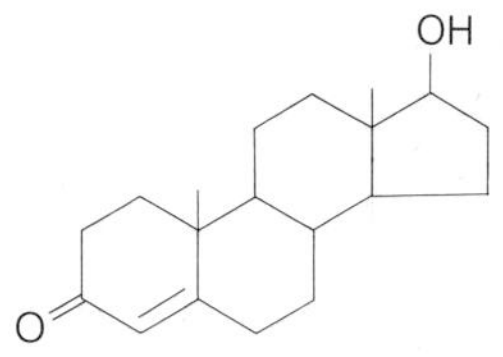 OH, O
A progestational hormone Progesterone	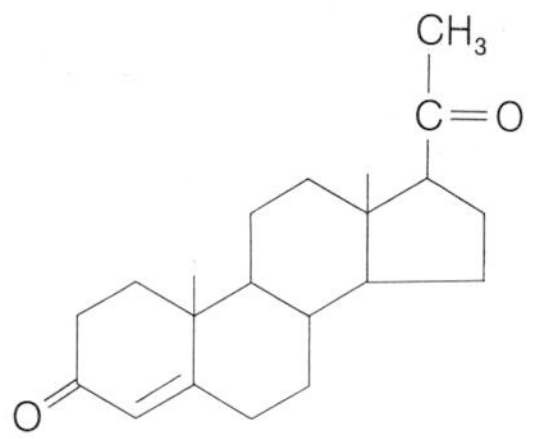 CH_3, C=O, O

FAT-SOLUBLE VITAMINS 20.7

Like other lipids, the fat-soluble vitamins are not soluble in water but will dissolve in organic solvents and lipid mixtures. **Vitamins** are specific organic compounds needed in the diet for maintenance of life and health. There are four fat-soluble vitamins required by human beings: A, D, E, and K. Each of these vitamins exists in several forms, which have slightly different chemical structures but identical physiological properties. Table 20.5 gives the structures of one form of each of these.

What is a vitamin?

TABLE 20.5 Fat Soluble Vitamins

Vitamin	Chemical structure
A (A_1)	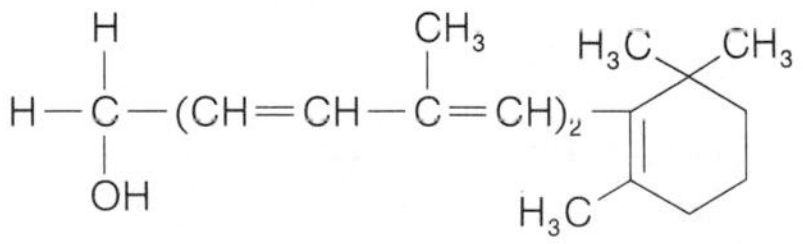 H, H—C(OH)—(CH=CH—C(CH_3)=CH)$_2$—, H_3C, CH_3, H_3C

TABLE 20.5 Fat Soluble Vitamins (Continued)

Vitamin		Chemical structure
K	(K_1)	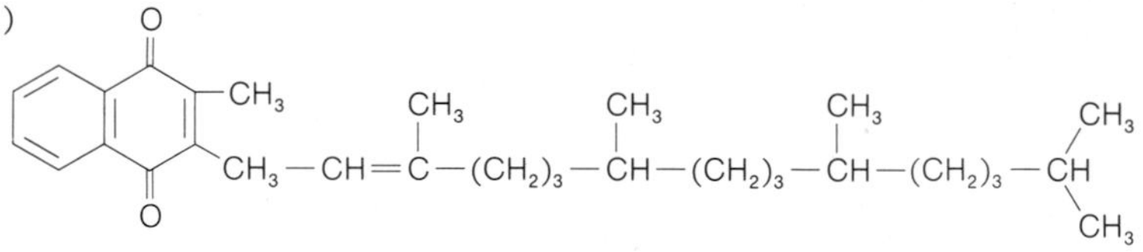 Naphthoquinone ring (O, O) with $-CH_3$ and $CH_3-CH=C(CH_3)-(CH_2)_3-CH(CH_3)-(CH_2)_3-CH(CH_3)-(CH_2)_3-CH(CH_3)_2$
D	(D_1)	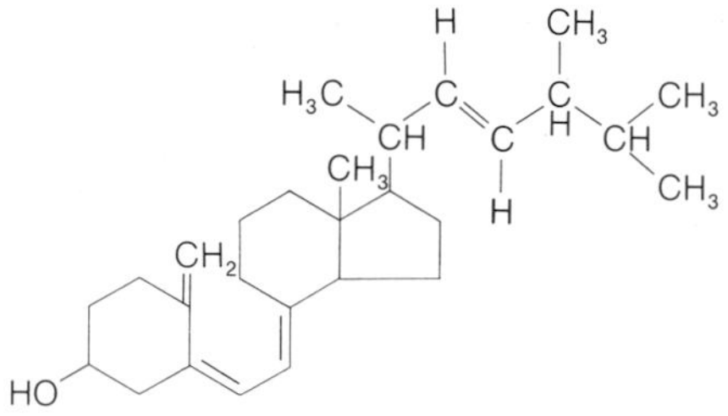 HO, CH_2, CH_3; side chain $H_3C-CH-C(H)=C(H)-C(H)(CH_3)-CH(CH_3)_2$
E		HO, H_3C, CH_3, CH_3, O; $-C(CH_3)-(CH_2)_3-CH(CH_3)-(CH_2)_3-CH(CH_3)-(CH_2)_3-CH(CH_3)_2$

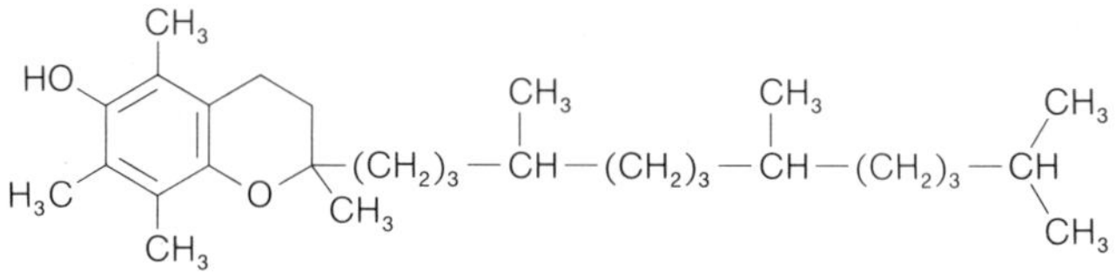

Source: Davies and Littlewood, *Elementary Biochemistry* (Englewood Cliffs, N.J.: Prentice Hall, 1979).

Large doses of fat-soluble vitamins may be toxic, but this is not true of the water-soluble vitamins (which we will discuss in Chapter 21). Fat-soluble materials are more difficult to excrete than water-soluble materials and may accumulate excessively in the membranes and fatty tissues of the body. In Chapter 23, we will discuss the functions of these vitamins and the food sources from which they are obtained.

20.8 PROSTAGLANDINS

How does aspirin work?

Prostaglandins are a recently discovered class of lipids that have a variety of biological and pharmacological actions. They are cyclic, oxygen-containing C_{20} fatty acids. Two examples are PGE_1 and $PGF_{2\alpha}$:

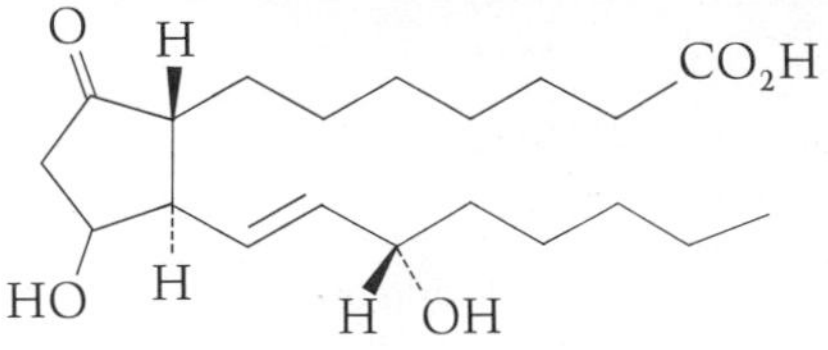

PGE_1 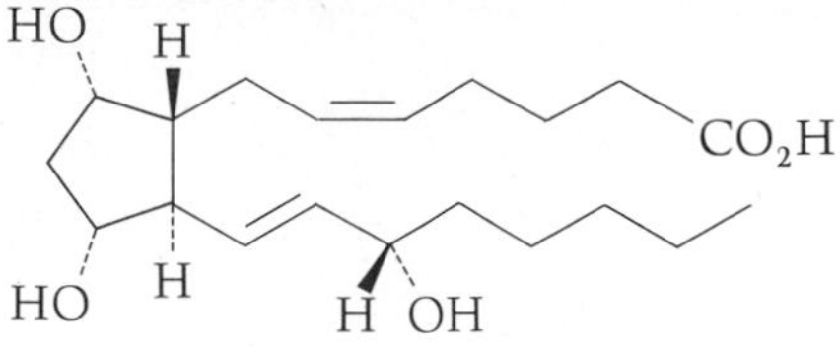 $PGF_{2\alpha}$

The prostaglandins have a variety of functions in the body. Some cause smooth muscle, such as that of blood vessels, to relax and thus reduce blood pressure. Others have the opposite effect, causing smooth muscle to contract. Prostaglandins also appear to be active in the induction of inflammatory and allergic responses, as well as in the relief of these processes. Both the inflammatory response and the immune response are normal body defense mechanisms, but if they are not properly regulated by the body, they can be harmful. Examples of such harmful results are rheumatoid arthritis, an inflammatory disease that affects millions of people, and asthma, hay fever, and other allergic responses.

Aspirin inhibits prostaglandin synthesis in body tissues; this may well be how it works to reduce inflammation and fever. For example, if aspirin could inhibit the synthesis of PGE_1, which is a potent **pyrogen,** or fever-inducing agent, it would reduce a fever.

SAPONIFICATION 20.9

How is soap made from fats?

A common way of classifying lipids is according to whether they contain fatty acids. Fats, oils, and phosphoglycerides, which contain fatty acids, are **complex lipids.** Steroids, which do not contain fatty acids, are **simple lipids.** Complex lipids are saponifiable; simple lipids are not. Saponification involves the solubilization of fats by hydrolysis with a base (Section 15.11).

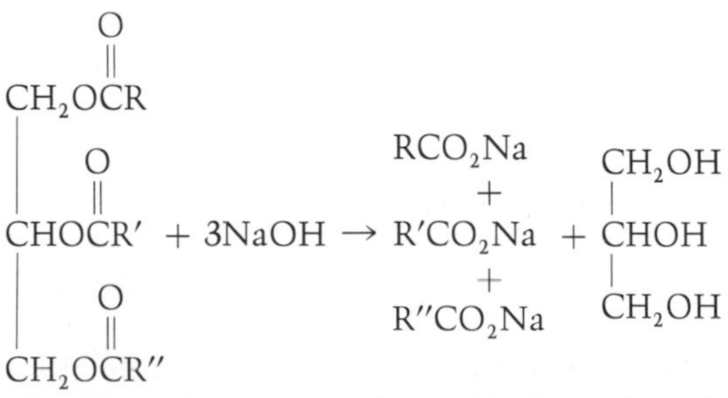

$$\begin{matrix} CH_2OC(=O)R \\ | \\ CHOC(=O)R' \\ | \\ CH_2OC(=O)R'' \end{matrix} + 3NaOH \rightarrow \begin{matrix} RCO_2Na \\ + \\ R'CO_2Na \\ + \\ R''CO_2Na \end{matrix} + \begin{matrix} CH_2OH \\ | \\ CHOH \\ | \\ CH_2OH \end{matrix}$$

a triacylglycerol   fatty acid salts   glycerol

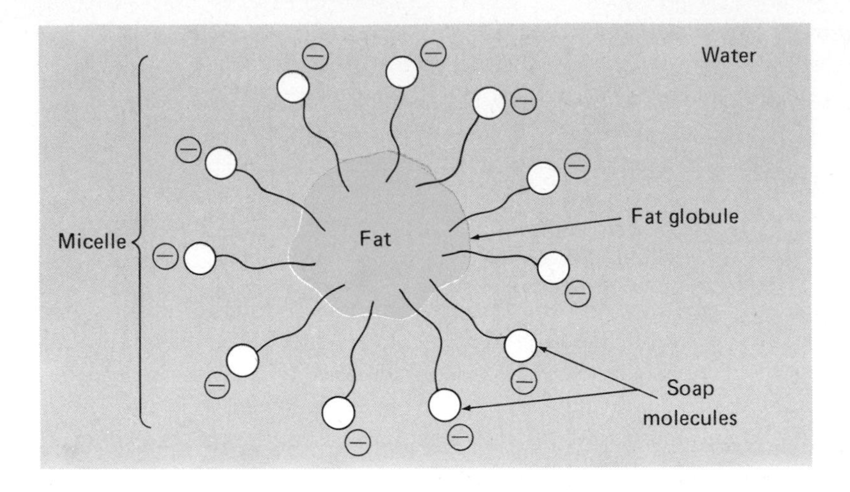

FIGURE 20.2 Solubilization of fat by soap.

The salts of the fatty acids are called soaps. Soaps are capable of interacting both with water (via their carboxylate groups) and with fatty materials (via their hydrophobic hydrocarbon chains). Because of this property, they can effectively disperse fats or oils into micelles in water. That is why soaps are good cleaning agents (Figure 20.2).

Problem 20.10 What functional group is found in all complex lipids?
Problem 20.11 Waxes are lipids that are esters of long chain saturated alcohols and long chain fatty acids. Are waxes saponifiable? Are they simple or complex lipids?

SUMMARY

Lipids are organic compounds that are usually insoluble in water but readily soluble in nonpolar solvents. Many diverse compounds fall into this class, including **fatty acids,** fats, **steroids, fat-soluble vitamins, phosphoglycerides, sphingolipids,** and **prostaglandins.** Fatty acids (general formula $R—CO_2H$) usually contain an unbranched alkyl chain 1 to 25 carbons in length, which may contain from zero to four *cis* double bonds. Fats and oils (**triacylglycerols**) are formed by the esterification of three fatty acid molecules to the trihydroxy compound glycerol. Phosphoglycerides, sphingolipids, and free fatty acids have both hydrophilic and hydrophobic character and aggregate in water to form **micelles, monolayers,** or **bilayers.** Cholesterol is the major animal steroid, and all other steroids are derived from it. Saponification provides a way of determining whether a lipid is **complex** (contains fatty acids that can be hydrolyzed) or **simple** (contains no fatty acids).

Key Terms

Bilayer (Section 20.4)
Cerebroside (Section 20.5)
Complex lipid (Section 20.9)
Essential fatty acid (Section 20.1)
Lipid (Introduction)
Micelle (Section 20.4)
Monolayer (Section 20.4)
Phosphatidic acid (Section 20.3)
Phosphoglyceride (Section 20.3)
Phospholipid (Section 20.5)
Polyunsaturated fatty acid (Section 20.1)
Prostaglandin (Section 20.8)
Simple lipid (Section 20.9)
Sphingolipid (Section 20.5)
Sphingomyelin (Section 20.5)
Steroid (Section 20.6)
Triacylglycerol (Section 20.2)
Vitamin (Section 20.7)

ADDITIONAL PROBLEMS

20.12 Draw the structure of a triacylglycerol containing palmitic acid, stearic acid, and linolenic acid. How many different triacylglycerols can you draw that have the same fatty acids?

20.13 Draw the structure of phosphatidyl choline.

20.14 Draw a *galacto*-cerebroside in which linoleic acid is in amide linkage with sphingosine and galactose is the monosaccharide present rather than glucose.

20.15 What does a bilayer or micelle of phospholipid molecules have in common with the three-dimensional structure of a globular protein? (*Hint:* What kind of interactions are involved in forming the three-dimensional structures?)

20.16 Name several lipids that are found in biological membranes.

20.17 Describe some of the functions of steroids in the body.

***20.18** If fat soluble vitamins are soluble in fats, as their name suggests, what kinds of foods would you expect to contain them? Explain why it is convenient to enrich milk with vitamin D and margarine with vitamin A. What vitamin is present in large amounts in cod liver oil?

20.19 Write an equation for the saponification of one of the triacylglycerols whose structural formula you drew for Problem 20.12.

20.20 Cholesterol, along with phospholipids, is present in most biological membranes. Where do you think the —OH group will be found—in association with polar head groups of phospholipids, or in the hydrocarbon core of the bilayer? Why?

Problem-Solving Hints

If you have trouble solving any of the additional problems, refer to the sections listed next to the problem numbers.

20.12 20.1, 20.2
20.13 20.3
20.14 20.5
20.15 20.4
20.16 20.3, 20.4, 20.5
20.17 20.6
20.18 20.7
20.19 20.9
20.20 20.6, 20.4

FOCUS 12 Atherosclerosis and Dietary Lipids

Atherosclerosis is characterized by the formation of thickened areas, or plaques, in the walls of major arteries. One result is elevated blood pressure (hypertension). A plaque in a coronary artery may grow large enough to block circulation to an area of the heart muscle, causing a myocardial infarction (heart attack). The plaque consists of an abnormal accumulation of muscle cells, accompanied by huge deposits of cholesteryl esters and other lipids. A cholesteryl ester contains a fatty acid in ester linkage with the hydroxyl group of cholesterol.

Atherosclerosis and myocardial infarction, its most common complication, are more common in people with elevated plasma lipid levels ("hyperlipemia") and, specifically, elevated plasma cholesterol ("hypercholesterolemia"). In general, however, the amount of cholesterol in the diet does not seem to affect the level of lipids in the blood. Instead, high plasma lipid levels (and atherosclerosis) appear to result from a relative deficiency in dietary polyunsaturated fatty acids.

When 20 to 60% of the total calories a person consumes are lipids (typical American diets contain 40% of calories as lipids), the blood

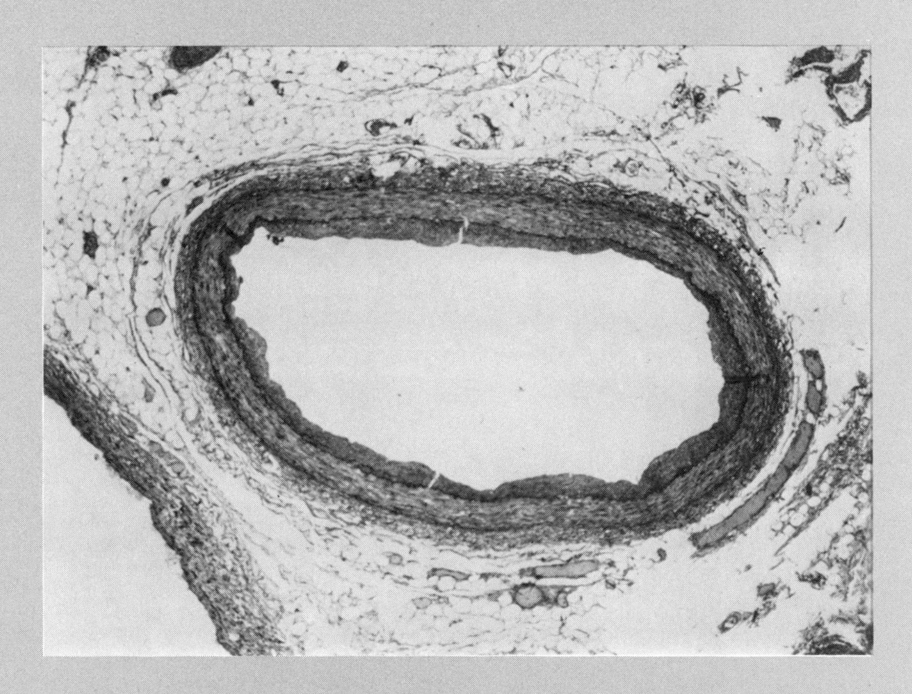

FIGURE 1 The coronary artery shown here has an average build-up of cholesterol. (Courtesy American Heart Association)

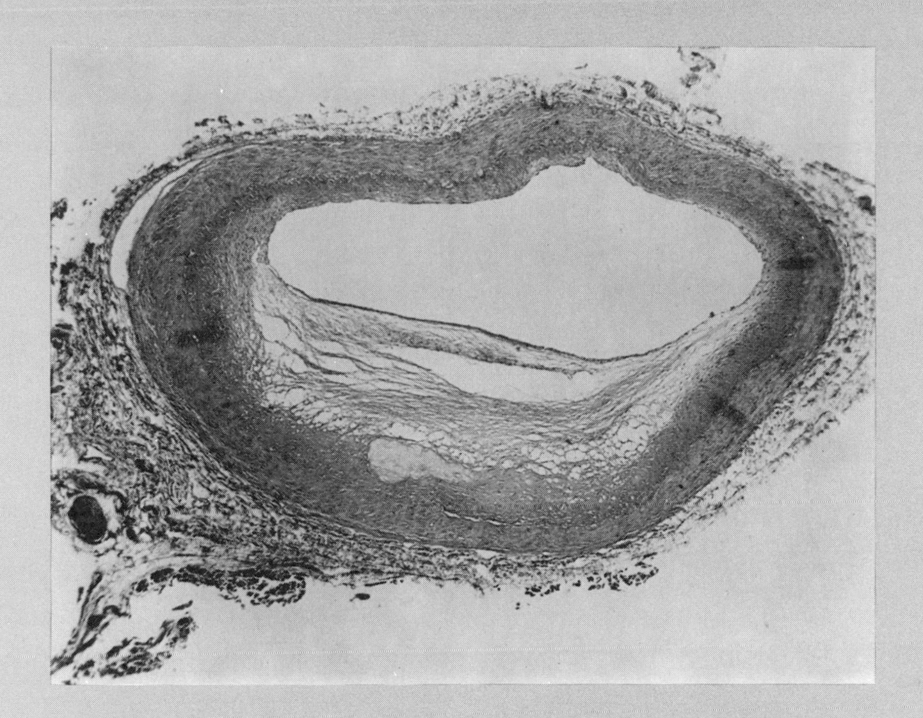

FIGURE 2 The excessive cholesterol deposits in this coronary artery are almost obstructing it. (Courtesy American Heart Association)

lipid levels are influenced by the nature of the dietary lipid. A person who consumes mostly animal fat—which is largely saturated and monounsaturated—may have a blood lipid level double that of someone whose diet is rich in polyunsaturated lipids. Studies with human subjects have shown that the blood lipid levels and the incidence of atherosclerotic disease are much lower in economically poorer areas of the world, where vegetable oils rather than animal products provide the major source of dietary lipid. We do not yet know exactly how fatty acid intake influences the blood lipid level.

Many studies have been performed to determine the ideal amount of dietary lipid. Experimental animals grow fastest and live longest when they eat a diet containing about 30% of total calories as lipids, with at least 1 to 2% of this being polyunsaturated lipids.

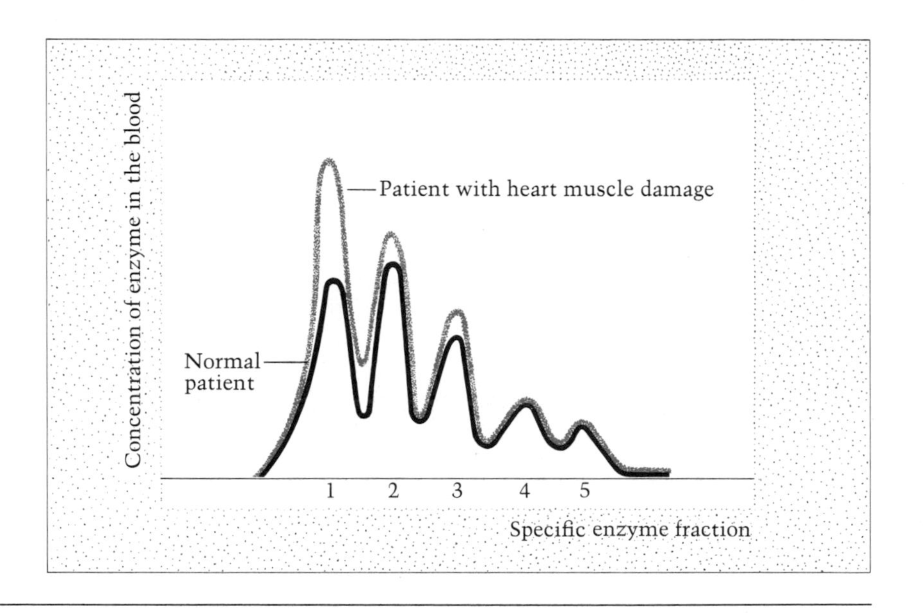

Lactate dehydrogenase (LDH) and creatine kinase (CK) are two very important enzymes found in the human blood. Lactate dehydrogenase appears in nearly all human tissues. However, its concentration is highest in the liver and skeletal muscle, as well as in the heart. On the other hand, we find creatine kinase not only in skeletal muscle and in the heart, but also in the brain.

We will learn in this chapter that the diagnosis and treatment of diseases has come to rely on measurements of enzyme levels. For example, the blood of a normal, healthy person has a fairly constant level of both lactate dehydrogenase and creatine kinase. The presence of these enzymes at normal levels in the blood is a result of muscle cell breakdown, which is a continuing process as the body builds up and breaks down tissue. An abnormal increase in either or both of the two enzymes reflects increased muscle cell breakdown, indicating some type of muscle damage. In the case of lactate dehydrogenase and creatine kinase, the muscles involved might be any of those that produce these particular enzymes. For instance, the blood of a patient with a recent heart attack would show a high level of lactate dehydrogenase and creatine kinase. The damage to the heart muscle would cause increased release of both enzymes.

21

BIOMOLECULES IN CELL STRUCTURE AND FUNCTION; ENZYMES

Most clinical laboratories are now able to offer two procedures: one that provides a measurement of the total level of lactate dehydrogenase and creatine kinase in the blood, and one that can detect the exact source of the increased level of either enzyme. The development of this valuable diagnostic tool represents a chemical milestone.

[Illustration. Damage to the heart muscle can be seen in laboratory testing by a rise of enzyme fraction 1 from its normal level (lower than enzyme fraction 2).]

BIOMOLECULES IN CELL STRUCTURE AND FUNCTION

An understanding of the structures and properties of biomolecules will enable you to understand cell structure and function in chemical terms. For example, it will help explain why cellulose makes a flexible but inelastic wall around a plant cell; why polar molecules cannot ordinarily pass through biological membranes (but nonpolar molecules can); and why a particular protein molecule binds to one sugar and catalyzes its reaction but does not bind to another sugar. It isn't possible to provide detailed answers to all these questions, but we want to leave you with the conviction that all these things can be explained and even predicted on the basis of ordinary chemical principles.

21.1 INTERACTIONS AMONG MOLECULES

How do molecules stick together?

A cell is made up of multitudes of molecules. Some of these molecules must stick together in some very specific and stable ways in order for the cell to be highly organized and durable. This particularly applies to the spontaneous adhesion of particular macromolecules to form organized structures such as chromosomes, biological membranes, and the contractile fibers of muscle cells. This spontaneous sticking together is often called self-assembly.

The interaction of individual small molecules with macromolecules is also fundamental to many biological reactions or processes. We usually call this interaction **binding:** the large molecule *binds* the small one. For example, a reactant molecule must be bound by an enzyme protein before a reaction can be catalyzed, and steroid hormones are bound by specific receptor proteins within cells. The binding of small molecules to larger ones usually concerns cell function rather than the formation of organized structures. Specific interactions among macromolecules (rather than between a macromolecule and a small molecule) also affect cell function. For instance, interactions among proteins are necessary in the contraction of muscle and in many other aspects of the control of cell processes.

By this time you may be wondering what it is that causes molecules to stick together. The attractive forces between molecules are nothing more than the familiar noncovalent interactions: hydrophobic interaction, hydrogen bonding, and ionic bonding (Section 17.2). However, since these forces are relatively weak, two molecules will stick tightly together only if their surfaces fit tightly together: the surfaces must be **complementary** to one another.

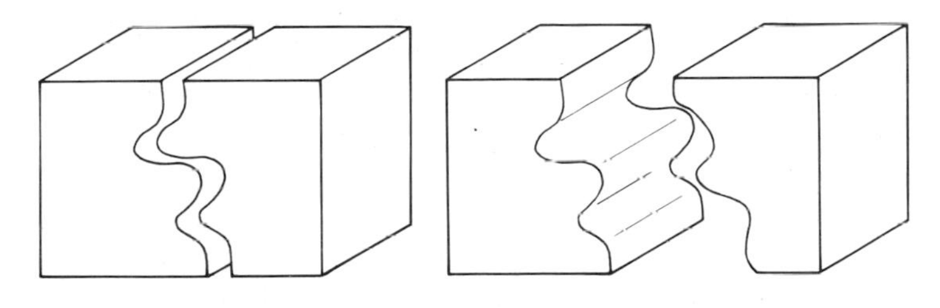

Complementary: tight binding   Not complementary: poor binding

The illustration shows complementarity in surface shape, but other kinds of complementarity are also important. For example, if there is a hydrophobic patch on one surface, there should be a matching hydrophobic patch on the other. And if there is a positively charged region on one surface, there should be a complementary area of negative charge on the other.

In general, then, molecules with complementary surfaces tend to stick together, and molecules without complementary surfaces do not. This spontaneous sticking together of complementary molecules is the basis for the self-assembly of cell structures like fibers and membranes. It is also the means for the specific pairing of bases in DNA and RNA (Section 18.6) and for the association of polypeptides in protein quaternary structures (Section 17.12).

The binding of small molecules by proteins also requires complementarity. However, we'll postpone further discussion of that topic until Section 21.9.

Problem 21.1 Write definitions in your own words for
(a) self-assembly (b) binding

21.2 BIOLOGICAL MEMBRANES

What is the most basic function of cellular membranes?

The limiting membrane of a cell, commonly called simply the **cell membrane,** is a thin, flexible structure that acts as a barrier to most substances on either side. It protects the integrity of the cell's structure, preventing the loss of materials from within the cell and preventing the entry of many (but not all) materials from outside the cell. Many cells have internal membranes as well, which divide the cell into distinct compartments and improve its ability to regulate its functions.

Membranes also take an active part in cell metabolism because they contain specific proteins. Some of these proteins are enzymes that catalyze important reactions; others are involved in the transport of ions or other materials across membranes. We'll discuss membrane proteins in more detail in Section 21.4.

21.3 THE LIPIDS OF MEMBRANES

Why is the cell membrane an effective barrier to polar molecules?

A discussion of the structure of the cell membrane—or indeed of any biological membrane—follows quite naturally from our discussion of lipid structures in the last chapter. A biological membrane is basically an extensive bilayer of phospholipid molecules—a lipid film. You probably will recall that bilayers and micelles can form spontaneously when phospholipids or detergents are put into water (Section 20.4). Membranes, too, are self-assembling structures.

The lipid film nature of a biological membrane is very important for membrane function. Since most of the materials dissolved in the cell water and in the fluid outside cells are charged or polar, what better way to keep the cell contents intact than to have a nonpolar lipid barrier—the membrane? The polar molecules on either side will be quite unwilling to leave their stable, hydrogen-bonded, aqueous surroundings to enter the unfriendly nonpolar interior of the membrane. For this reason, most dissolved materials do not readily cross biological membranes.

THE PROTEINS OF MEMBRANES 21.4

The modern view of membranes is embodied in the **fluid-mosaic model** of membrane structure (Figure 21.1). You have probably seen mosaics, which are pictures or designs formed by small stones or tiles set in mortar. In the fluid-mosaic model, protein molecules are like the stones, and the lipid bilayer is like the mortar. But whereas mortar is stiff and hard, the lipid bilayer into which the protein molecules are set or embedded is rather fluid. Thus we have a fluid mosaic, in which the protein molecules embedded in the lipid bilayer have some freedom to move—to diffuse along the membrane. The high content of unsaturated fatty acids in the membrane lipids is very important for the fluidity of the membrane bilayer.

Different kinds of cells have different proteins in their membranes. All membrane proteins have some function, whether as enzymes, trans-

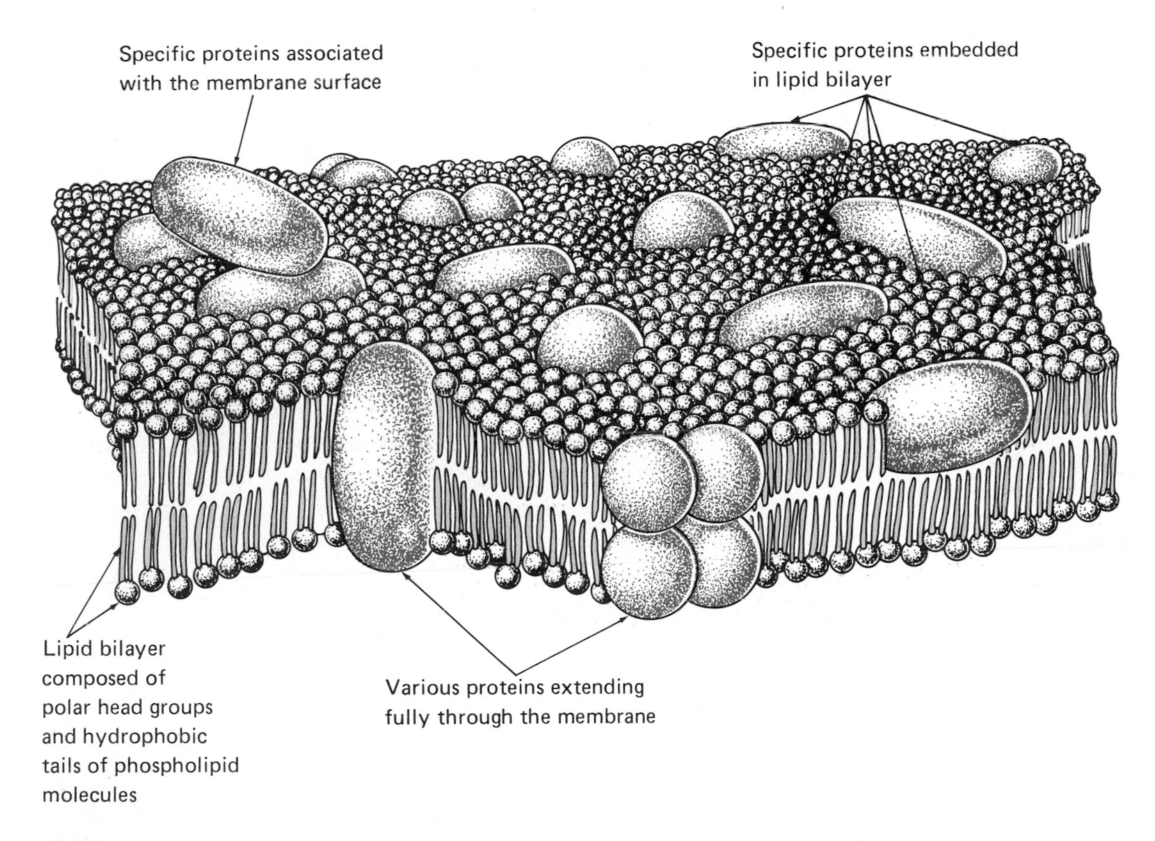

FIGURE 21.1 The fluid-mosaic model of membrane structure. (From *Experimental Biochemistry,* 2nd ed. by John M. Clark, Jr., and Robert L. Switzer, W. H. Freeman and Company. Copyright © 1977.)

port proteins (which serve to transport molecules through the membrane), binding sites for hormone molecules, or something else. Some membranes contain as little as 25% protein and 75% lipid, but others may contain as much as 80% protein and only 20% lipid. Certain membrane proteins are deeply inserted into the fluid lipid matrix. Others seem to be only weakly associated with the inner or outer surface of the membrane.

Problem 21.2 Proteins that are deeply embedded into the lipid matrix might be expected to contain a large percentage of what kind of amino acids? Explain.

Problem 21.3 Would you expect globular proteins that are weakly and noncovalently bonded to the surface of a membrane to have polar or nonpolar surfaces? Explain.

†21.5 STRUCTURE OF THE BACTERIAL CELL

Bacterial cells are rather simple in structure.

We begin our discussion of cell structure with bacterial cells because they tend to be simpler than the cells of higher plants and animals (see Figure 21.2). A bacterial cell is usually surrounded by a semirigid, polysaccharide **cell wall,** which protects the cell from mechanical damage and from swelling. Inside the cell wall is the limiting cell membrane, which contains many enzymes and transport proteins responsible for the uptake and oxidation of foodstuffs. The **cytoplasm,** the material enclosed by the limiting membrane, is a watery, soft gel; it contains many soluble proteins and many dissolved organic molecules and inorganic ions. The cell's proteins are made in the **ribosomes,** particles in the cytoplasm that contain both protein and nucleic acid. Granules of food storage materials, such as polysaccharides, are also found in the cytoplasm. A single **chromosome**—a single, circular molecule of DNA—lies in an area of the cell known as the nuclear region, or **nucleoid.**

†21.6 STRUCTURE OF THE PLANT CELL

How does the cell described here differ from the bacterial cell?

We will discuss a representative plant cell—a cell from the leaf of a green plant (see Figure 21.3). This cell is much larger than a bacterial cell. It too is surrounded by a limiting membrane, which is in turn surrounded by a cell wall. The cell wall provides mechanical support for the plant

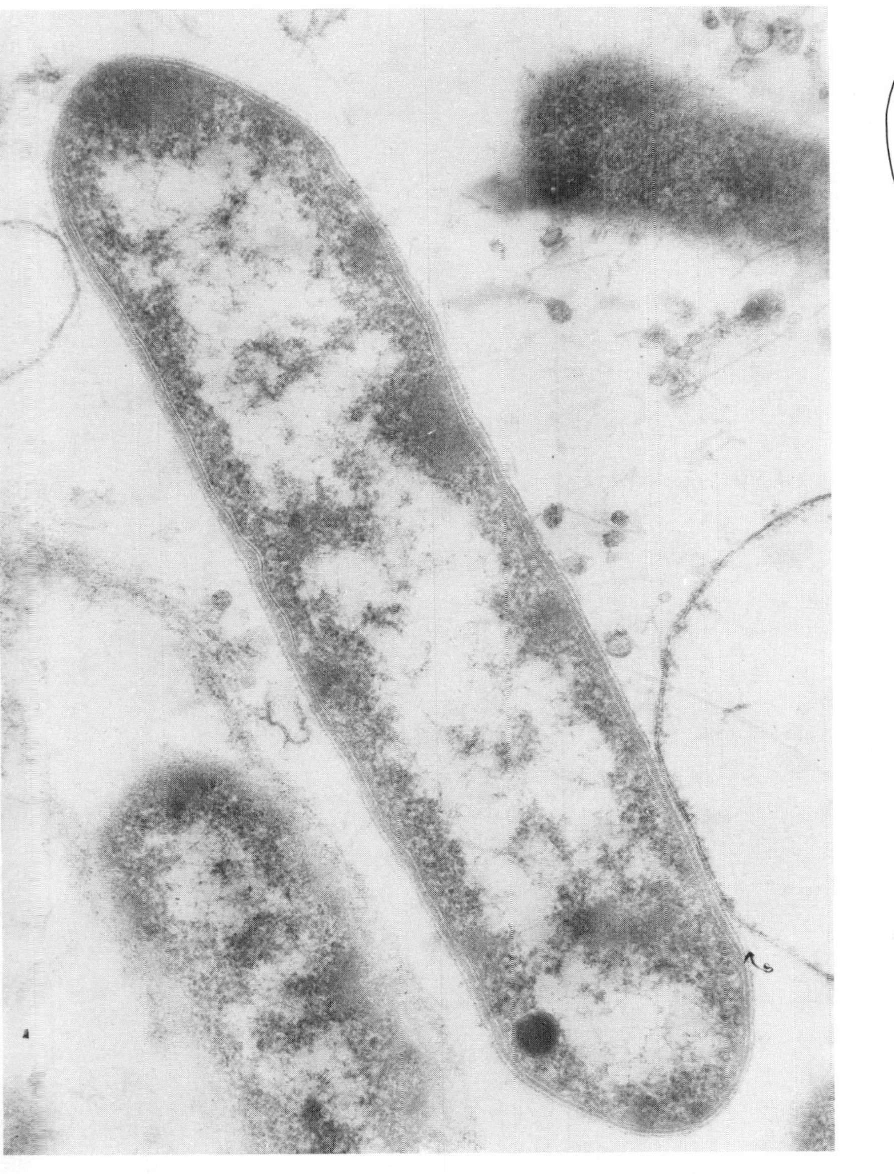

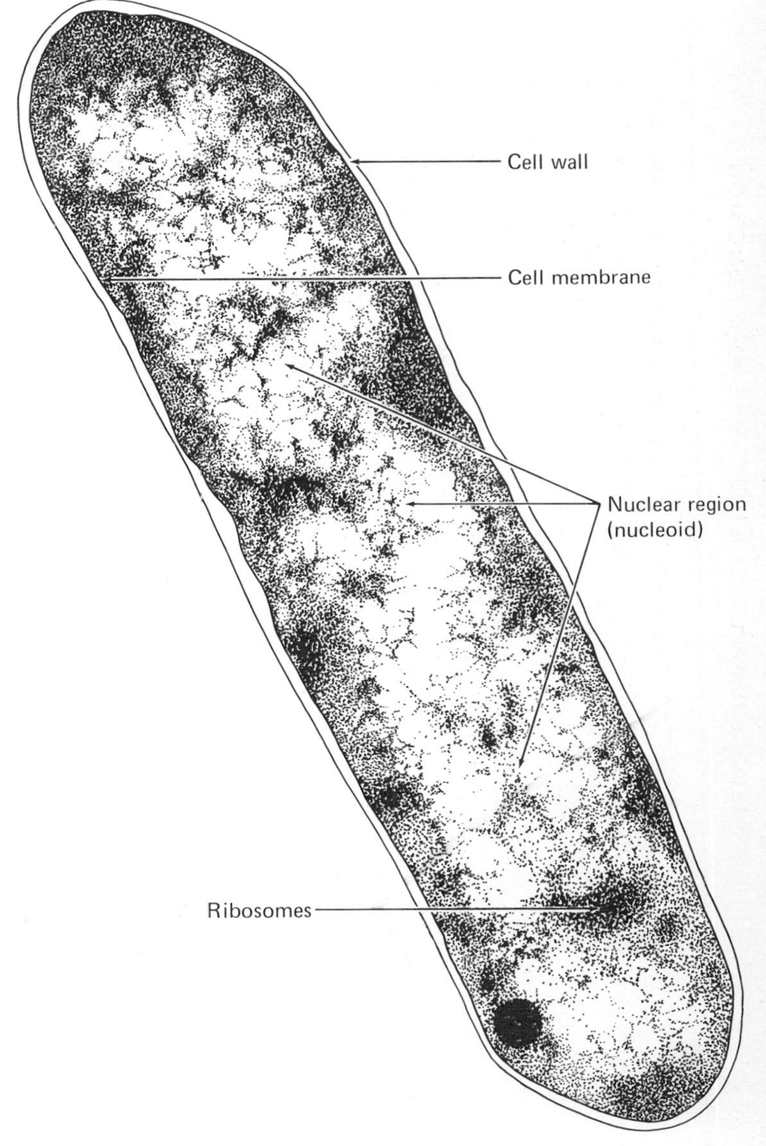

FIGURE 21.2 Structural features of a bacterial cell. (Courtesy of Paulette Curtis Furman)

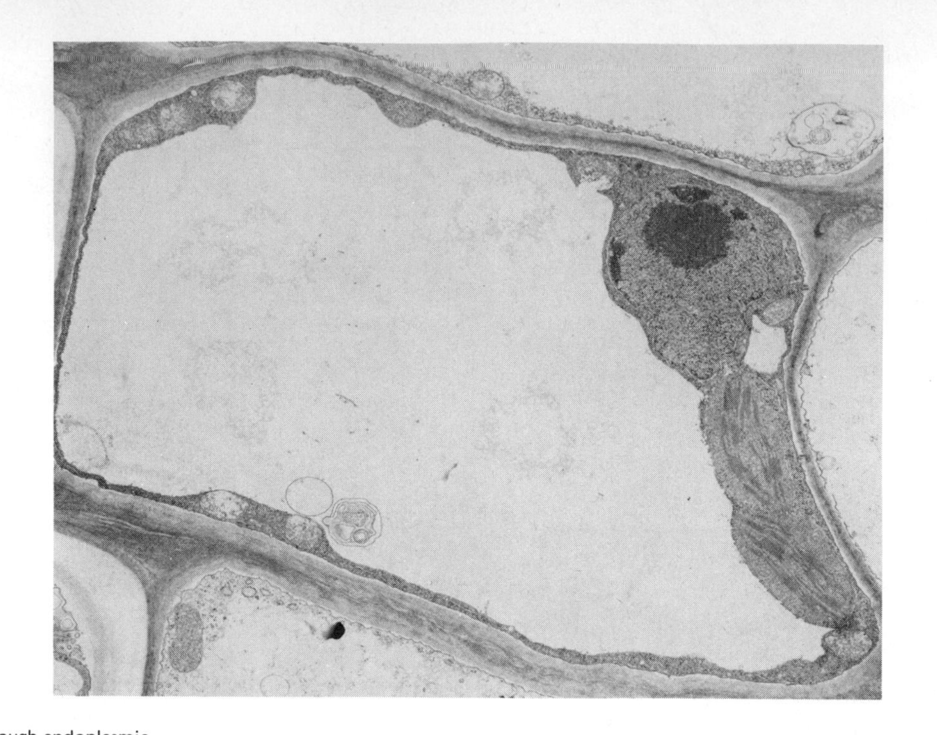

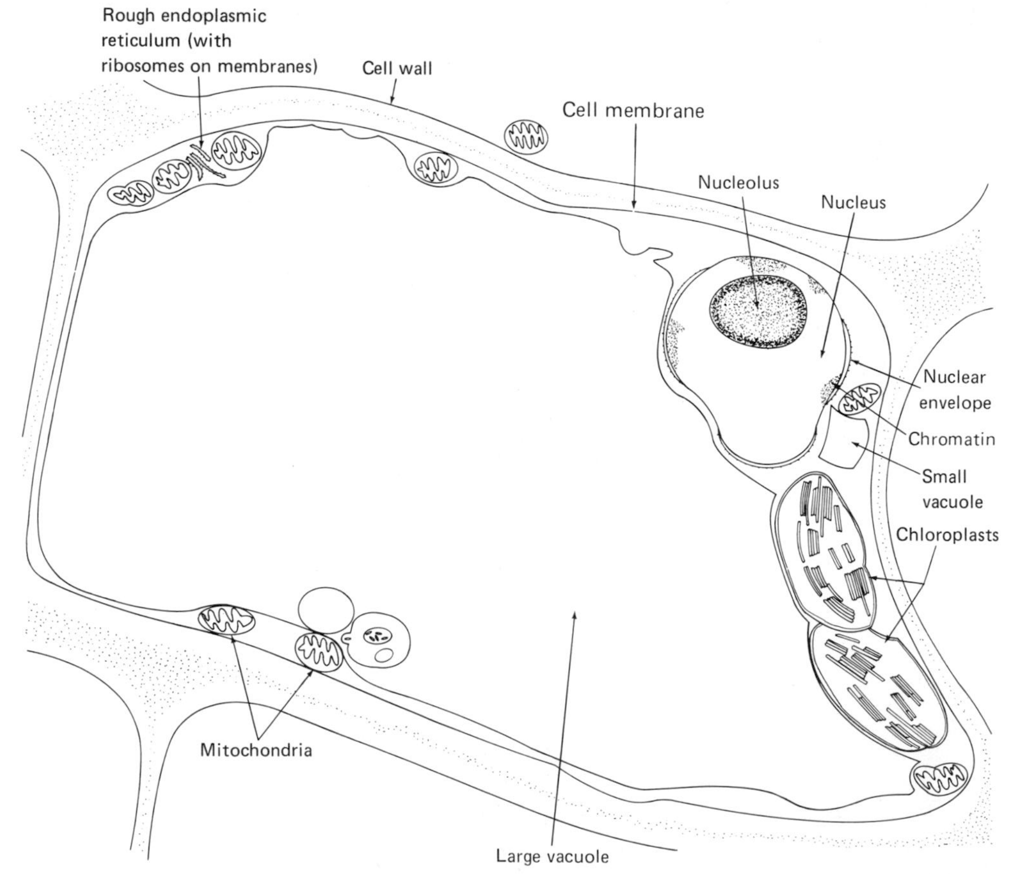

FIGURE 21.3 Major structural features of a plant cell. (Courtesy of Clinton Dawes, University of South Florida)

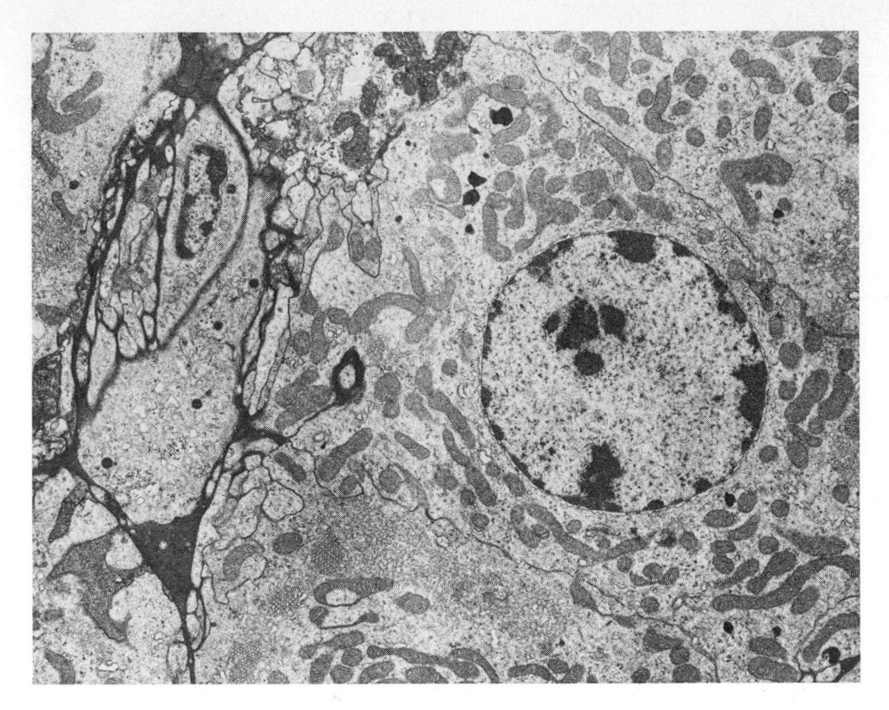

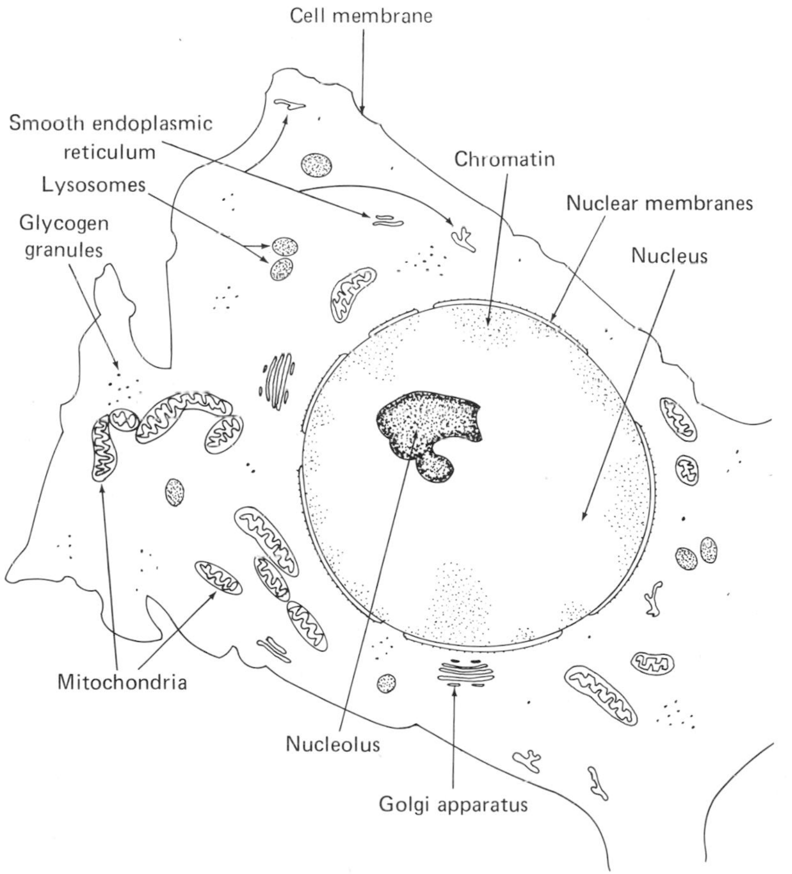

FIGURE 21.4 Major structural features of an animal cell. (Courtesy of Dr. G. W. Hinsch, University of South Florida)

tissues, and it is chiefly composed of fibers of cellulose. The cell has a large internal **vacuole,** which is also surrounded by a membrane. The vacuole contains dissolved salts, sugars, and a variety of other substances. The soluble portion of the cytoplasm (the **cytosol**) is like that of the bacterial cell, but the cytoplasm also contains many membranous organelles. **Organelles** are "little organs," that is, membrane-enclosed cell structures that have specific functions in the cell. The **nucleus** is the organelle that houses the chromosomes of the cell. The chromosomes appear as a dense material called **chromatin,** which contains both DNA and protein. The **chloroplasts** are photosynthetic organelles; they contain the green pigment chlorophyll and are responsible for the cell's energy production in daylight hours. In darkness, the **mitochondria** (singular, *mitochondrion*) take over the task of providing usable energy to the plant cell. The **endoplasmic reticulum** is a network of membranes formed into tubules and channels. In **rough endoplasmic reticulum,** ribosomes are attached to the membranes. Rough endoplasmic reticulum carries out the manufacture and intracellular transport of new proteins. These proteins are secreted from the cell or remain as membrane proteins. Other ribosomes dot the cytoplasm but are not attached to membranes. They synthesize proteins that are to be retained in the cytoplasm.

†21.7 STRUCTURE OF THE ANIMAL CELL

How do animal cells differ from plant cells?

An animal cell (see Figure 21.4 for a photomicrograph and a diagram) has a cell membrane, but it does not have a cell wall or vacuoles, as plant cells usually do. The cytoplasm of an animal cell contains roughly the same organelles as the plant cell does (with the exception of the chloroplasts). Many mitochondria are present, and so is a prominent nucleus. The endoplasmic reticulum, which appears in both rough and smooth forms, is very well developed. As in the plant cell, rough endoplasmic reticulum has numerous ribosomes dotting the surfaces of its membranes. **Smooth endoplasmic reticulum,** on the other hand, has no ribosomes, and it is involved in other processes, such as the synthesis of hormones and fats. Another important organelle is the **Golgi apparatus,** which helps package cell products that are to be secreted or exported from the cell. **Lysosomes** contain enzymes that can break down certain materials that come into the cell or that can digest the cell itself if severe injury has occurred. Granules of glycogen and droplets of fat may also be seen in the cytoplasm; these are food storage deposits that can be oxidized to provide energy.

Problem 21.4 Bacteria are called procaryotes and plant and animal cells are called eucaryotes. Why? (*Hint: Caryo* comes from the Greek word for "nucleus." *pro-* and *eu-* are common prefixes whose meanings you probably know; if you don't, consult a dictionary.)

ENZYMES, THE CELL'S MOST IMPORTANT MOLECULES 21.8

What are the important characteristics of enzymes?

An **enzyme** is a protein that is capable of acting as a catalyst. You will recall from Chapter 6 that a catalyst produces a dramatic increase in the *rate* of a reaction. Thus, a reaction that normally goes at a very slow or undetectable rate may proceed very rapidly in the presence of the proper catalyst. Enzymes and other catalysts do not alter the equilibrium constant (Section 6.8) of a reaction. Rather, they only decrease the time required for the reaction to reach equilibrium. Enzymes typically increase reaction rates a billion times or more. Almost every reaction occurring in a cell or in the body of a many-celled organism is an enzyme-catalyzed reaction.

Enzymes are much better catalysts than inorganic or other catalysts are. For one thing, they are usually very **specific.** This means that an enzyme will catalyze the reaction of only a small number of compounds. A particular enzyme may act on only amino acids, or even on only one or two certain amino acids. Enzymes are also specific for the type of reaction they catalyze. A given enzyme will usually catalyze only one type of reaction. For example, an enzyme might be capable of catalyzing only the hydrolysis of phosphate esters. Another enzyme might catalyze only the oxidation of alcohols to aldehydes.

Every enzyme can be classified according to the type of reaction it catalyzes and the type of compound it acts on (see Table 21.1). We will see examples of many of these enzymes and reactions in our study of biological chemistry. Looking at Table 21.1, you will note the characteristic ending *-ase* on the enzyme names. Virtually every time you see a word with the *-ase* suffix, that word will be the name of an enzyme.

One very important characteristic of enzymes is that their function as catalysts can be regulated. Enzyme proteins are marvelously effective catalysts. At the same time, however, they have the built-in ability to react or respond to changes in the cell's needs. In other words, the rate of an enzyme-catalyzed reaction—how fast the reactants are converted to products—can change to suit the cell's needs for the product of the reaction.

In the remainder of the book we will be talking a lot about the breakdown and synthesis of biomolecules in our continuing discussions of

TABLE 21.1 Enzyme Classification System

Class names	Types of reactions catalyzed
1 Oxido-reductases (oxidation-reduction reactions)	1.1 Acting on $\rangle$CH—OH
	1.2 Acting on $\rangle$C=O
	1.3 Acting on $\rangle$C=CH—
	1.4 Acting on $\rangle$CH—NH_2
	1.5 Acting on $\rangle$CH—NH—
	1.6 Acting on NADH; NADPH
2 Transferases (transfer of functional groups)	2.1 One carbon groups
	2.2 Aldehydic or ketonic groups
	2.3 Acyl groups
	2.4 Glycosyl groups
	2.7 Phosphate groups
	2.8 S-containing groups
3 Hydrolases (hydrolysis reactions)	3.1 Esters
	3.2 Glycosidic bonds
	3.4 Peptide bonds
	3.5 Other C—N bonds
	3.6 Acid anhydrides
4 Lyases (addition to double bonds)	4.1 $\rangle$C=C$\langle$
	4.2 $\rangle$C=O
	4.3 $\rangle$C=N—
5 Isomerases (isomerization reactions)	5.1 Racemases
6 Ligases (formation of bonds with ATP cleavage)	6.1 C—O
	6.2 C—S
	6.3 C—N
	6.4 C—C

carbohydrates, fats, proteins, and nucleic acids. Each breaking-down and building-up process is the result of a number of enzymes acting together in a highly coordinated and well-organized sequence. These sequences of enzyme-catalyzed reactions are called **metabolic pathways.**

Enzymes, then, are central to what occurs in the metabolic pathways, and metabolic pathways in turn are central to all life processes. Since enzymes have such a fundamental role, it is important to understand where and how they perform their functions and what factors influence their activity. The following chart outlines the ground we'll cover in the final sections of this chapter as we learn more about the characteristics of enzymes.

Enzyme function	Factors influencing enzyme activity
(21.9) The enzyme's active site	(21.11) Cofactors and coenzymes—enzyme partners in catalysis
(21.10) How does an enzyme catalyze a reaction?	(21.12) Coenzymes and the water soluble vitamins
(21.14) Enzymes in medical diagnosis	(21.13) The study of enzyme reaction rates
	(21.15) Regulation of enzyme activity
	(21.16) Sequences of enzyme-catalyzed reactions and their regulation

THE ENZYME'S ACTIVE SITE 21.9

How are substrate molecules bound on the enzyme surface?

The enzyme's **active site** is the small area of the surface of the folded protein molecule where the binding of reactants (which we call *substrates*) and the catalysis of the reaction actually take place (Figure 21.5). Let us take a hypothetical reaction:

$$A + B \overset{E}{\rightleftharpoons} P + Q$$

The enzyme first binds one reactant and then the other. For example:

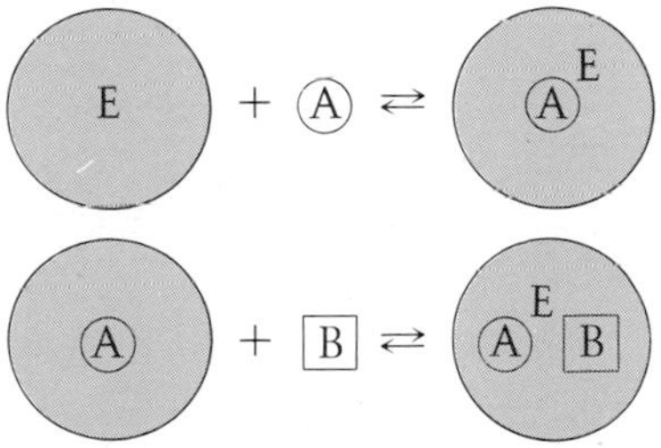

E + (A) ⇌ E(A)

E(A) + [B] ⇌ E(A)[B]

(The products P and Q are then formed and released, but we'll consider this later.) The binding process requires complementarity (Section 21.1) between the active site and the substrate molecules. Complementarity between an enzyme protein molecule and its substrate has been viewed in two different ways: (1) as a *lock and key* (a rigid complementarity) and (2) as an *induced fit* (an induced complementarity), where the

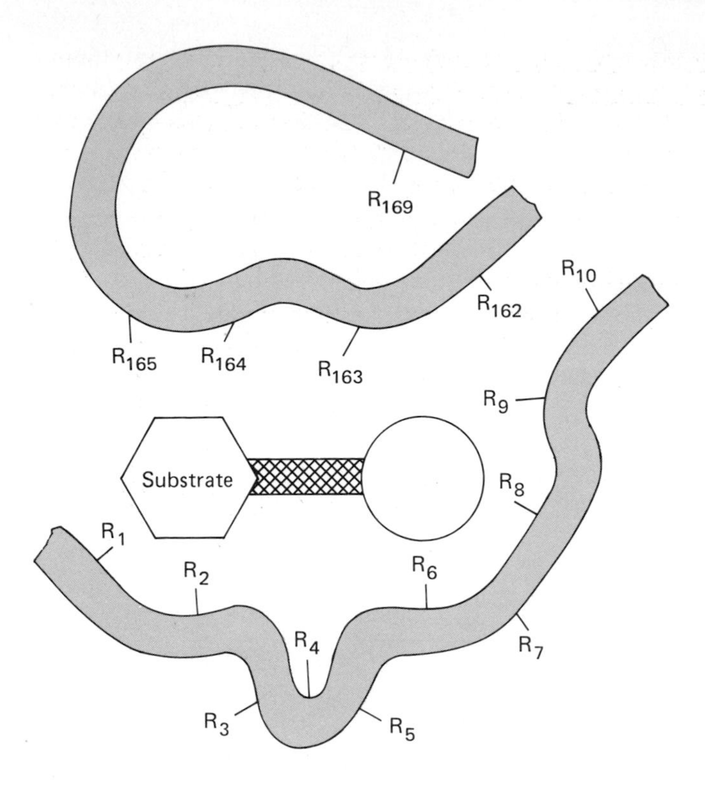

FIGURE 21.5 Schematic active site. The cross hatched area indicates a bond to be broken in the reaction. The R's represent some side chains in the active site and the colored line represents the backbone of two segments of the protein chain. The remainder of the protein is not shown. (From D. E. Koshland, Jr., *Advanced Enzymology 22,* 45 (1960).)

shape of the active site changes to match the substrate better as binding takes place.

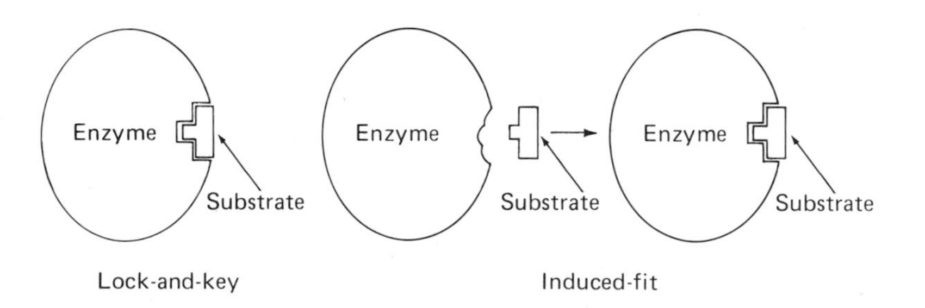

Since it is probably accurate to think of a globular protein's shape as being somewhat flexible rather than rigid, the induced-fit concept appears to be the more realistic one.

The binding of reactants by the enzyme is very specific. That is, only a limited number of different molecules can bind to the enzyme. Certain enzymes have what is called **absolute specificity;** this means that no molecule other than the normal reactant can be bound by the enzyme. An example is urease, a plant enzyme that catalyzes the hydrolysis of urea. It will bind and act on only urea. Other enzymes are less specific—they may bind several closely related molecules. For example, hexokinase, an enzyme that catalyzes a reaction of D-glucose, can also bind D-mannose, 2-deoxyglucose, and a couple of other similar sugars.

The substances bound and acted on by an enzyme are called the **substrates** of the enzyme. Thus, D-glucose and D-mannose are alternate substrates of hexokinase. The size and shape of the active site determine what molecules can be bound there. In addition, specific areas of nonpolar residues in the active site may attract hydrophobic substrate molecules, or a positively charged lysine residue may be in just the right spot to bind a negatively charged carboxyl group of a substrate (Figure 21.6). Size and shape, polarity, and charge are all features that contribute to complementarity between active site and substrate structures.

Problem 21.5 Name and draw the structural formulas of four amino acids whose R groups might be found in a hydrophobic active site.
Problem 21.6 Name and draw the structural formulas of basic amino acids whose R groups might bind a substrate carboxyl group.

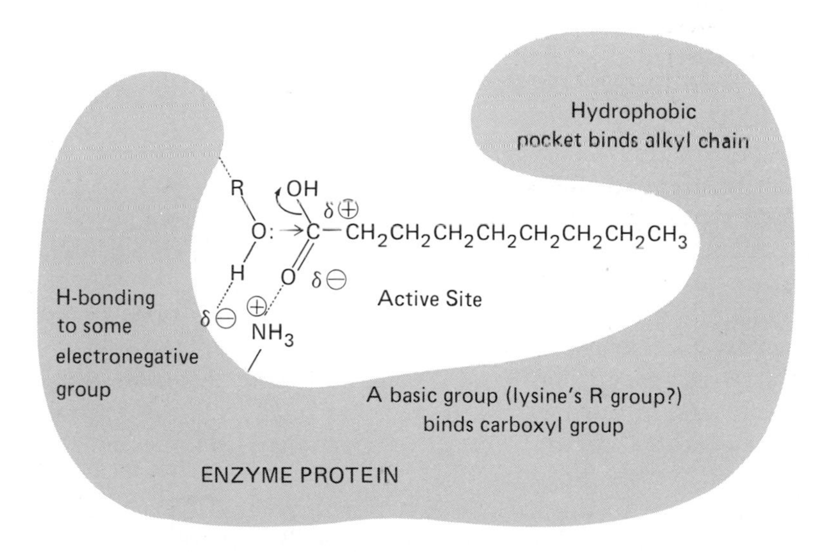

FIGURE 21.6 Orientation effect in catalysis. Enzyme binds substrates to place them in correct orientation for reaction. Reaction is ester formation involving a fatty acid and an alcohol.

21.10 HOW DOES AN ENZYME CATALYZE A REACTION?

Several factors account for an enzyme's ability to increase the rate of a reaction after binding of substrate(s) has occurred. The binding of two substrates in close proximity on the surface of an enzyme makes it more likely that they will react with one another. This is one way that a catalyst increases the rate of reaction. An additional factor in the rate increase is the enzyme's ability to bind the substrates so that they are correctly oriented to one another. A reaction will not occur unless the reactant molecules meet one another at a proper angle. Imagine, for example, the reaction between a fatty acid and an alcohol to form an ester. If the molecules collide in such a manner that the carboxyl and hydroxyl groups do not come close enough to one another, reaction is very unlikely to occur.

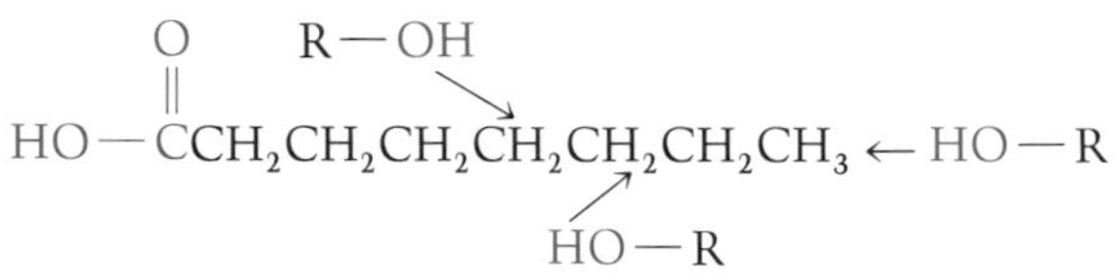

$$R-OH \searrow$$

$$HO-\overset{\overset{\large O}{\|}}{C}CH_2CH_2CH_2CH_2CH_2CH_2CH_3 \leftarrow HO-R$$

$$\nearrow HO-R$$

no reaction occurs

In free solution, the molecules will rarely hit just right. However, at the active site of the appropriate enzyme, binding takes place so that the two groups are brought together in the best orientation for reaction (see Figure 21.6).

Another factor in the enzyme's ability to increase the rate of reaction is that in binding the substrate, the enzyme may twist or pull the substrate into a conformation that facilitates the breaking of old covalent bonds and the making of new ones. This has sometimes been called the **rack mechanism** of enzyme catalysis (Figure 21.7).

Besides binding or bending substrate molecules, amino acid R groups in the active sites of enzymes often play active roles as donors or acceptors of hydrogen ions for acid or base catalysis. In organic chemistry, some examples of acid or base catalysis are the acid catalysis of the reaction between an alcohol and an acid to form an ester, and the acid- or base-catalyzed hydrolysis of esters.

Problem 21.7 Suggest some amino acids that might be capable of acting as hydrogen-ion donors or acceptors (that is, acids or bases) in enzyme catalysis. Remember that only the R groups are free for such purposes, since the carboxyl and alpha-amino groups are tied up in peptide linkages.

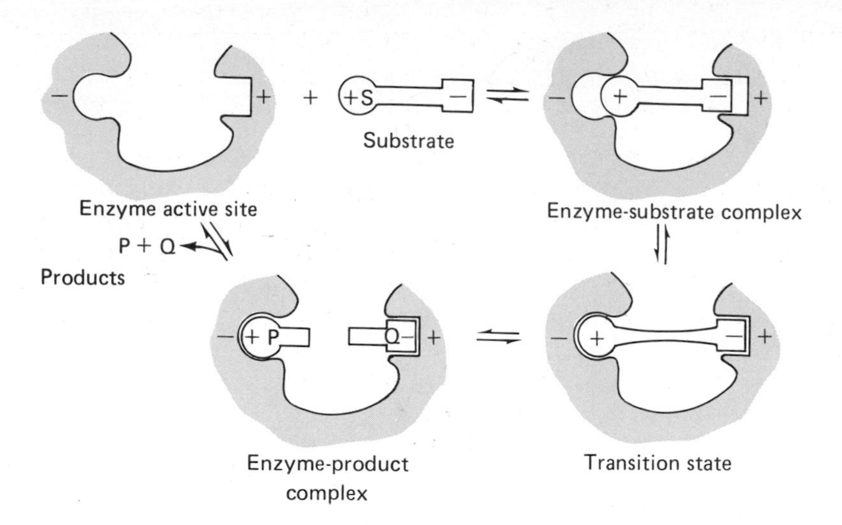

FIGURE 21.7 The rack mechanism, in which the enzyme distorts the substrate molecule to form the transition state from which product formation occurs. (From Segel, *Biochemical Calculations,* 2nd ed., 1976, p. 213, John Wiley and Sons, Inc.)

21.11 COFACTORS AND COENZYMES, PARTNERS OF ENZYMES IN CATALYSIS

Some enzymes contain metal ions (such as Fe^{2+}, Zn^{2+}, Mg^{2+}, Mn^{2+}, and K^{+}) or other non–amino acid constituents within their active sites. These constituents, which help to oxidize or reduce substrate molecules or assist in various other types of reactions, are called **cofactors** or **coenzymes.** Cofactors are usually metallic, and coenzymes are usually organic.

Many proteins require the presence of one or more cofactors or coenzymes in order to function. In many cases the coenzyme (or cofactor) is quite firmly bound to the protein. In others the binding is only temporary. That is, it occurs only during the reaction the enzyme catalyzes; after the reaction, the coenzyme dissociates from the enzyme protein. *Coenzyme* literally means "with enzyme," that is, something that works with an enzyme. Coenzymes themselves are *not* enzymes.

Complex organic molecules can act as cofactors. For example, the protein hemoglobin contains the compound heme as a cofactor. Heme is a complex organic molecule containing bound Fe^{2+}. Heme illustrates the difficulty of distinguishing coenzymes and cofactors just on the basis of whether they are organic. Although heme is organic, it cannot be considered a coenzyme of hemoglobin, because hemoglobin is not an enzyme.

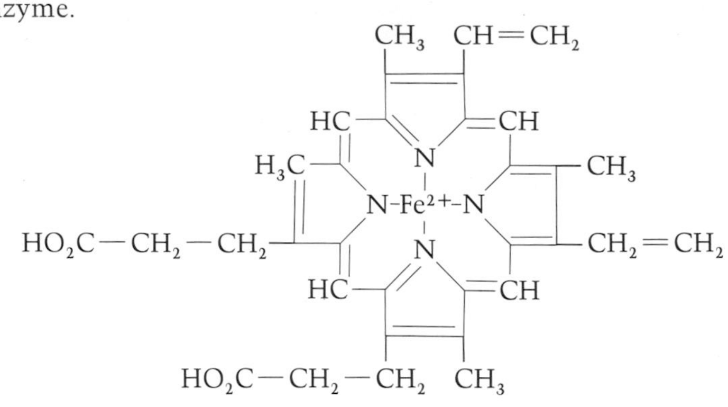

CH_3 $CH{=}CH_2$

HC— =CH

H_3C— N —CH_3

N-Fe^{2+}-N

$HO_2C{-}CH_2{-}CH_2$— N —$CH_2{=}CH_2$

HC— =CH

$HO_2C{-}CH_2{-}CH_2$ CH_3

heme

21.12 COENZYMES AND THE WATER SOLUBLE VITAMINS

Different enzymes involved in the catalysis of the same type of reaction frequently utilize the same coenzyme. For example, many enzymes involved in oxidation-reduction reactions in the breakdown of foodstuffs utilize the coenzyme *nicotinamide adenine dinucleotide* (NAD). Many enzymes that catalyze reactions of amino acids contain the coenzyme *pyridoxal phosphate.* And a number of enzymes that catalyze reactions of α-keto acids contain *thiamine pyrophosphate.* In later chapters we will consider some of these reactions and the specific roles of some of the coenzymes in more detail.

For now, let us focus on another important point. Each of the coenzymes NAD, pyridoxal phosphate, and thiamine pyrophosphate contains a different aromatic ring structure (shown in color in the structures in Figure 21.8) that humans and other animals cannot manufacture. This means that compounds containing those ring structures must be obtained in the diet. These compounds are **water soluble vitamins** (Table 21.2). (We considered the fat soluble vitamins in Section 20.7.) Vitamins are organic substances that are normally required in rather small quantities, but when adequate amounts are not consumed, a vitamin-deficiency disease develops. For example, thiamine deficiency leads to beriberi, and nicotinamide (also called niacinamide) deficiency leads to pellagra. Beriberi is seen in two forms, dry and wet, in humans; both are widespread in areas where polished rice is the main staple of the diet. Most of the thiamine in grains and seeds is found in the outer layers

nicotinamide

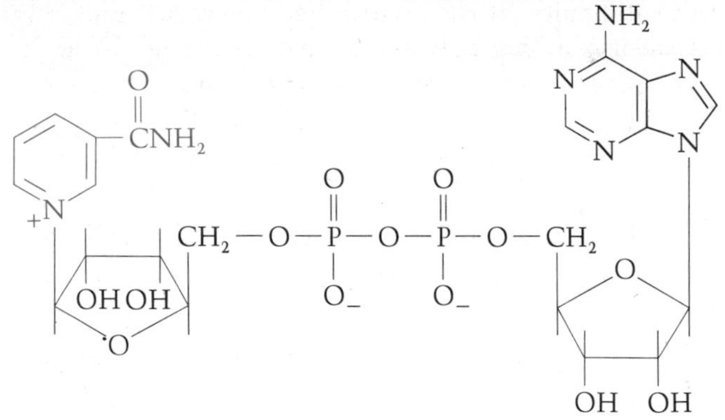

NAD

pyridoxal

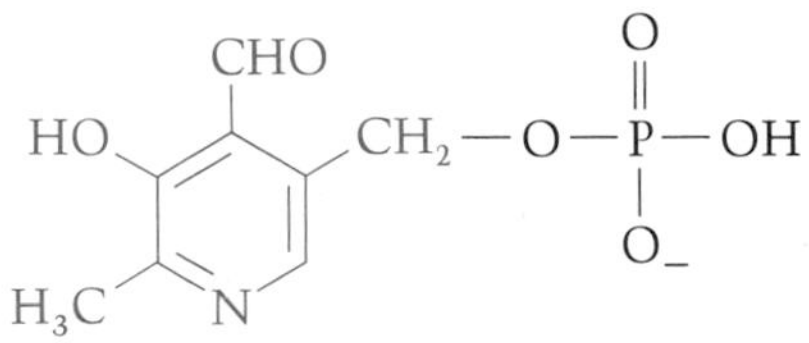

pyridoxal phosphate

thiamine

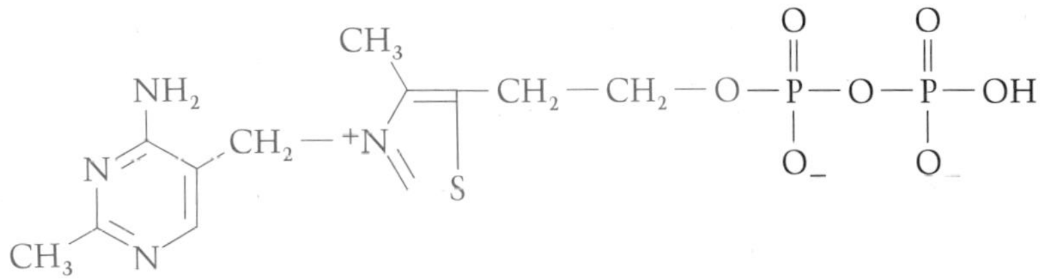

thiamine pyrophosphate

FIGURE 21.8 Some coenzymes and the water soluble vitamins (shown in color) in their structures.

of the rice, which are removed in the polishing or milling processes. The dry form of beriberi involves rapid weight loss, peripheral neuritis, muscle weakness, and wasting, along with mental confusion. In wet beriberi, edema (accumulation of excess fluid in the tissues) may mask the muscle wasting. Both forms respond rapidly to the administration of thiamine in the diet. Thiamine deficiency is not often seen in the Western world, except in chronic alcoholics who obtain most of their calories from alcohol.

Pellegra is somewhat less well defined than beriberi; it appears to involve a deficiency not only of niacin but also of other B vitamins and protein. The major symptoms are dermatitis of skin exposed to

TABLE 21.2 Water Soluble Vitamins and Their Coenzyme Forms

Vitamin	Coenzyme or active form	Function promoted
Thiamine (B_1)	Thiamine pyrophosphate (TPP)	Transfer of aldehyde groups
Riboflavin (B_2)	Flavin mononucleotide (FMN)	Oxidation-reduction reactions
	Flavin adenine dinucleotide (FAD)	Oxidation-reduction reactions
Nicotinic acid (Niacin)	Nicotinamide adenine dinucleotide (NAD)	Oxidation-reduction reactions
	Nicotinamide adenine dinucleotide phosphate (NADP)	Oxidation-reduction reactions
Pantothenic acid	Coenzyme A (CoA)	Transfer of acyl groups
Pyridoxine (B_6)	Pyridoxal phosphate	Transfer of amino groups
Biotin	Biotin	Carboxylation reactions
Folic acid	Tetrahydrofolic acid	Transfer of one-carbon groups
Vitamin B_{12}	Coenzyme B_{12}	Intramolecular rearrangements
Lipoic acid	Lipoyllysine	Oxidation-reduction and acyl group transfer
Ascorbic acid	—	Cofactor in hydroxylation reactions

sunlight, inflammation of the stomach lining, and impaired digestion with diarrhea.

Nicotinamide, pyridoxine, and thiamine are members of the B complex of water soluble vitamins. Table 21.2 lists the names of all the water soluble vitamins known today, along with brief descriptions of their functions in living organisms. We will encounter most of these vitamins in their coenzyme forms again in later chapters, and in Chapter 23 we'll consider some aspects of vitamins and nutrition.

†21.13 THE STUDY OF ENZYME REACTION RATES

What is saturation kinetics?

Enzyme function, or *activity*, depends on the concentrations of substrates, activators, and inhibitors. **Kinetics** is the study of reaction rates, and especially the way in which reaction rates depend on the concentrations of the reactants. For an ordinary chemical reaction involving one reactant, in which no catalyst is present, the rate at which the product is formed depends directly on the concentration of reactant. From Figure 21.9 you can see that at point B, where the reactant concentration is larger than it is at point A, the rate of product formation is greater.

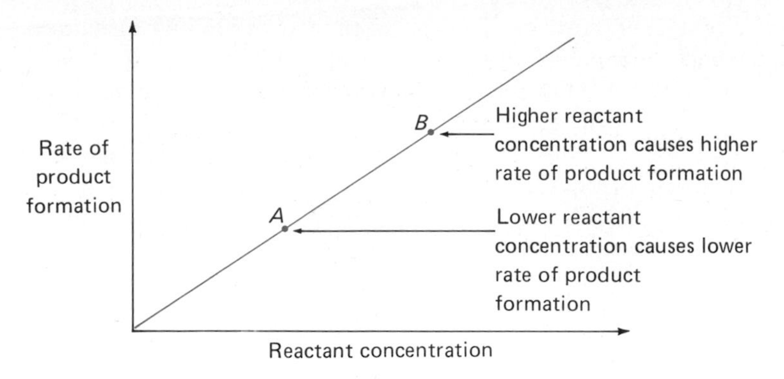

FIGURE 21.9 Dependence of the rate of formation of product on the reactant concentration in an ordinary chemical reaction.

Enzyme-catalyzed reactions behave differently. (The rate we are concerned with is the *initial* rate—the rate when the reaction begins.) Figure 21.10 shows that the rate of product formation increases steadily with increased substrate concentration up to a point, but the rate approaches a maximum value. The rate does not exceed this value no matter how large the substrate concentration might be. We can explain this behavior if we look at the way an enzyme works. An enzyme must bind its substrate molecule before products can form.

$$\mathrm{E} + \mathrm{S} \rightleftarrows \mathrm{ES} \qquad \text{substrate binding}$$
$$\mathrm{ES} \rightleftarrows \mathrm{P} + \mathrm{E} \qquad \text{product formation}$$

The rate of product formation really depends on the concentration of ES, which is known as the enzyme-substrate complex. The concentration of ES cannot be larger than the concentration of E, no matter how much

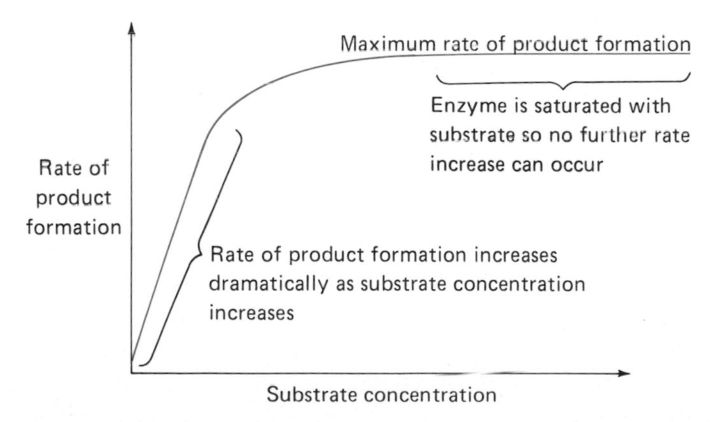

FIGURE 21.10 Dependence of the rate of formation of product on the substrate concentration in an enzyme-catalyzed reaction.

substrate might be present. In other words, the rate of product formation finally depends on the concentration of enzyme. Thus, whatever substrate concentration is enough to "saturate" the enzyme—to convert all E molecules to ES—will produce the maximum rate of the reaction. Enzyme-catalyzed reactions are thus said to exhibit **saturation kinetics.** The rate of the reaction attains its maximum value when the enzyme is saturated with substrate, and further increase in the concentration of substrate will not increase the rate of product formation.

Example 21.1 Consider the following two graphs:

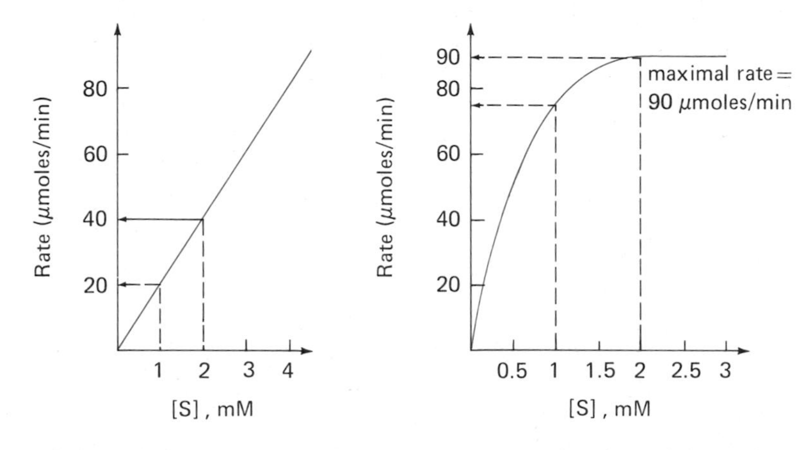

(a) If the substrate concentration [S] is 1 mM, the rate of the ordinary chemical reaction is, according to the graph on the left, 20 μmoles of product per minute. What is the approximate rate of the enzyme-catalyzed reaction at the same [S]?
(b) If [S] is doubled in both cases, from 1 mM to 2 mM, what happens to the rates of the two reactions?
Solution (a) Find the point at which the [S] = 1 mM line crosses the curve in the graph on the right. The rate is about 75 μmoles/min at [S] = 1 mM.
(b) The rate of the ordinary uncatalyzed reaction doubles to 40 μmoles/min when [S] = 2 mM. The rate of the enzyme-catalyzed reaction cannot double because it was already close to the maximum rate—it can only increase to equal the maximum rate (90 μmoles/min) when [S] = 2 mM.

21.14 ENZYMES IN MEDICAL DIAGNOSIS

The diagnosis and treatment of diseases of the internal organs has come to depend very heavily on measurements (known as assays) of enzyme levels in patients' blood samples. When a tissue is diseased or injured, cells of that tissue die and release their contents into the blood stream. If an enzyme known to be a normal component of a tissue other than blood is found in unusually high concentration in the blood, it indicates tissue damage. Table 21.3 lists a number of enzymes that aid in modern medical diagnosis.

What do enzymes have to do with medical practice?

TABLE 21.3 Enzyme Assays Useful in Diagnosis or Treatment of Disease

Enzyme	Organ or disease of interest
Commonly assayed	
Acid phosphatase	Prostatic carcinoma
Alkaline phosphatase	Liver, bone disease
Amylase	Pancreatic disease
Glutamate transaminase	Liver, heart disease
Aspartate transaminase	Liver, heart disease
Alanine transaminase	Liver, heart disease
Lactate dehydrogenase	Liver, heart, red blood cells
Creatine phosphokinase*	Heart, muscle, brain
Less commonly assayed	
Isocitrate dehydrogenase	Liver
Ceruloplasmin	Wilson's disease (liver)
Aldolase	Muscle, heart
Trypsin	Pancreas, intestine
Glucose 6-phosphate dehydrogenase	Red blood cells (genetic defect)
γ-Glutamyl transpeptidase	Liver disease
Guanase	Liver disease
Sorbitol dehydrogenase	Liver disease
Ornithine transcarbamylase	Liver disease
Pseudocholinesterase	Liver (poisonings, insecticides)
Leucine aminopeptidase	Liver, pancreas
5′-Nucleotidase	Liver disease
Pepsin	Stomach
Hexose isomerase	Liver disease
Hexose 1-phosphate-uridyl transferase	Galactosemia (genetic defect)
Malate dehydrogenase	Liver disease
Glutathione reductase	Anemia, cyanosis
Arginase	Liver disease
Lipoprotein lipase	Hyperlipoproteinemia
Elastase	Collagen diseases
Plasmin	Blood clotting disease

Source: Montgomery, Dryer, Conway, and Spector, *Biochemistry: A Case-Oriented Approach,* 2nd ed. (St. Louis: The C. V. Mosby Co., 1977), p. 139.

*Also commonly known as creatine kinase.

In some cases, several tissues are possible sources of an enzyme that is present in high serum concentrations. The choice can be narrowed if the doctor or technician looks for tissue-specific forms of enzymes. Different forms of an enzyme that catalyze the same reaction are known as **isoenzymes.** Some tissues contain unique isoenzyme forms. The different forms have different physical properties, which allow them to be distinguished from one another. Electrophoresis (Figure 17.8) is often used to make this distinction. Serum samples are treated to allow the separation and identification of the isoenzymes, and the tissue source of the enzyme can then be pinpointed.

Many inherited diseases, such as phenylketonuria (PKU), result simply from the body's inability to make a particular enzyme. We will encounter a number of examples in later chapters. Assays for the missing enzyme in blood or small tissue samples are routinely being used in the diagnosis of some of these diseases, as well as for the detection of carriers who may transmit the disease to their children.

21.15 REGULATION OF ENZYME ACTIVITY

For an organism to make the most economical use of energy and raw materials, enzyme activity must be regulated. Several ways of controlling enzyme activity are possible.

How are these highly efficient cell catalysts controlled?

Some enzymes are **allosteric;** this means that a small molecule (an **allosteric effector**) that does not resemble a normal substrate or product can bind to a special site on the enzyme to change its activity. The special site for binding such a molecule is called an **allosteric site.** The binding of an allosteric effector at the allosteric site causes a change in the tertiary or quarternary structure of the protein, leading to a change in the shape of the active site (Figure 21.11). This change in shape alters the properties of the active site. As a result, enzyme activity may be increased or decreased. In other words, allosteric effectors may inhibit or activate enzymes. Allosteric effectors often change the enzyme's ability to bind the substrate.

An **allosteric inhibitor** may decrease the binding of substrate so that much more substrate must be present to saturate the enzyme and produce the maximum reaction rate. This is shown by the lowermost curve in Figure 21.12. On the other hand, an **allosteric activator** will do just the opposite: increase the binding of the substrate. The topmost curve in Figure 21.12 shows that when the activator is present, the maximum rate is obtained with a much lower substrate concentration.

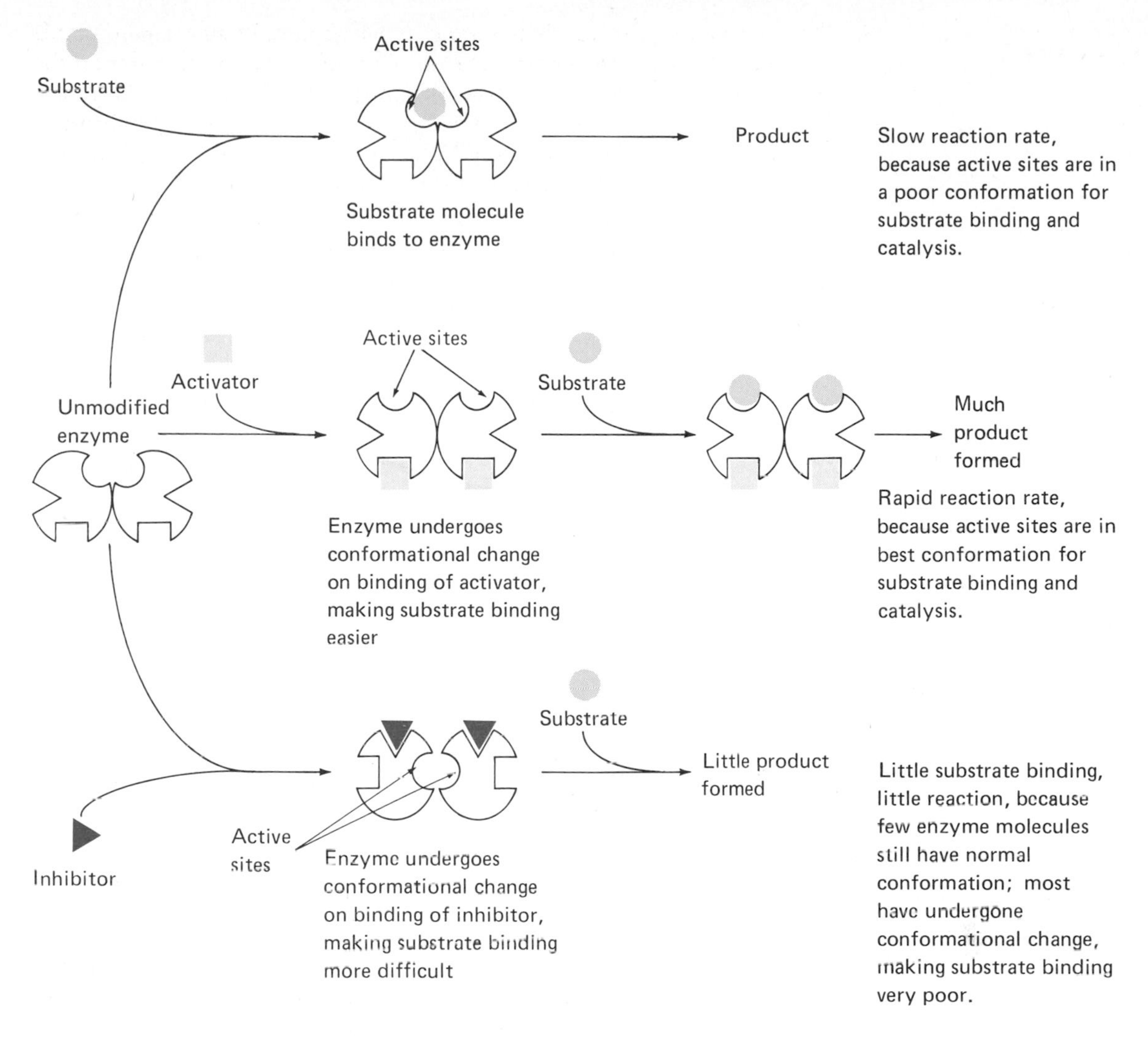

FIGURE 21.11 Interactions of a hypothetical allosteric enzyme with substrate, activator, and inhibitor molecules.

Problem 21.8 Refer to Figure 21.12 and estimate the rate of the reaction (as a fraction or percent of the maximum rate) if the actual substrate concentration is equal to 1 mM and
(a) the inhibitor is present, (b) the activator is present.

Problem 21.9 Assume that a particular allosteric enzyme is present in a liver cell. This enzyme controls the rate of synthesis of an essential cell material, say, an amino acid. In the absence of allosteric inhibitors or activators, the substrate concentration that gives half the maximum rate

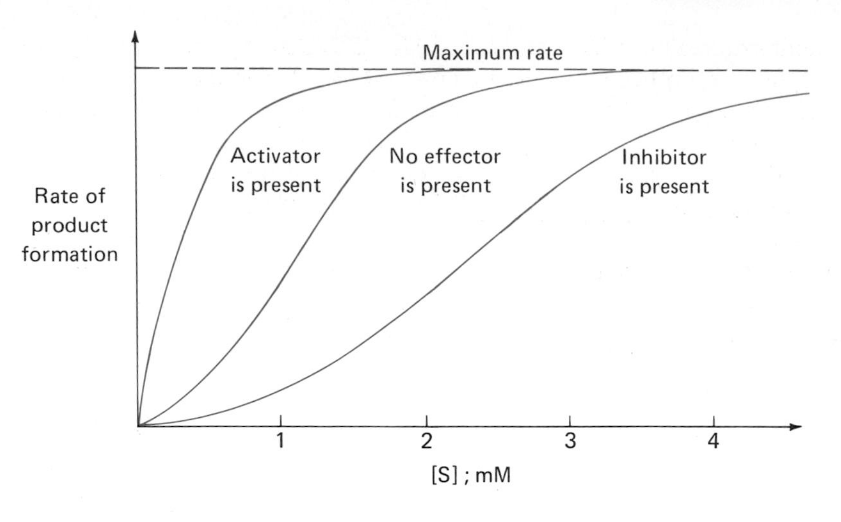

FIGURE 21.12 Kinetics of an allosteric enzyme.

of product formation (assumed to be an adequate rate under normal conditions) is equal to 1 mM. In the cell, however, the substrate concentration is only 0.1 mM. Describe how the cell might regulate enzyme activity to obtain an adequate rate of product formation.

21.16 SEQUENCES OF ENZYME-CATALYZED REACTIONS AND THEIR REGULATION

How are metabolic pathways regulated?

Few enzymes function in isolation. On the contrary, a number of enzymes usually act in sequence to carry out the breakdown or synthesis of a biomolecule. We mentioned earlier that such sequences of enzyme-catalyzed reactions are called metabolic pathways. We might present such a pathway like this:

$$A \xrightarrow{E_1} B \xrightarrow{E_2} C \xrightarrow{E_3} D \xrightarrow{E_4} P$$

where A, B, C, D, and P are reactants and products ("intermediates") of the pathway, and E_1, E_2, E_3, and E_4 are enzymes catalyzing the conversion of one intermediate into another.

Since living systems must be able to control their own functions, every metabolic pathway must be subject to regulation. That is, the rate at which A is broken down or P is formed must be controlled to suit the needs of the cell or organism. Perhaps P is an amino acid required for protein synthesis or a fatty acid needed for the manufacture of membrane phospholipid. When active protein or membrane formation is occurring, rapid formation of the amino acid or fatty acid is needed. When the processes are slow, rapid formation of P is not wanted and the activity of the pathway must be limited to save energy and raw materials.

In general, one enzyme controls the overall rate of operation of a pathway. The reaction catalyzed by this key enzyme is called the **rate-limiting step.** The rate-limiting step in a metabolic pathway determines the overall rate of flow of intermediates through the pathway, and therefore determines the rate of formation of the end product.

$$A \xrightarrow{E_1} B \rightarrow C \rightarrow D \rightarrow P$$

A common pattern of regulation is **feedback inhibition,** in which the end product of the pathway will inhibit an allosteric enzyme that catalyzes the rate-limiting step. When the concentration of P rises high enough, P begins to bind to E_1. Binding of the inhibitor P to the enzyme reduces the rate of formation of B, C, D, and P. Thus, the concentration of P in the cell can control the rate of formation of P. Often, as shown, the first enzyme in the pathway is the enzyme that is inhibited. This is the most economical way, since no extra B or C, and so on, will be made.

SUMMARY

Cells are made up of molecules that must stick together in specific ways to form cell structures or perform cell functions. Noncovalent interactions hold the molecules together. **Complementarity**—"fit" of surface features—between molecules is necessary for stable interactions. The structure of biological membranes is best visualized by the **fluid-mosaic model,** in which a phospholipid bilayer provides a matrix into which protein molecules are embedded, either deeply or superficially. A bacterial cell is rather simple in structure. It is enclosed by a **cell membrane,** which may be surrounded by a **cell wall;** it contains a single circular **chromosome,** numerous **ribosomes,** and many soluble **enzymes.** Plant and animal cells are much larger and contain numerous membrane-enclosed **organelles** within their **cytoplasm;** they have several noncircular chromosomes. Animal cells lack cell walls.

Enzymes are proteins that act as catalysts. They may be classified according to the types of reactions they catalyze. The substrate is bound at the enzyme **active site;** this binding requires complementarity and is very **specific.** Catalysis by enzymes is attributable to several factors: proximity and orientation effects, as well as bending or stretching of the substrate molecules (the rack mechanism). Acid or base catalysis by amino acid side chains is often involved as well. Many proteins require a nonprotein component—a **cofactor** or **coenzyme**—for activity. Enzyme-catalyzed reactions exhibit **saturation kinetics.** Enzyme measurements are used a great deal in modern medical diagnoses.

Enzyme activity is often regulated by **allosteric** mechanisms. Allosteric regulatory enzymes, which are generally proteins with quaternary structures, are affected by allosteric **inhibitors** or **activators.** Enzymes act in sequence in **metabolic pathways** to carry out the synthesis or degradation of biomolecules. Rates of such metabolic pathways are controlled by a regulatory enzyme that catalyzes the **rate-limiting step.** In **feedback inhibition,** a common pattern of regulation, an end product of the pathway serves as an allosteric inhibitor of the regulatory enzyme.

Key Terms

Absolute specificity (Section 21.9)
Active site (Section 21.9)
Allosteric (Section 21.15)
Allosteric activator (Section 21.15)
Allosteric effector (Section 21.15)
Allosteric inhibitor (Section 21.15)
Allosteric site (Section 21.15)
Binding (Section 21.1)
Cell membrane (Section 21.2)
Cell wall (Section 21.5)
Chloroplast (Section 21.6)
Chromatin (Section 21.6)
Chromosome (Section 21.5)
Coenzyme (Section 21.11)
Cofactor (Section 21.11)
Complementarity (Section 21.1)
Cytoplasm (Section 21.5)
Cytosol (Section 21.6)
Enzyme (Section 21.8)
Feedback inhibition (Section 21.16)
Fluid-mosaic model (Section 21.4)
Golgi apparatus (Section 21.7)
Isoenzyme (Section 21.14)
Kinetics (Section 21.13)
Lysosome (Section 21.7)
Metabolic pathway (Section 21.8)
Mitochondria (Section 21.6)
Nucleoid (Section 21.5)

Nucleus (Section 21.6)
Organelle (Section 21.6)
Rate-limiting step (Section 21.16)
Ribosome (Section 21.5)
Rough endoplasmic reticulum (Section 21.6)
Saturation kinetics (Section 21.13)
Smooth endoplasmic reticulum (Section 21.7)
Specific (Section 21.8)
Substrate (Section 21.9)
Vacuole (Section 21.6)
Water soluble vitamins (Section 21.12)

ADDITIONAL PROBLEMS

21.10 Describe the basic features of the fluid-mosaic model of membrane structure.
21.11 Why do nonpolar molecules penetrate biological membranes so much better than ionic and polar molecules?
21.12 Summarize in a sentence or two how bacterial cells differ from plant and animal cells.
21.13 Distinguish between rough and smooth endoplasmic reticulum.
21.14 What are the general types of reactions catalyzed by enzymes with the following code numbers? (See Table 21.1).

(a) 4.1 (b) 1.1
(c) 3.4 (d) 2.7
(e) 3.2

21.15 What is meant by *enzyme specificity*?
21.16 Name the coenzyme form(s) of:

(a) Vitamin B_1 (b) vitamin B_6
(c) niacin (d) riboflavin

21.17 Name several metal ions that often serve as enzyme cofactors.
21.18 Heme contains what metal ion?
21.19 Describe the basic difference between ordinary chemical kinetics and enzyme kinetics.

Problem-Solving Hints

If you have trouble solving any of the additional problems, refer to the sections listed next to the problem numbers.

21.10 21.4 **21.11** 21.2 **21.12** 21.5, 21.6, 21.7
21.13 21.6, 21.7 **21.14** 21.8 **21.15** 21.8, 21.9
21.16 21.12 **21.17** 21.11 **21.18** 21.11
21.19 21.13

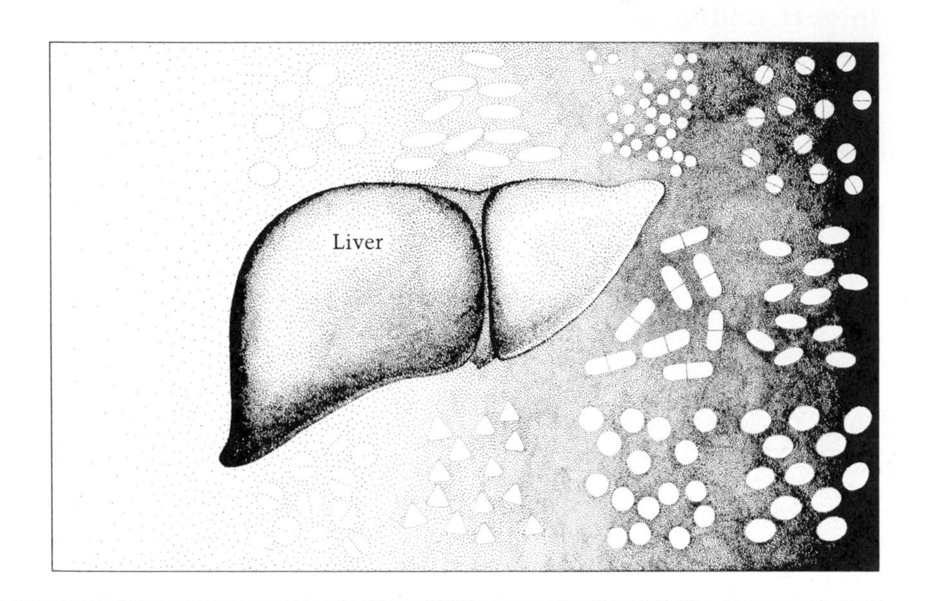

In some cases the function of the liver can be severely affected by certain drugs to the extent that hepatic (liver) failure or death results. The normal liver can disarm most toxic substances by making them more water soluble. Then they can be readily excreted in the urine. This process depends on certain enzymes that catalyze oxidation, reduction, and other chemical reactions necessary for the breakdown of the particular toxic substance. As we will find out later in this chapter, many chemicals are poisonous because they block electron transport systems or interfere with phosphorylating enzymes.

After repeated use, certain drugs (barbiturates, for example) can cause the liver enzymes that are involved in their breakdown to become more active. As a result, the drug is broken down into its nonreactive products too soon and is no longer effective.

If consumed, the laboratory reagent carbon tetrachloride can lead to liver damage. Jaundice, a condition in which the skin and the whites of the eyes take on a yellowish appearance, occurs shortly after ingestion of carbon tetrachloride. Damage to the liver is not always permanent in the patient who survives the toxic effects of this substance.

Tetracyclines are antibiotics used for the treatment of specific infections. However, these drugs also have the ability to inhibit cell metabolism. In large doses, they can cause hepatic failure by interfering with the breakdown reactions that are a normal part of routine liver function.

22

BIOLOGICAL ENERGY AND THE USES AND FORMATION OF ATP

Rapid, reliable tests are now available in the clinical laboratory for many drug analyses. If a toxic condition is suspected in a patient and a specific drug or chemical agent is thought to be responsible, the testing, diagnosis, and treatment are fairly straightforward. But in most cases the procedure becomes quite extensive. The laboratory results for a variety of tests must be examined as a group in the form of a so-called liver profile to confirm the presence of disease and determine the extent of the damage. One test alone is not sufficient to provide a positive diagnosis.

[Illustration: The human liver helps the body to metabolize certain toxic materials but may be overloaded by the toxic substances in some drugs, causing liver disease.]

Energy can take several forms: chemical energy, radiant energy, electrical energy, mechanical energy, and heat energy. All forms of energy are used by living organisms. Chemical energy is stored in carbohydrates and lipids, as well as in other molecules. In the nerve cells, this chemical energy is converted to electrical energy. In the muscles, chemical energy is converted to mechanical energy. The heat energy given off by chemical reactions is necessary to maintain body temperature.

The ultimate source of energy for living organisms is the radiant energy of the sun. This energy is captured by green plants, which transform it into the chemical energy stored in carbohydrates and other organic compounds. Animals eat the plants and convert these compounds to their own uses. They use the organic compounds in the plants as raw materials for the synthesis of body structures or as sources of energy. We also oxidize the organic molecules laid down in petroleum and coal to obtain the energy needed to run automobiles, airplanes, and industrial machinery; to heat our homes; and to generate electricity.

We oxidize organic molecules from plants and other animals to obtain the energy we need to operate our own bodies. When petroleum fuels are oxidized, large amounts of heat and light are given off. In contrast, when foodstuff molecules are oxidized in the cells of the body, the process is carried out in a tightly controlled, step-by-step fashion at constant temperature. The energy released during this oxidation is stored in a new chemical form: adenosine triphosphate (ATP), an energy-rich compound that serves as the cell's universal energy currency. The chemical energy in ATP can be converted into all the other forms of energy needed by the body.

CONSERVATION OF ENERGY 22.1

What fundamental idea is contained in the first law of thermodynamics?

All forms of energy are interconvertible (Section 1.11). Radiant energy may be converted into heat energy, as when sunlight strikes and warms your skin. Heat energy can be transformed into mechanical energy, as in a steam engine. An important concept in the study of energy is that energy cannot be created or destroyed, but only converted from one form to another. This is the familiar principle of the conservation of energy, which is called the first law of thermodynamics. The study of energy transformations is most readily approached by the measurement of heat changes; this is why the study of energy is called **thermodynamics** (*thermo*, "heat"; *dynamics*, "flow or change").

Problem 22.1 Suggest several other everyday examples of energy transformations. As a start, consider the different ways in which electricity is generated for sale by electric companies across the country. (*Hint:* Recall Problem 1.11.)

SPONTANEOUS AND NONSPONTANEOUS REACTIONS 22.2

What is a spontaneous reaction?

Another important idea we gain from thermodynamics is that (at a given temperature and pressure) a reaction will take place only if free energy is given off during the process. Free energy is given off if the products contain less energy than the starting materials do (Figure 22.1). Knowing this, we can predict whether a reaction will be spontaneous or not. Note

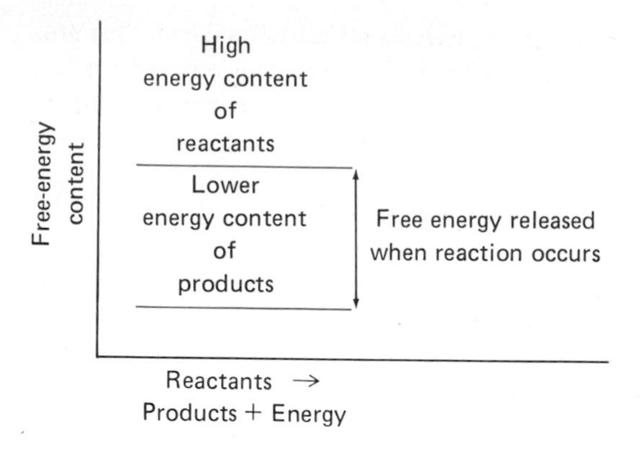

FIGURE 22.1 Free-energy change of a spontaneous reaction.

that we will use the word *spontaneous* in a special way. A **spontaneous reaction** is one that may proceed by itself because free energy is given off as the reaction occurs:

$$\text{reactants} \longrightarrow \text{products} + \text{free energy}$$

Reactions that do not release free energy are not spontaneous. (Note, however, that a reaction will not necessarily take place even though it may be classified as spontaneous. The *rate* of the reaction may be nearly zero unless a suitable catalyst is present.)

Since the idea of spontaneous and nonspontaneous reactions is important, let's look at it a little more closely. We will review and expand our understanding of free energy, a concept we first discussed in Chapter 1. All reactions and processes tend to go (spontaneously) to equilibrium. For example, if you start a rock rolling down a hill, the rock tends to continue rolling until it reaches a stable or equilibrium position at the bottom (assuming a round rock and a smooth hill). At equilibrium, the free-energy content of a system is at a minimum. In the case of the rock, we can say that the rock's potential energy (free-energy content) is at a minimum when it is at the bottom of the hill. Thus a reversible process

$$\text{rock at top of hill} \rightleftharpoons \text{rock at foot of hill}$$

will spontaneously proceed toward equilibrium (where it has its lowest energy content). Naturally, as it does so, energy is given off. The direction this particular process will take is not hard to predict: the rock will roll from the top of the hill to the bottom, losing potential energy as it does so.

$$\text{rock at top of hill} \longrightarrow \text{rock at foot of hill} + \text{energy}$$

The reverse of the downhill-rolling process is an uphill rolling. You can easily predict that the rock won't go by itself (spontaneously) from the bottom to the top of the hill, but you could carry it up there, using your own energy to drive the process:

$$\text{rock at foot of hill} + \text{your energy} \rightarrow \text{rock at top of hill}$$

Reactions that give off energy as they occur are called **exergonic reactions;** those that require an input of energy are called **endergonic reactions.** The useful energy given off or used by a reaction is called the **free-energy change** of the reaction, symbolized $\mathbf{\Delta G}$ ("delta G"). When free energy is released (in an exergonic reaction), ΔG is negative (the system *loses* energy). When free-energy input is needed (in an endergonic reaction), ΔG is positive (the system *gains* energy). Thus the downhill-rolling reaction has a negative ΔG, and the reverse reaction has a positive ΔG. We can summarize the results of the two kinds of reactions as follows:

Type of reaction	Example	What happens	Change in free energy ΔG
Exergonic	Rock rolling downhill	Energy is released	Energy is lost from the system; ΔG is negative ($-\Delta G$)
Endergonic	Rock rolling uphill	Energy is consumed	Energy is gained by the system; ΔG is positive ($+\Delta G$)

Problem 22.2 Does a spontaneous reaction have a negative or a positive free-energy change? Does a nonspontaneous reaction have a negative or a positive free-energy change?

22.3 A STANDARD FOR COMPARING THE FREE-ENERGY CHANGES OF REACTIONS

What is a handy way to predict whether a reaction is exergonic or endergonic?

Most chemical reactions do not go "all the way" like the rock on the hill. Instead a reaction such as

$$A \rightleftarrows P$$

will proceed until it reaches equilibrium where some of the reactant A and some of the product P are both present (Section 6.6). Different reactions will proceed to different extents. If we are going to compare the free-energy changes of different reactions, we need a standard of comparison. For this purpose we use the **standard free-energy change,** symbolized **ΔG^0** ("delta G zero"). The standard free-energy change of a reaction is the free-energy change that occurs when the reaction goes to equilibrium starting from standard conditions. These conditions are a temperature of 25°C, a pressure of 1 atm, a pH of 7, and all reactants and products at 1.0 *M* concentrations at the start. The standard free-energy changes of reactions give us a handy way of judging whether reactions are likely to be exergonic and spontaneous, or endergonic.

However, a note of caution! The standard free-energy change really applies only to standard conditions. The actual free-energy change is what really counts in determining which direction a reaction will go spontaneously. The actual free-energy change depends on both the standard free-energy change *and* the actual concentrations of reactants and products. If reactants are high in concentration and products are low, more energy will be released in going to equilibrium than if all were present at the same concentration (standard conditions). Therefore, the actual free-energy change will be more negative than the standard free-energy change.

An example of this situation is the ionization of a weak acid. If you put a little bit of acetic acid (CH_3CO_2H) into water—say, 1 mL of concentrated acetic acid in 1 L of water—and use pH paper to see whether the solution is acid, you'll find that it definitely is. The ionization reaction is

$$CH_3CO_2H + H_2O \rightleftarrows CH_3CO_2^- + H_3O^+$$

Would it surprise you to learn that the *standard* free-energy change of the ionization reaction is very positive? The reaction does occur to a small extent, however, because there is a high concentration of water, and a much higher concentration of CH_3CO_2H than the products $CH_3CO_2^-$ and H_3O^+, when you start the reaction by adding the concentrated acid to the water. Thus the actual ΔG, which depends on the concentration of products and reactants as well as on the standard free-energy change, is negative, and spontaneous ionization occurs, up to a point. Thus, H_3O^+ ions are formed and the solution becomes acidic.

Problem 22.3 If a reaction has a negative standard free-energy change and the products are in very high concentration compared with the reactants, are more products likely to be formed spontaneously? Explain.

BIOLOGICAL EXAMPLES OF ENERGY-YIELDING AND ENERGY-REQUIRING REACTIONS 22.4

Many exergonic, or energy-yielding, processes occur in biological systems. Examples are the oxidation of foodstuffs (carbohydrates, fats, and proteins), photosynthesis (in photosynthetic organisms only, of course), and fermentation (the breakdown of carbohydrate without oxygen). A great deal of energy is released in these processes. The endergonic, or energy-requiring, processes of biological systems are the synthesis of body materials, mechanical motion (as in muscle contraction), the storage of fat and starch, and the creation of concentration differences across membranes. Neither type of process occurs in isolation from the other. The free energy released in exergonic reactions is used to drive the endergonic reactions of living systems (Figure 22.2).

Is carbohydrate oxidation an exergonic reaction?

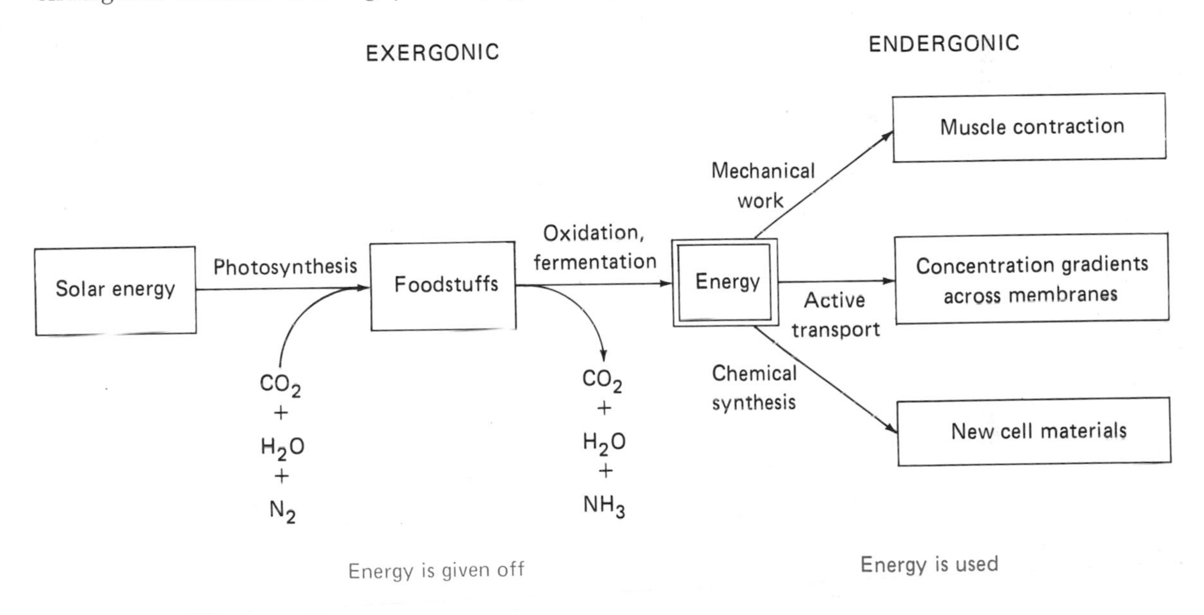

FIGURE 22.2 Energy-yielding and energy-requiring reactions within living systems.

THERMODYNAMICS AND METABOLIC PATHWAYS 22.5

If we look at the standard free-energy changes of the reactions involved, we find that metabolic pathways include three kinds of reactions: reactions with large negative values of ΔG^0, some reactions with positive values of ΔG^0, and many reactions that lie between these extremes.

Why are the first and last steps in a metabolic pathway often very exergonic?

A very exergonic reaction is often found right at the start of a metabolic pathway. As you know, a very exergonic reaction will proceed until most of the available starting material has been converted to product. Such reactions are essentially **irreversible;** that is, the back reaction scarcely occurs at all. In the cell, this has the effect of committing the starting material of the pathway to being consumed. An initial irreversible reaction in a metabolic pathway is often called a **committing step.**

A few reactions with quite large positive values of ΔG^0 occur in metabolic pathways. You might wonder how the product of such a reaction is ever formed. The answer is that product can be formed as long as its concentration is kept low enough for the actual free-energy change of the reaction (ΔG as opposed to ΔG^0) to be negative. In such cases the product concentration is kept low because the product is rapidly removed by later reactions in the metabolic pathway. This is one reason that we often find an irreversible (highly exergonic) reaction situated at the end of a metabolic pathway, as well as at the beginning.

The third class of reactions—those that have small ΔG^0s, positive or negative—is the most common. These reactions are readily reversible. If the reaction is $A \rightleftarrows B$, A may be converted to B, or B to A, depending on their relative concentrations in the cell. These reversible reactions are versatile, since they can easily be made to go in either direction, depending on the needs of the cell.

22.6 FLOW OF ENERGY IN THE BIOLOGICAL WORLD

What is the ultimate source of energy on earth?

The cell, the fundamental unit of the biological system, requires energy to maintain life. For example, energy is required to run the membrane transport processes that preserve the proper ionic environment inside the cell. Energy is also required for muscle contraction and for the manufacture of cell materials for growth and reproduction.

The ultimate energy source for all these biological processes is the sun. However, most cells cannot utilize solar energy directly. Only photosynthetic cells, chiefly those of green plants, can absorb the sun's radiant energy and transform it into chemical energy in the form of carbohydrate molecules. Other cells can take the chemical energy found in plants and transform it into mechanical or electrical energy, or into other forms of chemical energy. In effect, then, animals, most bacteria, and nonphotosynthetic plant life depend on the photosynthetic organisms for food. The green plants change simple inorganic molecules and solar energy into organic chemical compounds, or foodstuffs. All other organisms oxidize these foodstuffs to obtain energy. We will later examine how green plants carry out this process and how energy is released in the breakdown of foodstuff molecules.

Most of the solar energy reaching the earth is transformed into heat energy, and only a small fraction of it is turned into chemical energy by photosynthetic plants. When a living organism eats a plant, it receives chemical energy from the plant. When that organism is eaten, the energy is passed on to yet another organism. This sequence of eating and being eaten is called a **food chain.** Each food chain has at least three levels. The green plants (the producers) are on the first level. Plant eaters, or **herbivores,** are on the second; they are called the primary consumers. **Carnivores** (meat eaters), which eat the primary consumers, are on the third level; they are the secondary consumers. Carnivores that eat the secondary consumers are on a fourth level. See Figure 22.3 for diagrams of typical food chains.

Energy is lost at each step in the food chain. Only about 10% of the energy is conserved as cell material when a carnivore eats a herbivore, or when a herbivore eats a plant. Therefore, the shorter the food chain, the less total energy will be lost. This is economically important. When food is in short supply, it is advantageous for people to eat rice and other plant products, rather than feeding the plants to domestic animals and then eating the meat of the animals.

Problem 22.4 Wild cats are carnivorous, that is, they eat only meat. Are these animals dependent on photosynthetic plants? Explain.

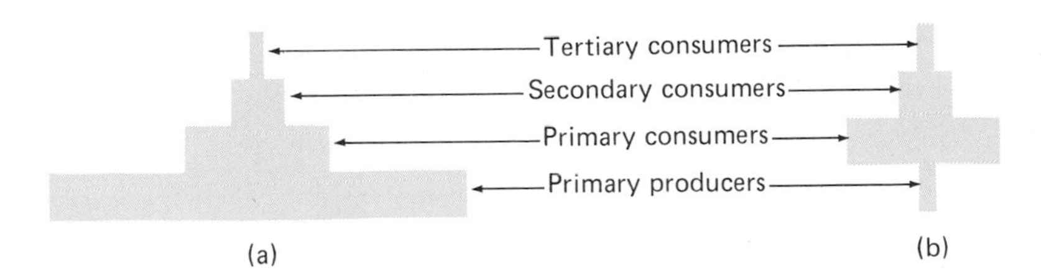

FIGURE 22.3 Relationships of producers and consumers in the major types of food chains found in nature. In (a), the primary producers are small (like single-celled marine organisms) and so a large quantity of them is required to support the herbivores that feed on them. In (b), the primary producers are large (for instance, trees) and an individual can support many herbivores. (From Curtis, H., *Biology*, 3rd ed., Worth Publishers, New York, 1979, p. 883.)

22.7 ATP: THE ENERGY CURRENCY OF CELLS

ATP molecules are the dollar bills of the cell's economic system.

We have mentioned that foodstuffs are oxidized to release energy that cells can use to fill their needs. Later we will look at the chemistry of these oxidations more closely; for the present, we will focus on the end product of these processes. This end product is a compound rich in chemical energy that is used to drive endergonic processes in cells. We will answer several questions: What is the chemical nature of this end product? Why is it so rich in energy? How is its energy harnessed to drive chemical reactions, muscle contraction, ion transport, and other energy-requiring processes?

The chief energy-rich product of virtually all foodstuff oxidations is **adenosine-5′-triphosphate,** or **ATP.**

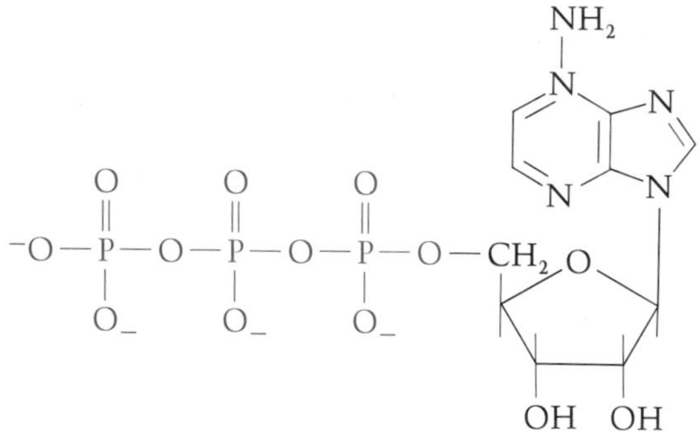

$$^{-}O-\overset{\overset{\displaystyle O}{\|}}{\underset{\underset{\displaystyle O^{-}}{|}}{P}}-O-\overset{\overset{\displaystyle O}{\|}}{\underset{\underset{\displaystyle O^{-}}{|}}{P}}-O-\overset{\overset{\displaystyle O}{\|}}{\underset{\underset{\displaystyle O^{-}}{|}}{P}}-O-CH_2-\text{(ribose, OH OH)}-\text{adenine }(NH_2)$$

adenosine-5′-triphosphate (ATP)

ATP differs from the nucleoside monophosphates like AMP (Section 18.4) only in that it contains three phosphoric acid groups rather than just one. These three groups are in anhydride linkage—that is, each phosphoric acid group has condensed with another (with the loss of a water molecule) to form a new phosphorus–oxygen bond.

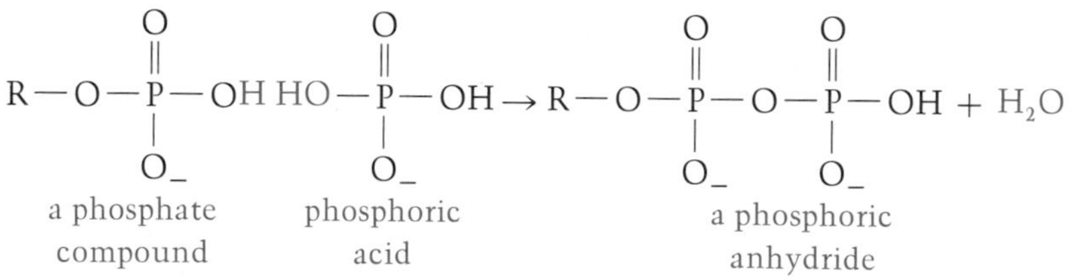

$$\underset{\text{a phosphate compound}}{R-O-\overset{\overset{\displaystyle O}{\|}}{\underset{\underset{\displaystyle O^{-}}{|}}{P}}-OH} \quad \underset{\text{phosphoric acid}}{HO-\overset{\overset{\displaystyle O}{\|}}{\underset{\underset{\displaystyle O^{-}}{|}}{P}}-OH} \rightarrow \underset{\text{a phosphoric anhydride}}{R-O-\overset{\overset{\displaystyle O}{\|}}{\underset{\underset{\displaystyle O^{-}}{|}}{P}}-O-\overset{\overset{\displaystyle O}{\|}}{\underset{\underset{\displaystyle O^{-}}{|}}{P}}-OH} + H_2O$$

Such phosphoric anhydrides (Section 15.13) are quite reactive compounds, much like the carboxylic anhydrides (Section 15.7)

Problem 22.5 Draw the structures of GTP, CTP, and UTP. (Refer to Section 18.4 if needed.)

HIGH-ENERGY COMPOUNDS 22.8

What is a high-energy compound?

ATP is often called a **high-energy compound.** Indeed, all the nucleoside-5′-triphosphates are high-energy compounds. Certain other biologically important compounds are also considered high-energy compounds; some of these are listed in Table 22.1.

When high-energy compounds are hydrolyzed, a great amount of energy is released. One way to measure the energy content of such a compound is to look at the standard free-energy change for the reaction in which the compound is broken down. The more negative the standard free-energy change (ΔG^0), the more energy is released. Thus, a high-energy compound is often defined as one whose standard free energy of hydrolysis (the breakdown reaction) is more negative than about -7 kcal/mol. The reaction for the hydrolysis of ATP is

$$ATP^{4-} + H_2O \rightleftarrows ADP^{3-} + HPO_4{}^{2-} + H^+$$

$HPO_4{}^{2-}$ is "inorganic phosphate" (often abbreviated $P_i{}^{2-}$), which at a pH near 7 exists as an anion. ADP (adenosine-5′-diphosphate) is adenosine coupled with two phosphates rather than three. (Note that we have written H^+ rather than H_3O^+ in our equation; this is a simplification that we will continue to use because many of the equations

TABLE 22.1 ΔG^0 Values for Hydrolysis of Some Important Metabolic Intermediates

Compound	ΔG^0 (kcal/mol)
Phosphoenolpyruvate	−14.8
1,3-Diphosphoglycerate	−11.8
Phosphocreatine	−10.3
Phosphoarginine	−7.7
ATP	−7.3
Glucose-1-phosphate	−5.0
Glucose-6-phosphate	−3.3
Glycerol-1-phosphate	−2.2

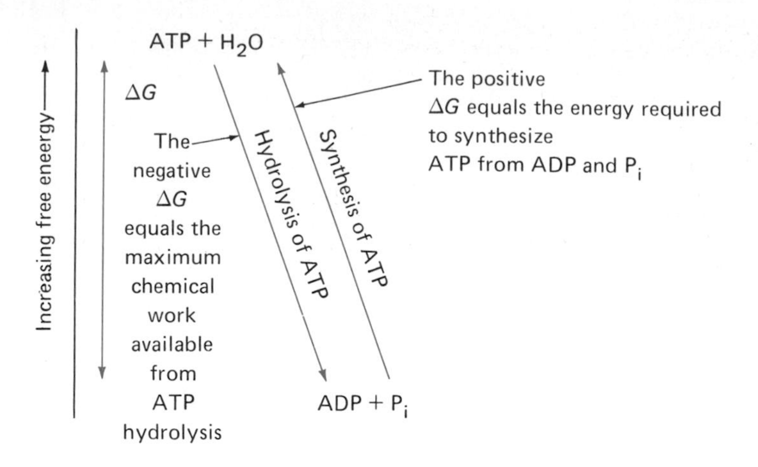

FIGURE 22.4 Free-energy changes associated with ATP hydrolysis and ATP synthesis.

will be somewhat long.) The ΔG^0 for this reaction is -7.3 kcal/mol, and therefore this reaction proceeds with a substantial release of free energy (Figure 22.4). In fact, it is the major exergonic reaction employed in the body to drive endergonic reactions.

On the other hand, it should be clear that it will take energy to drive the synthesis of ATP from ADP and phosphate. This is the reverse of the ATP hydrolysis reaction (Figure 22.4). The energy to drive ATP synthesis from ADP and phosphate comes from the exergonic reactions of photosynthesis and foodstuff oxidation. Later in this chapter we will study the major pathways for ATP synthesis.

Problem 22.6 In Figure 22.2 there is a box labeled "energy." Can you say something specific about the form in which energy is given off by the exergonic reactions and employed by the endergonic reactions?

22.9 COUPLING OF REACTIONS

Specifically how is energy supplied to drive endergonic reactions in cells?

We have seen that a reaction must have an actual negative free energy change if it is to proceed spontaneously. We have also seen that it is possible to supply energy to a reaction that has a positive free-energy change and thus to overcome this unfavorable energy situation. Let us see how this can be done with a chemical reaction. We'll use as an example a reaction that initiates the breakdown of glucose in a typical cell.

Glucose is phosphorylated to give glucose-6-phosphate.

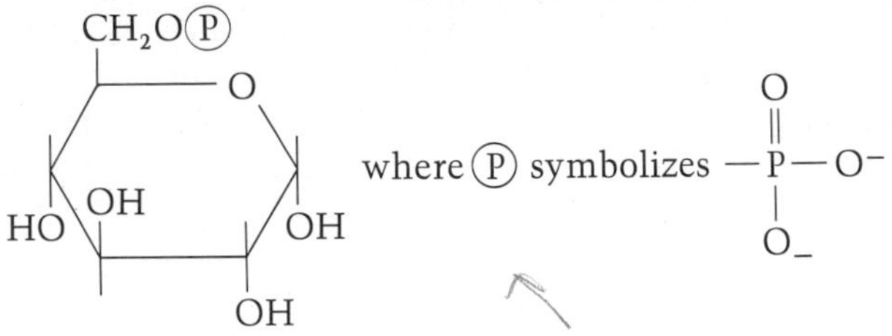

CH_2OⓅ — O — HO — OH — OH — OH

glucose-6-PO_4

where Ⓟ symbolizes $-\overset{\displaystyle O}{\overset{\|}{P}}\underset{\displaystyle O^-}{\underset{|}{}}-O^-$

One way we might imagine this product being formed from glucose is as follows:

$$\text{glucose} + HPO_4^{2-} \longrightarrow \text{glucose-6-}PO_4^{2-} + H_2O$$

The standard free-energy change for this reaction is approximately +3.9 kcal/mole. As the positive value of ΔG^0 indicates, this reaction would not proceed under standard conditions. Energy can be transferred to this endergonic reaction by **coupling** it with an exergonic reaction. As is often the case in cells, the exergonic reaction that serves this purpose is ATP hydrolysis. When we add the negative ΔG^0 of ATP hydrolysis to the positive ΔG^0 of the reaction between glucose and HPO_4^{2-}, we get a net negative ΔG^0:

$$ATP^{4-} + H_2O \longrightarrow ADP^{3-} + HPO_4^{2-} + H^+ \qquad \Delta G^0 = -7.3 \text{ kcal/mol}$$

$$\text{glucose} + HPO_4^{2-} \longrightarrow \text{glucose-6-}PO_4^{2-} + H_2O \qquad \Delta G^0 = +3.9 \text{ kcal/mol}$$

$$\text{glucose} + ATP^{4-} \longrightarrow \text{glucose-6-}PO_4^{2-} + ADP^{3-} + H^+ \qquad \Delta G^0 \text{ overall} = -3.4 \text{ kcal/mol}$$

You can see that the ΔG^0 values of coupled reactions are simply added together to give the overall standard free-energy change, just as the reactions are added together to give an overall reaction. This does not necessarily mean that the reactions take place separately, as written. The overall reaction does not involve HPO_4^{2-} at all.

The reaction of glucose phosphorylation is catalyzed in cells by the enzyme hexokinase. It occurs with a large negative free-energy change (which is even larger than the standard free-energy change since the glucose concentration will be high and the glucose-6-phosphate is kept low by later reactions). In other words, glucose-6-phosphate *can* be formed from glucose in the cell, provided the source of phosphate is ATP. The terminal phosphate of ATP is transferred directly to glucose in the active site of hexokinase.

In this example, the free energy content of the glucose-6-phosphate is greater than that of glucose and inorganic phosphate combined. Similarly, the free energy of ADP and inorganic phosphate is lower than that of ATP. The reaction of a high-energy compound (ATP) and a relatively low-energy compound (glucose) to give a different relatively high-energy compound (glucose-6-phosphate) and a relatively low-energy compound (ADP) is referred to as **coupling.**

We often speak of the energy-releasing and energy-consuming reactions of the cell, but in fact these are only half-reactions. They represent two aspects of one reaction. In order for the overall reaction to occur, there must be a net decrease in the free energy of the system involved in the reaction. The energy-releasing half and the energy-consuming half are inextricably bound—coupled—together.

In pathways in which carbohydrate, fat, protein, or other cell materials are made, many reactions that require the input of free energy are driven by being coupled with ATP hydrolysis in just this way. We will examine some of these pathways in more detail later.

There is another sort of reaction coupling that does not involve the hydrolysis of ATP. Here there are two separate reactions. The second reaction is highly exergonic, and it "pulls" the first reaction (which has a positive ΔG^0) by removing one of its products. For example, two reactions that are part of one of the major metabolic pathways, the citric acid cycle (Section 23.9), are coupled in this way:

$$\text{malate} + \text{NAD}^+ \longrightarrow \text{oxaloacetate} + \text{NADH} + \text{H}^+ \qquad \Delta G^0 = +6{:}1 \text{ kcal/mol}$$

$$\text{oxaloacetate} + \text{acetyl SCoA} \longrightarrow \text{citrate} + \text{CoASH} \qquad \Delta G^0 = -7.7 \text{ kcal/mol}$$

By keeping the concentration of oxaloacetate very low (through conversion to citrate) the malate can be siphoned off. The low oxaloacetate concentration ensures that ΔG of the first reaction is negative even though ΔG^0 is not. Thus, the conversion of malate to oxaloacetate, which under standard conditions is a very unfavorable reaction energetically, is made possible through the coupling with the very exergonic conversion of oxaloacetate to citrate. An exergonic reaction situated at the end of a metabolic pathway can often pull several preceding reactions.

Example 22.1 The hydrolysis of phosphoenolpyruvate has a ΔG^0 of -14.8 kcal/mole. Will this reaction provide enough energy to drive the synthesis of ATP from ADP in a coupled reaction?

Solution Note that the ΔG^0 for the reverse of the ATP hydrolysis reaction has the same numerical value as the ΔG^0 for ATP hydrolysis, but the opposite *sign:*

$$\text{phosphoenolpyruvate}^{3-} + H_2O \longrightarrow \text{pyruvate}^- + P_i^{2-} \qquad \Delta G^0 = -14.8 \text{ kcal/mol}$$

$$H^+ + ADP^{3-} + P_i^{2-} \rightarrow ATP^{4-} + H_2O \qquad \Delta G^0 = +7.3 \text{ kcal/mol}$$

$$net \text{ phosphoenolpyruvate}^{3-} + ADP^{3-} + H^+ \rightarrow \text{pyruvate}^- + ATP^{4-} \qquad \Delta G^0 = -7.5 \text{ kcal/mol}$$

The coupled reactions have a net standard free-energy change of −7.5 kcal/mole. That is, the overall reaction is very exergonic and plenty of energy is available to drive ATP synthesis.

Problem 22.7 A reaction in a metabolic pathway has a ΔG^0 of +3.2 kcal/mole. Will coupling with ATP hydrolysis provide enough energy to make it go under standard conditions?

ATP AND MUSCLE CONTRACTION 22.10†

What is muscle contraction on the molecular level?

The energy of ATP hydrolysis drives muscle contraction, which represents a transformation of chemical into mechanical energy. The contraction of your biceps muscle, causing your arm to bend at the elbow, is a matter of physiology, not chemistry, some would say. However, the contraction of the biceps muscle occurs because of the contraction or shortening of many individual muscle cells. The events that lead to the shortening of the muscle cells involve enzyme-catalyzed reactions, ATP hydrolysis, and changes in the conformations of proteins—all of which are certainly concerns of biochemistry. In fact, all physiological processes have their basis in chemistry.

A muscle consists of many muscle cells, each of which contains many **myofibrils** (Figure 22.5). Each myofibril is in turn made up of two kinds of protein filaments, the **thick and thin filaments.** The arrangement of the filaments gives muscle cells a striated (striped) pattern that can be seen under a microscope. The repeating unit of the striated pattern is called a **sarcomere.** In a relaxed myofibril, the filaments of each sarcomere present a characteristic pattern.

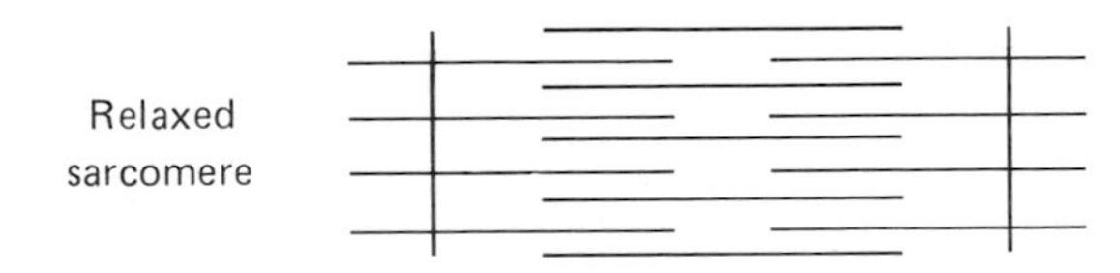

When contraction occurs, the pattern changes.

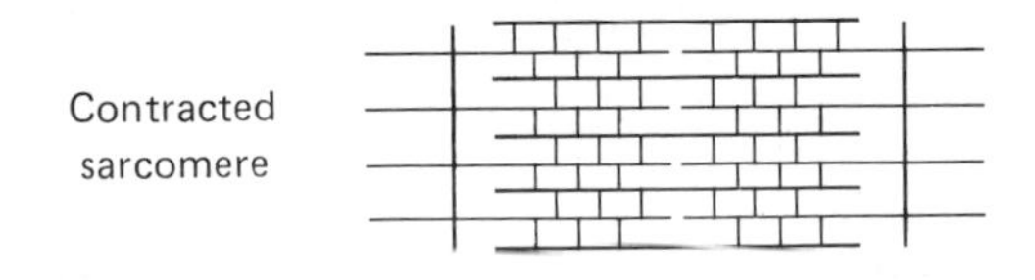

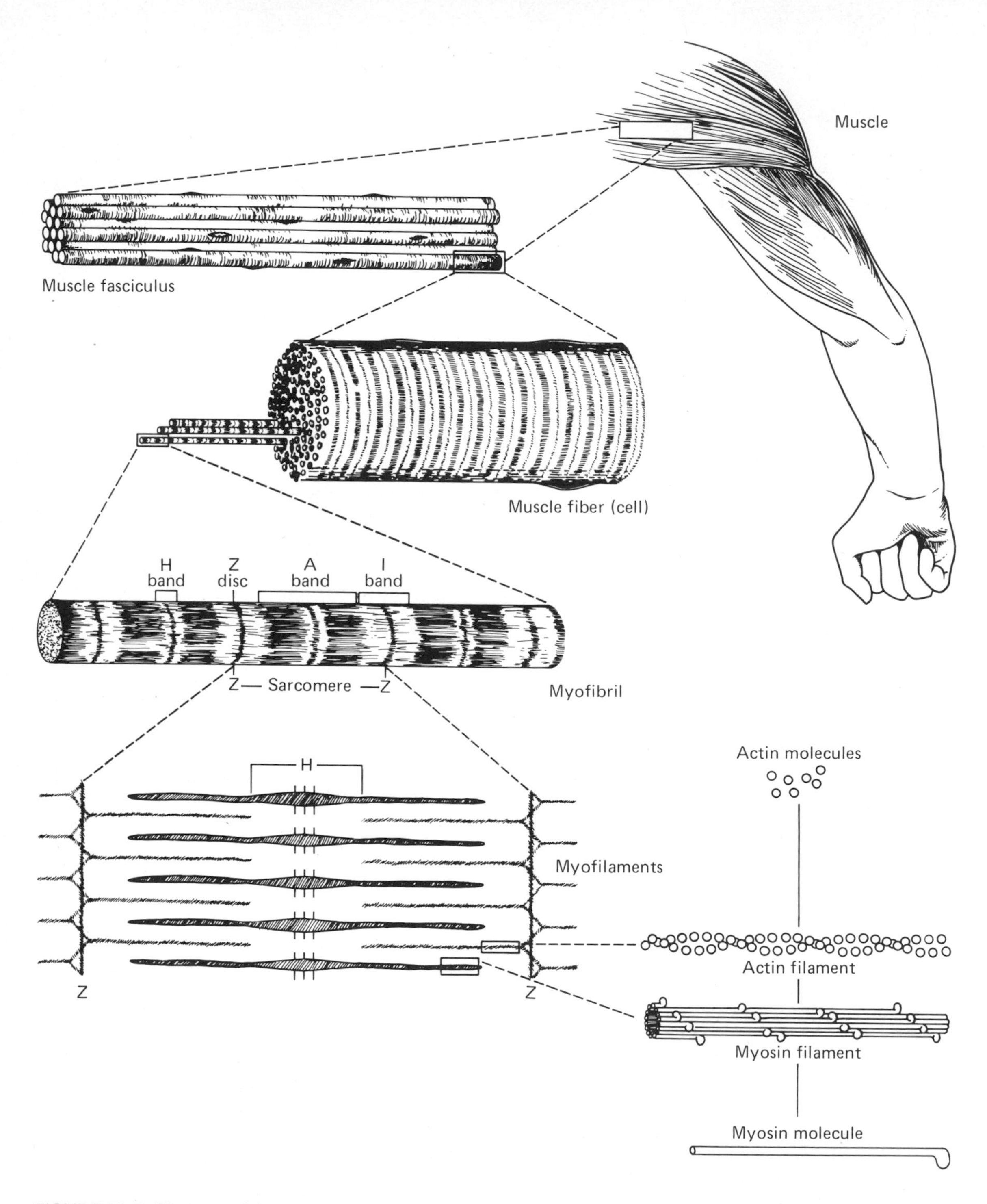

FIGURE 22.5 Diagram of the organization of skeletal muscle. (From Bloom and Fawcett, *Histology,* 10th ed., 1975, W. B. Saunders Company, p. 306.)

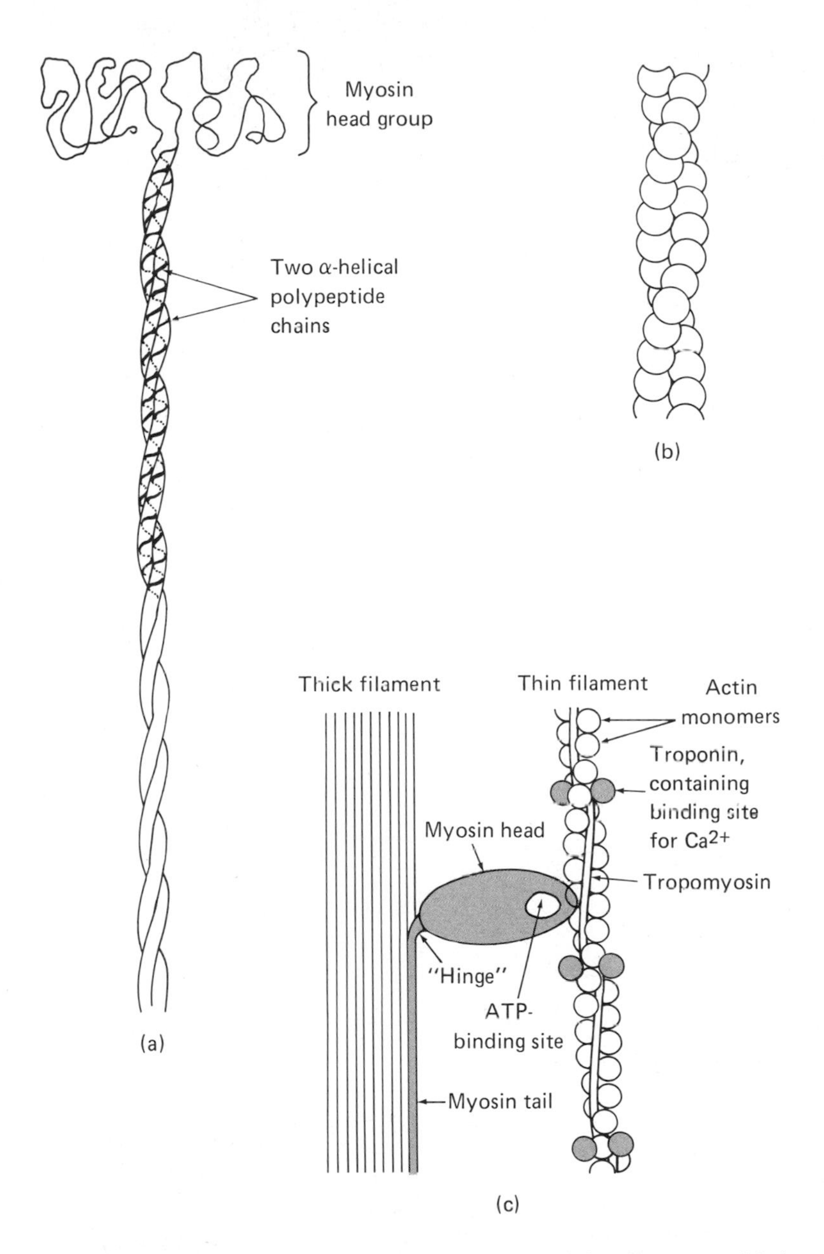

FIGURE 22.6 Molecular structure of the thick and thin filaments. (a) A myosin molecule of the thick filaments. (b) Actin strands in the thin filaments, each consisting of many globular actin subunits. The two strands are coiled around each other. (c) Arrangement of tropomyosin and troponin molecules in the thin filaments and their relationship to the myosin head groups. (From Lehninger, A. L., *Biochemistry,* 2nd ed., Worth Publishers, 1975, pp. 775(a), 760(b), 761(c).)

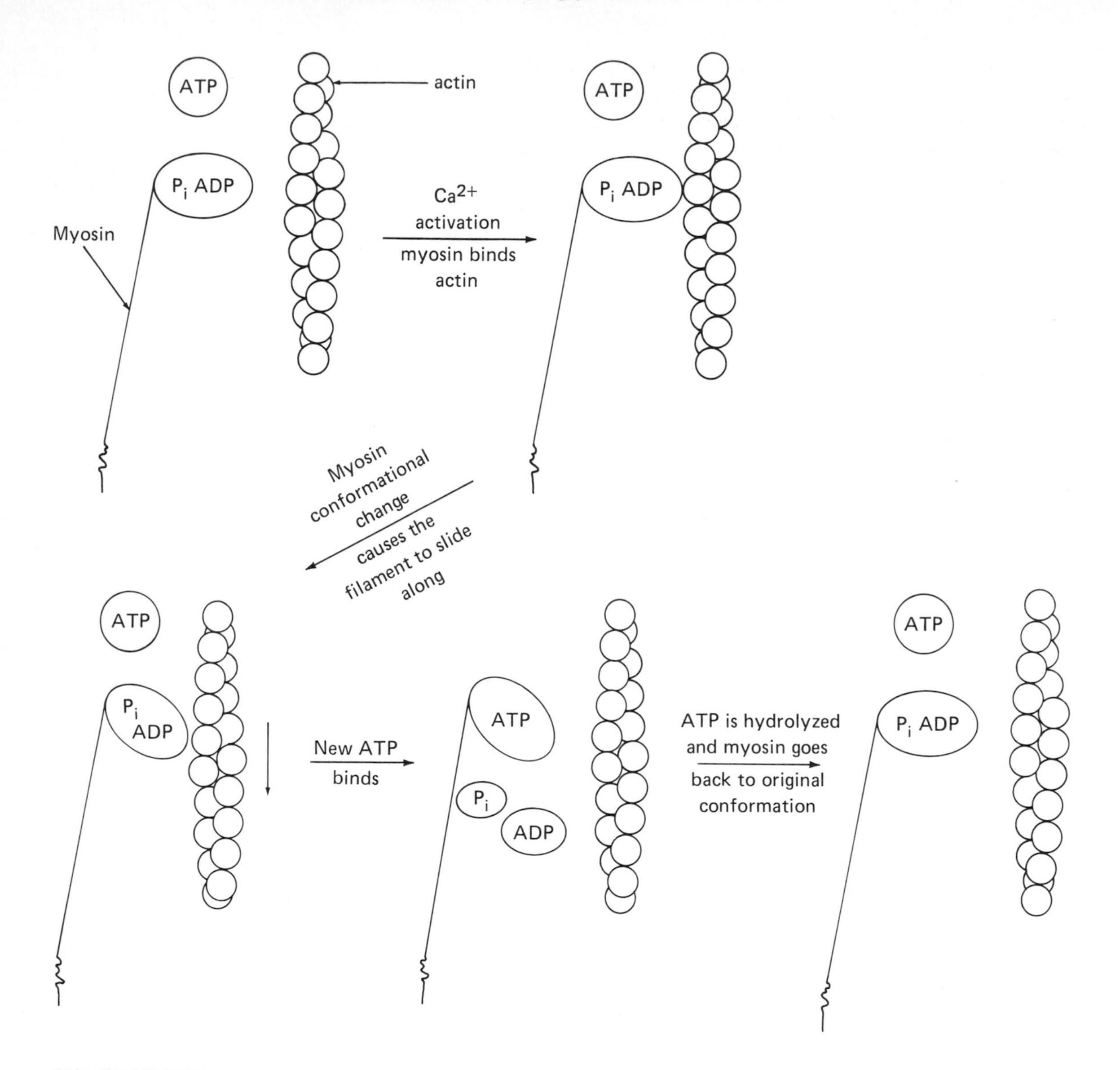

FIGURE 22.7 The myosin conformational change that pulls thin filaments along, shortening the sarcomere. (Adapted from H. G. Mannberg, J. B. Leigh, K. C. Holmes, and G. Rosenbaum. *Nature New Biology* 241:226 (1973).)

The change comes about because many bonds form between the thick and thin filaments, causing the filaments to slide along one another, shortening the sarcomeres and in turn shortening the muscle cell.

The thick filaments are made up of many molecules of a protein called **myosin.** The thin filaments consist of three proteins: **actin, troponin,** and **tropomyosin.** Figure 22.6 shows how the protein molecules are arranged in the two kinds of filaments.

The formation of bonds requires, first, the binding (and hydrolysis) of ATP by the myosin molecules (Figure 22.6a and c). A bond is then formed between a myosin head group (a protruding part of the myosin molecule) and an actin molecule of a thin filament. However, myosin molecules can form bonds with actin only if the actin molecules are exposed. In the relaxed state, the actin binding sites are hidden by tropomyosin molecules and no bonds can form. A change in Ca^{2+} concentration exposes these sites.

The Ca^{2+} concentration in the cell rises when a nerve impulse arrives at the muscle cell to signal contraction, and causes Ca^{2+} to be released from the endoplasmic reticulum (Section 21.6). This Ca^{2+} binds to molecules of troponin, causing a change in the shape of the thin filament that uncovers the actin binding sites. Now the myosin head groups can bind actin molecules (Figure 22.7). The binding of actin causes a change in the conformation of the myosin head group, which pulls the thin filament along toward the center of the sarcomere. Then the bond breaks. As long as the Ca^{2+} concentration remains high, however, new ATPs are bound, new bonds between myosin and actin form and break, and the sarcomere continues to shorten until it reaches its limit.

Relaxation occurs when the endoplasmic reticulum takes up the Ca^{2+} again and reduces the Ca^{2+} concentration enough to return the thin filaments to the relaxed state, in which actin binding sites are hidden again.

Thus in muscle contraction, ATP hydrolysis releases energy that induces conformational changes (movement) in protein molecules. ATP chemical energy is thereby transformed into the mechanical energy of muscle contraction.

Problem 22.8 Sketch the myofibrils of the sarcomere in relaxed and stretched forms.

Problem 22.9 In molecular terms, what is involved when a sarcomere shortens?

MEMBRANE TRANSPORT 22.11†

Living cells can transport a variety of ions and molecules across their membranes (Section 21.2). In some cases these molecules must be pumped in so that the concentration within the cell becomes higher than it is outside. This process, called **active transport,** requires energy, since molecules are being moved from a region of lower concentration to a region of higher concentration. Not all transport processes require energy. If the process simply involves equalizing the concentrations of the material on the two sides of the membrane, it is called **facilitated diffusion.** The diffusion of molecules from a region of higher concentration to a region of lower concentration is a spontaneous process.

Most transport processes appear to involve specific membrane proteins that behave much like enzymes. Figure 22.8 shows ways in which a transport protein can cause the transfer of a molecule across a membrane, either by creating a channel through which the molecule can move or by actually serving as a carrier that moves across the membrane and releases the molecule on the other side.

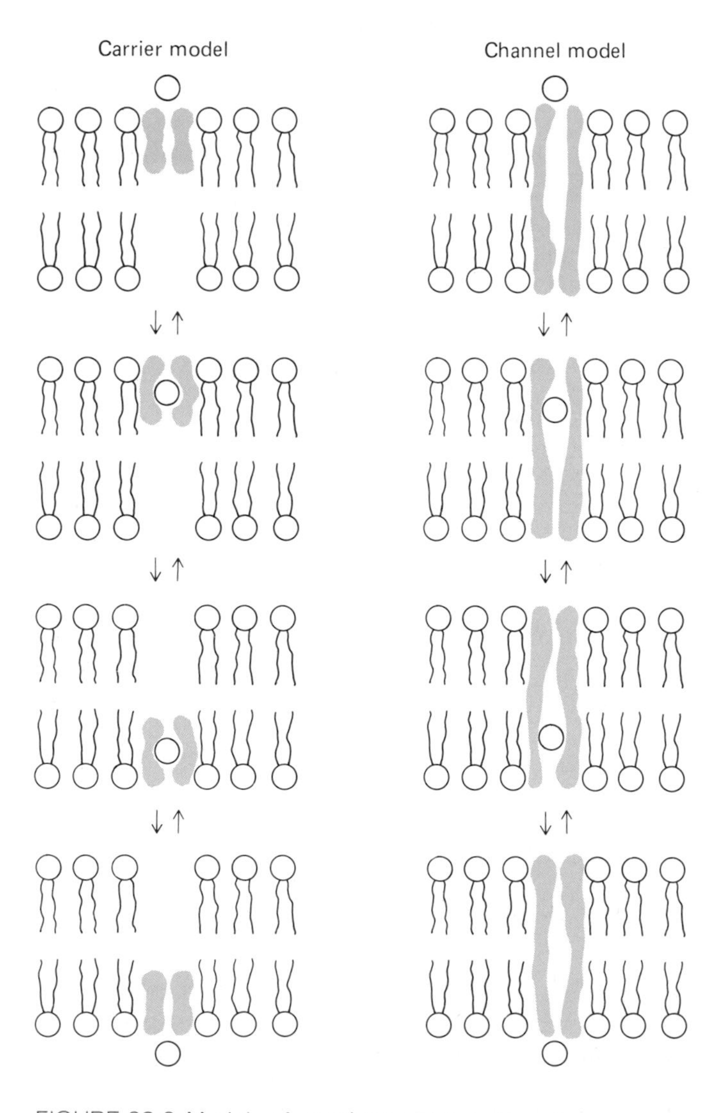

FIGURE 22.8 Models of membrane transport systems.

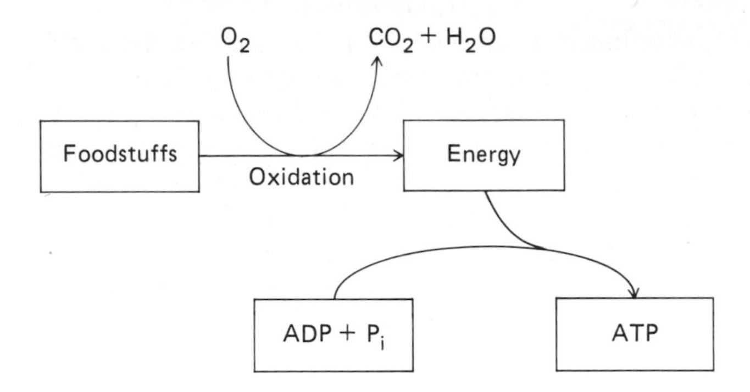

FIGURE 22.9 Energy for ATP formation is provided by the oxidation of foodstuffs.

In some active transport systems, coupling with ATP hydrolysis directly supplies the necessary energy. In other cases an indirect source of energy is used, in the form of an electrical potential across the membrane. The inside of a cell is always more negative than the outside—in other words, there is always a difference in the electrical potential across the cell membrane. Because, as you know, opposite charges attract, a positive ion will be attracted to the negative interior of a cell. Bringing a positive charge together with a negative charge is an exergonic, spontaneous reaction. Thus, bringing a positive ion into the negative interior of a cell releases energy. This energy can be used to pump the positive ion into the cell so that the concentration becomes higher than it is outside

Now that we have acquired a broad overview of biological energy concepts, we are ready to examine the specific ways in which most of the ATP is made in cells: the pathways of ATP synthesis. Oxidation of foodstuffs provides energy for making ATP, as we have seen (Figure 22.9). Since oxidation is such an important part of biological energy changes, it will be good to review briefly the nature of oxidation-reduction reactions.

OXIDATION-REDUCTION (REDOX) REACTIONS 22.12

What are two simple definitions of oxidation-reduction?

There are several ways of defining oxidation. It is useful, especially in biological chemistry, to define it in terms of the removal of hydrogen atoms (H·) from a compound. For example:

$$R\,CH_2CH_2R' \rightleftarrows RCH{=}CH\,R' + 2(H\cdot)$$

The reverse of this reaction is a reduction—the hydrogenation of (the addition of hydrogen atoms to) the unsaturated compound.

Another example of the same definition is the oxidation of an alcohol to an aldehyde:

$$R\,CH_2OH \rightleftharpoons R\overset{\overset{\displaystyle O}{\|}}{C}H + 2(H\cdot)$$

Again, the reverse reaction is a reduction.

We can also define oxidation in terms of the removal of electrons (Section 3.12) from an atom or ion. An example is the oxidation of the Fe(II) ion to the Fe(III) ion:

$$Fe^{2+} \rightleftharpoons Fe^{3+} + e^-$$

Again, the reverse reaction is a reduction: the Fe(III) ion gains an electron, yielding the more reduced Fe(II) ion. This type of reaction is involved in the ATP-synthetic pathways themselves, as we will see.

Each of the reactions just shown is a **half-reaction.** In other words, they do not show what is accepting or donating the hydrogen atoms or electrons. In fact, for every substance that is oxidized, something else must be simultaneously reduced: something must gain the hydrogen atoms or electrons that the substance being oxidized is losing. The cell contains special agents called **redox coenzymes,** which serve as carriers of hydrogen atoms or electrons in oxidation-reduction reactions.

Problem 22.10 Write the half-reaction for the reduction of an aldehyde to an alcohol.
Problem 22.11 Write the half-reaction for the oxidation of H_2O to O_2.

22.13 REDOX COENZYMES

When a substrate molecule is oxidized, what is usually simultaneously reduced?

A coenzyme (Section 21.11) is a nonprotein organic molecule that is part of, or acts jointly with, an enzyme catalyst in a biochemical reaction. Four redox coenzymes carry hydrogen atoms or electrons in biological oxidation-reduction reactions:

nicotinamide adenine dinucleotide (NAD)
nicotinamide adenine dinucleotide phosphate (NADP)
flavin mononucleotide (FMN)
flavin adenine dinucleotide (FAD)

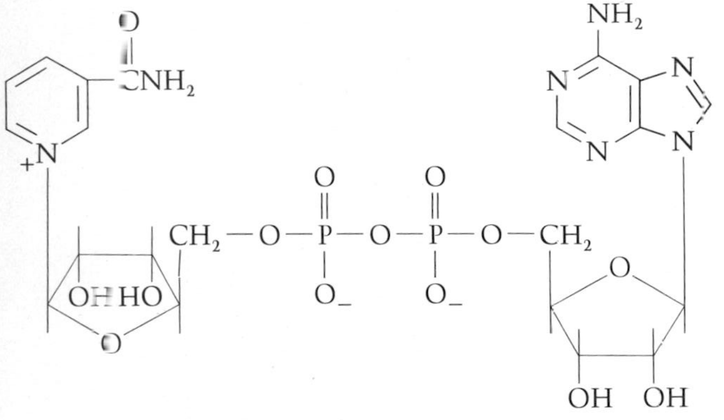

nicotinamide adenine dinucleotide
(NAD)

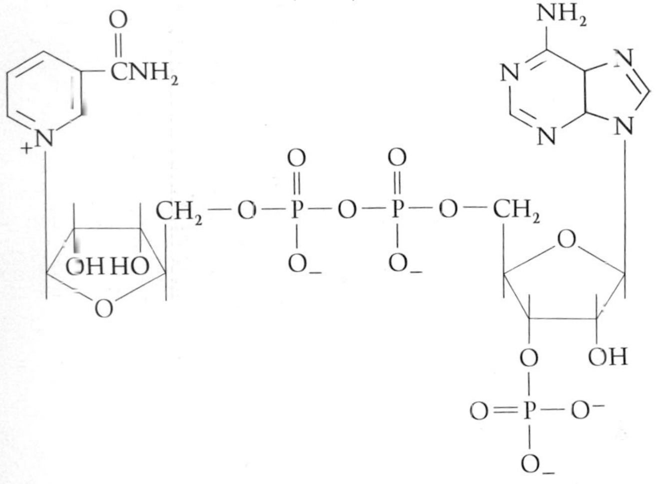

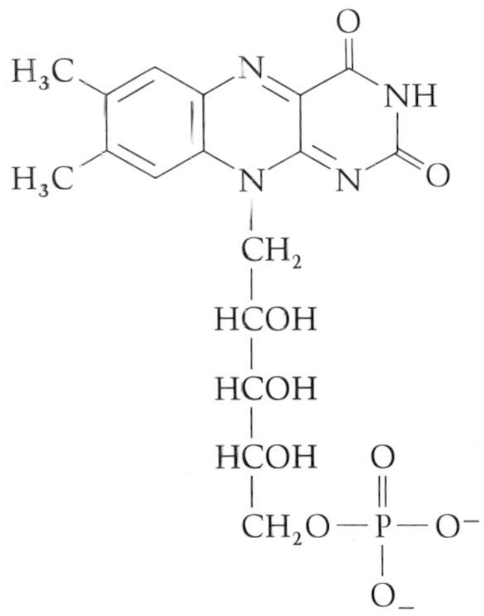

flavin mononucleotide
(FMN)

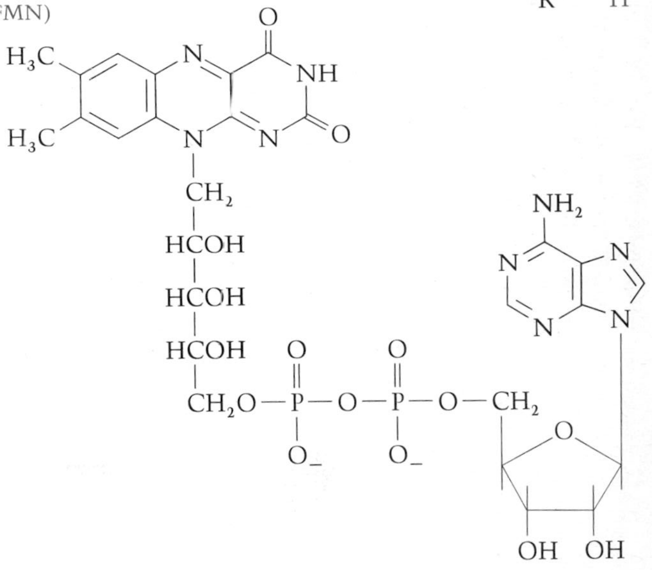

Reduced forms of NAD and NADP contain:

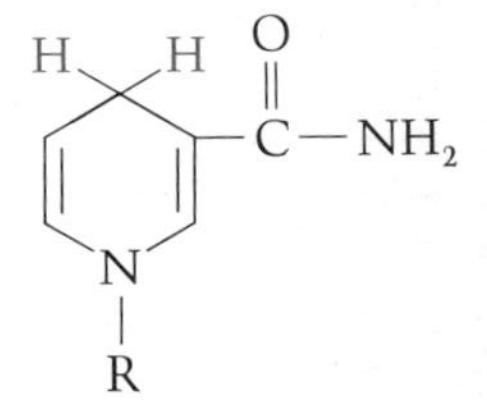

Reduced forms of FMN and FAD contain:

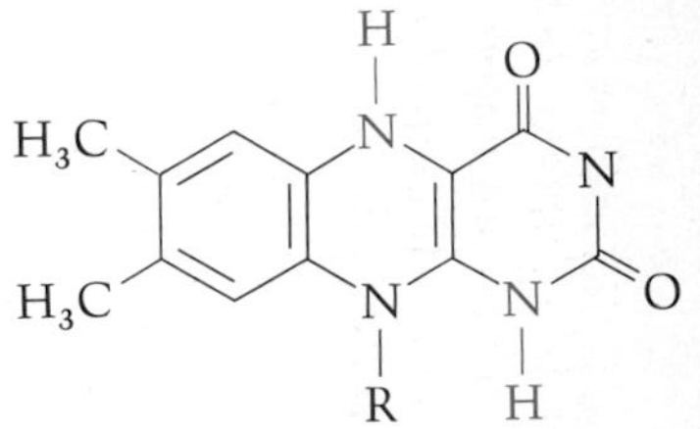

nicotinamide adenine dinucleotide phosphate
(NADP)

flavin adenine dinucleotide
(FAD)

FIGURE 22.10 Structural formulas of some important coenzymes.

In Figure 22.10, you will find the structural formulas of each of these coenzymes. Note that each contains a water soluble vitamin derivative, either nicotinamide or riboflavin (Table 21.2). In each case the coenzyme molecule can be reversibly oxidized and reduced as follows:

$$NAD^+ + 2(H\cdot) \rightleftarrows NADH + H^+$$

$$NADP^+ + 2(H\cdot) \rightleftarrows NADPH + H^+$$

$$FAD + 2(H\cdot) \rightleftarrows FADH_2$$

$$FMN + 2(H\cdot) \rightleftarrows FMNH_2$$

Let's look at a specific example. NAD is reduced when it accepts a hydrogen atom and an additional electron. A second hydrogen (as H^+) is also liberated in the reaction. The two hydrogen atoms come into the picture when they are removed from an organic molecule that is undergoing oxidation. If we let SH_2 represent the substrate of an enzyme that catalyzes such an oxidation-reduction reaction, then we can write a general reaction:

$$SH_2 + NAD^+ \xrightleftharpoons{\text{enzyme}} S + NADH + H^+$$

The reaction going to the right is a substrate oxidation. The reverse reaction is a substrate reduction, in which NADH is the donor of two reducing equivalents. A **reducing equivalent** is an electron or a hydrogen atom (a hydrogen atom is equivalent to an electron and a proton).

22.14 FREE-ENERGY CHANGES OF REDOX REACTIONS

Reduced coenzymes are energy-rich compounds.

In general, oxidations of organic substrates are energy-yielding reactions; free energy is released when they occur. Much of the free energy released in biological oxidations is trapped in the form of the reduced coenzymes (such as NADH or $FADH_2$). Some of the chemical energy of the substrate molecule is thus transformed into a *new* chemical form—the reduced coenzyme. These reduced coenzymes are energy-rich compounds. They serve as intermediaries carrying energy temporarily during the oxidation of foodstuffs. NADH and NADPH, in particular, can donate energy in many different ways, so they are a more useful source of energy than the substrate molecule is. Although they are not as universal an energy currency as ATP, the energy of NADH and other

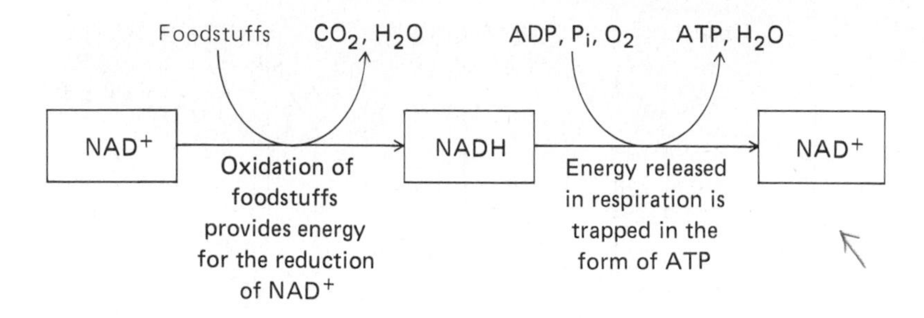

FIGURE 22.11 Oxidation of reduced coenzymes can drive ATP synthesis.

reduced coenzymes can be used to drive the synthesis of ATP in the mitochondria (Figure 22.11).

THE ELECTRON TRANSPORT SYSTEM OF MITOCHONDRIA 22.15

What is the final acceptor of reducing equivalents in respiration?

In animal and plant cells, foodstuffs are oxidized mainly in the mitochondria (Figures 22.12 and 22.13). The mitochondria are often said to be the power plants of the cell, because they produce large amounts of the high-energy compound ATP. Body tissues that depend heavily on a steady supply of ATP energy, such as the brain and the heart muscle,

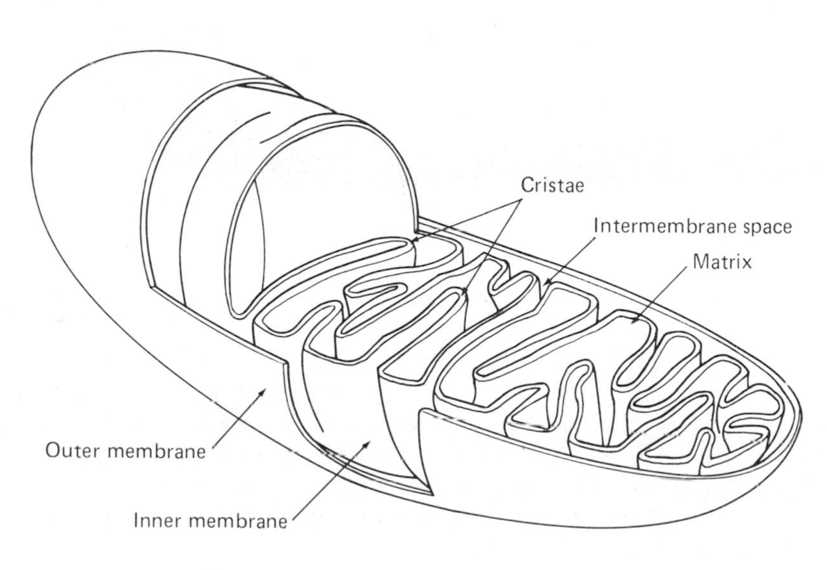

FIGURE 22.12 Diagram of the membranous structure and internal compartments of the mitochondrion.

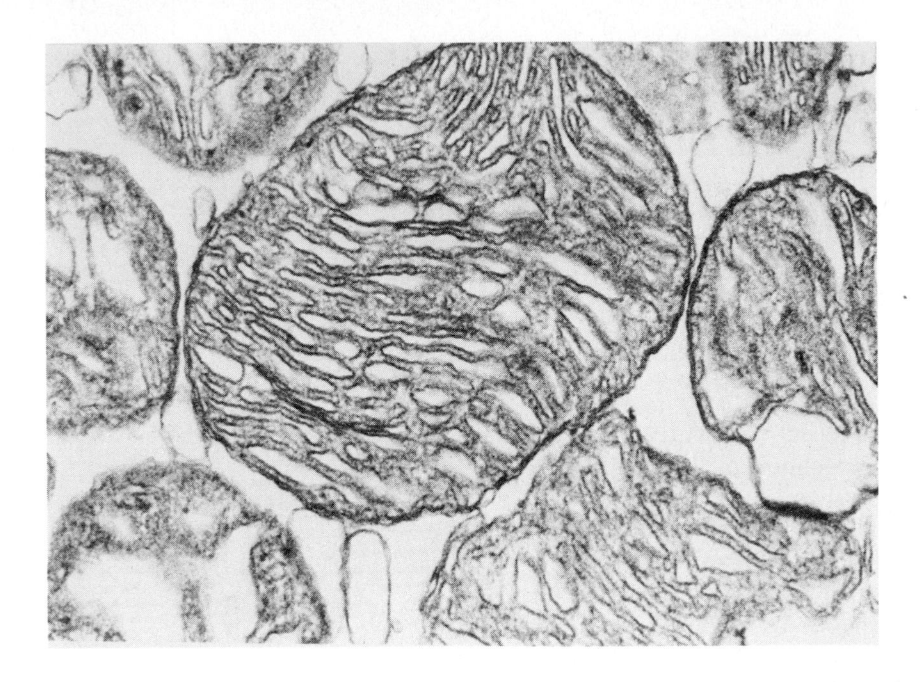

FIGURE 22.13 Electronmicrograph of mitochondria. (Photo by Margaret Bryant and Ronald Sans)

are especially rich in mitochondria. As the foodstuffs are oxidized, NADH is produced. NADH stores chemical energy obtained from the oxidation of foodstuffs until it can be used for ATP synthesis. How is ATP made in the mitochondria using the energy of NADH?

The inner mitochondrial membrane contains a number of special proteins that make up a system of hydrogen and electron carriers. In their role as carriers, these proteins catalyze the oxidation of NADH by oxygen:

$$NADH + H^+ + \frac{1}{2}O_2 \rightarrow NAD^+ + H_2O + \text{energy}$$

Since NADH is a very high-energy compound, its oxidation is very exergonic. So, although O_2 is not directly involved in the earlier stages of foodstuff oxidation, it is the final acceptor—the last stop—for the hydrogen atoms that are removed from substrates and transferred to redox coenzymes when foodstuff molecules are oxidized. A biological process in which oxygen is consumed is called **respiration. Cellular respiration** takes place in the mitochondria. It is responsible for nearly all the oxygen that the living organism consumes. In other words, respiration in the mitochondria is the reason we need oxygen.

Cellular respiration is catalyzed by several oxidation-reduction enzymes or redox "carriers" that make up the mitochondrial **electron**

transport system. These carriers act in a sequence of steps to pass reducing equivalents from NADH to oxygen. There are three types of carriers:

1. **flavoproteins** (proteins containing tightly bound FMN or FAD)
2. **coenzyme Q** (ubiquinone)
3. **cytochromes** (heme proteins)

All these carriers are localized in the mitochondrial inner membrane. We can represent the sequence in which the carriers act like this:

$$\begin{array}{l} \text{NADH} \rightarrow \text{FP}_1 \rightarrow \text{CoQ} \rightarrow \text{cyt } b \rightarrow \text{cyt } c_1 \rightarrow \text{cyt } c \rightarrow \text{cyt } (a + a_3) \rightarrow \text{O}_2 \\ \text{succinate} \rightarrow \text{FP}_2 \nearrow \end{array}$$

FP_1 and FP_2 stand for two different flavoproteins; CoQ represents coenzyme Q; and cyt *b*, cyt *c*, and so on represent different cytochromes. All these except coenzyme Q, which is *not* a protein, are respiratory enzymes. They catalyze the oxidation-reduction reactions that result in the step-by-step transfer of reducing equivalents from NADH to O_2. Each member of the sequence "carries" reducing equivalents from one member of the chain to the next. For example, cyt *b* carries electrons from CoQ to cyt c_1; cyt c_1 carries electrons from cyt *b* to cyt *c*; and so on.

A CLOSER LOOK AT THE REPIRATORY CARRIERS 22.16

What kinds of molecules serve as electron carriers in the electron transport system?

Because of the importance of the function they perform, the three types of redox carriers in the electron transport system are deserving of a more detailed consideration.

Flavoproteins

There are two flavoproteins in the electron transport system: NADH dehydrogenase (FP_1) and succinate dehydrogenase (FP_2). A **dehydrogenase** is an enzyme that catalyzes the removal of hydrogen atoms from some molecule. Each of these two dehydrogenases catalyzes in sequence the two general reactions:

$$SH_2 + FP \rightarrow S + FPH_2$$

$$FPH_2 + CoQ \rightarrow FP + CoQH_2$$

which add up to the overall reaction:

$$SH_2 + CoQ \rightarrow S + CoQH_2$$

SH_2 symbolizes the donor of reducing equivalents. The donor is NADH if the flavoprotein is FP_1 (NADH dehydrogenase) and succinate if the flavoprotein is FP_2 (succinate dehydrogenase). Succinate is a small molecule that is produced in the major metabolic pathway called the citric acid cycle (Section 23.9). The final acceptor of reducing equivalents in the overall reaction is coenzyme Q (CoQ). The flavin coenzyme, which is an integral part of the enzyme, first accepts and then passes on a pair of reducing equivalents. Thus the flavin coenzyme serves as an intermediate carrier of the reducing equivalents being passed from NADH or succinate to coenzyme Q.

Coenzyme Q

Coenzyme Q, also known as ubiquinone, is a lipid-soluble compound that is actually dissolved in the lipid bilayer of the mitochondrial inner membrane. In the structure, R represents a long polyunsaturated hydrocarbon chain. CoQ is capable of reversible oxidation-reduction, as shown:

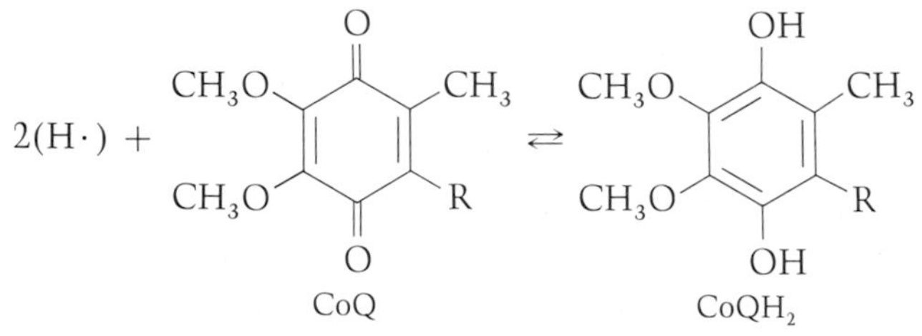

O
CH_3O CH_3 CH_3O OH CH_3
2(H·) + ⇌
CH_3O R CH_3O R
O OH
CoQ $CoQH_2$

Cytochromes

The cytochromes are hemoproteins, that is, proteins containing tightly bound heme groups (Section 21.11). The heme molecule contains an iron atom, which is capable of undergoing a reversible redox reaction:

$$\text{heme–Fe(III)} + e^- \rightleftharpoons \text{heme–Fe(II)}$$

This simple reaction describes the role of cytochromes *b*, c_1, and *c* in the mitochondrial electron transport chain. (Hemoglobin, unlike the cytochromes, does not normally undergo such redox reactions.) The *b*, c_1, and c cytochromes carry out the step-by-step transfer of electrons from $CoQH_2$ to cyt $(a + a_3)$. Cytochrome $(a + a_3)$, also called cytochrome c oxidase, differs from the other cytochromes in being able to bind O_2 as well as catalyze electron transfer reactions. It catalyzes the final transfer of electrons from cytochrome c to O_2.

ATP SYNTHESIS IN THE MITOCHONDRIA 22.17

The oxidation of NADH by O_2, which is catalyzed by the electron transport system, results in the overall release of a very large quantity of free energy. This energy is used to drive the energy-requiring reaction of ATP synthesis. In other words, electron transport from NADH to O_2 is coupled with the synthesis of ATP. The ATP is made by a process called **oxidative phosphorylation.**

What is the immediate source of energy for oxidative phosphorylation?

Careful measurements have shown that three specific reactions in electron transport are coupled to oxidative phosphorylation (Figure 22.14). Each of these three specific redox steps releases energy that is used to power the synthesis of an ATP molecule from ADP and phosphate. Note that the oxidation of a succinate molecule yields only two ATP molecules, whereas three ATPs are gained for each NADH oxidized. The enzyme that catalyzes ATP synthesis is ATP synthetase, a protein of the mitochondrial inner membrane. The overall reaction of mitochondrial electron transport and oxidative phosphorylation is

$$NADH + 4H^+ + \frac{1}{2}O_2 + 3ADP^{3-} + 3P_i^{2-} \rightleftharpoons NAD^+ + 4H_2O + 3ATP^{4-}$$

The big question still remaining is, How is electron transport coupled to oxidative phosphorylation? Much research has been devoted to trying to learn specifically how electron transport donates its energy to drive oxidative phosphorylation. Several hypotheses have been suggested, but as yet we do not entirely understand how the process occurs.

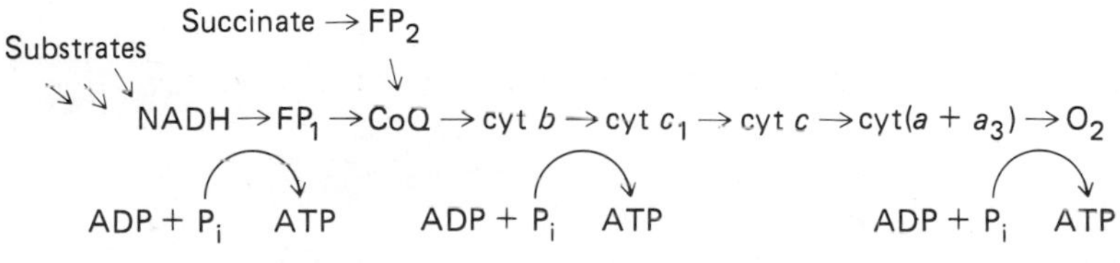

Substrates

Succinate $\rightarrow$ FP_2

$NADH \rightarrow FP_1 \rightarrow CoQ \rightarrow$ cyt $b \rightarrow$ cyt $c_1 \rightarrow$ cyt $c \rightarrow$ cyt$(a + a_3) \rightarrow O_2$

ADP + P_i ATP   ADP + P_i ATP   ADP + P_i ATP

FIGURE 22.14 The mitochondrial electron transport system and the sites of oxidative phosphorylation.

We do, however, know some important things about the way the system behaves. Normal mitochondrial electron transport is said to be tightly coupled with oxidative phosphorylation, because one does not occur unless the other occurs at the same time. Thus, if no oxygen is available to accept electrons from the electron transport system, no electron transport is possible, and no oxidative phosphorylation (ATP synthesis) can occur. This is why oxygen is so important to life. Likewise, if ADP or P_i concentrations are too low, ATP synthesis cannot take place. Therefore electron transport does not take place either, so NADH and succinate cannot be oxidized. The rate of cellular respiration is mainly controlled by the availability of ADP. We'll discuss this point again in Chapter 23, because it has important effects on the rate of oxidation of foodstuffs.

†22.18 CHLOROPHYLL, ELECTRON TRANSPORT, AND ATP SYNTHESIS IN CHLOROPLASTS

What do chloroplasts have in common with mitochondria?

Chloroplasts (Section 21.6) are special organelles possessed only by green plants—plants capable of photosynthesis (Figures 22.15 and 22.16). The chloroplasts contain enzymes of electron transport and ATP synthesis in their membranes, just as mitochondria do. The heart of the photosynthetic electron transport system is the green pigment **chlorophyll.**

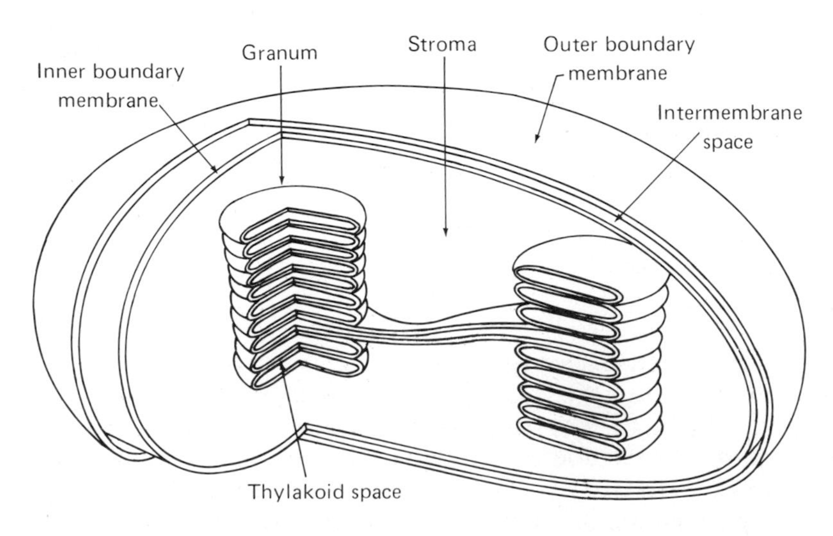

FIGURE 22.15 Diagram of the membranes and internal compartments of the chloroplast.

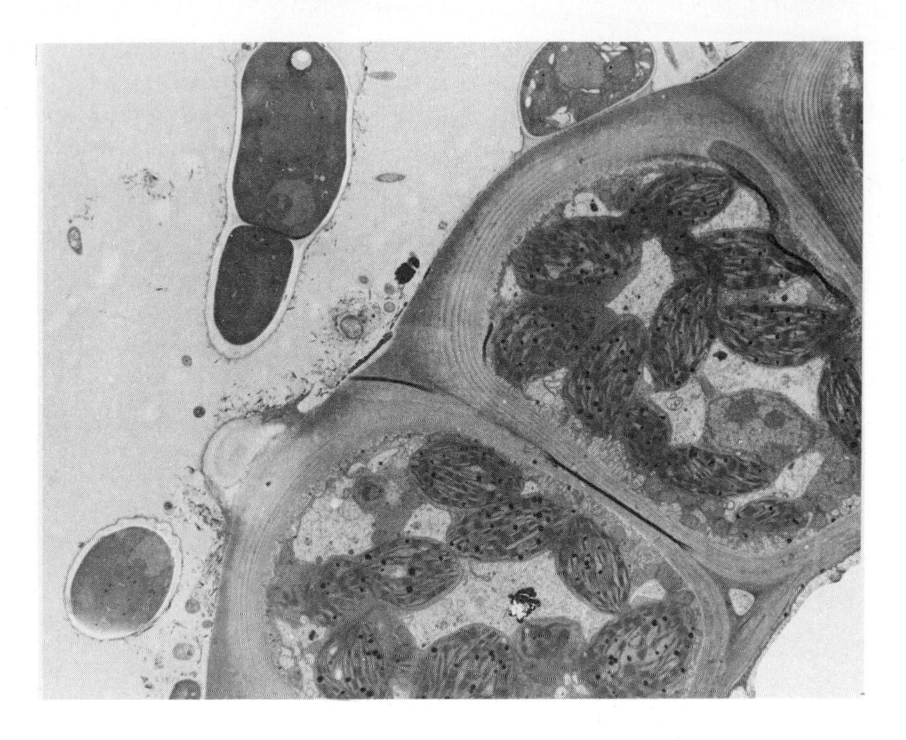

FIGURE 22.16 Electronmicrograph of chloroplasts. (Courtesy of Clinton Dawes, University of South Florida)

The most important of the several known forms of chlorophylls are proteins containing a structure that resembles heme, except that a Mg^{2+} ion replaces the Fe^{2+} of hemes (Figure 22.17). Like hemoproteins, chlorophylls undergo reversible oxidation-reduction reactions. In addition, they are capable of absorbing visible light, including the red light used for photosynthesis. After they absorb energy in the form of light, chlorophylls become very powerful redox compounds. Thus, they convert light energy into chemical energy. (Remember that a highly reduced compound is generally an energy-rich compound.) Since all life on earth depends on the chlorophyll–light reaction, it is rather sobering to see its relative simplicity.

The complete sequence of photosynthetic electron transport in the green plant involves two types of chlorophylls. The two types are found in two different **photosystems,** designated photosystem I and photosystem II. Together these photosystems catalyze chloroplast electron transport. The redox carriers involved, other than chlorophylls, are similar to those involved in mitochondrial electron transport: cytochromes (different from the ones in mitochondria), flavoproteins, a quinone (plastoquinone), and ferredoxin (an iron–sulfur protein). The sequence of carriers for photosynthetic electron transport is shown in Figure 22.18.

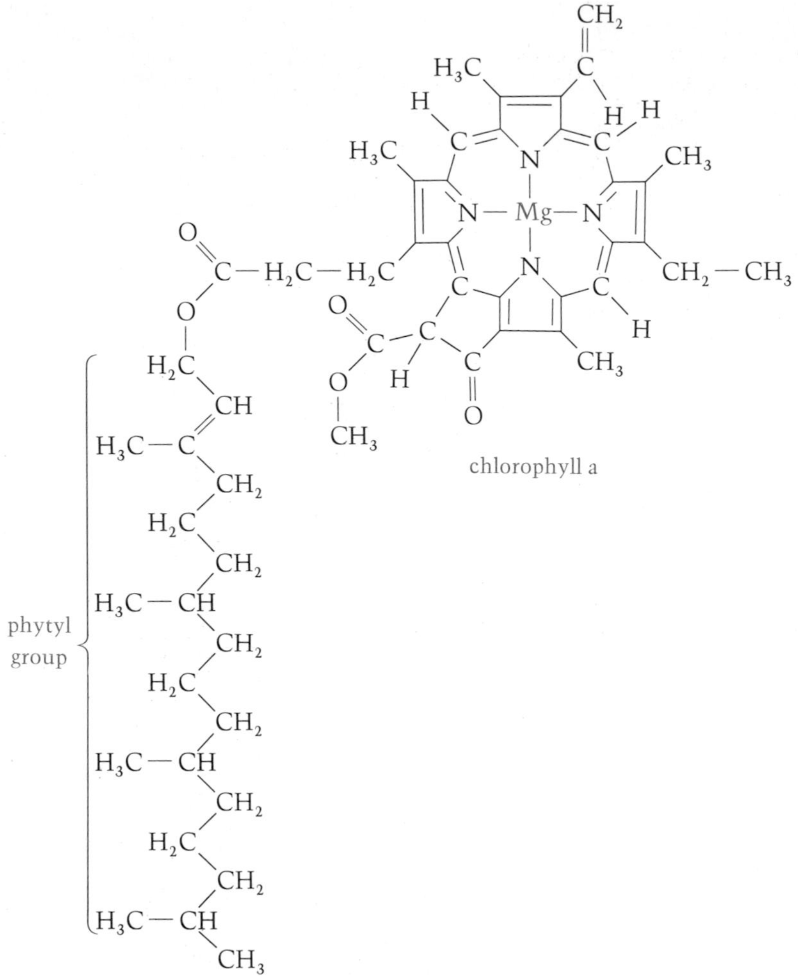

chlorophyll a

phytyl group

FIGURE 22.17 Chlorophyll a is a magnesium porphyrin with a long phytyl tail that anchors the molecules into the chloroplast membranes.

Phosphorylation is coupled with electron transport in the chloroplast as it is in mitochondrial electron transport. It appears that two ATPs are produced for each pair of reducing equivalents that pass down the chain. As you know, in mitochondrial oxidative phosphorylation, three ATPs are produced for each pair of reducing equivalents that pass from NADH to oxygen. However, part of the energy in photosynthesis is stored in the form of the energy-rich reduced coenzyme NADPH rather than ATP, so the energy yield is even higher than that of mitochondrial electron transport.

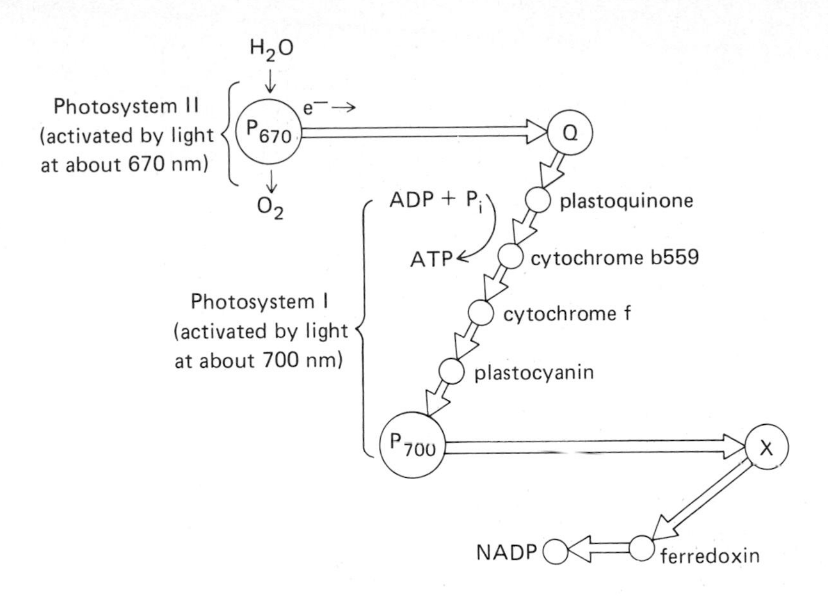

FIGURE 22.18 Z-diagram depicts the two photosystems involved in photosynthetic electron transport. P_{670} and P_{700} are special forms of chlorophyll that act as the electron donors in Photosystems II and I, respectively. The NADPH formed is later used to reduce CO_2 in reactions that do not require light. Although the diagram shows only one ATP-forming step, recent evidence indicates that there may be an additional ATP-forming (photophosphorylation) step. Its location in the scheme is still uncertain.

INHIBITORS OF ELECTRON TRANSPORT AND PHOSPHORYLATION ARE POISONS 22.19†

Can inhibition of electron transport in your mitochondria kill you?

A wide variety of chemicals interfere with electron transport systems or with phosphorylating enzymes, as shown in Figure 22.19. These chemicals make particularly effective poisons.

Cytochrome *c* oxidase (cytochrome $(a + a_3)$) is the site of action of several poisons. Normally, cytochrome *c* oxidase binds O_2 at its heme groups and catalyzes the reduction of O_2 to H_2O, using electrons it accepts from cytochrome *c* and H^+ ions it takes from the surroundings. However, hydrogen sulfide (H_2S), cyanide ion (CN^-), azide ion (N_3^-), and carbon monoxide (CO) are all capable of binding to cytochrome *c* oxidase in place of O_2. This prevents electron transport to O_2. Since the electron carriers cannot get rid of electrons coming into the electron

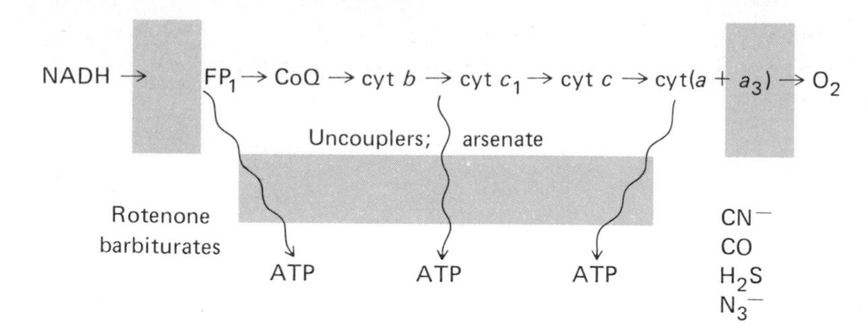

FIGURE 22.19 Sites of action of some poisons of mitochondrial electron transport and ATP synthesis.

transport system, the oxidation of NADH and succinate very quickly comes to a halt. In most cells this blocks the supply of ATP from oxidative phosphorylation, and the affected cells may die. Cyanide has been used in gas chamber executions. Carbon monoxide is a potentially deadly component of automobile exhaust gases and an important factor in air pollution (Section 4.5).

Rotenone, a common insecticide, blocks electron transport at the first phosphorylation site, which comes before the point where electrons from succinate enter the electron transport system. Thus it blocks the oxidation of NADH but not of succinate. Barbiturates (Seconal, Amytal, and others), used as sleep-inducing drugs, also block electron transport at this point. In normal dosages, however, they probably do not inhibit electron transport; rather, they appear to act in some unrelated way on the brain. Nevertheless, an overdose of barbiturates can produce significant inhibition of respiration. The tranquilizing drug chlorpromazine has a similar effect.

A variety of antibiotics are known to inhibit electron transport and oxidative phosphorylation. Certain toxic glycosides from plants have also been shown to act in this way. The herbicide DCMU (dichlorophenylmethylurea) specifically blocks chloroplast electron transport. This inhibits photosynthesis, and the plant eventually dies because it is unable to make enough NADPH and ATP to stay alive.

A special class of agents known as **uncouplers** act by preventing the use of electron transport energy for ATP synthesis. In the presence of an uncoupler, the breakdown of foodstuff molecules and electron transport proceeds rapidly, but the energy is released as heat, and little if any energy is trapped in the form of ATP. 2,4-Dinitrophenol (2,4-DNP) and a variety of other phenolic compounds are active in this way, both in mitochondrial and photosynthetic electron transport. Surprisingly, 2,4-DNP was used as a weight-reducing drug at one time, but establishing dosages was tricky and the treatment was so risky that it was abandoned.

Another type of uncoupler is arsenate, $HAsO_4^{2-}$, which behaves as an analog of phosphate (HPO_4^{2-}). Arsenate can substitute for phosphate in the reaction

$$\text{ADP} + \text{P}_i \xrightarrow{\text{energy}} \text{ATP} + \text{H}_2\text{O}$$

That is,

$$\text{ADP} + \text{As}_i \xrightarrow{\text{energy}} \underset{\substack{\text{unstable}\\ \text{intermediate}}}{\text{ADP} \cdot \text{As}_i} \xrightarrow{\text{hydrolysis}} \text{ADP} + \text{As}_i + \text{heat energy}$$

Thus ATP is not synthesized. This is a major reason why arsenic is toxic.

Problem 22.12 Explain why an uncoupler might be effective as a weight-reducing drug.

SUMMARY

A **spontaneous reaction** is a reaction that proceeds with a release of free energy. Such a reaction is an **exergonic reaction. Endergonic reactions** (reactions accompanied by an increase of free energy) are not spontaneous reactions. The **standard free-energy change ΔG^0** is the amount of free energy given off by a reaction proceeding to equilibrium under certain standard conditions. The actual free-energy change, **ΔG,** of a reaction depends on the ΔG^0 and the actual concentrations of reactants and products. A reaction is spontaneous if its actual ΔG is negative. A highly exergonic (**irreversible**) reaction is usually found at the start of a metabolic pathway. It serves as a **committing step.**

All biological energy derives from the sun; since only **photosynthetic cells** are capable of transforming solar energy into chemical energy, all other organisms depend on them for foodstuffs. Energy is passed along **food chains,** but the transfer is not very efficient.

Adenosine-5′-triphosphate (**ATP**) is the cell's energy currency. When foodstuffs are oxidized, their stored chemical energy is converted into another form of chemical energy, ATP, which is a **high-energy compound.** ATP energy can drive an endergonic reaction when that reaction is **coupled** with ATP hydrolysis. Conversely, ATP can be formed from ADP and phosphate when the reaction is coupled with a very exergonic one. ATP energy is used to drive the synthesis of cell materials, muscle contraction, and active transport. Muscle contraction, controlled by the concentration of Ca^{2+} in the cytoplasm, involves the interaction among **actin** of the **thin filaments, myosin** of the **thick filaments,** and ATP. Active transport—pumping of molecules or ions across membranes against a concentration difference—may be driven by ATP hydrolysis or by an electrical potential across the membrane.

Organic molecules contain chemical potential energy that is released when the molecules are oxidized. **Redox coenzymes** are the acceptors of reducing equivalents when substrates are oxidized. Energy released by the oxidation of the substrate is temporarily stored in the reduced coenzyme. Most of the reduced coenzyme formed during foodstuff oxidation arises in the mitochondria. The mitochondria make most of the cell's ATP while oxidizing NADH; they transfer reducing equivalents from the coenzyme to the final acceptor, O_2. The catalysts of this process are the electron carriers of the **electron transport system.** The carriers are of three types: **flavoproteins, coenzyme Q,** and the **cytochromes.** Electron transport from NADH to O_2 is very exergonic; the energy released is used to drive ATP synthesis, that is, **oxidative phosphorylation.** Electron transport and oxidative phosphorylation are very tightly coupled. The rate of cell respiration is controlled by the availability of ADP.

Photosynthetic electron transport takes place in the chloroplasts of green plants. The pigment **chlorophyll** traps light energy and converts it into chemical energy. Electron transport from these energy-rich reduced chlorophylls to NADP is coupled with phosphorylation. Electron transport in mitochondria and chloroplasts can be inhibited by a number of poisons. Poisons known as **uncouplers** act by preventing the use of electron transport energy for phosphorylation.

Key Terms

Actin (Section 22.10)
Active transport (Section 22.11)
Adenosine-5′-triphosphate (ATP) (Section 22.7)
Carnivore (Section 22.6)
Cellular respiration (Section 22.15)
Chlorophyll (Section 22.18)
Coenzyme Q (Section 22.15)
Committing step (Section 22.5)
Coupling (Section 22.9)
Cytochrome (Section 22.15)
Dehydrogenase (Section 22.16)
Electron transport system (Section 22.15)

Endergonic reaction (Section 22.2)
Exergonic reaction (Section 22.2)
Facilitated diffusion (Section 22.11)
Flavoprotein (Section 22.15)
Food chain (Section 22.6)
Half-reaction (Section 22.12)
Free-energy change (ΔG) (Section 22.2)
Herbivore (Section 22.6)
High-energy compound (Section 22.8)
Irreversible reaction (Section 22.5)
Myofibril (Section 22.10)
Myosin (Section 22.10)
Oxidative phosphorylation (Section 22.17)
Photosynthetic cells (Section 22.6)
Photosystem (Section 22.18)
Redox coenzyme (Section 22.12)
Reducing equivalent (Section 22.13)
Respiration (Section 22.15)
Sarcomere (Section 22.10)
Spontaneous reaction (Section 22.2)
Standard free-energy change (ΔG^0) (Section 22.3)
Thermodynamics (Section 22.1)
Thick filaments (Section 22.10)
Thin filaments (Section 22.10)
Troponin (Section 22.10)
Tropomyosin (Section 22.10)
Uncoupler (Section 22.19)

ADDITIONAL PROBLEMS

22.13 Describe the role of Ca^{2+}, on the molecular level, in controlling muscle contraction and relaxation.

22.14 In certain diseases of calcium metabolism, tissue calcium levels may be excessively high or low. The heart is the major tissue affected. Predict the effect of high or low calcium concentrations in the cells on the contractile function of the heart.

22.15 List and describe the uses of ATP in the cell.

22.16 Pick the compounds from Table 22.1 whose hydrolysis might be coupled to ATP synthesis from ADP and P_i. Explain why these are good choices.

22.17 Compare the types of enzymes found in the photosynthetic and mitochondrial electron transport systems.

22.18 Which of the enzymes you named in Problem 22.17 contain substances recognized as vitamins?

22.19 If you were poisoned with 2,4-DNP, would you expect your body temperature to rise? Why?

22.20 Why is ubiquinone found dissolved in the mitochondrial membrane?

22.21 Explain why cyanide is a highly effective poison.

22.22 Write the half-reaction for the reduction of NAD, using the full structural formulas of the oxidized and reduced forms.

22.23 The ΔG^0 for photosynthetic electron transport from H_2O to NADP is very positive. What provides energy to drive this endergonic reaction?

Problem Solving Hints

If you have trouble solving any of the additional problems, refer to the sections listed next to the problem numbers.

22.13 22.10
22.14 22.10
22.15 22.7
22.16 22.9
22.17 22.16, 22.18
22.18 22.16
22.19 22.19
22.20 22.16
22.21 22.19
22.22 22.13
22.23 22.18

FOCUS 13 The Energy of ATP

There are several reasons why ATP hydrolysis gives off so much free energy. We will look at two of them. First, note that the chain of three phosphate groups in ATP bears four full negative charges. Like charges tend to repel one another. This makes the compound very reactive to hydrolysis, because hydrolysis will relieve much of the strain of having so many negative charges close together. The products of the hydrolysis of ATP are thus more stable (have a lower free-energy content) than the reactants are (Figure 22.4).

An additional contributing factor is *opposed resonance.* You may recall that benzene has an unusually stable structure because the pi bonding electrons are shared by the six carbons in the ring (Section 10.14):

not this form (a) or this form (b) but this form (c)

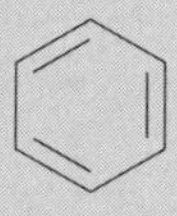

This phenomenon is known as resonance (Section 10.14). The electrons are able to occupy several alternate bonding positions. We show this by representing benzene with structure c. Such a compound has a lower energy content and hence is more stable than it would be if it had either of the structures represented by a or b. A phosphate group also has resonance, because several equivalent arrangements of the bonding electrons are possible:

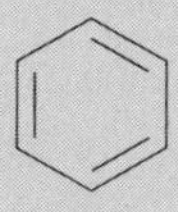

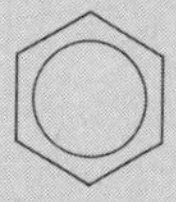

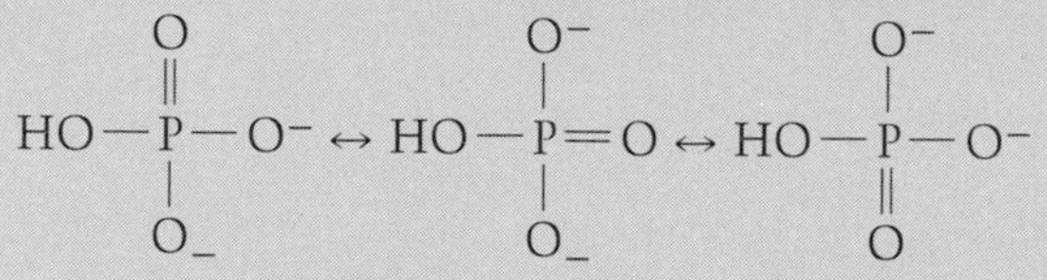

$$\begin{array}{ccccc} \mathrm{O} & & \mathrm{O^-} & & \mathrm{O^-} \\ \| & & | & & | \\ \mathrm{HO{-}P{-}O^-} & \leftrightarrow & \mathrm{HO{-}P{=}O} & \leftrightarrow & \mathrm{HO{-}P{-}O^-} \\ | & & | & & \| \\ \mathrm{O_-} & & \mathrm{O_-} & & \mathrm{O} \end{array}$$

The phosphoric anhydride groups of ATP have a smaller number of resonance forms than do ADP and P_i taken together. That is, the anhydrides oppose or prevent the greater resonance stabilization of non-anhydride phosphate groups. Thus the energy content of ATP is greater than that of ADP and P_i, and some of the negative ΔG^0 of ATP hydrolysis results from the formation of the more resonance-stabilized compounds ADP and P_i.

FIGURE 1 This athlete is using ATP as a source of energy. (Ellis Herwig/Stock, Boston)

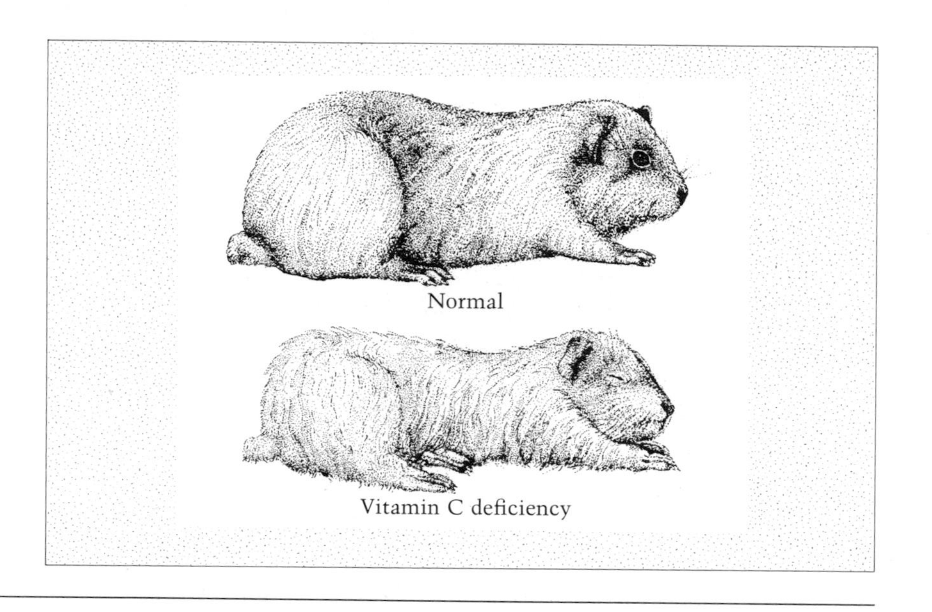

Vitamins are chemical compounds needed in small amounts by the body to maintain normal metabolism. We will stress the importance of these essential components of food in this chapter. Essential compounds (those that cannot be made in the body) must be obtained in our diet. Certain vitamins are fat soluble and can be stored in the cells of various organs. However, some of these stored vitamins are retained for only short periods of time and must be supplied regularly to avoid a vitamin deficiency. Also, other vitamins are water soluble and are excreted more readily than those that can be stored. Sometimes vitamin shortages can occur during strenuous exercise, pregnancy, and disease. The following case study should help us understand the consequences of a vitamin deficiency.

Brenda, a ten-month-old infant whose diet consisted mainly of milk and soup, developed dry, rough skin that formed blood blisters. Bleeding areas were found on her arms and legs, and a few blood spots were present on the whites of her eyes. She had swollen gums that bled easily, and she appeared to be weak and short of breath. When her mother moved Brenda's arms and legs, the baby appeared to be in pain.

The pediatrician diagnosed a severe vitamin C deficiency (scurvy), which had resulted from the baby's failure to take in proper amounts of fruits, meats, and vegetables. Vitamin C is important in the formation of connective tissue. Brenda's lack of vitamin C led her to have fragile skin and blood

23

DIGESTION AND METABOLISM: THE CITRIC ACID CYCLE

vessels that were easily injured. The pain in her arms and legs was caused by bleeding in the tissue that lined her bones.

With a change in her diet and the addition of a vitamin C supplement, Brenda was able to replenish her vitamin stores, and her condition improved rapidly.

[Illustration: Like humans, guinea pigs that are deprived of vitamin C will develop scurvy. In the guinea pig, this condition is characterized by enlarged joints and a general weakened appearance.]

DIGESTION AND METABOLISM: THE CITRIC ACID CYCLE

In the last chapter, we saw how adenosine-5′-triphosphate (ATP) is made in the mitochondria of all cells. We also looked at two major cell activities that consume ATP energy: muscle contraction and the active transport of materials across membranes. Now we are ready to begin studying the processes that supply energy and raw materials to the body. First, there is the digestion of foods, which makes glucose and other sugars, amino acids, and fatty acids available to the body cells. Then, the cells oxidize these substances, releasing energy that is used for making ATP. The breakdown of sugars, fatty acids, and amino acids in cells leads at the same time to the formation of certain small molecules that can be used for making new cell materials. The manufacture of new body materials requires energy from ATP hydrolysis. It is the third major cell activity that consumes ATP. Both foodstuff oxidation and the synthesis of new biomolecules are carefully controlled, step-by-step processes. They consist of a number of metabolic pathways crisscrossing in a sometimes bewildering manner. In this chapter we will consider the digestion of foods and some general features of metabolism. We will also introduce you to the important central metabolic pathway known as the citric acid cycle.

23.1 FOOD FOR THOUGHT

Let's say that you have just come back from lunch. Today you had a chicken sandwich on lightly buttered whole wheat bread and a glass of milk. How are these materials—meat and milk proteins, butterfat, wheat starch and protein, milk sugar, and so on—going to be put to work in your body to energize your afternoon activities and provide raw material

TABLE 23.1 Primary Components of Your Lunch

	Protein	Fat	Carbohydrate
Bread	✓		✓✓*
Chicken	✓✓*	✓	✓
Milk	✓	✓	✓
Butter		✓✓*	

*Two check marks indicate the major component.

for necessary cell repairs and such? In what new form, and where, will these raw materials end up? Let us first consider the handling of the materials now in your stomach, that is, the digestion and absorption of the various foodstuffs. We'll look at carbohydrate, fat, and protein digestion, in turn. Table 23.1 indicates which of your lunchtime foods will be involved in each of these kinds of digestion. We'll find out which of these nutrients is digested most quickly to provide you, quite literally, with the food your brain needs for thought. (The rest of the story will be told as we make our way through the next several chapters.)

DIGESTION OF CARBOHYDRATES 23.2

Where in the body does most carbohydrate digestion take place?

The major carbohydrate in the diet is starch, as in the bread of your chicken sandwich. Smaller amounts of other carbohydrates are also present in a normal diet—glycogen (in the chicken meat, for example) and the disaccharides lactose (milk sugar) and sucrose. The aim of carbohydrate digestion is to break the glycosidic bonds (Section 19.3) of these compounds. Hydrolysis releases the sugars glucose, fructose, and galactose, which can be absorbed directly from the intestine into the circulatory system and then into body cells.

Several enzymes contribute to the hydrolysis of starch and glycogen (Figure 23.1). **Amylase** (from amylose = starch) is found in saliva; it is also secreted by the pancreas into the duodenum. Amylase usually attacks polysaccharides at every other $\alpha(1\rightarrow4)$ glycosidic bond; however, it cannot attack a bond closer than two glucose units from a $1\rightarrow6$ branch. The major products of amylase action are the disaccharide maltose and the trisaccharide maltotriose. About 65% of the polysaccharide is broken down to these forms. Various short branched polymers (α-dextrins) make up about 30% of the products.

The brush-border membrane that lines the intestine contains additional enzymes capable of breaking down these short glucose polymers.

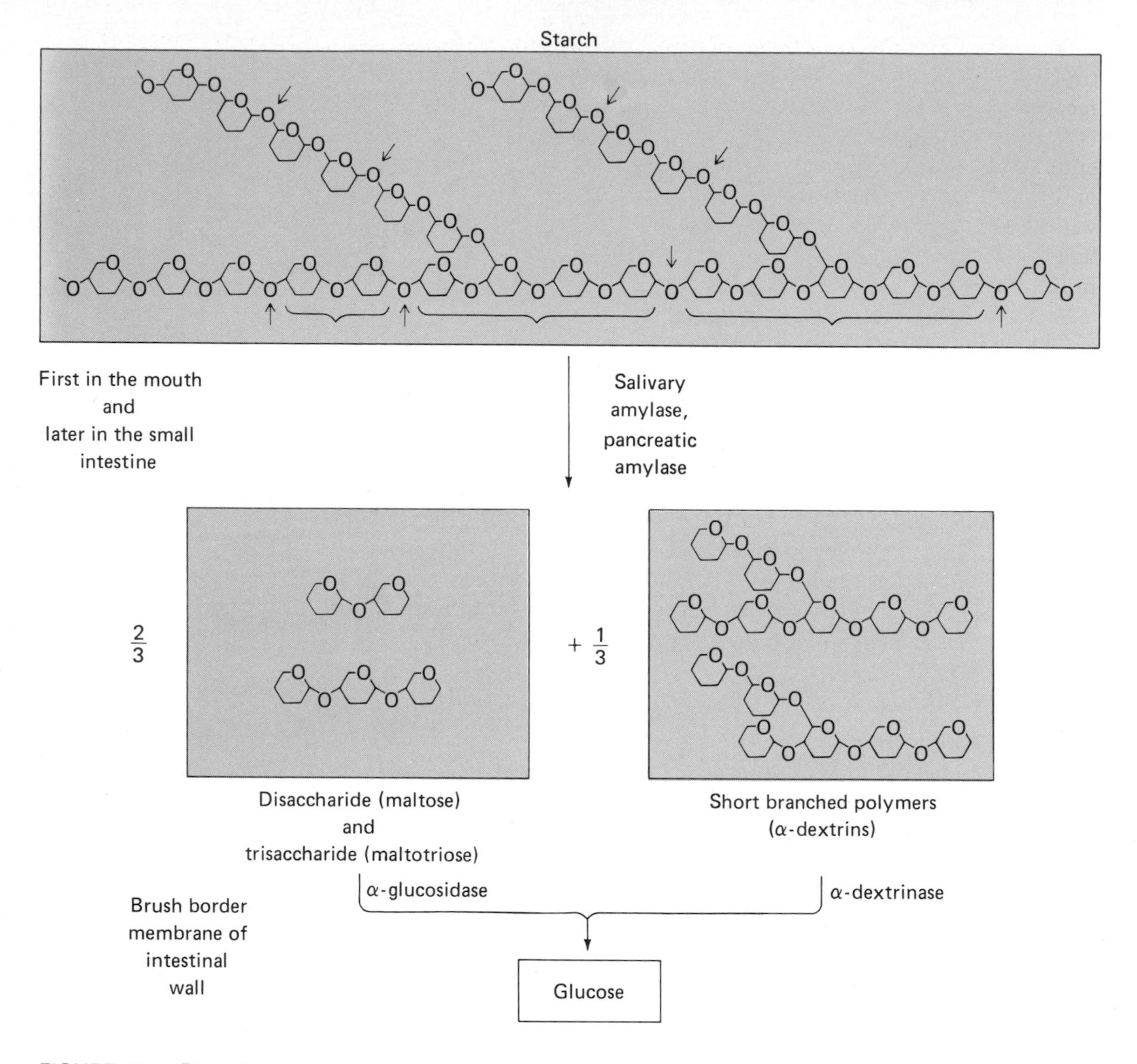

FIGURE 23.1 Digestion of starch.

One enzyme attacks the linear ones, cleaving one glucose unit at a time. A second enzyme, which can attack $\alpha(1\rightarrow6)$ as well as $\alpha(1\rightarrow4)$ glycosidic bonds, breaks the short branched polymers down to glucose. Glucose is then absorbed into the blood stream. The brush border contains a **sucrase** and a **lactase** as well, which release glucose, fructose, and galactose from the disaccharides sucrose and lactose. The sugars, and then the starches, are the foods that are most quickly digested and absorbed. This is an advantage because the blood glucose levels must be kept up so that the brain has enough glucose—its major fuel.

Some people would not choose to drink milk at lunch because sometimes they don't feel too well after doing so. It has been recognized only recently that many adults and some children cannot digest milk well because they produce little or no lactase. Apparently the enzyme is present during infancy and early childhood when milk is an important part of the diet, but later the body produces much less of the enzyme. People with this deficiency are said to be lactose-intolerant; that is, they experience digestive upsets if they ingest significant amounts of milk or milk products.

Problem 23.1 Write a chemical equation for the reaction catalyzed by sucrase. (Use names of the compounds only.)
Problem 23.2 Write a chemical equation for the reaction catalyzed by lactase. (Use structural formulas for the compounds.)

DIGESTION OF FATS 23.3

The lipids in your diet consist primarily of triacylglycerols such as the butterfat in your sandwich and milk. There are normally much smaller amounts of cholesterol and phosphoglycerides and other lipids that contain fatty acids. The first step in the digestion of fats is to break them up into many small droplets. This increases the surface area where the pancreatic **lipase** can work. The lipase is the enzyme that catalyzes hydrolysis of the ester bonds in the fats. The mechanical mixing provided by the stomach and intestine, as well as the detergent action of certain dietary substances and digestive juices, helps break up fats.

Are there natural detergents in the body?

The major detergents involved are bile salts and partly digested fats—fatty acid soaps (Section 20.9) and monoacylglycerols (see Figure 23.2). The bile salts are formed when bile acids, like cholic or deoxycholic acids (Table 20.4), form amide linkages with glycine, for example:

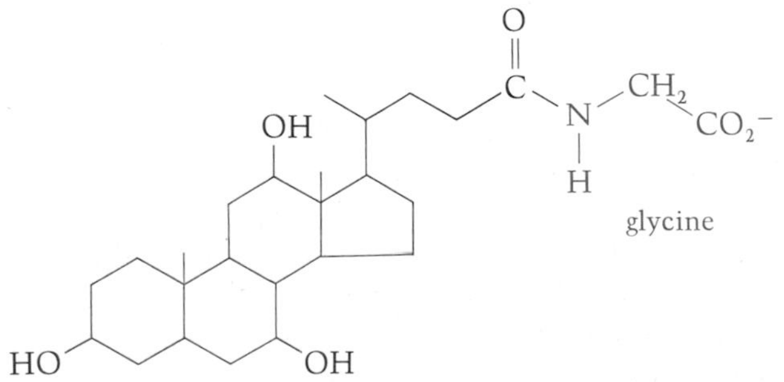

O
C N CH_2 CO_2^-
OH
H
glycine
HO OH

glycocholic acid
(a bile salt)

When fat enters the small intestine from the stomach, the gall bladder, which stores bile manufactured by the liver, discharges bile into the intestine. Fatty materials are broken up into small droplets, as just described, by the action of the bile salts and the other detergents. Lipase, secreted into the intestine in pancreatic juice, begins to attack the triacylglycerols. Usually two fatty acids are hydrolyzed, one at a time, by the action of the lipase. The products of this hydrolysis are a monoacylglycerol (Figure 23.2) and the pair of fatty acids. The fatty acids, monoacylglycerols, and bile salt molecules form micelles (Section 20.4) that

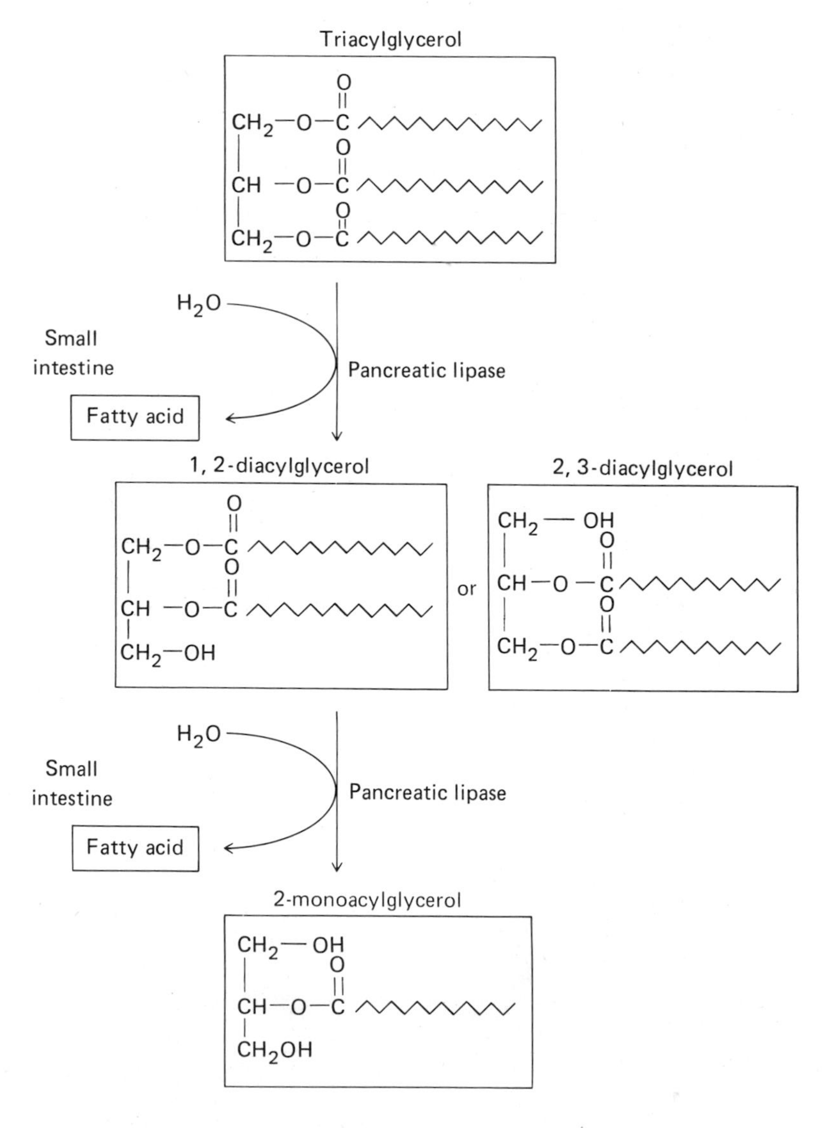

FIGURE 23.2 Digestion of fats.

diffuse to the brush-border membrane, where the digestion products are absorbed.

The bile salts are absorbed in the ileum portion of the small intestine. Later the liver recycles them. A deficiency of bile salts, which may be caused by decreased production or a blockage between the gall bladder and the intestine, causes poor absorption of lipids and fat soluble vitamins. The condition is called steatorrhea.

Problem 23.3 Write an overall reaction for the hydrolysis of a trioleoylglycerol (a triacylglycerol containing three oleic acid groups) as shown in Figure 23.2. Include complete structural formulas.

DIGESTION OF PROTEINS 23.4

Now we'll see how the proteins of your luncheon meal will be digested. This involves breaking proteins down into their component amino acids. It may seem simple enough, but the digestion of proteins poses a complicated set of problems. Among them are the following:

How many different enzymes are involved in protein digestion?

1. Since proteins contain a variety of amino acids, there are a great many different peptide bonds to be attacked—that is, a great many peptide bonds with different amino acids on either side of them.
2. Enzymes that are so nonspecific that they will attack any peptide bond, regardless of the amino acids involved, are rather poor catalysts. Large amounts of such enzymes have to be produced.
3. Highly specific enzymes, which will attack only one or a few combinations of amino acids, are excellent catalysts, but complete protein digestion would require the production of many different enzymes.
4. Since enzymes are proteins themselves, they too are digested when they pass through the digestive tract; thus they must be produced in some excess.

The actual scheme for protein digestion balances these factors against one another to arrive at a compromise between specificity and speed. The digestive juices contain six protein-degrading enzymes. Four of these enzymes cleave *internal* peptide bonds within protein chains, and the remaining two cleave peptide bonds at one or the other *end* of the protein chain. The combined action of the six enzymes acting first in the stomach and then in the small intestine produces a mixture of free amino acids and short peptides. These are absorbed by the intestine, and the peptides are further hydrolyzed inside the cells of the intestinal lining. As a result, essentially all dietary protein is completely broken

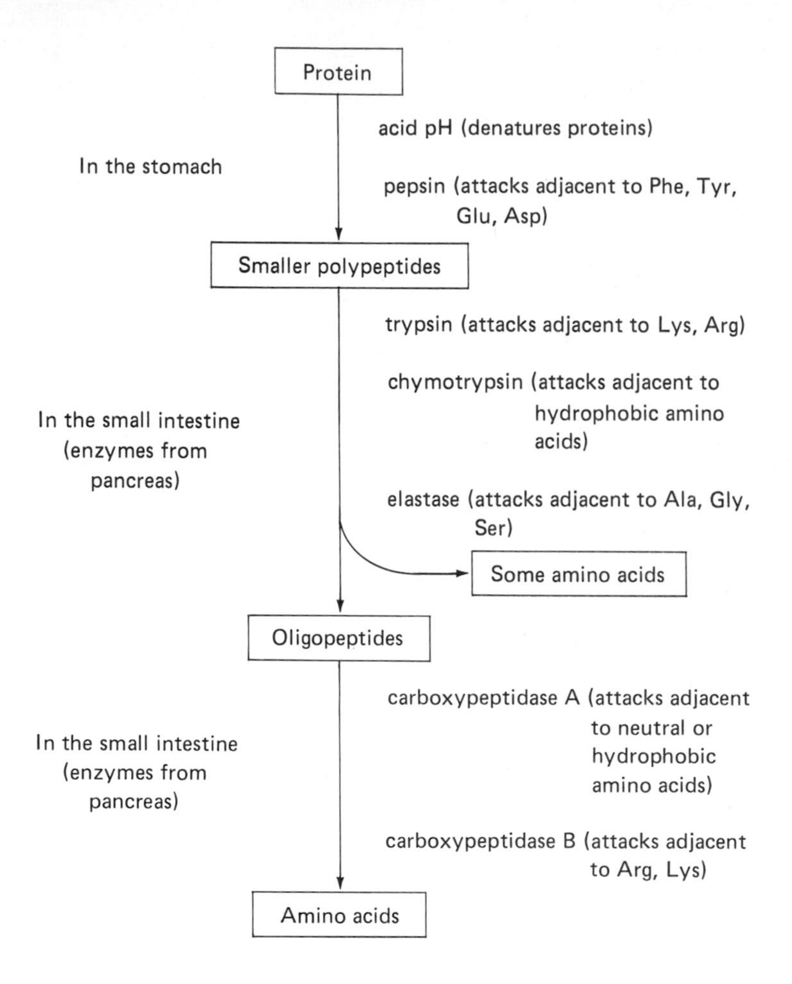

FIGURE 23.3 Digestion of protein.

down to amino acids before it reaches the blood. The sequence of events in protein digestion is shown in Figure 23.3.

Since proteins and fats are more slowly digested than carbohydrates, your lunch will continue to provide food materials to be absorbed for several hours, and you probably won't feel hungry again until close to dinner time.

Problem 23.4 Write an equation for the reaction in which a protein-degrading enzyme cleaves (catalyzes hydrolysis of) the C-terminal amino acid of the following peptide (include complete structural formulas):

asp–gly–lys

CALORIC VALUE AND ENERGY YIELD OF FOODS 23.5

While we are on the subject of the digestion and absorption of the major classes of foods, let us also look at some other factors in nutrition.

Carbohydrates, fats, and proteins can replace one another nutritionally to a major extent. All are oxidized in cells by a final common pathway that leads to ATP synthesis by mitochondria. The best measure of the energy yield of a foodstuff is the ATP production associated with its oxidation. Carbohydrates and fats yield about the same percentage of ATP—about 40% of the free energy of oxidation of these materials is converted to ATP. More protein calories are required to produce the same amount of ATP, however. Thus, a high-protein diet yields less total energy in terms of ATP synthesis.

ESSENTIAL FOOD COMPONENTS 23.6

What are the "essential" amino and fatty acids?

Protein is not needed primarily for energy. Rather, protein is the major dietary source of nitrogen, which is needed for the manufacture of many important biomolecules—nucleotides, various hormones, and nerve transmission substances, to name but a few.

Naturally, protein is also needed to supply amino acids and nitrogen for the synthesis of proteins and some of the amino acids in the body. However, the body cannot synthesize all 20 amino acids of proteins; some must be obtained from foods. Several amino acids are indispensable components of the diet—these are known as the **essential amino acids** (Table 23.2).

High-quality protein is protein that contains adequate amounts of the essential amino acids. Animal proteins are usually of high quality. Many plant proteins, however, lack one or more of the essential amino acids. Wheat protein is poor in lysine, for example, and corn protein is poor in both lysine and tryptophan. People whose protein intake is close

TABLE 23.2 Essential Amino Acids

Valine	Lysine	Methionine
Leucine	Phenylalanine	Threonine
Isoleucine	Tryptophan	Cysteine

TABLE 23.3 Minerals Required in the Diet

Sodium	Magnesium	Iron	Cobalt	Chloride
Potassium	Zinc	Copper	Phosphate	Iodide
Calcium		Manganese		Fluoride

to the minimum necessary (35 to 50 g per day) must take care to include high-quality proteins in their diet or they may not take in enough of the essential amino acids. Usually this means including some animal proteins—milk, meat, and eggs. A strict vegetarian diet can be dangerous unless one knows enough about the quality of the various plant proteins to make sure of taking in an adequate amount of the essential amino acids. This is especially true for children, who are still growing and have a greater demand for the synthesis of protein and other nitrogen-containing compounds.

Essential fatty acids, as their name implies, must also be components of the diet. Linoleic acid (Section 20.1) is the primary essential fatty acid. Generally, there is little difficulty in taking in adequate amounts of linoleic acid since most foods contain it.

Minerals and certain **trace elements** must also be obtained in the diet. The mineral elements necessary for the health of the animal organism are listed in Table 23.3.

A variety of other metallic elements and other trace elements are required in very, very small, or "trace," amounts. The diet usually contains enough of all of these minerals and trace elements. Deficiency symptoms sometimes do develop, however. Examples are anemia due to iron deficiency and goiter due to shortage of iodine in the diet.

23.7 VITAMINS

What are most vitamins actually needed for?

The vitamins are another essential component of foods. We have already discussed, in Chapters 20 and 21, some aspects of the structure and function of fat soluble and water soluble vitamins. As we mentioned previously, these materials cannot be made in the body and thus must be supplied by the diet. Only trace amounts are necessary: except for vitamin C (ascorbic acid), less than 20 mg per day of any vitamin are required. Most of the vitamins are converted to some coenzyme (Section 21.1) that is necessary for the normal functioning of food oxidation enzymes or enzymes involved in the synthesis of body materials. Vitamin-deficiency diseases are rare in modern society, because most people have adequate nutrition.

Table 23.4 gives information about the functions and dietary sources of all the fat soluble and water soluble vitamins. Note that these names give important clues to the likely food sources of the vitamins.

TABLE 23.4 Some Common Vitamins

Name	Minimum daily adult requirement	Function of vitamin	Deficiency disease or symptoms	Source of vitamin in diet
Fat soluble				
Vitamin A (retinol and dehydroretinol)	5000 I.U.* (1.5 mg)	Helps form visual pigments and maintains normal epithelial structure	Night blindness	Fish-liver oils, liver, egg yolk, green leafy or yellow vegetables
Vitamin D (ergocalciforol and cholecalciforol ·	400 I.U.	Needed for formation of good bones and teeth	Rickets, which result in defective bone formation	Fish-liver oil, butter, egg yolk, along with ultraviolet light (sunlight)
Vitamin E (tocopherol)	30 mg as synthetic *dl*-α-tocopherol acetate	Stabilizes biological membranes and acts as antioxidant	Red blood cells have greater tendency for hemolysis	Vegetable oil, wheat germ, leafy vegetables
Vitamin K (K_2-phylloquinone)	About 0.03 mg per kg of body weight	Needed for formation of prothrombin and for normal coagulation of blood	Hemorrhage	Leafy vegetables, vegetable oils; also produced by intestinal flora after fourth day of life
Water soluble				
Vitamin C (ascorbic acid)	60 mg	Needed for maintenance of connective tissues, vascular function, tissue respiration, and wound healing	Scurvy; loose teeth and bleeding gums	Citrus fruits, tomatoes, potatoes
Folic acid (folacin)	0.1 to 0.5 mg, depending on source	Needed for synthesis of purines and pyrimidines and for maturation of red blood cells	Anemia	Fresh leafy green vegetables, fruits, organ meats, liver, dried yeast
Niacin (nicotinic acid)	15–20 mg	Acts as a coenzyme in redox reactions, also used in metabolism of carbohydrates and tryptophan	Pellagra	Dried yeast, liver, meat, fish, whole grains
Vitamin B_1 (thiamine)	1–1.5 mg	Needed for metabolism of carbohydrates and for function of nerve cells	Beriberi (acute cardiac symptoms and heart failure, also weight loss and nerve inflammation)	Whole grains, pork, liver, nuts, enriched cereal products
Vitamin B_2 (riboflavin)	1.0–1.7 mg	Coenzyme in redox reactions; also involved in protein metabolism	Visual problems; skin fissures	Milk, cheese, liver, meat, eggs

*I.U. stands for international unit. One international unit equals 0.3 μg of retinol.

TABLE 23.4 Some Common Vitamins (Continued)

Name	Minimum daily adult requirement	Function of vitamin	Deficiency disease or symptoms	Source of vitamin in diet
Vitamin B_6 group (pyridoxine, pyridoxal, pyridoxamine)	2 mg	Needed for cellular function and for metabolism of certain amino and fatty acids	Convulsions (in infants); skin disorders (in adults)	Dried yeast, liver, organ meats, fish whole-grain cereals
Vitamin B_{12} (cyanocobalamine)	5 μg	Needed for DNA synthesis related to folate coenzymes, also for maturation of red blood cells	Pernicious anemia, and certain psychiatric syndromes	Liver, beef, pork, eggs, milk and milk products
Vitamin H (biotin)	0.15–0.30 mg	Needed for protein synthesis, transamination, and CO_2 fixation	Skin disorders	Liver, dried peas and lima beans, egg white, also produced by bacteria in alimentary canal
Pantothenic acid	10 mg	Forms part of coenzyme A	Fatigue, malaise, neuromotor and digestive disorders	Dried yeast, liver, eggs

Source: Sherman, Sherman, and Russikoff: *Basic Concepts of Chemistry,* 2nd edition. Copyright © 1980 by Houghton Mifflin Company. Used by permission.

23.8 INTRODUCTION TO METABOLISM

The fate of the products of digestion is our next concern. Sugars, amino acids, and fatty acids circulating in the bloodstream are absorbed into cells all over the body according to their needs. The cells, in turn, metabolize these materials. **Metabolism** involves all the metabolic pathways and the reactions that a cell or organism carries out in breaking down large molecules and synthesizing new body materials. Metabolism may be divided into two parts—catabolism and anabolism. **Catabolism** is the breaking down of large molecules into smaller ones. Catabolic pathways are oxidative and energy-yielding (exergonic) pathways. **Anabolism,** or **biosynthesis,** is the synthesis of larger molecules from smaller ones. Anabolic pathways are reductive and energy-requiring (endergonic) pathways.

Catabolism leads to the formation of ATP and reduced coenzymes, both of which are energy-rich. The energy released during catabolism is thus stored temporarily in these chemical forms. Biosynthetic (anabolic) pathways use the ATP and reduced coenzymes as their sources of energy and reducing power. Thus catabolism and anabolism are closely intertwined processes (Figure 23.4).

In addition, the pathways of catabolism of sugars, amino acids, and

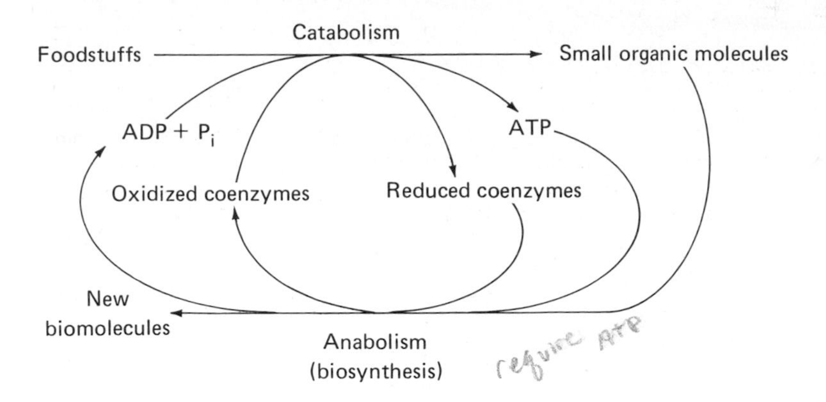

FIGURE 23.4 Coupling of catabolism and anabolism (biosynthesis).

fatty acids produce a few small, organic molecules. These are the raw materials from which new cell materials are synthesized in the biosynthetic pathways. Most of these molecules are intermediates in the central metabolic pathway known as the citric acid cycle, which we will discuss further in Section 23.9. The citric acid cycle therefore provides a concrete link between the pathways of catabolism and those of anabolism.

Figure 23.5 summarizes the catabolic and anabolic pathways for all three major types of foodstuffs. The largest, most complex molecules—namely, proteins, polysaccharides, and lipids—are placed at the very top of the diagram. These are shown being broken down to amino acids, glucose, and fatty acids, respectively. In addition, arrows pointing in the opposite direction show that these materials can be built back up into proteins, polysaccharides, and lipids. We can make the same sort of observation about other stages in the metabolism of these materials. For example, fatty acids are catabolized to the compound acetyl CoA. Acetyl CoA is also the raw material used to synthesize fatty acids, as shown by the oppositely directed arrow in the diagram. An important point that emerges from a close look at the diagram is that the pathways are separate and different, even though the reactions involved in each pathway might appear to be simply the reverse of one another.

The circular, or cyclic, pathway at the bottom of the diagram is the citric acid cycle. The close connection of this central pathway with all the other catabolic and anabolic pathways is clearly shown. Since the citric acid cycle is of such major importance in metabolism, it is the first metabolic pathway that we will study. However, in later chapters we will study the other catabolic and anabolic pathways shown in Figure 23.5.

Problem 23.5 Make a table that contrasts the substrates and products of anabolism and catabolism.

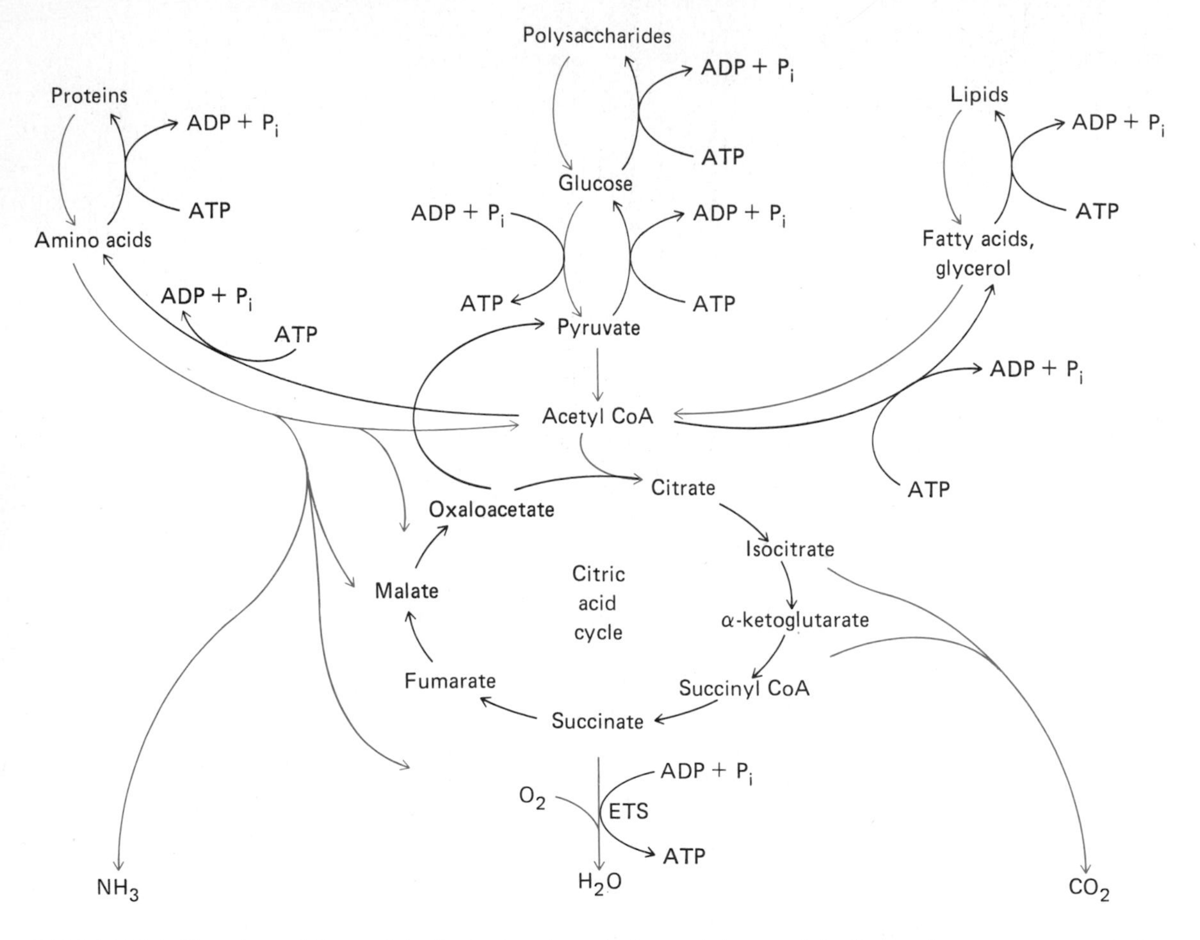

FIGURE 23.5 Overview of the metabolism of carbohydrate, protein, and fat. Colored arrows are catabolic pathways; black arrows are anabolic (biosynthetic) pathways.

23.9 THE CITRIC ACID CYCLE—THE HUB OF METABOLISM

What are the products of the citric acid cycle?

The **citric acid cycle** (Figure 23.6) is a series of eight enzyme-catalyzed reactions that has both catabolic and anabolic functions. Its catabolic function is to accomplish complete oxidation of the two-carbon acetyl group of acetyl CoA. Its anabolic function is to make available a pool of small molecules, which can be drawn on to supply raw material to the biosynthetic pathways. We sometimes call the citric acid cycle the *Krebs cycle* in honor of Sir Hans Krebs, who first demonstrated its existence.

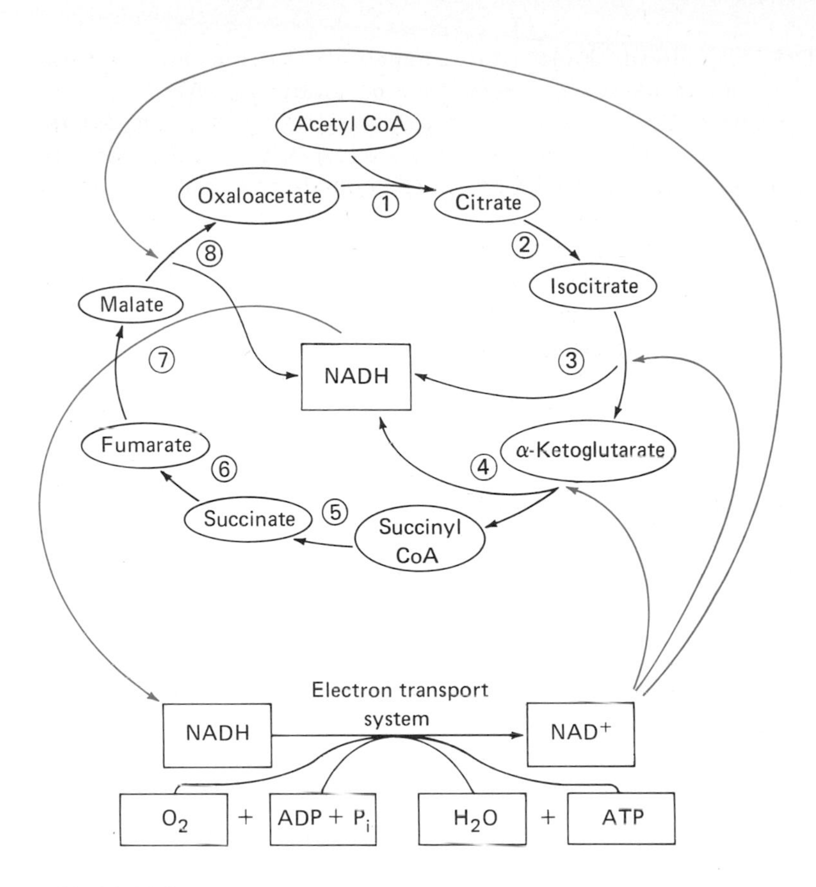

FIGURE 23.6 The citric acid cycle and its coupling to the electron transport system. The electron transport system must continuously oxidize NADH so the citric acid cycle can operate.

The citric acid cycle is the crucial metabolic pathway through which most of the energy stored in food molecules is released. Before entering the citric acid cycle, carbohydrates (such as glucose), fatty acids, and amino acids must first be degraded to yield acetyl CoA. (In Chapters 24 and 25 we will see how this occurs.) The two-carbon acetyl group of acetyl CoA is the fuel of the citric acid cycle. Just as gasoline is burned in a car or coal in a furnace, the acetyl group is burned (oxidized) in the citric acid cycle. Two molecules of CO_2 are formed, and energy is released. The energy released by the cycle is temporarily stored in the reduced coenzymes NADH and $FADH_2$.

These reduced coenzymes are formed as products of the four dehydrogenase (Section 22.16) reactions in the citric acid cycle. All the enzymes of the citric acid cycle are localized in the mitochondria. The three NAD-linked dehydrogenase reactions of the cycle can proceed only if enough NAD is available, so there must be continuous oxidation of

NADH. You already know from Chapter 22 that the mitochondrial electron transport system is responsible for oxidizing NADH in a process that uses O_2, ADP, and P_i and generates ATP. Therefore, for the citric acid cycle dehydrogenases to work effectively in oxidizing acetyl CoA, oxygen (and ADP) must be available for electron transport system activity.

Problem 23.6 Write an equation for a reaction in which a substrate molecule

$$\mathrm{R'{-}\underset{}{\overset{\displaystyle OH}{\overset{|}{C}H}}{-}R}$$

is oxidized by an NAD-linked dehydrogenase to yield a product and NADH. (Refer to Chapter 22, if you need to.)

Problem 23.7 Write an overall equation for the oxidation of the NADH formed by the electron transport chain in this reaction.

23.10 THE REACTIONS OF THE CITRIC ACID CYCLE

How many NADH molecules are formed per acetyl CoA molecule oxidized in the citric acid cycle?

Now we are ready to look at the eight individual reactions of the citric acid cycle. Figure 23.7 shows the structural formulas of the intermediates; we'll refer to them by name only as we go through the cycle. The reactions are numbered to key them to the reactions in Figures 23.6 and 23.7.

1. The first reaction of the citric acid cycle is an irreversible reaction—a commiting step (Section 22.5). The reaction is the condensation of acetyl CoA with the four-carbon dicarboxylic acid oxaloacetate to form the six-carbon tricarboxylic acid citrate. (The name of the enzyme appears over the reaction arrow.)

CH3 C(=O) SCoA

$$\text{acetyl—S—CoA} + \text{oxaloacetate}^{2-} \xrightarrow{\text{citrate synthase}} \text{citrate}^{3-} + \text{H—S—CoA} + \text{H}^{+}$$

The reaction is irreversible because it has a very large negative free-energy change—acetyl CoA is a high-energy compound and its high-energy thioester bond is hydrolyzed in the reaction.

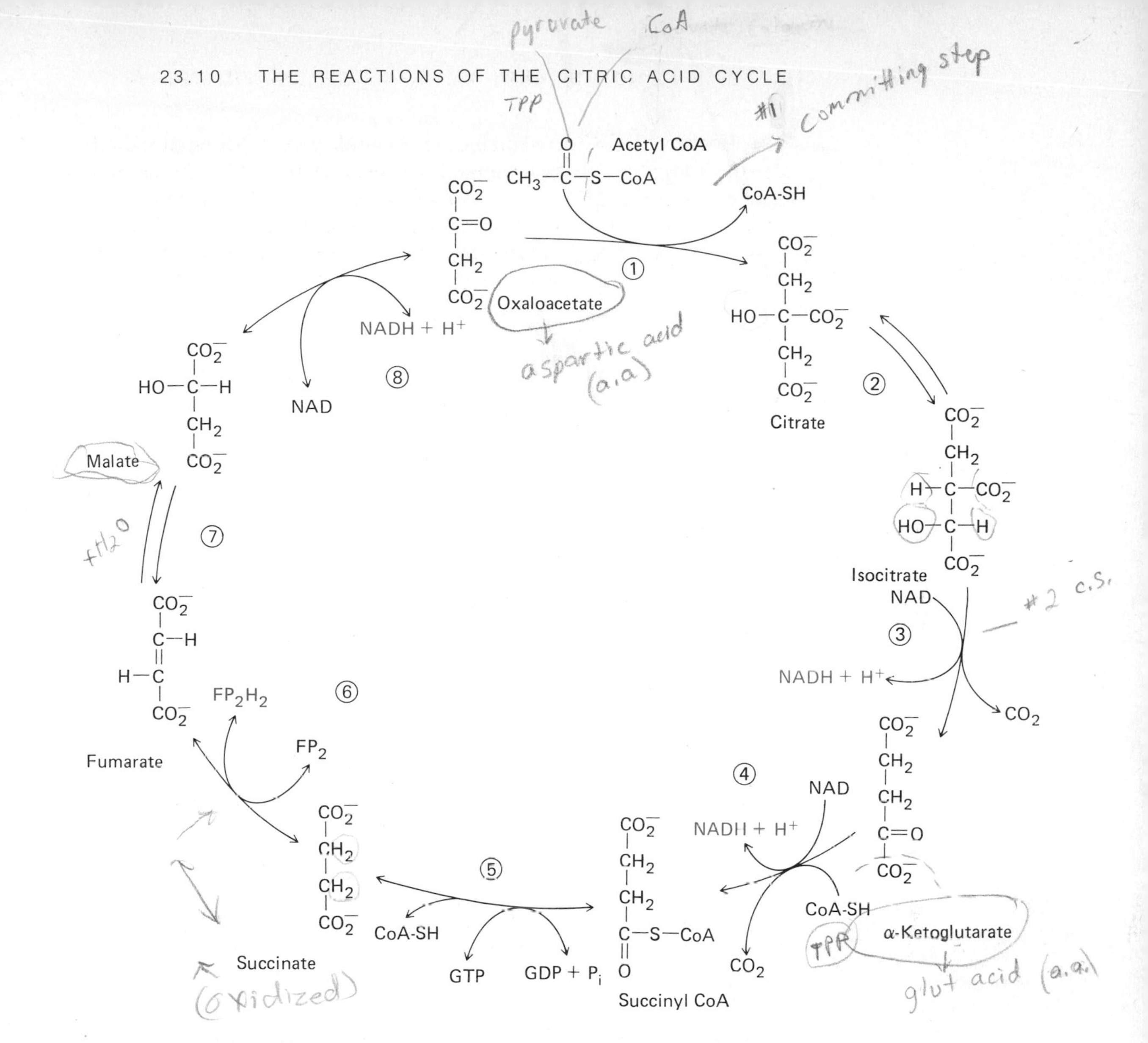

FIGURE 23.7 The reactions of the citric acid cycle. Reactions are numbered 1–8 to key them to the discussion in Section 23.10.

2. Citrate then undergoes an isomerization reaction:

$$\text{citrate}^{3-} \xrightleftharpoons{\text{aconitase}} \text{isocitrate}^{3-}$$

The enzyme aconitase contains an iron atom, which is an essential cofactor. A new tricarboxylic acid, isocitrate, is the product of this readily reversible reaction.

3. Isocitrate is now acted on by an NAD-linked dehydrogenase, which catalyzes both decarboxylation and oxidation of the alcohol group to a ketone:

$$\text{isocitrate}^{3-} + \text{NAD}^+ \xrightarrow[\text{Mg}^{2+}]{\text{isocitrate dehydrogenase}} \alpha\text{-ketoglutarate}^{2-} + \text{NADH} + \text{CO}_2$$

This reaction too is essentially irreversible, making it the second committing step in the citric acid cycle. Remember that NADH formed in this and other citric acid cycle reactions will be reoxidized by the electron transport system.

4. Next, α-ketoglutarate is oxidized and decarboxylated. This reaction, the second oxidative decarboxylation we have seen in the cycle, has effectively oxidized the two-carbon acetyl group to $2CO_2$. It is worth noting that most of the CO_2 that the body forms comes from these citric acid cycle reactions.

$$\alpha\text{-ketoglutarate}^{2-} + \text{NAD}^+ + \text{H—S—CoA} \xrightarrow{\alpha\text{-ketoglutarate dehydrogenase}} \text{succinyl—S—CoA}^- + \text{NADH} + \text{CO}_2$$

5. Succinyl CoA, like acetyl CoA, is a high-energy compound. The free energy of hydrolysis of the thioester linkage can now be employed to drive a phosphorylation reaction:

$$\text{succinyl—S—CoA}^- + \text{GDP}^{3-} + \text{P}_i^{2-} \xrightarrow{\text{succinyl thiokinase}} \text{succinate}^{2-} + \text{GTP}^{4-} + \text{H—S—CoA}$$

This reaction is an example of a **substrate-level phosphorylation.** Here the energy to drive the phosphorylation of GDP to GTP comes from the hydrolysis of the energy-rich substrate succinyl CoA.

6. Succinate is then oxidized by the action of another dehydrogenase, succinate dehydrogenase, which is a familiar flavoprotein (FP_2; see Section 22.15).

$$\text{succinate}^{2-} + \text{FP}_2 \rightleftharpoons \text{fumarate}^{2-} + \text{FP}_2\text{H}_2$$

The electron transport system will quickly reoxidize the reduced succinate dehydrogenase (FP_2H_2).

7. Fumarate then reacts with water:

$$\text{fumarate}^{2-} + H_2O \xrightarrow[\text{hydratase}]{\text{fumarate}} \text{L-malate}^{2-}$$

This enzyme is specific for formation of L-malate from fumarate; the D-isomer is never formed. Fumarate hydratase provides a good example of enzyme specificity.

8. Finally, L-malate is oxidized by the last of the citric acid cycle enzymes, malate dehydrogenase. This closes the circle: the dicarboxylic acid we began with, oxaloacetate, is regenerated:

$$\text{L-malate}^{2-} + NAD^+ \xrightleftharpoons[\text{dehydrogenase}]{\text{malate}} \text{oxaloacetate}^{2-} + NADH + H^+$$

This is a typical example of an alcohol dehydrogenase reaction, of which there are many in metabolism. Here the alcohol group of L-malate is oxidized to yield the keto group of oxaloacetate. Alcohol dehydrogenases are almost always NAD-linked enzymes. It is worth noting that because oxaloacetate is regenerated by normal citric acid cycle activity, the cycle does not need a continuous new supply of oxaloacetate. Small amounts are sufficient to support the rapid oxidation of acetyl CoA in the cell.

23.11 SUMMARY OF THE CITRIC ACID CYCLE PLUS ELECTRON TRANSPORT AND OXIDATIVE PHOSPHORYLATION

We can write an overall reaction for the oxidation of acetyl CoA in the citric acid cycle as follows:

$$\text{acetyl—S—CoA} + 3NAD^+ + FP_2 + GDP^{3-} + P_i^{2-} + 2H_2O \rightarrow$$
$$2CO_2 + 3NADH + FP_2H_2 + GTP^{4-} + \text{H—S—CoA} + 2H^+$$

Problem 23.8 Why doesn't oxaloacetate appear in the overall citric acid cycle reaction?

We can also write the overall reaction of the electron transport system

(Section 22.15), which oxidizes the reduced coenzymes generated in conjunction with the citric acid cycle:

$$2O_2 + 3NADH + FP_2H_2 + 11ADP^{3-} + 11P_i^{2-} + 14H^+ \rightarrow 3NAD^+ + FP_2 + 11ATP^{4-} + 15H_2O$$

In addition, since GTP and ATP are equivalent in energy content, we can substitute ATP for the GTP formed in the succinyl thiokinase reaction. We can then add the two reactions just shown and write the reaction for the complete oxidation of acetyl CoA:

$$\text{acetyl CoA} + 12ADP^{3-} + 12P_i^{2-} + 2O_2 + 12H^+ \rightarrow 2CO_2 + 12ATP^{4-} + 13H_2O + H{-}S{-}CoA$$

Thus the oxidation of a single acetyl group to CO_2 and H_2O provides enough energy to drive the synthesis of 12 ATP molecules.

Now let's return for a moment to your chicken sandwich. You'll see by taking another look at Figure 23.5 that all three major foods you obtained in your lunch are broken down to acetyl CoA. It is true that most of the amino acids do not break down to acetyl CoA as directly as sugars and fatty acids do; however, all are converted to something that can be oxidized in the citric acid cycle. In summary, therefore, we can say that all the major foods are ultimately oxidized in the citric acid cycle, where most of their chemical energy content is released. First, the energy is temporarily stored in NADH molecules. Then electron transport and oxidative phosphorylation oxidize NADH and use its energy for the formation of ATP. ATP is what your body really runs on!

You have probably noticed that we have said nothing further about using the small-molecule products of foodstuff breakdown for anabolic purposes. Your cells will not completely oxidize all these materials. Some small molecules will be used to make glycogen or fat for storage or for making other biomolecules. We will consider these topics in the next few chapters.

SUMMARY

The digestion of foods—which contain carbohydrates, lipids (fats), and proteins—takes place in the digestive system and makes sugars, fatty acids, and amino acids available to the body's cells. Carbohydrate digestion involves hydrolysis of the glycosidic bonds between sugars in starches and disaccharides. It requires **amylase, sucrase,** and **lactase,** as well as other enzymes. The sugars are absorbed into the bloodstream. Fats are broken up into small particles by the detergent actions of the **bile salts,** fatty acids, and monoacylglycerols. Then **lipase** can catalyze hydrolysis of the ester bonds to release free fatty acids, which are then absorbed and delivered to the blood stream. Protein digestion requires six enzymes secreted by the stomach and pancreas and results in the release of free amino acids, which are then absorbed. Carbohydrates, fats, and proteins can replace one another nutritionally to a large extent. All are oxidized to give rise to ATP, although protein yields less total ATP than fats and carbohydrates do.

Certain essential food components must be obtained in the diet. For example, sufficient high-quality protein must be taken in to ensure that the body receives an adequate supply of the **essential amino acids**—those the body cannot make itself. Linoleic acid—an essential fatty acid—and minerals such as calcium, magnesium, iron, phosphate, and iodine must also be present in the diet, as must **trace elements** (certain minerals needed in very small amounts) and vitamins.

Metabolism involves all the pathways concerned with the synthesis and degradation of biomolecules. **Catabolic** (degradative) pathways are typically oxidative and exergonic, whereas **anabolic (biosynthetic)** pathways are typically reductive and endergonic. Catabolism yields, and anabolism uses (1) ATP, (2) reduced coenzymes, and (3) certain small organic molecules. The **citric acid cycle** is a central pathway of metabolism that serves both catabolic and anabolic purposes. It involves eight enzyme-catalyzed reactions, which accomplish the complete oxidation of the two-carbon acetyl groups of acetyl CoA. Acetyl CoA is derived from the breakdown of carbohydrates, fatty acids, and amino acids. Most of the energy stored in food molecules of all kinds is released in the citric acid cycle reactions and temporarily stored in the form of NADH. NADH is then oxidized by the electron transport system, and the energy is used to make ATP.

Key Terms

Amylase (Section 23.2)
Anabolism (Section 23.8)
Bile salt (Section 23.3)
Biosynthesis (Section 23.8)
Catabolism (Section 23.8)
Citric acid cycle (Section 23.9)
Essential amino acid (Section 23.6)
Lactase (Section 23.2)
Lipase (Section 23.3)
Metabolism (Section 23.8)
Substrate-level phosphorylation (Section 23.10)
Sucrase (Section 23.2)
Trace elements (Section 23.6)

ADDITIONAL PROBLEMS

23.9 What would happen to acetyl CoA oxidation in the citric acid cycle if arsenite (a poison that inhibits α-ketoglutarate dehydrogenase) were present?

23.10 If oxaloacetate levels fall very low, how would the levels of acetyl CoA in the mitochondria be affected?

23.11 Write chemical equations for the reactions in this chapter that are examples of the alcohol dehydrogenase–type of reaction.

23.12 Write equations for the reactions in which most of the CO_2 the body produces as waste is formed.

23.13 What is the main function of the citric acid cycle? Which enzymes accomplish this function?

23.14 Do citric acid cycle reactions themselves make any ATP? If not, how is the energy released during these reactions temporarily stored?

Problem-Solving Hints

If you have trouble solving any of the additional problems, refer to the sections listed next to the problem numbers.

23.9 23.10 **23.10** 23.10 **23.11** 23.10
23.12 23.10 **23.13** 23.9, 23.10 **23.14** 23.8

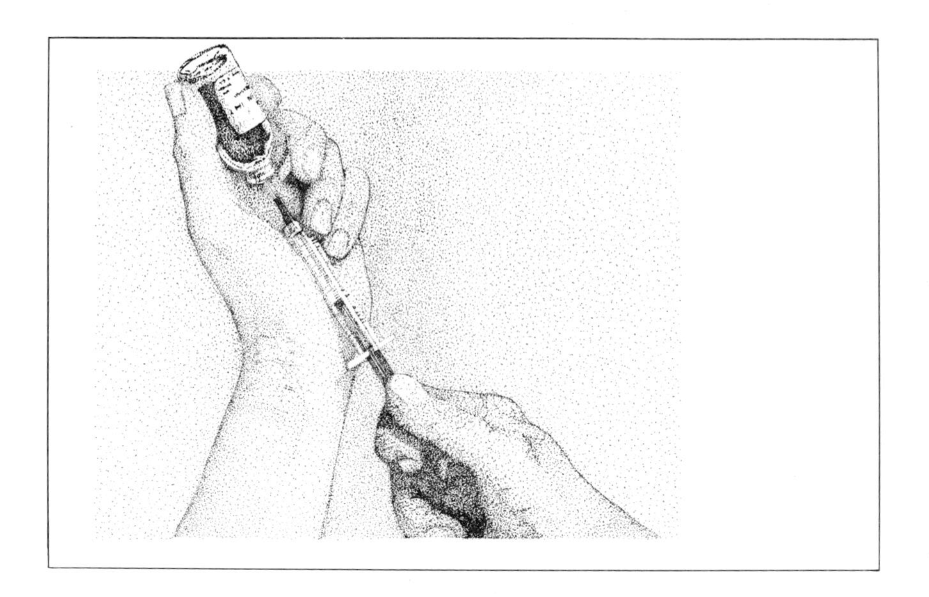

Diabetes is a disease that is characterized by a low level of active insulin, the hormone that most body tissues need for transporting and metabolizing glucose. Therefore a lack of insulin results in the inability of the body to metabolize glucose. One way of treating diabetes is through regular injections of commercially available (nonhuman) insulin.

Repeated injections of insulin may cause the production of massive amounts of insulin antibodies. These antibodies can bind to injected insulin and prevent it from initiating the active breakdown of glucose. Patients vary in the rate of antibody production, but some will produce enough antibodies to bind several thousand units of plasma insulin. This condition, which is known as insulin resistance, can render insulin totally ineffective. Insulin resistance is recognized when it takes tremendously large doses of insulin to keep the diabetic's glucose level within normal limits.

Bovine insulin (taken from the cow) is normally the type of insulin used to treat diabetes. When resistance to bovine insulin occurs, insulin from another source can be tried. Fish, turtle, hog, and whale insulins are also similar in structure and activity to bovine insulin but are difficult to obtain. Another thing that can be done to manage insulin resistance is to administer steroid chemicals. The steroids interfere with the binding of insulin by the antibodies, allowing the insulin to function effectively. In many cases, the insulin resistance may eventually disappear altogether, and then insulin treatment can continue as originally prescribed.

24

CARBOHYDRATE METABOLISM

[Illustration: The injection of insulin is the primary form of treatment of the diabetic.]

Now that you have been introduced to the purposes and general features of metabolism, you are ready to embark on an exploration of some specific metabolic pathways. In this chapter we are going to examine the major pathway responsible for the breakdown of carbohydrate foodstuffs. This pathway, called **glycolysis,** takes place in the cytoplasm of the cell and is responsible for breaking down glucose. We will also look at the important anabolic pathways for manufacturing carbohydrate for storage, and for making glucose to keep blood glucose levels up, when carbohydrate from the diet is not keeping up with the glucose demands of the brain.

24.1 CARBOHYDRATES FOR ENERGY: THE NEED FOR INSULIN

Carbohydrate metabolism is the most logical place to begin looking at the major pathways for degrading foodstuffs. Carbohydrate makes up the largest proportion of the normal diet. It is the most quickly digested and absorbed foodstuff; furthermore, the carbohydrate stored in cells for reserve energy is the most quickly mobilized energy-storage material.

Most people nowadays are aware that **insulin** is necessary for the normal metabolism of glucose in the body. Insulin is a polypeptide hormone produced in the pancreas. Almost everyone knows someone who is diabetic and must take insulin injections or keep careful watch over his or her diet. Diabetes (more properly, **diabetes mellitus**) actually occurs in two major forms: the juvenile-onset type (which, however, may not appear until adulthood), in which the pancreas does not manufacture a sufficient amount of insulin, and the adult-onset type, in which the body cells may not respond normally to insulin. The liver, red blood cells, and brain can take up glucose from the blood stream without the intervention of insulin. However, skeletal muscle and most of the other tissues require insulin in order to transport and metabolize glucose. We will have more to say later about the consequences of abnormal glucose metabolism in uncontrolled diabetes.

GLUCOSE CAN BE BROKEN DOWN WITH OR WITHOUT USING OXYGEN 24.2

What are the products of complete oxidation of glucose?

In glycolysis, which takes place in the cytoplasm of cells, glucose is degraded in a series of reactions that yield two molecules of a three-carbon compound, pyruvic acid, per molecule of glucose broken down. In most body cells, pyruvic acid (or pyruvate, since it is in salt form at physiological pH) is then converted to acetyl CoA, which in turn is oxidized in the citric acid cycle in the mitochondria (Figure 24.1). This is what happens when the cells need energy in the form of ATP. (If there is no need for ATP production at the moment, pyruvate or acetyl CoA can be used to make fat or other products.) The oxidation of pyruvate requires oxygen; that is, it is **aerobic.** Oxygen is required because, along with the citric acid cycle, the electron transport system is involved in the oxidation. The overall reaction looks like this:

$$\underset{\text{glucose}}{C_6H_{12}O_6} + \underset{\text{oxygen}}{6O_2} \rightarrow \underset{\text{carbon dioxide}}{6CO_2} + \underset{\text{water}}{6H_2O}$$

The complete oxidation of glucose to CO_2 and H_2O, carried out by the enzymes of glycolysis and those of the citric acid cycle and the electron

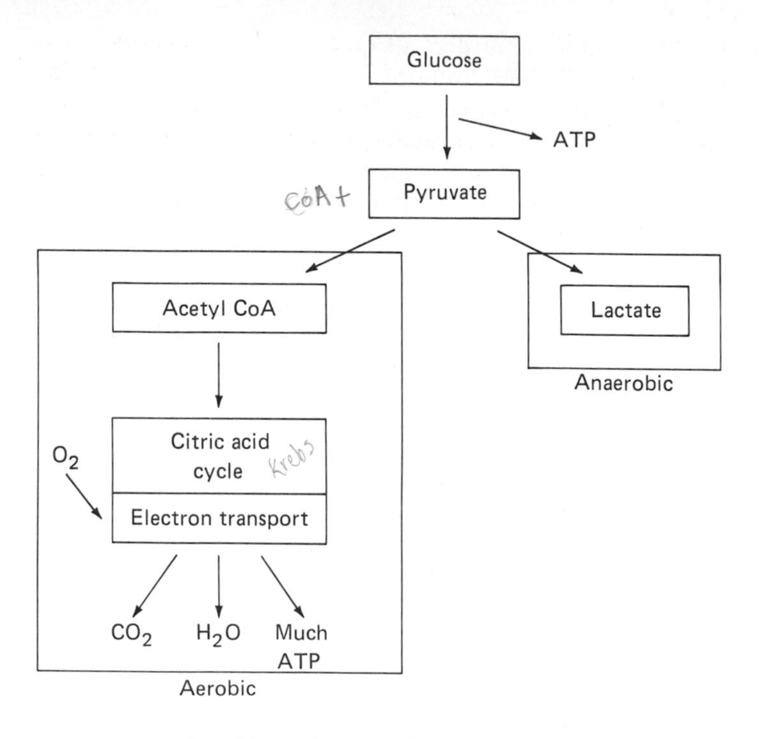

FIGURE 24.1 Aerobic and anaerobic glucose breakdown.

transport system, releases a great deal of energy. As you already know, much of this energy is used to power oxidative phosphorylation of ATP.

Problem 24.1 In which parts of the cell do the following take place?
(a) electron transport
(b) oxidative phosphorylation
(c) glycolysis
(d) citric acid cycle

In some types of cells, especially skeletal muscle and the red blood cells, pyruvate takes an alternate route that does not require oxygen, electron transport, or mitochondria. Because it does not require oxygen, the process is described as **anaerobic.** Glycolysis occurring under anaerobic conditions (anaerobic glycolysis) releases much less energy for every glucose molecule degraded. Glucose is not degraded completely to CO_2 and H_2O, but only to the level of the three-carbon compound, lactic acid (or lactate, since this too is in salt form at physiological pH).

$$\underset{\text{glucose}}{C_6H_{12}O_6} \rightarrow \underset{\text{lactate}}{2C_3H_6O_3}$$

ATP is made during anaerobic glycolysis, too, but much less ATP is obtained per glucose molecule. Anaerobic pathways of catabolism are often called **fermentations.** Yeasts carry out a process called alcoholic fermentation, which follows the same pathway as glycolysis except that pyruvate is converted to ethanol and CO_2. Alcoholic fermentation is the process that makes bread rise (carbon dioxide bubbles form in the dough). It is also the process responsible for alcohol production in the manufacture of beer, wine, and liquors. In anaerobic pathways like these, ATP synthesis occurs in enzyme-catalyzed reactions in the cell cytoplasm. The mitochondrial electron transport system is not involved.

Skeletal muscles do not always carry out glycolysis anaerobically. If exercise is taking place at a steady moderate rate, as in jogging, the muscle metabolism is largely aerobic, especially if the body is in good condition. However, during an intensive exercise period—the 100-yard dash, for example—anaerobic glycolysis is the main source of muscle ATP energy. Lactate builds up in the tissues and blood stream; it is in part responsible for the intense feeling of fatigue. The feeling of breathlessness that follows is caused by the need to obtain extra oxygen to oxidize the accumulated lactate.

Problem 24.2 What are the final products of the following:
(a) glycolysis in the presence of O_2
(b) glycolysis in the absence of O_2
(c) alcoholic fermentation in the absence of O_2

OVERVIEW OF GLYCOLYSIS 24.3

Let us now consider the breakdown of glucose to pyruvate. The first several reactions of glycolysis are called the **collecting phase,** because in this series of reactions all the common food sugars can be converted to the same three-carbon compound, glyceraldehyde-3-phosphate. In other words, the carbons of these different sugars are "collected" in the form of this one metabolic intermediate compound. The net reaction of the collecting phase looks like this:

$$\left.\begin{matrix}\text{glucose}\\ \text{galactose}\\ \text{or}\\ \text{fructose}\\ (C_6\text{ sugar})\end{matrix}\right\} \rightarrow \begin{matrix}2\text{ glyceraldehyde-3-phosphates}\\ (2\ C_3\text{ sugar phosphates})\end{matrix}$$

In the process, some ATP—although it is not shown here—must be used to supply phosphate and energy.

Next, glyceraldehyde-3-phosphate enters the second phase, which we will call the **ATP-synthetic phase** of glycolysis. In this phase,

ATP syn

several reactions convert glyceraldehyde-3-phosphate to pyruvate. The net reaction is

$$\underset{(2\ C_3\ \text{sugar phosphates})}{2\ \text{glyceraldehyde-3-phosphate}} \rightarrow \underset{(2\ C_3\ \alpha\text{-keto acids})}{2\ \text{pyruvate}}$$

There are two ATP-forming reactions and one NADH-forming reaction in this second phase of the pathway. Let us now look at the individual reactions of glycolysis.

24.4 REACTIONS OF THE COLLECTING PHASE OF GLYCOLYSIS

ATP is consumed in the collecting phase.

We will begin with glucose, and later we'll look at the entry of the other sugars into the pathway. Remember the chicken sandwich and milk we talked about in Chapter 23? The bread starch and lactose of milk are the main sources of the glucose that is absorbed into the blood and will now be oxidized in various body cells. Figure 24.2 shows the structural formulas of the metabolic intermediates that we will discuss.

The first of the five reactions of the collecting phase of glycolysis is the **phosphorylation** of glucose by ATP. In this reaction, which is catalyzed by hexokinase, a phosphate group is transferred from ATP to form an ester linkage with a particular hydroxyl group of glucose. The product of this phosphorylation reaction is glucose-6-phosphate.

$$\text{glucose} + \text{ATP}^{4-} \xrightarrow[\text{Mg}^{2+}]{\text{hexokinase}} \text{glucose-6-phosphate}^{2-} + \text{ADP}^{3-} + \text{H}^{+}$$

(reaction 1)

The Mg^{2+} placed under the reaction arrow means that the enzymatic reaction requires Mg^{2+} ions. All kinase reactions require Mg^{2+}. **Kinase** is the general name given to enzymes that catalyze phosphorylation reactions involving nucleoside triphosphates like ATP.

We previously used the hexokinase reaction as an example to illustrate the coupling of an exergonic reaction (ATP hydrolysis) and an endergonic reaction (glucose phosphorylation) (Section 22.9). In the cell, the reaction catalyzed by hexokinase is irreversible. This has the effect of trapping glucose in the cell, since the product, glucose-6-phosphate, cannot pass back across the cell membrane. Thus the hexose is committed to being metabolized inside the cell. Recall that many pathways begin with such an irreversible committing step (Section 22.5).

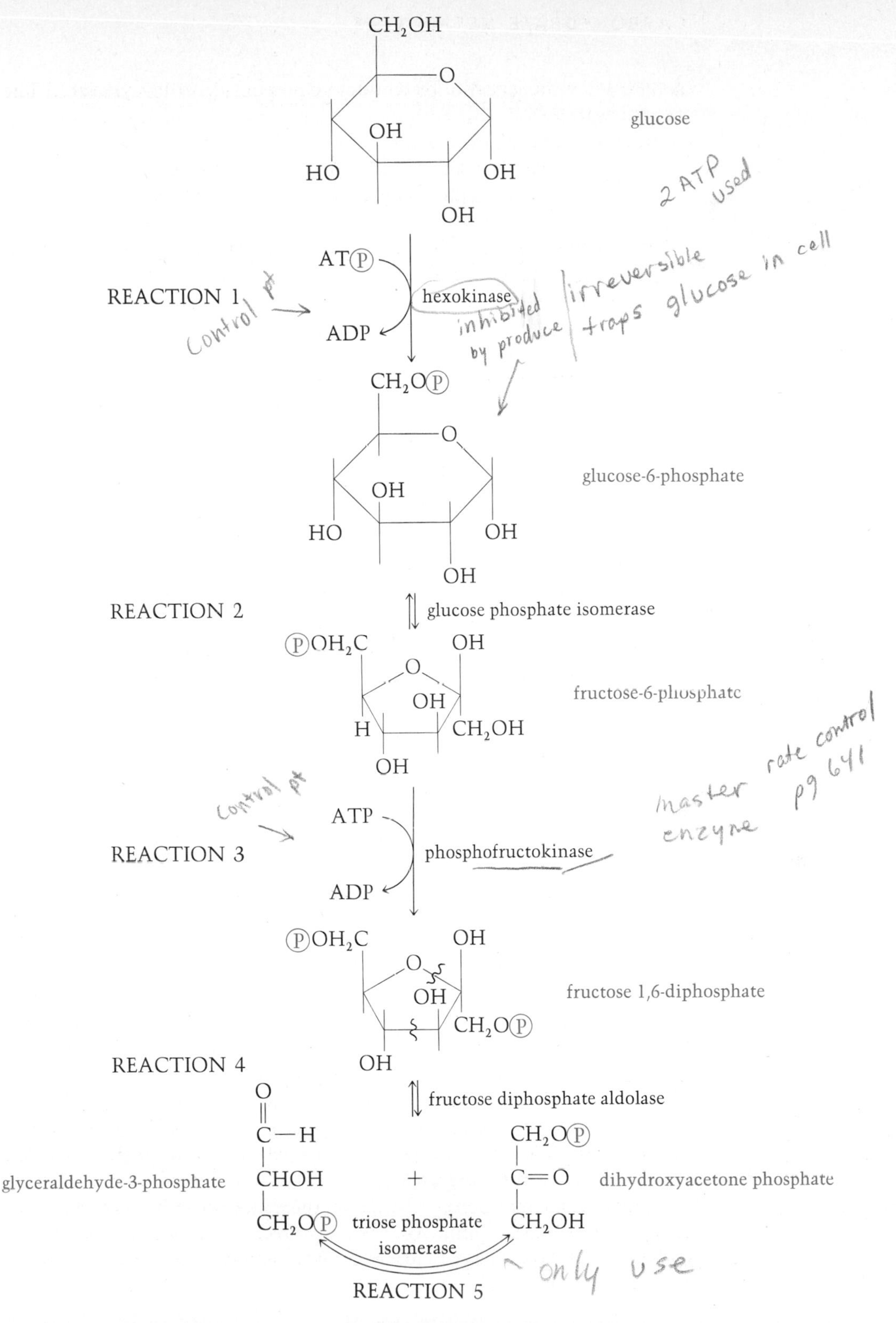

FIGURE 24.2 Collecting phase of glycolysis (Ⓟ represents a phosphate group in ester linkage).

The second reaction in glycolysis is simply a reversible isomerization. In this reaction, glucose-6-phosphate, an aldose phosphate, is converted to fructose-6-phosphate, a ketose phosphate.

$$\text{glucose-6-phosphate} \xrightleftharpoons{\text{glucose phosphate isomerase}} \text{fructose-6-phosphate}$$

(reaction 2)

Next, fructose 6-phosphate is phosphorylated by ATP in a reaction catalyzed by phosphofructokinase:

$$\text{fructose-6-phosphate}^{2-} + \text{ATP}^{4-} \xrightarrow[\text{Mg}^{2+}]{\text{phosphofructokinase}}$$

$$\text{fructose-1,6-diphosphate}^{4-} + \text{ADP}^{3-} + \text{H}^{+}$$

(reaction 3)

Like the hexokinase reaction, this one is irreversible in the cell. (Note the one-way arrows in this and the hexokinase reaction, which indicate boldly that they are irreversible.) The phosphofructokinase reaction commits the cell to breaking down glucose-6-phosphate or fructose-6-phosphate rather than converting it for storage or rearranging it to create a different sugar. Since this is an irreversible step, fructose-1,6-diphosphate has to be catabolized—it cannot go back. Phosphofructokinase catalyzes the *rate-limiting step* (Section 21.16) of glycolysis, and it is also the most important enzyme in the regulation of the pathway. More about this later.

In the next reaction, fructose-1,6-diphosphate is cleaved to yield two triose phosphates:

$$\text{fructose-1,6-diphosphate}^{4-} \xrightleftharpoons{\text{fructose diphosphate aldolase}}$$

$$\text{dihydroxyacetone phosphate}^{2-}\text{(DHAP)} + \text{glyceraldehyde-3-phosphate}^{2-}$$

(reaction 4)

The aldolase reaction is an example of a metabolic reaction with a large positive standard free-energy change (Section 22.5). However, because of exergonic (energy-releasing) reactions farther along in the pathway, its products are removed and kept at relatively low concentrations, so that the actual free-energy change of the reaction is negative and the reaction can proceed in the direction of triose phosphate formation.

One more isomerization reaction concludes the collecting phase of glycolysis. Dihydroxyacetone phosphate (DHAP) and glyceraldehyde-3-phosphate are interconverted in this reaction:

$$DHAP^{2-} \underset{}{\overset{\text{triose phosphate isomerase}}{\rightleftarrows}} \text{glyceraldehyde-3-phosphate}^{2-}$$

(reaction 5)

If we assume that all the DHAP formed at the aldolase-catalyzed step is converted to glyceraldehyde-3-phosphate, then we have collected all the glucose carbons in this form. Note that two ATPs were used up in this portion of the glycolytic pathway, so the process has not yet yielded any ATP.

REACTIONS OF THE ATP-SYNTHETIC PHASE 24.5

Which intermediates of the ATP-synthetic phase are high-energy compounds?

In the first of the five reactions of the second phase of glycolysis (Figure 24.3), glyceraldehyde-3-phosphate is oxidized from the aldehyde to the carboxylic-acid level, and then the acid is phosphorylated. Both steps are catalyzed by the same enzyme, which is an NAD-linked dehydrogenase. The overall reaction is

$$\text{glyceraldehyde-3-phosphate}^{2-} + NAD^+ + P_i^{2-} \overset{\text{glyceraldehyde-3-phosphate dehydrogenase}}{\rightleftarrows}$$

$$\text{1,3-diphosphoglycerate}^{4-} + NADH + H^+ \qquad \text{(reaction 1)}$$

Note that NAD accepts the pair of hydrogen atoms that the aldehyde loses when it is oxidized to a carboxylic acid.

In the next reaction of glycolysis, the high-energy anhydride bond between the carboxylic acid group and the 1-phosphate of 1,3-diphosphoglycerate is hydrolyzed. This hydrolysis releases energy that is used to drive the phosphorylation of ADP to yield ATP.

$$\text{1,3-diphosphoglycerate}^{4-} + ADP^{3-} \overset{\text{phosphoglycerate kinase}}{\rightleftarrows}$$

$$\text{3-phosphoglycerate}^{3-} + ATP^{4-} \qquad \text{(reaction 2)}$$

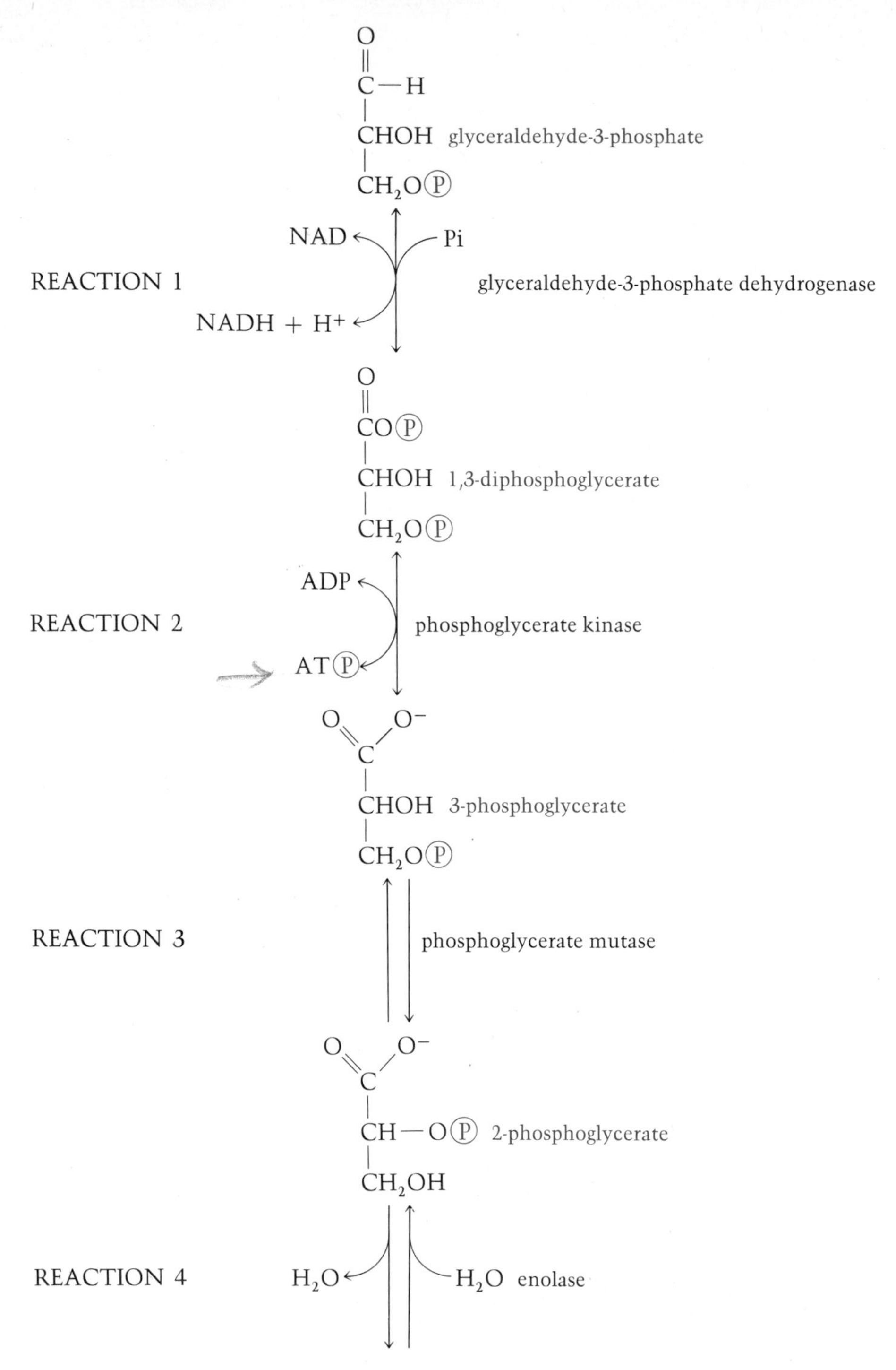

FIGURE 24.3 ATP: synthetic phase of glycolysis (Ⓟ represents a phosphate group in ester linkage).

produces 4 ATP
2 ATP used

REACTION 5

phosphoenolpyruvate

ADP → ATP

pyruvate

FIGURE 24.3 (Continued)

3-Phosphoglycerate is next rearranged to give 2-phosphoglycerate:

$$\text{3-phosphoglycerate}^{3-} \xrightleftharpoons{\text{phosphoglycerate mutase}} \text{2-phosphoglycerate}^{3-}$$

(reaction 3)

2-Phosphoglycerate then undergoes an unusual dehydration reaction forming a new high-energy compound called phosphoenolpyruvate or PEP for short (see Table 22.1):

$$\text{2-phosphoglycerate} \xrightleftharpoons[Mg^{2+}]{\text{enolase}} \text{phosphoenolpyruvate (PEP)} + H_2O$$

(reaction 4)

The standard free energy of hydrolysis of PEP is one of the most negative associated with any biological compound. Thus PEP is an excellent substrate for another ATP synthetic step:

$$PEP^{3-} + ADP^{3-} + H^+ \xrightarrow[Mg^{2+}]{\text{pyruvate kinase}} \text{pyruvate}^- + ATP^{-4}$$

(reaction 5)

The pyruvate kinase reaction is another irreversible step with a large negative free-energy change. (Again, the one-way arrow signals irreversibility.) Because of this, all the reactions occurring prior to it in the pathway are pulled toward product formation. We will consider pyruvate to be the three-carbon product of glycolysis. (If glycolysis occurs under anaerobic conditions, lactate will be the carbon product, but we will come back to this later.)

24.6 OVERALL REACTION OF GLYCOLYSIS

Does glycolysis produce acid?

We can obtain the overall reaction of glycolysis—taking pyruvate to be our carbon product—by adding the overall reactions of the two phases. The overall reaction of the collecting phase of glycolysis is

$$\text{glucose} + 2\text{ATP}^{4-} \longrightarrow 2\ \text{glyceraldehyde-3-phosphate}^{2-} + 2\text{ADP}^{3-} + 2\text{H}^{+}$$

The overall reaction of the ATP-synthetic phase of glycolysis is

$$2\ \text{glyceraldehyde-3-phosphate}^{2-} + 2\text{P}_{\text{i}}^{2-} + 4\text{ADP}^{3-} + 2\text{NAD}^{+} \longrightarrow 2\ \text{pyruvate}^{-} + 4\text{ATP}^{4-} + 2\text{NADH} + 2\text{H}_2\text{O}$$

The sum of the two reactions is then

$$\text{glucose} + 2\text{ADP}^{3-} + 2\text{P}_{\text{i}}^{2-} + 2\text{NAD}^{+} \longrightarrow 2\ \text{pyruvate}^{-} + 2\text{ATP}^{4-} + 2\text{H}^{+} + 2\text{NADH}$$

Thus we have a net yield of 2ATPs for each glucose molecule that is metabolized. In addition, we produce two NADH molecules (also energy-rich compounds) and two pyruvate molecules.

Only about one-third of the available energy is released when a cell breaks glucose down to pyruvate. The remainder of the free energy is released in the aerobic metabolism of pyruvate via the citric acid cycle and in the reoxidation of NADH. We'll pick up the further travels of pyruvate again after we consider the metabolism of galactose and fructose in glycolysis.

Problem 24.3 Write the equation for the overall reaction of glucose breakdown to pyruvate, including the structural formulas of glucose and pyruvate. Are all carbon, hydrogen, and oxygen atoms accounted for?

24.7† OXIDATION OF SOME MONOSACCHARIDES OTHER THAN GLUCOSE

Fructose, which is obtained largely from the digestion of the disaccharide sucrose, is oxidized in the usual glycolytic pathway. First it undergoes a phosphorylation reaction that is catalyzed by a special enzyme, fructokinase. The fructose-1-phosphate that results from this reaction is then cleaved by a second enzyme to give DHAP and glyceraldehyde (Figure 24.4). People with the genetic disease *fructosuria* lack either fructokinase or the second enzyme, fructose-1-phosphate aldolase. Since they cannot metabolize fructose, it accumulates in the body and some of it is excreted in the urine.

Fructose and galactose must be broken down via glycolysis, too.

Galactose is obtained mainly from the digestion of lactose, the major disaccharide of milk. Galactose is metabolized a bit differently from fructose or glucose since it is not simply phosphorylated before entering glycolysis. Two unique enzymes are involved. The pathway is shown in Figure 24.5.

Lactose provides a large part of an infant's total caloric intake. Genetic diseases in which one or the other of the enzymes necessary for galactose metabolism is missing or faulty can be fatal to infants. *Galactosemia* is the general name for disorders in which galactose is not metabolized and accumulates in the blood. The high levels of galactose cause clouding of the lens of the eye (cataracts) and mental disorders. The treatment is to eliminate milk and milk products from the diet.

Problem 24.4 Draw the structural formula of UDP-galactose. Using structural formulas, write the reaction that will convert UDP-glucose and galactose-1-phosphate to UDP-galactose and glucose-1-phosphate (the transferase reaction of Figure 24.5).

24.8 PYRUVATE OXIDATION IN THE MITOCHONDRIA

As we saw earlier in the chapter, pyruvate arises from glycolysis in the cytoplasm of the cell. It is then transported into the mitochondria to be oxidized by the pyruvate dehydrogenase system—a complicated reaction. The pyruvate dehydrogenase system is a complex of three enzymes, and several coenzymes derived from B vitamins are also involved (see Focus 14, "The Pyruvate Dehydrogenase System").

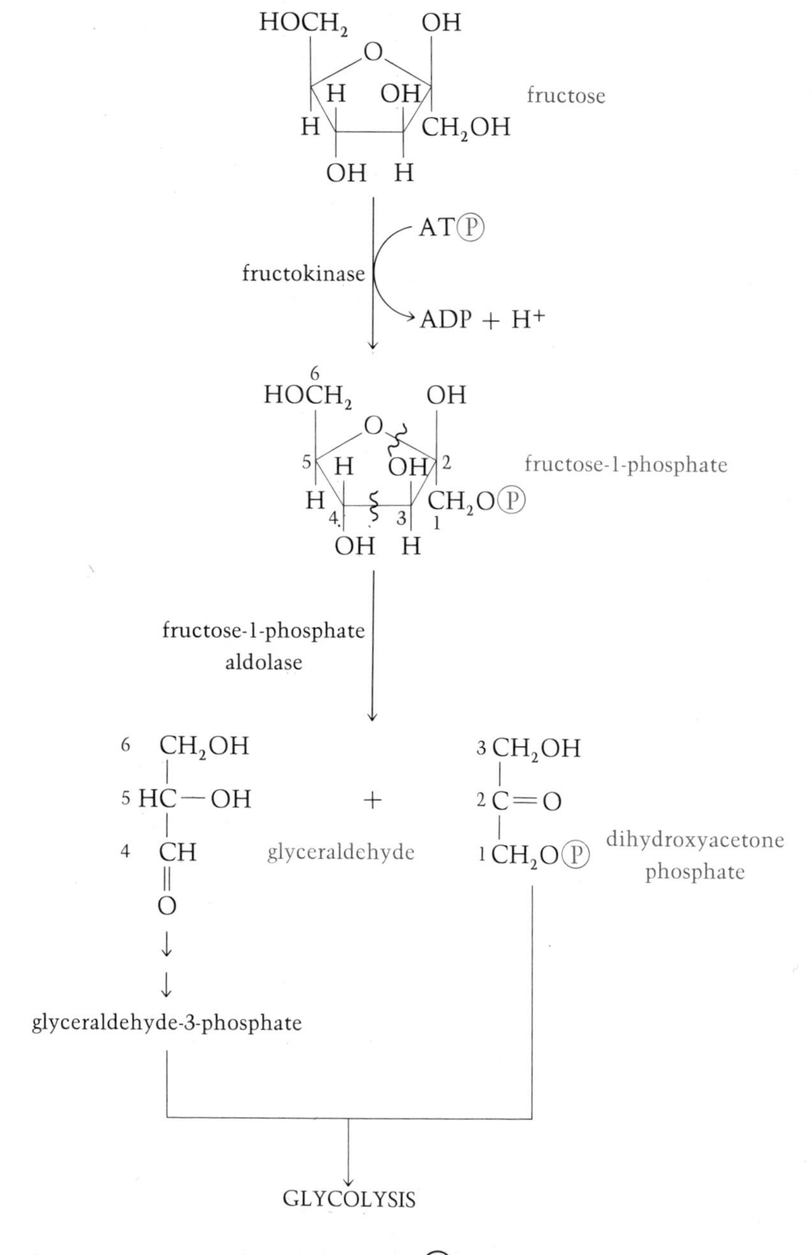

FIGURE 24.4 Fructose metabolism (Ⓟ represents a phosphate group in ester linkage).

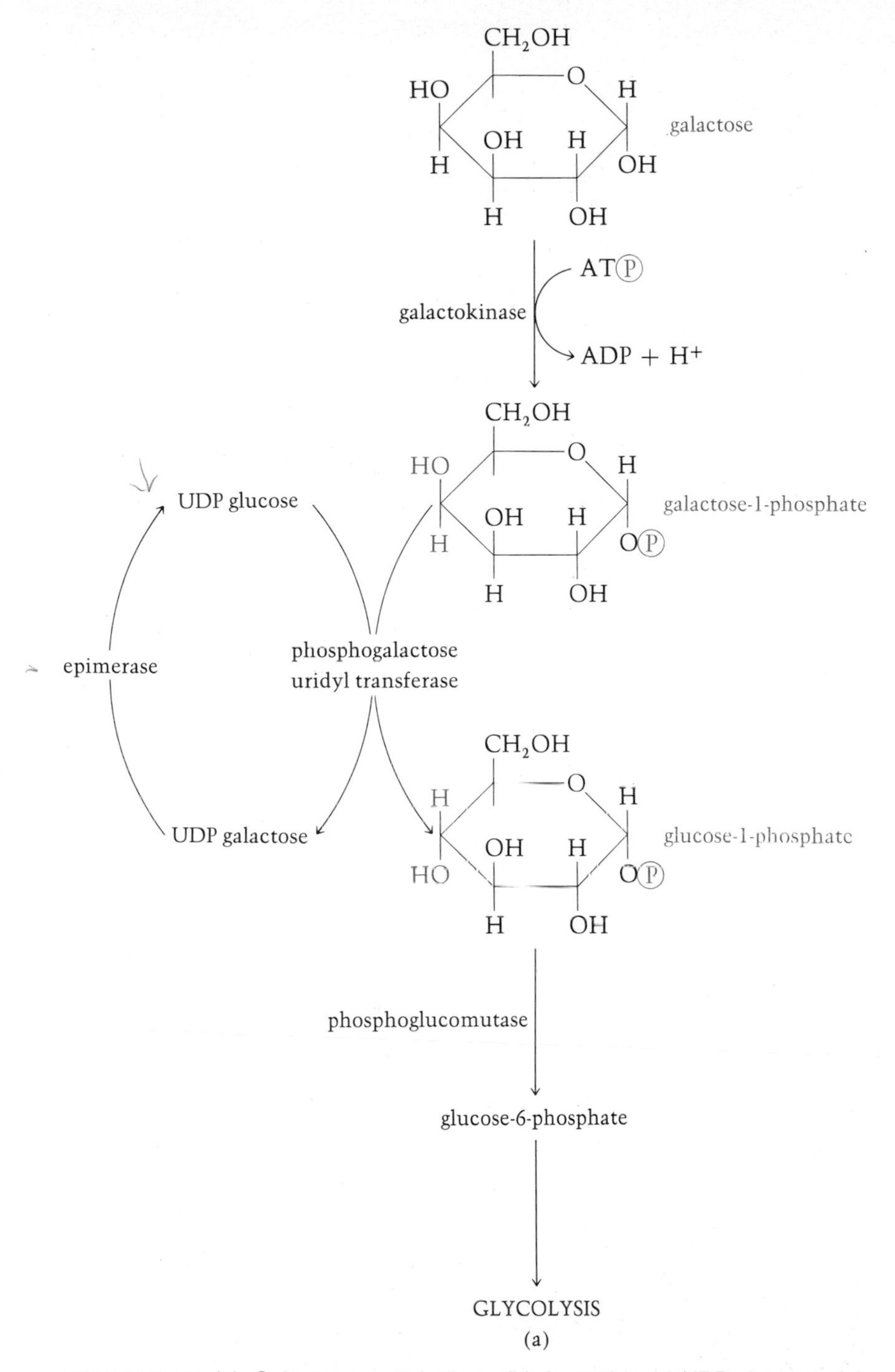

FIGURE 24.5 (a) Galactose metabolism; (b) formation of UDP-glucose; (c) complete structure of UDP-glucose (Ⓟ represents a phosphate group in ester linkage). (Continued on next page)

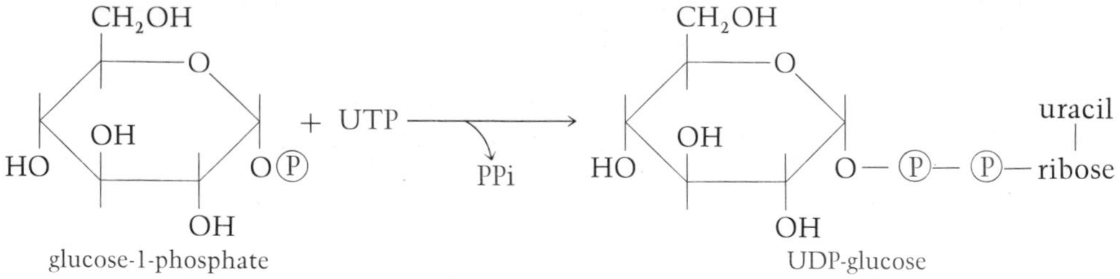

CH₂OH → written as: CH_2OH ; O ; OH ; HO ; O(P) ; + UTP → PPi ; CH_2OH ; O ; OH ; HO ; O—(P)—(P)—ribose ; uracil ; OH ; OH

glucose-1-phosphate   UDP-glucose

(b)

CH_2OH ; O ; H ; OH ; H ; HO ; H ; OH ; O ; O ; O—P—O—P—O—CH_2 ; O$_-$; O$_-$; O ; HN ; O ; N ; O ; H ; H ; H ; OH ; OH

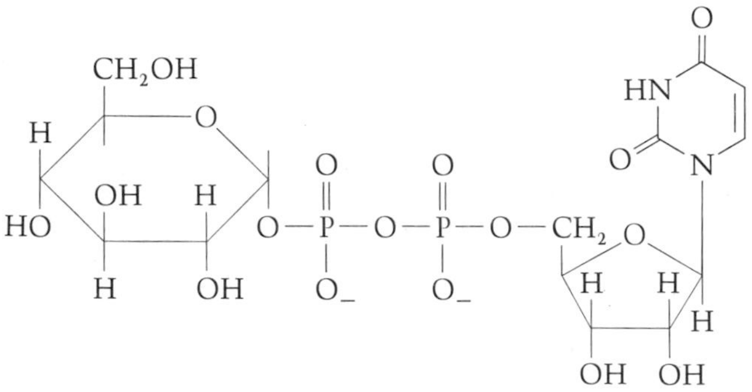

UDP-glucose

(c)

FIGURE 24.5 (Continued)

The overall reaction for the oxidation of pyruvate looks fairly simple, however:

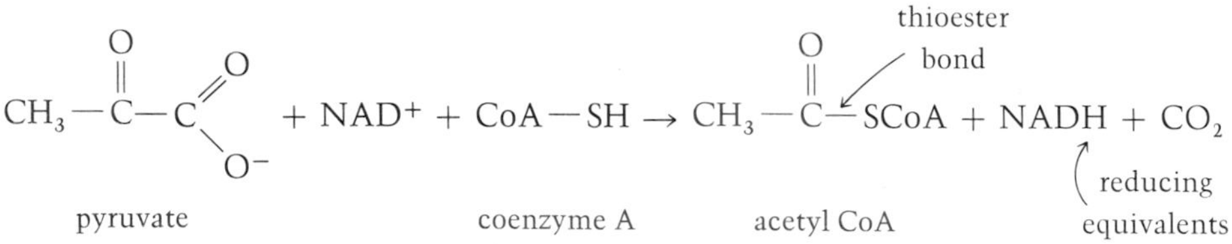

$$CH_3{-}\overset{\overset{O}{\|}}{C}{-}C\genfrac{}{}{0pt}{}{\nearrow O}{\searrow O^-} + NAD^+ + CoA{-}SH \rightarrow CH_3{-}\overset{\overset{O}{\|}}{C}{-}SCoA + NADH + CO_2$$

pyruvate   coenzyme A   acetyl CoA   thioester bond   reducing equivalents

In this reaction, a two-carbon fragment (the acetyl group) remains after the decarboxylation and oxidation of pyruvate. The acetyl group forms a thioester bond with the sulfhydryl (—SH) group of coenzyme A. (The structures of coenzyme A and acetyl CoA are shown in Figure 24.6.) Acetyl CoA is a high-energy compound by virtue of its thioester bond. NAD is the acceptor of the reducing equivalents (a hydrogen atom plus an electron) derived from the oxidation of pyruvate.

The pyruvate dehydrogenase reaction is also an irreversible committing step. In this case, the two-carbon acetyl group is committed to

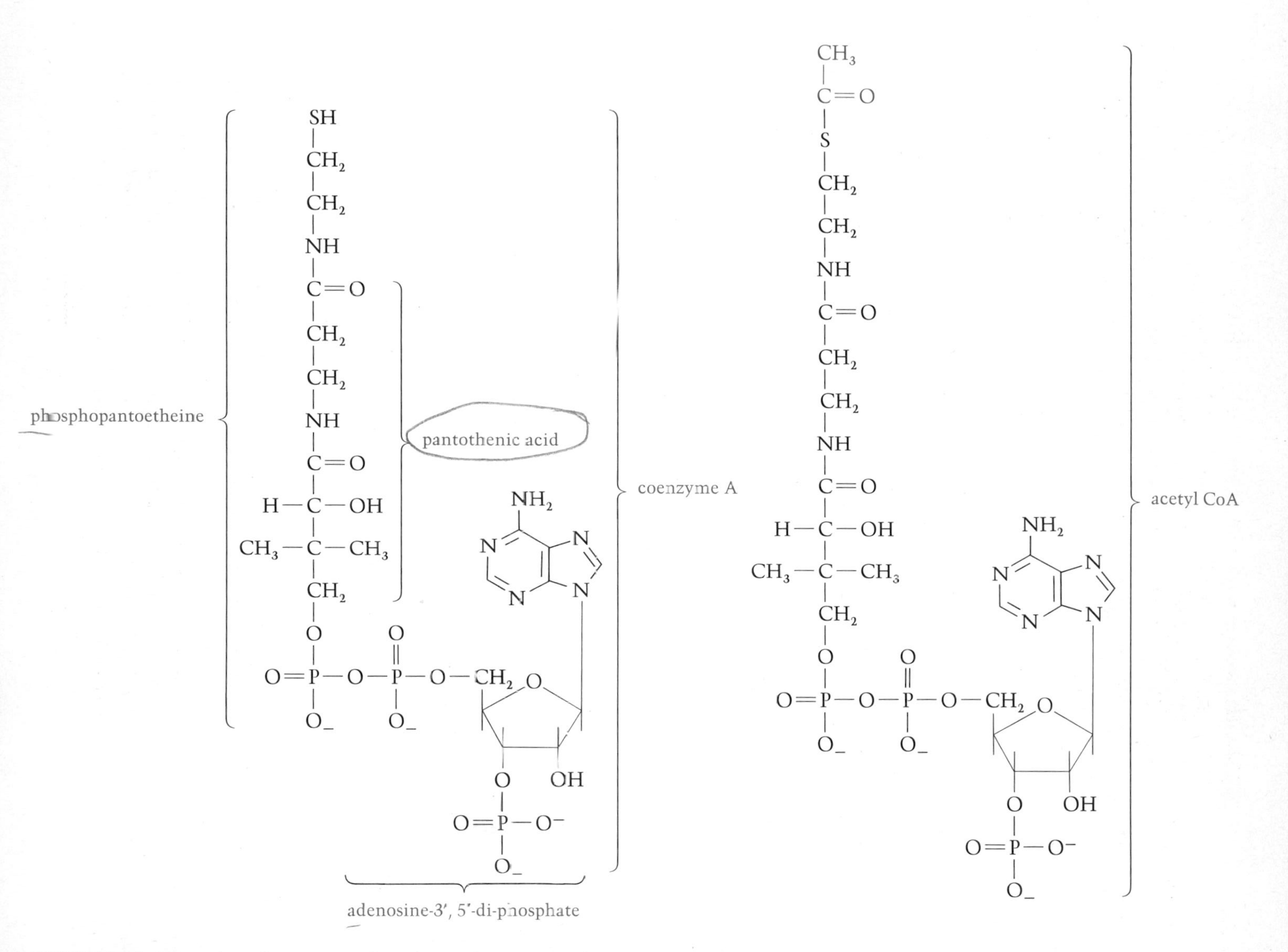

FIGURE 24.6 Structures of coenzyme A and acetyl coenzyme A.

metabolism within the mitochondria. Just as glucose-6-phosphate cannot pass across the cell membrane out of the cell, acetyl CoA cannot pass across the mitochondrial inner membrane out to the cytoplasm.

Since you are already familiar with the citric acid cycle, in which acetyl CoA will now be oxidized, we will just write overall reactions to describe the completion of pyruvate and glucose oxidation to CO_2 and H_2O:

$$\text{glucose} + 2ADP^{3-} + 2P_i^{2-} + 2NAD^+ \rightarrow$$
$$2\ \text{pyruvate}^- + 2ATP^{4-} + 2NADH + 2H^+$$

$$2\ \text{pyruvate}^- + 2NAD^+ + 2CoASH \rightarrow$$
$$2\ \text{acetyl CoA} + 2CO_2 + 2NADH$$

$$2\ \text{acetyl CoA} + 24ADP^{3-} + 24P_i^{2-} + 4O_2 + 24H^+ \rightarrow$$
$$4CO_2 + 24ATP^{4-} + 26H_2O + 2CoASH$$

To complete the picture, it is necessary to account for oxidation of the NADH that comes from the pyruvate dehydrogenase step, and the NADH from the glycolytic pathway itself. Since each NADH molecule oxidized by the mitochondria can drive the formation of 3ATPs, and since 4NADH molecules are formed for each glucose molecule, then 3×4, or 12, more ATPs can be formed. Thus 24ATPs + 12ATPs + 2ATPs (from glycolysis itself) gives a grand total of 38ATPs per glucose molecule oxidized completely to CO_2 and H_2O.

Problem 24.5 Write equations for the NADH oxidation we have just described and sum them with the three equations above to obtain a grand overall reaction for aerobic glucose oxidation. (Refer to Chapter 22 if necessary for the details of NADH oxidation by the electron transport system.)

†24.9 REGULATION OF AEROBIC CARBOHYDRATE BREAKDOWN

How is carbohydrate breakdown regulated by the cell's energy level?

We have mentioned before that the citric acid cycle can operate only if oxygen is available. The electron transport system will not work without oxygen (Section 22.15). However, even if plenty of oxygen is available, the electron transport system may be slow to oxidize NADH if the concentration of ATP is too much higher than the concentration of

ADP. Electron transport is coupled with oxidative phosphorylation. If the concentration of ADP is quite low (because most of the cell's adenine nucleotide is in the form of ATP), then oxidative phosphorylation cannot proceed and neither can electron transport. NADH will not be oxidized, it will pile up in the mitochondria, and activity in the citric acid cycle will slow down. When we consider that the primary purpose of oxidizing carbohydrates (or other foodstuffs) is to obtain ATP to drive the cell's energy-requiring reactions, this makes very good sense. If we already have a high level of ATP, why should we burn fuel to make more?

There are other important regulatory points in the pathway of aerobic carbohydrate catabolism. Various enzymes operate not only as catalysts that speed up the rate of specific reactions but also as regulators of metabolism that actually control when and to what extent a metabolic pathway will operate. One particular allosteric enzyme (Section 21.15) is of key importance in the regulation of glycolysis. This enzyme, phosphofructokinase, is the master rate-controlling enzyme of glycolysis. It is inhibited by a high concentration of ATP, and the inhibition is counteracted by a high concentration of adenosine monophosphate (AMP).

A high concentration of ATP signals that the breakdown of carbohydrates (or other foodstuffs) should be slowed or stopped. But a high concentration of ADP or AMP (the latter is an especially clear indication that the concentration of ATP is low) is a signal that more rapid breakdown is needed. The mechanism is called **acceptor control.** Thus if the energy level in your cells is high (that is, if the ATP concentration is high), the glucose you obtain from your lunchtime sandwich will not be oxidized completely. Instead, it will be used mainly to make fat to store food energy. We will consider how this is done in Chapter 25.

Going back even farther in the glycolysis pathway, to hexokinase, we find that there is a mechanism to block glucose from entering glycolysis right at the start. Hexokinase catalyzes the reaction that commits glucose to metabolism in the cell (Section 24.4). It is inhibited by its product, glucose-6-phosphate. Normally, most of the glucose-6-phosphate formed is removed by further metabolism. However, if ATP levels are high, phosphofructokinase (the rate-limiting enzyme of glycolysis), the citric acid cycle, and electron transport will move slowly. Glucose-6-phosphate will build up in the cell, inhibiting hexokinase and blocking the entry of more glucose into glycolysis.

Problem 24.6 Draw a diagram of the pathways of glycolysis and the citric acid cycle (use names or structures of compounds, as you please). Indicate the enzymes at which regulation of glucose catabolism occurs, and name the nucleotides or other compounds that influence their activities.

24.10 ANAEROBIC GLYCOLYSIS

What human cells or tissues carry out anaerobic glycolysis?

Anaerobic glycolysis is the breakdown of glucose without oxygen. In the aerobic pathway, which we have just discussed, the mitochondrial electron transport system directly utilizes oxygen. Citric acid cycle activity and the aerobic breakdown of glucose depend on the ability of electron transport to oxidize the NADH produced (Section 24.8). In contrast, tissues that carry out anaerobic glycolysis either do not possess mitochondria (such as the red blood cells), or their metabolism is adapted to making particularly rapid response to a call for action ("fast" skeletal muscle is the main example). In the latter case, anaerobic glycolysis is preferred because it can start providing ATP for muscle contraction much faster than the aerobic route can.

Anaerobic glycolysis differs in two specific ways from the aerobic glucose-degrading pathway already described. First, the fate of pyruvate is different, and second, cytoplasmic NADH must be reoxidized without using oxygen. In fact, the conversion of pyruvate to the final product (lactate) is coupled with the oxidation of NADH in a single reaction catalyzed by **lactate dehydrogenase:**

$$\underset{\text{pyruvate}}{CH_3-CO-COO^-} + NADH + H^+ \xrightleftharpoons{\text{lactate dehydrogenase}} \underset{\text{lactate}}{CH_3-CH(OH)-COO^-} + NAD^+$$

This is a typical alcohol dehydrogenase reaction. In anaerobic glycolysis, lactate is used as a "sink" (a holding tank) for reducing equivalents, that is, hydrogen atoms, enabling NAD to be regenerated right in the cytoplasm for use by glyceraldehyde phosphate dehydrogenase. This enzyme is the only dehydrogenase in the glycolytic pathway to pyruvate. It is very important to regenerate NAD, because if there were no mechanism for doing this, all the cytoplasmic NAD soon would be in reduced form. Glucose catabolism would then come to a halt, starving the cell of ATP.

The overall reaction of anaerobic glycolysis is

$$\text{glucose} + 2P_i^{2-} + 2ADP^{3-} \rightarrow 2\ \text{lactate}^- + 2ATP^{4-} + 2H_2O$$

A molecule of glucose yields much less ATP in the muscle cell or red blood cell where it is broken down anaerobically rather than aerobically (2ATPs versus 38ATPs per molecule of glucose). Thus much more glucose must be burned to get the same amount of ATP.

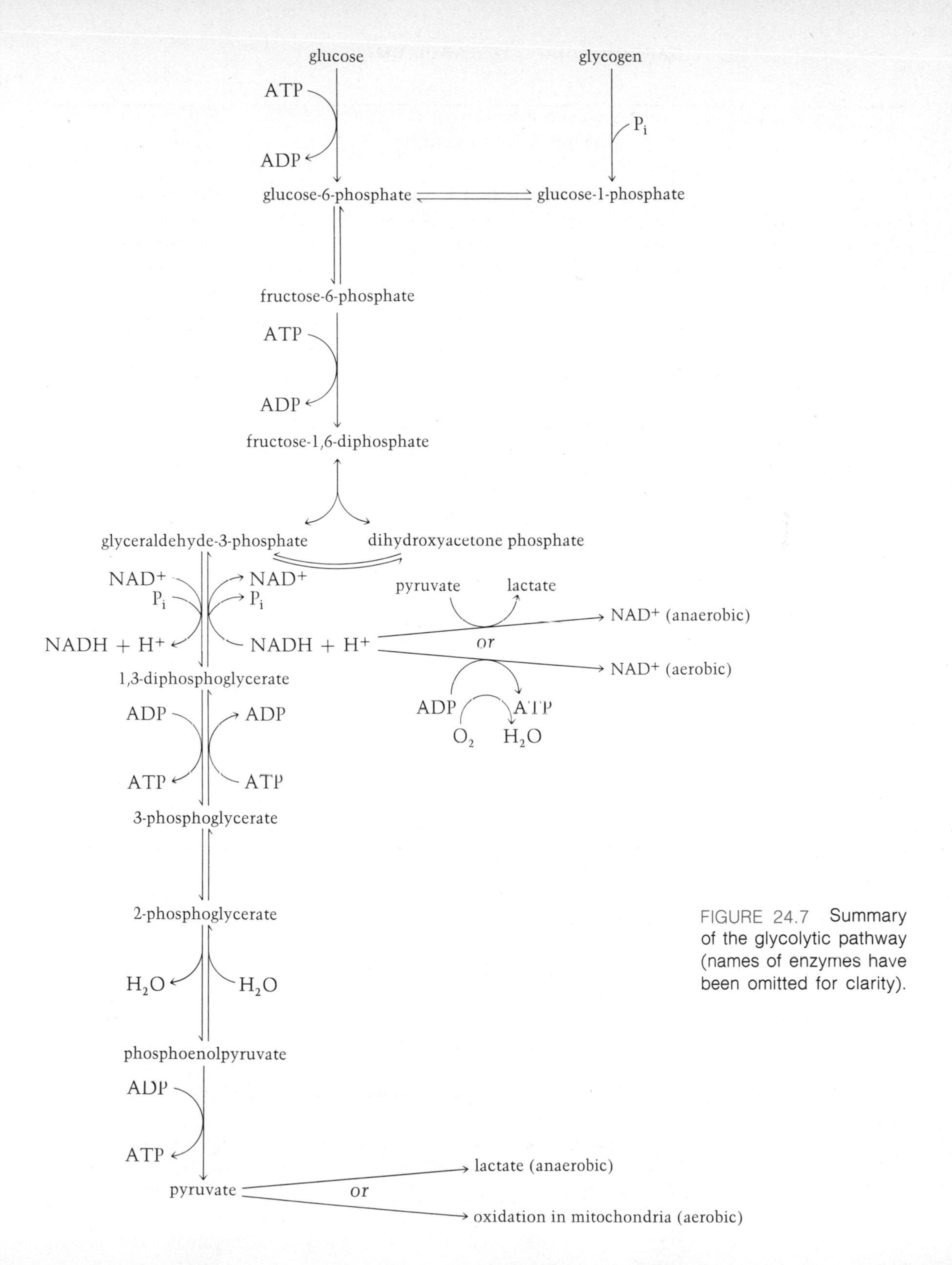

FIGURE 24.7 Summary of the glycolytic pathway (names of enzymes have been omitted for clarity).

Lactate cannot be further metabolized by the cells that do anaerobic glycolysis, and it diffuses out into the blood stream. Some of the lactate is absorbed by the liver, reoxidized to pyruvate by the action of liver lactate dehydrogenase, and used for glucose resynthesis. (We'll discuss this pathway shortly.) Some lactate is also used by "slow" skeletal muscles and heart muscles, which carry out aerobic metabolism. In these muscles, too, lactate is first converted to pyruvate, but then pyruvate is oxidized in the citric acid cycle to provide ATP energy to the cells. Since lactate is ultimately used by these other cells, the energy yield for glucose molecules is about the same whether they are initially metabolized under aerobic or anaerobic conditions. The pathways of aerobic and anaerobic carbohydrate breakdown are shown together in Figure 24.7.

24.11 NEW SYNTHESIS OF GLUCOSE: GLUCONEOGENESIS

Gluconeogenesis means the synthesis of glucose from smaller molecules (*gluco,* "glucose"; *neo,* "new"; *genesis,* "creation"). Only the liver (and, to a small extent, the kidney) are capable of carrying out glucose synthesis from these smaller molecules.

The liver is responsible for maintaining normal blood glucose levels when glucose is no longer being absorbed from the digestive system. Under most circumstances the liver has enough stored glycogen (Section 19.2) to supply glucose for this purpose. However, when fasting (even overnight) has depleted the liver's stock of glycogen, and in certain disease states, the liver cells find it necessary to make glucose from scratch.

24.12 CARBON SOURCES FOR LIVER GLUCONEOGENESIS

What raw materials are used for liver gluconeogenesis?

Most of the carbons used for gluconeogenesis come from amino acids that are being broken down to yield pyruvate or four-carbon citric acid cycle intermediates. Energy, in the form of ATP and NADH, comes from the oxidation of fatty acids. An alternative source of carbon atoms is lactate from skeletal muscle. During active exercise (when fast skeletal muscle carries out anaerobic glycolysis), the blood stream carries this lactate to the liver, where it is converted back to pyruvate (Figure 24.8).

Fatty acid carbons generally cannot be used for gluconeogenesis, for two reasons. First, acetyl CoA is the product of the breakdown of fatty

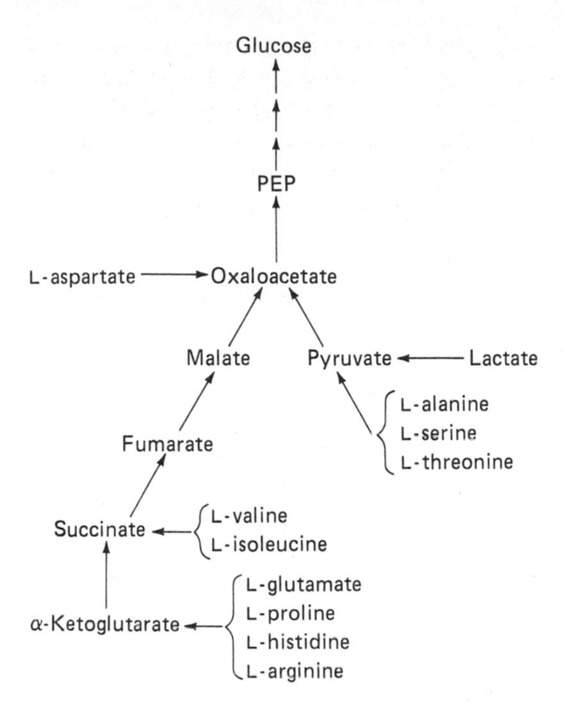

FIGURE 24.8 Carbon sources for gluconeogenesis.

acids. When acetyl CoA is oxidized in the citric acid cycle, two carbon atoms are lost in the form of CO_2 (Section 23.10). This is the same as saying that the acetyl group is completely oxidized to CO_2 in the cycle. Thus acetyl CoA does not undergo *net* conversion to four-carbon citric acid cycle intermediates that could be used in gluconeogenesis. Second, the pyruvate dehydrogenase reaction (Section 24.8) is irreversible, meaning that acetyl CoA cannot be converted to pyruvate. In effect, then, the road from fatty acids to glucose synthesis is simply blocked.

COMPARISON OF GLUCONEOGENESIS AND GLYCOLYSIS 24.13

Most of the reactions of gluconeogenesis are the same as those of glycolysis, except that they occur in reverse. Except for the first reaction of gluconeogenesis, all the reactions of both processes take place in the cytosol of the liver cell. The glycolytic enzymes can catalyze the reversible reactions in either direction. The direction of the reaction depends on whether the cell needs ATP or has a good supply of it, and on the signals received by the major regulatory enzymes.

But there are three reactions in the glycolytic pathway that are highly exergonic in the catabolic direction (Sections 24.4 and 24.5) and therefore irreversible. The reverse of these reactions must be carried out in a different way in order to make the biosynthetic pathway energetically possible. This is accomplished through the use of unique gluconeogenic enzymes and the avoidance of the irreversible reactions catalyzed by the glycolytic enzymes. See Figure 24.9 for a flow chart comparing both pathways.

†24.14 THE FOUR UNIQUE REACTIONS OF GLUCONEOGENESIS

Acetyl CoA is a regulator of gluconeogenesis.

We will start by examining the synthesis of glucose from pyruvate. The pyruvate comes either from lactate or from certain amino acids (Figure 24.8). The first synthetic reaction takes place in the mitochondria:

$$\text{pyruvate}^- + \text{ATP}^{4-} + CO_2 \underset{\text{(acetyl CoA)}}{\overset{\text{pyruvate carboxylase}}{\rightleftharpoons}} \text{oxaloacetate}^{2-} + \text{ADP}^{3-} + P_i^{2-} + 2H^+$$

c̄ Biotin

This carboxylation reaction is driven by ATP hydrolysis. The enzyme contains the coenzyme *biotin*—which, you may recall, is one of the water soluble vitamins (Section 21.12). The biotin coenzyme is carboxylated, and then it donates the carboxyl group to pyruvate to form oxaloacetate. Pyruvate carboxylase is an important regulatory point in the gluconeogenic pathway. It absolutely requires its allosteric activator, acetyl CoA. An accumulation of acetyl CoA in the mitochondria may signal either a shortage of oxaloacetate (for the citrate synthase reaction of the citric acid cycle), or a rapid oxidation of fatty acids. Rapid fatty acid oxidation in the liver cells indicates that there is a carbohydrate shortage and that gluconeogenesis is needed. The logical response is to increase oxaloacetate formation by pyruvate carboxylase, and this is just what an increase in acetyl CoA concentration in the mitochondria does.

In the next reaction, phosphoenolpyruvate (PEP) is formed from oxaloacetate:

$$\text{oxaloacetate}^{2-} + \text{GTP}^{4-} \underset{Mg^{2+}}{\overset{\text{PEP carboxykinase}}{\rightleftharpoons}} \text{PEP}^{3-} + CO_2 + \text{GDP}^{3-}$$

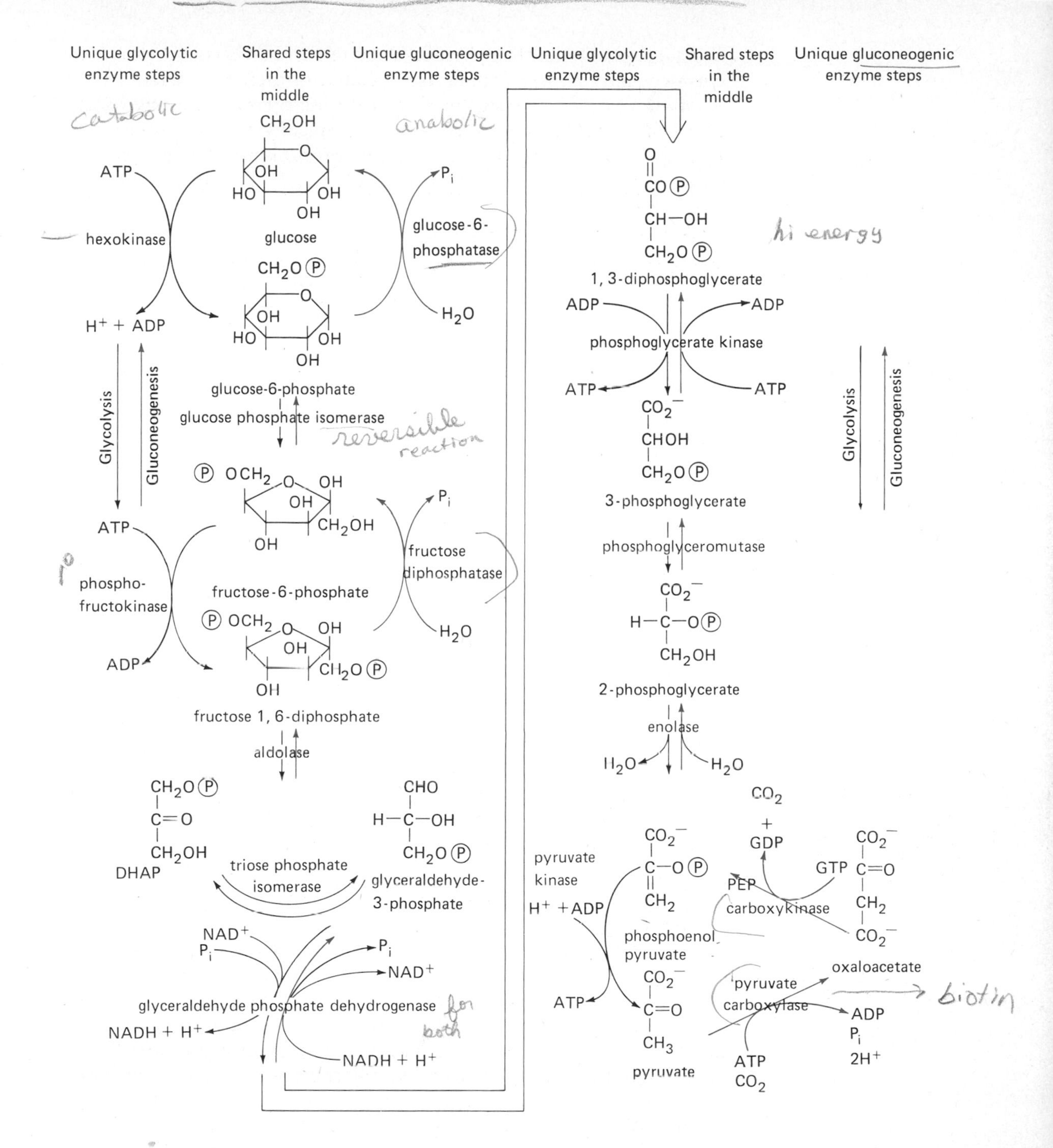

FIGURE 24.9 Flowchart comparing glycolysis and gluconeogenesis (Ⓟ symbolizes a phosphate group).

The same CO_2 that was attached by pyruvate carboxylase is released in this decarboxylation reaction. Note that with these two reactions the reaction catalyzed by pyruvate kinase during glycolysis (Section 24.5) has been reversed. That is, pyruvate has been converted to phosphoenolpyruvate. Note also that whereas the pyruvate kinase reaction occurs with the formation of one ATP molecule (*one* high-energy phosphate bond) for each pyruvate formed, the formation of PEP from pyruvate requires the hydrolysis of one GTP and one ATP molecule (*two* high-energy phosphate bonds). It is the energy obtained from the hydrolysis of these bonds that makes it energetically possible for PEP to be formed from pyruvate in this way. That is, the very exergonic reactions of ATP and GTP hydrolysis drive the very endergonic reaction of forming PEP from pyruvate.

The oxaloacetate used for the formation of PEP may also come from the citric acid cycle. Most of the amino acids yield one or another citric acid cycle intermediate when they are being broken down. Thus, the carbon skeletons of many amino acids can be turned into glucose without the participation of pyruvate carboxylase (Figure 24.8).

Problem 24.7 Write a series of reactions, starting with succinyl CoA, that result in the formation of PEP. Include all necessary coenzymes, nucleotides, and other reactants.

Once phosphoenolpyruvate is formed, several glycolytic enzymes operate in reverse to convert it to fructose-1,6-diphosphate (Figure 24.7). Then comes the problem of reversing the irreversible reaction catalyzed by phosphofructokinase in glycolysis, that is, how to make fructose-6-phosphate from fructose-1,6-diphosphate. The gluconeogenic enzyme fructose-1,6-diphosphatase catalyzes a simple hydrolysis of the 1-phosphate ester bond:

$$\text{fructose-1,6-diphosphate}^{4-} + H_2O \xrightarrow{\text{fructose-1,6-diphosphatase}} \text{fructose-6-phosphate}^{2-} + P_i^{2-}$$

Since fructose-1,6-diphosphate is a medium-energy compound (recall that its formation during glycolysis required ATP), its hydrolysis releases a fair amount of energy and the reaction is irreversible.

Both fructose diphosphatase and phosphofructokinase are allosteric enzymes (Section 21.15). They are regulated very closely to ensure that they are not active simultaneously. Phosphofructokinase is *inhibited* by a high concentration of ATP, which signals that the cell contains enough ATP and that no more carbohydrate breakdown is needed at the moment. Conversely, fructose diphosphatase is *activated* by ATP, which stimulates glucose synthesis while ATP energy is readily available. So when the concentration of ATP is high in the liver cell, phosphofructokinase is turned off and fructose diphosphatase may be turned on. (However, there are some other regulatory mechanisms that we cannot pursue here.)

When the energy level of the cell is low, the concentration of ATP is low and those of ADP and AMP are high. ADP and AMP activate phosphofructokinase (turning on glucose catabolism, which will provide energy to make ATP). At the same time, they inhibit fructose diphosphatase, ensuring that the two enzymes are not working at cross-purposes.

We should note that the pyruvate kinase reaction and the pair of gluconeogenic steps that "reverse" it are also closely controlled to prevent their acting at the same time.

Fructose-6-phosphate is converted to glucose-6-phosphate by the glycolytic enzyme glucose-6-phosphate isomerase. The liver cell can now form glycogen (see below) or it can release free glucose into the blood stream:

$$\text{glucose-6-phosphate}^{2-} + H_2O \xrightarrow{\text{glucose-6-phosphatase}} \text{glucose} + P_i^{2-}$$

This step represents the reversal of the reaction catalyzed by hexokinase—glucose is formed from glucose-6-phosphate.

Problem 24.8 How many high-energy phosphate bonds are used in the synthesis of one D-glucose from pyruvate? from oxaloacetate?

Problem 24.9 Write equations for the reactions that will synthesize glucose from lactate. Include all necessary coenzymes, nucleotides, and other reactants.

STORAGE OF GLYCOGEN AND STARCH 24.15

Glycogen, the storage carbohydrate of animal cells, and starch, the storage carbohydrate of plant cells, are very similar in structure (Section 19.2). All animal cells maintain stores of glycogen, laid down when blood glucose is plentiful. Glycogen is the most quickly mobilized energy storage material. The glycogen polymer is simply deposited in the form of small granules in the cell cytoplasm. The amount of glycogen that can be stored in animal cells is relatively small and intended only for short-term purposes. In animals, fat is the long-term energy storage form. On the other hand, certain plant cells are capable of storing large quantities of starch. Consider, for example, the cells of the potato tuber, or the seeds of grains such as wheat; these are nothing more than specialized cells packed with stored food molecules, most of which are polysaccharides. The carbohydrate stores in both plant and animal cells are readily broken down and oxidized when energy is needed within the cell.

24.16 THE SPECIAL ROLE OF LIVER GLYCOGEN

How does the body maintain blood glucose at about 5 mM?

In animals, the glycogen stored in liver cells has an additional very important function beyond that of an energy reserve for the liver cell itself. Certain body cells, particularly the cells of the brain and the red blood cells, are almost completely dependent on a supply of glucose from the blood stream to satisfy their energy requirements. (Most other cells can equally well use, or even prefer to use, fatty acids or lactate as their source of energy.) Therefore, the blood glucose must be maintained at a certain level (about 5 mM) to ensure an adequate supply of energy to these important glucose-dependent tissues. Whenever the blood glucose content falls below the normal value, the liver has the responsibility of breaking down its cellular glycogen stores to release free glucose into the blood stream.

The level of glucose in the blood is regulated by the pancreatic hormones **glucagon** and insulin. These hormones have opposing effects on the liver and other cells. Glucagon acts on the liver cells to promote the release of glucose and raise blood glucose levels; it does this by activating glycogen breakdown and also gluconeogenesis. Insulin acts on

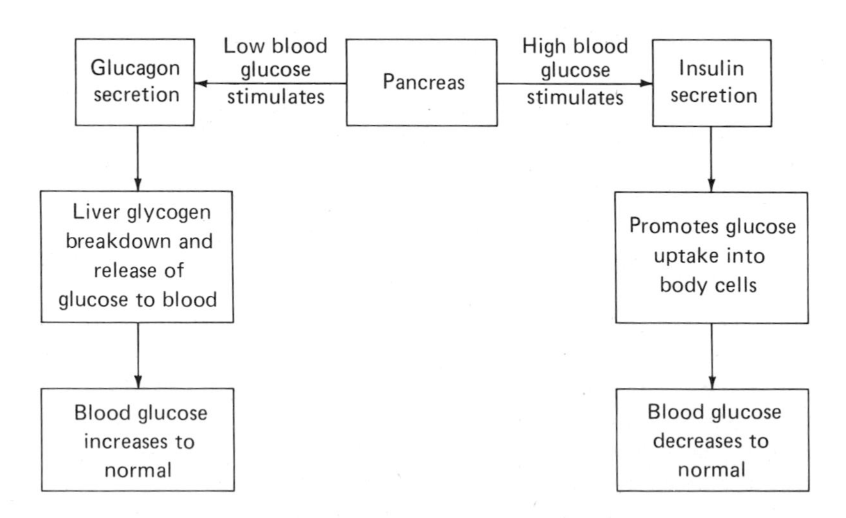

FIGURE 24.10 Control of blood glucose by pancreatic hormones insulin and glucagon.

many other cells, especially muscle, to promote the uptake of glucose, which reduces blood glucose levels. The secretion of the hormones is controlled in turn by the blood glucose levels themselves. High blood glucose (**hyperglycemia**) elicits insulin secretion, whereas low blood glucose (**hypoglycemia**) elicits glucagon secretion (see Figure 24.10). The glucose tolerance test is given to suspected diabetics, among others, to measure the ability of the pancreas to respond to glucose in the blood. After the patient is given a large "dose" of glucose, usually in a cola drink, his or her blood glucose is measured several times in the hours following to see if glucose disappears from the blood in the normal time.

Problem 24.10 If you eat a meal containing a substantial amount of carbohydrate, will your pancreas secrete insulin or glucagon?

BREAKDOWN OF GLYCOGEN AND ITS REGULATION 24.17†

Glycogen is a polysaccharide of $\alpha(1\rightarrow4)$ linked glucose units with frequent $\alpha(1\rightarrow6)$ branches. Glycogen phosphorylase is the enzyme that catalyzes the step-by-step cleavage of single glucose molecules from the glycogen polymer.

$$\text{glycogen} + P_i \xrightarrow{\text{glycogen phosphorylase}} \text{glucose-1-phosphate} + \text{glycogen (minus 1 glucose unit)}$$

In liver cells, glycogen phosphorylase is activated by the hormone glucagon, leading to the release of glucose into the blood. In muscle cells, **epinephrine** (adrenalin) activates phosphorylase, which leads to the release of glucose phosphate in the cell itself to supply energy through glycolysis. In either case, the activation takes place by a series of reactions called an **activation cascade.** The activation cascade for phosphorylase activation is shown in Figure 24.11.

The series of reactions involved in activation is complicated, but at least one major thing should be noted: the role of **cyclic AMP** (Figure 24.12). When either glucagon or epinephrine binds to its "target" cells by binding to specific proteins in the cell membrane, this causes the activation of a membrane enzyme that catalyzes the formation of cyclic AMP (abbreviated cAMP) inside the cell. Cyclic AMP then sets into motion the other events that lead to the formation of active phosphorylase molecules and the release of glucose phosphate units from glycogen. Cyclic AMP is involved in a similar way in the action of

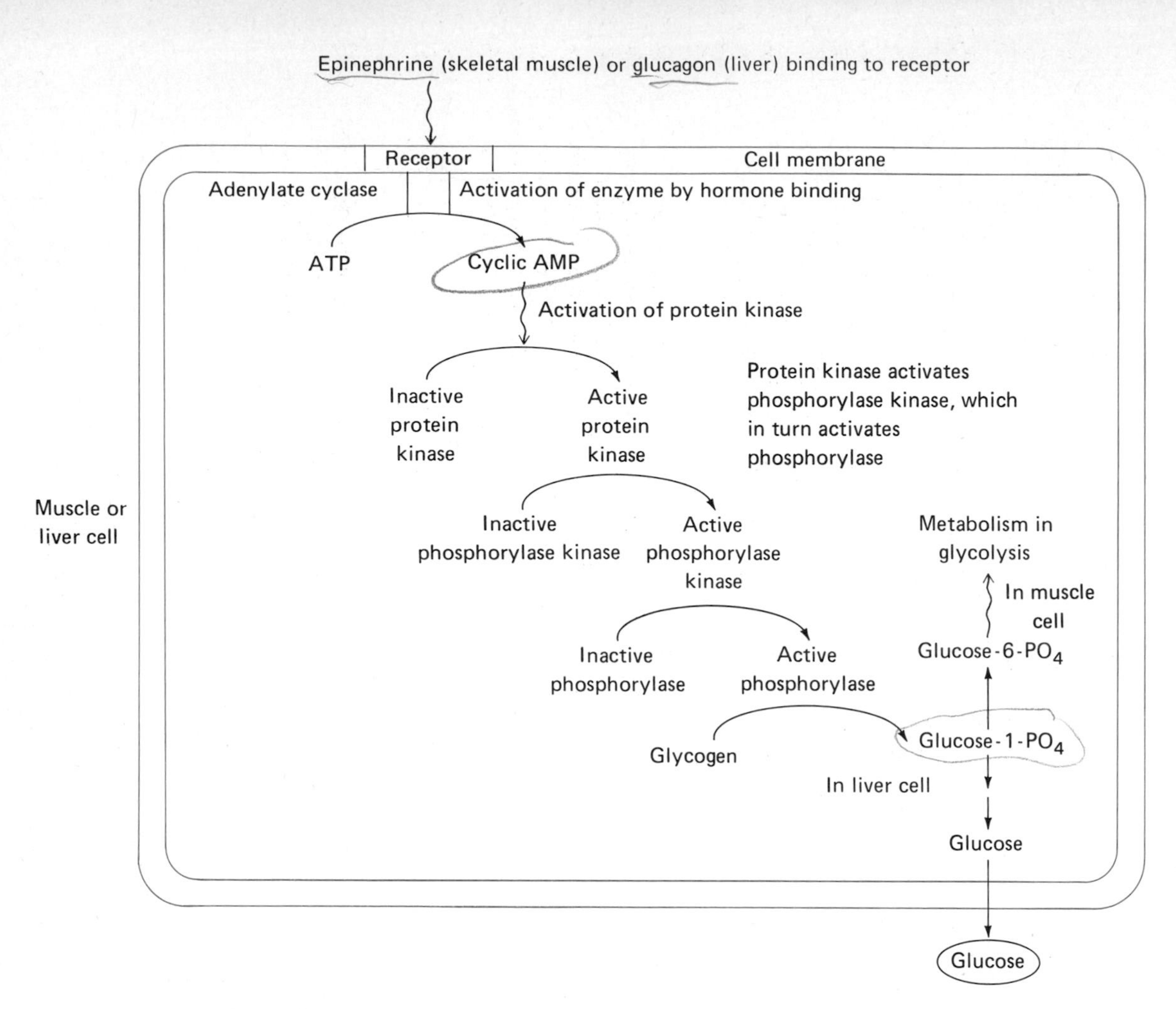

FIGURE 24.11 The activation cascade that turns on the breakdown of glycogen in liver and muscle cells.

NH_2
N
N
N
N
O—CH_2 O
O=P——O OH
$^-$O

FIGURE 24.12 Cyclic AMP (cAMP).

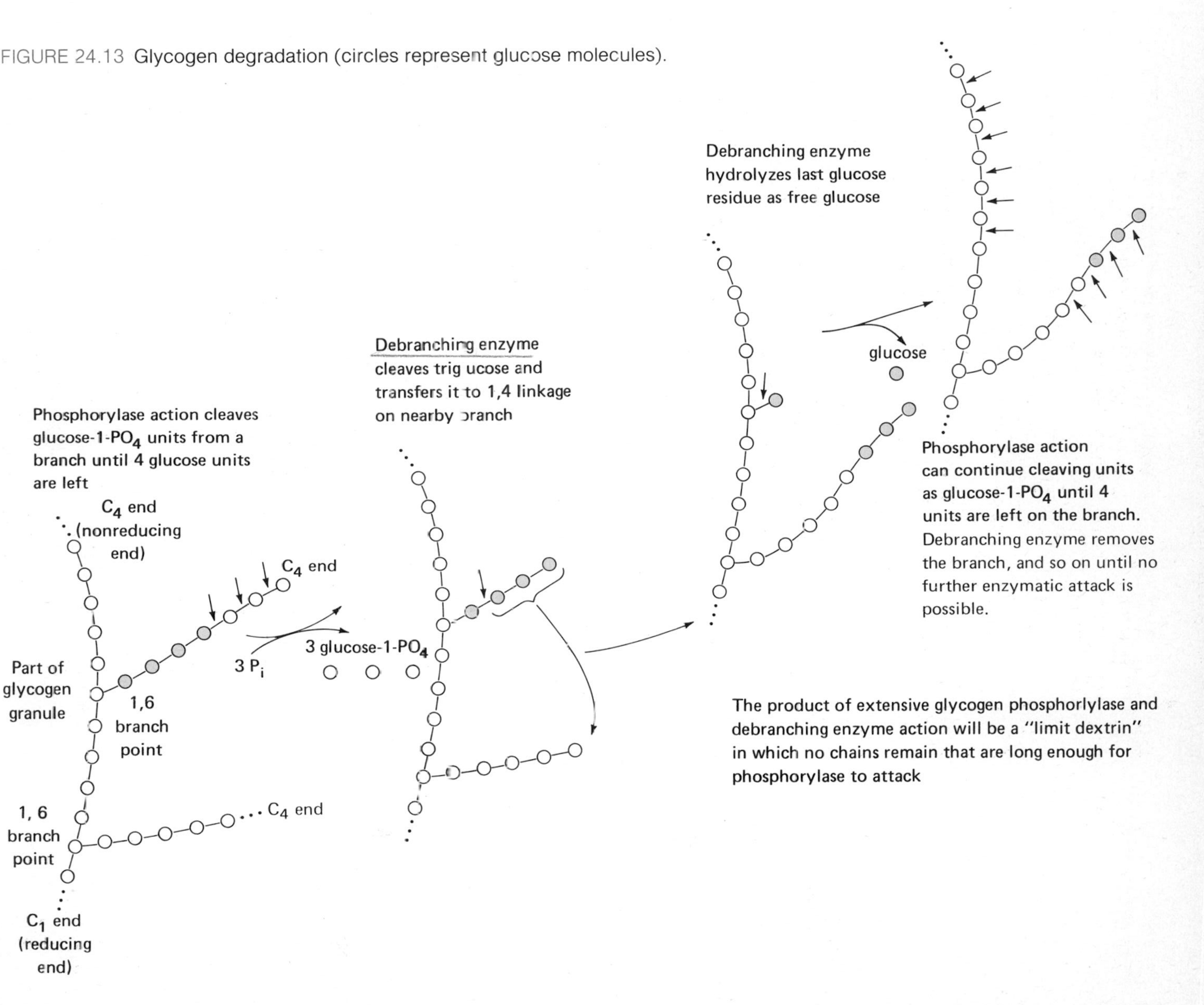
Phosphorylase action cleaves glucose-1-PO4 units from a branch until 4 glucose units are left
C4 end (nonreducing end)
C4 end
Part of glycogen granule
1,6 branch point
1, 6 branch point
C4 end
C1 end (reducing end)
3 Pi
3 glucose-1-PO4
Debranching enzyme cleaves trig ucose and transfers it to 1,4 linkage on nearby branch
Debranching enzyme hydrolyzes last glucose residue as free glucose
glucose
Phosphorylase action can continue cleaving units as glucose-1-PO4 until 4 units are left on the branch. Debranching enzyme removes the branch, and so on until no further enzymatic attack is possible.
The product of extensive glycogen phosphorlylase and debranching enzyme action will be a "limit dextrin" in which no chains remain that are long enough for phosphorylase to attack

FIGURE 24.13 Glycogen degradation (circles represent glucose molecules).

various other hormones on other cells, so its role in hormone action is an important one.

Problem 24.11 Glucagon does not affect muscle glycogen breakdown. Why is it that some tissues respond to particular hormones and others do not? (*Hint:* How does a hormone bind to its target cells?)

Glycogen breakdown can also be turned on by allosteric mechanisms that do not involve hormones. You will recall that muscle contraction occurs when the Ca^{2+} concentration rises in the cytoplasm of muscle cells (Section 22.10). Contraction requires energy, of course, and in contracting skeletal muscle glycolysis is the major source of ATP. Thus it is appropriate that the rise in Ca^{2+} concentration also stimulates glycogen breakdown by activating phosphorylase kinase, immediately providing glucose phosphate molecules to fuel glycolysis.

The starch phosphorylase found in plants is activated differently. In the potato tuber, phosphorylase is activated by storage near 0°C. Have you ever noticed the sweetness of potatoes that have been touched by frost or stored in an unheated room during very cold weather? The release of glucose into the cytoplasm protects the cells from freezing by lowering the freezing point of the cell water.

Phosphorylases cannot bind the glycogen or starch substrates closer than four residues from a branch point. A debranching enzyme is necessary, in addition to phosphorylase, for extensive degradation of glycogen or amylopectin to occur (see Figure 24.13). The active site of the debranching enzyme is suitable for binding glycogen at branches, unlike that of phosphorylase.

24.18 FATE OF GLUCOSE-1-PHOSPHATE

Once phosphorylase action releases glucose-1-phosphate molecules from glycogen, glucose-6-phosphate is formed.

$$\text{glucose-1-phosphate}^{2-} \xrightleftharpoons{\text{phosphoglucomutase}} \text{glucose-6-phosphate}^{2-}$$

Glucose-6-phosphate can be broken down by glycolysis if the cell needs energy. This breakdown will surely occur in the muscle cell (as indicated in Figure 24.11). In the liver cell, glucose can be released from glucose-6-phosphate in a simple hydrolysis reaction catalyzed by glucose-6-phosphatase:

$$\text{glucose-6-}PO_4{}^{2-} + H_2O \xrightarrow{\text{glucose-6-phosphatase}} \text{glucose} + P_i{}^{2-}$$

Glucose is then released into the blood stream to raise the blood glucose levels. Cells in other animal tissues do not have this enzyme at all, so they cannot make free glucose. Only the liver is capable of exporting glucose for use by other tissues.

GLYCOGEN STORAGE DISEASES 24.19

A number of inherited diseases of glycogen metabolism have been described by physicians. In some cases biochemists have positively identified the specific enzyme whose deficiency is responsible for the symptoms. Table 24.1 lists inherited diseases that involve flaws in the ability to degrade glycogen. Aside from those, certain other diseases are known that involve abnormalities in enzymes of glycogen synthesis.

TABLE 24.1 Glycogen Storage Diseases

Name	Type	Principal tissue affected	Enzyme deficiency	Symptoms
von Gierke's disease	I	Liver	Glucose-6-phosphatase	Hypoglycemia; weakness; sweating; pallor; enlarged liver with excessive glycogen deposits in cells
Limit dextrinosis	III	All tissues	Debranching enzyme	Excessive glycogen deposits; enlarged liver
McArdle's disease	V	Skeletal muscle	Phosphorylase	Muscle weakness; muscle pain after exercise; excessive glycogen in muscle cells
Hers' disease	VI	Liver	Phosphorylase	Similar to type I
	VII	Muscle	Phosphofructokinase	Similar to type V
	VIII	Liver	Phosphorylase kinase	Similar to type I

GLYCOGEN SYNTHESIS FROM GLUCOSE PHOSPHATES 24.20

All animal cells can make glycogen from glucose derivatives. Beginning with glucose-1-phosphate, the first step is the formation of an activated glucose derivative called UDP-glucose (Figure 24.5). UDP-glucose is a high-energy compound that will readily transfer a glucose molecule to an appropriate acceptor. The acceptor in this case is a glycogen polymer already present in tissue:

UDP-glucose + glycogen $\xrightleftharpoons{\text{glycogen synthase}}$

UDP + glycogen (lengthened by 1 glucose unit)

The reaction is essentially the reverse of that catalyzed by glycogen phosphorylase (Section 24.17). Furthermore, the reaction is tightly regulated by a cascade mechanism similar to the one that regulates glycogen phosphorylase, but it is oppositely directed.

Glycogen has $\alpha(1\rightarrow6)$ branches in addition to $\alpha(1\rightarrow4)$ glycosidic bonds (Section 19.3). An additional enzyme, a branching enzyme, is responsible for transferring short lengths of the chain (about seven glucose units) into $\alpha(1\rightarrow6)$ linkages (Figure 24.14). This is how the branched glycogen polymer is formed.

SUMMARY

Glycolysis is the major route for catabolism of the sugars that are derived from starches and other carbohydrates in the diet. Glycolysis results in the breakdown of glucose to pyruvate (which is then completely oxidized to CO_2 and H_2O under **aerobic** conditions) or lactate (formed from pyruvate under **anaerobic** conditions). Two ATP molecules (net) are formed in conjunction with the reaction. Under aerobic conditions, pyruvate is converted to acetyl CoA, which is completely oxidized in the citric acid cycle. In this cycle, reduced coenzymes are

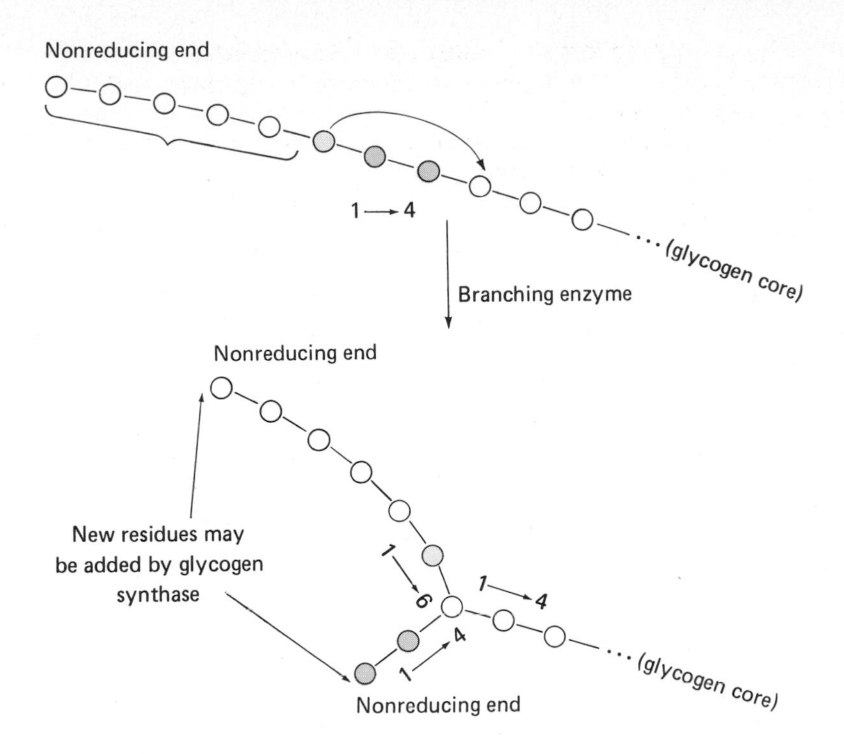

FIGURE 24.14 Action of glycogen branching enzyme.

formed that are then oxidized by the electron transport system, yielding ATP. The total ATP yield from the complete oxidation of glucose is 38 ATP. Carbohydrate catabolism is regulated by the cell's energy level (ATP level). In humans, anaerobic glycolysis occurs mainly in red blood cells and fast skeletal muscle. Pyruvate is reduced to lactate by NADH in an anaerobic reaction catalyzed by *lactate dehydrogenase (LDH)*. This reaction regenerates NAD so that glycolysis can continue. The overall ATP yield of anaerobic glycolysis is two ATP molecules per glucose molecule metabolized.

Glycogen and starch may be broken down to be oxidized for energy. Liver glycogen also provides a source of glucose for the maintenance of blood glucose levels, which are regulated by **insulin** and **glucagon.** Glycogen phosphorylase is the enzyme that catalyzes the release of glucose-1-phosphate molecules from glycogen. Phosphorylase is activated by means of an **activation cascade** involving glucagon (in liver cells), or **epinephrine** (in muscle cells), **cyclic AMP,** and several enzymes. In the liver, glucose-6-phosphatase may release free glucose to the blood stream.

Gluconeogenesis takes place only in the liver and kidney. The liver can use this pathway to supply blood glucose if liver glycogen is depleted. The carbon sources are lactate, amino acids, and four-carbon citric acid cycle intermediates. Fatty acids cannot be used for gluconeogenesis. Gluconeogenesis and glycolysis use the same enzymes, except for four unique gluconeogenic enzymes that bypass irreversible steps of glycolysis. Gluconeogenesis and glycolysis are regulated so that both are not active at the same time. All cells can synthesize glycogen from glucose via glycogen synthase.

Key Terms

Acceptor control (Section 24.9)
Activation cascade (Section 24.17)
Aerobic (Section 24.2)
Anaerobic (Section 24.2)
ATP-synthetic phase (Section 24.3)
Collecting phase (Section 24.3)
Cyclic AMP (Section 24.17)
Diabetes mellitus (Section 24.1)
Epinephrine (Section 24.17)
Fermentation (Section 24.2)
Fructosuria (Section 24.7)
Galactosemia (Section 24.7)
Glucagon (Section 24.16)
Gluconeogenesis (Section 24.11)
Glycolysis (Introduction)
Hyperglycemia (Section 24.16)
Hypoglycemia (Section 24.16)
Insulin (Section 24.1)
Kinase (Section 24.4)
Phosphorylation (Section 24.4)

ADDITIONAL PROBLEMS

24.12 What is the main purpose of glycolysis?
24.13 Glycolysis is considered to be an irreversible process. Why?
24.14 If triose phosphate isomerase is not present, can glycolysis occur?
24.15 Under aerobic conditions, little lactate is formed when carbohydrate is being oxidized. Why?
24.16 Which enzyme in the citric acid cycle is very similar to the pyruvate dehydrogenase system?
24.17 Epinephrine (adrenalin) has been called the fight-or-flight hormone. How does this relate to the action of the hormone on the skeletal muscle cells?

24.18 Describe the effects of cyclic AMP, glucagon, and epinephrine on glycogen metabolism.

24.19 Write reactions to show how dihydroxyacetone phosphate might be used for net glucose synthesis in the liver cell.

24.20 What is the message transmitted when acetyl CoA accumulates inside the liver mitochondria? What enzyme responds and with what result?

24.21 Write the equations for and name the enzymes that catalyze the unique gluconeogenic reactions. Do the same for the glycolytic reactions that the unique gluconeogenic enzymes bypass.

24.22 Describe the conversion of glucose-6-phosphate to glycogen.

24.23 Write a series of reactions, as well as the overall reaction, for the complete oxidation of

(a) DHAP
(b) PEP
(c) lactate

24.24 Write a series of reactions and the overall reaction for the anaerobic glycolysis of sucrose. What is the ATP yield per sucrose molecule? (Assume simple hydrolysis of sucrose to glucose and fructose).

Problem-Solving Hints

If you have trouble solving any of the additional problems, refer to the sections listed next to the problem numbers.

24.12 24.3
24.13 24.6
24.14 24.5
24.15 24.8
24.16 24.8
24.17 24.17
24.18 24.17
24.19 24.14
24.20 24.14
24.21 24.14
24.22 24.15
24.23 24.4, 24.5, 24.8
24.24 24.4, 24.5, 24.7

FOCUS 14 The Pyruvate Dehydrogenase System

The pyruvate dehydrogenase system catalyzes the reaction that turns pyruvate into acetyl CoA, and thus allows the citric acid cycle to get started. This reaction is of particular interest because it illustrates the functions of several coenzymes. The pyruvate dehydrogenase system contains many polypeptide subunits. It is a very large protein molecule (molecular weight 7×10^6), which exhibits three distinct enzyme activities. Each of the three enzymes contains a tightly bound coenzyme, and two additional coenzymes (coenzyme A and NAD) participate in the complete reaction. The enzymes and the reactions they catalyze are as follows:

1. *Pyruvate dehydrogenase* (coenzyme: thiamine pyrophosphate), abbreviated $E_1 \cdot TPP$, where E_1 stands for pyruvate dehydrogenase and TPP for thiamine pyrophosphate. $E_1 \cdot TPP$ catalyzes the oxidative decarboxylation of pyruvate (an α-keto acid) and transfers the resulting two-carbon acetyl group into thioester linkage with lipoic acid (the cofactor of the second enzyme):

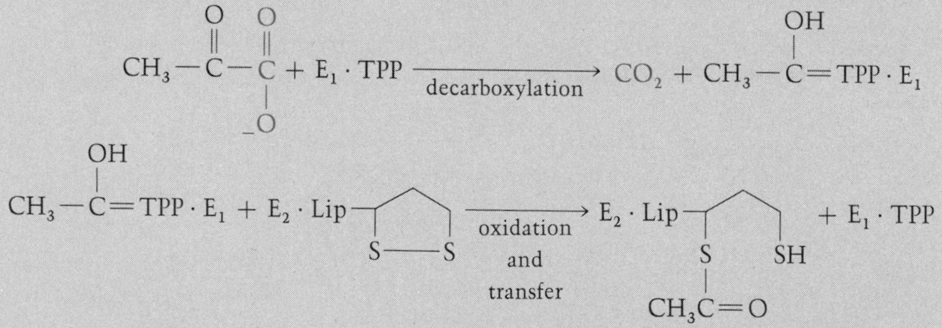

$$CH_3{-}\overset{O}{\overset{\|}{C}}{-}\underset{\underset{{}^-O}{|}}{\overset{O}{\overset{\|}{C}}} + E_1 \cdot TPP \xrightarrow[\text{decarboxylation}]{} CO_2 + CH_3{-}\overset{OH}{\overset{|}{C}}{=}TPP \cdot E_1$$

$$CH_3{-}\overset{OH}{\overset{|}{C}}{=}TPP \cdot E_1 + E_2 \cdot Lip\text{(S—S ring)} \xrightarrow[\text{oxidation and transfer}]{} E_2 \cdot Lip\text{(S–}C(=O)CH_3\text{, SH)} + E_1 \cdot TPP$$

Note that the product of the decarboxylation step is an enol (unsaturated alcohol), which is at the same oxidation level as an aldehyde or ketone. Oxidation brings it to the oxidation level of a carboxylic acid (acetic acid), although here it is in the form of acetyl thioester.

2. *Lipoate acetyltransferase* (cofactor: lipoic acid). In its oxidized form this enzyme is represented as

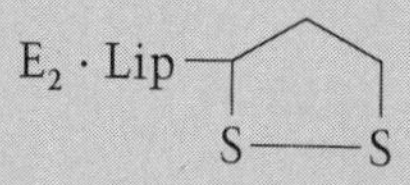

$$E_2 \cdot Lip\text{(S—S ring)}$$

In its reduced form it may be represented by the formula

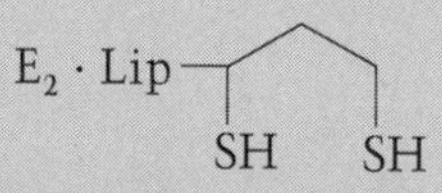

$$E_2 \cdot Lip\text{(SH, SH)}$$

The enzyme catalyzes the transfer of the acetyl group from lipoate to coenzyme A:

$$E_2 \cdot \text{Lip}(\text{S}\text{-}\text{C(=O)CH}_3)(\text{SH}) + \text{CoASH} \xrightarrow[\text{acetyl transfer to CoA}]{} CH_3\overset{\text{O}}{\overset{\|}{\text{C}}}SCoA + E_2 \cdot \text{Lip}(\text{SH})(\text{SH})$$

Note that the enzyme (more specifically, its coenzyme) is left in fully reduced form. If it is to be able to catalyze the reaction again with another acetyl group, it must be reoxidized. That is the role of the third enzyme of the system.

3. *Dihydrolipoate dehydrogenase* (coenzyme: FAD), represented as $E_3 \cdot FAD$ (oxidized form) and $E_3 \cdot FADH_2$ (reduced form). $E_3 \cdot FAD$ catalyzes the oxidation of dihydrolipoate, using NAD as the coenzyme:

$$E_2 \cdot \text{Lip}(\text{SH})(\text{SH}) + E_3 \cdot FAD \xrightarrow[\text{oxidation of dihydrolipoate}]{} E_2 \cdot \text{Lip}(\text{S—S}) + E_3 \cdot FADH_2$$

The reduced $E_3 \cdot FADH_2$ is then reoxidized by NAD:

$$E_3 \cdot FADH_2 + NAD^+ \xrightarrow[\text{oxidation of } E_3 \cdot FADH_2]{} E_3 \cdot FAD + NADH + H^+$$

Note that both steps are catalyzed by dihydrolipoate dehydrogenase ($E_3 \cdot FAD$).

If we add the five reactions for the enzymes of the complex, we obtain the overall reaction for the dehydrogenation of pyruvate to form acetyl CoA (see Section 24.8). Acetyl CoA may be oxidized in the citric acid cycle, whereas NADH is oxidized by the electron transport system, giving rise to ATP through oxidative phosphorylation.

Since the pyruvate dehydrogenase reaction is central to the metabolism of carbohydrates, and since carbohydrates make up the major part of most people's diets, it is not surprising that thiamine deficiency (Section 19.12) is reflected in dramatic weight loss and wasting of the body. No thiamine, no thiamine pyrophosphate. No thiamine pyrophosphate, no acetyl CoA and no citric acid cycle. No citric acid cycle, no ATP and no usable energy. Neurological symptoms are also observed in people with thiamine deficiencies, since nerve tissues depend heavily on aerobic glucose oxidation for their energy supply.

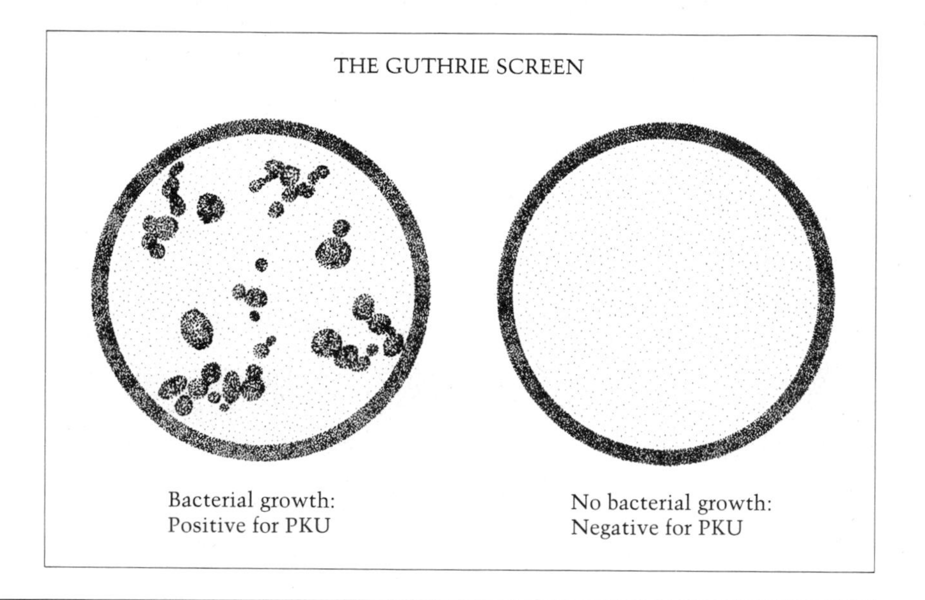

The condition known as phenylketonuria (PKU), which will be discussed in this chapter, affects newborn infants. The condition is characterized by an increased amount of the amino acid phenylalanine in the blood.

In a hospital environment, a federal regulation requires that all newborn infants be tested for phenylketonuria. A procedure called the Guthrie screen is the commonest test used to detect PKU. This test is a microbiological procedure that involves a special strain of bacterial organism. The organism requires phenylalanine to grow. First the organism is placed in an environment with thienylalanine, which prevents growth in the organism. Blood drawn from the infant's heel from the fourth to sixth day of life is then added to the mixture. Only if the level of phenylalanine in the blood exceeds that of the thienylalanine will the bacteria grow. In other words, the action of the thienylalanine can be overcome by appropriately high levels of phenylalanine, which strongly suggest the presence of PKU.

Recently, it has come to the attention of hospitals involved that newborns are spending less time in the hospital and are being released as early as the second day of life. This trend has a direct effect on the validity of the Guthrie screen, since the infant is usually given four days in which to ingest phenylalanine. After four days, the level of the amino acid in the blood becomes detectable, and any increase of phenylalanine above the normal level (such as would occur in a PKU baby) is apparent. However, on the

25

METABOLISM OF FATS AND AMINO ACIDS

second day of life, when phenylalanine might not yet be present in the infant's diet, a PKU baby could falsely project a normal level.

New techniques have been developed that are now in use in many hospital laboratories. Instead of a screen, which can only detect a level above a normal diet level, a quantitative technique is used. This quantitative method measures the actual blood level of phenylalanine and is more accurate than the bacterial procedure.

[Illustration: The Guthrie screen provides an accurate and fairly rapid method for the diagnosis of PKU, but it must be used on serum obtained from newborns shortly after birth.]

METABOLISM OF FATS AND AMINO ACIDS

Whereas carbohydrate breakdown produces energy most quickly, fats and amino acids provide a longer-term and steadier energy supply for most cells. We've discussed how carbohydrates (like the bread of a chicken sandwich) are digested and absorbed more quickly, and how glycogen stored in muscle and other cells provides an immediate source of glucose phosphates for energy when needed. Fats and proteins, like the butter and chicken in the sandwich, are digested and absorbed rather slowly. They stay in the digestive system longer and provide a continuing supply of foodstuff molecules to the blood stream for some time after a meal is over. Similarly, fats are stored in the body in relatively large amounts and are made available (mobilized) when needed to supply energy and to keep the body running smoothly. In this chapter we will see how fats and fatty acids are manufactured to make fat for storage as well. Finally, we will see how the amino acids we obtain from dietary proteins are degraded and how excess nitrogen is excreted. An important feature of the breakdown of both fatty acids and amino acids is that, like glucose, they are broken down to yield acetyl CoA, which is then oxidized in the citric acid cycle to provide energy for ATP synthesis in oxidative phosphorylation.

FATS AS FUELS 25.1

How does the capacity for fat storage compare with the capacity for glycogen storage?

Fats are the second major source of food calories (or energy) after the carbohydrates. They compose about 40% of the average American diet. Fats, or triacylglycerols (Section 20.2), are also energy-storage compounds, like the polysaccharides. Indeed, animals have a much greater capacity for fat storage than for glycogen storage. Once the limited glycogen stores are filled, all the excess food molecules an animal takes in (those left over after its immediate requirements for energy glycogen, protein synthesis, and so forth), are converted to fat for storage. Many of us have had distressing personal experiences with the body's marvelous ability to store fat!

There is a second important difference between fats and polysaccharides as energy-storage materials. Glycogen and starch are fuels for short-term storage, whereas fats are fuels for long-term storage. A major advantage of fat as a stored fuel is that it weighs less than would an energy-equivalent amount of stored polysaccharide.

Specialized fat-storage tissues have evolved to accommodate large amounts of stored fat in many animals. These are called **adipose tissues** (Figure 25.1).

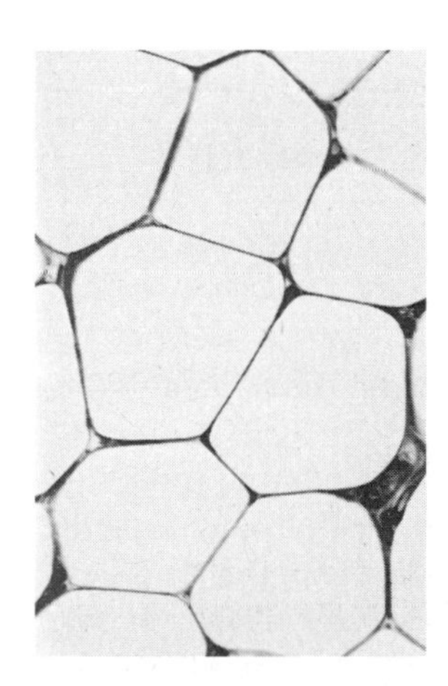

FIGURE 25.1 Photomicrograph of a thin section of adipose tissue. (From Bloom and Fawcett, *Histology*, © 1968, W. B. Saunders Company. By permission of D. W. Fawcett and W. B. Saunders Co.)

Plants store triacylglycerols mostly in fruits and seeds, where the growing embryo can use them for energy and raw materials. Many commercially important plant oils are obtained by the pressing of fruits or seeds—olive, corn, peanut, soybean, and safflower oils, for example. These plant products are usually oils rather than solid fats because they contain a greater proportion of polyunsaturated fatty acid residues (Section 20.2).

25.2 BREAKING DOWN STORED FATS TO FATTY ACIDS AND GLYCEROL

What reaction is catalyzed by a lipase?

Fats may be stored in one of two places—within the cell that is to oxidize them or in the fat-storage tissue (adipose tissue). In either case, the first step in degrading a fat is breaking the ester linkages between the glycerol backbone and the fatty acids. This reaction is catalyzed by a class of enzymes known as **lipases.** The many kinds of lipases differ in their location in the body and the way they are regulated, but they all catalyze the same general reaction:

$$\underset{\text{a triacylglycerol}}{\begin{array}{l} CH_2O-\overset{O}{\overset{\|}{C}}-R \\ R'-\overset{O}{\overset{\|}{C}}-OCH \\ CH_2O-\overset{O}{\overset{\|}{C}}-R'' \end{array}} + 3H_2O \xrightarrow{\text{lipase}} \underset{\text{glycerol}}{\begin{array}{c} CH_2OH \\ | \\ CHOH \\ | \\ CH_2OH \end{array}} + \underset{\text{fatty acids}}{\begin{array}{c} R-\overset{O}{\overset{\|}{C}}-OH \\ + \\ R'-\overset{O}{\overset{\|}{C}}-OH \\ + \\ R''-\overset{O}{\overset{\|}{C}}-OH \end{array}}$$

(This is an overall reaction; hydrolysis of the three fatty acids takes place in three separate steps.)

Glycerol is converted to the glycolytic intermediate compound, dihydroxyacetone phosphate (DHAP) (Section 24.5), which can then be metabolized through glycolysis. The fatty acids are metabolized separately. The process begins when adipose-tissue fats are hydrolyzed. The free fatty acids that are released are transported by **serum albumin** (a blood protein) to other parts of the body. This release of free fatty acids from adipose tissue is called **fatty acid mobilization.**

WILL A TISSUE OXIDIZE FATS OR CARBOHYDRATES FOR ENERGY? 25.3

Fatty acids are lipids and weak acids that exist partly in uncharged (protonated) form, so they cross the nonpolar lipid bilayer of cellular membranes easily. Unlike glucose, the fatty acids do not need insulin or membrane transport systems in order to enter cells. This means that many cells will choose to oxidize fatty acids if they are given a choice. Important exceptions are the red blood cells and brain cells. Red blood cells can catabolize only glucose for energy because they have no mitochondria, and fatty acid oxidation is a mitochondrial process. Brain cells do not receive any fatty acids because they do not obtain their nutrients directly from the blood stream, but rather from the cerebrospinal fluid. The so-called **blood–brain barrier** permits glucose and many other substances to reach the cerebrospinal fluid but keeps the albumin-bound fatty acids out.

Fatty acids are mobilized from adipose tissue during fasting; that's why you lose weight when you go on a diet. However, anxiety and physical exercise also cause extensive fatty acid mobilization. Thus, more fatty acids are made available to body tissues when there is a high demand for energy. Before mobilization can occur, lipases must first break down the fats to release free fatty acids. Various hormones are responsible for activating the triacylglycerol lipase of adipose cells by means of a cascade mechanism (Section 24.3) involving cyclic AMP. The mechanism is very similar to the one that activates glycogen phosphorylase in liver or muscle. The triacylglycerol lipase of adipose cells is activated by epinephrine, norepinephrine, cortisol, and glucagon.

In uncontrolled diabetes, most of the body's tissues are unable to use glucose. Instead, they are forced to oxidize large amounts of fatty acids. As we continue, we'll see some of the consequences of such an imbalance in the ratio of carbohydrate metabolism to fat metabolism.

PREPARATION OF FATTY ACIDS FOR MITOCHONDRIAL FATTY ACID OXIDATION 25.4

What is the committing step in fatty acid oxidation?

Fatty acids can reach the cytoplasm by diffusing across the cell membrane from the blood. They can also be released from stored fats by lipase activity within the cell. The first step in the breakdown of fatty acids, whether they came from the butter you ate today or were mobilized from stored fats, is the formation of a CoA ester of the fatty acid:

$$R{-}CH_2CH_2CH_2\overset{\overset{\displaystyle O}{\|}}{C}{-}O^- + ATP^{4-} + H{-}S{-}CoA \rightarrow$$

$$R{-}CH_2CH_2CH_2\overset{\overset{\displaystyle O}{\|}}{C}{-}S{-}CoA + AMP^{2-} + PP_i^{3-}$$

$$PP_i^{3-} + H_2O \rightarrow 2P_i^{2-}$$

Because many different fatty acids occur in the diet and are made in the body, we use a general formula to represent the fatty acid. Typically, the fatty acid will have 16 or 18 carbons, so here R represents a 12- or 14-carbon chain.

The formation of the CoA ester of a fatty acid—which is called a fatty acyl CoA—is a committing step. For one thing, the formation of fatty acyl CoA is irreversible because of the large negative free-energy change that results from ATP hydrolysis and then the hydrolysis of PP_i (inorganic pyrophosphate) to two P_i molecules. But also, the large CoA group (Figure 24.6) prevents the fatty acid (fatty acyl group) from leaving the cell.

The large CoA group also prevents fatty acyl CoA molecules from being transported intact into the mitochondria, where oxidation of the fatty group will take place. However, the fatty acyl groups themselves are transferred across to be picked up again by other coenzyme A molecules inside the mitochondria.

Problem 25.1 Compare the initial committing steps of fatty acid and glucose catabolism.

25.5 OVERVIEW OF FATTY ACID OXIDATION

When fatty acyl CoA arrives in the mitochondria, the four enzymes of fatty acid oxidation take over. Fatty acid oxidation is called **beta oxidation** because the scene of oxidative action is the beta carbon, which is just two carbons away from the carboxyl carbon.

β carbon → carboxyl carbon

$$RCH_2CH_2CH_2\overset{\overset{\displaystyle O}{\|}}{C}SCoA$$

Four reactions must occur to remove one two-carbon group from the carboxyl end of the fatty acyl CoA molecule. The sum of these four reactions gives the overall reaction

$$RCH_2CH_2CH_2\overset{\overset{\displaystyle O}{\|}}{C}SCoA \xrightarrow[HSCoA]{} RCH_2\overset{\overset{\displaystyle O}{\|}}{C}SCoA + CH_3\overset{\overset{\displaystyle O}{\|}}{C}SCoA$$

The oxidation and cleavage of an acetyl CoA molecule from the fatty acid chain thus shorten the fatty acid chain by two carbons. The four reactions will be repeated again and again until the entire chain has been broken down to acetyl CoA units. Equations for each of the four reactions are given in Table 25.1.

The first reaction is the oxidation of the $-CH_2-CH_2-$ group at the beta position. As a result, a carbon–carbon double bond forms between this group and the adjacent alpha carbon (which is one carbon away from the carboxyl carbon). In the second reaction, the addition of water across the double bond (a hydration) causes the beta carbon to take on a hydroxyl group:

$$-\overset{\overset{\displaystyle OH}{|}}{C}H-$$

TABLE 25.1 The Four Reactions of Fatty Acid Beta Oxidation

Type of reaction	Equation for the reaction
Dehydrogenation	$RCH_2CH_2CH_2\overset{\overset{O}{\|}}{C}-S-CoA + FP \rightarrow RCH_2CH{=}CH\overset{\overset{O}{\|}}{C}-S-CoA + FPH_2$
Hydration	$RCH_2CH{=}CH\overset{\overset{O}{\|}}{C}SCoA + H_2O \rightarrow RCH_2\overset{\overset{OH}{|}}{C}HCH_2\overset{\overset{O}{\|}}{C}SCoA$
Dehydrogenation	$RCH_2\overset{\overset{OH}{|}}{C}HCH_2\overset{\overset{O}{\|}}{C}-S-CoA + NAD^+ \rightarrow RCH_2\overset{\overset{O}{\|}}{C}CH_2\overset{\overset{O}{\|}}{C}-S-CoA + NADH + H^+$
Cleavage of carbon–carbon bond	$RCH_2\overset{\overset{O}{\|}}{C}CH_2\overset{\overset{O}{\|}}{C}-S-CoA + HS-CoA \rightarrow RCH_2\overset{\overset{O}{\|}}{C}-S-CoA + CH_3\overset{\overset{O}{\|}}{C}-S-CoA$

The third reaction, which is another dehydrogenation or oxidation reaction, results in the formation of a keto group:

$$\begin{array}{c} \mathrm{O} \\ \| \\ -\mathrm{C}- \end{array}$$

Finally, in the fourth reaction, a molecule of coenzyme A reacts with this group, cleaving the chain at the beta carbon and forming a new fatty acyl CoA molecule, which is two carbons shorter than at the start. The other product is a molecule of acetyl CoA. This completes one cycle of beta oxidation. Each of these four reactions is catalyzed by a separate enzyme.

The new, shorter fatty acyl CoA then goes through the same series of reactions. This produces another acetyl CoA molecule and another new fatty acyl CoA that is two carbons shorter than before. The same cycle is repeated until all 16 or 18 (or more) carbons of the fatty acyl chain form acetyl CoA molecules. Since energy-rich reduced coenzymes (NADH and a reduced flavoprotein) are also products of these reactions, many energy-rich molecules are formed and can now be oxidized to yield a great deal of ATP.

Problem 25.2 Write the complete sequence of reactions necessary for the degradation of hexanoic acid to acetyl CoA.

25.6 FATE OF THE PRODUCTS OF FATTY ACID OXIDATION

What is the ATP yield for complete oxidation of palmitate?

The normal products of mitochondrial oxidation of fatty acids are acetyl CoA, NADH, and reduced flavoprotein. These are oxidized by the citric acid cycle and the electron transport system to drive the production of ATP. Each acetyl CoA molecule gives rise to the formation of 12 ATPs (Section 23.11); the oxidation of each NADH molecule yields 3 ATPs; and the oxidation of each reduced flavoprotein molecule yields 2 ATPs. Thus, for each cycle of beta oxidation, 17 ATPs are formed.

For example, palmitic acid, a 16-carbon fatty acid, yields 8 acetyl CoA units and 7 molecules each of NADH and reduced flavoprotein:

$$\mathrm{C_{16}H_{32}O_2 + 8HSCoA + 7NAD^+ + 7FP \longrightarrow}$$
$$\underset{(\rightarrow 12\mathrm{ATP})}{\text{8 acetyl CoA}} + \underset{(\rightarrow 3\mathrm{ATP})}{\mathrm{7NADH}} + \mathrm{7H^+} + \underset{(\rightarrow 2\mathrm{ATP})}{\mathrm{7FPH_2}}$$

Therefore, a total of $(8 \times 12) + (7 \times 3) + (7 \times 2) = 131$ ATPs will ultimately be produced. However, since two high-energy phosphate bonds are broken ($ATP \rightarrow AMP + PP_i \rightarrow 2P_i$; see Section 25.4) in activating the fatty acid, the net yield will be $131 - 2 = 129$ ATPs per palmitic acid molecule oxidized. As you see, the ATP energy yield of fatty acid oxidation is very high.

Problem 25.3 Calculate the ATP yield for the catabolism of the 18-carbon fatty acid stearic acid.

Problem 25.4 Write the complete overall equation for the oxidation of palmitate to CO_2 and H_2O. (Refer to Chapter 23 as needed for details about the oxidation of acetyl CoA in the citric acid cycle.)

OXIDATION OF UNSATURATED, ODD-NUMBERED, AND BRANCHED FATTY ACIDS 25.7†

Why are certain detergents non-biodegradable?

The oxidation of unsaturated fatty acids is basically the same as that of saturated fatty acids. One or two additional enzymes are required, depending on where the double bonds are located in the fatty acid chain. The energy yield is slightly lower because an unsaturated chain is already partly oxidized, in comparison with a fully saturated chain:

$$\underset{\text{saturated}}{CH_3CH_2CH_2CH_2CH_2\overset{O}{\overset{\|}{C}}OH} \xrightarrow[\text{oxidation}]{} \underset{\text{unsaturated}}{CH_3CH_2CH{=}CHCH_2\overset{O}{\overset{\|}{C}}OH} + 2(H\cdot)$$

Branched-chain fatty acids (the branches are usually methyl groups) are found in certain bacteria and in plant waxes. If methyl groups occur only at even-numbered carbons, beta oxidation occurs normally, yielding the three-carbon compound propionyl CoA along with acetyl CoA.

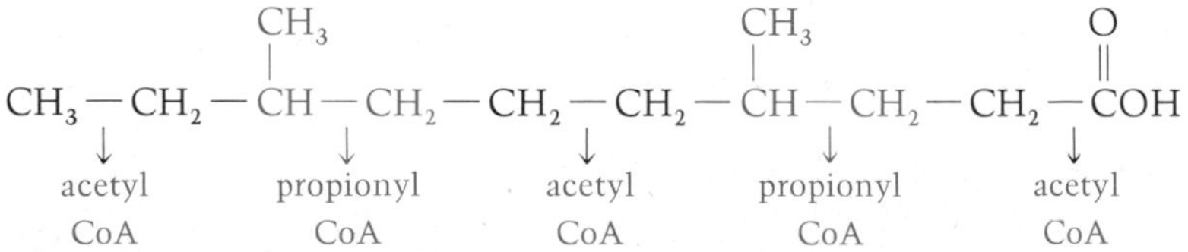

$$\underset{\substack{\downarrow \\ \text{acetyl} \\ \text{CoA}}}{CH_3-CH_2}-\underset{\substack{\downarrow \\ \text{propionyl} \\ \text{CoA}}}{\overset{CH_3}{\overset{|}{CH}}-CH_2}-\underset{\substack{\downarrow \\ \text{acetyl} \\ \text{CoA}}}{CH_2-CH_2}-\underset{\substack{\downarrow \\ \text{propionyl} \\ \text{CoA}}}{\overset{CH_3}{\overset{|}{CH}}-CH_2}-\underset{\substack{\downarrow \\ \text{acetyl} \\ \text{CoA}}}{CH_2-\overset{O}{\overset{\|}{C}}OH}$$

If branches occur at odd-numbered positions, however, beta oxidation is blocked. Synthetic detergents in use up to 1966 contained branched hydrocarbon chains with more or less random distribution of methyl groups. That is, some were at odd-numbered positions and some at even-numbered positions. Beta oxidation of these chains was usually blocked, and bacteria and other organisms could not degrade them. The result was foamy pollution in many rivers and lakes. Since 1966, only straight-chain **biodegradable** detergents have been manufactured, and this source of environmental pollution has been largely eliminated.

25.8 COMPARISON OF ENERGY YIELDS OF FAT AND CARBOHYDRATE OXIDATIONS

Ounce for ounce and gram for gram, fats yield almost twice the energy carbohydrates do. We can illustrate this in part by comparing the ATP yield of the oxidation of a six-carbon sugar, of glucose, and of a six-carbon fatty acid, hexanoic acid (hexanoate).

Both 6C

Glucose, as we saw in Chapter 24, yields 36 ATPs in its complete oxidation to CO_2 and H_2O. Hexanoate yields 44 ATPs. The fundamental reason for this difference is that hexanoate is more reduced than glucose; it has more of the higher-energy C—H bonds and fewer of the lower-energy C—O bonds. This illustrates the fact that more highly reduced compounds have greater energy content: as a molecule becomes increasingly oxidized, it gives up energy from its chemical bonds. Conversely, the synthesis of a more highly reduced compound from less highly reduced materials will usually require energy.

Problem 25.5 The 44 ATPs obtained from hexanoate oxidation are considerably less than twice the 38 ATPs obtained from glucose oxidation. What other factor is responsible for the difference in energy content of equal weights of glucose and hexanoic acid?

25.9 STARVATION, DIABETES, AND KETONE BODIES

What are ketone bodies?

Someone who is starving or fasting is obviously not taking in any carbohydrate. Since the body's glycogen-storage capacity is small (most of the liver glycogen is consumed overnight), very little carbohydrate is available for maintaining blood glucose levels. As we saw in Chapter 24, new carbohydrate can be made in the liver from small molecules in the process called gluconeogenesis. However, fatty acid oxidation must provide the bulk of the energy to most of the body's cells, thereby sparing glucose for red blood cells and brain cells.

In uncontrolled diabetes mellitus, even if the patient's diet contains a normal amount of carbohydrate, glucose may not enter cells normally because of inadequate insulin production or poor cell response to insulin. Glucose levels in the blood are high (hyperglycemia), and glucose may appear in the urine (glucosuria), but most of the body's cells can't use the glucose and must depend on oxidation of fatty acids mobilized from adipose tissue for energy.

When these conditions of limited carbohydrate availability and extensive fatty acid mobilization exist, the liver oxidizes many fatty acids. Acetyl CoA is produced by beta oxidation in the liver even faster than it can be oxidized in the citric acid cycle. It therefore builds up in the mitochondria. To dispose of the excess acetyl CoA, a new sequence of reactions takes place, which leads to the formation of **ketone bodies**—acetoacetate, β-hydroxybutyrate, and acetone.

$$CH_3\overset{\overset{O}{\|}}{C}CH_2CO_2^- \qquad CH_3\overset{\overset{OH}{|}}{C}HCH_2CO_2^- \qquad CH_3\overset{\overset{O}{\|}}{C}CH_3$$

acetoacetate — β-hydroxybutyrate — acetone

Acetone is formed only when the acetoacetate concentration is rather high—it is a waste product that cannot be further metabolized. However, both acetoacetate and β-hydroxybutyrate diffuse out of the liver cells and are carried by the blood to other tissues that can oxidize them. Even the brain can use ketone bodies when necessary. Ketone-body oxidation thus spares much of the limited glucose the body has available, and in this way the formation of ketone bodies can be beneficial.

In extreme starvation or diabetes, however, the ketone-body levels may rise enough to produce **ketoacidosis.** In this disorder, blood and tissue pH fall to less than 7.35 because of the large amounts of acetoacetic and β-hydroxybutyric acids that are produced. Acetone levels also rise, and the distinctive odor of acetone may be detectable on the patient's breath. If the disease is not controlled, the patient becomes severely dehydrated because the kidneys excrete too much water trying to get rid of the excess blood acids. Coma and death may result. Some fad diets restrict carbohydrates to force the body to oxidize fatty acids. Such diets can be somewhat dangerous because of the tendency to develop ketoacidosis.

Diabetes is controlled by insulin, which allows cells to take up glucose and reduces the body's excessive dependence on fat catabolism. An overdose of insulin can cause the blood glucose levels to fall excessively low (hypoglycemia); symptoms such as trembling, pallor, and weakness may occur. If hypoglycemia is severe, the diabetic may become unconscious because the brain is starving for glucose (insulin shock). Diabetic patients on insulin often keep a convenient supply of candy or other sweets to counteract the first symptoms of hypoglycemia.

Problem 25.6 The brain does not oxidize fatty acids, but it can oxidize ketone bodies. Why?

25.10 FATTY ACID SYNTHESIS: WHEN AND WHERE?

Organisms need to synthesize various lipids: fatty acids, triacylglycerols, various types of phospholipids, and steroids. The most heavily traveled pathways are those that lead to the synthesis of fatty acids and triacylglycerols.

Fatty acids and fats are synthesized when more food molecules are present in the organism than are necessary to satisfy energy needs. Most of this synthesis takes place in the liver and the adipose tissue, but all cells are capable of some fatty acid synthesis. The mammary glands of the breasts are especially active in fatty acid synthesis during lactation (milk production).

The main enzymes of fatty acid synthesis are localized in the cytoplasm of the cell. (Remember that fatty acid oxidation takes place in mitochondria.) Additional enzymes that introduce double bonds or elongate the fatty acid chain are found in the membranes of the endoplasmic reticulum.

25.11 OVERVIEW OF FATTY ACID SYNTHESIS

Fatty acid synthesis and fatty acid oxidation are quite similar.

You will remember that fatty acids are broken down step by step in the beta-oxidation cycle (Section 25.5). At each step a two-carbon fragment, in the form of acetyl CoA, is removed. Similarly, in fatty acid biosynthesis, fatty acids are built up from two-carbon units that come from acetyl CoA.

A carbon–carbon bond formation, two reduction steps, and one dehydration step occur in each cycle of fatty acid synthesis. This is just the reverse of the carbon–carbon bond cleavage, two oxidation steps, and one hydration step encountered in beta oxidation. Thus fatty acid biosynthesis closely resembles fatty acid oxidation in reverse. However, we will see that the enzymes and location of the process in the cell are completely different.

PROVISION OF ACETYL CoA FOR FATTY ACID SYNTHESIS 25.12

Where does acetyl CoA for fatty acid synthesis come from?

When more carbohydrates have been taken into the body than are needed for energy and glycogen synthesis, the excess carbohydrates are broken down via glycolysis. The product of the degradation is pyruvate (Section 24.3), which is turned into acetyl CoA in the mitochondria (Section 24.9). If the cell's energy level (ATP level) is adequate, citric acid cycle and electron transport activity will be inhibited by the high ATP and NADH levels (Section 24.14). In that case, the acetyl CoA produced will not usually be oxidized. Instead, only the first reaction of the citric acid cycle (an irreversible committing step) will occur:

$$\text{acetyl CoA} + \text{oxaloacetate} + H_2O \rightarrow \text{citrate} + \text{HSCoA}$$

The excess citrate formed in this way is transported from the mitochondria to the cytoplasm, where the citrate is cleaved to yield acetyl CoA and oxaloacetate again. Only in this way can mitochondrial acetyl CoA be made available for fatty acid synthesis in the cytoplasm, since acetyl CoA itself cannot pass across the mitochondrial membrane.

The acetyl CoA made available is then carboxylated in an ATP-driven reaction to make malonyl CoA. This reaction is catalyzed by another biotin-containing enzyme:

$$\underset{\text{acetyl CoA}}{H_2O + CH_3\overset{\overset{\displaystyle O}{\|}}{C}SCoA} + CO_2 + ATP^{4-} \xrightarrow[\text{(citrate)}]{\text{acetyl CoA carboxylase}}$$

$$\underset{\text{malonyl CoA}}{{}^{-}O_2CCH_2\overset{\overset{\displaystyle O}{\|}}{C}SCoA} + ADP^{3-} + P_i^{2-} + 2H^+$$

Acetyl CoA carboxylase is activated by citrate (an allosteric mechanism; see Section 21.15). The arrival of citrate in the cytoplasm tells this enzyme to start fatty acid synthesis. Acetyl CoA carboxylase is the most important regulatory enzyme of the pathway. It plays the same regulatory role as pyruvate carboxylase does in gluconeogenesis (Section 24.14).

In gluconeogenesis, an accumulation of acetyl CoA in the mitochondria signals pyruvate carboxylase to start glucose synthesis. Similarly, in fatty acid synthesis, an accumulation of citrate in the cytoplasm tells acetyl CoA carboxylase to begin synthesizing fatty acids. Acetyl CoA carboxylase is also subject to feedback inhibition (Section 21.17) by the end products of fatty acid biosynthesis—specifically palmitoyl CoA and other long-chain fatty acyl CoAs.

Problem 25.7 Draw a diagram that shows the cell compartments (mitochondria, cytoplasm) involved in the reactions above.

25.13 OUTLINE OF FATTY ACID SYNTHESIS

Palmitic acid ($CH_3(CH_2)_{14}CO_2H$) is the major product of fatty acid synthesis. The formation of palmitate from acetyl CoA and malonyl CoA takes place on the surface of a large protein, **fatty acid synthase,** that contains several enzymes. The overall reaction catalyzed by the enzymes of fatty acid synthase is very similar to the overall reaction of a fatty acid beta-oxidation cycle, except in reverse:

$$CH_3\overset{O}{\overset{\|}{C}}SACP + {}^{-}O_2CCH_2\overset{O}{\overset{\|}{C}}SACP \xrightarrow[CO_2 \quad H-SACP]{} CH_3CH_2CH_2\overset{O}{\overset{\|}{C}}SACP$$

The acetyl group, the malonyl group, and all other intermediates are carried by a protein that is part of the fatty acid synthase. It is called **ACP** for **acyl carrier protein.** Again there are four steps in the cycle, as in beta-oxidation. In this case we have a carbon–carbon bond formation, a reduction, a dehydration, and then a second reduction (Table 25.2). Again a number of cycles are necessary for completion of the process. In fatty acid oxidation, one cycle of four reactions is necessary to *remove* each pair of carbons from the fatty acid chain. In fatty acid synthesis, one cycle of four reactions is responsible for *adding* each pair of carbons to the growing chain. In each cycle a new malonyl group comes in to be decarboxylated and condensed with the growing chain.

An important feature of fatty acid synthesis is the use of the reduced coenzyme NADPH as the source of the reducing equivalents in both reduction reactions. Except in gluconeogenesis, which uses NADH, anabolic or biosynthetic pathways use NADPH for their reduction steps.

TABLE 25.2 The Four Reactions of Fatty Acid Synthesis

Reaction type	Reaction
Carbon–carbon bond formation	$CH_3C(=O)S{-}ACP + {}^{-}O_2CCH_2C(=O)S{-}ACP \rightarrow CH_3C(=O)CH_2C(=O)S{-}ACP + CO_2$ (+ H—SACP)
Reduction	$CH_3C(=O)CH_2C(=O)S{-}ACP + NADPH + H^+ \rightarrow CH_3CH(OH)CH_2C(=O)S{-}ACP + NADP^+$
Dehydration	$CH_3CH(OH)CH_2C(=O)S{-}ACP \rightarrow CH_3CH{=}CHC(=O)S{-}ACP + H_2O$
Reduction	$CH_3CH{=}CHC(=O)S{-}ACP + NADPH + H^+ \rightarrow CH_3CH_2CH_2C(=O)S{-}ACP + NADP^+$

Note: In fatty acid synthesis, the intermediates are carried by ACP rather than by coenzyme A as in fatty acid oxidation. Acyl carrier protein is part of the fatty acid synthase molecule.

The overall reaction for the formation of palmitate from acetyl CoA is

$$8 \text{ acetyl CoA} + 14NADPH + 7H^+ + 7ATP^{4-} + H_2O \rightarrow C_{16}H_{32}O_2 + 8CoASH + 14NADP^+ + 7ADP^{3-} + 7P_i^{2-}$$

You can see that a great deal of energy, in the form of ATP and reduced coenzyme, is necessary to drive the biosynthesis of a fatty acid molecule.

Problem 25.8 With reference to the overall reaction for fatty acid synthesis:
(a) How many of the eight acetyl CoAs are used to make malonyl CoA?
(b) What specific reaction does ATP hydrolysis drive?

Problem 25.9 Make a list of differences between the pathways of fatty acid synthesis and beta oxidation. Include, for example, cellular location of enzymes, coenzymes, energy changes, and so on. What is the major advantage of using different pathways instead of simply using fatty acid beta oxidation in reverse?

25.14† SYNTHESIS OF TRIACYLGLYCEROLS

Where is fat made in the body?

The body obtains its fatty acids from the digestion of dietary fats, as well as from fatty acid synthesis. Unless they are needed for membrane lipids or other materials, the fatty acids are synthesized into storage fat—triacylglycerols. Adipose tissue, liver, and lactating mammary glands are the major sites of triacylglycerol synthesis in the body.

Triacylglycerols are triesters of glycerol and three fatty acids. The glycerol comes from a glycolytic intermediate, dihydroxyacetone phosphate (DHAP). The availability of glycerol (which really implies the availability of blood glucose, from which DHAP is formed) is an important determinant of the rate of fat synthesis, especially in adipose tissue. Insulin promotes glucose uptake into adipose cells and enhances fat synthesis while it inhibits fatty acid mobilization. On the other hand, when carbohydrate is in short supply, the synthesis and storage of fat in adipose tissue are inhibited, and fatty acid mobilization (Section 25.3) is enhanced. The synthesis of fat in lactating mammary glands also depends on the presence of the appropriate hormones, as you might expect.

The fatty acyl groups are added to the glycerol backbone one at a time in reactions involving three different enzymes. These enzymes are quite specific: they usually ensure that a saturated fatty acid is esterified first, and then two unsaturated ones.

Triacylglycerols are stored in adipose tissue cells as fat droplets in the cytoplasm (Figure 25.1). In many cases, almost the entire volume of the cell seems to be taken up by a huge fat droplet. There is virtually no limit to the amount of fat that can be stored in an animal's adipose tissue.

†25.15 SYNTHESIS OF OTHER LIPIDS

Phosphoglycerides are synthesized from phosphatidic acid (Section 20.3) and alcohols (serine, choline, or inositol, for example). Activated CDP (cytidine diphosphate) derivatives of the alcohols are used, much as UDP derivatives are used to donate glucose groups in glycogen formation. Cholesterol (Section 20.6) is derived from acetyl CoA. Cholesterol itself is the raw material from which all other steroids, such as the sex hormones, are made. Bile acids are also produced from cholesterol by the liver. They are used to make bile salts that are secreted into the small intestine to aid in fat digestion (Section 23.3).

25.16 AMINO ACIDS AS FUELS

What are amino acids used for in the body?

Proteins obtained in the diet (like the chicken in our luncheon sandwich in Chapter 23) are broken down to their amino acid components during digestion. The essential amino acids (Section 23.6) are usually needed for protein synthesis in the body, even in an adult animal that is no longer growing. Amino acids are also used for synthesizing many other cell materials. If the intake of protein is higher than necessary for these purposes, however, the amino acids are oxidized for energy, used for carbohydrate synthesis, or converted to fat for storage.

Amino acid degradation must be considered in two parts. One part is the fate of the nitrogen of the α-amino group. The other is the fate of the carbon skeleton. We'll consider these one by one.

NITROGEN EXCRETION 25.17

When amino acids are broken down, the nitrogen that is removed becomes ammonia. Ammonia is highly toxic, so much so that the human body normally maintains blood ammonia at a level barely detectable by ordinary analytical methods. The body has an efficient mechanism for converting ammonia to a less toxic waste product, urea. Urea is made from ammonia in the liver and is excreted by the kidney into the urine.

Not all organisms excrete nitrogen as urea. Some, mostly marine animals, excrete free ammonia, which is rapidly diluted to nontoxic levels in their watery environment. Others, such as birds and reptiles, excrete nitrogen in the form of a more complex molecule, uric acid, which is structurally related to the purines (Section 18.2). The **ureotelic** (urea-excreting) organisms, including humans, metabolize amino groups as described in the next two sections.

FORMATION OF AMMONIA 25.18

The major route for the removal of ammonia from amino acids begins with enzymes known as **transaminases,** which catalyze the transfer of an amino group from an amino acid to an α-keto acid:

$$\underset{\text{L-amino acid}}{\overset{\displaystyle NH_3^+ \atop |}{RCHCO_2^-}} + \underset{\alpha\text{-keto acid}}{\overset{\displaystyle O \atop \|}{R'CCO_2^-}} \rightleftharpoons \underset{\text{new }\alpha\text{-keto acid}}{\overset{\displaystyle O \atop \|}{RCCO_2^-}} + \underset{\text{new L-amino acid}}{\overset{\displaystyle NH_3^+ \atop |}{R'CHCO_2^-}}$$

During the reaction, the amino group of the amino acid substrate is transferred to the pyridoxal phosphate cofactor (Section 21.12) of the transaminase. The α-carbon of the amino acid becomes a keto group, and a new α-keto acid is formed. Then the cofactor transfers the amino group to the α-keto acid substrate, to give the second product—a new α-amino acid. In many such reactions, α-ketoglutarate is the α-keto acid that accepts the amino group from the amino acid, and glutamate is formed:

$$\alpha\text{-ketoglutarate} + \text{amino acid} \xrightleftharpoons{\text{transaminase}} \text{glutamate} + \alpha\text{-keto acid}$$

Thus, many amino groups may be collected in the cell in the form of a pool of glutamate molecules.

Glutamate may now be transferred to the mitochondria. (In fact, it may have been formed there by mitochondrial transaminases). Then glutamate is oxidized and NH_4^+ is released, forming α-ketoglutarate again. Ammonia is now available for conversion to urea. And the α-ketoglutarate may participate in a new transamination reaction, or it may enter the citric acid cycle.

25.19 THE UREA CYCLE

Where in the body does urea formation take place?

The **urea cycle** is a series of five reactions that convert ammonia to urea. The reactions take place only in the liver of the urea-excreting organism. The first two reactions are catalyzed by mitochondrial enzymes, and the last three occur in the cytosol (Figure 25.2). Note that the basis of the cycle is the formation and regeneration of L-ornithine.

Urea, once formed, readily diffuses out of the liver cells into the circulation. The kidneys then excrete it into the urine.

25.20 CARBON CATABOLISM OF AMINO ACIDS

Each of the 20 common amino acids follows a specific pathway of carbon skeleton breakdown after the α-NH_2 group is removed. We will not examine these pathways in detail, but we do wish to point out that the breakdown of all these amino acids gives rise to the same few central

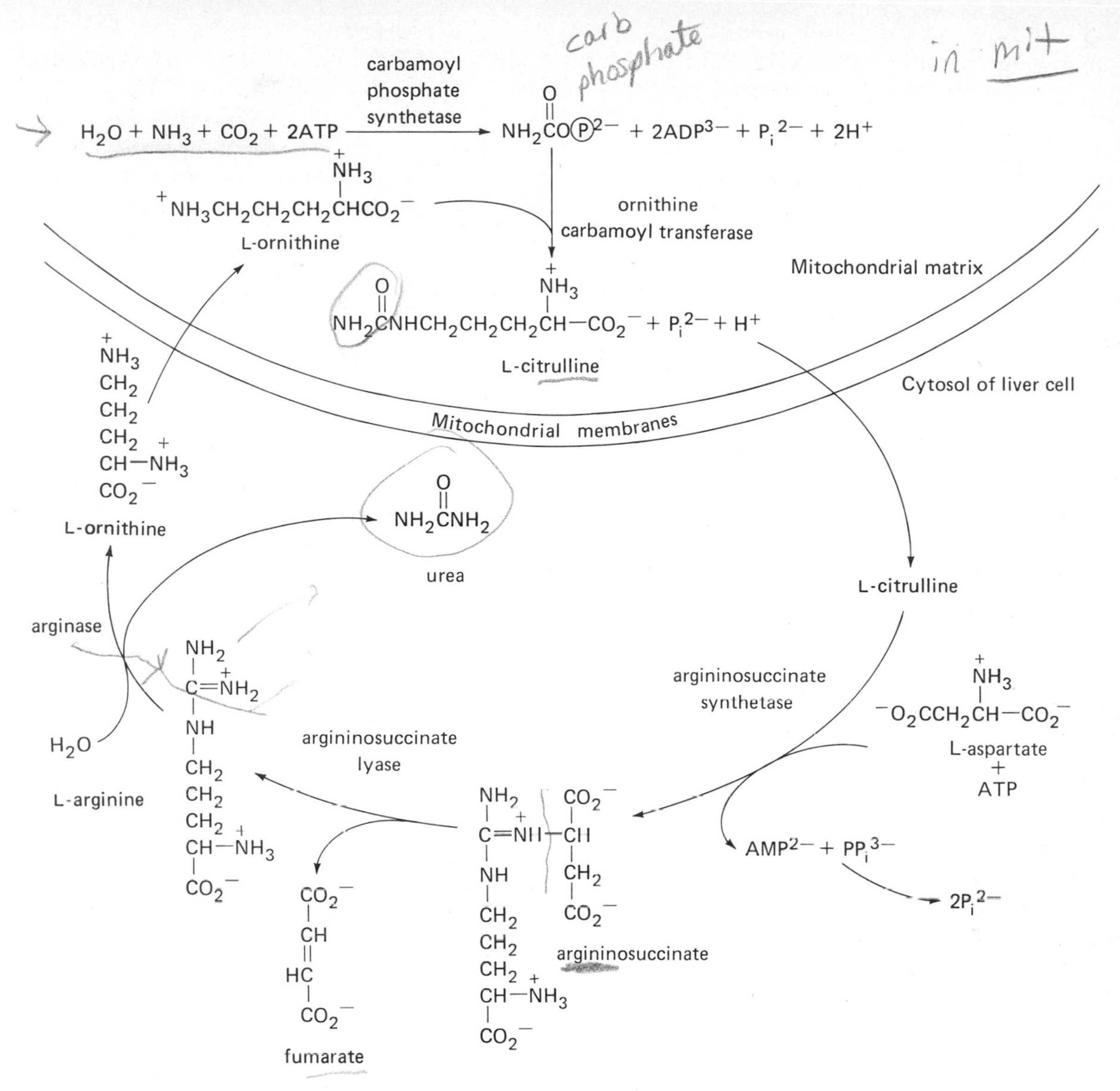

FIGURE 25.2 The urea cycle for disposal of ammonia as urea in the liver cell.

metabolites. These metabolites can be oxidized in the citric acid cycle to yield NADH and thus ATP. The amino acids can be divided into six families on the basis of which of the metabolites they yield: pyruvate, α-ketoglutarate, acetyl CoA, fumarate, oxaloacetate, and succinyl CoA. Figure 25.3 shows the members of the various families in relation to the route of their entry into the citric acid cycle.

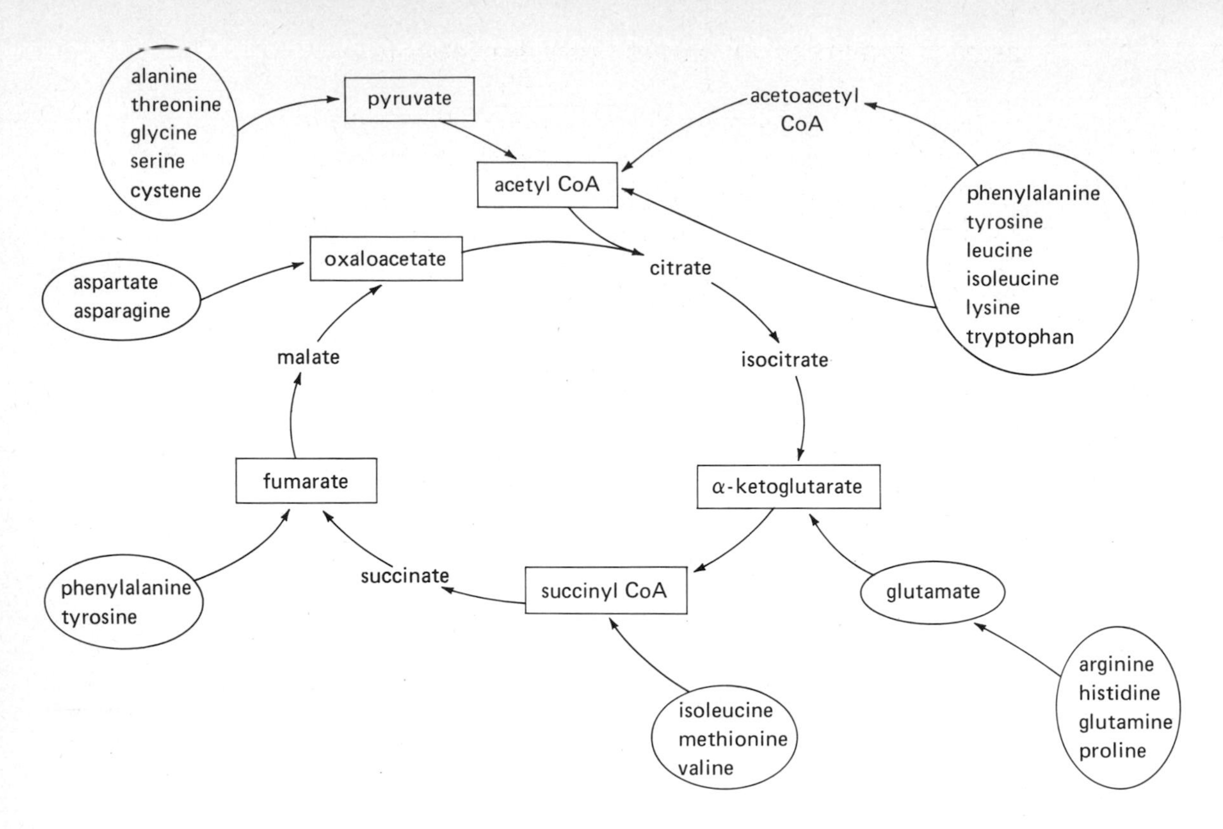

FIGURE 25.3 The amino acid families, indicating the common metabolites formed during catabolism of the amino acids. Names of families are in boxes. Note that the larger amino acids are likely to give rise to more than one important metabolite.

Here are some particularly straightforward examples of reactions leading to the formation of these central metabolites:

$$\underset{\text{L-glutamate}}{{}^{+}H_3N-CH(CO_2^-)-CH_2-CH_2-CO_2^-} \xrightarrow{\text{transamination}} \underset{\alpha\text{-ketoglutarate}}{O{=}C(CO_2^-)-CH_2-CH_2-CO_2^-} + (-NH_2)$$

$$\underset{\text{L-alanine}}{{}^{+}H_3N-C(CO_2^-)(H)-CH_3} \xrightarrow{\text{transamination}} \underset{\text{pyruvate}}{O{=}C(CO_2^-)-CH_3} + (-NH_2)$$

$$
\begin{array}{ccc}
\begin{array}{r}
CO_2^- \\
| \\
CH_2 \\
| \\
{}^+H_3N-CH \\
| \\
CO_2^-
\end{array}
& \xrightarrow{\text{transamination}} &
\begin{array}{l}
CO_2^- \\
| \\
CH_2 \\
| \\
C{=}O + (-NH_2) \\
| \\
CO_2^-
\end{array}
\end{array}
$$

L-aspartate → oxaloacetate

Some of the other pathways are long and complex, but in the end each of the fundamental amino acids can be degraded to one or more small, easily oxidized metabolites, as shown in Figure 25.3. Thus amino acids may be oxidized for energy. It is important to remember also that some of these central metabolites are important sources of carbon atoms for gluconeogenesis (Section 24.12); therefore, amino acids can be used for carbohydrate synthesis in the liver when the body requires it.

25.21 DISEASES OF AMINO ACID METABOLISM

Several inherited diseases are known to result from abnormalities of urea cycle enzymes. In one type of **hyperammonemia** (high blood ammonia), carbamyl phosphate synthetase is the defective enzyme; in another type, ornithine transcarbamylase is at fault. The treatment is restriction of protein in the diet.

Flaws in the metabolism of the aromatic amino acids phenylalanine and tyrosine are involved in **phenylketonuria (PKU)** and **alkaptonuria.** In PKU, the enzyme phenylalanine hydroxylase is defective. This enzyme is normally responsible for converting phenylalanine to tyrosine, which is the first step in the degradation of phenylalanine. As a result of the disease, phenylalanine and certain abnormal breakdown products, such as phenylpyruvate, accumulate in the body and are found in the urine. Phenylpyruvate is an inhibitor of the mitochondrial pyruvate transport protein that is necessary to move pyruvate into the mitochondria for oxidation. In the presence of high levels of phenylpyruvate, aerobic carbohydrate metabolism is inhibited. Since the brain depends on aerobic carbohydrate oxidation for most of its energy supply, this may account for the mental retardation associated with the disease. PKU, the most common of the diseases of amino acid metabolism (occurring in 1 of 10,000 births), can be treated by a low-phenylalanine diet. Newborn infants are routinely tested for PKU. The test is very simple and inexpensive and has allowed detection of the disease early enough to prevent mental retardation in many young PKU patients.

Alkaptonuria (black urine disease) results from a defect in an enzyme, homogentisate oxidase, in tyrosine breakdown. In its absence, homogentisic acid accumulates and is found in the urine. Black oxidation products form when the urine is exposed to air. The symptoms are rather mild compared with those of other genetic diseases; arthritis occurs in middle age, and black pigment accumulates in the connective tissues.

25.22 INTEGRATION OF CARBOHYDRATE, FATTY ACID, AND PROTEIN METABOLISM

How are the metabolisms of different foods related?

Let's stand back for a moment and try to develop an integrated picture of the pathways we have studied so far. We have seen how brain cells and red blood cells depend mainly or exclusively on glucose oxidation for energy, and how most other tissues prefer fatty acids or ketone bodies if they are available. We have also seen how the liver exports glucose to keep blood glucose levels within normal limits so that brain and red blood cells function normally. Usually, the liver gets its export glucose from the breakdown of glycogen.

But when the glycogen stores in the liver are low because of fasting, the liver must carry out gluconeogenesis, making glucose from lactate, citric acid cycle intermediates, and amino acids. (Let us emphasize once again that fatty acid carbons *cannot* be used for gluconeogenesis.) When the carbohydrate shortage is extreme, normal tissue proteins must be degraded to provide the raw material for glucose synthesis. In later stages of starvation, not only will body fats be totally depleted (because the body depends almost completely on fat oxidation for energy), but the muscles and other tissues will be extensively wasted to provide carbons for gluconeogenesis. Both the mobilization of amino acids from tissue proteins and amino acid degradation are stimulated by hydrocortisone (cortisol) (Section 20.6), a steroid hormone made in the adrenal glands, and by other hormones secreted in response to stress (starvation is an example of stress). The same hormones promote fatty acid mobilization from adipose cells and therefore send increased amounts of both amino acids and fatty acids into pathways designed to help the organism survive the carbohydrate shortage.

In uncontrolled diabetes, the insulin insufficiency causes an increase in activity of the key gluconeogenic enzymes in the liver. Protein synthesis is depressed and protein catabolism increased, so more amino acids are channeled into liver gluconeogenesis. (These effects are caused by various hormonal and allosteric controls that we will not go into.) At the same time, glucagon levels are usually elevated, so the liver is unable to retain much glycogen and continually sends more glucose into the blood stream. A great deal of glucose may be lost in the urine. The outcome is that the body is actively degrading muscle and other body proteins, and the resulting amino acids are being used to make glucose, which is largely going to waste. This is one reason that undiagnosed diabetics usually lose weight. Another reason is that their body fat is broken down to supply the fatty acids necessary to meet the energy needs that glucose cannot be used to fulfill.

Problem 25.10 Insulin is necessary for transporting certain amino acids (as well as glucose) in tissues such as muscle. This activity contributes still further to the derangement of metabolism in uncontrolled diabetes. Explain.

SUMMARY

Fats are hydrolyzed by lipases to yield fatty acids and glycerol. This hydrolysis can take place in **adipose tissue** or other tissues. Most tissues except the brain and red blood cells will oxidize available fatty acids in preference to glucose. Hormones promote the **mobilization** of fatty acids from adipose tissue in fasting, stress, physical exertion, and diabetes. Fatty acids are degraded, after conversion to **fatty acyl CoA,** in a mitochondrial process called **beta oxidation.** The total energy yield from fatty acid oxidation exceeds that from the oxidation of an equal weight of carbohydrate. When carbohydrates are unavailable, so much fatty acid oxidation takes place in the liver that **ketone bodies** are formed. Most tissues can oxidize them, sparing glucose for use by the red blood cells and brain. However, the formation of excessive ketone bodies, as in uncontrolled diabetes, causes **ketoacidosis.** In this condition, the blood pH falls below normal, and dehydration, coma, and death may occur.

Fatty acids are synthesized from acetyl CoA that arrives in the cytoplasm by way of citrate exported from mitochondria. The **fatty acid synthase** is a multienzyme protein that carries out all the enzyme activities necessary to form palmitate. **Acyl carrier protein** carries the fatty acid chain while four reactions similar to those of beta oxidation (but in reverse) add each two-carbon unit to the growing fatty acid chain. Triacylglycerols are formed by esterification of three fatty acids to glycerol.

Amino acids that are not needed for synthesizing new proteins or other cell components may be oxidized for energy. There are two aspects of amino acid catabolism: nitrogen excretion and carbon chain catabolism.

Nitrogen from the amino acid amino groups becomes ammonia, which is then converted to urea, a less toxic excretion product, in a series of reactions called the **urea cycle.** The carbon skeletons of each of the amino acids can be degraded to give one or more small, central metabolites, which can be oxidized by the citric acid cycle. Many inherited diseases of amino acid metabolism, such as **phenylketonuria, hyperammonemia,** and **alkaptonuria,** are known.

Key Terms

Acyl carrier protein (ACP) (Section 25.13)
Adipose tissues (Section 25.1)
Alkaptonuria (Section 25.21)
Beta oxidation (Section 25.5)
Biodegradable (Section 25.7)
Blood–brain barrier (Section 25.3)
Fatty acid mobilization (Section 25.2)
Fatty acid synthase (Section 25.13)
Hyperammonemia (Section 25.21)
Ketoacidosis (Section 25.9)
Ketone bodies (Section 25.9)
Lipase (Section 25.2)
Phenylketonuria (PKU) (Section 25.21)
Serum albumin (Section 25.2)
Transaminase (Section 25.18)
Urea cycle (Section 25.19)
Ureotelic (Section 25.17)

ADDITIONAL PROBLEMS

25.11 What is the function of serum albumin?

25.12 Explain why the brain cannot use fatty acids as fuel.

25.13 Write the overall reaction for the complete oxidation of the fatty acid $CH_3(CH_2)_{10}CO_2H$.

25.14 What is meant by *fatty acid mobilization?*

25.15 What do a starving person and an untreated diabetic have in common?

25.16 Can fatty acids be oxidized for energy in (anaerobic) fast skeletal muscle? Explain.

25.17 Explain why branched-chain fatty acids with branches at odd-numbered carbons are not biodegradable.

25.18 Compare the two dehydrogenases of the fatty acid beta-oxidation pathway.

25.19 What citric acid cycle reaction is similar to the hydration reaction in beta oxidation?

25.20 Why is citrate required for fatty acid biosynthesis?

25.21 What fatty acid is the primary product of fatty acid biosynthesis?

25.22 What is biotin? What types of enzymes use biotin as cofactor? Give two examples of such enzymes.

25.23 The overall reaction for the biosynthesis of palmitate can be expressed as follows:

$$\text{acetyl CoA} + 7\text{ malonyl CoA}^- + 14\text{NADPH} + 20\text{H}^+ \rightarrow \text{palmitate}^- + 7\text{CO}_2 + 8\text{CoA—SH} + 14\text{NADP}^+ + 6\text{H}_2\text{O}$$

Describe the source(s) of acetyl CoA and malonyl CoA. Stress the immediate reactions involved in the generation of the substrates.

25.24 Assume more carbohydrate is available in the diet than is needed for immediate energy demands. In general terms, describe the process for converting the carbohydrate into fat for storage.

25.25 Write equations for all the reactions necessary for the complete degradation of the amino acid alanine to CO_2, H_2O, and NH_3.

25.26 Compare the structural formulas of phenylalanine and tyrosine. What is the structural formula of phenylpyruvate? (*Hint:* It is formed from phenylalanine by transamination.)

25.27 Write the overall reaction for the operation of the urea cycle.

Problem-Solving Hints

If you have trouble solving any of the additional problems, refer to the sections listed next to the problem numbers.

25.11 25.2
25.12 25.2
25.13 25.5, 25.6
25.14 25.2
25.15 25.9
25.16 25.3
25.17 25.7
25.18 25.5
25.19 25.5
25.20 25.12
25.21 25.13
25.22 25.12, 25.13
25.23 25.12
25.24 25.10
25.25 25.18, 25.19, 25.20
25.26 25.18
25.27 25.19

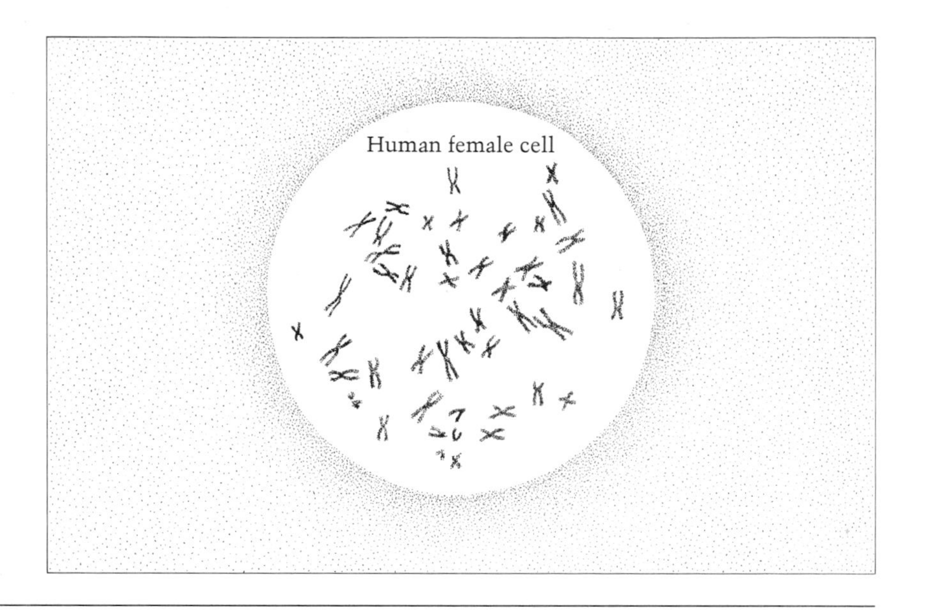

DNA (deoxyribonucleic acid) is the hereditary substance responsible for each individual's genetic characteristics. As we will discover in this chapter, each cell in the body (except sex cells) contains 46 chromosomes. The DNA in the chromosomes has all the information necessary to manufacture a new cell that will have exactly the same chromosomal make-up and contain the same genetic material. The chromosomes can be arranged by size, in 23 pairs, the smallest of which are the sex chromosomes.

Certain genetic disorders can be detected by abnormalities in chromosome structure. For example, children suffering from mongolism display distinct chromosome mutations (changes). Likewise, strikingly obvious defects in chromosomes can be seen after the influence of irradiation and certain drugs.

Identification and classification of chromosomes is performed by a technique called chromosome karyotyping. Until the development of this procedure, human beings were thought to have 48 chromosomes. Further refinement of the technique and further study of its results have led to the discovery of important chromosomal abnormalities in certain hereditary diseases.

In chromosome karyotyping, the chromosomes are studied only at one particular stage of cell division, mitotic metaphase. This is the stage at which the chromosomes are most clearly defined. Bone marrow cells taken from a patient will begin dividing when placed in a special tissue culture

26

BIOSYNTHESIS OF NUCLEIC ACIDS AND PROTEINS

solution in a test tube. Blood cells can also be used if they are incubated in a chemical called colchicine, which stops cell division at metaphase.

Once the cells in the test tube have reached the proper stage of cell division, they are examined on a slide under a microscope. A cell with mitotic activity can be photographed with a special lens and camera attached to the microscope. Then the photograph is enlarged to make it easier to match up each pair of characteristically sized chromosomes.

The final result is called a chromosome karyotype. This technique of culturing cells is used to confirm suspected genetic disorders.

[Illustration: The chromosome karyotype is constructed from a photograph of the chromosomes in the human cell.]

BIOSYNTHESIS OF NUCLEIC ACIDS AND PROTEINS

Fats, carbohydrates, and other molecules are essential to the functioning of an organism, as we have seen. But the proteins actually determine the kind of organism, and the nucleic acids determine the kinds of proteins that are available.

The DNA (deoxyribonucleic acid) molecules found in the nuclei of an organism's cells carry all the information needed to manufacture that particular organism. The DNA tells what proteins to make, and the specific proteins determine the organism's characteristics. In addition, the DNA carries from generation to generation the information needed to manufacture more organisms of that type.

DNA, as you probably know, is found in **genes.** The study of how inherited characteristics are passed from generation to generation is called **genetics.** When we look at how genetics operates at the molecular level, as we will be doing in this chapter, we are dealing with **molecular genetics.**

DNA is not the only nucleic acid that plays a role in genetics. The information contained in the DNA is translated into proteins by other nucleic acids, called RNAs (ribonucleic acids). The DNA stores the information, and the RNAs put the information to work. Some viruses do not have any DNA, so RNA does the whole job—both transmitting the information and putting together the proteins.

The nucleic acids and the proteins illustrate vividly the effect that the structure of molecules has on their chemical behavior, which in turn affects the biology of the whole organism. These molecules are able to play the role they do because of a fascinating property they have: base pairing. We discussed base pairing earlier (in Chapter 18), but now we will discover its true importance.

GENETIC INFORMATION 26.1

What is genetic information?

By **genetic information,** we mean the "recipes" for the synthesis of all the proteins that each cell can produce. Some proteins are structural, such as the proteins of hair and nails, bones, and connective tissue. Other proteins serve as enzymes, transport proteins, hormones, or receptors. A cell cannot synthesize a protein for which it does not possess the recipe. By the same token, the proteins a cell can make completely determine the form and behavior of that cell. The recipe for synthesis of a protein is called a **gene.** A gene is a small segment of a DNA molecule, a portion of its nucleotide sequence.

The individual DNA molecules within a cell are usually called **chromosomes** (Section 21.5). The small, single chromosome of a very simple bacterial cell may contain the recipes for only a few thousand proteins. Many more proteins are specified by the 23 *pairs* of large human chromosomes. (In organisms that reproduce sexually, each body cell contains two of each type of chromosome: one is derived from each parent when the egg and sperm cells unite to form the unique cell from which the new individual develops.) Perhaps we should add, however, that it is now known that the DNA in eucaryotic cells (cells with nuclei) contains many repeated sequences, as well as many sequences that do not seem to be represented in finished protein or RNA products. So the 46 chromosomes of the human cell may contain less genetic information than the large amount of DNA would suggest.

TRANSFER OF GENETIC INFORMATION 26.2

There are two quite separate aspects of genetic information transfer. First, the parent cell must make copies of its chromosomes to transfer to the daughter cell during cell division. This process, in which new, identical DNA molecules are synthesized, is called **replication.** Replication ensures that the daughter cell will contain information identical to that in the parent cell.

The second aspect is the transfer of the information (the protein recipe) from a gene in the nucleus to the cytoplasm, where **protein synthesis** takes place on the ribosomes. (We will cover this topic in Section 26.6.) Only some of the genetic information in a cell is used at any one time, although some proteins are being made all the time.

26.3 STRUCTURE OF DNA

What are the complementary bases?

A native DNA molecule consists of two polynucleotide strands connected by base pairing along their entire lengths (Section 18.6). The two strands are complementary; that is, where one strand has the base sequence 5′-ATCGTA-3′, for example, the other strand must have the sequence 3′-TAGCAT-5′ since A (adenine) pairs with T (thymine) and C (cytosine) pairs with G (guanine). Note that the arrangement is antiparallel: the strands run in opposite directions, with the 5′ positions at opposite ends.

Strand 1   5′-ATCGTG-3′

Strand 2   3′-TAGCAC-5′

The double-stranded structure is tightly coiled into a double helix that might be visualized as a twisted ladder. (See Section 18.6.)

Problem 26.1 Write the complement of 5′-GCGCATGC-3′. (Refer to Section 18.6, if necessary, for a discussion of the complementary bases.)

26.4 DNA SYNTHESIS, OR REPLICATION

The two strands of a native DNA molecule are complementary copies because of specific base pairing; G pairs only with C, and A pairs only with T. Because the strands are complementary, a marvelous system exists for replicating a DNA molecule.

First, the two strands are gradually separated. A supply of deoxyribonucleoside triphosphates (dATP, dTTP, dGTP, dCTP) must be available to provide the nucleotides needed for the new strands. Then enzymes can copy each of the strands to form a complete new complementary partner. Thus two complete double-stranded DNA molecules exist where only one existed before:

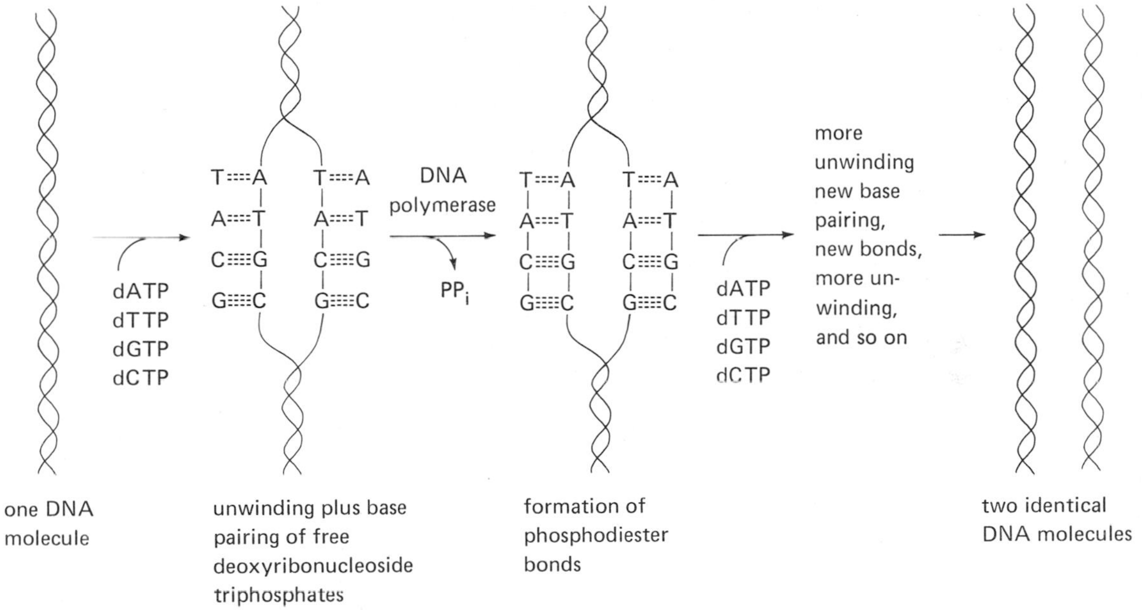

dATP
dTTP
dGTP
dCTP

T::::A T::::A
A::::T A::::T
C::::G C::::G
G::::C G::::C

DNA polymerase
PP_i

T::::A T::::A
A::::T A::::T
C::::G C::::G
G::::C G::::C

dATP
dTTP
dGTP
dCTP

more unwinding
new base pairing,
new bonds,
more unwinding,
and so on

one DNA molecule

unwinding plus base pairing of free deoxyribonucleoside triphosphates

formation of phosphodiester bonds

two identical DNA molecules

Each of the new molecules will contain one "old" strand and one new strand. This is called **semiconservative replication** (Figure 26.1): one old strand is conserved in each of the new molecules.

Energy to drive the biosynthesis of DNA comes from the hydrolysis of the deoxyribonucleoside triphosphates being linked into the chain. All the nucleoside triphosphates are high-energy compounds like ATP. Two high-energy phosphate bonds are cleaved for each mononucleotide inserted into a growing chain. When a mononucleotide is linked into the chain, one bond is cleaved, yielding inorganic pyrophosphate (PP_i). Then PP_i is cleaved to two P_i's by the action of the enzyme inorganic pyrophosphatase. Thus, a good deal of energy is made available.

The factors that determine when and in what cells replication and cell division occur are complex and not yet well understood, especially in cells of animals and plants. Different animal tissues can have very different normal timetables for division. For example, adult muscle cells very seldom divide. On the other hand, cells in the gastrointestinal tract are almost continuously dividing. There have to be powerful controls on these processes in normal tissues.

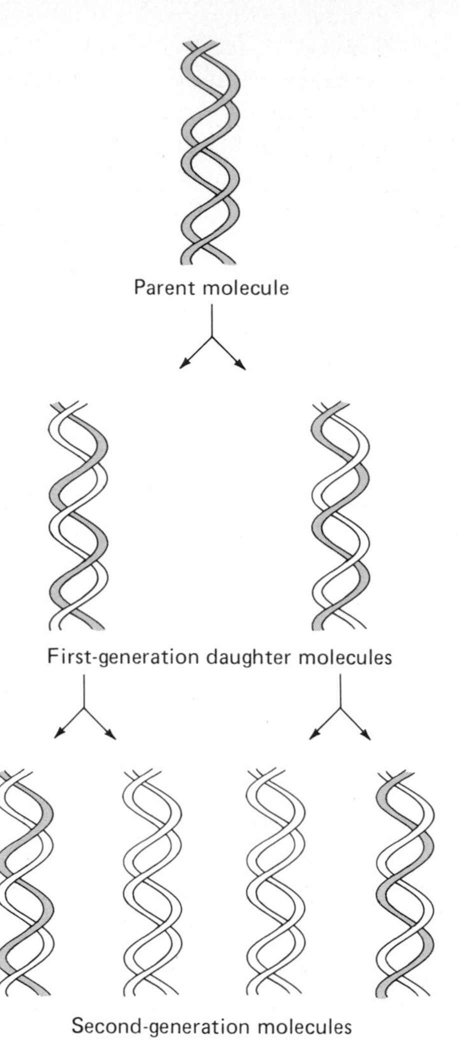

FIGURE 26.1 A schematic diagram of semiconservative DNA replication. Colored strands are "old" (parental) DNA; white strands are newly synthesized DNA.

26.5 ANTICANCER DRUGS AND DNA REPLICATION

How does cancer chemotherapy work?

Cancer cells metabolize rapidly, and for unknown reasons they do not respond to the controls that apparently restrict DNA replication and cell division in normal tissues. Certain chemicals, however, do inhibit

DNA replication (and therefore cell division) in cancer cells. The strategy of chemotherapy (that is, treatment with drugs or chemicals) often revolves around the inhibition of thymine production, because thymine is a base employed only in DNA synthesis and not in RNA synthesis. A treatment that focuses on DNA synthesis minimizes the number of normal body cells that the drug treatment affects. Most normal cells divide less frequently than the cancer cells do, so they replicate DNA less frequently. All cells carry out RNA synthesis regularly, though; therefore, chemicals that inhibit RNA synthesis would be toxic to normal and cancer cells alike.

Two anticancer drugs that inhibit thymine production are 5-fluorouracil and methotrexate (amethopterin). Most cancer chemotherapeutic agents have unfortunate side effects because they do affect normal body tissues in which cells divide relatively frequently, such as the lining of the gastrointestinal tract and the hair follicles. Diarrhea and hair loss are common side effects of anticancer drugs.

TRANSFER OF THE GENETIC MESSAGE TO THE RIBOSOME 26.6

How is genetic information transferred to the cytoplasm where proteins are made?

The second important aspect of genetic information transfer is the transfer of information from the chromosome (specifically, from a particular gene) to the ribosomes, which are the protein-synthesizing sites in the cell. Recall that a gene is a segment of a DNA molecule. Information is contained in this DNA segment by virtue of its specific nucleotide sequence. The specific nucleotide sequence of a gene, in turn, contains all the information necessary to generate a particular amino acid sequence and thus a unique protein. In other words, the nucleotide sequence is what translates into the amino acid sequence of a specific protein. And the unique character of a particular protein is completely dictated by its amino acid sequence.

In animal and plant cells, the chromosomes are in the nucleus. The ribosomes either float free in the cytoplasm or are attached to the rough endoplasmic reticulum. Some messenger clearly is necessary to carry information from the nucleus to the ribosome, since the chromosome itself does not leave the nucleus. That messenger is **messenger RNA (*m*RNA).** Messenger RNAs are single-stranded polyribonucleotides that are made in the nucleus in a manner that closely resembles DNA replication. A portion of the DNA double helix unwinds, and then free nucleotides pair with the exposed bases of one DNA strand and become linked. The enzyme that catalyzes formation of the phosphodiester bonds of the growing RNA chain is called **RNA polymerase.**

The synthesis of RNA differs from that of DNA in that the nucleotides that base-pair to the unwound segment of DNA are *ribonucleoside* triphosphates (ATP, GTP, CTP, UTP), not deoxyribonucleotides. The most important difference, however, is that only a specific segment of the DNA—a particular gene or set of genes—is copied. Furthermore, only *one* of the DNA strands in a given gene is copied. The term we use to describe the copying of an RNA sequence from a gene is **transcription.** In sum, transcription is RNA synthesis directed by DNA and catalyzed by RNA polymerase. We should note that all forms of RNA, not only *m*RNA, are transcribed on the DNA in a similar way. Therefore, some genes contain recipes for synthesis of the other forms of RNA, transfer RNA or ribosomal RNA, rather than recipes for proteins. These RNAs play other roles in protein synthesis, as we will see.

One of the most important and interesting questions in science today concerns the way transcription is regulated. There is abundant evidence that all of an individual's body cells contain the same genetic information. Yet it is obvious that not all genes in all cells are copied into RNA (and then protein); if they were, all the cells of a multicellular organism would be the same, and they are not at all the same. For example, skeletal muscle cells are packed with protein fibers that are involved in contraction, and most other cells do not synthesize these proteins. Liver cells have gluconeogenic and urea cycle enzymes that no other cells contain. The evidence suggests that there are specific proteins (repressors or activators) that bind to DNA or to RNA polymerase, to block, or to induce, the transcription of certain genes. Such regulatory proteins can control which genes are transcribed or "expressed" in a particular cell, and when.

26.7 THE GENETIC CODE

What purpose is served by degeneracy in the genetic code?

In the preceding section we outlined how information contained in the DNA nucleotide sequence is transcribed into the nucleotide sequence of *m*RNA. The *m*RNA nucleotide sequence is then translated into an amino acid sequence during polypeptide biosynthesis. Before we look at this process, let us investigate the nature of the code that is contained in the nucleotide sequence.

RNA contains three-nucleotide sequences known as **triplet codons,** or "code words" that specify particular amino acids. The four bases (A, G, C, and U) can be arranged in 64 different sequences of three, so there are 64 different triplet codons (Table 26.1). Each triplet codon specifies a particular amino acid (refer to Section 17.4 to refresh your memory about amino acids). For example, 5′-AUG-3′ codes for methionine. Note that no thymine is present in these codons since

TABLE 26.1 The Genetic Code

	U		C		A		G	
U	UUU	Phe	UCU	Ser	UAU	Tyr	UGU	Cys
	UUC	Phe	UCC	Ser	UAC	Tyr	UGC	Cys
	UUA	Leu	UCA	Ser	UAA	End	UGA	End
	UUG	Leu	UCG	Ser	UAG	End	UGG	Trp
C	CUU	Leu	CCU	Pro	CAU	His	CGU	Arg
	CUC	Leu	CCC	Pro	CAC	His	CGC	Arg
	CUA	Leu	CCA	Pro	CAA	Gln	CGA	Arg
	CUG	Leu	CCG	Pro	CAG	Gln	CGG	Arg
A	AUU	Ile	ACU	Thr	AAU	Asn	AGU	Ser
	AUC	Ile	ACC	Thr	AAC	Asn	AGC	Ser
	AUA	Ile	ACA	Thr	AAA	Lys	AGA	Arg
	AUG	Met	ACG	Thr	AAG	Lys	AGG	Arg
G	GUU	Val	GCU	Ala	GAU	Asp	GGU	Gly
	GUC	Val	GCC	Ala	GAC	Asp	GGC	Gly
	GUA	Val	GCA	Ala	GAA	Glu	GGA	Gly
	GUG	Val	GCG	Ala	GAG	Glu	GGG	Gly

these are RNA base sequences; uracil replaces thymine in all RNAs. Adenine and uracil are complementary bases in RNA, just as adenine and thymine are complementary bases in the DNA double helix. Three codons (UAA, UAG, and UGA) do not specify amino acids, but rather signal the end of the message. When one of these **termination codons** appears in the coded message, it announces the end of the amino acid sequence for that particular polypeptide.

Problem 26.2 If a protein has a molecular weight of 50,000, and if the average amino acid residue has a molecular weight of 120, how many nucleotides will the *m*RNA coding for this protein contain? How many nucleotides will the gene contain?

You will also note from Table 26.1 that there are numerous instances in which two or more different codons specify one amino acid. Since there are 61 code words that specify amino acids and only 20 fundamental amino acids, this is not surprising. This duplication is referred to as the **degeneracy** of the genetic code. It results mainly from variations in the third position of the codon. It is thought that this "redundancy" ensures that fewer errors occur in information transfer.

Perhaps you are wondering how the code words were deciphered. The earliest experiments were conducted using synthetic *m*RNAs made with only one kind of nucleotide. For example, if only adenine nucleotides were used, the result was a polyadenine chain:

AAAAAAAAAAAAAAAAA

When polyadenine was used instead of normal *m*RNA in a cell-free bacterial system for protein synthesis, a polypeptide chain containing only lysine (polylysine) could be isolated. (Inspection of Table 26.1 will show that AAA is indeed the code word for lysine). This experiment was done for each of the four possible polynucleotides containing only one base. It was shown that CCC codes for proline, GGG for glycine, and UUU for phenylalanine.

Of course, most of the codons contain two or three different bases, and these were more difficult to decipher. One approach was to make synthetic polynucleotides in which two nucleotides were present in a known ratio—for example, two parts uracil to one part guanine. Most of the codons would then be UUG, UGU, or GUU. Polypeptides synthesized in the bacterial system using this message contained predominantly cysteine, valine, and leucine. However, not until 1964 was a technique developed for specifically determining which codon codes for which amino acid. Description of those experiments is beyond the scope of this book, but they supplied the code word dictionary given in Table 26.1.

26.8 OVERVIEW OF PROTEIN BIOSYNTHESIS

We are now ready to consider protein biosynthesis. As shown in Figure 26.2, the process takes place in several steps. It begins with **activation** of the amino acids, yielding aminoacyl-*t*RNA molecules (*t*RNA mole-

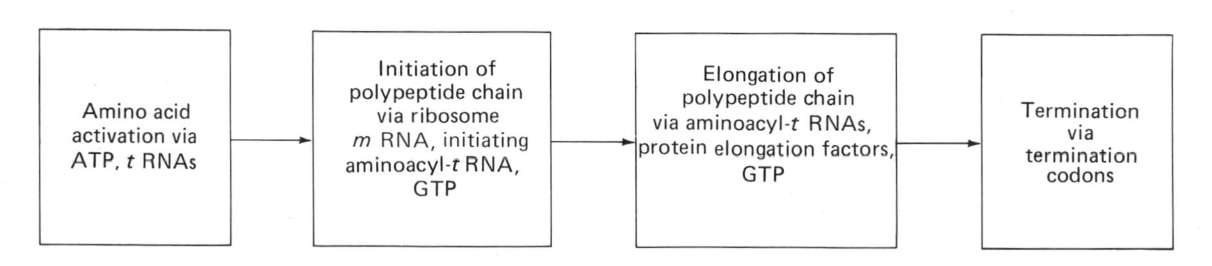

FIGURE 26.2 Overview of protein biosynthesis.

cules with amino acids attached). Then the polypeptide chain is initiated. This requires a ribosome, an *m*RNA, and a particular aminoacyl-*t*RNA. The third part of the process is elongation of the chain—amino acids are linked with the chain according to the message given by the *m*RNA. Finally, the chain is ended when a termination codon appears in the message. We will discuss each of these steps in turn.

AMINO ACID ACTIVATION 26.9

Amino acids are attached to specific tRNA molecules.

As usual, our biosynthetic starting materials must be activated in order to make the process of joining them in long chains more energetically favorable. Amino acids are activated by linking them to AMP; ATP is the source of the AMP and of the energy required for the reaction:

$$\underset{\text{amino acid}}{R-\overset{NH_2}{\overset{|}{C}H}-CO_2H} + \underset{\text{ATP}}{^{-}O-\overset{O}{\overset{\|}{\underset{O_-}{\underset{|}{P}}}}-O-\overset{O}{\overset{\|}{\underset{O_-}{\underset{|}{P}}}}-O-\overset{O}{\overset{\|}{\underset{O_-}{\underset{|}{P}}}}-O-CH_2\text{-ribose(OH, OH)-adenine}} \rightarrow$$

$$\underset{\text{an aminoacyl adenylate}}{R-\overset{NH_2}{\overset{|}{C}H}-\underset{O}{\underset{\|}{C}}-O-\overset{O}{\overset{\|}{\underset{O_-}{\underset{|}{P}}}}-O-CH_2\text{-ribose(OH, OH)-adenine}} + PP_i \xrightarrow[\text{inorganic pyrophosphatase}]{H_2O} 2P_i$$

Next, the aminoacyl group is transferred to a specific transfer RNA (*t*RNA) molecule:

$$\underset{\text{an aminoacyl adenylate}}{R-\overset{NH_2}{\overset{|}{C}H}-\underset{O}{\underset{\|}{C}}-O-\overset{O}{\overset{\|}{\underset{O_-}{\underset{|}{P}}}}-OCH_2\text{-ribose(OH, OH)-adenine}} + \underset{3'}{HO}-tRNA-5' \rightarrow \underset{\text{an aminoacyl-}t\text{RNA}}{R-\overset{NH_2}{\overset{|}{C}H}-\underset{O}{\underset{\|}{C}}-\underset{3'}{O}-tRNA-5'}$$

In Section 26.10 we will discuss *t*RNA in more detail.

The entire reaction (formation of the aminoacyl adenylate and then formation of the aminoacyl-*t*RNA) is catalyzed by the same enzyme, an aminoacyl-*t*RNA synthetase. There are 20 such enzymes in the cell cytoplasm, each of which is specific for *one* particular amino acid and one (or more) particular *t*RNA, which will carry that particular amino acid.

Problem 26.3 The amino acid activation reaction is rather similar to what reaction in fatty acid metabolism?

Problem 26.4 What purpose is served by the inorganic pyrophosphatase-catalyzed reaction $PPi + H_2O \rightarrow 2Pi$?

Problem 26.5 Are the starting materials of replication and transcription activated? Explain.

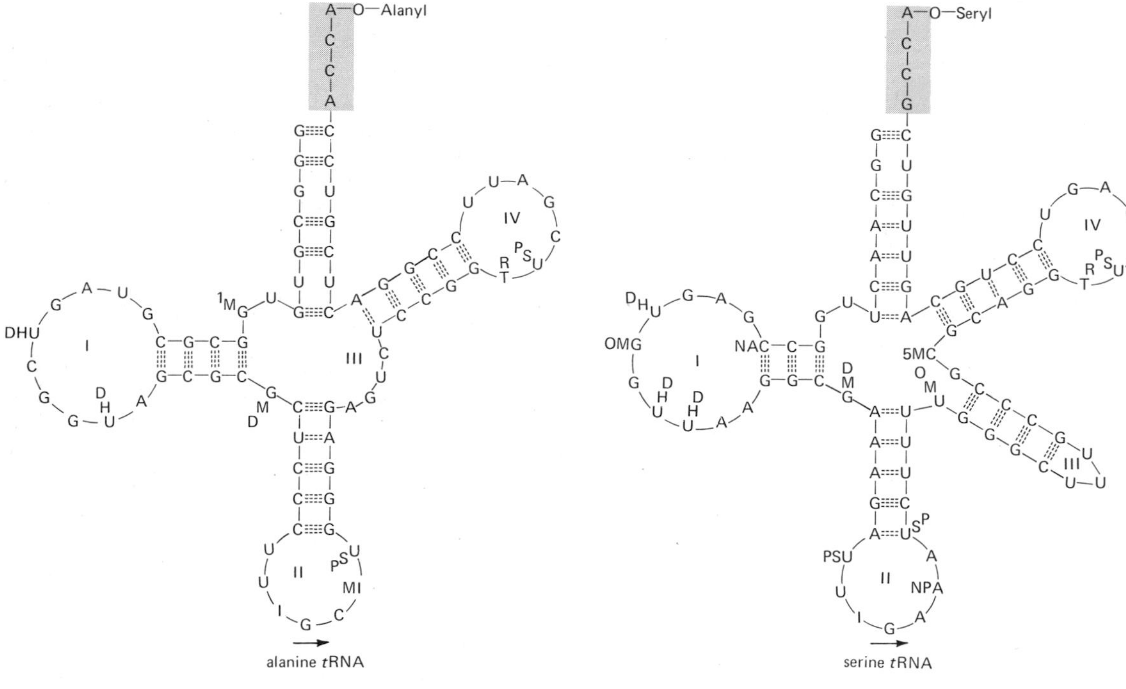

alanine *t*RNA

serine *t*RNA

FIGURE 26.3 Comparison of the nucleotide sequences for two *t*RNA molecules obtained from baker's yeast. The molecules are shown in the cloverleaf pattern with an indication of the suggested base pairing. In each case the anticodon is the sequence of three bases just over the arrow at the bottom of the molecule. The symbols DHU, PSU, I, DMG, and so on, represent certain rare modified nucleotides found only in transfer RNAs. (Adapted from M. O. Dayhoff, *Atlas of Protein Sequence and Structure,* vol. 4, National Biomedical Research Foundation, Silver Spring, Md., 1969.)

TRANSFER RNA 26.10

What do all tRNAs have in common?

Transfer RNAs (*t*RNAs) are small polyribonucleotide chains that contain about 75 to 90 nucleotide units. There are some 60 different *t*RNA species in the cell. Although each has a different sequence, all have certain structural features in common. All *t*RNAs have a special cloverleaf structure, formed by the single polynucleotide strand folding back on itself and forming base pairs in four regions (Figure 26.3). Figure 26.4 is a three-dimensional representation of a *t*RNA molecule.

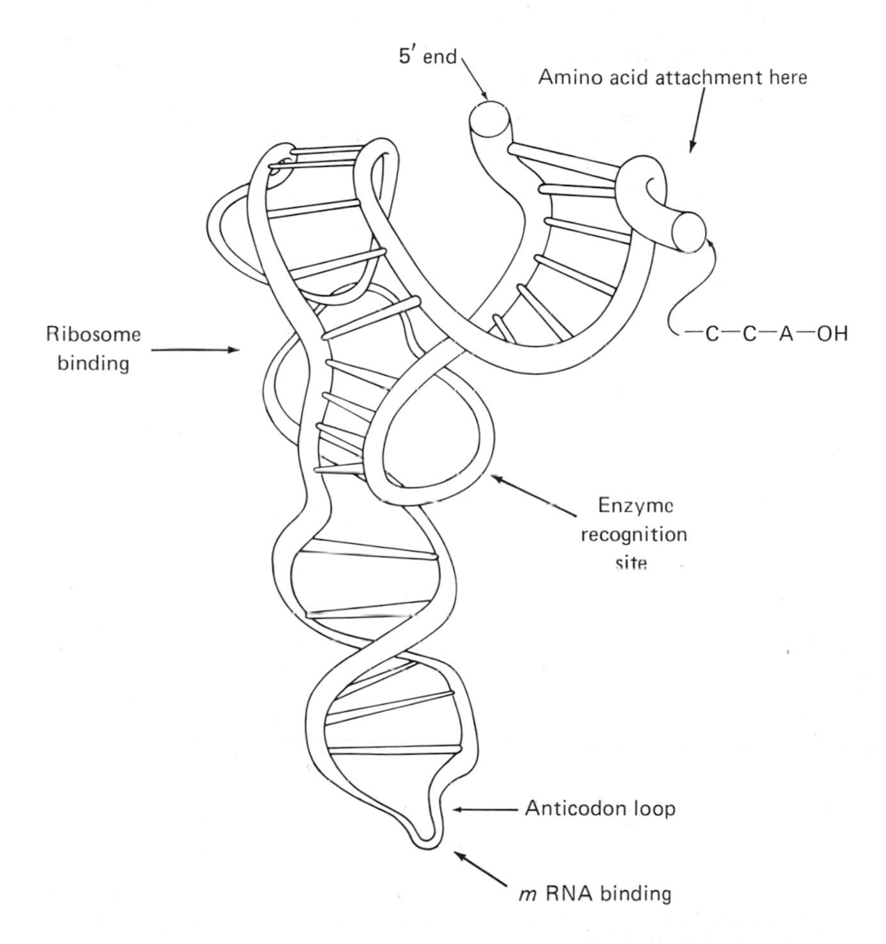

FIGURE 26.4 The three-dimensional arrangement of a transfer RNA (*t*RNA) molecule. The molecule is composed of a single polynucleotide chain that is folded back on itself to form bonds, indicated by cross-bars, between complementary base pairs in different portions of the chain. Note that the site of amino acid attachment and the anticodon are at opposite ends of the molecule. (Adapted from *Biochemical Concepts*, Robert W. McGillvery. Copyright © 1975 by W. B. Saunders Company. Reprinted by permission of Holt, Rinehart & Winston.)

Two other structural features are of particular importance. First, all *t*RNAs have the same sequence at the 3′ end, —CCA—3′ (Figure 26.3). This is the part of the molecule that accepts the aminoacyl group in the reaction catalyzed by an aminoacyl-*t*RNA synthetase. In addition, all *t*RNAs have an **anticodon** loop that contains a triplet sequence—an anticodon—that is complementary to the sequence of a particular *m*RNA codon.

The anticodon is the most important part of a *t*RNA molecule or an aminoacyl-*t*RNA molecule. The reason is that base pairing between an *m*RNA codon and the anticodon of the proper aminoacyl-*t*RNA is the sole means of properly positioning an amino acid in the sequence during protein synthesis. In other words, codon–anticodon binding is the key to translating the *m*RNA code into an amino acid sequence. It is clearly very important that an amino acid be attached to the proper *t*RNA during amino acid activation. Otherwise the amino acid and the anticodon would not match, and the wrong amino acid would be inserted into the polypeptide chain. Ensuring the accuracy of the matches between *t*RNAs and amino acids is the job of the very specific aminoacyl-*t*RNA synthetases.

26.11 INITIATION OF THE POLYPEPTIDE

What is needed to begin translation?

The process of synthesizing a polypeptide according to the instructions contained in an *m*RNA codon sequence is called **translation.** For the translation of the *m*RNA nucleotide sequence to begin, the following must come together: a messenger RNA, a ribosome, certain protein factors, and a particular initiating *t*RNA with its attached amino acid. In all cells, the particular aminoacyl-*t*RNA required for the initiation of a polypeptide is a certain methionyl-*t*RNA^{met}. (The superscript *met* designates that the *t*RNA is specific for methionine. The prefix *methionyl-* indicates that the *t*RNA is covalently bonded to methionine.)

In an animal or plant cell, each ribosome contains three molecules of **ribosomal RNA (*r*RNA)** of different molecular weights, as well as 90 or so different proteins. These are organized so that the ribosome consists of a larger and a smaller subunit (Figure 26.5). The ribosome binds *m*RNA, aminoacyl-*t*RNAs, and various protein factors that are necessary for protein synthesis. In addition, the ribosome contains the enzyme that catalyzes formation of the peptide bonds that hold the amino acids together. However, the specific functions of most of the ribosomal proteins and the *r*RNAs are not yet known.

The *m*RNA becomes bound to the ribosome (see Figure 26.6 for a diagram detailing the initiation process). The initiating codon of the

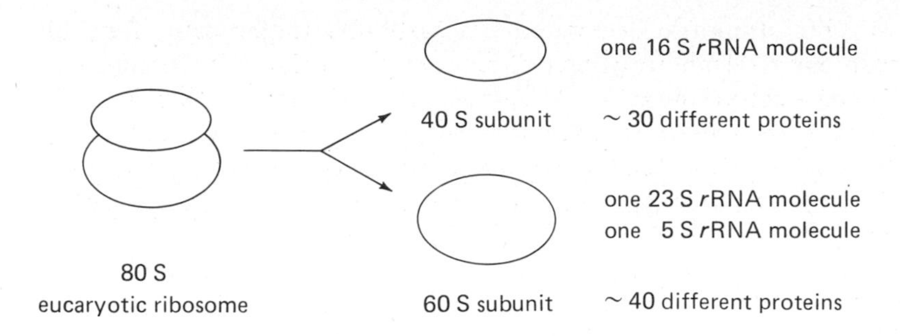

FIGURE 26.5 Structure of the cytoplasmic ribosomes of animals and plants.

*m*RNA is then available for base pairing with the initiating methionyl-*t*RNA^{met}. The initiating codon of the *m*RNA is always a 5′-AUG-3′ codon that is some distance from the 5′ end of the *m*RNA. The only *t*RNA with the complementary anticodon (5′-CAU-3′) is *t*RNA^{met}. Methionyl-*t*RNA^{met} becomes bound to *m*RNA when the initiating codon and the anticodon are base-paired. Other parts of the *t*RNA

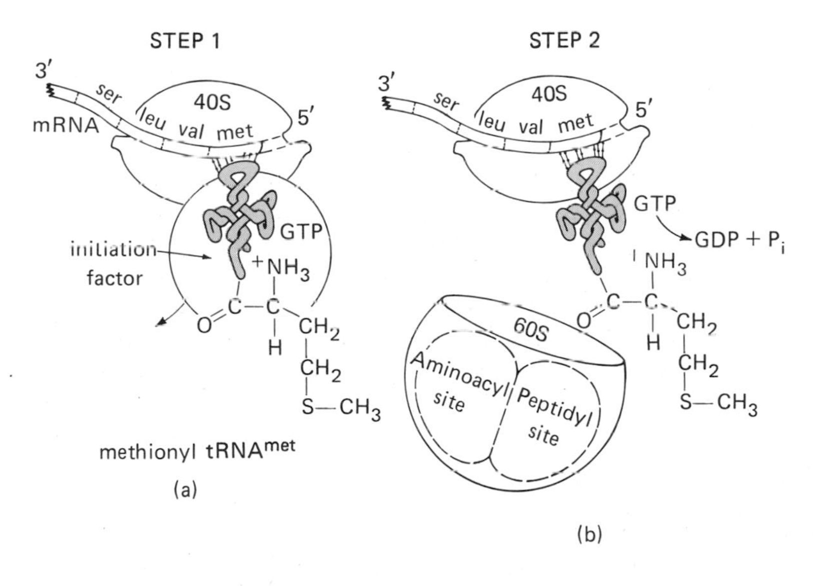

FIGURE 26.6 Initiation of polypeptide synthesis. (a) Step 1: a complex forms between a special initiator protein, an initiator *t*RNA carrying a methionyl residue, and a molecule of guanosine triphosphate (GTP). The complex then binds to the initial methionine codon (AUG) on a molecule of *m*RNA attached to a small ribosomal subunit. (b) Step 2: a complete ribosome is then formed by binding of a large subunit. The formation of the complete ribosome places the initiator *t*RNA with its attached methionyl residue on the peptidyl site. (Adapted from *Biochemical Concepts*, Robert W. McGillvery. Copyright © 1975 by W. B. Saunders Company. Reprinted by permission of Holt, Rinehart & Winston.)

molecule also bind to specific proteins or RNAs of the ribosome. The site on the ribosome at which methionyl-*t*RNA^{met} binds is called the **peptidyl site (P-site).**

26.12 ELONGATION OF THE POLYPEPTIDE

What are the three steps of elongation?

After initiation, the main requirements for polypeptide synthesis are ample supplies of aminoacyl-*t*RNAs and of guanosine triphosphate (GTP). (GTP is an energy-rich molecule, like ATP, which is hydrolyzed to provide energy.) The second *m*RNA codon to be translated (shown as "val" in step 1 of Figure 26.7) lies in the **aminoacyl site (A-site)** of the ribosome. The aminoacyl-*t*RNA with the complementary anticodon binds there (step 1 of Figure 26.7). GTP is also needed. Now, peptidyl synthetase (an enzyme that is part of the larger ribosome subunit) catalyzes formation of a peptide bond between the new amino acid (valine) and the initial methionine. It does this by attaching the methionine to the second amino acid, which is still attached to the second *t*RNA (see step 2 of Figure 26.7).

At this point in the elongation process, the step called **translocation** occurs (step 3 of Figure 26.7). First, the empty *t*RNA^{met} leaves the P-site. Then the second *m*RNA codon and the second *t*RNA (now with both its own amino acid and the initiating methionine) are moved ("translocated") to fill the P-site. This requires the energy of GTP hydrolysis.

The A-site is now open for the third *m*RNA codon to be translated, and the stage is set for a repeat of the elongation process. A new complementary aminoacyl *t*RNA binds to the third codon (step 4 of Figure 26.7), and peptide-bond formation and translocation (steps 1, 2, and 3) occur again. The process is repeated for addition of the next amino acid to the chain, and so on.

Problem 26.6 If the second amino acid in the polypeptide is to be phenylalanine, what *m*RNA codon will appear at the A-site during the first elongation step? What might be the *t*RNA anticodon?

26.13 TERMINATION OF THE POLYPEPTIDE

The polypeptide gets longer and longer until an *m*RNA termination codon (UAA, UAG, or UGA) appears at the A-site. No *t*RNAs have

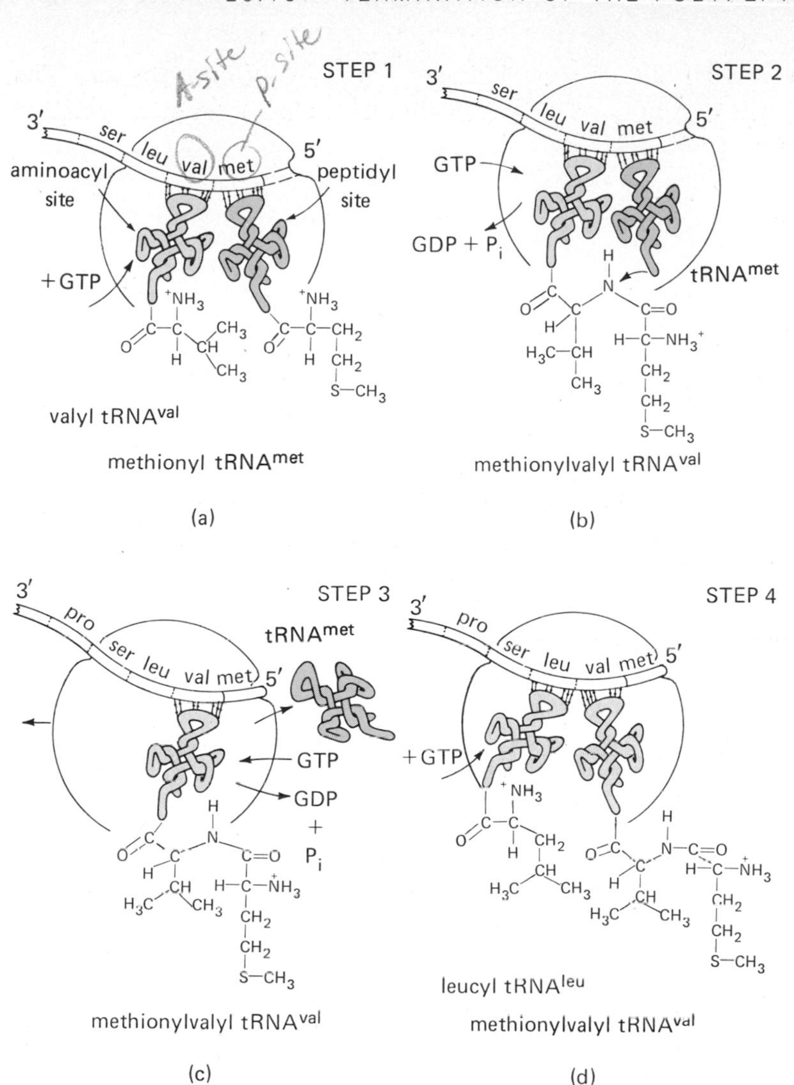

FIGURE 26.7 Polypeptide elongation. (a) Step 1: now a valine-specific *t*RNA binds at the aminoacyl site via codon–anticodon base pairing. GTP is also bound. (b) Step 2: peptide synthetase (a protein of the ribosome) catalyzes peptide bond formation between met and val (the met is transferred to give methionylvalyl-*t*RNA^{val}). (c) Step 3: another ribosomal enzyme, a translocase, causes the ribosome to shift along the attached *m*RNA by the space of one triplet codon. This mechanical work is driven by GTP hydrolysis. The now free initiator *t*RNA dissociates as methionylvalyl-*t*RNA^{val} takes its place on the peptidyl site. (d) Step 4: the next codon, specifying leu, has been brought into position at the aminoacyl site, and now a leucyl-specific *t*RNA binds by codon–anticodon pairing. GTP binds as well. Peptide bond formation and translocation occur, and cycles of elongation continue. (Adapted from *Biochemical Concepts,* Robert W. McGillvery. Copyright © 1975 by W. B. Saunders Company. Reprinted by permission of Holt, Rinehart & Winston.)

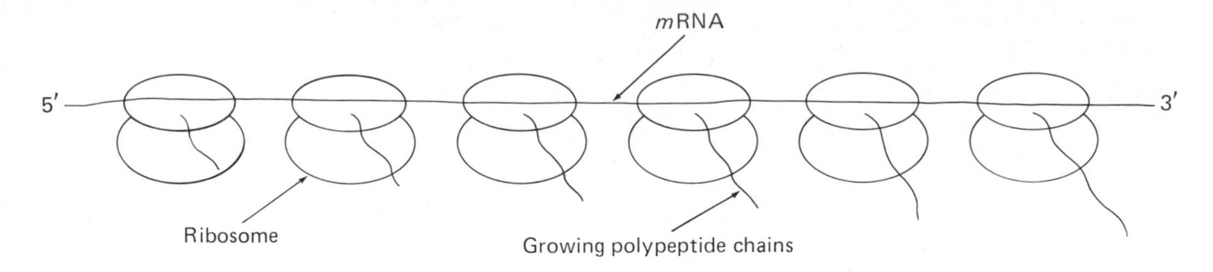

FIGURE 26.8 A polyribosome.

What happens when a termination codon arrives at the ribosome?

anticodons that are complementary to these termination codons. Thus, when the completed polypeptide is hydrolyzed from the last *t*RNA, the *m*RNA dissociates from the ribosome. If the ribosome or the *m*RNA is to continue in protein synthesis, a new initiation routine must take place. The completed polypeptide, which has already folded into its native conformation, is ready to take its place in cell function.

In the overall process, an amount of energy equivalent to 3ATPs is used for the synthesis of each peptide bond. Generally, many ribosomes will simultaneously be translating a single *m*RNA molecule, each ribosome bearing its own growing polypeptide chain (Figure 26.8). This complex of ribosomes and *m*RNA is called a **polyribosome,** or **polysome.**

Sometimes a single *m*RNA contains the information for several polypeptides. The different messages are separated by termination signals, and initiation codons are present at the beginning of each new message.

Problem 26.7 If a protein molecule has a molecular weight of 50,000, how many GTPs are necessary for its synthesis? (Assume the molecular weight of an average amino acid residue to be 120).

26.14 GENETIC ERRORS

What is a mutation?

Many human genetic diseases result from faulty proteins. The fault is caused by a mistake in the genetic information transmitted from parent to offspring. In many of these diseases, only a single amino acid is different from normal, and yet the function of the protein is profoundly altered.

In sickle cell anemia and some of the **thalassemias,** hemoglobin has undergone a single amino acid substitution. Hemoglobin, as you may know, is the red blood cell protein that transports oxygen. In sickle cell hemoglobin, a particular glutamic acid residue has been replaced by a valine residue. This results in a marked reduction in the protein's attraction for oxygen. It also changes its physical properties, and that in turn changes the shape of the red blood cell. Instead of a normal doughnut shape, a sickled cell has a distorted crescent shape (hence its name). This shape causes sickled cells to jam up more often during circulation through the small blood vessels. This jamming leads to a failure in the delivery of oxygen to the cells and causes the patient intense pain.

The substitution of valine for glutamic acid results from a **point mutation** in the gene for one of the two types of polypeptides in hemoglobin. A **mutation** is a change in the nucleotide sequence (the information content) of a gene. The actual mutation—possibly caused by a mistake in replication—probably occurred many generations ago, but the error has been perpetuated in DNA replications since then. All the polypeptides synthesized using the *m*RNA transcribed from this gene would contain the error, resulting in the synthesis of sickle cell hemoglobin chains.

ANTIBIOTICS AND PROTEIN SYNTHESIS 26.15

Many antibiotics are effective in treating disease because they block protein synthesis on bacterial ribosomes but not on the eucaryotic ribosomes of animals. There are several differences in the details of translation on bacterial 70S and higher plant or animal 80S ribosomes that result in their different susceptibilities to antibiotics. (70S and 80S refer to measures of size of these two types of ribosomes; the 80S ribosomes are larger.) When bacteria infect a host organism, they rapidly grow and divide in the host's tissues. Antibiotics that block protein synthesis block the growth and proliferation of the bacteria and thus limit the infection.

One of the many known antibiotics, streptomycin, binds to bacterial ribosomes and apparently causes a misreading of the *m*RNA code, resulting in the synthesis of faulty bacterial proteins. Tetracyclines inhibit the binding of aminoacyl-*t*RNAs at the A-site of the bacterial ribosomes. Chloramphenicol inhibits peptidyl synthetase and blocks peptide bond formation. Erythromycin inhibits the translocation step of bacterial protein synthesis. In each case, the blockage of protein synthesis in the bacterial cells severely inhibits their ability to grow and divide. The body's natural defense mechanisms then have a better chance to destroy the crippled bacterial cells. The usefulness of antibiotics in treating disease depends on their being more toxic to the infecting organism than they are to the host.

SUMMARY

Genetic information is stored in the nucleotide sequence of a cell's DNA, which is contained in chromosomes in the cell nucleus. This information consists of "recipes" for synthesizing all the proteins that cell can produce. Genetic information is transferred from generation to generation by **semiconservative replication** of DNA molecules. **Protein synthesis** takes place on ribosomes in the cytoplasm of the cell. The genetic information reaches the cytoplasm by the **transcription** of a segment of the DNA nucleotide sequence, a **gene,** into **messenger RNA (*m*RNA).** The messenger RNA can move from the nucleus to the cytoplasm.

The **genetic code** consists of 64 **triplet codons,** 61 of which specify particular amino acids and 3 of which are **termination codons,** marking the end of a polypeptide chain. Before protein synthesis can begin, amino acids must be **activated** as aminoacyl-transfer RNAs. **Transfer RNAs (*t*RNAs)** are small polynucleotides containing anticodons that base pair with the codons of *m*RNA. The initiation of protein synthesis requires the formation of a complex consisting of a ribosome, an *m*RNA molecule, the initiating aminoacyl-*t*RNA (which is methionyl-*t*RNA^{met}), and certain initiation proteins. Elongation of the polypeptide requires a supply of aminoacyl *t*RNAs, GTP, and various protein factors. The synthesis of a protein is terminated when a termination codon arrives at the ribosome's **aminoacyl site (A-site).**

A **mutation** is a mistake in replication. Proteins made from a gene in which a mutation occurs may be faulty or nonfunctional. Antibiotics often block protein synthesis on the smaller ribosomes found in bacteria, but they do not affect protein synthesis on a human's larger ribosomes.

Key Terms

Amino acid activation (Section 26.8)
Aminoacyl site (A-site) (Section 26.12)
Anticodon (Section 26.10)
Chromosome (Section 26.1)
Degeneracy (Section 26.7)
Gene (Section 26.1)

Genetic information (Section 26.1)
Genetics (Introduction)
Messenger RNA (*m*RNA) (Section 26.6)
Molecular genetics (Introduction)
Mutation (Section 26.14)
Peptidyl site (P-site) (Section 26.11)
Point mutation (Section 26.14)
Polyribosome (polysome) (Section 26.13)
Protein synthesis (Section 26.2)
Replication (Section 26.2)
Ribosomal RNA (rRNA) (Section 26.11)
RNA polymerase (Section 26.6)
Semiconservative replication (Section 26.4)
Termination codon (Section 26.7)
Thalassemia (Section 26.14)
Transcription (Section 26.6)
Transfer RNA (tRNA) (Section 26.10)
Translation (Section 26.11)
Translocation (Section 26.12)
Triplet codons (Section 26.7)

ADDITIONAL PROBLEMS

26.8 Predict the amino acid sequences that will be formed by a ribosome using the following messenger RNAs. Assume that each chain begins with the triplet codon at the left.

(a) AAA UUU CGA AGG GGG AGC

(b) GUA UCC AGC UCG GCU AGU CCC

(c) AUG GCA AAA UAC

(d) CCA CGA CUA

26.9 One DNA strand contains the following sequence:
5′-TCAGGTAAGCCATAG-3′

(a) Write the sequence of the other strand of the DNA.
(b) Write the sequence of bases in the *m*RNA transcribed from the first DNA strand.
(c) Write the amino acid sequence that *m*RNA codes for.
(d) If the A in the middle of the first DNA strand is deleted, what amino acid sequence will be coded for?

26.10 What are termination codons? What would happen if a mutation caused a termination codon to appear in the middle of an *m*RNA message?

26.11 In the laboratory you have made a synthetic polynucleotide consisting of alternating uridine and cytidine nucleotides (UCUCUCU . . .). What would be the sequence of the polypeptide produced if this synthetic message were used by ribosomes?

26.12 If you were to synthesize in the laboratory a polynucleotide containing one part adenine and two parts guanine, what codons would be most likely to form? What amino acids would predominate in a polypeptide made using the synthetic message?

26.13 If there is a severe deficiency of even one amino acid in the diet, protein synthesis in general is very much depressed. Explain.

26.14 In all organisms, the relative molar base composition of DNA obeys what is called Chargaff's rule: adenine = thymine, guanine = cytosine. Explain why this should be.

26.15 What are the specific requirements for the initiation of a polypeptide chain?

26.16 The great specificity of the aminoacyl-*t*RNA synthetases is extremely important in keeping mistakes in protein synthesis to a minimum. Explain.

***26.17** In general terms, how might the transcription of a gene be regulated?

***26.18** Can you think of ways in which translation might be regulated?

Problem-Solving Hints

If you have trouble solving any of the additional problems, refer to the sections listed next to the problem numbers.

26.8	26.7	**26.9**	26.3, 26.7	**26.10**	26.13
26.11	26.7	**26.12**	26.7	**26.13**	26.8, 26.11
26.14	26.4	**26.15**	26.10	**26.16**	26.10
26.17	26.6	**26.18**	26.11		

FOCUS 15 Mutagens and Carcinogens

A mutation is a change in the nucleotide sequence of a gene that changes the gene's information content. The change may be caused by a mistake in replication. Apparently, these spontaneous mutations occur fairly often without major effects. However, some mutations do have drastic effects, and various inherited human diseases result from mutations. Mutagenic chemicals and physical agents such as ultraviolet light and x rays greatly increase the rate of mutation. A virus may be able to cause a mutation by inserting its own DNA (or RNA) into a

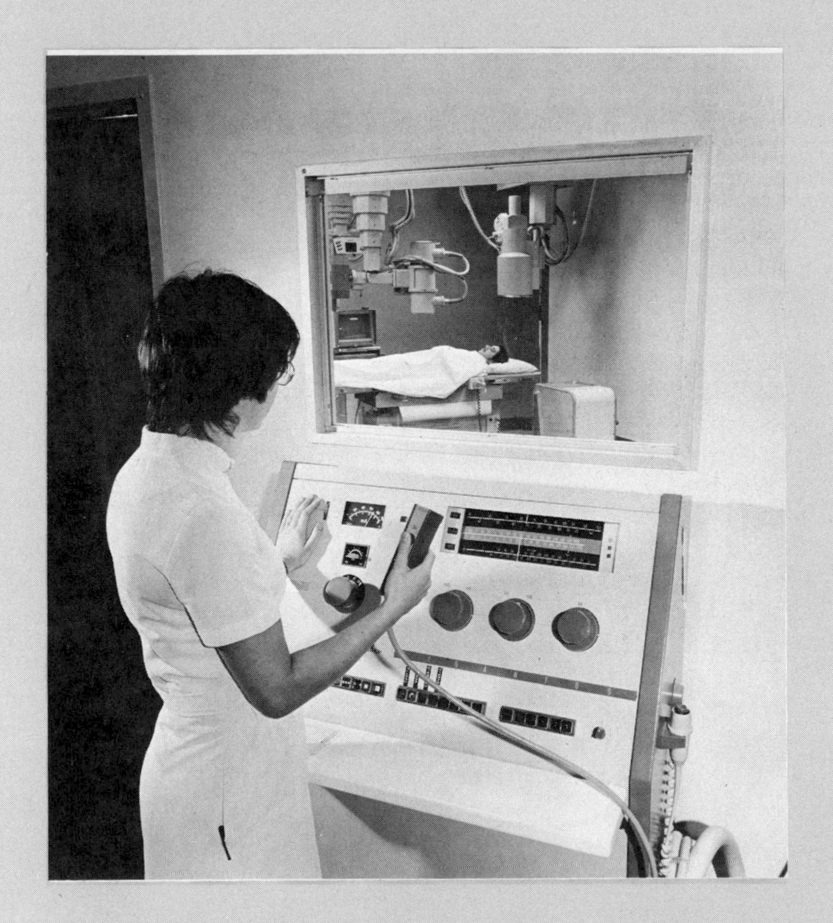

FIGURE 1 Repeated exposures to x rays greatly increase the rate of mutation. Thus, patients are given x rays only when necessary, while medical personnel remain behind protective barriers. (Ellis Herwig/Stock, Boston)

cell's chromosomes. The high incidence of cancer in the elderly is thought to be at least partly the result of the accumulation of somatic mutations, mutations that occur in ordinary body cells. The longer a cell lives, the greater the probability that some mutagenic agent will affect it. The mutated body cell may divide many times to form a tumor, but the mutation will not be transmitted to offspring.

Mutagenic chemicals may act in several ways. One class, the alkylating agents, alkylate susceptible nitrogen or oxygen atoms of the pyrimidine or purine bases (Section 18.2) of DNA, changing the structure of the bases and causing pairing mistakes to occur. Among the most potent alkylating mutagens are the nitrosamines (Section 13.12):

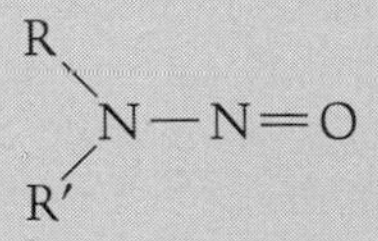

$$\begin{matrix} R \\ & \diagdown \\ & & N{-}N{=}O \\ & \diagup \\ R' \end{matrix}$$

Nitrosamines are highly carcinogenic (cancer-causing) and are thought to be a major cause of human cancer. They are formed when secondary amines react with nitrous acid (HNO_2):

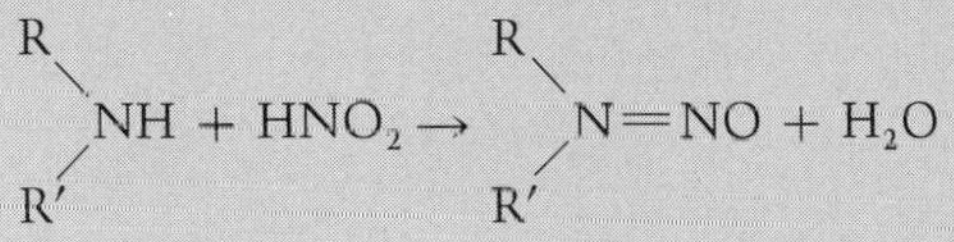

$$\begin{matrix} R \\ & \diagdown \\ & & NH + HNO_2 \rightarrow \\ & \diagup \\ R' \end{matrix} \quad \begin{matrix} R \\ & \diagdown \\ & & N{=}NO + H_2O \\ & \diagup \\ R' \end{matrix}$$

Many foods are excellent sources of both these reactants. Plants contain nitrate, which can be reduced to nitrite. Cured meats, like bacon, are often preserved by the addition of nitrites and nitrates. The nitrosamine-forming reaction can occur in the stomach, and the products are absorbed in the blood and carried throughout the body. Cells exposed to these nitrosamines may undergo somatic mutations leading to cancer. Recent evidence seems to suggest, however, that nitrites in foods are not a major cause of human cancers.

Another class of mutagen is itself incorporated directly into the DNA structure. For example, 5-bromodeoxyuridine can be incorporated into DNA in place of the normal thymidine nucleotide. This type of mutation, in which a base is replaced, is called a base-substitution mutation.

Still other mutagens act as intercalating agents. Many antibiotics, drugs, dyes, and other substances have a flat, aromatic-ring structure. This structure enables them to intercalate (insert) between the normal stacked base pairs in the DNA double helix. This intercalation can induce frame-shift mutations. The meaning of this term can be appreciated from the following diagram:

one strand of DNA	ACT CAC TAG GTT AGC CCG CAT normal codons
same strand but with inserted X molecule	ACT C X A CTA GGT TAG CCC GCA T frame-shift mutation codons

Note that every codon following the intercalating agent is altered, as well as the codon that actually contains the inserted agent. Imagine the effect on the structure of a protein molecule whose gene has undergone such a frame-shift mutation. Many of the amino acids inserted during protein synthesis would be totally wrong, and the protein might be completely useless. Or a frame-shift mutation might occur in a region of the chromosome that is involved with the regulation of gene expression or replication. Normal regulatory controls—such as controls over cell division—might then be lost.

Genetic Engineering FOCUS 16

Technology is now available for introducing DNA from virtually any organism into the genetic material of bacterial cells. Recently, for example, the gene for human growth hormone was incorporated into the

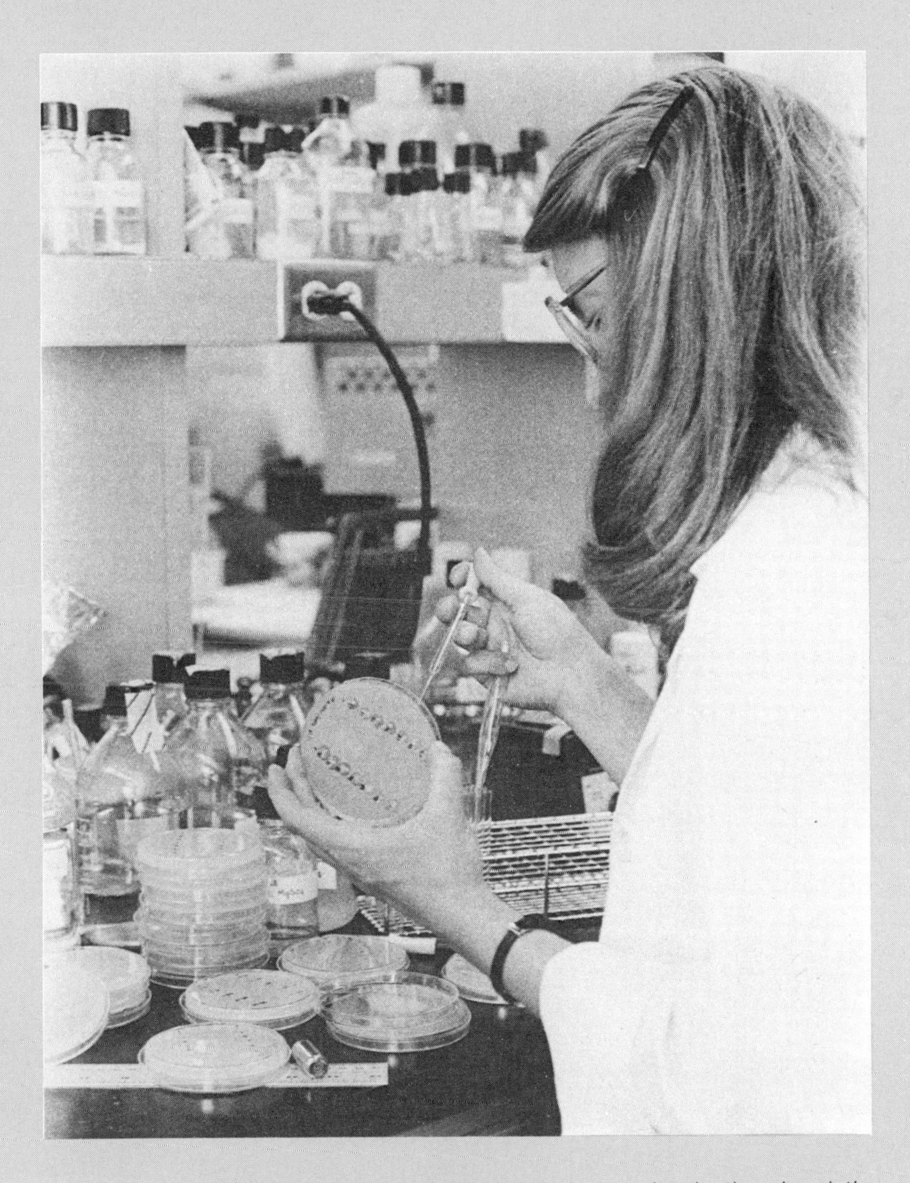

FIGURE 1 Genetic engineering is being carried on today in the chemist's laboratory. (Martha Morgan, University of Utah)

DNA of a strain of Escherichia coli (*E. coli*), a common bacterial species that has been used as experimental material for much research in biochemical genetics. The bacteria then produced the hormone. It was possible to isolate quantities of the human growth hormone synthesized by cultures of *E. coli*. Growth hormone is used to treat children who are suffering from dwarfism because their pituitary glands do not produce adequate amounts of the hormone. The bacteria-produced hormone has not yet been approved for medical use, but it has been shown to be identical in structure to normal human growth hormone.

This manipulation of genetic material has been called recombinant DNA research, or genetic engineering. The technique holds great promise for the production of medically important products such as hormones and other proteins. Many commercial applications can be imagined, too, such as the synthesis of large amounts of enzymes for use in the chemical industry. One day some form of DNA recombination may be used to treat genetic diseases in humans.

We will summarize one procedure that has been used to introduce foreign genes into bacteria. Bacteria contain small, circular DNA molecules that are separate from the larger circular bacterial chromosome. These small DNA molecules, called plasmids, can be isolated from broken *E. coli* cells. The plasmid DNA is then cleaved at one particular location by a specific endonuclease, an enzyme isolated from another strain of *E. coli*. An endonuclease catalyzes hydrolysis of internal 3′,5′-phosphodiester linkages. The specific endonucleases used hydrolyze only phosphodiester bonds located within specific and unusual sequences of several nucleotides. Since the circular plasmid DNA is clipped open at only one spot, it assumes a linear shape. The linear plasmid DNA is then mixed with isolated DNA containing a desired gene, such as a growth-hormone gene. The plasmid incorporates the foreign gene and then forms itself into a circle again.

In the next step, *E. coli* cells are treated with calcium chloride to make them more permeable. They are then exposed to the new plasmid DNA, which enters some of the bacterial cells and is replicated along with the normal chromosome as the cells grow and divide. The bacteria are plated on a growth medium containing an antibiotic to which the cells containing the new plasmid are resistant. Only the resistant cells reproduce, forming visible colonies on the plate that can be picked out and cultured on a larger scale.

A large culture of these bacteria will contain many plasmids that have the foreign gene. The gene is said to have been cloned; that is, a large number of identical copies of the gene have been made. Cloning is also useful for obtaining large quantities of a gene for nucleotide sequencing.

The protein product of the foreign gene is synthesized in the culture along with the native proteins of the *E. coli* cells. The foreign protein may then be separated from the others and purified.

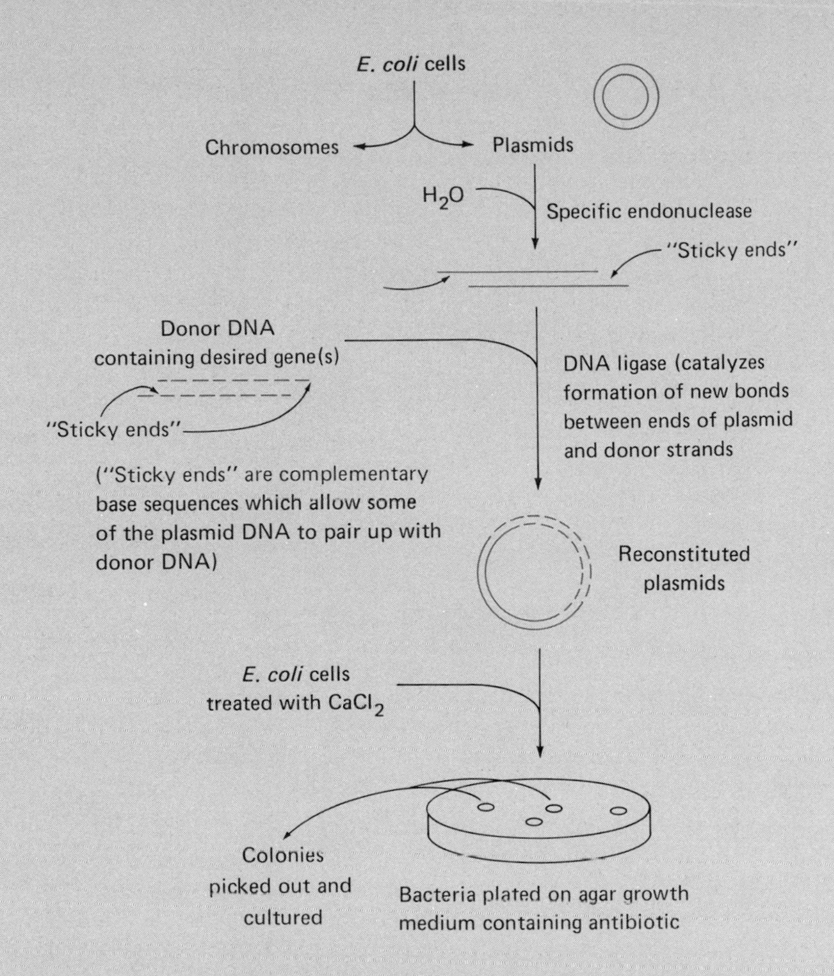
E. coli cells
Chromosomes
Plasmids
H_2O
Specific endonuclease
"Sticky ends"
Donor DNA
containing desired gene(s)
"Sticky ends"
("Sticky ends" are complementary
base sequences which allow some
of the plasmid DNA to pair up with
donor DNA)
DNA ligase (catalyzes
formation of new bonds
between ends of plasmid
and donor strands
Reconstituted
plasmids
E. coli cells
treated with $CaCl_2$
Colonies
picked out and
cultured
Bacteria plated on agar growth
medium containing antibiotic

```
                 ST. ELIZABETH HOSPITAL MEDICAL CENTER
             DEPARTMENT OF PATHOLOGY AND LABORATORY MEDICINE
                  SECTION OF DEVELOPMENTAL RESEARCH

             DIAGNOSTIC TESTING RESULTS--SERUM ELECTROLYTES
DATE: 1/1/81
PATIENT NAME: DOE, JOHN
TEST NAME        TEST RESULT      NORMAL RANGE     INTERPRETATION  STD. DEV. FROM MEAN
SODIUM            133             135-145          LOW                 1.3
POTASSIUM         3.1             3.5-5.0          LOW                 4.0
CHLORIDE          98              97-107           NORMAL              0.0
CARBON DIOXIDE    24.8            22-32            NORMAL              0.0
ANION GAP         13.3            11.0-19.5        NORMAL
**THE ABOVE PATTERN OF ELECTROLYTE CHANGES SUGGESTS THE FOLLOWING DIAGNOSTIC CONSIDERATIONS**
FLUID OVERLOAD
DIURETIC DRUG PRESENT
CONGESTIVE HEART FAILURE
*LABORATORY TESTING SUGGESTIONS AND ADDITIONAL DIAGNOSTIC CONSIDERATIONS BASED ON
             THE ANALYSIS OF INDIVIDUAL ELECTROLYTE RESULTS*
CONGESTIVE HEART FAILURE--CARDIAC PROFILE
SPOT URINE SODIUM--IF <10, POSSIBLE DILUTIONAL HYPONATREMIA EXISTS.
                   IF >20, POSSIBLE URINE SODIUM LOSS DUE TO RENAL DISEASE, DIURETICS,
                   DEFICIENCY OF MINERALOCORTICOID, OR INAPPROPRIATE ADH.
SPOT URINE POTASSIUM--IF <10, POSSIBLE GI TRACT LOSS OF POTASSIUM.
                      IF >20, POSSIBLE RENAL LOSS OF POTASSIUM.
PLEASE NOTE:
THESE DIAGNOSTIC CONSIDERATIONS HAVE BEEN PRIORITIZED BY SEVERITY OF THE PATIENT'S TEST RESULTS.
```

Physicians order tests of serum electrolyte levels frequently and routinely as a basis for determining the origin of certain symptoms of disease. Maintaining electrical neutrality is of great importance in the human body. Electrical imbalance can point to a variety of conditions. Sodium (Na^+) and potassium (K^+) are both positively charged ions (cations), while chloride (Cl^-) and bicarbonate (HCO_3^-) are negatively charged ions (anions). To maintain an equal balance of positive and negative ions, there must not be too much of either one.

The anion gap is a measurement of the difference between positive and negative ions present in the blood. The following electrolyte levels might be found in the blood serum of a normal, healthy person:

$$\begin{array}{rr} \text{sodium} & \left.\begin{array}{r}136\\4\end{array}\right\}140 \\ \text{potassium} & \\ \text{chloride} & \left.\begin{array}{r}98\\25\end{array}\right\}123 \\ \text{bicarbonate} & \end{array}$$

To calculate the anion gap, we first add the sodium and potassium cation levels. In our example, the result would be 140. Similarly, we can add the anion levels and get 123. The difference, $140 - 123 = 17$, is referred to as the anion gap. (What actually accounts for the anion gap are the other ions

27

BODY FLUIDS: TRANSPORT AND THE INTEGRATION OF METABOLISM

present in the blood in low concentrations that are simply not figured in the final result.)

A very low or very high anion gap indicates a problem to the physician. If the value is low, supplementary ions can be given. If it is high, supplementary ions can be extracted, such as in the use of dialysis to lower the potassium ion level in the blood. The anion gap can also be useful in conjunction with measurements of specific electrolyte levels. When the anion gap is high (because of an increase in unmeasured anion) and the chloride level is normal, we have a condition of normochloremic acidosis. On the other hand, if the anion gap is normal but the bicarbonate level is decreased and the chloride level is high, we have hyperchloremic acidosis. Certain specific diagnoses are indicated by either of these two findings. Any form of acidosis can be a direct indication of a blood pH imbalance that can result in tissue damage. If left uncorrected, the imbalance may result in death.

[Illustration: Analytical research in clinical chemistry has produced rapid computer methods for providing possible diagnostic considerations with such test series as the serum electrolytes.]

Every cell carries on its own metabolic activities. That means nutrients must be brought into the cell, and wastes must be removed from the cell. In a single-celled organism, this is no problem. Nutrients are absorbed directly from the environment, and wastes are released back into the environment. But in large, multicelled organisms, most cells are far from the surface. If those cells were to simply dump their wastes, they would be dumping them on other cells, and all cells would have a difficult time searching through their environment for nutrients that they could absorb. Thus, a complicated circulatory system is necessary to bring nutrients to the cells, to remove wastes from the cells (and eventually from the body), and to carry messages among cells.

The overall circulatory system involves various body fluids that have different chemical compositions and perform different functions. The blood is the best-known body fluid, and we will look at its transport activities and composition first. Then we will investigate the other body-fluid compartments and the nature of certain specialized body fluids.

27.1 COMPOSITION OF BLOOD

What is the difference between serum and plasma?

If blood is drawn from a vein (and clotting is prevented), the blood cells can be separated from the blood **plasma.** The plasma is normally a slightly yellow, clear liquid. The blood cells usually constitute 40 to 50% of the blood volume (the exact percentage is called the **hematocrit**); the liquid plasma accounts for the remaining 50 to 60% of the volume.

TABLE 27.1 Cellular Elements of the Blood

Cell type	Number per mm³ of blood	Size (μm)	Nucleus?	Where formed	Function
Erythrocytes	4,500,000–6,000,000	6–9	No	Red bone marrow	O_2 transport
Leukocytes					
Neutrophils	3000–7000	10–15	Yes	Bone marrow	Phagocytosis and destruction of foreign materials such as bacterial cells
Eosinophils	50–500	10–15	Yes	Bone marrow	
Basophils	0–50	10–15	Yes	Bone marrow	
Monocytes	100–600	12–20	Yes	Bone marrow	
Lymphocytes	1000–3000	10–20	Yes	Spleen and lymph nodes	Formation of antibodies (gamma globulins)
Platelets	200,000–400,000	1.8	No	Bone marrow, spleen	Aid in control of bleeding

If drawn blood is allowed to clot, a clear, yellowish fluid—the blood **serum**—separates from the clot. The serum consists of the plasma minus the protein **fibrinogen.** Fibrinogen forms another protein, called **fibrin,** when blood clots. The clot consists of the cellular elements of the blood entrapped in a network of fibrous strands of fibrin.

The cellular elements of the blood are the **erythrocytes** (red blood cells), the **leukocytes** (white blood cells), and the **platelets** (Table 27.1). The erythrocytes are responsible for oxygen transport in the blood. The leukocytes, of which there are several types, are part of the body's defense system against disease. Platelets participate in the clotting of the blood, another of the body's defense mechanisms.

Problem 27.1 What are the functions of red and white blood cells?

COMPOSITION OF BLOOD PLASMA 27.2

Plasma is about 90% water, about 9% organic constituents, and about 1% inorganic salts. Table 27.2 gives a partial listing of plasma components and their concentrations.

Proteins make up 6 to 8% of the plasma. Numerous plasma proteins have been identified and more or less well characterized (Table 27.3). Some particularly important plasma proteins are serum albumin, which is involved in maintaining osmotic-pressure balance and transporting

TABLE 27.2 Composition of Plasma

Organic constituents (9%)		Inorganic salts (0.9%)		Special substances (90%)	
	g/100 mL		mEq/L‡		Vol. (%)
Proteins	6–8	Sodium	142	Water	90
Albumin	5	Potassium	4	Blood gases	
Globulins	2	Calcium	5	Nitrogen	1
Alpha	0.7	Magnesium	2	Oxygen	0.3
Beta	0.7	Chlorides	103	Carbon dioxide	2.5
Gamma	0.6	Phosphates	3		
Fibrinogen	0.3	Sulfates	1	Enzymes	
		Bicarbonates	28	Transaminase	
Nutrients	mg/100 mL			SGOT (aspartate)	5–100 I.U.*
Glucose	80–115		g/100 mL	SGPT (alanine)	4–13 I.U.*
Lactic acid	5–20	Iron	80–200	Lactic dehydrogenase	60–250 I.U.*
Cerebrosides	15	Iodine	0.5	Acid phosphatase	1.1 Bod. units†
Amino acids	5			Alkaline phosphatase	4.0 Bod. units†
Fatty acids	370			Vitamins	
Phospholipids	6–13			Vitamin A	29–64 g/100 mL
Cholesterol	150–280				
	mg/100 mL			Hormones	
Nonprotein nitrogeneous waste (NPN)	18–35			Protein-bound iodine (PBI) (thyroid hormone)	3–8 g/100 mL
Urea (BUN)	8–25				
Uric acid	3–7				
Creatine	0.4				
Creatinine	1.2				
Ammonium salts	0.2				

Source: King and Showers, *Human Anatomy and Physiology*, 6th ed. (Philadelphia: W. B. Saunders, 1969), p. 262.
*International units.
‡milliequivalents/L. (An ionic concentration expressed in mEq/L is equal to the concentration expressed in mM/L multiplied by the charge on the ion.)

fatty acids (Section 25.2) and certain other substances; the **gamma globulins** (γ globulins), which are the antibodies of the body's immune system; and the **beta globulins** (β globulins), certain of which are involved in transporting cholesterol, iron, and other materials.

Among the numerous other substances dissolved in the plasma are glucose, amino acids, oxygen, carbon dioxide, and salts of sodium, potassium, and calcium. A major function of the blood and plasma is the transport of dissolved nutrients, oxygen, and many other materials about the body.

TABLE 27.3 Major Protein Components of Human Blood Plasma

	Concentration (mg/100 mL)	Approximate molecular weight	Other components	Function
Serum albumin	3,000–4,500	68,000		Osmotic regulation, transport of fatty acids, bilirubin, etc.
α_1 Globulins	100	40,000–55,000	Carbohydrate	Not known
α_1 Lipoproteins	350–450	200,000–400,000	40–70% lipid	Lipid transport
α_2 Globulins	400–900			
α_2 Glycoproteins		up to 800,000	Carbohydrate	Unknown
Ceruloplasmin	30	150,000	Carbohydrate	Copper transport
Prothrombin		63,000	Carbohydrate	Blood clotting
β Globulins	600–1,200			
β_1 Lipoproteins	350–450	3–20 million	80–90 % lipid	Lipid transport
Transferrin	40	85,000	Carbohydrate	Iron transport
Plasminogen		90,000		Precursor of fibrinolysin
γ Globulins	700–1,500	150,000	Carbohydrate	Antibodies
Fibrinogen	300	340,000	Carbohydrate	Blood clotting

Source: Lehninger, A. L., *Biochemistry,* (New York: Worth Publishers, 1976), p. 830.
NOTE: Some 50 minor protein components, including a number of enzymes, have also been found in plasma.

BODY DEFENSES AND THE BLOOD 27.3

What is the main protein in a blood clot?

Clotting of blood, when it spills out of the blood vessels, is an important defense mechanism of the body. Numerous protein factors participate in the clotting process (Figure 27.1). In the figure, we have simply used Roman numerals to refer to the numerous proteins involved in the cascade mechanism (Section 24.17) of clotting. Clotting may be set off by factors derived from injured tissue, or by factors of the blood itself, chiefly the platelets. In either case the final result is the formation of an insoluble clot consisting of a polymer of many fibrin molecules along with trapped blood cells.

Some details about the clotting process are still not definitely understood. The well-known bleeding condition called **hemophilia** is caused by the absence or dysfunction of some plasma factor. In hemophilia A, it is factor VIII and in hemophilia B, factor IX (see Figure 27.1). Deficiency in (or inactivity of) these factors delays or prevents the final formation of fibrin polymer—the clot.

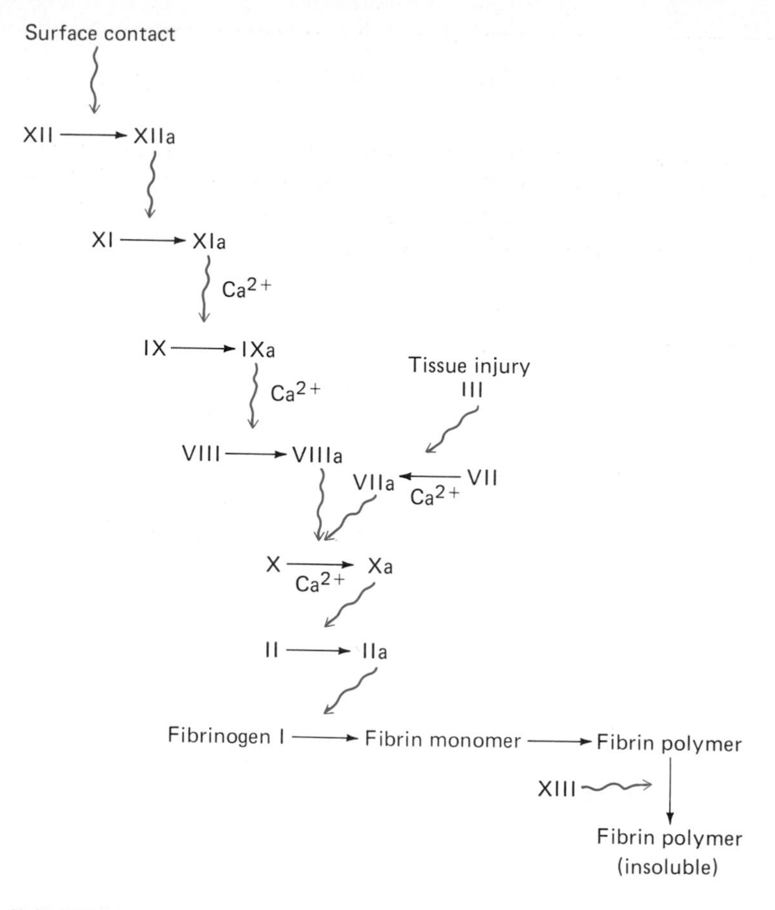

FIGURE 27.1 The blood clotting cascade. Each of the Roman numerals represents a protein factor which is activated (indicated by ⟿) by some other protein factor.

A second important body defense is the immune system. All the various types of white blood cells participate in one way or another in protecting the body against foreign invaders such as bacteria and viruses. Some of the cells actively envelop and consume bacterial cells or other foreign materials, using the hydrolytic enzymes of the white cells' lysosomes. Others are responsible for making antibodies, also known as gamma globulins. These proteins can bind specific foreign molecules and cause them to aggregate into clumps that other white cells can envelop and break down.

Problem 27.2 Where are the gamma globulins made? (Consult Table 27.1)

CONTROL OF PLASMA pH 27.4

The pH of the blood is normally between 7.35 and 7.45. Slight changes in the hydrogen-ion concentration of the blood will have marked effects on body metabolism. For example, **acidosis** (arterial blood pH below 7.35) leads to coma, and **alkalosis** (blood pH above 7.45) leads to convulsions. The body has both long-term and short-term mechanisms for controlling hydrogen-ion concentration. The kidneys and the lungs are involved in the long-term regulation of blood pH. In addition, all body fluids are supplied with **buffer systems,** which can respond within a fraction of a second to rectify alterations in hydrogen-ion concentration (Section 7.10).

The most important buffer system in the blood plasma is the bicarbonate buffer system. Carbon dioxide is continually formed by body cells, largely as a result of citric acid cycle activity (Section 23.9). Carbon dioxide reacts with water to yield carbonic acid:

$$CO_2 + H_2O \rightleftarrows H_2CO_3$$

The reaction will occur spontaneously, but a blood enzyme also catalyzes the reversible reaction.

Carbonic acid undergoes an acid–base reaction with H_2O to yield hydrogen (or hydronium) ion and bicarbonate ion:

$$H_2O + H_2CO_3 \rightleftarrows H_3O^+ + HCO_3^-$$

A mixture of a weak acid (H_2CO_3) and its salt (HCO_3^-) is a buffer.

The HCO_3^-/H_2CO_3 buffer system responds to a rise in H_3O^+ ion concentration by consuming H_3O^+ ions in the reaction

$$H_3O^+ + HCO_3^- \rightarrow H_2CO_3 + H_2O$$

and returning the pH toward normal. Conversely, the buffer system can respond to a decrease in H_3O^+ ion concentration by the dissociation

$$H_2O + H_2CO_3 \rightarrow H_3O^+ + HCO_3^-$$

which again pushes pH back toward normal.

At a pH of 7.4 in the blood, the ratio of $[HCO_3^-]$ to $[H_2CO_3]$ is about 20 to 1. Most of the time the body must consume hydrogen ions being produced by metabolism. To buffer these ions, it is good for a great deal of HCO_3^- to be available at normal pH. In general, though, buffering works best when the buffer's acid and conjugate base components are present in equal concentrations. When we use a buffer—in the laboratory, for example—we want the buffered solution to be able to resist a rise or a fall in pH equally well. In the blood, though, the pH would be too low if carbonic acid and bicarbonate were present in the same concentration.

Problem 27.3 When you hyperventilate, you blow off too much CO_2 from your lungs. What effect would this have on your blood's buffering power?

27.5 TRANSPORT OF NUTRIENTS TO TISSUES

Unlike many simpler organisms, humans eat periodically rather than feeding continuously. When food is being digested and nutrients absorbed (the **absorptive state**), plenty of foodstuff molecules are made immediately available to the various body tissues. On the other hand, when little or no food is present in the digestive tract (the **postabsorptive state**), special mechanisms are required to maintain sufficient nutrition for all body cells. We'll consider several nutrient transport patterns: (1) transport from the digestive tract to the liver and other tissues during the absorptive state; (2) transport from the liver to other tissues; and (3) transport from adipose tissue to the liver and other tissues in the postabsorptive state.

27.6 NUTRIENT TRANSPORT IN THE ABSORPTIVE STATE

What organ receives absorbed nutrients first?

When a recent meal is actively being digested, sugars, amino acids, fatty acids, and fats are being absorbed, mainly from the small intestine, into the blood of the **portal system.** The portal system is composed of veins that lead from the intestines to the liver, and then through the liver into the vena cava (the large vein that returns blood to the heart). The system's function is to pass the blood from the intestines through the liver, which removes a large proportion of the dissolved nutrients, before it reenters general circulation. The liver cells remove approximately two-thirds of the glucose and one-half of the amino acids absorbed into the portal blood from the intestines. These nutrients are retained in the liver (the glucose is used to make glycogen and the amino acids used for synthesis of plasma proteins and other materials) until the body needs them elsewhere. The absorbed materials that remain in the blood are available to other body tissues, which take up glucose and amino acids as dictated by insulin levels and their needs for energy and raw materials.

After a meal, triacylglycerols are delivered mainly to the liver and adipose tissue, but other tissues also receive some. Most triacylglycerols are transported in the blood and lymph. During transport, they are usually bound to other molecules. A particular class of blood lipoprotein (lipid-protein complex; see Table 27.4), called **very-low-density lipoprotein (VLDL),** is responsible for much triacylglycerol transport. In addition, after a fatty meal, triacylglycerols travel in the blood and lymph in the form of **chylomicrons.** These fatty globules are larger than the VLDL particles and contain a high fat-to-protein ratio. Shortly after a meal, chylomicrons disappear from the plasma; the triacylglycerols have been hydrolyzed in the blood vessels and the fatty acids absorbed by body cells. Table 27.4 summarizes the characteristics of the blood lipoproteins. Note that other lipids, such as cholesterol, are also transported in the blood by specific lipoproteins.

TABLE 27.4 Lipoproteins of Human Plasma

Property	Chylomicrons	Very low density (VLDL)	Low density (LDL)	High density (HDL)	Very high density (VHDL)
Density	<0.95	0.95–1.006	1.006–1.063	1.063–1.210	>1.21
Diameter (Å)	300–5,000	300–750	200–250	100–150	100
Amount (mg/100 mL plasma)	100–250	130–200	210–400	50–130	290–400
Approximate composition (%)					
Protein	2	9	21	33	57
Phosphoglyceride	7	18	22	29	21
Cholesterol:					
Free	2	7	8	7	3
Ester	6	15	38	23	14
Triacylglycerol	83	50	10	8	5
Fatty acids	—	1	1		
Lipid characteristic	Mainly triacylglycerol	Mainly triacylglycerol; phosphatidylcholine and sphingomyelin main phospholipid components	High in cholesteryl linoleate	High in phosphatidylcholine and cholesteryl linoleate	

Source: Adapted from White, Handler, et al. *Principles of Biochemistry,* 6th ed. (New York: McGraw-Hill, 1979), p. 573.

Problem 27.4 In hypercholesterolemia, which plasma lipoproteins would you expect to be most affected? (Consult Table 27.4 for information.)

27.7 NUTRIENT TRANSPORT IN THE POSTABSORPTIVE STATE

The liver maintains adequate blood glucose levels when carbohydrate is no longer being absorbed from the digestive tract (Section 24.16). Under the influence of the hormone glucagon, as you may recall, the liver breaks glycogen down to yield free glucose for export to the blood stream. Glucagon is secreted by the pancreas when plasma glucose drops below normal (Section 24.16). Thus, a mechanism is available to keep blood glucose levels high enough to satisfy the needs of brain cells and red blood cells for energy. The liver can also release amino acids to the blood stream when circulating amino acid levels fall.

The triacylglycerol stored in adipose tissue then constitutes the major reservoir of fuel for the oxidative metabolism of most tissues in the postabsorptive state. Glucagon and a variety of other hormones, such as epinephrine (adrenalin) and cortisol (hydrocortisone), stimulate the breakdown of adipose-tissue fats (Section 25.2). Hydrolysis of these triacylglycerols is catalyzed by a special lipase in the adipose tissue. The hydrolysis results in the release of free fatty acids into the blood stream. They are promptly bound by circulating serum albumin molecules and transported in this form to the liver and other tissues. They may then be oxidized for energy or, possibly in small amounts, used for synthesizing phospholipids or other lipids.

Heart and skeletal muscle, among other tissues, prefer to oxidize fatty acids instead of glucose in the postabsorptive state (Section 25.3). As a result, more glucose is available for tissues that must have this fuel.

27.8 TRANSPORT OF OXYGEN

Small, primitive organisms obtain the oxygen they require for metabolism by simple diffusion of the gas from their environment into their cells. Large, complex organisms must transport oxygen to tissues that may be far from the lungs, where oxygen initially enters. Again, the body fluids, particularly the blood stream, are responsible for this transport. Oxygen is transported in two different ways. Dissolved oxygen is carried in the plasma, but since it is not very soluble (only 0.3 mL O_2/100 mL blood), this method of transport would not by itself be adequate to maintain body functions. The second mode of transport involves the chemical combination of oxygen with the hemoglobin of red blood cells (Section 17.12).

HEMOGLOBIN AND THE ERYTHROCYTE 27.9

What special characteristics does the erythrocyte possess?

The human erythrocyte is a peculiar cell. Its cytoplasm contains a large amount of hemoglobin but relatively few other proteins, and it has no nucleus, mitochondria, ribosomes, or other organelles. It depends exclusively on glycolysis for energy production. Thus, the erythrocyte is a highly specialized cell, whose sole function, essentially, is the transport of O_2 and, to a lesser extent, CO_2.

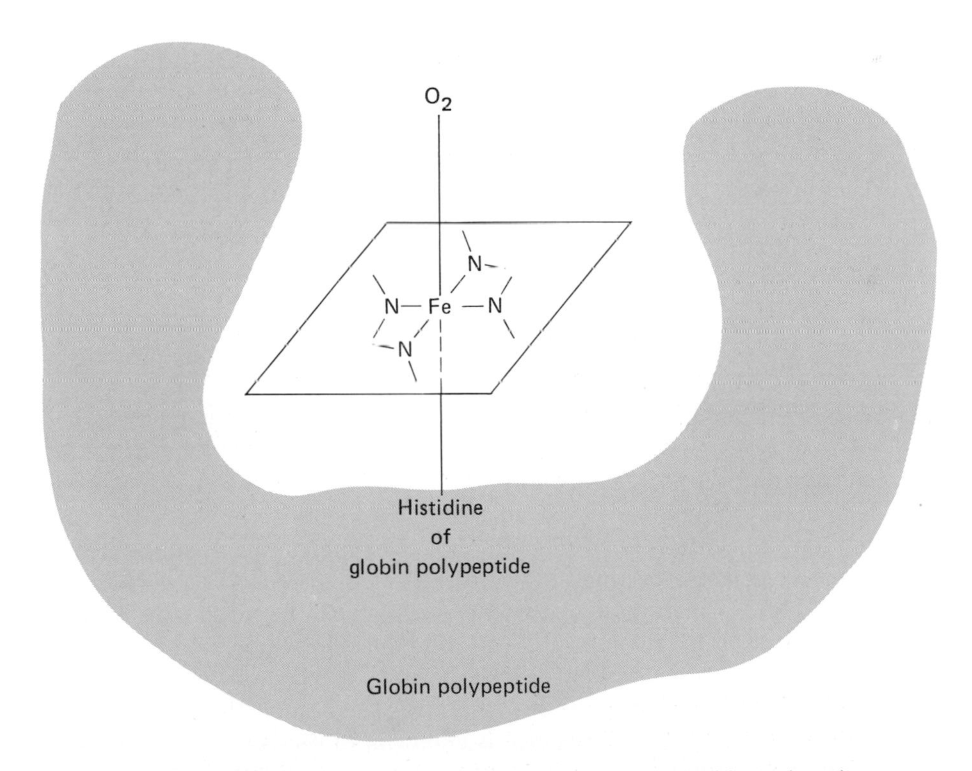

FIGURE 27.2 The plane of the heme ring system within a globin active site. Each of the four nitrogens is part of one of heme's four imidazole rings (5-membered N-containing rings) coordinated to Fe^{2+}. Oxygen and histidine are bonded perpendicular to the plane of the heme ring system.

Hemoglobin (Hb) is composed of four subunits, each consisting of a protein component, globin, in combination with iron protoporphyrin (heme) (Section 21.12). Heme has a highly conjugated bonding system (numerous alternating double and single bonds), which enables it to absorb light in the visible region of the spectrum (Section 10.7). This absorption gives a red color to hemoglobin and thus to blood. The Fe^{2+} of heme forms four bonds with the nitrogens of the porphyrin ring system, and it can also form two additional bonds, for a total of six. In hemoglobin, a specific histidine residue of globin forms the fifth Fe^{2+}—N bond. The sixth bond may be formed between Fe^{2+} and O_2, H_2O, carbon monoxide, or cyanide ion (CN^-).

Problem 27.5 Carbon monoxide (CO) is a deadly poison, mainly because it interferes with O_2 transport by hemoglobin. People who have been exposed to CO and have lost consciousness are sometimes treated with hyperbaric therapy. That is, they are given air with much higher than normal O_2 content. Why should this work?

27.10 BINDING AND RELEASE OF OXYGEN BY HEMOGLOBIN

Why does hemoglobin bind O_2 in the lungs and release it in other tissues?

Each hemoglobin molecule consists of four polypeptides, each of which has its own heme molecule (Section 17.9). Thus every hemoglobin molecule can bind and transport four molecules of oxygen. (See Figure 17.6 for a model of the three-dimensional structure of hemoglobin.)

The relationship between hemoglobin's binding of oxygen and the concentration of oxygen is very important in the body. The binding of oxygen may be visualized as a saturation curve (Figure 27.3). In the curve, the percentage to which hemoglobin is saturated with oxygen is plotted against the pressure of oxygen (P_{O_2}, measured in mmHg) in equilibrium with O_2 dissolved in solution. The curve is S-shaped, or sigmoidal. When the blood circulates through the lungs, where the oxygen concentration is high (P_{O_2} = 100mmHg), the hemoglobin in the red blood cells becomes fully saturated with oxygen. Then, when the blood reaches the muscles and other working tissues, where the oxygen concentration is low (P_{O_2} = 20–40 mmHg), the hemoglobin molecules release much of the oxygen. See Figure 27.3, which shows that when the oxygen concentration measures 20 mmHg, the hemoglobin saturation is only about 20 to 25%. This means hemoglobin binds oxygen much less effectively when the oxygen concentration is low, and therefore oxygen is released from hemoglobin and diffuses out of the red blood cells and into the tissues. In the tissues, it can be used in the oxidation of fuels to generate ATP energy.

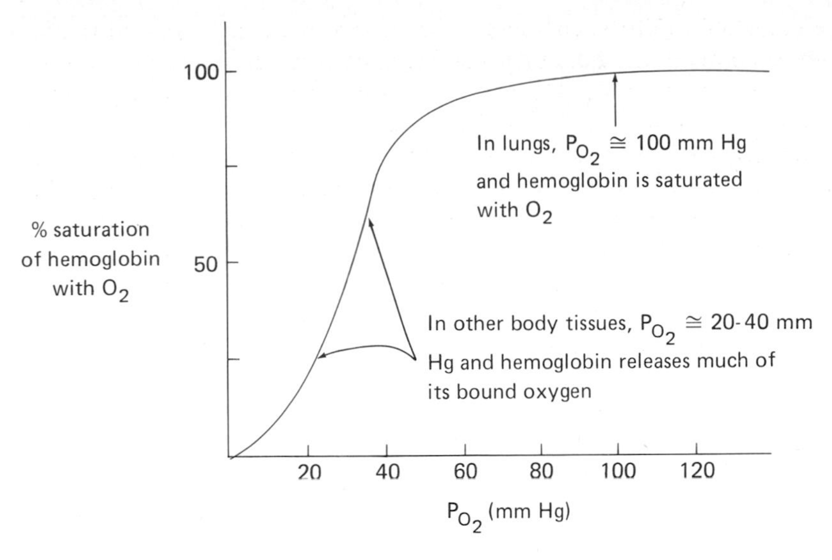
100
50
% saturation
of hemoglobin
with O_2
In lungs, $P_{O_2} \cong 100$ mm Hg
and hemoglobin is saturated
with O_2
In other body tissues, $P_{O_2} \cong$ 20-40 mm
Hg and hemoglobin releases much of
its bound oxygen
20
40
60
80
100
120
P_{O_2} (mm Hg)

FIGURE 27.3 O_2 saturation curve of hemoglobin.

Other factors lead to an even more complete dissociation of O_2 from hemoglobin in the tissues. For example, active oxidative metabolism results in higher CO_2 levels and reduced pH. These two conditions, in turn, promote the release of oxygen from the red blood cells, thus ensuring that cells with the greatest need of oxygen receive a greater supply. Another factor influencing oxygen transport is 2,3-diphosphoglycerate (DPG), a normal byproduct of glycolysis in red blood cells that promotes the dissociation of O_2 from hemoglobin. DPG binds more tightly to deoxyhemoglobin (Hb) than to oxyhemoglobin (HbO_2), which stimulates the release of oxygen from hemoglobin.

BODY FLUID COMPARTMENTS 27.11

Approximately 50% of an adult human's body weight consists of **intracellular fluid,** the fluid located inside cells. **Extracellular fluid,** which represents all the body fluid not located in cells, accounts for about 25%. Extracellular fluid exists in several compartments: the blood plasma represents nearly 5% of the body weight, and the **interstitial fluid,** the fluid that directly bathes most cells, represents 15 to 20%. Specialized fluids—such as cerebrospinal fluid, lymph, aqueous humor, and synovial fluid—are present in smaller amounts.

27.12 IONIC COMPOSITION OF BODY FLUIDS

What are the main differences in ionic composition of intracellular and extracellular fluids?

Figure 27.4 contrasts the ionic compositions of extracellular fluid (blood plasma and interstitial fluid) and intracellular fluid. Note that the major extracellular metal cation is Na^+ ion and that the major intracellular cation is K^+ ion. In addition, there is much more Mg^{2+} ion inside cells than outside of them. Also, whereas Ca^{2+} ion is present in significant amounts in extracellular fluid, its intracellular concentration is so low that it is off the scale in Figure 27.4.

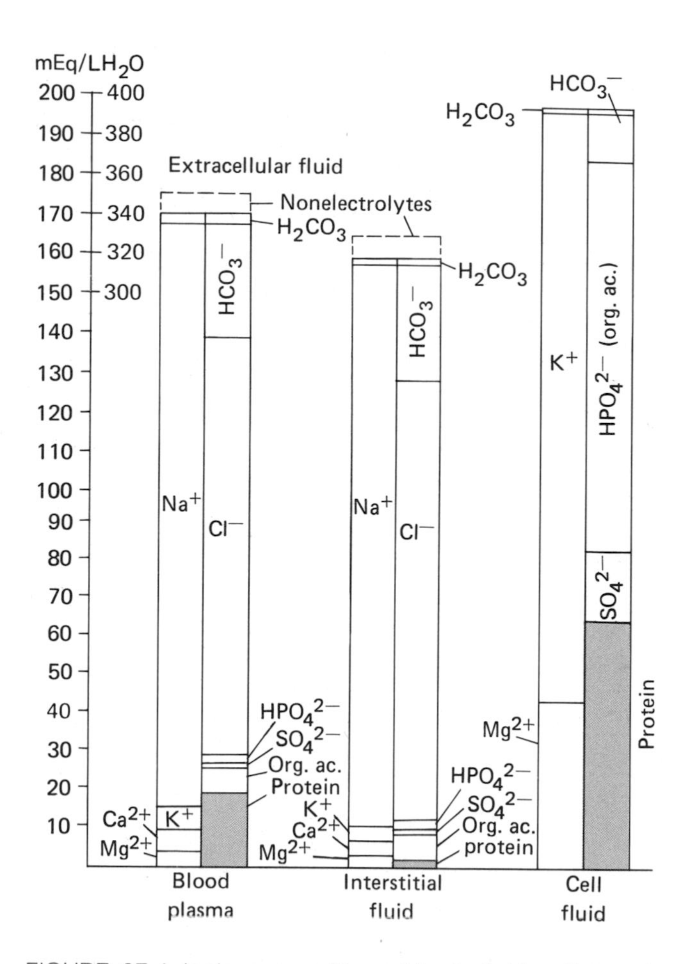

FIGURE 27.4 Ionic composition of body fluids. (Adapted from Gamble, J. L., *Chemical Anatomy, Physiology and Pathology of Extracellular Fluids.* Harvard University Press, Cambridge, Mass., 1954.)

The major anion (negative ion) in plasma and interstitial fluids is chloride; bicarbonate also makes a significant contribution. Phosphate and sulfate are present in smaller amounts. Inside the cells, chloride is almost completely replaced by phosphate and organic anions such as pyruvate, citrate, and α-ketoglutarate. The differences between intracellular fluid and extracellular fluid are maintained through the active transport of various ions across the cell membrane, the selective permeability of cell membranes, and the cellular metabolism itself.

As we have already seen, bicarbonate is of major importance in the pH buffering of the plasma. Phosphate plays a similar role in cells, being the chief intracellular buffer. The relevant acid–base reaction at physiological pH is

$$H_2O + H_2PO_4^- \rightleftarrows H_3O^+ + HPO_4^{2-}$$

The pK_a (Section 7.9) for $H_2PO_4^-$ is 7.2. These $H_2PO_4^-$ and HPO_4^{2-} ions are the main phosphate species present in cells (at a pH around 7.4). They are present in roughly equivalent concentrations, providing very effective buffering.

Problem 27.6 Write the reaction the phosphate buffer system will undergo if the pH rises inside the cell.

Problem 27.7 Write the reaction the phosphate buffer system will undergo if the pH falls inside the cell.

PROTEIN CONCENTRATION IN BODY FLUIDS 27.13

The protein concentrations of the three fluids are also strikingly different. The protein constituents have been placed in the anion column of the bar graphs of Figure 27.4 because most proteins have a net negative charge at physiological pH. The cell fluid contains the largest proportion of protein anions, as we might expect. The proteins of plasma are mainly serum albumins, transport lipoproteins, and gamma globulins, as previously discussed. The interstitial fluid has much less protein than the plasma, because the protein molecules are too large to diffuse, as the smaller ions do, from the blood across the cells lining the capillaries. The capillaries are the smallest blood vessels, and it is through them that materials are exchanged between the blood and tissues.

27.14 ROLE OF THE ENDOCRINE SYSTEM IN METABOLIC INTEGRATION

What are target tissues?

The **endocrine glands** are glands that release their secretions (hormones) directly into the blood, which then transports these hormones throughout the body. Table 27.5 lists the principal endocrine glands and their hormones. Note that in many cases only certain **target tissues** respond to a hormone. For example, glucagon's target tissue is the liver (Section 24.17).

The endocrines are responsible for the overall control of metabolism, including the assimilation of foodstuffs, the storage of fat and glycogen, and energy metabolism. Many aspects of metabolism are under the control and integration of more than one endocrine gland.

Hormones regulate the *rate* at which cell functions and metabolism occur. They act indirectly, usually by influencing the activities of enzymes or transport proteins, or the rate of synthesis of enzymes.

Recall how glucagon acts on liver cells (or epinephrine on muscle cells). It binds to specific cell membrane receptors, which in turn causes activation of an enzyme that catalyzes cyclic AMP formation, which initiates the activation cascade leading to increased activity of glycogen phosphorylase (Section 24.17). Mechanisms of this sort, which involve cyclic AMP, seem to be typical of polypeptide hormones (and certain other hormones).

Steroid hormones—such as estrogens, cortisol, and testosterone (Table 20.4)—act by means of a different mechanism. These lipid molecules can readily cross cell membranes and are bound by specific receptor proteins after they arrive in the cytoplasm (Figure 27.5). The steroid-receptor complex then enters the nucleus, where it increases the rate of transcription (Section 26.6) of specific *m*RNAs that code for specific proteins. As a result, these proteins are synthesized at a greater rate, and their concentrations in the cell increase. This leads to increased activity of whatever metabolic pathways these proteins catalyze. Thus, particular steroids stimulate particular metabolic pathways. For example, cortisol enhances the synthesis of carbohydrate from protein (amino acids) (Table 27.5). It does this by increasing the concentration of various transaminases (involved in amino acid catabolism) (Section 25.18) and certain enzymes of gluconeogenesis (Section 24.11) in liver cells. These enzymes increase the rate of protein utilization for the purpose of carbohydrate synthesis.

Thus, hormones provide a sort of super control of body metabolism that is superimposed on the many controls that exist within individual cells. Because of hormones, the body can organize and control its overall responses to different physiological events and nutritional conditions.

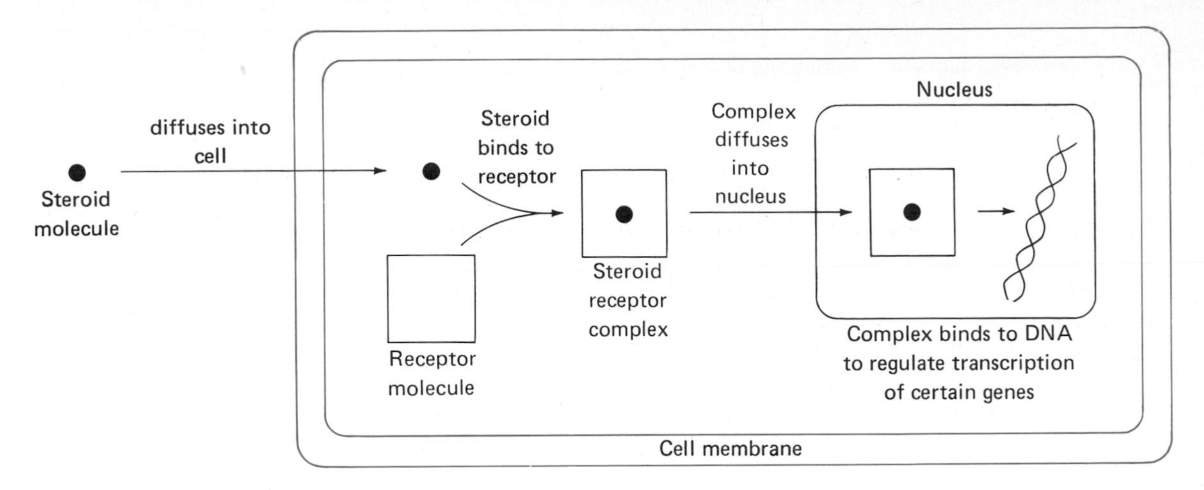

FIGURE 27.5 Target cell of steroid hormone and how the hormone acts.

TABLE 27.5 Secretions and Functions of Major Endocrine Glands

Endocrine gland and hormones	Influence on metabolic functions or water and salt balance (assimilation, storage, energy, metabolism; retention, excretion)	Target tissue
Anterior pituitary		
Adrenocorticotrophin (ACTH)	Stimulates secretion of steroid hormones	Endocrine system, adrenal cortex
Thyrotrophin (TSH)	Stimulates thyroid hormone production	Endocrine system, thyroid
Gonadotrophins		
Follicle stimulating (FSH)	Necessary for ovulation and sperm formation	Endocrine system, ovaries, testes
Luteinizing (LH or ICSH)	Stimulates ovulation; stimulates testosterone production	Endocrine system, ovaries, testes
Prolactin	Stimulation of milk production	Endocrine system, ovaries, mammary glands
Somatotrophin (GH or STH)	Growth; carbohydrate, protein, lipid metabolism; glycogen storage; sustained lactation	All body cells
Posterior pituitary		
Antidiuretic hormone (ADH), or vasopressin	Water retention	Urinary system, kidney
Oxytocin	Ejection of milk from mammaries	Mammary glands, uterus
Adrenal cortex		
Cortisone, hydrocortisone	Conversion of protein to carbohydrate	Liver
Aldosterone	Sodium retention, potassium excretion	Kidney, all cells

TABLE 27.5 Secretions and Functions of Major Endocrine Glands (Continued)

Endocrine gland and hormones	Influence on metabolic functions or water and salt balance (assimilation, storage, energy, metabolism; retention, excretion)	Target tissue
Adrenal medulla		
Epinephrine	Carbohydrate catabolism; glycogenolysis; mobilization of fat	Cardiovascular system, muscle
Norepinephrine	Carbohydrate catabolism; glycogenolysis	Sympathetic nervous system, cardiovascular system, liver
Thyroid		
Thyroxine and triiodothyronine	Increase of metabolic rate; O_2 consumption; growth; support of lactation	Heart, all cells
Calcitonin	Mineral metabolism; reduction of Ca^{2+} resorption; reduction of blood Ca^{2+}	Skeletal system
Parathyroid		
Parathormone	Mineral metabolism, Ca^{2+}, P_i; increase of blood Ca^{2+}; increase of P_i excretion	Skeletal system, muscular system
Pancreatic islets		
Insulin	Increase of carbohydrate utilization; lowering of blood sugar	Digestive system, liver, muscle
Glucagon	Promotes glycogenolysis of liver glycogen; elevates blood sugar	Digestive system, liver
Testes		
Testosterone	Promotion of protein anabolism and nitrogen, salt, water retention	Reproductive system, testes
Ovaries		
Estrogens	Protein anabolism; nitrogen, salt, water retention	Reproductive system, ovaries
Progestins	Preparation for and maintenance of pregnancy	Reproductive system, uterus, mammary glands
Relaxin		Reproductive system, cervix, symphysis pubis
Placenta		
Estrogens	Same as for ovaries	Reproductive system, ovaries
Progesterone		Reproductive system, uterus, mammary glands
Relaxin		Reproductive system, cervix, symphysis pubis

Source: King and Showers; *Human Anatomy and Physiology,* 6th ed. (Philadelphia: W. B. Saunders, 1969), p. 136.

SUMMARY

If clotting is prevented, blood can be separated into blood cells and blood **plasma.** If clotting does occur, the fluid that separates is the blood **serum.** There are three kinds of blood cells: the **erythrocytes,** several types of **leukocytes,** and the **platelets.** Plasma contains water, salts, and organic constituents, including plasma proteins. Among those proteins are serum albumin, **gamma globulins,** and **beta globulins.** The pH of plasma is controlled mainly by the bicarbonate **buffer system.** A major function of the blood is to transport nutrients in the **absorptive** and **postabsorptive states.** Another major function is oxygen transport, which is accomplished mainly by the red blood cell, the erythrocyte. The hemoglobin in the erythrocyte binds oxygen in a characteristic fashion that allows it to be delivered to the tissues.

The various body-fluid compartments—**intracellular fluid,** blood plasma, and **interstitial fluid**—differ in ionic composition because of active transport, selective membrane permeability, and metabolism. Phosphate is the major intracellular buffer.

The **endocrine glands** release hormones into the blood, which transports them to their **target tissues.** The endocrines are responsible for the overall control of metabolism, growth, and reproduction.

Key Terms

Absorptive state (Section 27.5)
Acidosis (Section 27.4)
Alkalosis (Section 27.4)
Beta globulins (Section 27.2)
Buffer system (Section 27.4)
Chylomicrons (Section 27.6)
Endocrine gland (Section 27.14)
Erythrocyte (Section 27.1)
Extracellular fluid (Section 27.11)
Fibrin (Section 27.1)
Fibrinogen (Section 27.1)
Gamma globulins (Section 27.2)
Hematocrit (Section 27.1)
Hemophilia (Section 27.3)
Interstitial fluid (Section 27.11)
Intracellular fluid (Section 27.11)
Leukocyte (Section 27.1)

Plasma (Section 27.1)
Platelet (Section 27.1)
Portal system (Section 27.6)
Postabsorptive state (Section 27.5)
Serum (Section 27.1)
Target tissue (Section 27.14)
Very-low-density lipoproteins (VLDL) (Section 27.6)

ADDITIONAL PROBLEMS

27.8 What tissue(s) are responsible for the formation of red blood cells?
27.9 Are all types of white blood cells formed in the same tissue(s)?
27.10 What feature do most plasma proteins have in common?
27.11 Compare the buffer systems of intracellular and extracellular fluids.
27.12 Compare the erythrocyte with the liver cell in terms of its functions and structural features.
27.13 How many heme molecules does a molecule of hemoglobin contain? How many oxygen molecules can a molecule of hemoglobin bind?
27.14 Draw a hypothetical curve showing the effect of reduced pH or of an increased amount of diphosphoglycerate on the oxygen saturation of hemoglobin.
27.15 Name as many body-fluid compartments as you can.
27.16 Compare the Na^+ and K^+ concentrations (in mEq/L H_2O) of plasma and intracellular fluid.
27.17 Compare the protein contents of interstitial fluid and intracellular fluid.
27.18 Which hormones in Table 27.5 do you recognize as steroids? Which endocrine glands manufacture and secrete steroids?
27.19 Which hormone in Table 27.5 seems to have the most wide-ranging effect on body metabolism?

Problem-Solving Hints

If you have trouble solving any of the additional problems, refer to the sections listed next to the problem numbers.

27.8 27.1
27.9 27.1
27.10 27.2
27.11 27.4, 27.12
27.12 27.9
27.13 27.9, 27.10
27.14 27.10
27.15 27.11
27.16 27.12
27.17 27.13
27.18 27.14
27.19 27.14

SELF-TEST

Chapters 1–7 (General Chemistry)

1. Write the chemical symbol for each of the following elements.
(a) Potassium (b) Sodium
(c) Sulfur (d) Chlorine
(e) Titanium

2. Write the name of each of the following elements.
(a) Si (b) Ca
(c) Fe (d) Mn
(e) Mg

3. The volume of an object can be determined be measuring the volume of water displaced when the object is totally submerged. A student determined the volume of a small irregular object by observing the displacement of water in a graduated cylinder.

$$\text{volume of water in cylinder with object} = 1.68 \text{ mL}$$
$$\text{volume of water in cylinder without object} = 1.00 \text{ mL}$$

The student next determined the weight of the object by weighing it in a small beaker.

$$\text{weight of beaker with object} = 23.463 \text{ g}$$
$$\text{weight of beaker alone} = 22.102 \text{ g}$$

Use these data to calculate the density of the object.

4. A completely discharged storage battery is recharged, and the change in free energy for the complete recharge is $\Delta G = 5.0 \times 10^2$ kcal.
(a) What is ΔG for the chemical reaction in the battery when the battery completely discharges?
(b) What is the maximum amount of electrical energy that could be obtained from this battery when it goes from a completely charged to discharged state?

5. Balance the following equations.
(a) $ClNO + O_3 \rightarrow ClNO_2$
(b) $SO_2 + O_2 \rightarrow SO_3$
(c) $PF_5 + H_2O \rightarrow H_3PO_4 + HF$

6. Write the electron configurations for the following atoms using the $1s^22s^2$ . . . notation.
(a) Ne (atomic number = 10)
(b) Si (atomic number = 14)

7. Fill in the proper spaces with arrows to indicate the filling of orbitals and the electron spins for the following atoms.

3p ___ ___ ___

3s ___

2p ___ ___ ___

2s ___

(a) C 1s ___
(Atomic number = 6)

3p ___ ___ ___

3s ___

2p ___ ___ ___

2s ___

(b) S 1s ___
(Atomic number = 16)

8. Write Lewis dot structures for the following atoms.

(a) Mg (group II)
(b) C (group IV)
(c) F (group VII)
(d) Ar (group VIII)
(e) Ba (group II)
(f) Se (group VI)
(g) Sn (group IV)
(h) I (group VII)
(i) Xe (group VIII)
(j) Cs (group I)

9. Complete and balance the following nuclear decay reactions.

(a) ${}^{14}_{6}C \rightarrow {}^{14}_{7}N +$
(b) ${}^{225}_{89}Ac \rightarrow {}^{221}_{87}Fr +$

10. Classify the following as pure substance, solution, or heterogeneous mixture.

(a) air
(b) water
(c) a diamond ring
(d) a slice of orange
(e) solid carbon dioxide (dry ice)

11. For each of the following pairs of atoms, indicate whether the compound they form is ionic or covalent, give the formula of the compound, and draw Lewis dot structures for the covalent compounds. (Refer to a periodic table in answering this question.)

(a) Ba and F
(b) P and Br
(c) C and H
(d) Na and S

12. Write balanced chemical equations that show the formation of each of the compounds in Problem 11 from the elements.

13. Name the following compounds.

(a) NaBr
(b) MgF_2
(c) H_2S
(d) K_3PO_4
(e) CCl_4

14. Write formulas for the following compounds.

(a) Calcium carbonate
(b) Sodium sulfate
(c) Chlorine dioxide
(d) Nitrogen tribromide
(e) Sulfur hexafluoride

15. Calculate the number of grams of H_2 and Cl_2 needed to react to form 3.65 g of HCl (Atomic weights: H = 1.0, Cl = 35.5).

16. A sample of helium gas is in a container of fixed volume. Indicate how each property of the gas given in the following table would be affected by the changes listed. (Use a plus sign for increase, a minus sign for decrease, and a zero for no change.) The effects on some of the properties are already entered in the table.

Change	Temperature of gas	Pressure of gas	Average kinetic energy of He atoms	Number of collisions per second between He atoms and walls of the container	Average speed of He atoms
More gas is added	0				
The gas is cooled					
Some gas is removed and the remaining gas is heated				+	
More gas is added and the gas is cooled		+			

17. Start with the ideal gas law and explain under what conditions it becomes the combined gas law. Also show under what conditions the ideal gas law becomes Boyle's and Charles's laws.

18. Liquid benzene and gaseous benzene are in equilibrium at a pressure of 1 atm. For each of the following, tell whether the liquid state value is less than, greater than, or the same as the gaseous state value.

(a) temperature
(b) average kinetic energy of molecules
(c) energy per gram of benzene

19. Classify each of the following as amorphous, ionic crystal, molecular crystal, or network covalent crystal.

(a) NaF
(b) diamond
(c) soot
(d) solid methane (CH_4)

20. Why are the melting and boiling points of water exceptionally high for a substance with a relatively low molecular weight?

21. A student makes a solution by mixing 18.0 g of glucose with enough water to make 400 mL of solution. The molecular weight of glucose is 180.

(a) Calculate the weight-volume percent of this solution.
(b) Calculate the molar concentration of this solution.

22. A 0.15 *M* NaCl solution is isotonic with the solution in red blood cells.
(a) What will happen if red blood cells are placed in 0.15 *M* NaCl?
(b) What will happen if red blood cells are placed in 1 *M* NaCl?
(c) What will happen if red blood cells are placed in pure water?
23. Assume that the following chemical equilibrium has been reached in a container at a fixed temperature.

$$2SO_2 + O_2 \rightleftharpoons 2SO_3$$

(a) Write the expression for the equilibrium constant K.
(b) Indicate how the changes given in the following table would affect the various listed quantities after equilibrium is re-established in each case at the same temperature. (Use a plus sign for increase, a minus sign for decrease, a zero for no change, L for left, and R for right.)

Change	$[SO_2]$	$[O_2]$	$[SO_3]$	Position of equilibrium	K	ΔG	$\Delta G^{\ddagger}$
Some SO_2 is added							
Some SO_3 is removed							
Some O_2 is added							
Some SO_2 and some O_2 are removed							
Some V_2O_5, a surface catalyst, is added							

24. In each of the following chemical equations, circle all acids and underline all bases.
(a) $HNO_3 + H_2O \rightarrow H_3O^+ + NO_3^-$
(b) $NH_4^+ + HPO_4^{2-} \rightleftharpoons NH_3 + H_2PO_4^-$
(c) $N_2 + O_2 \rightleftharpoons 2NO$
25. Write chemical equations for the acid-base reaction in each of the following solutions.
(a) 0.1 *M* HCl reacts with 0.1 *M* NaOH
(b) Vinegar (remember vinegar is an aqueous 5% acetic acid solution)
(c) Soda water (dissolved CO_2 forms a solution of carbonic acid in water)

26. When 15.0 mL of an unknown HNO_3 solution is titrated with 0.150 *M* NaOH, the pH of the solution reaches 7 after 20.0 mL of the standard solution has been added. Calculate the concentration of the HNO_3 solution.

27. Fill in the missing values (assume 25°C).

Solution	$[H_3O^+]$	$[OH^-]$	pH
0.10 *M* HCl			
0.10 *M* NaOH			
0.10 *M* NaCl			
An $NH_4^+—NH_3$ buffer			10.00
0.10 *M* solution of an unknown weak acid		1.0×10^{-11}	

Chapters 8–16 (Organic Chemistry)

28. Supply the appropriate information in each of the spaces below.

Compound name	Lewis dot formula	Three-dimensional formula	Name of functional group
ethane		$CH_2{=}CH$–C(CH_3)(H)(CH_3)	
3-methyl-1-butene			
acetone		H–C(H)(H)–C(=O)–C(H)(H)–H	
propyne			

29. An unknown compound has the molecular formula C_6H_{12}. It does not react with Br_2 in the dark, but it does react slowly when placed in the direct sunlight. Give a possible structural formula for the original hydrocarbon.

30. Circle and name each of the functional groups in the compound below. Do not include alkyl groups.

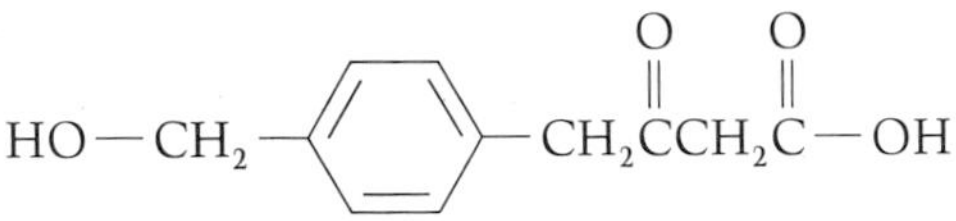

$$HO-CH_2-C_6H_4-CH_2\overset{O}{\overset{\|}{C}}CH_2\overset{O}{\overset{\|}{C}}-OH$$

31. Give the appropriate structural formula of the missing reactants or products. The equations are not balanced. (Write N.R. if no reaction occurs.)

(a) $CH_3CH_2Br + Na \rightarrow ?$

(b) $CH_3CH{=}CHCH_3 + ? \xrightarrow{Ni} CH_3CH_2CH_2CH_3$

(c) $CH_3CH_3(\text{excess}) + Br_2 \xrightarrow{\text{light}} ?$

(d) 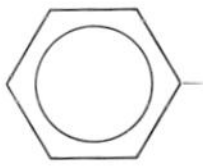 $C_6H_5-Cl + KOH \xrightarrow{\text{alcohol}} ?$

(e) $CH_3CH{=}CHCH_3 + H_2O \xrightarrow{H_2SO_4} ?$

(f) $CH_3CH{=}CHCH_3 + HBr \rightarrow ?$

(g) 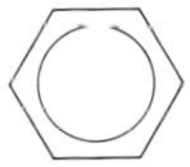 $C_6H_6 + SO_3 \xrightarrow{H_2SO_4} ?$ (h) 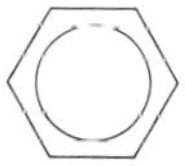 $C_6H_6 + Br_2 \xrightarrow{Fe} ?$

32. Give an acceptable name (common or IUPAC) for each of the following compounds.

(a) $CH_2{=}\overset{CH_3}{\overset{|}{C}}CH_2CH_2CH{=}CH_2$

(b) 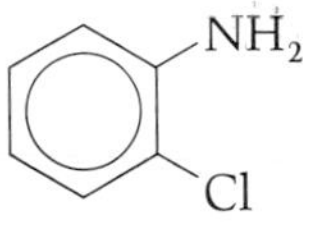 benzene ring with NH_2 and, on the adjacent carbon, Cl

(c) 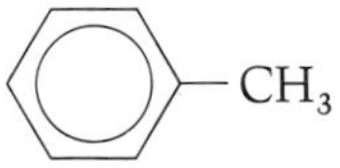 $C_6H_5-CH_3$

(d) 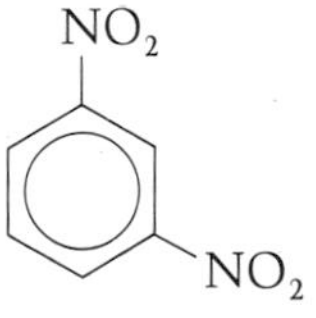 benzene ring with NO_2 and, meta to it, NO_2

(e) $CH_3CH_2C{\equiv}CCH_3$

33. Give the formula of the missing reactant or major organic product in each of the following reactions. If no appreciable amount of product is formed, write N.R. (no reaction).

(a) CH_4 (excess) + $Cl_2 \xrightarrow{\text{light}}$?

(b) $CH_4 + Cl_2$ (excess) $\xrightarrow{\text{light}}$?

(c) $CH_3CH_2\overset{\overset{\displaystyle OH}{|}}{C}HCH_3 + ? \rightarrow CH_3CH_2\overset{\overset{\displaystyle O}{\|}}{C}CH_3$

(d) $CH_3OH + C_6H_5-\overset{\overset{\displaystyle O}{\|}}{C}-OH \xrightarrow{H_2SO_4}$?

(e) $CH_3OH + Na \rightarrow$?

(f) $C_6H_5-CH_2OH + HBr \rightarrow$?

(g) $C_6H_5-OH + NaOH \rightarrow$?

(h) $CH_3CH_2-O-CH_2CH_3 + HI$ (excess) $\rightarrow$?

(i) $CH_3CH_2CH_2OH + NaOH \rightarrow$?

(j) $CH_3CH_2CH_2SH + NaOH \rightarrow$?

(k) $CH_3NH_2 + HCl \rightarrow$?

(l) $CH_3NH_2 + CH_3Cl \rightarrow$?

(m) $H_2NCH_2CO_2H \xrightarrow[\text{in } H_2O]{\text{dissolve}}$

34. Which compound in each of the following pairs has (1) the higher boiling point and (2) the higher solubility in water?

(a) CH_3F and CH_3OH

(b) CH_3OH and CH_3NH_2

(c) $CH_3CH_2CH_2NH_2$ and $(CH_3)_3N$

(d) $CH_3CH_2-O-CH_3$ and $CH_3CH_2CH_2OH$

35. Which compound in each of the following pairs is more acidic?

(a) CH_3CH_2OH and CH_3CH_2SH

(b) $C_6H_4(CH_3)-OH$ (ring bearing CH_3 ortho to OH) and $C_6H_5-CH_2OH$

(c) CH_3CH_2OH and NH_4Cl

36. Give the structural formula of each of the following.

(a) a tertiary alcohol whose molecular formula is $C_5H_{12}O$

(b) a secondary alcohol whose molecular formula is $C_5H_{12}O$

(c) a primary amine whose molecular formula is $C_5H_{13}N$

37. Give an acceptable name for each of the following compounds.

(a) $CH_3CH_2-O-C_6H_5$

(b) $CH_3CH_2\overset{\overset{\displaystyle CH_3}{|}}{C}H\overset{\overset{\displaystyle OH}{|}}{\underset{\underset{\displaystyle CH_3}{|}}{C}}HCHCH_2-C_6H_5$

(c) 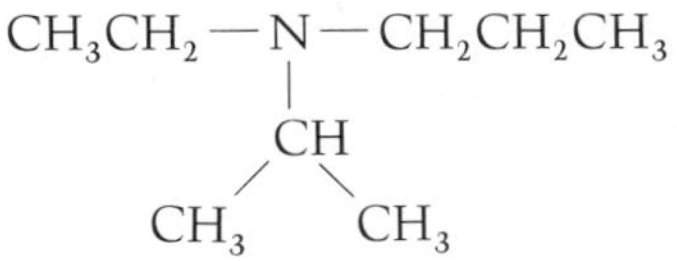 $CH_3CH_2-N(CH(CH_3)_2)-CH_2CH_2CH_3$

(d) 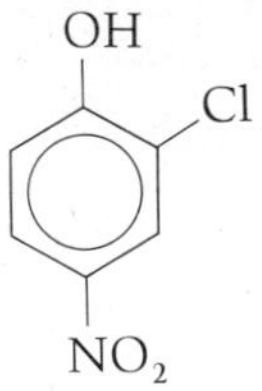 2-chloro-4-nitrophenol ring structure: OH, Cl, NO_2 on benzene ring

(e) 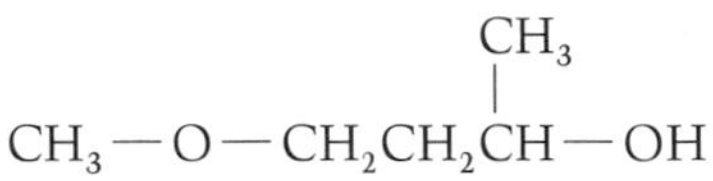 $CH_3-O-CH_2CH_2CH(CH_3)-OH$

(f) 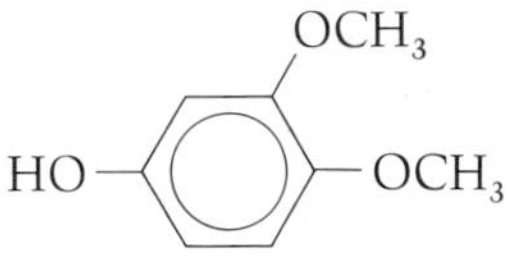 HO–(benzene ring)–OCH_3, with a second OCH_3 on the ring

(g) $CH_3CH_2CH_2SH$

(h) 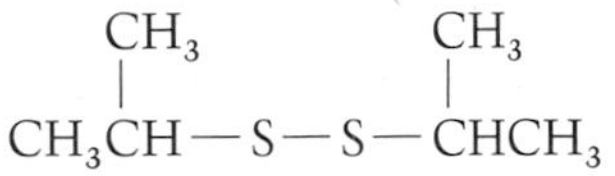 $CH_3CH(CH_3)-S-S-CH(CH_3)CH_3$

(i) 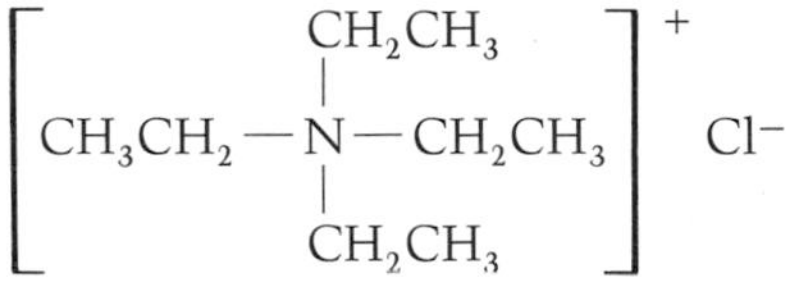 $\left[CH_3CH_2-N(CH_2CH_3)_2-CH_2CH_3\right]^+ Cl^-$

38. Give the structural formula of each of the following compounds.

(a) propanal (b) 2-pentanone
(c) benzamide (d) ethanoic anhydride
(e) formaldehyde (f) methyl methanoate

39. Give the structural formula of the major organic product of each of the above reactions. If the reaction does not occur to any appreciable extent, write N.R.

(a) 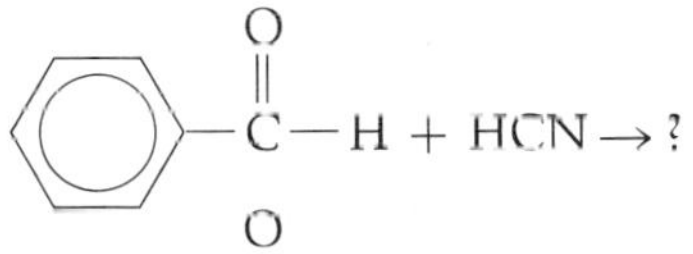 $C_6H_5-\overset{O}{\overset{\|}{C}}-H + HCN \rightarrow ?$

(b) $C_6H_5-\overset{O}{\overset{\|}{C}}-H + Ag(NH_3)_2{}^+ \rightarrow ?$

(c) $C_6H_5-\overset{O}{\overset{\|}{C}}-H + CH_3OH \xrightleftharpoons{H^+} ?$

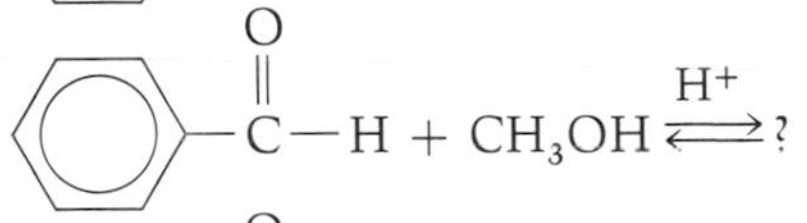

(d) 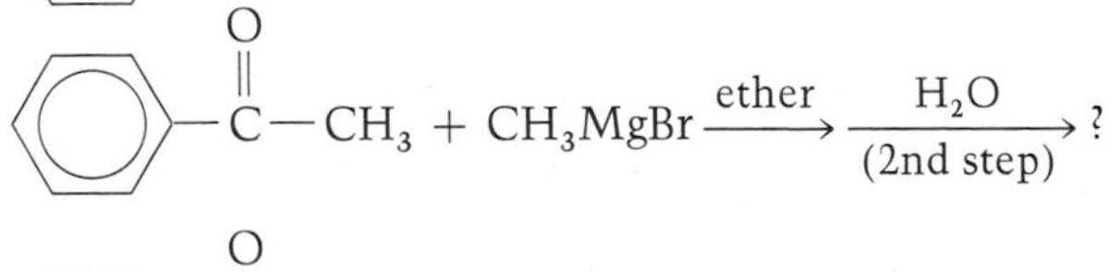 $C_6H_5-\overset{O}{\overset{\|}{C}}-CH_3 + CH_3MgBr \xrightarrow{\text{ether}} \xrightarrow[\text{(2nd step)}]{H_2O} ?$

(e) 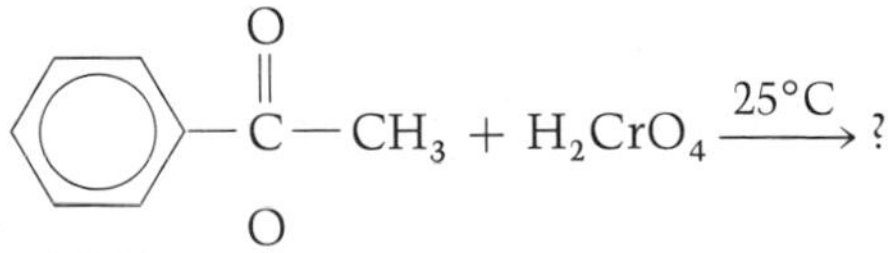 $C_6H_5-\overset{O}{\overset{\|}{C}}-CH_3 + H_2CrO_4 \xrightarrow{25°C} ?$

(f) 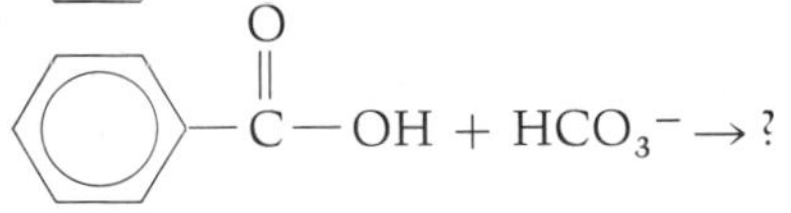 $C_6H_5-\overset{O}{\overset{\|}{C}}-OH + HCO_3{}^- \rightarrow ?$

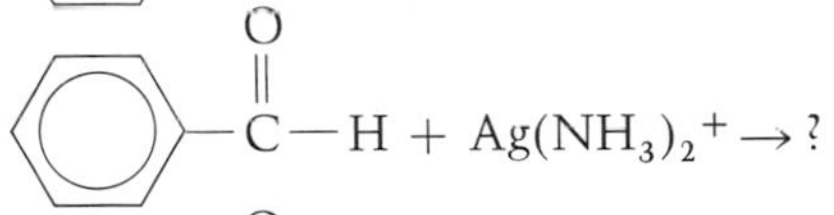

(g) $C_6H_5-\overset{O}{\overset{\|}{C}}-OCH_3 + NH_3 \rightarrow ?$

(h) $C_6H_5-\overset{O}{\overset{\|}{C}}-OH + CH_3NH_2 \rightarrow ?$

(i) $C_6H_5-\overset{O}{\overset{\|}{C}}-OCH_3 + OH^- \rightarrow ?$

(j) $\left(\overset{O}{\overset{\|}{C}}-C_6H_4-\overset{O}{\overset{\|}{C}}-O-CH_2-C_6H_{10}-CH_2-O\right)_n + \underset{\text{(excess)}}{H_2O} \xrightarrow{H^+} ?$ (Kodel)

(k) $\left(\overset{H}{\overset{|}{N}}-CH_2-\overset{O}{\overset{\|}{C}}\right)_n + \underset{\text{(excess)}}{H_2O} \xrightarrow{H^+} ?$ (silk)

40.
(a) A certain carbohydrate $C_6H_{12}O_6$, is crystalline, very soluble in water, and very sweet to the taste. It does not react when heated in aqueous HCl solution. It has four chiral carbon atoms. Give a one word description of this sugar.
(b) The above carbohydrate is one of how many stereoisomers?
(c) Give the Fischer projection formula of one of the D-stereoisomers.
(d) Give the Haworth formula of the six-membered cyclic hemiacetal form of the above compound.
(e) What is this cyclic form called?
(f) Use asterisks (*) to designate the chiral carbon atoms in the cyclic formula in (d).
(g) Is your cyclic form an α- or β- anomer?
(h) Draw the Fisher projection formula of the enantiomier of (c)?

41. Explain why carboxylic acids are more acidic than alcohols.

42. Many of the reactions of alkenes involve addition of an acidic reagent (HCl, HBr, HI, H_3O^+, and H_2SO_4). Explain why alkenes are reactive toward acids.

43. Explain why (in general) primary and secondary amines have lower boiling points and lower water solubilities than alcohols of comparable molecular weights.

44. Explain why thiols are stronger acids than alcohols.

45. Give the structural formula and describe the common uses of each of the following chemicals.
(a) chloroform
(b) freon-12
(c) acetone
(d) formaldehyde

(e) ethanol (f) diethyl ether
(g) ethyl acetate

46. Which of the following molecules are chiral? Draw the line-dash-wedge formula of both enantiomers of each.

(a) CH_2ClF

(b) 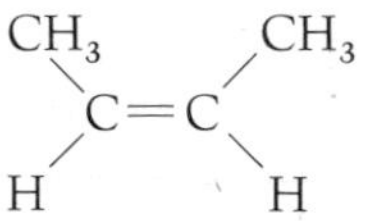 $CH_3(H)C{=}C(CH_3)H$

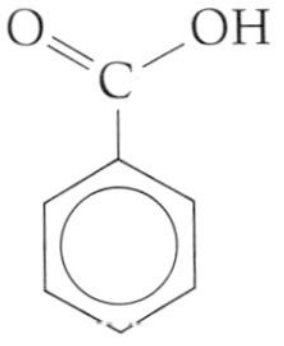

(c) $C_6H_5{-}C(=O)OH$

(d) $CH_3CH(Br)CH_2CH_3$

(e) $CH_2{=}CHCH(CH_3)CH_2CH_3$

47. A pair of enantiomers would differ in which of the following properties: (1) boiling point, (2) solubility in water, (3) density, (4) rotation of plane polarized light, and (5) reactivity toward strong chemical agents such as H_2SO_4, Br_2, and NaOH?

48. One resonance structure of acetamide is $CH_3{-}C(=\ddot{O}{:}){-}\ddot{N}(H){-}H$. Draw another resonance structure of this compound.

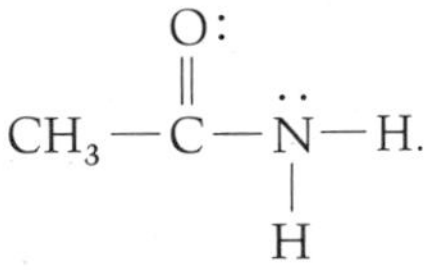

49. Give the appropriate structural formula of each of the following.

(a) the enantiomer of C(H)(F)(Br)(CH_3) [H up, F left, Br dashed, CH_3 wedged]

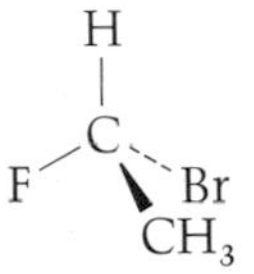

(b) a primary alcohol containing three carbon atoms
(c) a tertiary amine derived from aniline
(d) an L-ketopentose
(e) an acetal containing four carbon atoms

50. A certain compound $C_4H_8O_2$ does not neutralize NaOH but reacts with Na metal to produce H_2 gas. It also reacts with Br_2 to form a colorless product ($C_4H_8Br_2O_2$). Give a reasonable structural formula of the original compound and write equations for the reactions described.

51. A compound $C_4H_8O_2$ reacts with an aqueous solution of $NaHCO_3$ to produce bubbles. It does not react with Br_2. Give a reasonable structural formula for the original compound and write the equation of its reaction with $NaHCO_3$.

52. Which of the following compounds might be suspected of being explosive?

(a) $CH_3CH_2CH_2CH_2CH_2CH_2NO_2$ (b) CH_3NO_2

(c) CH_3NH_2 (d) OH, HO, OH

(e)

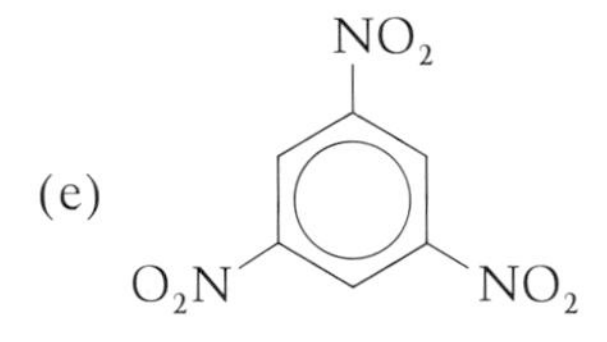

Chapters 17–27 (Biochemistry)

53. The following structure is the nitrogenous base adenine. Add to the drawing whatever is necessary to convert the structure to that of AMP.

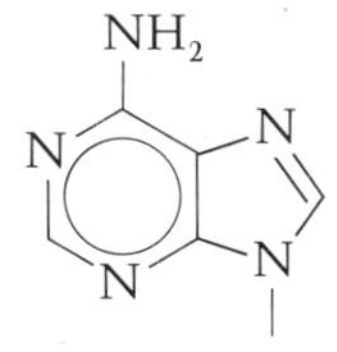

54. Which bond or bonds (covalent, ionic, hydrogen, or hydrophobic) is (are) involved in:
(a) protein 1° structure (b) protein 2° structure
(c) protein 3° structure

55.
(a) Draw the structure of a fatty acid.
(b) Name two types of lipids that contain esterified fatty acids in their structures.

56. Give the name of the covalent backbone linkage in
(a) proteins (b) nucleic acids
(c) polysaccharides

57. Sketch the graph usually obtained when the velocity of an enzyme-catalyzed reaction is plotted against the substrate concentration (v versus [S] plot). Label on the graph the v value that represents the maximum velocity.

58.
(a) Enzymes ________ reaction rates.
(b) Enzymes ________ reaction activation energies.
(c) Name two of the factors thought to be responsible for the ability of enzymes to do the above.

59. State one important difference between bacterial cells and those of animals and plants.

60. Give a brief description of biological membrane structure. (A labeled diagram will be acceptable.)

61. Identify the class of each of the following lipids.

(a)

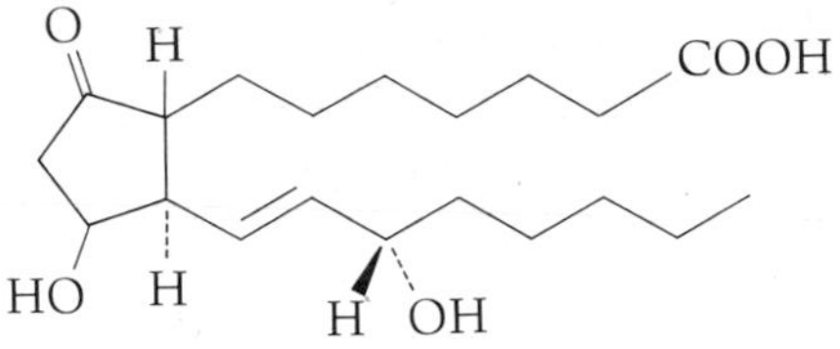

O H COOH

HO H H OH

(b)

CH_2OH

C=O

HO

O

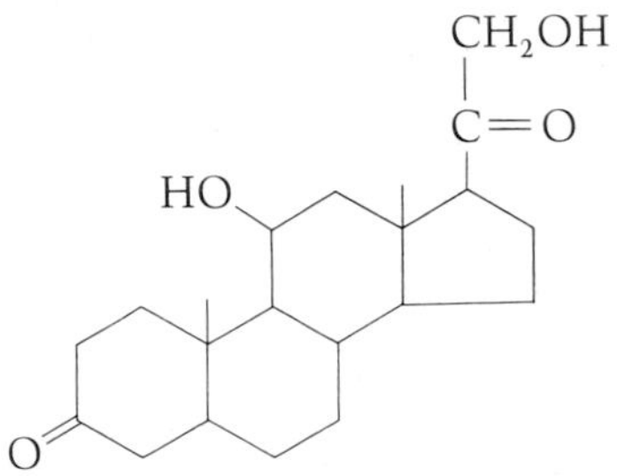

62. The complementary base pairs in DNA are ________ and ________.

63. When a protein loses its native structure and activity by being heated, treated with acid, and so on, it is said to be ________.

64.

(a) Draw the general structure of an α-amino acid.

(b) If an amino acid is called an acidic amino acid, what important functional group does its R group or side chain possess?

65. Add what is necessary to complete the following reactions:

(a) 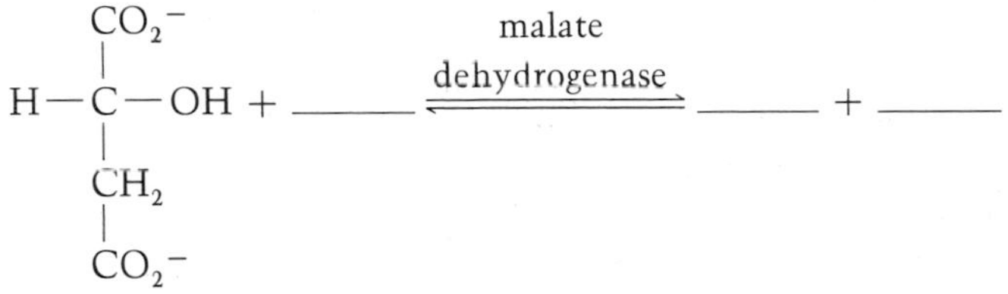 CO_2^- | H—C—OH | CH_2 | CO_2^- + ______ $\xrightleftharpoons{\text{malate dehydrogenase}}$ ______ + ______

(b) 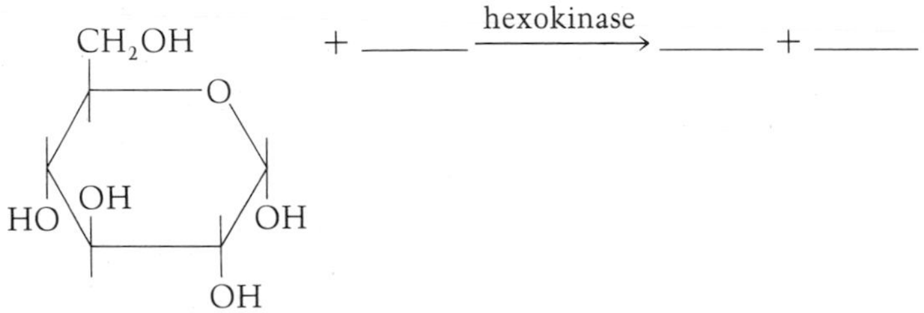 CH_2OH (glucose ring: O, HO, OH, OH, OH) + ______ $\xrightarrow{\text{hexokinase}}$ ______ + ______

(c) CO_2^- | C—OⓅ ‖ CH_2 + ______ $\xrightarrow{\text{pyruvate kinase}}$ ______ + ______

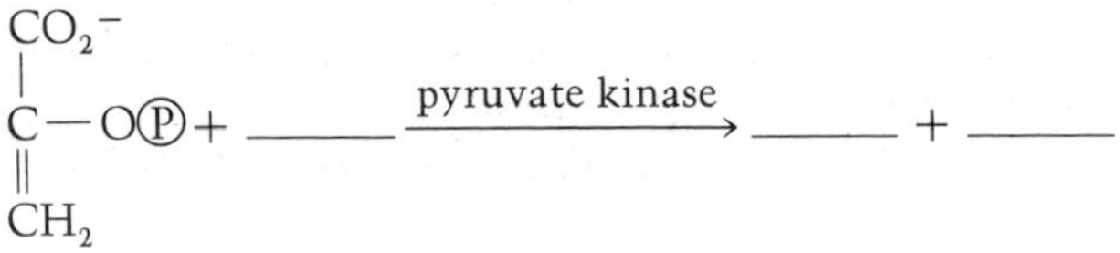

(d) $^{-}O-\overset{O}{\overset{\|}{P}}(O^{-})-O-\overset{O}{\overset{\|}{P}}(O^{-})-O-\text{ribose}(\text{adenine}) + {}^{-}O-\overset{O}{\overset{\|}{P}}(O^{-})-OH \xrightarrow{\text{ATP synthetase}}$

______ + ______

66. The $\Delta G°$ of the reaction $A \rightarrow B$ is $+5$ kcal/mole; the $\Delta G°$ of the reaction $C \rightarrow D$ is -8 kcal/mole.
(a) What is the $\Delta G°$ of the reaction $A + C \rightarrow B + D$?
(b) What is the $\Delta G°$ of the reverse reaction?
67. What is the primary purpose of both aerobic and anaerobic glycolysis?
68. How does the energy level (or ATP level) of the cell regulate the complete oxidation of glucose?
69. In terms of energy changes (in general), oxidation reactions are __________ and reduction reactions are __________.
70. ATP is the primary energy "currency" in cells. In plant cells the process called __________ that takes place in the __________ is primarily responsible for ATP production. In animal cells, and in plant cells in the darkness, the __________ synthesize ATP in the process called __________.
71. In each of the following pairs of compounds, which one is more energy-rich?
(a) ATP or AMP
(b) acetyl CoA or citrate
(c) NAD or NADH
(d) NADPH or NADP
72. Glucose, galactose, and fructose are all converted to __________ by the reactions of the collecting phase of glycolysis (plus certain special reactions in which galactose and fructose are phosphorylated).
73. What is meant by a "committing step" in metabolism?
74. What are the products of the citric acid cycle?
75. Specifically, what is ATP used for in cells?
76. What is the *purpose* of the lactate dehydrogenase reaction occurring in anaerobic glycolysis?
77. Indicate how each of the following affects carbohydrate (glucose or glycogen) metabolism; where possible, note the tissue(s) affected.
(a) insulin
(b) glucagon
(c) epinephrine (adrenaline)
(d) cyclic AMP
78. List three ways in which the pathways of fatty acid β-oxidation and fatty acid biosynthesis *differ* from one another.
79.
(a) What is ketoacidosis?
(b) Give the structure of one of the compounds which is associated with the development of ketoacidosis.
80.
(a) Pyridoxal phosphate is to __________ as FAD is to fatty acyl CoA dehydrogenase.

(b) __________ is to the urea cycle as oxaloacetate is to the citric acid cycle.
(c) Alanine is to __________ as glutamate is to α-ketoglutarate.

81.
(a) Name three metabolites that can provide carbons for gluconeogenesis.
(b) In what tissue does gluconeogenesis occur?

82. Add whatever is necessary to complete the following reactions:

(a) $CH_3\overset{\overset{O}{\|}}{C}CO_2^- + ATP + \_\_\_\_ \xrightleftharpoons{\text{pyruvate carboxylase}} \_\_\_\_ + ADP + P_i$

(b) $\_\_\_\_ + GTP \xrightleftharpoons{\text{PEP carboxykinase}} \_\_\_\_ + \_\_\_\_ + GDP$

(c) This pair of reactions represents the "reversal" of what glycolytic reaction?

83.
(a) Where does acetyl CoA for fatty acid biosynthesis come from?
(b) How is malonyl CoA made?

84. *Briefly* describe the role of each of the following in protein synthesis
(a) *m*RNA (b) *t*RNA (c) *r*RNA

85. What is a polyribosome or polysome?

86. What is the energy source for protein biosynthesis?

87. What is a termination codon?

88. What are the main functions of the blood?

89. The primary blood buffer system is the __________ buffer system.

90. What is meant by the term *post-absorptive state?*

91. Describe briefly how hemoglobin performs its job of transporting O_2 from the lungs to the body tissues.

92.
(a) In general, what are the functions of hormones?
(b) Contrast the general modes of action of steroid and polypeptide hormones.

ANSWERS TO SELF-TEST

1. (a) K (b) Na (c) S (d) Cl (e) Ti

2. (a) silicon (b) calcium (c) iron (d) manganese (e) magnesium

3. $\text{density} = \dfrac{\text{mass}}{\text{volume}}$

volume = 1.68 mL − 1.00 mL = 0.68 mL

mass = 23.463 g − 22.102 g = 1.361 g

$\text{density} = \dfrac{1.361\ \text{g}}{0.68\ \text{mL}} = 2.0\ \text{g/mL}$

4. (a) $\Delta G = -5.0 \times 10^2$ kcal. The reaction in the discharging battery is the reverse of the reaction when the battery is charged. (b) The maximum energy that can be obtained is equal to the loss in free energy, so the electrical energy = 5.0×10^2 kcal.

5. (a) $3ClNO + O_3 \rightarrow 3ClNO_2$ (b) $2SO_2 + O_2 \rightarrow 2SO_3$
(c) $PF_5 + 4H_2O \rightarrow H_3PO_4 + 5HF$

6. (a) $1s^22s^22p^6$ (b) $1s^22s^22p^63s^23p^2$

7. (a) C 1*s* ⇅ 2*s* ⇅ 2*p* ↑ ↑ __ 3*s* __ 3*p* __ __ __

(b) S 1*s* ⇅ 2*s* ⇅ 2*p* ⇅ ⇅ ⇅ 3*s* ⇅ 3*p* ⇅ ↑ ↑

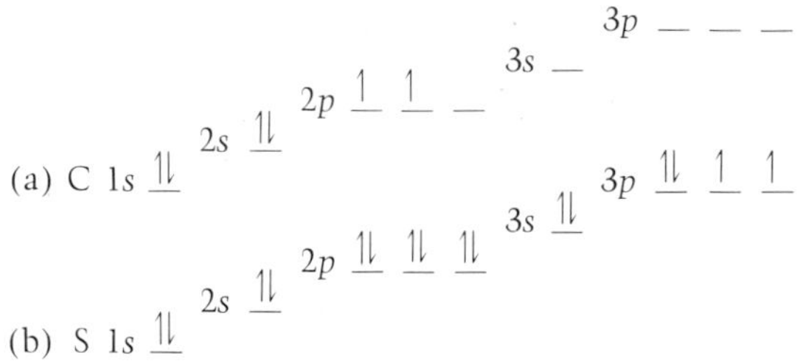

8. (a) Mg: (b) ·C̈·
(c) :F̤̈· (d) :Ar̤̈:
(e) Ba: (f) ·Sё:
(g) S̈n· (h) ·Ï̤:
(i) :Xë̤: (j) Cs·

9. (a) ${}^{14}_{6}C \rightarrow {}^{14}_{7}N + {}^{0}_{-1}e$ (β decay)
(b) ${}^{225}_{89}Ac \rightarrow {}^{221}_{87}Fr + {}^{4}_{2}He$ (α decay)

10. (a) solution (b) pure substance (c) heterogeneous mixture
(d) heterogeneous mixture (e) pure substance

11. (a) Ba is a metal (group II) and F is a nonmetal (group VII). A metal and a nonmetal react to form an ionic compound. The ions are Ba^{2+} and F^-, so the compound is BaF_2. (b) P is a nonmetal (group V), as is Br (group VII). Two nonmetals react to form a covalent compound.

·P̈· ·B̈r̤: shows that three Br atoms are required for one P atom to complete the octets.

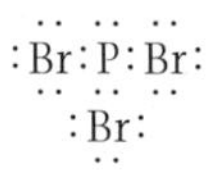

:B̤̈r:P̤̈:B̤̈r:
:B̤̈r:

(c) C is a nonmetal (group IV) as is H. (Therefore C and H will form a covalent compound.

$$
\begin{array}{c} H \\ \cdot\cdot \\ H:C:H \\ \cdot\cdot \\ H \end{array}
$$

(d) Na (group I) is a metal and S (group VI) is a nonmetal, so they will form an ionic compound, Na_2S.

12. (a) $Ba + F_2 \rightarrow BaF_2$ (b) $2P + 3Br_2 \rightarrow 2PBr_3$ (c) $C + 2H_2 \rightarrow CH_4$
(d) $2Na + S \rightarrow Na_2S$

13. (a) sodium bromide (b) magnesium fluoride (c) hydrogen sulfide
(d) potassium phosphate (e) carbon tetrachloride

14. (a) $CaCO_3$ (b) Na_2SO_4 (c) ClO_2 (d) NBr_3 (e) SF_6

15.

$$H_2 + Cl_2 \rightarrow 2HCl$$

$$\text{molecular weight HCl} = 1.0 + 35.5 = 36.5$$

$$3.65\ \cancel{g} \times \frac{1\ \text{mol}}{36.5\ \cancel{g}} = 0.100\ \text{mol HCl}$$

$$0.100\ \cancel{\text{mol HCl}} \times \frac{1\ \text{mol}\ H_2\ \text{or}\ Cl_2}{2\ \cancel{\text{mol HCl}}} = 0.0500\ \text{mol}\ H_2\ \text{or}\ Cl_2\ \text{reacted}$$

$$0.0500\ \cancel{\text{mol}\ H_2} \times \frac{2.0\ \text{g}\ H_2}{1\ \cancel{\text{mol}\ H_2}} = 0.10\ \text{g}\ H_2\ \text{reacted}$$

$$0.0500\ \cancel{\text{mol}\ Cl_2} \times \frac{71.0\ \text{g}\ Cl_2}{1\ \cancel{\text{mol}\ Cl_2}} = 3.55\ \text{g}\ Cl_2\ \text{reacted}$$

16.

Change	Temperature of gas	Pressure of gas	Average kinetic energy of He atoms	Number of collisions per second between He atoms and walls of the container	Average speed of He atoms
More gas is added	0	+	0	+	0
The gas is cooled	–	–	–	–	–
Some gas is removed and the remaining gas is heated	+	+	+	+	+
More gas is added and the gas is cooled	–	+	–	+	–

17. $PV = nRT$
If the amount (moles) of gas is constant, n is constant. Therefore, $PV/T = nR$ = constant, and the combined gas law results. If moles and temperature are both constant, $PV = nRT$ = constant, and Boyle's law results. If moles and pressure are constant, $V/T = nR/P$ = constant; and Charles's law results.

18. (a) same (b) same (c) Energy per gram of liquid benzene is less than the energy per gram of gaseous benzene by the heat of vaporization.
19. (a) ionic crystal (b) network covalent crystal (c) amorphous (d) molecular crystal
20. The strong intermolecular attraction among water molecules accounts for the relatively high melting and boiling points. Hydrogen bonding and the dipolar attraction between water molecules are the cause of this strong intermolecular attraction.
21. (a) $\frac{18.0\text{ g}}{400\text{ mL}} \times \frac{100}{100} = \frac{4.50\text{ g}}{100\text{ mL}} = 4.50\%$

(b) $18.0\,\cancel{\text{g}} \times \frac{1\text{ mol}}{180\,\cancel{\text{g}}} = 0.100\text{ mol}; \frac{0.100\text{ mol}}{0.400\text{ L}} = 0.250\ M$

22. (a) Nothing. There is no net transfer of materials across a membrane separating isotonic solutions. (b) The red blood cells will undergo crenation (shrivel) as water passes out of the cells into the more concentrated (hypertonic) solution. (c) They will swell and undergo hemolysis (rupture) as water passes into the more concentrated solution within the cells.
23. (a) $K = \frac{[SO_3]^2}{[SO_2]^2[O_2]}$

(b)

Change	$[SO_2]$	$[O_2]$	$[SO_3]$	Position of equilibrium	K	ΔG	$\Delta G^‡$
Some SO_2 is added	+	−	+	R	0	0	0
Some SO_3 is removed	−	−	−	R	0	0	0
Some O_2 is added	−	+	+	R	0	0	0
Some SO_2 and some O_2 are removed	−	−	−	L	0	0	0
Some V_2O_5, a surface catalyst, is added	0	0	0	0	0	0	−

24. (a) $HNO_3 + H_2O \rightarrow H_3O^+ + NO_3^-$

(b) $NH_4^+ + HPO_4^{2-} \rightleftarrows NH_3 + H_2PO_4^-$

(c) This is not an acid-base reaction. It is a redox reaction.
25. (a) $H_3O^+ + OH^- \rightarrow 2H_2O$ (Remember that 0.1 M HCl contains only H_3O^+ and Cl^- ions in solution.)
(b) $CH_3CO_2H + H_2O \rightleftarrows H_3O^+ + CH_3CO_2^-$
(c) $H_2CO_3 + H_2O \rightleftarrows H_3O^+ + HCO_3^-$
26. The reaction is $H_3O^+ + OH^- \rightarrow 2H_2O$. At pH of 7, moles H_3O^+ originally present = moles OH^- added.

$$0.150 \frac{\text{mol}}{\cancel{\text{L}}} \times 0.0200\ \cancel{\text{L}} = 0.00300\ \text{mol}$$

$$\frac{0.00300\ \text{mol}}{0.0150\ \text{L}} = 0.200\ M\ \text{HNO}_3$$

27.

Solution	$[H_3O^+]$	$[OH^-]$	pH
0.10 *M* HCl	1.0×10^{-1}	1.0×10^{-13}	1.00
0.10 *M* NaOH	1.0×10^{-13}	1.0×10^{-1}	13.00
0.10 *M* NaCl	1.0×10^{-7}	1.0×10^{-7}	7.00
An NH_4^+—NH_3 buffer	1.0×10^{-10}	1.0×10^{-4}	10.00
0.10 *M* solution of an unknown weak acid	1.0×10^{-3}	1.0×10^{-11}	3.00

28.

Compound name	Lewis dot formula	Three-dimensional formula	Name of functional group
ethane	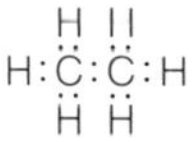 H H H:C:C:H H H	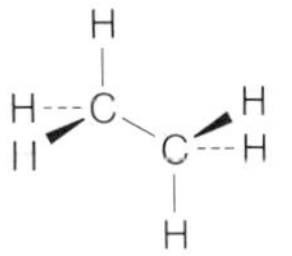 H H---C H H C---H H	
3-methyl-1-butene	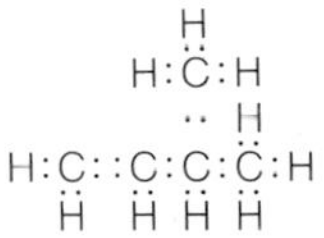 H H:C:H H H:C::C:C:C:H H H H H	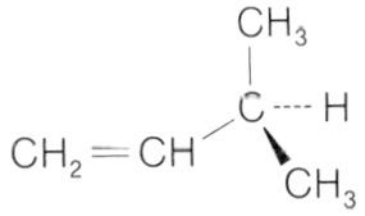 CH_3 C---H CH_2═CH CH_3	alkene
acetone	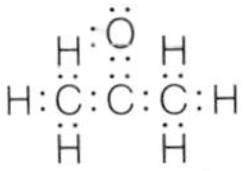 H :O: H H:C:C:C:H H H	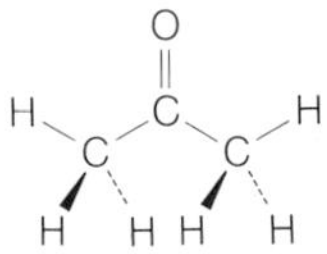 O ‖ H C H C C H H H H	keto
propyne	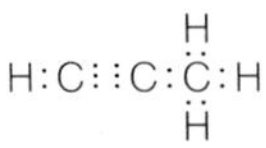 H H:C:::C:C:H H	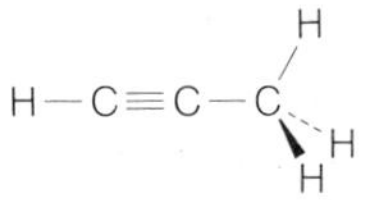 H H—C≡C—C H H	alkyne

29. Any cycloalkane (for example cyclohexane or cyclopentane—CH_3) except cyclopropane derivatives.

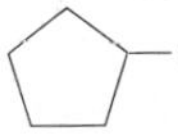

30. alcohol hydroxyl, benzene ring, keto, carboxyl

31. (a) $CH_3CH_2CH_2CH_3 + NaBr$ (b) H_2
(c) $CH_3CH_2Br + HBr$ (d) N.R.
(e) $CH_3CH(OH)CH_2CH_3$ (f) $CH_3CH(Br)CH_2CH_3$
(g) C_6H_5—SO_3H (h) C_6H_5—$Br + HBr$

32. (a) 2-methyl-1,5-hexadiene (b) 2-aminochlorobenzene (c) toluene
(d) o-dinitrobenzene (e) 2-pentyne

33. (a) $CH_3Cl + HCl$ (b) $CCl_4 + HCl$
(c) H_2CrO_4 (d) C_6H_5—$C(=O)$—O—$CH_3 + H_2O$
(e) $CH_3ONa + H_2$ (f) C_6H_5—$CH_2Br + H_2O$
(g) C_6H_5—$ONa + H_2O$ (h) $CH_3CH_2I + H_2O$
(i) N.R. (j) $CH_3CH_2CH_2SNa + H_2O$
(k) $CH_3NH_3^+Cl^-$ (l) $(CH_3)_2NH_2^+Cl^-$
(m) $H_3\overset{+}{N}CH_2C(=O)$—$O^-$

34.

	Higher boiling point	Higher solubility
(a)	CH_3OH	CH_3OH
(b)	CH_3OH	CH_3OH
(c)	$CH_3CH_2CH_2NH_2$	similar
(d)	$CH_3CH_2CH_2OH$	similar

35. (a) CH_3CH_2SH (b) 2-methylphenol ($C_6H_4(CH_3)$—OH, CH_3 ortho to OH)
(c) NH_4Cl (compare the conjugate bases)

36. (a) $CH_3CH_2C(CH_3)_2$—OH (b) $CH_3CH_2CH_2CH(CH_3)$—OH
(must have —C(—C—)—CHOH)
(c) $CH_3CH_2CH(CH_3)CH_2NH_2$
(must have —C—NH_2)

37. (a) ethyl phenyl ether or ethoxybenzene
(b) 3,4-dimethyl-1-phenyl-2-hexanol (c) ethyl propyl isopropyl amine
(d) 2-chloro-4-nitrophenol (e) 4-methoxy-2-butanol

(f) 3,4-dimethoxyphenol (g) propanethiol or propyl mercaptan
(h) diisopropyl disulfide (i) tetraethylammonium chloride

38. (a) $CH_3CH_2\overset{O}{\overset{\|}{C}}H$ (b) $CH_3\overset{O}{\overset{\|}{C}}CH_2CH_2CH_3$

(c) 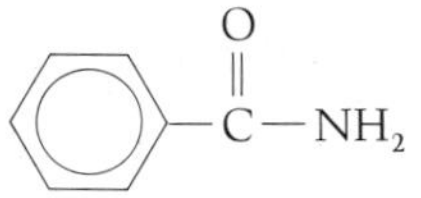 $C_6H_5-\overset{O}{\overset{\|}{C}}-NH_2$ (d) $CH_3\overset{O}{\overset{\|}{C}}-O-\overset{O}{\overset{\|}{C}}CH_3$

(e) $H\overset{O}{\overset{\|}{C}}H$ (f) $CH_3-O-\overset{O}{\overset{\|}{C}}-H$

39. (a) 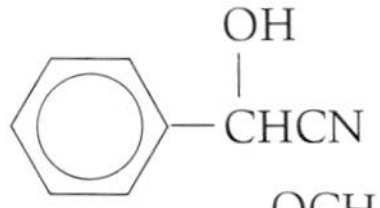 $C_6H_5-\overset{OH}{\overset{|}{C}}HCN$ (b) $C_6H_5-\overset{O}{\overset{\|}{C}}-O^-$

(c) $C_6H_5-CH(OCH_3)_2$ (d) 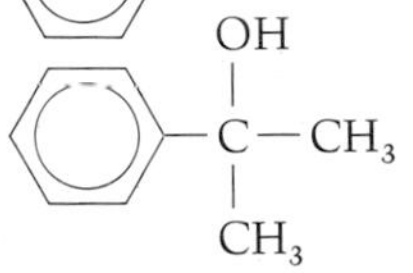 $C_6H_5-\overset{OH}{\overset{|}{C}}(CH_3)-CH_3$

(e) N.R. (f) 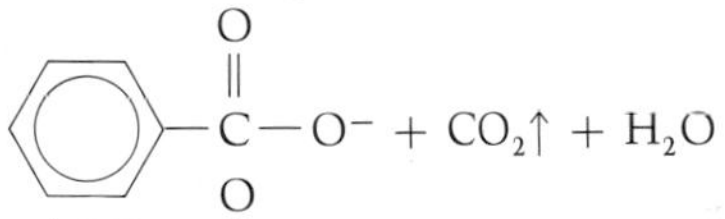 $C_6H_5-\overset{O}{\overset{\|}{C}}-O^- + CO_2\uparrow + H_2O$

(g) $C_6H_5-\overset{O}{\overset{\|}{C}}-NH_3 + CH_3OH$ (h) $C_6H_5-\overset{O}{\overset{\|}{C}}-O^-\quad CH_3NH_3^+$

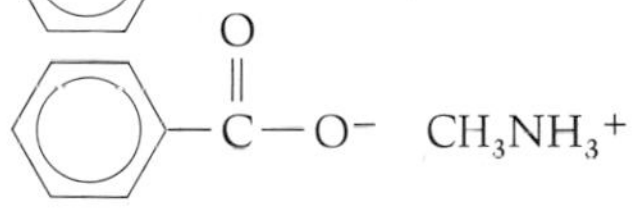

(i) 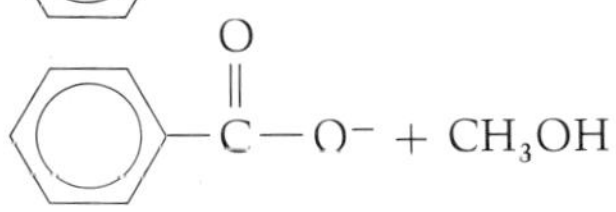 $C_6H_5-\overset{O}{\overset{\|}{C}}-O^- + CH_3OH$

(j)  $HO-\overset{O}{\overset{\|}{C}}-C_6H_4-\overset{O}{\overset{\|}{C}}-OH + HOCH_2-C_6H_{10}-CH_2OH$

(k) $H_3\overset{+}{N}-CH_2-\overset{O}{\overset{\|}{C}}-OH$

40. (a) aldohexose (b) $2^4 = 16$

(c)

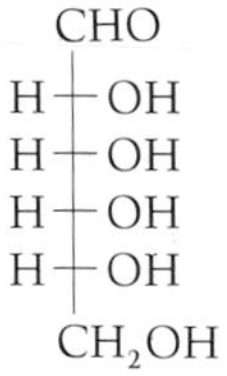

```
    CHO
  H-+-OH
  H-+-OH
  H-+-OH
  H-+-OH
    CH2OH
```

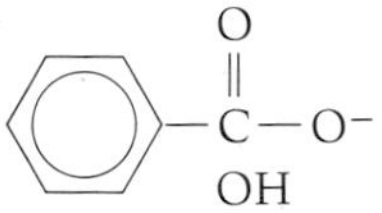

(d) 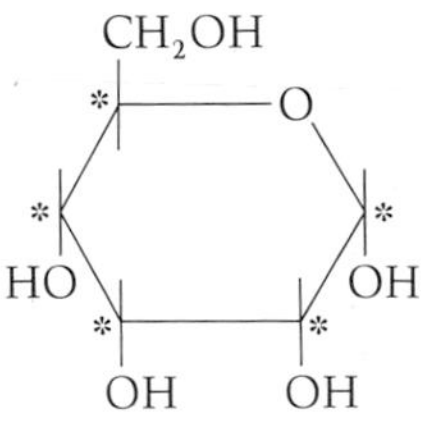 pyranose ring: ring O; ring carbons marked *; CH_2OH on C-5; HO, OH, OH, OH substituents on the remaining ring carbons.

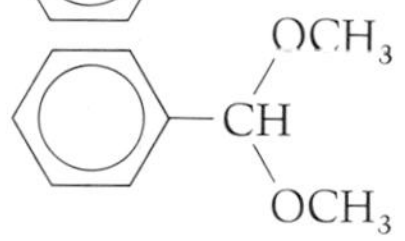

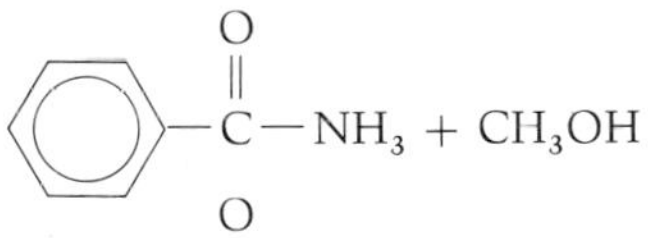

(e) pyranose (f) see (d)

(g) α (h)

```
    CHO
 HO-+-H
 HO-+-H
 HO-+-H
 HO-+-H
    CH2OH
```

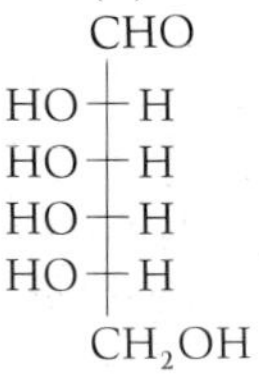

41. Their conjugate bases are stabilized by resonance:

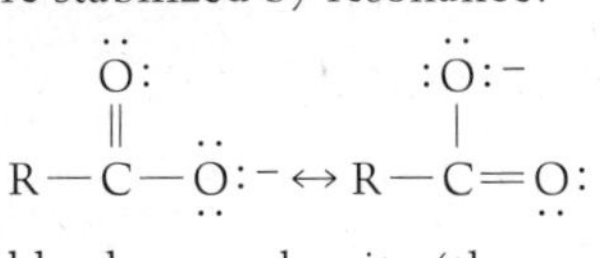

$$R-\overset{\overset{\displaystyle \ddot{O}:}{\|}}{C}-\ddot{\underset{..}{O}}:^- \leftrightarrow R-\overset{\overset{\displaystyle :\ddot{O}:^-}{|}}{C}=\underset{..}{O}:$$

42. They have a high available electron density (the π orbital).
43. N is less electronegative than O, so O—H····O hydrogen bonds are stronger than N—H····O or N—H····N.
44. S is a larger atom so the conjugate base R-S⁻ has its negative charge spread over a larger space.
45. (a) $CHCl_3$; anesthetic, solvent (b) CF_2Cl_2; refrigerant
(c) $CH_3\overset{\overset{O}{\|}}{C}CH_3$; solvent (d) $H-\overset{\overset{O}{\|}}{C}-H$; embalming agent
(e) beverage, solvent (f) anesthesia, solvent
(g) solvent

46. (d) Br above C; C bonded to CH_3, H (dashed), CH_2CH_3 (wedge) — and mirror image: Br above C; C bonded to H (dashed), CH_3CH_2 (wedge), CH_3

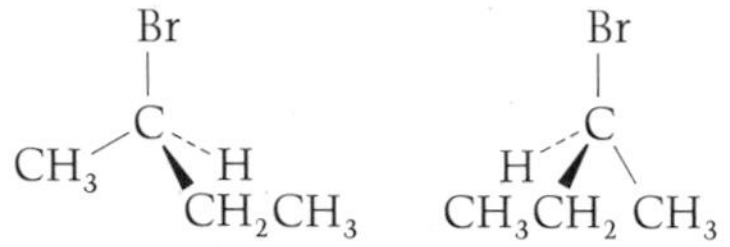

(e) CH_3 above C; C bonded to CH_3CH_2, H (wedge), $CH=CH_2$ (dashed) — and mirror image: CH_3 above C; C bonded to $CH_2=CH$ (dashed), H (wedge), CH_2CH_3

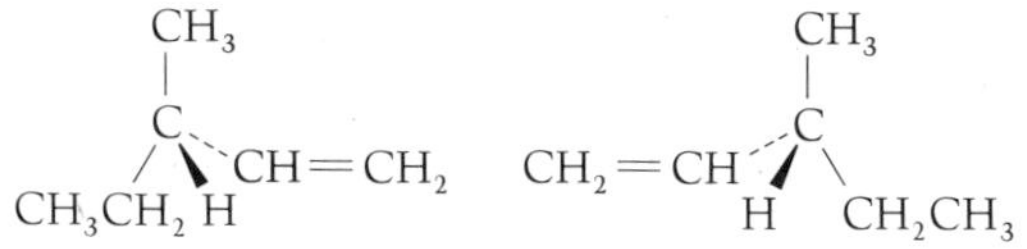

47. Only (4)

48. $CH_3-\overset{\overset{\displaystyle :\ddot{O}:^-}{|}}{C}=\underset{\underset{\displaystyle H}{|}}{\overset{+}{N}}-H$

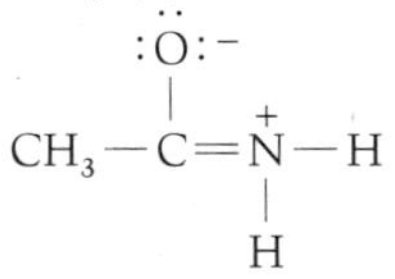

49. (a) H above C; C bonded to Br (dashed), CH_3 (wedge), F (b) $CH_3CH_2CH_2OH$

(c) $C_6H_5-N(CH_3)_2$ (d) CH_2OH / $C=O$ / $H-C-OH$ / $HO-C-H$ / CH_2OH (Fischer projection)

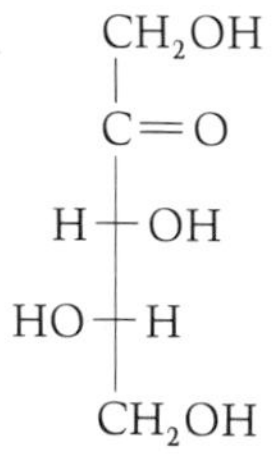

(e) $CH_3CH(OCH_3)_2$

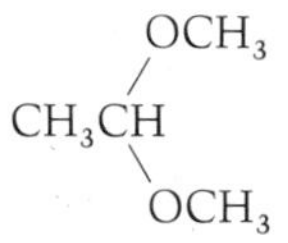

50.

$$HOCH_2CH=CHCH_2OH + 2Na \rightarrow {}^-O-CH_2CH=CHCH_2-O^- + 2Na^+ + H_2$$

$$HOCH_2CH=CHCH_2OH \xrightarrow{Br_2} HOCH_2\underset{\underset{\displaystyle Br}{|}}{C}H\overset{\overset{\displaystyle Br}{|}}{C}HCH_2OH$$

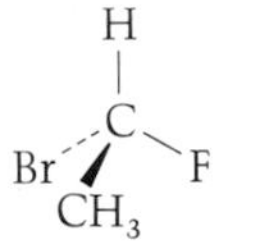

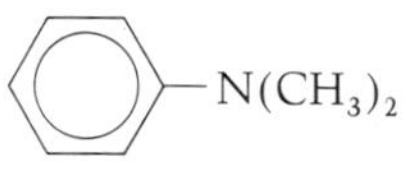

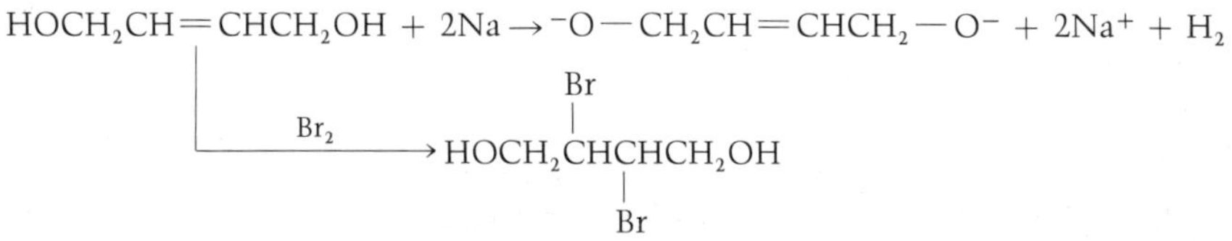

51. $CH_3CH_2CH_2\overset{\overset{\displaystyle O}{\|}}{C}{-}OH + NaHCO_3 \rightarrow CH_3CH_2CH_2\overset{\overset{\displaystyle O}{\|}}{C}{-}ONa + CO_2\uparrow + H_2O$

(or $CH_3\underset{}{\overset{\overset{\displaystyle CH_3}{|}}{C}H}\ \overset{\overset{\displaystyle O}{\|}}{C}{-}OH$)

52. (b), (e) because a high proportion of their molecules are composed of nitro groups ($-NO_2$), which provide O_2 for combustion and also form N_2 gas on combustion.

53. $^{-}O{-}\overset{\overset{\displaystyle O}{\|}}{\underset{\underset{\displaystyle OH}{\|}}{P}}{-}O{-}CH_2$ (attached to a five-membered ring containing O, with OH OH on the two lower ring carbons)

54. (a) covalent (b) hydrogen (c) all except covalent

55. (a) example: $CH_3CH_2CH_2CH_2CH_2CH_2CH_2CH_2CH_2CH_2CH_2CO_2H$
(b) triacylglycerols, phosphoglycerides

56. (a) peptide bond (b) 3′,5′-phosphodiester bond
(c) (α or β) glycosidic bond

57.

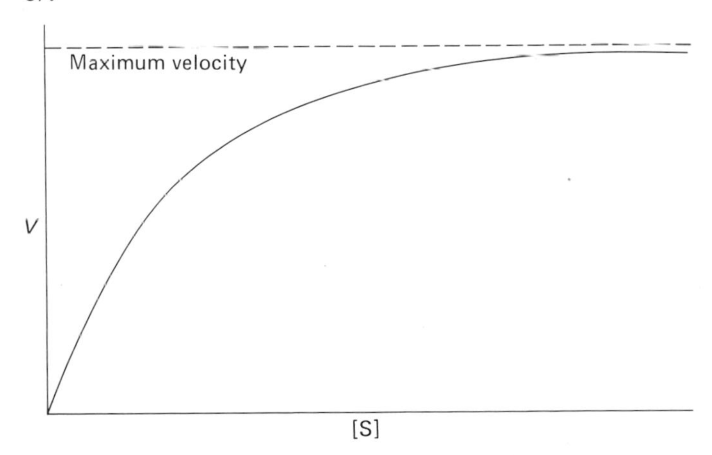

58. (a) increase (b) decrease (c) proximity, orientation, acid-base catalysis

59. Bacterial cells have no nuclei or other organelles, unlike animal and plant cells. Bacterial cells have only one (circular) chromosome; animal and plant cells have several or many (linear) chromosomes.

60. A biological membrane consists of a bilayer of phospholipid molecules (and cholesterol) in which are embedded many protein molecules, some deeply, some only superficially.

61. (a) prostaglandin (b) steroid

62. adenine-thymine; guanine-cytosine

63. denatured

64. (a) $R{-}\overset{\overset{\displaystyle NH_3^{+}}{|}}{C}H{-}CO_2^{-}$ (b) a carboxyl group (CO_2H)

65. (a) NAD, $\begin{matrix} CO_2^- \\ | \\ C{=}O \\ | \\ CH_2 \\ | \\ CO_2^- \end{matrix}$, NADH + H⁺ (b) ATP, [glucose-6-phosphate ring: CH_2OⓅ, O, HO, OH, OH, OH], ADP

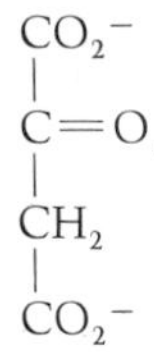

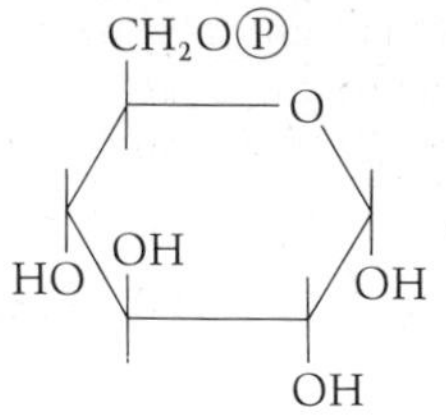

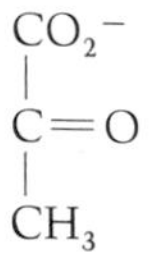

(c) ADP, $\begin{matrix} CO_2^- \\ | \\ C{=}O \\ | \\ CH_3 \end{matrix}$, ATP (d) 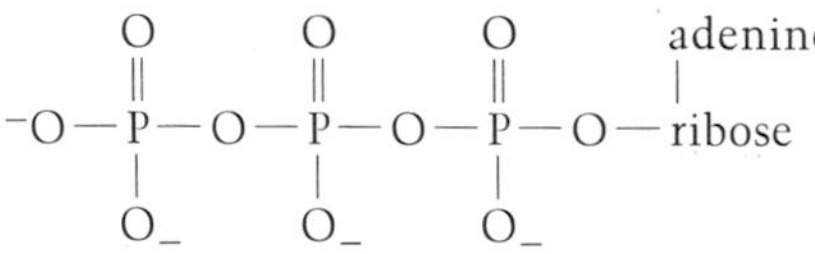 $^-O-\overset{O}{\overset{\|}{\underset{O^-}{\underset{|}{P}}}}-O-\overset{O}{\overset{\|}{\underset{O^-}{\underset{|}{P}}}}-O-\overset{O}{\overset{\|}{\underset{O^-}{\underset{|}{P}}}}-O-\overset{\text{adenine}}{\overset{|}{\text{ribose}}}$, H_2O

66. (a) −3 kcal/mole (b) +3 kcal/mole

67. ATP formation

68. The ATP level regulates the rate of electron transport to oxygen, because normal coupled electron transport cannot occur unless adequate ADP is available (when ATP is high, ADP is low, and vice versa). Electron transport is necessary to keep oxidizing NADH to NAD so the citric acid cycle dehydrogenases can function in oxidation of the acetyl CoA derived from breakdown of glucose in glycolysis (and conversion of pyruvate to acetyl CoA).

69. energy-yielding (exergonic), energy-requiring (endergonic)

70. photophosphorylation, chloroplasts, mitochondria, oxidative phosphorylation

71. (a) ATP (b) acetyl CoA (c) NADH (d) NADPH

72. glyceraldehyde-3-phosphate

73. A step at or near the beginning of a pathway that irreversibly converts the starting material to the first intermediate of the pathway; this "commits" the starting material to being metabolized—it can't go back.

74. CO_2, GTP, NADH, CoASH

75. ATP is used to drive energy requiring processes including biosynthesis, muscle contraction, and active transport of material across membranes.

76. Reoxidation of NADH, making NAD available for glyceraldehyde-3-phosphate dehydrogenase activity.

77. (a) promotes glucose uptake into cells (also glycogen synthesis); muscle, especially, but most tissues other than liver, red blood cells, and brain. (b) promotes glycogen breakdown, releasing glucose; liver (c) promotes glycogen breakdown; muscle and other tissues (d) promotes glycogen breakdown in tissues affected by glucagon and epinephine.

78. different enzymes; different cellular localizations: synthesis is in cytoplasm, oxidation in mitochondria; different coenzymes: synthesis—NADPH, oxidation—NAD and FAD; different acyl group carrier: coenzyme A in oxidation, acyl carrier protein (ACP) in synthesis.

79. (a) An excessive level of ketone bodies (acetoacetic acid, beta-hydroxy butyric acid, and acetone) in the bloodstream, leading to fall in blood pH.

(b) 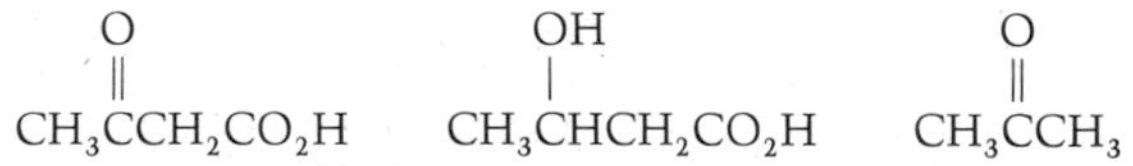 $CH_3\overset{O}{\overset{\|}{C}}CH_2CO_2H$ acetoacetic acid $CH_3\overset{OH}{\overset{|}{C}}HCH_2CO_2H$ beta-hydroxy butyric acid $CH_3\overset{O}{\overset{\|}{C}}CH_3$ acetone

80. (a) a transaminase (b) L-ornithine (c) pyruvate

81. lactate, pyruvate, amino acids, oxaloacetate, and so on (b) primarily the liver; small amount in kidney

82. (a) CO_2, $^-O_2CCH_2\overset{\overset{\large O}{\|}}{C}CO_2^-$

(b) $^-O_2CCH_2\overset{\overset{\large O}{\|}}{C}CO_2^-$, $CH_2{=}\overset{\overset{\large OⓅ}{|}}{C}{-}CO_2^-$, CO_2

(c) Reaction catalyzed by pyruvate kinase

83. (a) From cleavage of citrate exported from the mitochondria (b) By carboxylation of acetyl CoA, catalyzed by acetyl CoA carboxylase

84. (a) *m*RNA contains a transcribed copy of the nucleotide sequence of a DNA gene—this sequence represents the information necessary to specify the amino acid sequence of a polypeptide or protein. (b) *t*RNA molecules are connected to specific amino acids by amino acyl *t*RNA synthetases, and each contains a particular three-nucleotide sequence (called an anticodon) that is complementary to a particular *m*RNA codon. Base-pairing between codon and anticodon automatically brings the correct amino acid to the proper site on the ribosome where it can be linked onto the growing polypeptide chain. (c) *r*RNA molecules of three sizes contribute to the ribosome structure; they probably also function in some ways in binding *m*RNA and *t*RNAs to the ribosome, but little is yet known about these functions.

85. A group of ribosomes connected together by a molecule of *m*RNA—they are all simultaneously translating the *m*RNA message into polypeptides.

86. GTP and the activated amino acyl *t*RNAs themselves—when peptide bonds are formed, the high energy bond between the amino acyl group and the *t*RNA is also broken.

87. One of the three codons that does not code for a particular amino acid but rather signifies the end of the *m*RNA message.

88. Transport of nutrients, oxygen, hormones, and so on, defense of the body against invaders (the immune system), communication between tissues.

89. H_2CO_3/HCO_3^-

90. The nutritional state of the body when there is no food in the digestive tract being absorbed.

91. Hemoglobin becomes saturated with O_2 in the lungs where the oxygen pressure or concentration is high. When the circulation carries the blood cells containing HbO_2 to other tissues where the O_2 pressure is much lower, O_2 dissociates from Hb (the equilibrium shifts to the Hb + O_2 side) and O_2 diffuses into the tissues where it is needed.

92. (a) Regulation and integration of metabolism, communication between different tissues, regulation of growth and development. (b) Polypeptide hormones generally bind to receptors in the cell membrane and activate formation of cyclic AMP, which then sets into motion a cascade of enzyme activation reactions culminating in the activation of a key enzyme in a pathway. Steroid hormones can cross cell membranes and bind to cytoplasmic receptor proteins; then the hormone-receptor complex enters the nucleus and interacts with the DNA to increase or decrease the rate of transcription of *m*RNAs for particular proteins.

Unit Analysis APPENDIX 1

Quantities can be measured in various units. Just using the metric system, we can measure mass in grams, kilograms, or other units, length in millimeters, centimeters, meters. But when we add or subtract quantities, their units must be identical. For example, if we wish to add 0.50 m and 42 cm, we must first change either meters to centimeters or change centimeters to meters.

We can change a given quantity from one unit to a different unit by using an appropriate conversion factor. First, we write an equation that shows the relation between the different units. For the simple problem above, the appropriate equation (see Table 1.3) is

$$100 \text{ cm} = 1 \text{ m}$$

We can divide both sides of this equation by one meter and obtain

$$\frac{100 \text{ cm}}{1 \text{ m}} = 1$$

The right-hand side of this equation is unity—that is, it has the value of 1. Therefore, the left-hand side is also unity, because the two sides are equal. We can multiply any quantity by unity without affecting the quantity itself, so we can multiply our 0.50 m by 100 cm/1 m and not affect the actual distance expressed by 0.50 m. However, we do change the units of the distance from meters to centimeters, because when the numerator and denominator of a fraction have identical units, those units cancel.

To summarize, a ratio such as 100 cm/1 m, which expresses the same quantity in different units, is called a *conversion factor.* A conversion factor does not change the value of a quantity, it only changes the units in which the quantity is expressed.

To continue with our example:

these units cancel each other

$$0.50 \cancel{\text{m}} \times \frac{100 \text{ cm}}{1 \cancel{\text{m}}} = 50 \text{ cm}$$

We have thus changed meters to centimeters.

In order to change centimeters to meters, we will go back to our original equation.

$$100 \text{ cm} = 1 \text{ m}$$

We can divide both sides of the equation by 100 cm and obtain a new conversion factor.

$$1 = \frac{1\ \text{m}}{100\ \text{cm}}$$

Each side of this equation is unity, so we can multiply a quantity by the factor 1 m/100 cm without changing the actual quantity.

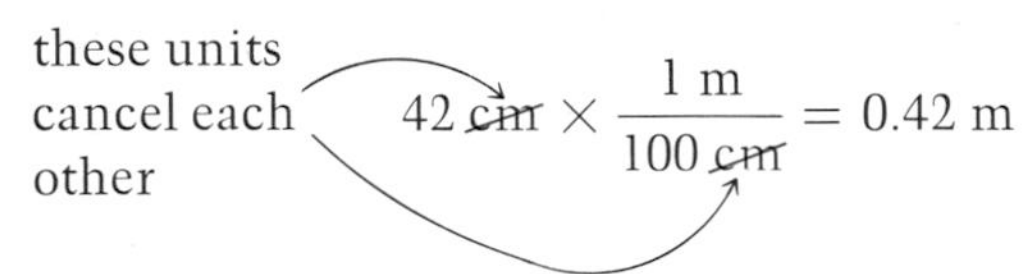

So, to change a quantity from one unit of measurement to another, we first need a conversion factor that has the original units in its denominator and the new units in its numerator. We then multiply the quantity by the conversion factor. The old units cancel out, and the quantity is expressed in the new units.

To avoid mistakes in unit analysis, it is very important to use the correct conversion factor—that is, to have the right units in the numerator and in the denominator. For example, what if we multiply 0.50 m by 1 m/100 cm?

these units do not cancel

$$0.50\ \text{m} \times \frac{1\ \text{m}}{100\ \text{cm}} = 0.0050\,\frac{\text{m}^2}{\text{cm}}$$

We wind up with a quantity expressed in square meters per centimeter. These units are not meaningful, so we realize that we made a mistake.

In some cases, we have to use more than one conversion factor. Let us consider the following.

Example Change a speed of 0.50 miles per hour to centimeters per second.

Solution Since "per" means "divided by," we can restate the problem as

$$\text{Change } 0.50\,\frac{\text{mi}}{\text{h}} \text{ to } \frac{\text{cm}}{\text{s}}$$

Here we have units in both the numerator (mi) and denominator (h) that must be changed. We can work on them one at a time. Let's start by changing miles to centimeters. According to Table 1.2,

$$1\ \text{km} = 0.6214\ \text{mi}$$

or

$$\frac{1\ \text{km}}{0.6214\ \text{mi}} = 1$$

therefore $$\frac{0.50\ \cancel{\text{mi}}}{1\ \text{h}} \times \frac{1\ \text{km}}{0.6214\ \cancel{\text{mi}}} = \frac{0.80\ \text{km}}{1\ \text{h}}$$

We see from Tables 1.3 and 1.4

$$1\ \text{km} = 1000\ \text{m}$$

$$1\ \text{m} = 100\ \text{cm}$$

Therefore $$\frac{1000\ \text{m}}{1\ \text{km}} = 1$$

and, $$\frac{100\ \text{cm}}{1\ \text{m}} = 1$$

We can now use these conversion factors to change kilometers to centimeters.

$$\frac{0.80\ \cancel{\text{km}}}{1\ \text{h}} \times \frac{1000\ \text{m}}{1\ \cancel{\text{km}}} = \frac{800\ \text{m}}{1\ \text{h}}$$

$$\frac{800\ \cancel{\text{m}}}{1\ \text{h}} \times \frac{100\ \text{cm}}{1\ \cancel{\text{m}}} = \frac{80000\ \text{cm}}{1\ \text{h}}$$

We have finally introduced centimeters. Now we need to change hour to seconds. We know some conversion factors from everyday experience.

$$60\ \text{min} = 1\ \text{h}$$

$$60\ \text{s} = 1\ \text{min}$$

From these we get

$$\frac{1\ \text{h}}{60\ \text{min}} = 1$$

$$\frac{1\ \text{min}}{60\ \text{s}} = 1$$

and we can use these conversion factors to change hour to minutes and minutes to seconds.

$$\frac{80000\ \text{cm}}{1\ \cancel{\text{h}}} \times \frac{1\ \cancel{\text{h}}}{60\ \text{min}} = \frac{1333\ \text{cm}}{1\ \text{min}}$$

$$\frac{1333\ \text{cm}}{1\ \cancel{\text{min}}} \times \frac{1\ \cancel{\text{min}}}{60\ \text{s}} = 22\ \frac{\text{cm}}{\text{s}}$$

In solving a problem that requires the use of several conversion factors, we would not ordinarily work out each answer separately, as we have done here. We would simply set up the problem as a continuous fraction and then use a calculator to find the final answer.

$$\frac{0.50 \text{ mi}}{1 \text{ h}} \times \frac{1 \text{ km}}{0.6214 \text{ mi}} \times \frac{1000 \text{ m}}{1 \text{ km}} \times \frac{100 \text{ cm}}{1 \text{ m}} \times \frac{1 \text{ h}}{60 \text{ min}} \times \frac{1 \text{ min}}{60 \text{ s}} = 22 \frac{\text{cm}}{\text{s}}$$

(Note: In these calculations, some of the separate answers were not expressed to the correct number of significant figures. However, the final answer is correctly expressed.

Many of the problems you will encounter in chemistry will require converting a quantity from one unit to another. Careful attention should always be paid to units, and numbers should always be accompanied by proper units. You can partially check the accuracy of a mathematical operation by checking whether or not the result is expressed in the desired units. You can make an arithmetical mistake and still get correct units. However, if the final result carries improper units, you can be quite sure the procedure was faulty.

SI Units* APPENDIX 2

Seven fundamental SI units have been defined. These are called base units. All other SI units are derived from these base units. Table A.1 gives the seven base units, and Table A.2 lists several of the derived

TABLE A.1 SI Base Units

Quantity	Unit	Symbol	Definition
Length	meter	m	The length equal to 1,650,763.73 wavelengths, in vacuum, of the orange-red line of the emission spectrum of krypton-86.
Mass	kilogram	kg	The mass of a standard platinum-iridium alloy cylinder located at the International Bureau of Weights and Measures in France.
Time	second	s	The duration of 9,192,631,770 cycles of radiation associated with a certain energy transition in the cesium-133 atom.
Temperature	kelvin	K	$\frac{1}{273.16}$ of the temperature of the triple point of water (the condition under which solid, liquid, and gaseous water exist in equilibrium).
Electric current	ampere	A	The quantity of current in two parallel wires placed 1 meter apart, in vacuum, that produces between these conductors a force of 2×10^{-7} newton per meter of length.
Luminous intensity	candela	cd	The luminous intensity of $\frac{1}{600,000}$ of a square meter of a radiating cavity at the freezing point of platinum.
Amount of substance	mole	mol	The amount of substance of a system that contains as many elementary entities as there are atoms in exactly 0.012 kg of carbon-12.

*SI is the abbreviation for the International System of Units.

units occasionally encountered in chemistry. Table A.3 provides some conversion factors for SI units and some other units commonly used in chemistry.

TABLE A.2 Derived SI Units

Quantity	Unit	Symbol	Definition
Force	newton	N	The force that accelerates a mass of one kilogram one meter per second per second. $N = kg \cdot m/s^2$
Pressure	pascal	Pa	A force of one newton acting on one square meter. $Pa = N/m^2$
Energy	joule	J	A force of one newton acting through one meter. $J = N \cdot m$
Power	watt	W	An expenditure of energy at the rate of one joule per second. $W = J/s$
Charge	coulomb	C	The electrical charge carried by a current of one ampere flowing in a conductor for one second. $C = A \cdot s$
Voltage	volt	V	The energy of electrical charge such that one watt of power arises per ampere of current. $V = W/A$; $V = J/C$

TABLE A.3 Conversion Factors

Quantity	SI Unit	Common chemical unit
Force	1 newton	10^5 dynes
Pressure	1.01×10^5 Pa	1 atmosphere
	133 Pa	1 mm of Hg
Energy	1 joule	10^7 ergs
	4.18 J	1 calorie

Diagnostic Test and Mathematics Review APPENDIX 3

We use a few special mathematical concepts in this book. These concepts are explained fully when the need arises: exponentials and scientific notation (Section 1.8), significant figures (Section 1.9 and 1.10), and logarithms (Section 7.8).

An understanding of basic arithmetic and algebraic operations, however, is assumed. Some students may need a review of these operations. Quiz yourself with the following short, multiple-choice, diagnostic test. Do not use a calculator; you may use paper and pencil. Circle your choice for the correct answer to each question. After you have completed the test, check your answers against the correct answers provided.

DIAGNOSTIC TEST

Subtraction of a negative number

1 $6 - (-10) = ?$

(a) -4 (b) 4 (c) 16 (d) 60

Multiplication

2 $(4 + 1)(6 - 8) = ?$

(a) -10 (b) 3 (c) 16 (d) 22

Division

3 $\dfrac{(6 - 2)}{(1 + 3)} = ?$

(a) 4 (b) $5\frac{1}{3}$ (c) $\frac{3}{4}$ (d) 1

Multiplication with a two-term factor

4 $(x - 4)(1 - 4) = ?$

(a) 12 (b) $x - 12$ (c) $x + 12$ (d) $12 - 3x$

Multiplication and division

5 $\dfrac{(-6 + 2)(1 + 2)}{(10 - 12)} = ?$

(a) 12 (b) 6 (c) -6 (d) 3

Equation solving

6 $-3 + x = -1$ (solve for x)

(a) $x = 2$ (b) $x = -2$ (c) $x = 0$ (d) $x = 4$

7 $-6 = -y + 1$ (Solve for y)
(a) $y = -7$ (b) $y = 7$ (c) $y = 6$ (d) $y = 1$

8 $d = \frac{10}{V}$ (solve for V)

(a) $V = 10d$ (b) $V = 10 - d$ (c) $V = \frac{10}{d}$ (d) $V = \frac{d}{10}$

9 $15 = \frac{Z}{2} + 12$ (solve for Z)

(a) $Z = 1.5$ (b) $Z = 18$ (c) $Z = -9$ (d) $Z = 6$

10 $\frac{4}{T} = \frac{2}{y}$ (solve for y)

(a) $y = \frac{T}{2}$ (b) $y = 4T$ (c) $y = 8T$ (d) $y = \frac{8}{T}$

Answers

1 (c)
2 (a)
3 (d)
4 (d)
5 (b)
6 (a)
7 (b)
8 (c)
9 (d)
10 (a)

If you have answered all ten questions correctly, your basic mathematical skills should be adequate for you to understand the mathematical concepts presented in this book. If you missed one or more questions, you may experience some difficulty. You should study the following review of basic mathematical skills. The review is keyed by number to the test question that illustrates a particular skill.

REVIEW OF BASIC MATHEMATICAL SKILLS

Subtraction of a negative number

1 Subtraction of a negative number results in addition.

$$6 - (-10) = 6 + 10 = 16$$

The two negatives can be replaced by the plus sign $(+)$. Any even number of negative signs can be replaced by a plus sign, but an odd number of negative signs remains negative. For example,

$$-(-4) = +4$$
$$-[-(-4)] = -[+4] = -4$$
$$-[-\{-(-4)\}] = -[-\{+4\}] = -[-4] = +4$$

Multiplication

2 The symbols (   )(   ) mean that whatever appears inside the first set of parentheses is multiplied by whatever is inside the second set of parentheses. As many terms as desired can be multiplied—(   )(   ) (   )(   ) . . .—and the meaning is the same. All the quantities inside the parentheses are to be multiplied together. Usually, it is easier to carry out any operations inside the parentheses before multiplication.

$$(4 + 1)(6 - 8) = (5)(-2) = -10$$

The answer is -10 because a positive number (5) multiplied by a negative number (-2) gives a negative answer. If the problem had been stated as $(-4 - 1)(6 - 8)$, then combining inside the sets of parentheses followed by multiplication gives $(-5)(-2) - 10$. The answer is positive because two negative numbers multiplied together gives a positive answer.

Division

3 The symbols $\frac{(\quad)}{(\quad)}$ mean that whatever appears inside the parentheses on top (numerator) is divided by whatever appears inside the bottom parentheses (denominator). Usually, it is easier to carry out all operations inside each set of parentheses before dividing the numerator by the denominator.

$$\frac{(6 - 2)}{(1 + 3)} = \frac{(4)}{(4)} = 1$$

Multiplication with a two-term factor

4 Here we cannot combine the terms in the first parentheses; however, we can combine terms within the second parentheses.

$$(x - 4)(1 - 4) = (x - 4)(-3)$$

We must now multiply each term in the first set of parentheses by the -3.

$$(x - 4)(-3) = -3x - (4)(-3) = -3x + 12$$

The 12 is positive because multiplication of -4 and -3 produces a positive number. Of course, $-3x + 12$ is exactly the same as $12 - 3x$ (a plus sign is understood when no sign is written).

Multiplication and division

5 Once again, we should carry out the operations within each set of parentheses first.

$$\frac{(-6 + 2)(1 + 2)}{(10 - 12)} = \frac{(-4)(3)}{(-2)}$$

Now, we can multiply the terms in the numerator and get $-12/-2$ and then divide: $-12/-2 = 6$. Note that a negative number divided by a negative number gives a positive result. Alternatively, we could have divided either one of the terms in the numerator by -2 first and then multiplied by the other number.

$$\frac{(-4)(3)}{(-2)} = (2)(3) = 6 \qquad \text{or} \qquad \frac{(-4)(3)}{(-2)} = (-4)(-3/2) = \frac{12}{2} = 6$$

The order of multiplication and division makes no difference.

Equation solving

The last five questions all deal with solving an equation for some desired quantity. The problem we face in solving an equation is to isolate the desired quantity on one side of the equation. The following rules are useful: (1) we can add or subtract the same number to both sides of an equation; (2) we can multiply both sides of an equation by the same number (except we cannot multiply by zero); (3) we can divide every term on both sides of an equation by the same number (except we cannot divide by zero).

6 $-3 + x = 1$
We can isolate x if we add 3 to both sides of the equation

$$+3 - 3 + x = -1 + 3$$

$$x = 2$$

7 $-6 = -y + 1$

We can isolate $-y$ if we subtract 1 from both sides of the equation.

$$-1 - 6 = -y + 1 - 1$$

$$-7 = -y$$

We have solved for $-y$. To get $+y$, we can multiply both sides of the equation by -1.

$$(-1)(-7) = (-1)(-y)$$

$$7 = y$$

We can equally well express this result as $y = 7$.

8 $d = \dfrac{10}{V}$

First multiply both sides of the equation by V.

$$(d)(V) = \frac{(10)}{(V)}(V)$$

Notice that the V's cancel on the right-hand side.

$$(d)(V) = \frac{(10)}{(\cancel{V})}(\cancel{V})$$

If we now divide both sides by d, we can isolate V.

$$\frac{(\cancel{d})(V)}{(\cancel{d})} = \frac{(10)}{(d)} \qquad \text{or} \qquad V = \frac{10}{d}$$

9 $15 = \dfrac{Z}{2} + 12$

We can subtract 12 from both sides.

$$15 - 12 = \frac{Z}{2} + 12 - 12$$

$$3 = \frac{Z}{2}$$

Now, we simply multiply both sides by 2.

$$6 = Z$$

Note that we could have started out by multiplying through by 2.

$$2(15) = 2\left(\frac{Z}{2}\right) + 2(12)$$

$$30 = Z + 24$$

Then, we find Z by subtracting 24 from both sides.

$$6 = Z$$

10 $\frac{4}{T} = \frac{2}{y}$

We must get y out of the denominator and into the numerator. We can accomplish both by multiplying both sides by y.

$$(y)\left(\frac{4}{T}\right) = \left(\frac{2}{\cancel{y}}\right)(\cancel{y})$$

Now we multiply both sides by $T/4$ so we can cancel the factor multiplied by y.

$$(y)\left(\frac{4}{T}\right)\left(\frac{T}{4}\right) = (2)\left(\frac{T}{4}\right)$$

Canceling terms and dividing 4 into 2 we get

$$(y)\left(\frac{\cancel{4}}{\cancel{T}}\right)\left(\frac{\cancel{T}}{\cancel{4}}\right) = (\cancel{2})\left(\frac{T}{\underset{2}{\cancel{4}}}\right) \qquad \text{or} \qquad y = \frac{T}{2}$$

Here are a few more worked-out examples.

11 $6 + \frac{x}{3} = ab$ (solve for x)

Subtract 6 from both sides.

$$\frac{x}{3} = ab - 6$$

Multiply through by 3.

$$x = 3(ab - 6)$$

12 $-3 = \frac{(x - 1)}{(x + 3)}$ (solve for x)

Multiply through by $(x + 3)$

$$-3(x + 3) = x - 1$$

$$-3x - 9 = x - 1$$

Subtract x from both sides.

$$-4x - 9 = -1$$

Add 9 to both sides.

$$-4x = 8$$

Divide through by -4.

$$x = -2$$

13 $10 = \frac{10}{x + 1}$ (solve for x)

Multiply through by $x + 1$.

$$10(x + 1) = 10$$

$$10x + 10 = 10$$

Subtract 10 from both sides.

$$10x = 0$$

Divide both sides by 10.

$$x = \frac{0}{10} \quad \text{or} \quad x = 0$$

14 $5 - x = \frac{x}{3} + 1$ (solve for x)

Subtract $\frac{x}{3} + 5$ from both sides.

$$5 - x - \left(\frac{x}{3} + 5\right) = \frac{x}{3} + 1 - \left(\frac{x}{3} + 5\right)$$

Expand and cancel terms.

$$5 - x - \frac{x}{3} - 5 = \frac{x}{3} + 1 - \frac{x}{3} - 5$$

or

$$-x - \frac{x}{3} = -4$$

Multiply through by -3.

$$-3(-x) - 3\left(-\frac{x}{3}\right) = -3(-4)$$

or

$$3x + x = 12$$

and

$$4x = 12$$

Divide through by 4.

$$x = 3$$

Glossary APPENDIX 4

absolute configuration—the exact configuration of a given isomer. (*Section 16.7*)

absolute specificity—characteristic of an enzyme that can act on only a single substance. (*Section 21.9*)

absolute temperature—temperature expressed on the kelvin scale. (*Section 4.2*)

absorptive state—period after eating when food molecules are being absorbed from the digestive tract into the blood. (*Section 27.5*)

acetal—$R-\overset{\displaystyle OR'}{\underset{\displaystyle OR'}{\overset{|}{\underset{|}{C}}}}-H$ in which R may be alkyl or aryl, R′ is alkyl. Acetals are prepared by reacting an aldehyde, RCHO, with an alcohol, R′—OH, in the presence of a strong acid. (*Section 14.7*)

acid—a substance capable of giving up hydrogen ions (protons). (*Section 7.1*)

acid-base indicator—a weak acid with a color different from its conjugate base; the color changes over a specific pH range that depends on the K_a of the indicator. (*Section 7.11*)

acid-base titration—the addition of a measured amount of an acid or base of known concentration to a base or acid of unknown concentration until the reaction is complete as shown by an indicator or the pH of the solution. (*Section 7.12*)

acidic solution—a solution in which $[H_3O^+]$ is greater than $[OH^-]$. (*Section 7.6*)

acidosis—blood pH below 7.35. (*Section 27.4*)

actin—a major protein of the thin filaments of myofibrils in muscle. (*Section 22.10*)

actinide series—the elements with atomic numbers 90–103. (*Sections 2.10, 2.15*)

activation cascade—a sequence of reactions initiated by hormone binding at cell membrane and terminating in activation of a key enzyme; for example, glycogen phosphorylase activation by glucagon in liver cells. (*Section 24.17*)

active site—small region of an enzyme's surface where substrates bind and catalysis takes place. (*Section 21.9*)

active transport—transport of molecules or ions across membranes against the concentration gradient; requires energy in the form of ATP or a membrane potential. (*Section 22.11*)

acyl carrier protein (ACP)—protein containing 4′-phosphopantetheine that is part of the fatty acid synthase complex and carries the growing acyl chain during fatty acid synthesis. (*Section 25.13*)

acyl group—$R-\overset{\displaystyle O}{\overset{\|}{C}}-$, where R is alkyl, aryl, or hydrogen. (*Section 15.7*)

acyl halide—$R-\overset{\displaystyle O}{\overset{\|}{C}}-X$, where R may be alkyl, aryl, or hydrogen, and X = Cl, Br, or I. (*Section 15.7*)

addition—the chemical combination of a small molecule and an alkene or other unsaturated molecule. (*Section 10.6*)

adenosine-5′-triphosphate (ATP)—high energy compound whose hydrolysis drives most endergonic reactions in cells. (*Section 22.7*)

adipose tissues—specialized connective tissues in which fat is stored. (*Section 25.1*)

aerobic—oxygen-requiring. (*Section 24.2*)

alcohol—a compound in which an —OH group is bonded to an aliphatic carbon atom. (*Section 12.1*)

aldehyde—$R-\overset{\displaystyle O}{\overset{\|}{C}}-H$, where R is alkyl, aryl, or H. (*Section 14.3*)

aldose—an aldehyde sugar. (*Section 16.3*)

aliphatic hydrocarbon—all hydrocarbons except aromatic hydrocarbons: alkanes, alkenes, alkynes, and their cyclic derivatives. (*Section 10.12*)

alkaloids—amines of plant origin that are alkalilike and that usually exhibit physiological activity. (*Section 13.10*)

alkalosis—blood pH above 7.45. (*Section 27.4*)

alkanal—the general IUPAC name for aldehydes. (*Section 14.4*)

alkane (C_nH_{2n+2})—a noncyclic hydrocarbon that has only single bonds. (*Section 9.1*)

alkanoic acid—the general IUPAC name for carboxylic acids. (*Section 15.3*)

alkanol—the general IUPAC name for alcohols. (*Section 12.2*)

alkanone—the general IUPAC name for ketones. (*Section 14.9*)

alkaptonuria—genetic disease of amino acid metabolism in which homogentisic acid is excreted into the urine. (*Section 25.21*)

alkoxide—the conjugate base (RO^-) of an alcohol. (*Section 12.4*)

alkoxy—the IUPAC name for the ether functional group, —OR. (*Section 12.11*)

alkylation—reaction involving the addition of an alkyl group to a molecule. The reaction of ammonia with an alkyl halide to give an amine is an example of an alkylation. (*Section 13.7*)

alkyl group—a hydrocarbon group containing one hydrogen atom less than the alkane for which it is named. (*Section 9.3*)

$$\underset{\text{ethane}}{CH_3CH_2-H} \qquad \underset{\text{ethyl group}}{CH_3CH_2-}$$

alkyne—a hydrocarbon that has a carbon-carbon triple bond. (*Section 10.8*)

allosteric—meaning "other place" or "other site," as in allosteric site. (*Section 21.15*)

allosteric effectors—small molecules that may bind to special sites on allosteric enzymes to alter the conformation of the protein and thus its catalytic activity. (*Section 21.15*)

allosteric enzyme—a regulatory enzyme whose activity may be altered by allosteric effectors. (*Section 21.15*)

allosteric site—special site (different from the active site) where an allosteric effector may bind to increase or decrease the activity of an enzyme. (*Section 21.15*)

alpha helix (α-helix)—coiled secondary structure of protein. (*Section 17.10*)

alpha particle (α or ^4_2He)—the nucleus of a helium atom; two neutrons and two protons. (*Section 2.15*)

amide—$R-\overset{\overset{\large O}{\|}}{C}-NR'_2$, in which R and R′ may be alkyl, aryl, or hydrogen. (*Section 15.7*)

amine—derivative of ammonia in which one or more hydrogen atoms have been replaced by alkyl or aryl groups. (*Section 13.1*)

amine salt—an ionic compound containing an R_4N^+ ion in which at least one R group is alkyl or aryl and the others are hydrogen. (*Section 13.6*)

amino acid—a molecule that has *both* an amine group and a carboxylic acid group. (*Section 13.11*)

$$H_2N\underset{\underset{\large R}{|}}{C}H\overset{\overset{\large O}{\|}}{C}-OH$$

amino acid activation—formation of an aminoacyl transfer RNA, catalyzed by an aminoacyl *t*RNA synthetase. (*Section 26.8*)

aminoacyl site (A-site)—ribosomal site that contains an *m*RNA codon to be translated (other than the initiating codon) and to which an aminoacyl *t*RNA binds by codon–anticodon pairing. (*Section 26.12*)

ammonium ion (NH_4^+)—the conjugate acid of ammonia (NH_3). In substituted ammonium ions one or more of the hydrogen atoms are replaced by alkyl or aryl groups. (See also *quartenary ammonium ion.*) (*Section 13.5*)

amphoteric substance—a substance that can function as an acid or base. (*Section 7.2*)

amorphous solid—a noncrystalline solid; the units of an amorphous solid have no regular arrangement. (*Section 4.11*)

amylase—an enzyme that catalyzes hydrolysis of starches. (*Section 23.2*)

amylopectin—one of the polysaccharides in starch; a branched glucose polymer. (*Section 19.3*)

amylose—one of the polysaccharides in starch; a straight chain glucose polymer. (*Section 19.3*)

anabolism—biosynthesis. (*Section 23.8*)

anaerobic—occurring in the absence of O_2. (*Section 24.2*)

anhydrous—without water; an anhydrous salt is an ionic compound that has no water of hydration. (*Section 5.3*)

anion—a negative ion. (*Section 2.12*)

anomeric carbon—the carbon that bonds to two oxygen atoms in the hemiacetal form. (*Section 16.11*)

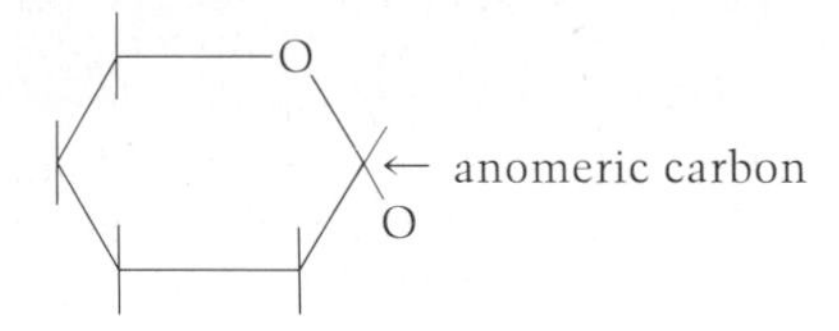

anomers—hemiacetal forms of a sugar that differ only in the configuration of the anomeric carbon. (*Section 16.11*)

anticodon—a three-nucleotide sequence in a *t*RNA that is complementary to a particular *m*RNA codon. (*Section 26.10*)

antioxidant—a compound that reacts with atmospheric oxygen to form harmless, nonoxidizing products. (*Focus 11*)

aromatic hydrocarbon—a cyclic, conjugated hydrocarbon that undergoes substitution rather than addition. The most common aromatic hydrocarbons have the benzene ring. (*Section 10.12*)

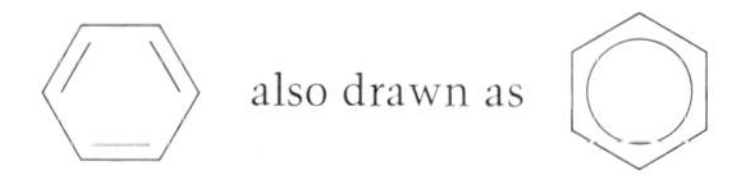

asymmetric carbon atom—a carbon atom bonded to four different groups or atoms. (*Section 16.5*)

atmosphere—a standard unit of pressure; a pressure that will support a column of mercury 760 millimeters high. (*Section 4.1*)

atmospheric pressure—the weight of air pushing on one unit of area at the surface of the earth. (*Section 4.1*)

atom—the smallest particle of chemical material that makes up an element. (*Section 2.1*)

atomic number—the number of protons in the nucleus of an atom. (*Section 2.6*)

atomic weight—the average mass of the atoms of a given element as they exist in nature. (*Section 2.13*)

ATP (synthetic phase of glycolysis)—second phase of glycolysis in which glyceraldehyde-3-phosphate is converted to pyruvate in several reactions that include two ATP-forming steps. (*Section 24.3*)

Avogadro's number—6.02×10^{23}. (*Section 3.14*)

barometer—an instrument designed to measure atmospheric pressure. (*Section 4.1*)

base—a substance capable of combining with hydrogen ions (protons). (*Section 7.1*)

base pairing—specific hydrogen bonding between adenine and thymine, or adenine and uracil, and cytosine and guanine, which is responsible for important features of structure and function of nucleic acids. (*Section 17.21*)

basic solution—a solution in which $[OH^-]$ is greater than $[H_3O^+]$. (*Section 7.6*)

Benedict's test—a test for aldehydes in which a copper(II) citrate complex oxidizes aldehydes to the carboxyl group and brown Cu_2O precipitates. (*Section 14.6*)

beta globulins—plasma proteins responsible for transporting lipids and various other materials. (*Section 27.2*)

beta oxidation—pathway for oxidation of fatty acyl groups in the mitochondria. (*Section 25.5*)

beta particle (β or ${}_{-1}^{0}e$)—a fast-moving electron that has been ejected from the nucleus of an atom undergoing beta decay. (Section 2.15)

beta sheet (β sheet)—extended secondary structure of protein. (*Section 17.10*)

bilayer—film of phospholipid molecules two molecules in thickness, with hydrophobic fatty acyl chains in the interior and hydrophilic head groups oriented to the surfaces of the film. (*Section 20.4*)

bile salt—a cholesterol derivative that is important in solubilizing fats for digestion. (*Section 23.3*)

binding—the specific interaction between a protein and a small molecule, for example. (*Section 21.1*)

biodegradable—able to be degraded by microorganisms or other organisms; for example detergents that can be degraded by microorganisms in soil or water. (*Section 25.7*)

biosynthesis—synthesis of organic compounds in cells. (*Section 22.1*)

biotin—a vitamin and coenzyme involved in carboxylation reactions such as that catalyzed by pyruvate carboxylase. (*Section 24.4*)

blood-brain barrier—membranes separating cerebrospinal fluid and bloodstream that prevent certain materials such as fatty acids from being utilized by brain tissue. (*Section 25.3*)

boiling point elevation—a colligative property of a solution due to a nonvolatile solute. (*Section 5.10*)

Boyle's law—a gas law that relates pressure and volume at constant amount and temperature; PV = constant. (*Section 4.4*)

branched hydrocarbon—a hydrocarbon chain that has one or more hydrocarbon groups in place of hydrogen atoms on the carbon chain. (*Section 9.2*)

Brownian motion—the random motion of colloidal particles caused by the collisions of solvent molecules with the particles. (*Section 5.11*)

buffer—a weak acid-conjugate base pair that counteracts the effect of added acids or bases to maintain nearly constant pH. (*Section 7.10*)

buffer system—a solution of a weak acid or weak base and its salt that is capable of resisting pH change when acid or base is generated in or added to the solution; for example, the bicarbonate buffer system of the blood. (*Section 27.4*)

calorie (cal)—a quantity of energy sufficient to raise the temperature of 1 gram of water 1°C; the calorie is defined as 4.184 joules. (*Section 1.16*)

carbohydrates—naturally occurring compounds that have the molecular formula $C_n(H_2O)_m$. (*Section 16.1*)

carbonyl group—the $-\overset{\displaystyle O}{\overset{\|}{C}}-$ functional group. (*Section 14.1*)

carboxyl group—the carboxylic acid functional group, $-\overset{\displaystyle O}{\overset{\|}{C}}-OH$. (*Section 14.1, 15.1*)

carboxylate ion—$R-\overset{\displaystyle O}{\overset{\|}{C}}-O^-$, the anion of a carboxylic acid in which R may be alkyl, aryl, or hydrogen. (*Section 15.5*)

carboxylic acid—$R-\overset{\displaystyle O}{\overset{\|}{C}}-OH$, in which R may be alkyl, aryl, or hydrogen. (*Section 15.2*)

carboxylic anhydride—$R-\overset{\displaystyle O}{\overset{\|}{C}}-O-\overset{\displaystyle O}{\overset{\|}{C}}-R$, in which R may be alkyl, aryl, or hydrogen. (*Section 15.7*)

carnivore—a meat eater. (*Section 22.6*)

catabolism—degradation of food molecules. (*Section 23.8*)

catalyst—any substance that speeds up a reaction but remains unchanged at the end of the reaction. (*Section 6.5*)

cation—a positive ion. (*Section 2.12*)

cell membrane—the limiting membrane that surrounds a cell. (*Section 21.2*)

cell wall—a (mainly) polysaccharide structure that surrounds plant and bacterial cells and helps protect them from swelling and other mechanical damage. (*Section 21.5*)

cellular respiration—O_2 consumption due to electron transport in mitochondria. (*Section 22.5*)

cellulose—fibrous polysaccharide consisting of glucose molecules in $\beta(1 \rightarrow 4)$ glycosidic linkage. (*Section 19.3*)

celsius scale (°C)—a temperature scale on which water freezes at 0°C and boils at 100°C. (*Section 1.14*)

cerebroside—a carbohydrate-containing sphingolipid. (*Section 20.5*)

change in free energy (ΔG)—the amount of useful energy released or absorbed when a chemical reaction takes place at constant temperature and pressure. (*Sections 1.15, 22.2*)

Charles's and Gay-Lussac's law—a gas law that relates volume and absolute temperature at constant amount and pressure; V/T = constant. (*Section 4.4*)

chemical change—a change in the composition as well as the form of a substance. (*Section 1.3*)

chemical energy—energy stored in substances that appears when the substances undergo chemical change. (*Section 1.15*)

chemical equilibrium—the situation when the rates of the forward and reverse reactions of a reversible chemical reaction become equal. (*Section 6.6*)

chemical reaction—a specific chemical change. (*Section 1.3*)

chemical symbol—a capital letter or a capital letter followed by a lowercase letter that stands for a particular element. (*Section 1.5*)

chiral—not superposable on the mirror image. (*Section 16.5*)

chitin—polysaccharide found in the insect exoskeleton. (*Section 19.4*)

chlorinated hydrocarbons—compounds containing only carbon, chlorine, and usually hydrogen. (*Section 11.1*)

chlorinated methanes—the organic products of the chlorination of methane: CH_3Cl, CH_2Cl_2, $CHCl_3$, and CCl_4. (*Section 11.1*)

chlorophyll—a green pigment of chloroplasts responsible for trapping and utilizing light energy to drive synthesis of ATP and NADPH. (*Section 22.18*)

chloroplast—a specialized organelle of green plants containing the enzymes and pigments of photosynthesis. (*Section 21.6*)

chondroitins—polysaccharides found in animal connective tissues. (*Section 19.4*)

chromatin—a complex of DNA and protein that is the substance of chromosomes. (*Section 21.6*)

chromosome—a DNA molecule with or without bound proteins. (*Section 21.5*)

chylomicrons—largest of blood lipoprotein particles, containing primarily triglyceride, little protein; only found in plasma soon after a fatty meal. (*Section 27.6*)

***cis-trans* isomers**—stereoisomers that differ in the orientation of two groups about a bond that cannot rotate freely such as a double bond. In the *cis* isomer, two similar groups X on different atoms occur on the same side of the double bond. In the *trans* isomer, the two groups occur on opposite sides. (*Section 8.10*)

```
X       X     X       H
 \     /       \     /
  C = C         C - C
 /     \       /     \
H       H     H       X
 cis isomer   trans isomer
```

citric acid cycle—cyclic pathway of eight enzyme catalyzed reactions responsible for complete oxidation of acetyl groups and provision of small molecules for biosynthesis. (*Section 23.9*)

cleavage—a reaction that splits a molecule into two parts. (*Section 12.13*)

codon—three nucleotide sequence of messenger RNA that specifies either an amino acid or termination of a genetic message. (*Section 26.7*)

coenzyme—an organic molecule that works with an enzyme to assist in catalysis or to carry a functional group or pair of reducing equivalents during a reaction. (*Section 21.11*)

coenzyme A (CoA or CoASH)—coenzyme containing within its structure the vitamin pantothenic acid, and responsible for carrying fatty acyl and certain other acyl groups in metabolism. (*Section 24.8*)

coenzyme Q—ubiquinone; a lipid-soluble compound found dissolved in the mitochondrial inner membrane; it is one of the members of the electron transport system. (*Section 22.15*)

cofactor—a metal ion or other molecule that is necessary for activity of an enzyme. (*Section 21.11*)

collecting phase of glycolysis—initial phase of glycolysis which allows various hexoses to be converted to glyceraldehyde-3-phosphate for oxidation in the ATP-synthetic phase of the pathway. (*Section 24.3*)

colligative property—any property that depends only on the relative numbers of particles present and not on their identity. (*Section 5.10*)

colloid—a particle between 10 and 1000 times as large as most molecules. (*Section 5.11*)

colloidal dispersion—a mixture of colloidal particles and a solvent such as water. (*Section 5.11*)

combined gas law—a gas law that relates pressure, volume, and absolute temperature at constant amount; $\frac{PV}{T}$ = constant. (*Section 4.4*)

combustion—the reaction of a substance with oxygen. (*Section 9.6*)

committing step—an enzymatic step that is irreversible in the cell and commits a metabolite to being metabolized in a particular pathway. (*Section 22.5*)

complementarity—the "fit" between molecules that will interact specifically with one another. (*Section 21.1*)

complementary sequence—a polynucleotide sequence capable of exact base-pairing with another polynucleotide sequence when the two are antiparallel to one another. (*Section 17.21*)

complex lipid—a lipid containing saponifiable fatty acyl groups. (*Section 20.9*)

compound—a substance that always consists of the same elements combined in a fixed and definite proportion by weight and that can be broken down into the elements of which it is composed. (*Section 1.4*)

condensation—the change of a gas into a liquid. (*Section 4.6*)

condensed formula—a formula that shows each carbon in a chain with its hydrogens or other atoms as CH_3, CH_2, CH, CCl_2, $CHCl_2$, and so forth. An example is $CH_3CH_2CH_3$. (*Section 8.4*)

conformations—arrangements of the same molecule that differ only by rotation about a single bond. (*Section 8.11*)

conjugated protein—a protein that in its functional form contains some nonprotein component such as lipid, polysaccharide, a metal or other cofactor, etc. (*Section 17.3*)

conjugate acid—the new species formed when a base combines with an H^+ ion. (*Section 7.1*)

conjugate base—the fragment remaining after an acid has lost an H^+ ion. (*Section 7.1*)

conversion factor—an equation that expresses the same quantity in different units. (*Appendix 1*)

coupling—(1) the simultaneous occurrence of a reaction with a negative $\Delta G°$ and a reaction with a positive $\Delta G°$, both catalyzed by the same enzyme. The overall $\Delta G°$ is negative, and products of both reactions are formed. (2) Occurrence of two or more reactions in sequence, such that the product of a reaction with a positive $\Delta G°$ is removed by a second reaction having a negative $\Delta G°$. The concentration of the product of the first reaction is kept so low that its actual ΔG is negative. (*Section 22.9*)

covalent bonding—bonding resulting from the sharing of electrons by atoms. (*Section 3.7*)

covalent compound—a compound in which the atoms share electrons; a compound of nonmetal atoms. (*Section 3.7*)

cracking—the process of heating a hydrocarbon mixture to between 500 and 1000°C in the absence of air. In this process, molecules break apart and reform to give mixtures of different composition. (*Focus 6*)

crenation—the shriveling of red blood cells as water passes out of the cells into a hypertonic solution. (*Section 5.12*)

crystal lattice—the orderly three-dimensional pattern of the atoms, molecules, or ions that make up a crystalline solid. (*Section 4.11*)

crystalline solid—a solid whose units (atoms, ions, or molecules) are arranged in an orderly, three-dimensional pattern. (*Section 4.11*)

Curie (Ci)—a specified rate of nuclear decay equal to 3.7×10^{10} nuclear disintegrations per second. (*Section 2.19*)

curved arrow notation—notation for expressing a series of chemical transformations in which the product of one reaction is the reactant for the next. Reagents and products that are not involved in the next reaction in the sequence are shown entering and leaving the reaction sequence via curved arrows. (*Section 14.11*)

cyanohydrin—the product of addition of HCN to an aldehyde or ketone. (*Section 14.7*)

$$R-\underset{\displaystyle CN}{\overset{\displaystyle OH}{\underset{|}{\overset{|}{C}}}}-R' \qquad (\text{R and R}' = \text{alkyl, aryl, or hydrogen})$$

cyclic AMP—a special nucleotide that is involved in regulation of metabolism; often referred to as a second messenger. (*Section 24.17*)

cyclic ether—a ring compound in which one atom of the ring is an oxygen atom. (*Section 12.14*)

cycloalkanes (C_nH_{2n})—cyclic hydrocarbons that have only single bonds. (*Section 9.7*)

cytochrome—one of several different heme proteins that are electron transport enzymes in mitochondria or chloroplasts. (*Section 22.15*)

cytoplasm—the cell substance between the cell membrane and the nuclear membrane. (*Section 21.5*)

cytosol—the soluble portion of the cell cytoplasm. (*Section 21.3*)

daughter isotope—the isotope produced by the radioactive decay of an atom. (*Section 2.16*)

DDT—chlorinated hydrocarbon pesticide whose formula is (*Section 11.3*)

$$Cl-C_6H_4-\underset{}{\overset{\displaystyle CCl_3}{\overset{|}{CH}}}-C_6H_4-Cl$$

degeneracy of genetic code—existence of more than one codon to specify a single amino acid. (*Section 26.7*)

dehydration reaction—the elimination of H_2O from a molecule. (*Section 10.5*)

dehydrogenase—an enzyme that catalyzes dehydrogenation of a substrate molecule and transfers the pair of reducing equivalents to a coenzyme acceptor such as NAD. (*Section 22.16*)

dehydrohalogenation reaction—the elimination of HCl, HBr, or HI from a molecule. (*Section 10.5*)

denaturation—loss of the native functional structure of a protein. (*Section 17.11*)

density—the mass of a sample of material per unit volume. (*Section 1.7*)

deoxyribonucleic acid (DNA)—polymer of deoxyribonucleotides generally consisting of two complementary chains that are base-paired along their entire lengths. (*Section 17.15*)

destructive distillation—the process of decomposing a substance by heating it in the absence of oxygen. (*Section 10.18*)

diabetes mellitus—disease in which insulin production is inadequate (or cells do not respond normally to insulin) resulting in poor carbohydrate utilization and excessive dependence on fatty acids for energy. (*Section 24.1*)

dialyzing membrane—a membrane that allows water molecules, ions, and small molecules to pass but holds back large molecules and colloids. (*Section 5.12*)

dipole—polar bonds possess bond dipoles due to unequal sharing of electrons in a covalent bond. Some molecules with polar bonds are polar (such as H_2O) and others are nonpolar, due to cancellation of the bond dipoles (such as CO_2). (*Section 5.1*)

disaccharide—a carbohydrate that can be hydrolyzed to two monosaccharide units. *Trisaccharides, tetrasaccharides,* and *polysaccharides,* are hydrolyzed to three, four, and many monosaccharide units. *Oligosaccharides* are small polysaccharides. (*Section 16.1*)

displacement (substitution)—a reaction in which one ion or group is displaced by another. In the alkylation of amines by alkyl halides, the amine molecule displaces a halide ion. (*Section 13.7*)

disulfide—a compound that has the —S—S— functional group. (*Sections 12.19, 17.9*)

disulfide linkage—covalent linkage (—S—S—) between cysteine residues of a protein that contributes to maintenance of its three-dimensional structure. (*Section 17.9*)

double bond—a covalent bond composed of two pairs of shared electrons (C::C; also shown as C=C). (*Section 8.7*)

double helix—the native three-dimensional structure of DNA in which two polynucleotide strands coil about one another. (*Section 18.7*)

dynamite—a high explosive made by impregnating an inert, porous solid with glyceryl trinitrate. Dynamite is more shock-resistant than glyceryl trinitrate. (*Section 11.7*)

electron—a negatively charged, subatomic particle of low mass; the smallest known negative charge. (*Section 2.5*)

electron transport system—series of redox enzymes in mitochondrial inner membrane; responsible for oxidation of succinate and NADH using O_2 and coupled to ATP synthesis (oxidative phosphorylation). (*Section 22.15*)

electronegativity—the relative attraction for electrons by an atom in a molecule. (*Section 3.9*)

element—a substance that cannot be subdivided into simpler chemical substances. (*Section 1.4*)

elimination reaction—a reaction in which a small molecule is lost from a larger one and in the process a double or triple bond is formed. (*Section 10.5*)

elongation of a polypeptide—stage of polypeptide synthesis following initiation, in which amino acids are linked together as the messenger RNA codons are translated. (*Section 26.11*)

emulsifier—a substance that causes two mutually insoluble liquids to remain in an emulsion. (*Focus 11*)

emulsion—a colloidal dispersion of two or more mutually insoluble liquids. (*Focus 11*)

enantiomers—two objects that are related as an object and its nonsuperposable mirror image. Also called *mirror image isomers.* (*Section 16.5*)

endergonic reaction—an energy-requiring reaction, one that consumes free energy and thus has a positive ΔG. (*Section 22.2*)

endocrine gland—tissue that manufactures and releases one or more hormones directly into the blood. (*Section 27.14*)

endothermic reaction—a chemical reaction that absorbs heat as it occurs. (*Section 1.13*)

energy—the capacity to alter substances. (*Section 1.1*)

enzymes—protein molecules that catalyze certain biochemical reactions. (*Sections 6.5, 17.3, 21.8*)

epinephrine (adrenaline)—a hormone that stimulates glycogen breakdown in muscle by an activation cascade involving cyclic AMP as the "second messenger." (*Section 24.17*)

epoxide—a cyclic ether in which the oxygen atom is part of a three-membered ring. (*Section 12.14*)

equilibrium constant—the ratio of the concentrations of the products and reactants raised to the power given by the coefficients in the balanced equation. (*Section 6.7*)

equilibrium state—state in which the rates of the forward and reverse reactions of a reversible process are equal. (*Section 4.7*)

erg—an energy unit; 10^7 erg = 1 J. (*Section 2.19*)

erythrocyte—red blood cell; responsible for oxygen transport; contains hemoglobin. (*Section 27.1*)

essential amino acid—an amino acid that the body cannot make and which must be obtained from dietary proteins. (*Section 23.6*)

essential fatty acid—a polyunsaturated fatty acid (linoleic acid) that the human body cannot make, and which must be obtained in the diet. (*Section 20.1*)

ester—$R-\overset{\overset{\displaystyle O}{\|}}{C}-OR'$, where R may be alkyl, aryl, or hydrogen, and R′ may be alkyl or aryl. (*Section 15.7*)

esterification—the reaction of an alcohol with a carboxylic acid to produce an ester. (*Section 12.4*)

ether—a compound in which an oxygen atom is bonded to two hydrocarbon groups. (Section 12.10)

exergonic reaction—a chemical reaction that gives off free energy as it occurs; ΔG for the reaction is negative. (*Section 22.2*)

exothermic reaction—a chemical reaction that releases heat as it occurs. (*Section 1.13*)

extracellular fluid—all fluid in body that is not contained in cells. (*Section 27.11*)

facilitated diffusion—transport of a molecule or ion across a membrane by a carrier molecule, in a process that does not require energy because it occurs with, rather than against, the concentration gradient. (*Section 22.11*)

Fahrenheit scale (°F)—a temperature scale on which water freezes at 32°F and boils at 212°F. (*Section 1.7*)

fat—an ester of glycerol and long-chain carboxylic acids that is a solid at room temperature. (*Section 12.9*)

fat-soluble vitamin—one of the several vitamins (A, D, E, and K) that are nonpolar by nature and thus fat-soluble rather than water-soluble. (*Section 20.7*)

feedback inhibition—inhibition of an enzyme that occurs near the beginning of a metabolic pathway by an end-product of the pathway. (*Section 21.16*)

Fehling's test—a test for complex aldehydes in which copper(II) tartrate complex oxidizes the aldehyde to a carboxylic acid, and brown Cu_2O precipitates. (*Section 14.6*)

fermentation—any pathway for degradation of carbohydrate or other organic substance, which occurs in the absence of oxygen, for example, anaerobic glycolysis. (*Section 24.2*)

fibrin—insoluble fibrous protein of blood clot, formed from fibrinogen. (*Section 27.1*)

fibrinogen—soluble blood protein that gives rise to fibrin when clotting occurs. (*Section 27.1*)

fibrous protein—protein that takes a fibrous structure in its functional form, such as the alpha keratins, collagen, and fibroin. (*Section 17.7*)

first law of thermodynamics—states that energy cannot be created or destroyed but only converted from one form to another. (*Section 20.1*)

Fischer projection formulas—the two-dimensional projection of a three-dimensional formula in the following form. (*Section 16.9*)

```
      A               A
      |               ¦
C ----+---- D  =  C ►-C-◄ D
      |               ¦
      B               B
```

flavoprotein—a redox enzyme containing tightly bound FAD or FMN. (*Section 22.15*)

flavor enhancer—a substance added to foods to improve their natural flavor. (*Focus 11*)

flavoring agent—a substance added to foods to impart flavor. (Focus 11)

fluid-mosaic model—current model of biological membrane structure in which a phospholipid bilayer provides a fluid matrix in which numerous protein molecules are embedded as in a mosaic. (*Section 21.4*)

fluorocarbons–compounds that contain only the elements carbon and fluorine. (*Section 11.6*)

food additive–any substance intentionally added to food for a functional purpose. (*Focus 11*)

food chain–sequence of organisms that feed upon one another. (*Section 22.6*)

food coloring–a substance added to foods to impart color. (*Focus 11*)

formula unit–the number of ions of an ionic compound indicated by the formula of the compound. (*Section 3.13*)

formula weight–the sum of the atomic weights of all atoms in the formula unit. (*Section 3.14*)

fractional distillation–separation of components of a mixture that have different boiling points by heating of the mixture and collection of the condensed vapor in different fractions. (*Focus 6*)

free energy (G)–a measure of the chemical energy stored in a substance. (See also *change in free energy.*) (*Section 1.15*)

free energy of activation–the energy barrier between the reactants and products of a chemical reaction. (*Section 6.4*)

freezing point lowering–a colligative property of a solution. (*Section 5.10*)

freons–compounds of carbon, fluorine, and chlorine used as refrigerants and aerosol propellants. (*Section 11.6*)

fructosuria–appearance of fructose in urine due to inability of body to metabolize it. (*Section 24.7*)

functional group–an atomic arrangement or grouping that occurs commonly and that behaves the same regardless of the molecule of which it is a part. (*Section 8.12*)

functional structure–the normal three-dimensional structure of a protein, which possesses biological activity. (*Section 17.2*)

furanose–the five-membered ring of the hemiacetal or acetal form of a monosaccharide. (*Section 16.11*)

galactosemia–appearance of galactose in the blood due to inability to metabolize it. (*Section 24.7*)

gamma globulins–plasma proteins that are antibodies against foreign substances. (*Section 27.2*)

gamma ray (γ)–electromagnetic radiation often emitted by nuclei undergoing radioactive decay; similar to but more energetic and penetrating than x rays. (*Section 2.15*)

gas–that physical state of a substance in which the substance assumes the volume and shape of its container. (*Section 2.3*)

gas laws–quantitative relationships for gases involving two or more of the properties of pressure, volume, absolute temperature, and amount. (*Section 4.4*)

gene–a segment of a DNA molecule containing instructions for synthesis of a protein or RNA molecule. (*Section 26.1*)

genetics–study of the transmission of physical and other characteristics from generation to generation. (*Section 26 Intro.*)

globular protein–a protein that exists in a globular or roughly spherical shape; most proteins are globular proteins. (*Section 17.7*)

glucagon–a hormone important in regulation of plasma glucose levels; acts on liver cells to promote glycogen breakdown by an activation cascade; result is export of glucose to the bloodstream. (*Section 24.16*)

gluconeogenesis–synthesis of glucose from small molecules such as lactate, amino acids, and pyruvate. (*Section 24.11*)

glucosuria–appearance of glucose in the urine; a symptom of diabetes. (*Section 24.16*)

glycogen–animal storage polysaccharide consisting of glucose in $\alpha(1 \rightarrow 4)$ and $\alpha(1 \rightarrow 6)$ linkages. (*Section 19.3*)

glycolysis–metabolic pathway in which glucose (or other sugars) may be degraded to pyruvate, occurring with synthesis of ATP. (*Section 24 Intro.*)

glycoside–acetal form of a carbohydrate. (*Section 16.12*)

glycosidic bond–the bond between the anomeric carbon of a sugar and a hydroxyl oxygen of another sugar or a nitrogen atom of an organic base. (*Section 18.3*)

golgi apparatus–organelle consisting of membrane channels and vesicles usually located adjacent to the nucleus and responsible for packaging certain cell products into membrane sacs. (*Section 21.7*)

Grignard reagent—RMgX, where R = alkyl or aryl, and X = Cl, Br, or I. Grignard reagents are prepared by the reaction of an alkyl or aryl halide with Mg in ether solution. (*Section 14.7*)

gutta percha—the natural rubber in which all the double bonds are *trans.* (*Focus 8*)

half-life—the time required for one half of a given number of radioactive atoms to decay. (*Section 2.17*)

halogenation—the reaction of a compound with halogen (Cl_2 or Br_2). (*Section 9.6*)

Haworth formula—a hexagonal representation of a cyclic structure. (*Section 16.11*)

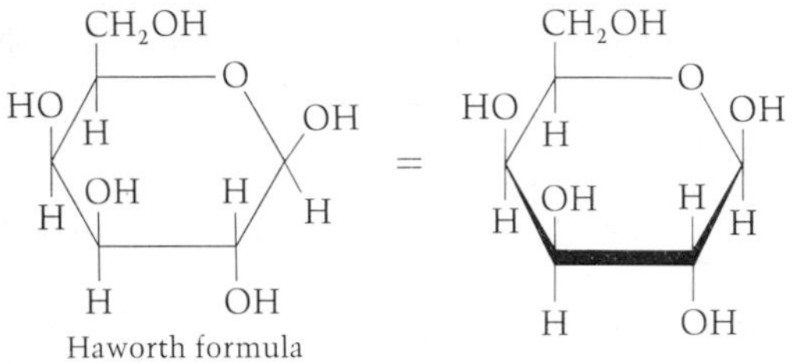

Haworth formula

heat—energy that transfers naturally from a hotter material to a cooler material. (*Section 1.13*)

heat of fusion—the amount of heat per gram that must be added or removed to melt or freeze a substance at its freezing temperature. (*Section 4.10*)

heat of vaporization—the amount of heat per gram that must be added or removed to vaporize or condense a substance at its boiling point. (*Section 4.8*)

hematocrit—the percent of blood volume contributed by the blood cells. (*Section 27.1*)

hemiacetal—$R-\overset{\displaystyle OH}{\underset{\displaystyle OR'}{C}}-H$, in which R may be alkyl or aryl, and R′ is alkyl. Hemiacetals exist in equilibrium with the aldehyde and the alcohol from which they are prepared. (*Section 14.7*)

hemiketal—$R-\overset{\displaystyle OR''}{\underset{\displaystyle OR''}{C}}-R'$, in which R and R′ may be alkyl or aryl, and R″ is alkyl. Hemiketals exist in equilibrium with the ketone and the alcohol from which they are prepared. (*Section 14.11*)

hemolysis—the rupture of red blood cells as water passes from a hypotonic solution into the cells. (*Section 5.12*)

hemophilia—genetic disease in which blood does not clot normally. (*Section 27.3*)

herbivore—a plant eater. (*Section 22.6*)

heterocyclic compounds—cyclic molecules in which the ring contains at least one atom that is not carbon. (*Section 13.9*)

heterogeneous catalyst—a surface catalyst; a catalyst that is in a different physical state from that of the reactants. (*Section 6.5*)

heterogeneous mixture—a mixture in which two or more substances are visually distinguishable from one another. (*Section 2.4*)

hevea—the natural rubber in which all the double bonds are *cis.* (*Focus 8*)

high-energy compound—a compound with a standard free energy of hydrolysis more negative than −7 kcal/mole. (*Section 22.8*)

homogeneous catalyst—a catalyst that is in the same physical state as the reactants. (*Section 6.5*)

homogeneous mixture—a mixture which is visually uniform and in which the substances are uniformly intermingled. (*Section 2.4*)

humectant—a substance that aids in retaining moisture. (*Section 12.9*)

hyaluronic acid—a polysaccharide found in animal connective tissues. (*Section 19.4*)

hydration of ions—the attraction between a positive ion and the negative end of the water molecule or between a negative ion and the positive end of the water molecule. (*Section 5.3*)

hydration reaction—the chemical combina- of a compound with water. (*Section 10.6*)

hydrogenation reaction—the chemical combination of a compound with hydrogen. (*Section 10.6*)

hydrogen bonding—the attraction between a partially positive hydrogen atom that is bonded to an electronegative atom and an adjacent electronegative atom. (*Section 5.4*)

hydrohalogenation reaction—the addition of HCl, HBr, or HI to a double bond. (*Section 10.6*)

hydrophobic interaction—noncovalent interaction between hydrophobic molecules or parts of molecules that tends to segregate hydrophobic and polar entities, as in formation of micelles or in globular protein structure. (*Section 17.2*)

hydroxyl group—the —OH group. (*Section 12.1*)

hyperglycemia—elevated blood glucose levels. (*Section 24.16*)

hyperammonemia—elevated ammonia levels in the blood. (*Section 25.21*)

hypertonic solution—a solution with a higher solute particle concentration than another solution. (*Section 5.11*)

hypoglycemia—excessively low blood glucose levels. (*Section 24.16*)

hypotonic solution—a solution with a lower solute particle concentration than another solution. (*Section 5.11*)

ideal gas law—a gas law that relates pressure, volume, absolute temperature, and moles; $PV = nRT$, where R is the universal gas constant. (*Section 4.4*)

initiation of a polypeptide—process in which intact ribosome is assembled with *m*RNA, initiation factors, and methionyl $t\text{RNA}^{\text{met}}$ to begin polypeptide synthesis. (*Section 26.10*)

insulin—polypeptide hormone secreted by the pancreas and required for normal metabolism of glucose. (*Section 24.1*)

intermolecular attractions—attractions between molecules; in water the attractions stem from dipolar interactions and hydrogen bonding. (*Section 5.6*)

interstitial fluid—fraction of extracellular fluid that is found outside of the blood vessels directly bathing cells of the tissues. (*Section 27.11*)

intracellular fluid—total volume of fluid contained within the body's cells. (*Section 27.11*)

ionic compound—a compound that consists of positive and negative ions. (*Section 3.4*)

ionization energy—the minimum energy required to remove an electron from a gaseous atom. (*Section 2.11*)

irreversible reaction—a biochemical reaction with a large negative standard free energy change in which the equilibrium lies very far to the side of the products and very little back reaction occurs. (*Section 22.5*)

isoelectric pH—the pH at which the net charge on a molecule (such as a protein or amino acid) is equal to zero. (*Section 17.5*)

isoenzyme (isozyme)—one of two or more forms of an enzyme that catalyze the same reaction. (*Section 21.14*)

isomers—different compounds that have the same molecular formula. (*Section 8.8*)

isotonic solution—a solution with a solute particle concentration the same as another solution. (*Section 5.11*)

isotopes—atoms of the same element with different mass numbers, that is, atoms with the same atomic number but different numbers of neutrons in their nuclei. (*Section 2.13*)

joule (J)—the derived SI unit of energy. (See Appendix 2.) (*Section 1.16*)

K_a—an acid constant; an equilibrium constant that indicates the strength of a weak acid. (*Section 7.4*)

K_b—a base constant; the equilibrium constant that indicates the strength of a weak base. (*Section 7.4*)

kelvin (K)—the scale used to designate absolute temperature and related to the celsius scale by the equation, $K = °C + 273$. (*Section 4.2*)

ketal—$\mathrm{R{-}\overset{\overset{\large OR''}{|}}{\underset{\underset{\large OR''}{|}}{C}}{-}R'}$, in which R and R′ may be alkyl or aryl, and R″ is alkyl. Ketals are prepared by reacting a ketone, $\mathrm{R{-}\overset{\overset{\large O}{\|}}{C}{-}R'}$, with an alcohol, R″—OH, in the presence of a strong acid. (*Section 14.11*)

ketone—$\mathrm{R{-}\overset{\overset{\large O}{\|}}{C}{-}R'}$, where R and R′ may be alkyl or aryl. (*Section 14.8*)

ketoacidosis—presence of excessive amounts of ketone bodies in the blood leading to lowered blood pH. (*Section 25.9*)

ketone bodies—acetoacetic acid, beta hydroxybutyric acid (and acetone). (*Section 25.9*)

ketose—a ketone sugar. (*Section 16.3*)

kinase—an enzyme that catalyzes the phosphorylation of a substrate using ATP as source of the phosphate group (or the reverse reaction). (*Section 24.4*)

kinetic energy—energy of motion. (*Section 1.12*)

kinetics—the study of reaction rates. (*Section 21.13*)

K_w—the water constant; $K_w = [H_3O^+][OH^-] = 1.0 \times 10^{-14}$ at 25°C. (*Section 7.7*)

lactase—an enzyme that catalyzes hydrolysis of lactose. (*Section 23.2*)

lactose—"milk sugar"; disaccharide consisting of glucose and galactose in $\alpha(1\rightarrow4)$ glycosidic linkage. (*Section 22.2*)

lanthanide series—the elements with atomic numbers 58–71. (*Sections 2.10, 2.14*)

Le Chatelier's Principle—a statement applicable to an equilibrium; equilibrium position will shift where possible so as to oppose a disturbance. (*Section 6.9*)

leukocyte—white blood cell, of which there are several types; all are involved in defending the body against infection. (*Section 27.1*)

Lewis dot structure—a representation of the electrons in the valence shell of an atom by means of dots placed around the chemical symbol for the atom. (*Section 2.9*)

line-dash-wedge formula—expresses the three-dimensional nature of bonds. A wedge (➤) signifies a bond coming out of the page toward the reader. A dash (----) signifies a bond directed back, behind the page away from the reader. A line (—) signifies a bond in the plane of the page. (*Section 8.5*)

line formula—formulas in which lines represent bonds. Carbons and hydrogens are not shown; the meeting of two lines or the end of a line represents a carbon atom with enough hydrogen atoms to satisfy the four values of carbon. (*Section 9.7*) For example,

$$CH_3CH_2CH(CH_3)CH_3 =$$

lipase—an enzyme that catalyzes hydrolysis of fatty acids from triacylglycerol molecules. (*Section 23.3, 25.2*)

lipid—one of a diverse class of biomolecules that are soluble in organic solvents and rather insoluble in water. (*Section 20 Intro.*)

liquid—that physical state of a substance in which the substance has its own volume but assumes the shape of its container. (*Section 2.3*)

lysosome—organelle containing a variety of hydrolytic enzymes. (*Section 21.7*)

macromolecule—a very large polymeric molecule such as a protein, polysaccharide, or nucleic acid. (*Section 17 Intro.*)

maltose—disaccharide consisting of two molecules of glucose in $\alpha(1\rightarrow4)$ glycosidic linkage. (*Section 23.2*)

mass—the quantity of matter in a material. (*Section 1.1*)

mass number—the sum of the number of protons and neutrons in a nucleus of an atom. (*Section 2.13*)

mass spectrograph—an instrument that separates chemical species according to their masses; useful in determining the masses of isotopes. (*Section 2.13*)

matter—the physical stuff that makes up the universe. (*Section 1.1*)

membrane transport system—a protein or proteins that can carry or provide a channel for passage of certain molecules or ions across a biological membrane. (*Sections 21.4, 22.11*)

messenger RNA—single stranded polyribonucleotide containing a nucleotide sequence transcribed from a gene; this sequence will be translated into the amino acid sequence of a polypeptide. (*Section 26.6*)

metabolism—the sum total of all the anabolic and catabolic reactions a cell organism can perform. (*Section 23.8*)

metabolic pathway—a sequence of enzyme-catalyzed reactions that accomplish the synthesis or degradation of some biomolecule. (*Section 21.8*)

metal—a sample of matter that is a good conductor of heat and electricity and that will bend without breaking. (*Section 2.10*)

meta-substitution—substitution at the 1,3-positions of benzene. (*Section 10.15*)

metric system—a measurement system based on powers of 10, composed of base units, prefixes, and derived units. (*Section 1.6*)

micelle—a spherical aggregate of phospholipid or fatty acid molecules oriented so as to bury the hydrophobic portions of the molecules in the interior and expose the polar groups on the surface. (*Section 20.4*)

millimeter of mercury (mmHg)—a unit of pressure that refers to the height in millimeters of a column of mercury that can be supported by that pressure. (*Section 4.1*)

mitochondria—organelles that are responsible for most of a cell's energy production. (*Section 21.6*)

mobilization of fatty acids—release of fatty acids into the bloodstream following hydrolysis of adipose tissue fat molecules. (*Section 25.2*)

molarity (M)—a concentration scale defined as moles of solute per liter of solution. (*Section 5.9*)

mole—Avogadro's number of units. (*Section 3.14*)

mole ratio—the coefficient of one reactant or product in a balanced chemical equation divided by the coefficient of one of the other reactants or products. (*Section 3.14*)

molecular formula—a formula that shows only the atomic composition; for example, C_2H_6. (*Section 8.4*)

molecular genetics—the study of the biochemistry of genetics. (*Section 26 Intro.*)

molecular orbitals orbitals formed by the overlapping of an atomic orbital of one atom with an atomic orbital of another atom. (*Section 10.3*)

molecular weight—the sum of the atomic weights of all the atoms in a molecule of a covalent compound. (*Section 3.14*)

molecule—the smallest particle of chemical material that makes up a covalent compound. (*Section 2.1*)

monolayer—a monomolecular film of phospholipid or fatty acid molecules. (*Section 20.4*)

monosaccharide—a simple, nonhydrolyzable carbohydrate. (*Section 16.1*)

mucoprotein—a protein with polysaccharide conjugated to it; component of animal connective tissues. (*Section 19.4*)

mutarotation—the conversion of either hemiacetal form into a mixture of both forms. (*Section 16.11*)

mutase—an enzyme that catalyzes intramolecular rearrangement of some functional group, such as a phosphate group. (*Section 22.6*)

mutation—a change in the nucleotide sequence of a gene that changes its information content. (*Section 26.14*)

myofibril—the contractile unit of the muscle cell, made up of thick and thin protein filaments. (*Section 22.10*)

myosin—the protein of the thick filaments of the myofibril. (*Section 22.10*)

NAD-linked dehydrogenase—a redox enzyme that employs NAD as coenzyme. (*Section 21.2*)

native conformation—the naturally occurring three-dimensional structure of a protein in which it exhibits its normal biological activity. (*Section 17.11*)

negative ion—an atom, group of atoms, or molecule that has gained one or more electrons and hence bears a negative electrical charge. (*Section 2.12*)

neutral solution—a solution in which $[H_3O^+]$ and $[OH^-]$ are equal. (*Section 7.6*)

neutron—an electrically neutral particle; a constituent of the nuclei of all atoms *except* the most common isotope of hydrogen. (*Section 2.5*)

nitration reaction—the reaction of an organic compound with nitric acid to produce a nitro compound, in which a carbon atom is bonded to the nitrogen of an NO_2 group. An example is the nitration of benzene and its derivatives with nitric acid in the presence of H_2SO_4 to produce nitrobenzene. (*Section 10.16*)

$$C_6H_6 + HONO_2 \xrightarrow{H_2SO_4} C_6H_5NO_2 + H_2O$$

nitroglycerine (glyceryl trinitrate)—nitrate ester of glycerol; a high explosive. (*Section 11.7*)

N-nitrosamines—($R_2N-N=O$) carcinogenic compounds produced by the reaction of inorganic nitrites with amines in the presence of acids. N-Nitrosamine may result in the human stomach when inorganic nitrites are ingested. (*Section 11.9*)

noble gases—the elements helium, neon, argon, krypton, xenon, and radon. (*Section 2.10*)

nonmetal—a material that is not metal; nonmetals are typically poor conductors of heat and electricity. (*Section 2.10*)

nonspontaneous reaction—reaction for which ΔG is positive. (*Section 1.15*)

nucleic acid—a polynucleotide: DNA or RNA. (*Section 18 Intro.*)

nucleoid—region of bacterial cell cytoplasm where chromosome is found. (*Section 21.5*)

nucleoside—a compound containing a heterocyclic base in beta-glycosidic linkage with a ribose or deoxyribose sugar. (*Section 18.3*)

nucleotide—a nucleoside phosphate, most commonly a nucleoside-5′-phosphate. (*Section 18.1*)

nucleus—(1) the tiny central part of an atom that accounts for most of the mass of the atom. (*Section 2.5*) (2) Organelle with double membrane that contains the chromosomes of a plant or animal cell. (*Section 21.6*)

octet rule—the tendency for atoms to achieve a valence shell of eight electrons when they react. The octet rule is followed most often by the representative elements. (*Section 3.3*)

optical activity—the rotation of the plane of polarization of plane-polarized light by a chiral substance. (*Section 16.7*)

orbitals—the various regions of space around the nucleus of an atom in which electrons are likely to be found. (*Section 2.7*)

orbital hybridization—the combination of atomic orbitals (s and p) to form *hybrid orbitals* such as sp^3, sp^2, and sp. (*Section 8.6*)

organelle—"little organ"; membrane-surrounded structure of cells with nuclei that serves a specialized function; examples are mitochondria, lysosomes, etc. (*Section 21.6*)

organic compounds—compounds of carbon; they are also defined as the hydrocarbons and their derivatives. (*Section 8.1*)

ortho substitution—substitution at the 1,2-positions of benzene. (*Section 10.15*)

osmotic pressure—the pressure that develops across a semipermeable membrane separating a solution from pure solvent. (*Section 5.10*)

oxidation—a real or apparent loss of electrons. (*Section 3.12*)

oxidative phosphorylation—ATP synthesis occurring in the mitochondria and coupled to electron transport to O_2. (*Section 22.17*)

para substitution—substitution at the 1,4-positions of benzene. (*Section 10.15*)

paraffin—an alkane. (*Section 9.6*)

peptide—a molecule containing two or more amino acids linked by peptide bonds. (*Section 17.1*)

peptide bond—the amide $-\overset{\displaystyle O}{\overset{\|}{C}}-N-$ bond in proteins. (*Section 15.16*)

peptidyl site (Psite)—ribosomal site at which initiating *t*RNA (methionyl-*t*RNA^{met}) binds; also occupied by growing polypeptide chain (bound to a *t*RNA) during an elongation cycle in protein biosynthesis. (*Section 26.11*)

periodic table—an arrangement of elements according to increasing atomic number such that elements with similar properties are grouped together. (*Section 2.10*)

pH—defined as $-\log [H_3O^+]$. (*Section 7.8*)

pH titration curve—a plot of pH *vs* volume of added standard solution in a titration. (*Section 7.13*)

phenol—compound in which a hydroxyl group is bonded to a carbon atom that is part of an aromatic ring. (*Section 12.5*)

phenoxide—the conjugate base of a phenol, ArO^-. (*Section 12.8*)

phenylketonuria (PKU)—genetic disease in which phenylalanine is not metabolized normally; its breakdown products accumulate in the blood and urine; may lead to mental retardation if not detected and treated in early infancy. (*Section 25.21*)

phosphatidic acid—a glycerol triester containing two fatty acid chains and a phosphate group. (*Section 20.3*)

phosphodiester linkage—linkage between the 5′-phosphate group of one nucleotide and the 3′-hydroxyl group of another. (*Section 18.5*)

phosphoglyceride—one of a class of compounds containing two fatty acids and a phosphate esterified to glycerol, with an alcohol esterified in turn to the phosphate group. (*Section 20.3*)

phospholipid—a phosphate-containing lipid: a phosphoglyceride or a sphingomyelin. (*Section 20.5*)

phosphorylation—formation of a phosphate ester by transfer of a phosphate group from a donor to an acceptor molecule. (*Section 24.4*)

photophosphorylation—ATP synthesis occurring in conjunction with photosynthetic electron transport in the chloroplast. (*Section 22.18*)

photosynthesis—process of converting light energy into chemical energy carried out by the chloroplasts of green plants. (*Section 22.18*)

photosystem—one of two connected electron transport systems in the chloroplast, each employing a different chlorophyll pigment; Photosystem I contains P_{700} and Photosystem II contains P_{670} (*Section 22.18*)

physical change—a change in the form but not the composition of a substance. (*Section 1.2*)

Pi (π) orbital—The molecular orbital formed by the overlap of *p*-type orbitals so that the bonding electrons lie above and below the plane of the atoms. (*Section 10.3*)

pK_a—the negative logarithm of K_a. (*Section 7.9*)

pK_b—the negative logarithm of K_b. (*Section 7.9*)

plane-polarized light—light that has been filtered so that all of its waves vibrate in one plane. (*Section 16.6*)

plasma—the liquid portion of the blood in which the blood cells are suspended. (*Section 27.1*)

platelet—cellular element of the blood involved in clotting. (*Section 27.1*)

point mutation—mutation in which only a single nucleotide in the sequence of a gene is altered. (*Section 26.14*)

polar covalent bond—a covalent bond in which the electrons are shared unequally by atoms. (*Section 3.10*)

polarimeter—instrument used to measure the angle of rotation of plane-polarized light by an optically active sample. (*Section 16.7*)

polyamide—a polymer in which the repeating units are connected through amide bonds. (*Section 15.17*)

polyester—a polymer in which the repeating units are connected through ester linkages. (*Section 15.12*)

polymerization—the combination of many small molecules (usually all the same) to form long molecular chains. (*Section 10.6*)

polynuclear aromatic hydrocarbon—an aromatic hydrocarbon that has multiple six-membered conjugated rings, such as naphthalene. (*Section 10.17*)

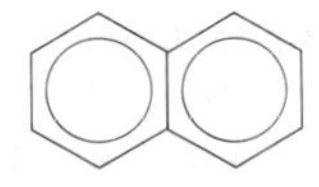

polynucleotide—polymer of nucleotides linked by 3′,5′-phosphodiester bonds. (*Section 18.5*)

polypeptide—polymer of many amino acids linked by peptide bonds. (*Sections 15.17, 17.1*)

polyprotic acid—an acid with more than one ionizable hydrogen atom; H_2SO_4. (*Section 7.14*)

polyprotic base—a base capable of combining with more than one hydrogen ion; for example, CO_3^{2-}. (*Section 7.14*)

polyribosome (polysome)—several ribosomes simultaneously translating the same messenger RNA. (*Section 26.13*)

polysaccharide—a polymer of many monosaccharides linked by glycosidic bonds. (*Section 19.1*)

polyunsaturated fatty acid—fatty acid containing two or more double bonds. (*Section 20.1*)

polyvinyl chloride (PVC)—polymer $\text{-}(CH_2CHCl)_n\text{-}$, prepared from vinyl chloride, $CH_2{=}CHCl$. (*Section 11.5*)

portal system—system of blood vessels that carries blood with absorbed nutrients from the intestines to the liver. (*Section 27.6*)

positive ion—an atom, group of atoms, or molecule that has lost one or more electrons and hence bears a positive electrical charge. (*Section 2.12*)

postabsorptive state—between-meal period, when nutrients are not being absorbed from the digestive tract. (*Section 27.5*)

potential energy—energy that is stored. (*Section 1.12*)

preservative—any chemical that inhibits the growth of microorganisms in foods. (*Focus 11*)

pressure—force divided by the area over which the force is exerted. (*Section 4.1*)

primary (1°) alcohol—an alcohol in which the hydroxyl-bearing carbon atom is bonded to one other carbon atom. RCH_2OH. (*Section 12.1*)

primary (1°) amine—an amine that has one hydrocarbon group bonded to nitrogen: RNH_2. (*Section 13.2*)

primary structure—the sequence of amino acids in a polypeptide or protein. (*Section 17.9*)

proof—a number that gives the alcohol content of beverages; twice the percent of alcohol. (*Section 12.9*)

prostaglandin—one of a class of lipid compounds derived from polyunsaturated fatty acids that have hormonelike activities. (*Section 20.8*)

protein—a polypeptide, or two or more polypeptides acting jointly, that possess biological activity. (*Section 17.3*)

proton—a particle that carries unit positive electrical charge; a constituent of the nuclei of all atoms. (*Section 2.5*)

pure substance—a sample of matter that consists of only one kind of molecule. (*Section 2.4*)

purine—parent compound of the nucleotide bases adenine and guanine. (*Section 18.2*)

pyranose—the six-membered ring of the hemiacetal or acetal form of a monosaccharide. (*Section 16.11*)

pyridoxal phosphate—coenzyme derived from vitamin B_6; essential coenzyme of transaminases. (*Section 25.18*)

pyrimidine—parent compound of the nucleotide bases uracil, cytosine, and thymine. (*Section 18.2*)

quantum level—a particular energy level around the nucleus of an atom into which orbitals of similar energy are grouped. (*Section 2.7*)

quaternary ammonium ion—R_4N^+ in which all four R groups are alkyl or aryl. (*Section 13.8*)

quaternary structure—the level of protein structure that involves the association of two or more polypeptide chains. (*Section 17.12*)

rack mechanism—model of enzyme action in which the enzyme is thought to stretch or twist the bonds of a substrate molecule, making reaction more likely to occur. (*Section 21.10*)

Rad (radiation absorbed dose)—an amount of radiation that leads to the absorption of 100 ergs (2.39×10^{-6} cal) by one gram of any material. (*Section 2.19*)

radioactive element—an element whose atoms have unstable nuclei that undergo decay. (*Section 2.15*)

radioisotope—an isotope that is radioactive. (*Section 2.16*)

random coil—a polypeptide, or segment of a polypeptide, that does not possess any specific secondary structure. (*Section 17.10*)

rare earth elements—an element of the Lanthanide series. (*Section 2.14*)

rate limiting step—the slowest step in a pathway, which determines the overall rate of the pathway. (*Section 21.16*)

redox—short for reduction-oxidation. (*Section 3.12*)

redox coenzyme—molecule such as NAD, NADP, FAD, or FMN that can be reversibly reduced and oxidized and is used as carrier of reducing equivalents when a dehydrogenase catalyzes oxidation of a substrate molecule. (*Section 22.12*)

reducing equivalent—hydrogen atom or electron. (*Section 22.13*)

reduction—a real or apparent gain of electrons. (*Section 3.12*)

Rem (radiation equivalent man)—an amount of radiation that causes damage to tissues equal to that caused by one roentgen. (*Section 2.19*)

replication—synthesis of new DNA. (*Section 26.2*)

representative elements—the elements not included in the transition elements, noble gases, lanthanide series, or actinide series. (*Section 2.10*)

resonance—the method of representing a molecular structure for which more than one valid structure can be drawn. The actual structure of the molecule is a hybrid of all the valid structures that can be drawn that differ only in the positions of electrons. (*Section 10.14*)

respiration—an oxygen-consuming process; cellular respiration is synonymous with mitochondrial electron transport. (*Section 22.15*)

reversible chemical reaction—a chemical reaction that can take place in forward and reverse directions. (*Section 6.6*)

ribonucleic acid (RNA)—polymer of ribonucleotides. (*Section 18 Intro.*)

ribosome—particle containing RNAs and proteins; site of protein synthesis in the cytoplasm. (*Sections 21.5, 26.6*)

ribosomal RNA (rRNA)—one of the three RNAs found in the ribosome. (*Section 26.11*)

roentgen (r)—the quantity of X or gamma radiation that leads to the absorption of 83.8 ergs (2.00×10^{-6} cal) in one gram of air. (*Section 2.19*)

rough endoplasmic reticulum—system of membrane channels studded with ribosomes on their exterior surfaces and involved in synthesis of membrane proteins and proteins to be exported from cells. (*Section 21.6*)

salt—a general term for an ionic compound. (*Section 3.4*)

saponification—hydrolysis of an ester with aqueous NaOH or KOH to produce a carboxylate salt and an alcohol. (*Section 15.11*)

sarcomere—the contractile unit of the myofibril. (*Section 22.10*)

saturated hydrocarbons—hydrocarbons that have only single bonds. (*Section 9.1*)

saturated solution—a solution that contains the maximum amount of solute at a given temperature and pressure. (*Section 5.2*)

saturation kinetics—the rate behavior of typical enzyme-catalyzed reactions. (*Section 21.13*)

scientific notation—use of powers of ten to express the magnitude of numbers. (*Section 1.8*)

secondary (2°) alcohol—an alcohol in which the hydroxyl-bearing carbon is bonded to two other carbon atoms, RCHOH (with R on the carbon). (*Section 12.1*)

secondary (2°) amine—an amine that has two hydrocarbon groups bonded to nitrogen: R_2NH. (*Section 13.2*)

secondary structure—three-dimensional polypeptide structures established by hydrogen-bonding between peptide groups; alpha-helical structure and beta structure are the two types of secondary structure. (*Section 17.10*)

semiconservative replication—normal mode of replication, in which each new DNA molecule contains one new strand and one old strand of polydeoxyribonucleotide. (*Section 26.4*)

semipermeable membrane—a membrane that allows only solvent molecules to pass. (*Section 5.10*)

serum—liquid that remains after blood has clotted. (*Section 27.1*)

serum albumin—a blood protein responsible for fatty acid transport. (*Section 25.18*)

SI units—a system of units that constitutes the modernized metric system and includes the kilogram, meter, second as the base units of mass, distance, and time. (*Section 1.6*)

sickle cell anemia—inherited disease caused by a mutation that gives rise to substitution of one amino acid in the beta chains of hemoglobin. (*Sections 17.9, 26.13*)

sigma orbital—the molecular orbital formed by the overlap of two atomic orbitals so that the bonding electrons lie in the region between the atoms. (*Section 10.3*)

significant figures—digits in a number that reflect the precision to which the quantity expressed by that number is known. (*Section 1.9*)

simple lipid—a lipid that cannot by hydrolyzed to give two or more components. (*Section 20.8*)

simple protein—a protein that in its functional form does not contain any nonprotein components. (*Section 17.3*)

smooth endoplasmic reticulum—system of membrane tubules and channels containing many enzymes of lipid metabolism and other specialized pathways. (*Section 21.7*)

solid—that physical state of a material in which the material has its own volume and shape. (*Section 2.3*)

solute—the substance dissolved in a solvent. (*Section 5.2*)

solution—a homogeneous mixture of two or more substances. (*Section 2.4*)

solvent—the medium in which solutes are dissolved to make a solution; water is the most important solvent. (*Section 5.2*)

specific gravity—the density of a substance divided by the density of water at the same temperature. (*Section 1.7*)

specificity—property exhibited by most enzymes, selectivity of substrate structure and type of reaction catalyzed. (*Section 21.8*)

sphingolipid—type of lipid similar to phosphoglycerides but containing sphingosine rather than glycerol. (*Section 20.5*)

sphingomyelin—a sphingolipid containing an alcohol esterified to phosphate. (*Section 20.5*)

sphygmomanometer—a device that uses air pressure to measure blood pressure as the heart contracts (systolic) and also during the relaxation period (diastolic). (*Section 4.3*)

spin—a property of an electron; two electrons in the same orbital must have opposite spins. (*Section 2.7*)

spontaneous reaction—a reaction that can occur because free energy is given off as it proceeds; the rate of such a reaction may be zero if the activation energy barrier is high, however. (*Sections 1.15, 22.2*)

standard free energy change ($\Delta G°$)—the free energy change associated with a reaction as it proceeds to equilibrium starting from standard conditions (all reactants and products at 1.0 *M*, 25°C, and pH 7.0, for biochemical systems). (*Section 27.3*)

starch—storage polysaccharide of plants; starch is a mixture of amylose and amylopectin. (*Section 19.3*)

stereoisomers—isomers that have the same sequential atomic arrangement but differ only in the orientation of their atoms in space. (*Section 8.10*)

steroid—one of a class of compounds derived from cholesterol that function as hormones. (*Section 20.6*)

strong acid—an acid that reacts 100 percent with water in solutions of 1 *M* or less to produce H_3O^+ and its conjugate base. (*Section 7.3*)

strong base—a base that reacts 100 percent with water in solutions of 1 *M* or less to produce OH^- and its conjugate acid; in practical terms, OH^- is the strongest base normally encountered. (*Section 7.3*)

structural formula—a formula in which covalent bonds are represented by lines joining the bonded atoms. (*Section 8.4*)

structural isomers—isomers that differ in the sequence in which their atoms are bonded. (*Section 8.9*)

substituent—an atom or atomic grouping attached to a molecular chain or ring. (*Section 9.3*)

substitution—a reaction in which one atom or group displaces another atom or group. (*Section 9.6*)

substrate—molecule acted upon by an enzyme. (*Section 21.9*)

substrate-level phosphorylation—phosphorylation driven by coupling with oxidation of a substrate molecule; to be contrasted with oxidative phosphorylation that occurs in conjunction with mitochondrial electron transport. (*Section 23.10*)

sucrase—an enzyme that catalyzes hydrolysis of sucrose. (*Section 23.2*)

sulfhydryl—(thiol), the —SH functional group. (*Section 12.17*)

sulfide—a thioether. (*Section 12.17*)

sulfonation reaction—the reaction of an organic compound with SO_3 and H_2SO_4 to produce a sulfonic acid. An example is the sulfonation of benzene or its derivatives with SO_3 and H_2SO_4 to produce a benzenesulfonic acid. (*Section 10.16*)

$$C_6H_6 + SO_3 \xrightarrow{H_2SO_4} C_6H_5SO_3H$$

superposable—capable of being placed ideally in the space occupied by another object so that the two objects coincide exactly. (*Section 16.5*)

surface active agent—a wetting agent. (*Section 5.6*)

surface catalyst—a heterogeneous catalyst; a catalyst that offers a surface upon which the reaction occurs more easily. (*Section 6.5*)

surface tension—a measure of the tendency of a liquid to resist the expansion of its surface area. (*Section 5.6*)

surfactant—short for surface active agent. (*Section 5.6*)

suspension—a temporary dispersal of a material in a solvent. (*Section 5.10*)

target tissue—a tissue affected by a particular hormone. (*Section 27.14*)

teflon—polymer, $\text{-}(CF_2CF_2)_n\text{-}$, prepared from tetrafluoroethene, $CF_2{=}CF_2$. (*Section 11.6*)

temperature—a measure of relative hotness and coldness. (*Section 1.13*)

termination codon—one of the three codons (UAA, UAG, and UGA) that signifies the end of an *m*RNA message and the end of a polypeptide. (*Section 26.7*)

tertiary (3°) alcohol—an alcohol in which the hydroxyl-bearing carbon atom is bonded to three other carbon atoms, R_3COH. (*Section 12.1*)

tertiary (3°) amine—an amine that has three hydrocarbon groups bonded to nitrogen: R_3N. (*Section 13.2*)

tertiary structure—three-dimensional structure of a polypeptide, established by noncovalent bonding and disulfide linkages. (*Section 17.11*)

tetrahedral geometry—the geometrical shape that results when four bonds are directed toward the corners of a tetrahedron. (*Section 8.5*)

thalassemias—a group of several inherited diseases in which a faulty hemoglobin is produced. (*Section 26.14*)

thermodynamics—the study of energy changes. (*Section 22.1*)

thick filaments—protein fibers made of myosin molecules that are part of the contractile system of muscle. (*Section 22.10*)

thin filaments—protein fibers made of actin, troponin, and tropomyosin that are part of the contractile system of muscle. (*Section 22.10*)

thioether—the sulfur analog of an ether: R—S—R; also called a sulfide. (*Section 12.17*)

thiol—the sulfur analog of an alcohol: R—S—H. (*Section 12.17*)

thiophenol—the sulfur analog of a phenol: Ar—SH. (*Section 12.17*)

titration—the addition of precisely measured amounts of a reagent of known concentration to a solution of unknown concentration until enough has been added to react with all the material of unknown concentration. (*Section 7.12*)

TNT—(2,4,6-trinitrotoluene) a highly explosive nitro compound. (*Section 11.7*)

Tollen's test—test for aldehydes in which the $Ag(NH_3)_2^+$ ion oxidizes the aldehyde group to a carboxyl group and a silver mirror is formed. (*Section 14.6*)

torr—a pressure unit equal to 1/760 of an atmosphere; one torr is identical with one millimeter of mercury. (*Section 4.1*)

trace elements—elements required by the body in only very small amounts. (*Section 23.6*)

transaminase—one of a class of enzymes that catalyzes transfer of the alpha-amino group of an amino acid to an alpha-keto acid, leading to formation of a new alpha keto acid and a new amino acid. (*Section 25.18*)

transcription—synthesis of RNA using the information in a segment of a DNA molecule. (*Section 26.6*)

transfer RNA (tRNA)—type of RNA that carries amino acids and directs insertion of correct amino acids during protein synthesis, by base-pairing between *t*RNA anticodon and codon of messenger RNA. (*Section 26.10*)

transition elements—elements with atomic numbers 21–30, 39–48, 57, 72–80, 89, 104, 105. (*Section 2.10*)

translation—process of protein or polypeptide synthesis on a ribosome, using directions in an *m*RNA molecule. (*Section 26.10*)

translocation—movement of *m*RNA and ribosome during elongation phase of protein synthesis. (*Section 26.12*)

triacylglycerol—neutral fat; glycerol with three esterified fatty acids. (*Section 20.2*)

triple bond—a covalent bond composed of three pairs of electrons (C⋮⋮C; also shown as C≡C). (*Section 8.7*)

triplet codon—three-nucleotide sequence in *m*RNA that specifies a particular amino acid. (*Section 26.7*)

tropomyosin—one of the proteins of the thin filaments of the myofibril. (*Section 22.10*)

troponin—one of the proteins of thin filaments of the myofibril. (*Section 22.10*)

Tyndall effect—the scattering of light by colloidal particles such that a light beam becomes visible as it passes through a colloidal dispersion. (*Section 5.11*)

uncoupler—a chemical that destroys the coupling of electron transport and oxidative phosphorylation (or photophosphorylation) permitting electron transport to occur without conservation of energy in the form of ATP. (*Section 22.19*)

unified atomic mass unit (u)—a mass equal to 1/12 the mass of one atom of the isotope $^{12}_{6}C$. (*Section 2.13*)

unit cell—the minimum number of atoms, ions, or molecules necessary to display the overall pattern of the crystal structure of a substance. (*Section 4.12*)

unsaturated hydrocarbon—a hydrocarbon that has one or more double or triple bonds. (*Chapter 10 Introduction*)

urea cycle—pathway responsible for converting ammonia derived from degradation of amino acids to urea for excretion. (*Section 25.19*)

ureotelic—urea-excreting. (*Section 25.17*)

vacuole—a fluid-filled membrane sac found in many plant cells. (*Section 21.6*)

valence shell—in an atom, the outermost shell (quantum level) that contains electrons. (*Section 2.9*)

vaporization—the change of a liquid into a gas. (*Section 4.7*)

vasodilator—a substance such as isoamyl nitrite that relaxes the smooth muscles of the blood vessels and lowers the blood pressure. (*Section 11.8*)

very low density lipoproteins (VLDL)—blood lipoproteins responsible for much triglyceride and cholesteryl ester transport. (*Section 27.6*)

vitamin—one of a diverse group of compounds that the body needs as coenzymes (or needs in order to make the coenzymes) but cannot manufacture, and which must be obtained in the diet. (*Section 20.7*)

volume-volume percent—a concentration scale defined as the volume of solute per 100 volumes of solution. (*Section 5.9*)

water-soluble vitamin—one of the large group of vitamins that are water-soluble rather than fat-soluble. (*Section 21.12*)

weak acid—an acid that reacts only partially with water and establishes an equilibrium. (*Section 7.4*)

weak base—a base that reacts only partially with water and establishes an equilibrium. (*Section 7.4*)

weight—the force of gravity exerted on a mass; often used synonymously with mass. (*Section 1.1*)

weight-volume percent—a concentration scale used in clinical laboratories and defined as the grams of solute per 100 mL of solution. (*Section 5.9*)

wetting agent—a substance that lowers the surface tension of water and thereby causes water to wet materials more readily. (*Section 5.7*)

Wurtz reaction—the coupling of two alkyl groups by the reaction of an alkyl halide with sodium. (*Section 9.5*)

$$2R{-}X + 2Na \xrightarrow{\text{heat}} R{-}R + 2NaX$$

zwitterion—a single species with a positively charged part and a negatively charged part, that is, it is both a cation and an anion. (*Section 17.5*)

Chapter 1

1.1 (a) physical change (b) physical change (c) chemical change (d) chemical change (e) physical change **1.2** SI indicates a sulfur–iodine combination. Si is the symbol for silicon. **1.3** 6.00 mg **1.4** 0.200 L **1.5** 8.89 cm **1.6** 5.00 kg **1.7** (a) 0.70 g/mL (b) 1.4×10^3 mL **1.8** (a) 2 (b) 2 (c) 3 (d) 4 (e) 5 (f) 3 (g) 4 (h) 1 (i) 2 **1.9** (a) 6.351×10^3 (b) 5.01×10^{-3} (c) 6.128503×10^3 (d) 1.040×10^{-5} (e) 2×10^{-3} (f) 1.02×10^2 (g) 5.654940×10^6 (h) 3.4×10^{-2} **1.10** (a) 2.632×10^3 (b) 1.024×10^{-3} (c) 9.5×10^7 (d) 1.8 (e) 2×10^9 (f) 11.4 **1.11** Heat, light, mechanical energy, electrical energy. **1.12** All kinetic while running; kinetic to potential while vaulting; all potential at top of vault; potential to kinetic while falling to ground. **1.13** Heat removed from refrigerator interior is rejected into kitchen. No. **1.14** Kinetic energy changes to work of displacing the water already at the bottom of the falls. **1.15** (a) 21°C (b) −40°F **1.16** 1°C is a larger temperature change than 1°F. **1.17** Loss; reaction is spontaneous. **1.18** (a) Reaction is spontaneous. (b) $\Delta G = +3 \times 10^3$ kcal **1.19** $\Delta G = -6 \times 10^3$ kcal **1.20** (a) Physical change (b) chemical change (c) physical change (d) chemical change (e) chemical change (*Sections 1.2, 1.3*) **1.24** (a) 1.00 cm (b) 30.5 cm (c) 91.4 cm (*Section 1.6*) **1.25** (a) 0.17 m (b) 0.305 m (c) 0.914 m (*Section 1.6*) **1.27** $0.57 per liter (*Section 1.6*) **1.29** (a) 2 (b) 3 (c) 3 (d) 4 (e) 4 (f) 4 (g) 1 (h) 3 (*Section 1.9*) **1.30** (a) 5.3×10^{-2} kg (b) 4.5×10^{-3} g (c) 5×10^2 g (d) 6.5 mg (e) 49 kg (*Sections 1.6, 1.8*) **1.33** Specific gravity: mercury = 13.5; lead = 11.2 (*Section 1.7*) **1.35** (a) 8.6×10^4 cal/h (b) 86 kcal/h (*Section 1.6*) **1.37** (a) $\Delta G = -1.4$ kcal/g (b) $\Delta G = +1.4$ kcal/g (*Sections 1.15, 1.16*) **1.38** (a) 4°C (b) 38°C (c) 121°C (d) −29°C (e) −273°C (*Section 1.14*) **1.39** (a) 32°F (b) 72°F (c) 212°F (d) 14°F (e) −459°F (*Section 1.14*) **1.43** 177 g of apple and 73 g of cheese (*Section 1.16*) **1.44** 7 days (*Section 1.16*) **1.46** 6.58×10^{-3} kiloyazoo; 6.58×10^3 milliyazoos (*Section 1.6*)

Chapter 2

2.1 (a) $2H_2 + O_2 \rightarrow 2H_2O$ (b) $CaC_2 + 2H_2O \rightarrow C_2H_2 + Ca(OH)_2$ (c) $2K + H_2S \rightarrow K_2S + H_2$ (d) $3H_2 + N_2 \rightarrow 2NH_3$ (e) $PCl_3 + 3H_2O \rightarrow H_3PO_4 + 3HCl$ (f) $4Al + 3O_2 \rightarrow 2Al_2O_3$
2.2 (a) pure substance (b) solution (c) heterogeneous mixture (d) heterogeneous mixture (e) pure substance (f) solution **2.3** 72
2.4 (a) (b)

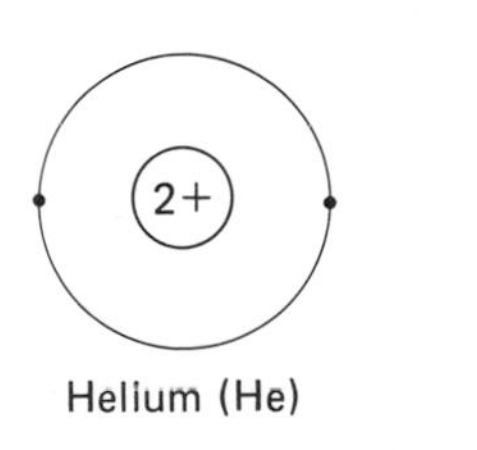

Helium (He)

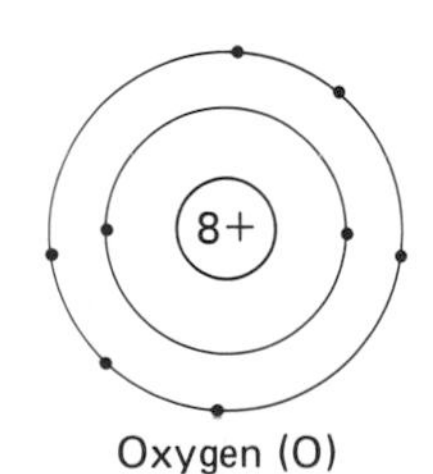

Oxygen (O)

(c)

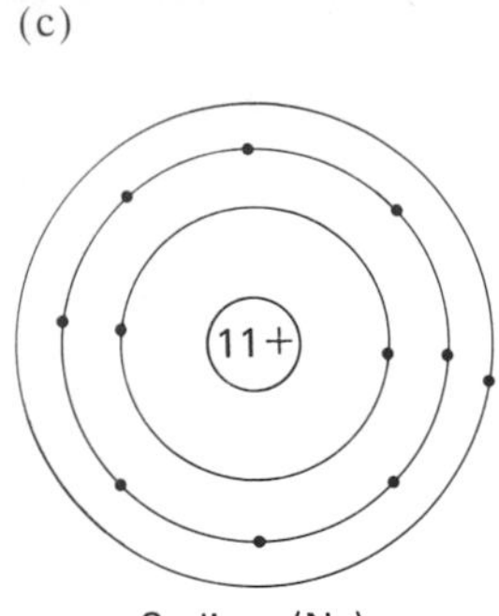

Sodium (Na)

(d)

29+

Copper (Cu)

2.5 2; 2 **2.6** (a) $1s^1$ (b) $1s^2 2s^1$ (c) $1s^2 2s^2 2p^1$ (d) $1s^2 2s^2 2p^3$ (e) $1s^2 2s^2 2p^4$ (f) $1s^2 2s^2 2p^6$

2.7

(a) $1s$ ↑

(b) $1s$ ↑↓ $2s$ ↑

(c) $1s$ ↑↓ $2s$ ↑↓ $2p_x$ ↑ $2p_y$ — $2p_z$ —

(d) $1s$ ↑↓ $2s$ ↑↓ $2p_x$ ↑ $2p_y$ ↑ $2p_z$ ↑

(e) $1s$ ↑↓ $2s$ ↑↓ $2p_x$ ↑↓ $2p_y$ ↑ $2p_z$ ↑

(f) $1s$ ↑↓ $2s$ ↑↓ $2p_x$ ↑↓ $2p_y$ ↑↓ $2p_z$ ↑↓

2.8 (a) H· (b) Li· (c) ·B: (d) :N· (e) ·O: (f) :Ne:

2.9 (a) K· (b) ·Sn: (c) ·Ga: (d) ·P· (e) Ca: (f) :Br:

2.10 (a) He; Na (b) F (c) S (d) N (e) Be **2.11** Cations: Li, Mg anions: N, S, H

2.12 (a) An element is a substance that consists of atoms with the same atomic number. (b) An isotope of an element has the same number of protons but a different number of neutrons than another isotope of that element. (c) The mass number of an atom is the sum of its number of protons and neutrons.

2.13 20.18 **2.14** ${}^{228}_{88}\text{Ra} \rightarrow {}^{0}_{-1}\text{e} + {}^{228}_{89}\text{Ac}$ ${}^{234}_{91}\text{Pa} \rightarrow {}^{0}_{-1}\text{e} + {}^{234}_{92}\text{U}$ **2.15** (a) ${}^{210}_{83}\text{Bi}$ (b) ${}^{4}_{2}\text{He}$ (c) ${}^{225}_{88}\text{Ra}$ (d) ${}^{231}_{90}\text{Th}$ **2.16** (a) beta decay; (b), (c), and (d) alpha decay **2.17** It forms 0.50 g of ${}^{234}_{91}\text{Pa}$.

2.18 1.5 mg **2.19** ${}^{24}_{11}\text{Na} \rightarrow {}^{0}_{-1}\text{e} + {}^{24}_{12}\text{Mg}$ ${}^{131}_{53}\text{I} \rightarrow {}^{0}_{-1}\text{e} + {}^{131}_{54}\text{Xe}$ **2.24** (a) $1s^2 2s^2$ (b) $1s^2 2s^2 2p^2$ (c) $1s^2 2s^2 2p^5$ (d) $1s^2 2s^2 2p^6 3s^2$ (e) $1s^2 2s^2 2p^6 3s^2 3p^1$ (f) $1s^2 2s^2 2p^6 3s^2 3p^4$ (g) $1s^2 2s^2 2p^6 3s^2 3p^6$ (*Section 2.8*)

2.25

(a) $1s$ ↑↓ $2s$ ↑↓

(b) $1s$ ↑↓ $2s$ ↑↓ $2p_x$ ↑ $2p_y$ ↑ $2p_z$ —

(c) $1s$ ↑↓ $2s$ ↑↓ $2p_x$ ↑↓ $2p_y$ ↑↓ $2p_z$ ↑

(d) $1s$ ↑↓ $2s$ ↑↓ $2p_x$ ↑↓ $2p_y$ ↑↓ $2p_z$ ↑↓ $3s$ ↑↓

(e) $1s$ ↑↓ $2s$ ↑↓ $2p_y$ ↑↓ $2p_y$ ↑↓ $2p_z$ ↑↓ $3s$ ↑↓ $3p_x$ ↑ $3p_y$ — $3p_z$ —

(f) $1s$ ⥮ $2s$ ⥮ $2p_x$ ⥮ $2p_y$ ⥮ $2p_z$ ⥮ $3s$ ⥮ $3p_x$ ⥮ $3p_y$ ↿ $3p_z$ ↿

(g) $1s$ ⥮ $2s$ ⥮ $2p_x$ ⥮ $2p_y$ ⥮ $2p_z$ ⥮ $3s$ ⥮ $3p_x$ ⥮ $3p_y$ ⥮ $3p_z$ ⥮

(*Section 2.8*)

2.26 (a) Be: (b) ·C̈· (c) :F̤̈· (d) Mg: (e) :Al· (f) :S̤̈· (g) :Är̤: (*Section 2.9*)

2.30

Species	Protons	Neutrons	Mass number
$^{19}_{9}F$	9	10	19
$^{12}_{6}C$	6	6	12
$^{16}_{8}O$	8	8	16
$^{32}_{16}S$	16	16	32
$^{13}_{6}C$	6	7	13
$^{17}_{8}O$	8	9	17
$^{32}_{15}P$	15	17	32
$^{18}_{8}O$	8	10	18
$^{31}_{15}P$	15	16	31
$^{14}_{6}C$	6	8	14

(*Section 2.13*)

2.33 25 $^{37}_{17}Cl$ atoms and 75 $^{35}_{17}Cl$ atoms (*Section 2.13*) **2.34** (a) $^{40}_{20}Ca$ (b) $^{32}_{15}P$ (c) $_{-1}^{0}e$ (d) $^{220}_{86}Rn$ (*Section 2.16*) **2.35** One year (*Section 2.17*) **2.38** 1.0×10^6 ergs (*Section 2.19*) **2.42** 8 alpha particles and 6 beta particles

Chapter 3

3.1 The properties of helium (a gaseous, unreactive, nonatomic element) clearly indicate that it must be included in the noble gas family. **3.2** Ionization energies increase in moving from bottom to top in a given group. The very high ionization energies of helium, neon, and argon serve to make removal of electrons extremely difficult. **3.3** The Na^+ ion has the electronic configuration $1s^22s^22p^6$. The Cl^- ion has the electronic configuration $1s^22s^22p^63s^23p^6$. Electrons in the third shell are somewhat farther from the nucleus than electrons in the second shell. **3.4** (a) NaF (b) MgO (c) LiCl (d) AlF_3 **3.5** (a) $4Al + 3O_2 \rightarrow 2Al_2O_3$ (b) $6Li + N_2 \rightarrow 2Li_3N$ (c) $2Na + F_2 \rightarrow 2NaF$ **3.6** (a) calcium sulfate (b) ammonium carbonate (c) aluminum oxide (d) sodium bromide (e) magnesium hydroxide (f) lithium sulfate (g) aluminum nitrate (h) potassium iodide (i) magnesium bromide (j) calcium sulfide (k) ammonium nitrate (l) lithium oxide (m) sodium hydroxide (n) sodium phosphate **3.7** iron(II) oxide and iron(III) oxide **3.8** $CoBr_2$ and CoF_3 **3.9** $N_2 + 3F_2 \rightarrow 2NF_3$

3.10 (a) H:F̤̈: (b) :C̤̈l:N̤̈:C̤̈l: with :C̤̈l: below N (c) :S̤̈:H with H below S

3.11 (a) $H_2 + F_2 \rightarrow 2HF$ (b) $N_2 + 3Cl_2 \rightarrow 2NCl_3$ (c) $N_2 + 3H_2 \rightarrow 2NH_3$

3.12 (a) polar H ⇸ F̤̈: (b) polar H ⇸ N̈ ⇷ H, with H ⇸ N below (c) nonpolar (d) nonpolar (e) polar :F̤̈ ⇷ N̈ ⇸ F̤̈:, with :F̤̈: below N

3.13 (a) sulfur dioxide (b) sulfur trioxide (c) hydrogen bromide (d) phosphorus pentachloride (e) nitrogen trichloride (f) dinitrogen trioxide (g) phosphorus trifluoride (h) oxygen difluoride (i) ammonia (j) hydrogen iodide **3.14** (a) Mg oxidized, Cl reduced (b) Na oxidized, F reduced (c) Li oxidized, O reduced **3.15** (a) H oxidized, Cl reduced (b) Cu oxidized, O reduced (c) H oxidized, F reduced (d) S oxidized, O reduced (e) H (in H_2) oxidized, C reduced
3.16 6.0×10^{20} Na atoms and 3.0×10^{20} Cl_2 molecules **3.17** (a) 36.5 g (b) 63.5 g (c) 58.1 g (d) 98.1 g (e) 342 g **3.18** (a) 3.6 g (b) 6.4 g (c) 5.8 g (d) 9.8 g (e) 34 g
3.19 $230 \text{ g Na} \times \dfrac{1 \text{ mol Na}}{23.0 \text{ g Na}} = 10.0 \text{ mol Na}$ $355 \text{ g } Cl_2 \times \dfrac{1 \text{ mol } Cl_2}{71.0 \text{ g } Cl_2} = 5.00 \text{ mol } Cl_2$

$585 \text{ g NaCl} \times \dfrac{1 \text{ mol NaCl}}{58.5 \text{ g NaCl}} = 10.0 \text{ mol NaCl}$

3.20 20 mol H_2; 20 mol H_2O **3.21** 90 g H_2O; 80 g O_2 **3.22** 20 g O_2 **3.23** (a) Li_2O (b) MgF_2 (c) NaCl (*Section 3.5*) **3.24** (a) $4Li + O_2 \rightarrow 2Li_2O$ (b) $Mg + F_2 \rightarrow MgF_2$ (c) $2Na + Cl_2 \rightarrow 2NaCl$ (*Section 3.5*) **3.25** (a) ClF (b) Cl_2O (c) H_2S (*Section 3.8*) **3.27** (a) $Cl_2 + F_2 \rightarrow 2ClF$ (b) $2Cl_2 + O_2 \rightarrow 2Cl_2O$ (c) $H_2 + S \rightarrow H_2S$ (*Section 3.8*)
3.28 Problem 3.24: (a) Li oxidized, O reduced (b) Mg oxidized, F reduced (c) Na oxidized, Cl reduced Problem 3.27: (a) Cl oxidized, F reduced (b) Cl oxidized, O reduced (c) H oxidized, S reduced (*Section 3.12*)
3.30 (a) calcium oxide (b) carbon disulfide (c) magnesium nitrate (d) dinitrogen pentoxide (e) aluminum sulfate (f) potassium hydroxide (g) hydrogen fluoride (h) iodine monofluoride (*Sections 3.6, 3.11*) **3.31** (a) copper(I) oxide; copper(II) oxide (b) iron(II) sulfate; iron(III) sulfate (*Section 3.6*) **3.33** (a) 2 mol H_2 (b) 0.5 mol H_2O (c) 1 mol Li (d) 4.99 mol H_2S (e) 0.25 mol MgO (*Section 3.14*) **3.34** (a) 34 g (b) 55 g (c) 52 g (d) 61 g (e) 1.7×10^2 g (*Section 3.14*)
3.35 (a) 0.334 mol HCl (b) 0.167 mol H_2 and 0.167 mol Cl_2 (c) 0.337 g H_2 and 11.8 g Cl_2 (d) 0.4 g Cl_2 (*Section 3.14*) **3.39** 721 g H_2O and 19 g H_2 (*Section 3.14*)

Chapter 4

4.1 1.5×10^2 cm, 1.5×10^3 mm, 1.5×10^3 torr **4.2** 0.500 atm **4.3** 298 K **4.4** 100°C
4.5 The volume of the balloon would decrease. **4.6** 43°C **4.7** 9.9×10^3 L **4.8** 1.94×10^{-2} mol
4.9 27.1 **4.10** 2.38 atm **4.11** (a) They are equal. (b) Total energy of molecules in the gaseous state is greater than that of molecules in the liquid state by 94 cal (the heat of vaporization). (c) Molecules in the gaseous state are relatively more separated than are the molecules in the liquid state. **4.12** (a) No. (b) The energy of a gram of liquid water is greater than a gram of ice by 80 cal (heat of fusion).
4.13 (a) 7.6×10^2 mmHg (b) 3.8×10^2 mmHg (c) 20 mmHg (d) 700 mmHg (*Section 4.1*)
4.14 (a) 293 K (b) 323 K (c) 258 K (d) 255 K (e) 366 K (*Section 4.2*) **4.19** 2.1 L (*Section 4.4*)
4.21 2.9 atm (*Section 4.4*) **4.23** 0°C (*Section 4.4*) **4.24** 0.206 g (*Section 4.4*) **4.28** 105 cal (*Section 4.10*) **4.33** 29.0 **4.34** 0.5 cal/(g °C)

Chapter 5

5.1 0.15 g **5.2** Dissolve 37.5 g of KCl in sufficient water to give a final volume of 250 mL.
5.3 Dissolve 44 g of NaCl in sufficient water to give a final volume of 500 mL. **5.4** 2.5×10^{-2} mol; 3.0 g
5.5 Dilute 5.0 mL of 16 *M* NH_3 with sufficient water to give a final volume of 800 mL.
5.6 0.05 *M* $MgCl_2$; the total concentration of particles is greatest for 0.05 *M* $MgCl_2$, namely 0.15 *M* (Mg^{2+} and $2Cl^-$). **5.7** 0.10 *M* sucrose **5.8** Crenation. A hypertonic solution has solute particles at a higher concentration; consequently water will pass out of the cell into the surrounding solution.

5.18 2.5 g (*Section 5.9*) **5.19** 3.43% (*Section 5.9*) **5.20** 0.200 *M* (*Section 5.9*) **5.21** 0.010 *M* (*Section 5.9*) **5.24** (a) and (d); (b) and (c) (*Section 5.12*) **5.26** $CaCl_2 \cdot 6H_2O$ **5.27** 15.7 *M*

Chapter 6

6.1 In general, increasing the concentrations of reactants leads to more frequent collisions among the reacting molecules, which leads to an increase in the rate of the reaction. An increase in temperature increases the average kinetic energy of reacting molecules, thereby causing a larger number of collisions to be effective in overcoming the free energy of activation barrier. The rate of the reaction thus increases.
6.2 A catalyst is a substance that speeds up a reaction but remains unchanged at the end of the reaction. A surface catalyst is a solid that speeds up a reaction between reactants in the liquid or gaseous states. The surface catalyst offers a suitable place for the reacting molecules to come together. A homogeneous catalyst is in the same physical state as the reactants. The homogeneous catalyst enters into chemical combination with one or more of the reactants but is regenerated in a later step.
6.3

$$H_2O_2 + I^- \rightarrow H_2O + IO^-$$
$$IO^- + H_2O_2 \rightarrow H_2O + O_2 + I^-$$
$$2H_2O_2 + \cancel{I^-} + \cancel{IO^-} \rightarrow 2H_2O + O_2 + \cancel{IO^-} + \cancel{I^-}$$
$$2H_2O_2 \rightarrow 2H_2O + O_2$$

6.4 ΔG is the same for both reactions. $\Delta G^\ddagger$ is greater for the uncatalyzed reaction. **6.5** When the rates of the forward and reverse reactions are equal (equilibrium), there can be no change in concentrations. **6.6** The concentration of SO_2 is 0.25 moles per liter. **6.7** No. The actual equilibrium concentrations of products and reactants does depend on initial concentrations, but the ratio of equilibrium concentrations (equilibrium constant) does not depend on initial concentrations. **6.8** Yes

6.9 $K = \dfrac{[SO_3]^2}{[SO_2]^2[O_2]}$

6.10 (a) right (b) right (c) left (d) left (e) right

6.17 (a) $K = \dfrac{[H_2O][CH_3OCHO]}{[CH_3OH][HCO_2H]}$ (b) $K = \dfrac{[HCl]^2}{[H_2][Cl_2]}$

(c) $K = \dfrac{[NO_2]^2}{[N_2O_4]}$ (d) $K = \dfrac{[C_2H_5Br][H_2O]}{[C_2H_5OH][HBr]}$ (*Section 6.7*)

6.19

	$[SO_2]$	$[O_2]$	$[SO_3]$
(a)	decrease	increase	increase
(b)	decrease	increase	decrease
(c)	increase	increase	increase
(d)	decrease	decrease	decrease

(*Section 6.9*)

6.21 2.0×10^{-2} 6.22 [HI] = 5.0

Chapter 7

7.1

NH_3	+	HSO_4^-	$\rightarrow$	NH_4^+	+	SO_4^{2-}
base		acid		acid (conjugate of NH_3)		base (conjugate of HSO_4^-)

7.2 (a) acid (b) acid (c) amphoteric (d) base (e) acid (f) amphoteric **7.3** (a) Cl^- (b) $CH_3CO_2^-$ (c) conjugate acid, H_2SO_4; conjugate base, SO_4^{2-} (d) NH_4^+ (e) NH_3 (f) conjugate acid, H_3O^+; conjugate base, OH^- **7.4** $H_2CO_3 + H_2O \rightleftarrows H_3O^+ + HCO_3^-$

7.5 $CO_3^{2-} + H_2O \rightleftarrows HCO_3^- + OH^-$ **7.6** Problem 7.4: acids, H_2CO_3 and H_3O^+; bases, H_2O and HCO_3^- Problem 7.5: acids, H_2O and HCO_3^-; bases, CO_3^{2-} and OH^- **7.7** Weak acids: strongest, acetic; weakest, hydrogen carbonate ion. Weak bases: strongest, carbonate ion; weakest, acetate ion.

7.8 Weak acids

Acetic $K_a = \frac{[H_3O^+][CH_3CO_2^-]}{[CH_3CO_2H]}$

Carbonic $K_a = \frac{[H_3O^+][HCO_3^-]}{[H_2CO_3]}$

Boric $K_a = \frac{[H_3O^+][H_2BO_3^-]}{[H_3BO_3]}$

Hydrogen carbonate ion $K_a = \frac{[H_3O^+][CO_3^{2-}]}{[HCO_3^-]}$

Weak bases

Carbonate ion $K_b = \frac{[HCO_3^-][OH^-]}{[CO_3^{2-}]}$

Ammonia $K_b = \frac{[NH_4^+][OH^-]}{[NH_3]}$

Hydrogen carbonate ion $K_b = \frac{[H_2CO_3][OH^-]}{[HCO_3^-]}$

Acetate ion $K_b = \frac{[CH_3CO_2H][OH^-]}{[CH_3CO_2^-]}$

7.9 The K_a for a strong acid is indefinitely large.
7.10 Conjugate base of boric acid is $H_2BO_3^-$. $H_2BO_3^- + H_2O \rightleftarrows H_3BO_3 + OH^-$
7.11 (a) acidic (b) basic (c) neutral (d) acidic (e) acidic (f) neutral (g) basic
7.12 (c) $2H_2O \rightleftarrows H_3O^+ + OH^-$ (d) $CH_3CO_2H + H_2O \rightleftarrows H_3O^+ + CH_3CO_2^-$
(e) $NH_4^+ H_2O \rightleftarrows NH_3 + H_3O^+$ (f) $2H_2O \rightleftarrows H_3O^+ + OH^-$ (g) $CH_3CO_2^- + H_2O \rightleftarrows OH^- + CH_3CO_2H$
7.13 $[OH^-] = 0.010$ $[H_3O^+] = 1.0 \times 10^{-12}$ **7.14** $[OH^-] = 1.0 \times 10^{-10}$ **7.15** (a) acidic (b) basic (c) basic (d) neutral (e) basic **7.16** (a) basic (b) basic (c) neutral (d) acidic (e) acidic **7.17** (a) 1.00 (b) 13.00 (c) 5.30 (d) 8.70 **7.18** (a) $[H_3O^+] = 5.0 \times 10^{-2}$ (b) $[H_3O^+] = 1.0 \times 10^{-2}$ (c) $[H_3O^+] = 1.8 \times 10^{-4}$ (d) $[H_3O^+] = 2.9 \times 10^{-13}$
7.19 (a) $[OH^-] = 2.0 \times 10^{-13}$ (b) $[OH^-] = 1.0 \times 10^{-12}$ (c) $[OH^-] = 5.6 \times 10^{-11}$ (d) $[OH^-] = 3.4 \times 10^{-2}$

7.20

Weak acids	Weak bases
Acetic, $pK_a = 4.74$	Carbonate ion, $pK_b = 3.68$
Carbonic, $pK_a = 6.35$	Ammonia, $pK_b = 4.74$
Boric, $pK_a = 9.10$	Hydrogen carbonate ion, $pK_b = 7.66$
Hydrogen carbonate ion, $pK_a = 10.33$	Acetate ion, $pK_b = 9.25$

7.21 Monohydrogen phosphate ion, HPO_4^{2-} **7.22** Acetic acid, CH_3CO_2H **7.23** $[H_3O^+] \sim 10^{-4}$
7.24 pH ~ 5–6 **7.25** 0.0900 *M* HNO_3 **7.26** 0.0586 *M* NH_3 **7.27** 30.0 mL
7.29 $H_3O^+ + OH^- \rightarrow 2H_2O$ (*Section 7.3*) **7.30** (a) HCl (b) H_2O (c) OH^- (d) O^{2-} (*Section 7.1*)
7.32 (a) $[H_3O^+] = 1.0 \times 10^{-7}$ (b) $[H_3O^+] = 1.0 \times 10^{-2}$ (c) $[H_3O^+] = 1.0 \times 10^{-13}$ (*Section 7.7*)
7.33 (a) $[OH^-] = 1.0 \times 10^{-7}$ (b) $[OH^-] = 1.0 \times 10^{-12}$ (c) $[OH^-] = 0.10$ (*Section 7.7*)
7.34 (a) pH = 7.00 (b) pH = 2.00 (c) pH = 13.00 (*Section 7.8*) **7.37** $K_a = 4.0 \times 10^{-5}$ $pK_a = 4.40$ (*Sections 7.4, 7.9*) **7.44** (a) 24.6 mL (b) 20.2 mL (*Sections 7.12, 7.14*) **7.50** pH = 11.70

Chapter 8

8.1 (a), (b), and (f) are inorganic; (c), (d), and (e) are organic.

8.2

```
(a)     H   H   H   H   H
        |   |   |   |   |
    H — C — C — C — C — C — H
        |   |   |   |   |
        H   H   H   H   H

(b)     H   H   H   H
        |   |   |   |
    H — C — C — C — C — H
        |   |   |   |
        H   H   |   H
            H — C — H
                |
                H

(c)             H
                |
            H — C — H
        H       |       H
        |       |       |
    H — C ——— C ——— C — H
        |       |       |
        H       |       H
            H — C — H
                |
                H
```

8.3

	Lewis dot formula	Molecular formula	Condensed formula
(a)	H H H H H / H:C:C:C:C:C:H / H H H H H	C_5H_{12}	$CH_3CH_2CH_2CH_2CH_3$
(b)	H H H H / H:C:C:C:C:H / H H ·· H / H:C:H / H	C_5H_{12}	$CH_3CH_2CHCH_3$ with CH_3 on the CH
(c)	H / H:C:H / H ·· H / H:C : C : C:H / H ·· H / H:C:H / H	C_5H_{12}	CH_3CCH_3 with CH_3 above and CH_3 below the central C

8.4

```
(a)  Cl          (b)  Cl          (c)  Cl
     |                |                |
     C···H            C···H            C···H
    / \              / \              / \
   H   H            Cl  H            Cl  Cl
```

8.5 (a) tetrahedral (b) tetrahedral (c) planar, 120° angles (d) linear

8.6 (a) sp^3 (b) sp^3 (c) sp^2

8.7

```
(a)  Cl        Cl       (b)  H         Cl       (c) H—C≡N
       \      /               \       /
        C == C                 C == C
       /      \               /       \
     Cl        Cl            H         Cl
```

8.8 (a) same (b) same. Note that you can convert one into the other by rotating the molecule in (a) and by rotating half of the molecule about the middle C–C bond in (b). **8.9** (a) structural isomers (b) not isomers (c) structural isomers **8.10** (a) same (b) stereoisomers (they are *cis-trans* isomers) (c) structural isomers (the two carbons have different atomic sequences) (d) not isomers (they have different molecular formulas) (e) conformations (f) structural isomers (g) conformations.
8.11 (a) ether (b) keto group (c) aldehyde (d) disulfide, aromatic ring (e) keto group, aromatic ring (f) ester, aromatic ring (g) peroxide, aromatic ring (h) phenol (i) alcohol, aromatic ring (j) phenol (k) amine, aromatic ring (l) carboxyl, aromatic ring **8.12** (a) CH_3OH (b) Cl—(benzene ring)—OH

(c) $C_6H_5{-}SH$ (d) $CH_3{-}O{-}CH_3$ (e) $CH_3{-}O{-}O{-}CH_3$ (f) $CH_3{-}S{-}S{-}CH_3$ (g) CH_3NH_2

(h) CH_3SH (i) $CH_3CH_2\overset{O}{\overset{\|}{C}}H$ (j) $CH_3CH{=}CHCH_3$ (k) $CH_3C{\equiv}CH$ (l) $C_6H_5{-}\overset{O}{\overset{\|}{C}}{-}C_6H_5$

(m) $CH_3CH_2\overset{O}{\overset{\|}{C}}{-}OH$ (n) $C_6H_5{-}\overset{O}{\overset{\|}{C}}{-}O{-}CH_3$ (o) $CH_3\overset{O}{\overset{\|}{C}}{-}NH_2$ **8.14** (a), (c) are inorganic, (e) is organic (*Section 8.1*) **8.16** (a) NaCl (*Section 8.2*) **8.17** (a) NaCl (*Section 8.2*)

8.18 (b) $H{-}\underset{H}{\overset{H}{C}}{-}\overset{H}{C}{=}\overset{H}{C}{-}\underset{H}{\overset{H}{C}}{-}H$ (*Section 8.4*)

8.19 (b) C_4H_8, $H{:}\underset{H}{\overset{H}{C}}{:}\overset{H}{C}{::}\overset{H}{C}{:}\underset{H}{\overset{H}{C}}{:}H$, $CH_3CH{=}CHCH_3$ (*Section 8.4*)

8.20 (a) $H{:}\underset{H}{\overset{H}{C}}{:}\overset{:\ddot{O}:}{C}{:}H$ (c) $H{:}\underset{H}{\overset{H}{C}}{:}\ddot{\underset{..}{O}}{:}H$ (*Section 8.4*)

8.21 (a) $CH_3CH{=}CHCH_3$ or $CH_3CH_2CH_2CH_3$ (c) $H{-}\underset{H}{\overset{H}{C}}{-}\underset{H}{\overset{H}{C}}{-}OH$ (*Section 8.4*)

8.22 $CH_3CH_2CH_2Cl$, $CH_3\overset{Cl}{\overset{|}{C}}HCH_3$ (*Section 8.4*)

8.23 (a) tetrahedral C with H (top), H, H, OH (*Section 8.5*) **8.25** (a) sp^2 (*Section 8.6*)

8.26 (c) $H_2C{=}CH{-}CH{=}CH_2$ (planar; all H shown) (*Section 8.7*)

8.27 (a), (c), (e) are the same; (g) different compounds that are not isomers. (*Sections 8.8–8.11*)

8.28 (a) $CH_3{-}O{-}CH_3$ (c) $CH_3{-}\underset{H}{\underset{|}{N}}{-}CH_3$ (*Section 8.9*)

8.29 (a) $C_6H_5(H)C{=}C(H)CH_3$ (*Section 8.10*) **8.30** (a) $FH_2C{-}CH_2F$ (*Section 8.11*)

8.31 (a) aromatic ring, keto group, alcohol (c) ether, aromatic ring, keto, carboxyl (e) amide (*Section 8.12*)

Chapter 9

9.1

```
    H   H   H   H   H   H          H   H   H   H   H            H   H   H   H   H
    |   |   |   |   |   |          |   |   |   |   |            |   |   |   |   |
H — C — C — C — C — C — C — H  H — C — C — C — C — C — H    H — C — C — C — C — C — H
    |   |   |   |   |   |          |   |   |   |   |            |   |   |   |   |
    H   H   H   H   H   H          H   H   H   |   H            H   H   |   H   H
                                           H — C — H                H — C — H
                                               |                        |
                                               H                        H

              H                            H
              |                            |
          H — C — H                    H — C — H
    H   H     |     H                  H   |   H   H
    |   |     |     |                  |   |   |   |
H — C — C ——— C ——— C — H          H — C — C — C — C — H
    |   |     |     |                  |   |   |   |
    H   H     |     H                  H   H   |   H
          H — C — H                        H — C — H
              |                                |
              H                                H
```

9.2

```
         CH3                                  CH3
         |                                    |
(a) CH3 — C — CH2CH2CH3          (b) CH3CH2 — C — CH2CH3
         |                                    |
         CH3                                  CH3

           CH2CH3                                     CH2CH2CH3
           |                                          |
(c) CH3CH2CHCH2CH2CH2CH2CH3      (d) CH3CH2CH2CHCH2CH2CH2CH2CH2CH3
```

9.3 (a) 2-methylpropane (b) pentane (c) 2,3,4-trimethylheptane (d) 3,6-dimethyloctane (e) 3,3-diethylpentane **9.4** (a) hexadecane (b) heptane (c) hexadecane

9.5 (a) $2CH_3CH_2CH_2Br + 2Na \xrightarrow{\text{heat}} CH_3CH_2CH_2CH_2CH_2CH_3 + 2NaBr$

```
         CH3                     CH3   CH3
         |                       |     |
(b) 2CH3CHBr + 2Na --heat--> CH3CH — CHCH3 + 2NaBr
```

(c) $CH_3CH_2CH{=}CHCH_3 + H_2 \xrightarrow{\text{Ni}} CH_3CH_2CH_2CH_2CH_3$

```
        CH3                        CH3
        |                          |
        C                          CH
      /   \\                      /    \
(d) CH2     CH + H2 --Ni--> CH2      CH2
     \     /                   \      /
    CH2 — CH2                  CH2 — CH2
```

9.6 (a) $2CH_3CH_2Cl + 2Na \rightarrow CH_3CH_2CH_2CH_3 + 2NaCl$

(b) $CH_3CH_2CH{=}CH_2 + H_2 \xrightarrow{\text{Ni}} CH_3CH_2CH_2CH_3$

9.7 Excess Cl_2 favors CCl_4; excess CH_4 favors CH_3Cl.

9.8

```
        CH2                             CH2 — CH2
      /     \                          /         \
   CH2       CH2                    CH2           CH2
   |           \                    |              |
   CH2          CH2                 CH2           CH2
     \         /                       \          /
      CH2 — CH2                         CH2 — CH2
```

cycloheptane
C_7H_{14}

cyclooctane
C_8H_{16}

9.9 (a) chlorocyclohexane (b) 1,1-dichlorocyclohexane (c) 1,3-dimethylcyclopentane (d) 1-chloro-1,3-dimethylcyclobutane

9.10 (a)  [cyclobutane with two Br on one carbon] (b) 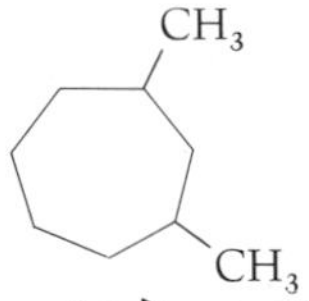 [cycloheptane with two CH_3 groups]

9.11 (a) [cyclopropane] + HBr → $CH_3CH_2CH_2Br$ (b) [cyclopropane] + Cl_2 → $ClCH_2CH_2CH_2Cl$ **9.12** (a) C_7H_{16} (*Section 9.2*)

9.13 (a) $CH_3CH(CH_3)CH_2CH_3$ (c) $CH_3C(CH_3)_2CH_2CH(CH_3)CH_3$ (e)  [cyclobutane with two CH_2CH_3 groups on one carbon] (*Section 9.3*)

9.14 (a) 2-methylpentane (c) 1,1,3,3-tetramethylcyclopentane **9.16** (a) no (*Section 9.4*)

9.17 (a) $CH_3CH{=}CH_2 + H_2 \xrightarrow{Ni} CH_3CH_2CH_3$ (*Section 9.5*)
(c) $CH_3CH_2CH_3 + 5O_2 \rightarrow 3CO_2 + 4H_2O$ (*Section 9.6*)

9.18 (a)  [propene skeletal] → [propane skeletal] (c) 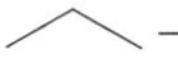 [propane skeletal] → O=C=O

9.19 $2CH_3CH(CH_3)Br + 2Na \xrightarrow{heat} CH_3CH(CH_3)CH(CH_3)CH_3 + 2NaBr$

9.20 (b) [cyclopropane] + HBr → $CH_3CH_2CH_2Br$ (*Section 9.9*)
$2CH_3CH_2CH_2Br + 2Na \xrightarrow{heat} CH_3CH_2CH_2CH_2CH_2CH_3 + 2NaBr$ (*Section 9.5*)

9.22 (a) 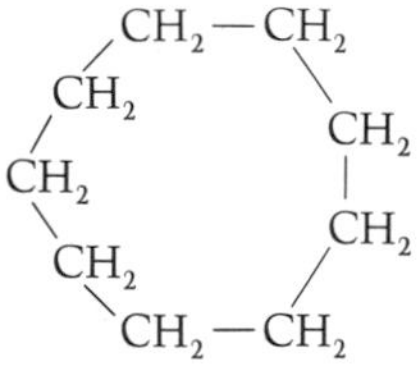 [ring of eight CH_2 groups] [cyclooctane skeletal] (*Section 9.7*)

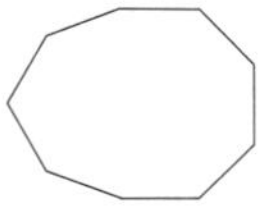

9.24 [cyclopropane] + HI → $CH_3CH_2CH_2I$ (*Section 9.9*) **9.26** (a) $CH_4 + 4Br_2 \xrightarrow{light} CBr_4 + 4HBr$ (*Section 9.6*)

9.28 (a) 2-methylbutane (*Section 9.3*) (c) chlorocyclohexane (*Section 9.8*)

Chapter 10

10.1 (a) propene (the number is unnecessary because there is no ambiguity without it.)
(b) 2-butene (c) 2-methyl-2-butene (d) 1-methylcyclohexene

10.2 (a) $CH_3CH{=}C(CH_3)CH_2CH_3$ (b) $CH_3CH(CH_3)CH{=}CHCH_2CH_3$ (c) 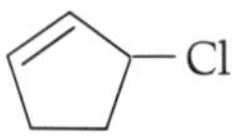 [cyclopentene with Cl]

10.3

pi bonds
sigma bonds

H, CH3, C, C, H, CH3

cis-2-butene

CH3, H, C, C, H, CH3

trans-2-butene

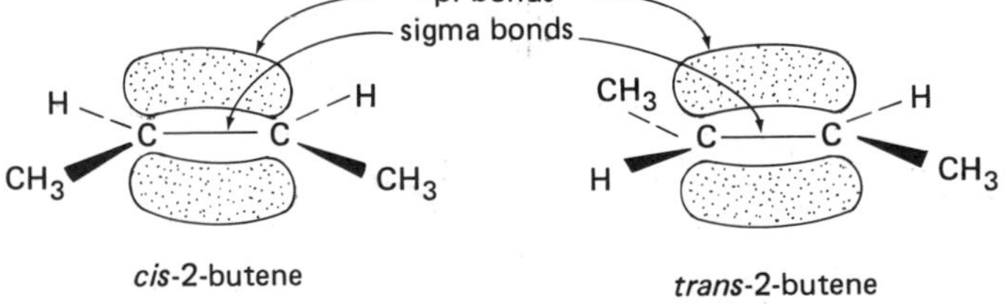

10.4 (a) $CH_3CH{=}CH_2$ (b) $CH_3CH{=}CH_2$ (c) $CH_3CH{=}CH_2$ (d)

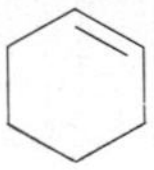

(e) $CH_3C(CH_3){=}CHCH_2CH_3$ (major) + $CH_3CH(CH_3)CH{=}CHCH_3$ (f) 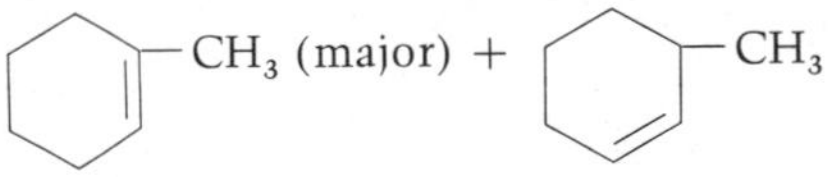 —CH_3 (major) + —CH_3

10.5 (a) $CH_3CH(OH){-}CH(OH)CH_3$ (b) $CH_3CH(Br){-}CH(Br)CH_3$ (c) $CH_3CH(OH){-}CH_2CH_3$ (d) $CH_3CH(Cl){-}CH_2CH_3$

10.6 (a) 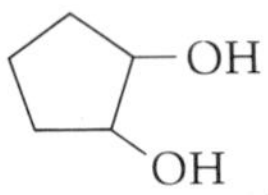 —OH, OH (b) 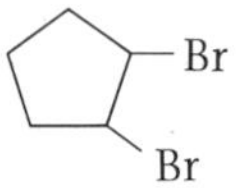 —Br, Br (c) 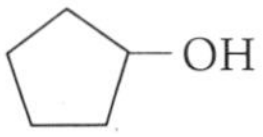 —OH (d) 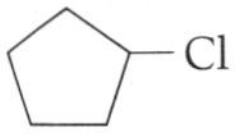 —Cl

10.7 (a) $HOCH_2{-}CH(OH)CH_2CH_3$ (b) $BrCH_2CH(Br)CH_2CH_3$ (c) $CH_3CH(OH)CH_2CH_3$ (d) $CH_3CH(Cl)CH_2CH_3$

10.8 1-hexene would decolorize Br_2 solution; cyclohexane would not. You could also use $KMnO_4$ solution, in which case 1-hexene would decolorize the solution, and cyclohexane would not.

$$CH_2{=}CHCH_2CH_2CH_2CH_3 + Br_2 \rightarrow CH_2(Br){-}CH(Br)CH_2CH_2CH_2CH_3$$

$$CH_2{=}CHCH_2CH_2CH_2CH_3 + KMnO_4 \rightarrow CH_2(OH){-}CH(OH)CH_2CH_2CH_2CH_3$$

(b) Ethylene would decolorize Br_2 solution or $KMnO_4$ solution. Polyethylene has no double bonds, so it would not react with these reagents.

$$CH_2{=}CH_2 + Br_2 \rightarrow CH_2(Br){-}CH_2(Br) \qquad CH_2{=}CH_2 + KMnO_4 \rightarrow CH_2(OH){-}CH_2(OH)$$

10.9 The β-carotene molecule has a cyclohexene ring at each end. In lycopene, that part of the molecule is not cyclic.

10.10 CH_3 CH_3 CH_3 CH_3 CH_3 CH_3 CH_3 CH_3 CH_3CH_3 CH_3

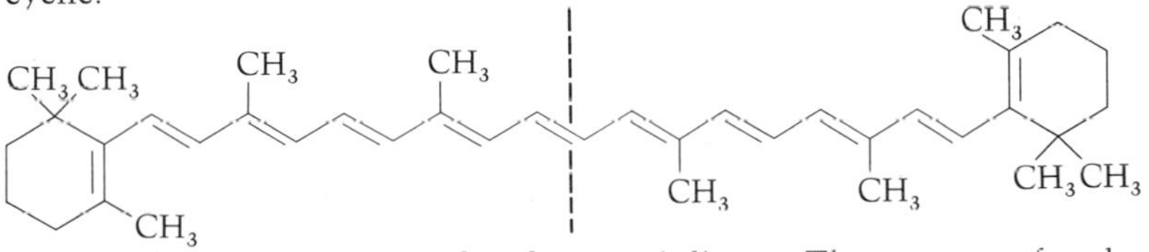

10.11 *sp* **10.12** The $-C{\equiv}C-$ bond system is linear. The presence of such a bond system in a six-membered ring would lead to a large bond angle strain.

10.13 (a) 4-methyl-2-pentyne (b) 1-cyclohexyl-5,5-dimethyl-1-heptyne (c) 2,5-heptadiyne

10.14 (a) $CH_3C{\equiv}CCH_3 + Cl_2 \rightarrow CH_3C(Cl){=}C(Cl)CH_3 \xrightarrow{Cl_2} CH_3CCl_2{-}CCl_2CH_3$

(b) $CH_3C{\equiv}CCH_3 \xrightarrow{H_2/Ni} CH_3CH{=}CHCH_3 \xrightarrow{H_2/Ni} CH_3CH_2CH_2CH_3$

10.15 Only (a). An aromatic compound is *cyclic* and is conjugated all the way around the ring.

10.16

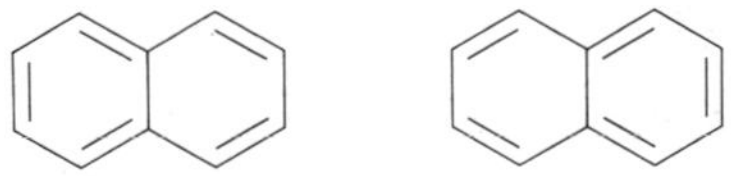

10.17 (a) p-nitrotoluene (b) 3,4-dichloroethylbenzene (c) 3,4,5-trinitrotoluene (d) o-nitroethylbenzene

10.18 (a) $C_6H_6 + Br_2 \xrightarrow{Fe} C_6H_5Br + HBr$

(b) $C_6H_6 + SO_3 \xrightarrow{H_2SO_4} C_6H_5SO_3H$

(c) $C_6H_6 + HONO_2 \xrightarrow{H_2SO_4} C_6H_5NO_2 + H_2O$

10.19 (a) [anthracene resonance structures] (b) [phenanthrene resonance structures]

10.21 (a) $CH_3C(CH_3){=}CHCH_3$ (*Section 10.2*) (c) p-dinitrobenzene, NO_2 (*Section 10.15*) (e) m-fluorotoluene, CH_3, F (*Section 10.15*)

(g) 1,5-dimethylcyclohexene, CH_3, CH_3 (*Section 10.2*) (i) Cl–cyclohexane–Cl (*Section 9.8*) (k) $ClCH{=}CHCH_2CH_3$ (*Section 10.2*)

10.22 (a) 5-ethyl-4-methyl-2-nonene (*Section 10.2*) (c) 2-methyl-1-butene (*Section 10.2*) (e) 3-nitrochlorobenzene (*Section 10.15*) (g) benzoic acid (*Section 10.15*) (i) 1-pentyne (*Section 10.9*)

10.23 (a) $CH_3CCl_2{-}CCl_2CH_3$ (*Section 10.11*) (e) $CH_3C(CH_3){=}CH_2$ (*Section 10.5*) (g) 1-chloro-1-methylcyclohexane, CH_3, Cl (*Section 10.6*)

(i) $\mathrm{+CH_2CH(CH_3)+_n}$ (*Section 10.6*) (k) 1-methylcyclohexanol, CH_3, OH (*Section 10.6*)

10.25

[polycyclic aromatic structure] (*Section 10.17*)

10.26 Shorter, because the extra electrons exert greater attraction between the two carbon nuclei. (*Section 10.8*)

10.28 (a) 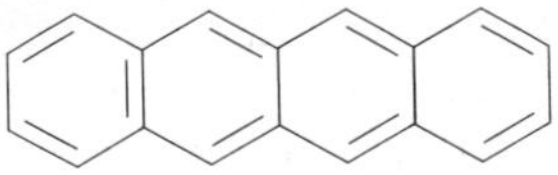 [tetracene structure] (*Section 10.17*)

10.31 $-C\equiv$ has only two atoms or groups bonded to it, so their repulsion is least when the molecule is linear. (*Section 10.8*) **10.33** Cyclooctyne, because the bond angles are greater (closer to 180°). (*Section 10.8*) **10.34** (a) and (e) (*Section 10.13*) **10.36** Addition would destroy the aromaticity. (*Section 10.13*) **10.38** cyclopropane, which undergoes addition reactions (*Section 9.9*)

10.40 (a) $CH_2{=}CHCH_2CH_2CH_3 \xrightarrow{Br_2} BrCH_2CH(Br)CH_2CH_2CH_3$. Red color disappears. Cyclopentane does not react. (*Section 10.6*)

Chapter 11

11.1 CH_3CH_2Cl (chloroethane) CH_3CHCl_2 (1,1-dichloroethane) CH_3CCl_3 (1,1,1-trichloroethane) CH_2ClCH_2Cl (1,2-dichloroethane) $CH_2ClCHCl_2$ (1,1,2-trichloroethane) CH_2ClCCl_3 (1,1,1,2-tetrachloroethane) $CHCl_2CHCl_2$ (1,1,2,2-tetrachloroethane) $CHCl_2CCl_3$ (1,1,1,2,2-pentachloroethane) CCl_3CCl_3 (hexachloroethane)
11.2 Fluorine-containing compounds are relatively inert and have low toxicities. Many are narcotics at high concentrations. **11.3** Fluorine-containing compounds are colorless, odorless, and low-boiling liquids or gases.
11.4 $CH_3CH(CH_3)CH_2CH_2OH$
11.5 Chlorinated methanes are volatile liquids except methyl chloride, which is a gas. They are insoluble in water and have relatively high densities because of their high molecular weights and small sizes. (*Sections 11.1, 11.2*) **11.7** photochemical (decomposition under the influence of sunlight); reaction with atmospheric oxygen; biodegradation (breakdown by living organisms) (*Section 11.3*)
11.9 Chlorine-containing compounds are much more toxic than fluorine-containing compounds. (*Tables 11.2, 11.3*) **11.11** high oxygen and nitrogen content, release of large amounts of energy upon decomposition, formation of gaseous products, and rapid decomposition (*Section 11.7*)

Chapter 12

12.1 (a) 1° (b) 1° (c) 3° (d) 2°
12.2 (a) $CH_3-CH_2-C(CH_3)_2-CH_2OH$ (b) $CH_3CH(CH_3)OH$ (c) 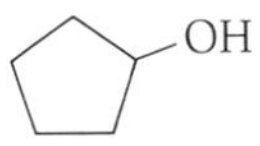 [cyclopentane ring]–OH (d) 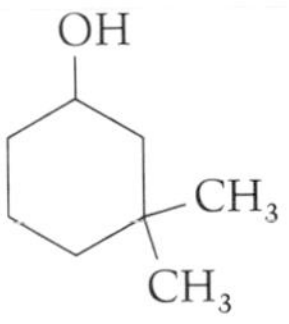 [cyclohexane ring with OH, and CH₃, CH₃ on one carbon]
12.3 (a) 1,2-cyclopentanediol (b) 3-methyl-3-ethyl-1-pentanol (c) 1-decanol (d) 1,8-octanediol
12.4 Butanol; it resembles water more than octanol does. **12.5** CH_3OH (acid) and CH_3O^- (base); OH^- (base) and H_2O (acid)
12.6 (a) $CH_3C(CH_3){=}CH_2$ (b) $CH_3CH_2CH_2CH_2Br$ (c) $CH_3CH(CH_3)-O-C({=}O)-C_6H_5$ (d) no reaction

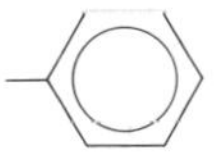

(e) $CH_3CH_2CH_2\overset{\displaystyle CH_3}{\overset{|}{C}}=O$ (f) $CH_3\overset{\displaystyle CH_3}{\overset{|}{C}H}-\overset{\displaystyle O}{\overset{\|}{C}H}$ **12.7** (a) 2-chlorophenol (b) 4-ethylphenol (c) 2,4,6-trinitrophenol (or picric acid) (d) 2-chloro-5-bromophenol

12.8 [structure: benzene ring with OH and ortho NO_2] = 2-nitrophenol; [structure: benzene ring with OH and para NO_2] = 4-nitrophenol

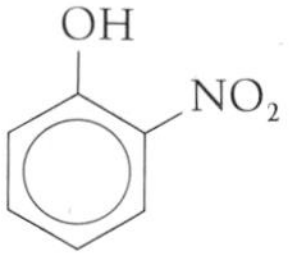

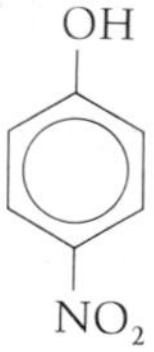

12.9

(a) 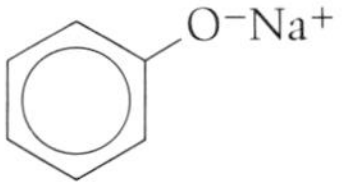 [structure: benzene ring with O^-Na^+] + H_2 (b) 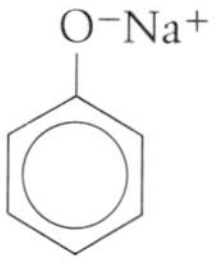 [structure: benzene ring with O^-Na^+] + H_2O

(c) equilibrium with a slight amount of [structure: benzene ring with $\bar{O}$] $+ H_3O^+$ (d) 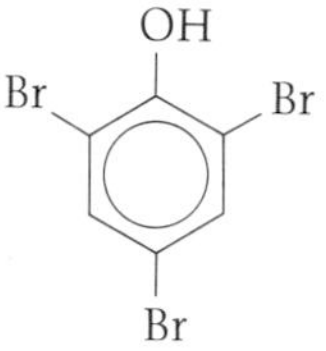 [structure: benzene ring with OH, Br at 2, 4, 6 positions] + 3HBr

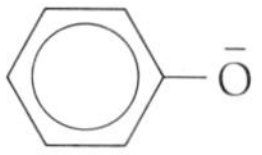

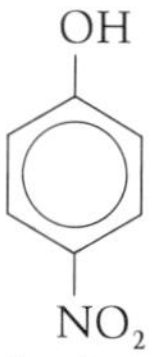

(e) 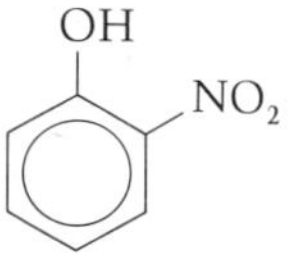 [structure: benzene ring with OH and ortho NO_2] + [structure: benzene ring with OH and para NO_2] + H_2O (f) 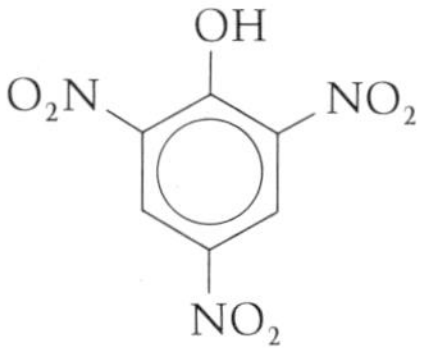 [structure: benzene ring with OH, O_2N and NO_2 ortho, NO_2 para] + $3H_2O$ (g) 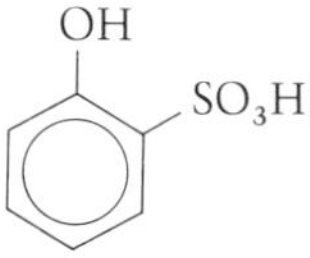 [structure: benzene ring with OH and ortho SO_3H] + [structure: benzene ring with OH and para SO_3H]

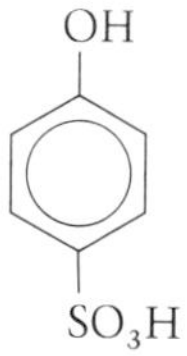

12.10 (a) ethyl propyl ether or 1-ethoxypropane (b) isopropyl phenyl ether or 2-phenoxypropane (c) diisopropyl ether or 2-isopropoxypropane **12.11** (a) Dipropyl ether is more soluble and has the higher boiling point. (b) 1-hexanol is more soluble and has the higher boiling point. (c) 1-hexanol has the higher boiling point; they have similar solubilities. **12.12** (a) $CH_3CH_2I + CH_3I + H_2O$

(b) CH_3CH_2O-[structure: para-disubstituted benzene ring]$-Cl$ + ortho isomer + HCl **12.13** (a) $CH_3\underset{\displaystyle OH}{CH}-\underset{\displaystyle OH}{CH}CH_3$ (b) $CH_3\underset{\displaystyle OH}{CH}-\underset{\displaystyle NH_2}{CH}CH_3$

12.14 Tetrahydrocannabinol has a phenol, an alkene, and an ether group. Morphine has a phenol, an ether, an alcohol, an alkene, and a 3° amine (Chapter 13).

12.15 (a) $CH_3CH_2CH_2-S-CH_2CH_2CH_3$ (b) 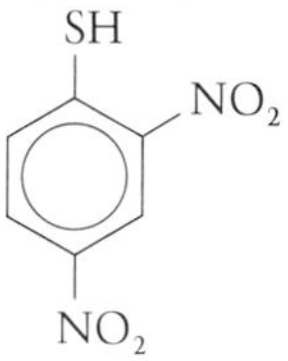 [structure: benzene ring with SH, ortho NO_2, para NO_2] (c) 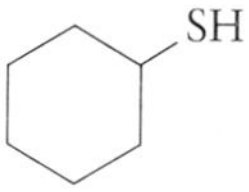 [structure: cyclohexane ring with SH]

12.16 (a) ethanethiol or ethyl mercaptan (b) 2,5-dimethyl-1-hexanethiol (c) 4-nitrothiophenol (d) ethyl phenyl sulfide (e) diphenyl sulfide **12.17** diethyl ether **12.18** Equilibrium favors the products.

12.19 (a) $CH_3CH_2\overset{\displaystyle CH_3}{\underset{\displaystyle OH}{C}}CH_2CH_3$ (c) $CH_3\overset{\displaystyle CH_3}{CH}-O-\overset{\displaystyle CH_3}{CH}CH_3$ (e) $CH_3\overset{\displaystyle CH_3}{\underset{\displaystyle SH}{C}}CH_2\overset{\displaystyle CH_3}{CH}CH_2CH_3$

(g) $CH_3O-C_6H_4-OCH_3$ (i) $CH_3CH(OH)CH_2CH(OH)CH_3$ (k) $CH_3CH_2C(CH_3)_2CH_2CH_2OH$ (*Section 12.2*)

12.22

Compound	Mol. wt.	B.P., °C	Density, g/mL
(a) 1-butanol	74	118	0.81
pentane	72	36	0.63
(c) 1-hexanol	102	157	0.82
heptane	100	98	0.68

(*Section 12.3*)

12.24 (a) aldehyde (*Section 12.4*) (c) no reaction (*Section 12.4*) (e) $CH_3\overset{O}{\overset{\|}{C}}CH_3$ (*Section 12.4*)

(g) $C_6H_5-\overset{O}{\overset{\|}{C}}-OCH_3$ (*Section 12.4*) (i) $C_6H_5-O^-Na^+ + H_2$ (*Section 12.8*)

(k) o-$HOC_6H_4SO_3H$ + p-$HOC_6H_4SO_3H$ (*Section 12.8*) (m) $CH_3CH_2Br + CH_3Br + H_2O$ (*Section 12.13*)

(o) $C_6H_5CH_2-S-S-CH_2C_6H_5 + 2H_2O$ (*Section 12.19*)

12.25 (a) $CH_2{=}CH_2 + H_2O \xrightarrow{H^+} CH_3CH_2OH$ (b) $\underset{\text{(glucose)}}{C_6H_{12}O_6} \xrightarrow{\text{yeast}} 2CH_3CH_2OH + 2CO_2$ (*Section 12.9*)

12.29 with acids: $R-O-R + H_2SO_4 \rightarrow R-\overset{H^+}{\overset{|}{O}}-R + HSO_4^-$ (*Section 12.13*) **12.31** The bond angles in epoxides are strained. (*Section 12.14*) **12.32** (a) Alcohols hydrogen bond; thiols do not. (*Section 12.18*) **12.33** Sulfur is less electronegative than oxygen. (*Section 12.18*)

Chapter 13

13.1 1° amines: methylamine, aniline; 2° amines: dimethylamine, piperidine; 3° amines: trimethylamine, methylethylaniline **13.2** (a) 1° (b) 2° (c) 3° **13.3** With amines, the classifications 1°, 2°, 3° refer to the number of hydrocarbon groups attached to nitrogen. With alcohols, the classification refers to the number of hydrocarbon groups attached to the carbon that bears the OH.

13.4 (a) $CH_3CH_2CH_2-N(CH_2CH_2CH_3)-CH_2CH_2CH_3$ b) $CH_3CH_2CH_2-N(CH_3)-CH_2CH_2CH_2CH_3$ (c) $F-C_6H_4-NH_2$

(d) $C_6H_5-NHCH_2CH_2CH_2CH_3$ (e) $CH_3CH_2-C_6H_4-NH_2$ (f) $C_6H_5-N(CH_3)(CH_2CH_3)$

13.5 (a) ethylpropylamine (b) propylamine (c) diphenylamine (d) *p*-nitroaniline

13.6

	B.P., °C	**Water solubility**
methylamine (molecular weight = 31)	−6.5	Very soluble
methanol (molecular weight = 32)	65	∞
ethane (molecular weight = 30)	−88.5	insoluble

13.7 1° amines: methylamine, ethylamine; 2° amines: dimethylamine, diethylamine, piperidine; 3° amines: trimethylamine, triethylamine **13.8** (a) butylamine: $CH_3CH_2CH_2CH_2NH_2$; diethylamine: $CH_3CH_2-NH-CH_2CH_3$ (b) Triethylamine cannot form hydrogen bonds to itself; dipropylamine can. **13.9** (a) propylamine (b) aniline **13.10** (a) tripropylamine (b) diphenylamine **13.11** The oxygen atom in morpholine can hydrogen bond to the N—H of adjacent molecules. **13.12** the anion X^- of the acid HX from which H_3O^+ was produced

13.13 (a) $\underset{\text{base 1}}{(CH_3)_2NH} + \underset{\text{acid 2}}{H_3O^+} \rightleftarrows \underset{\text{acid 1}}{(CH_3)_2NH_2^+} + \underset{\text{base 2}}{H_2O}$ (b) $\underset{\text{base 1}}{(CH_3)_3N} + \underset{\text{acid 2}}{H_3O^+} \rightleftarrows \underset{\text{acid 1}}{(CH_3)_3NH^+} + \underset{\text{base 2}}{H_2O}$

(c) $\underset{\text{base 1}}{(CH_3)_2NH} + \underset{\text{acid 2}}{H_2O} \rightleftarrows \underset{\text{acid 1}}{(CH_3)_2NH_2^+} + \underset{\text{base 2}}{OH^-}$ (d) $\underset{\text{base 1}}{(CH_3)_3N} + \underset{\text{acid 2}}{H_2O} \rightleftarrows \underset{\text{acid 1}}{(CH_3)_3NH^+} + \underset{\text{base 2}}{OH^-}$

13.14 (a) $CH_3CH_2-\overset{H}{\underset{CH_2CH_3}{N^+}}-CH_2CH_3 + Cl^-$ triethylammonium chloride

(b) $CH_3CH_2CH_2CH_2NH_3^+ + I^-$ butylammonium iodide

13.15 (a) $(CH_3)_2NH + CH_3Br \rightarrow (CH_3)_3\overset{+}{N}H + Br^-$
(b) $NH_3 + CH_3CH_2CH_2CH_2I \rightarrow CH_3CH_2CH_2CH_2\overset{+}{N}H_3 + I^-$
(c) $(CH_3CH_2)_3N + CH_3CH_2Cl \rightarrow (CH_3CH_2)_4N^+ + Cl^-$

13.16 [piperidinium ring: N⁺ bearing H and CH_3] Br^- **13.17** Purine has the ring systems of both pyrimidine and imidazole.

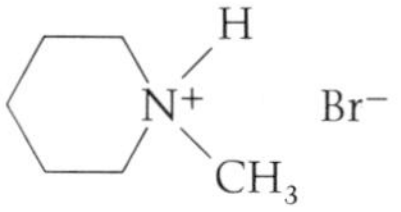

13.18 [imidazole ring: N ← pyridine; N–H ← pyrrole] **13.19** [imidazole ring: N ← 3°; N–H ← 2°] **13.20** [structure: CH_3O, O, HO, N, CH_3]

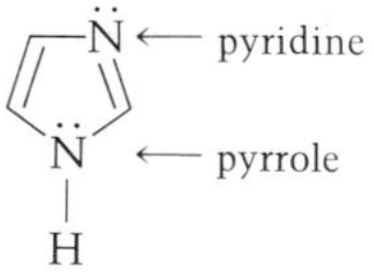

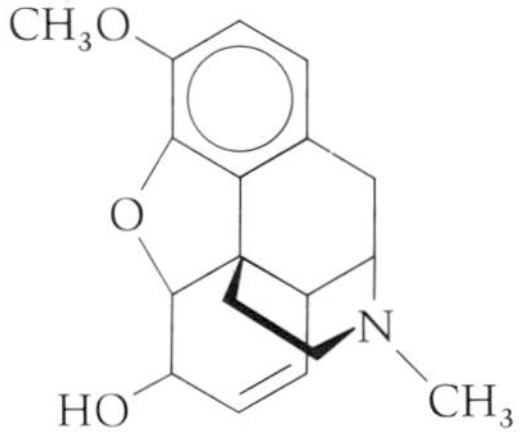

13.21 nicotine: two 3° amines; nicotinic acid: one 3° amine; atropine: one 3° amine; strychnine: one 3° amine (one N is part of an amide group—see Table 8.2: [ring with O=C–N]; morphine: one 3° amine; reserpine: one 2° and one 3° amine; ergotamine

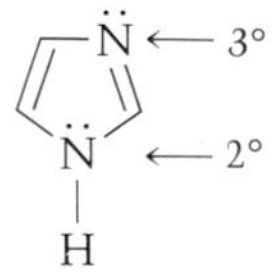

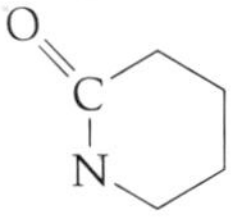

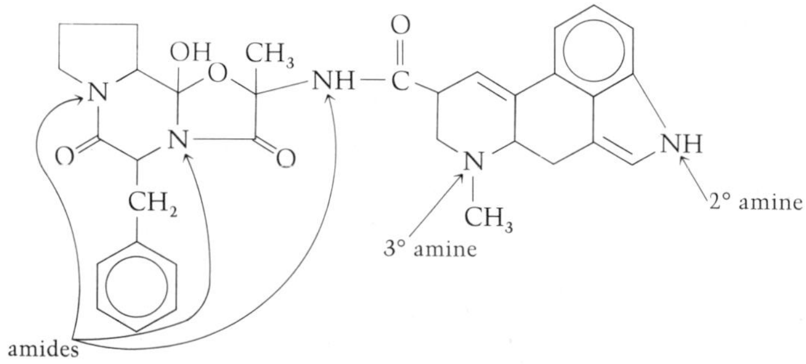

[ergotamine structure: OH, CH_3, O, NH—C(=O), N, N, O, O, CH_2, N–CH_3, NH; labels: amides, 3° amine, 2° amine]

13.22 $H_3\overset{+}{N}-CH_2-\overset{O}{\overset{\|}{C}}-O^-$
13.23 (a) $(CH_3CH_2)_2NH$ 2° (c) $CH_3CH_2-NH-\underset{CH_3}{\underset{|}{CHCH_3}}$ 2°

(e) $C_6H_5-NH_2$ 1°, (*Sections 13.1, 13.3*) **13.24** see 12.23 (*Section 13.2*) **13.25** Hydrazine has more H atoms and can form more hydrogen bonds. (*Section 13.4*)
13.26 (a) $CH_3CH_2NH_2 + H_3O^+ \rightleftarrows CH_3CH_2NH_3^+ + H_2O$

(c) $CH_3\underset{}{\overset{CH_3}{\overset{|}{C}}}HCH_2NH_2 + HBr \rightarrow CH_3\overset{CH_3}{\overset{|}{C}}HCH_2NH_3^+Br^-$

(g) $C_5H_5N: + H_3O^+ \rightleftarrows C_5H_5N-H^+ + H_2O$ (*Section 13.5*)

13.27 (a) reactants (*Section 13.5*) **13.28** (a) $CH_3CH_2CH_2CH_2CH_2-NH_3^+I^-$ pentylammonium iodide (*Section 13.6*) **13.30** Tetramethylammonium hydroxide is the stronger base. (*Section 13.8*)
13.32

phenol
HO
ether → O
N—CH_3
HO
3° amine
alkene
2° alcohol

(*Sections 8.12, 13.2*)

***13.34** (b) $C_6H_5-NH_3^+$ (*Section 13.5*)

Chapter 14

14.1 electronegativity and resonance **14.2** (a) aldehyde (b) carboxylic acid (c) ketone (d) ester (e) ketone (f) acid anhydride (g) amide (h) acyl bromide (i) aldehyde

14.3 (a) $CH_3\overset{OH}{\overset{|}{C}}HCH_2\overset{O}{\overset{\|}{C}}H$ (b) $CH_3CH_2\underset{CH_3}{\underset{|}{C}}H\overset{CH_3}{\overset{|}{C}}H\overset{O}{\overset{\|}{C}}H$ (c) $O_2N-C_6H_4-\overset{O}{\overset{\|}{C}}-H$

14.4 (a) 2,5,5-trimethylhexanal (b) 2-chloropropanal (c) 2-butenal **14.5** (a), (b), (e)
14.6 Both tests are

$$\underset{OH}{\underset{|}{CH_2}}\underset{OH}{\underset{|}{CH}}\underset{OH}{\underset{\backslash}{CH}}\underset{OH}{\underset{\backslash}{CH}}\underset{OH}{\underset{\backslash}{CH}}\overset{O}{\overset{\|}{C}}H + 2Cu(II)\text{ complex} + 4OH \rightarrow \underset{OH}{\underset{|}{CH_2}}\underset{OH}{\underset{|}{CH}}\underset{OH}{\underset{\backslash}{CH}}\underset{OH}{\underset{\backslash}{CH}}\underset{OH}{\underset{\backslash}{CH}}\overset{O}{\overset{\|}{C}}-OH + Cu_2O + 2H_2O.$$

14.7 Benedict's reagent uses sodium citrate to keep the Cu^{2+} ions in solution. Fehling's reagent uses sodium tartrate. Benedict's reagent is preferred because it has a longer shelf life.

14.8 (a) $C_6H_5-\overset{OH}{\overset{|}{C}}H-CN$ (b) C_6H_5-MgCl (c) $C_6H_5-\overset{OMgCl}{\overset{|}{C}}H-CH_3$

(d) $C_6H_5-CH(OH)CH_3$ (e) $C_6H_5-CH_2OMgCl$ (f) $C_6H_5-CH_2OH$

(g) $CH_3CH_2CH(OH)-OCH_3$ (h) $CH_3CH_2CH(OCH_3)_2$ (i) cyclic: $CH_2-CH_2-CH_2-CH(OH)-O-$ (ring: CH_2, CH_2, CH_2, $O-CH$; CH bears OH)

14.9 (c), (d)

14.10 (a) $CH_3CH_2-C(=O)-CH_2CH_3$ (b) $C_6H_5-C(=O)-CH_2CH_3$ (c) $CH_3-C(=O)-CH(CH_3)CH_3$

(d) cyclopentanone with CH_3, CH_3 on adjacent carbons (3,4-dimethylcyclopentanone) (e) $CH_3CH_2C(=O)CH_2CH_2CH_3$ (f) $O=$ cyclohexane $=O$ (1,4-cyclohexanedione)

14.11 (a) 4-heptanone (dipropyl ketone) (b) 3-methylhexanal (c) cyclobutanone
(d) 2,3-dichlorocyclohexanone (e) 7-ethyl-4-nonanone

14.12 (a) $CH_3C(OH)(H)CH_3 \xrightarrow{H_2CrO_4 \;\to\; Cr_2O_3 + H_2O} CH_3C(=O)CH_3$ (b) $CH_3C(=O)CH_3 \xrightarrow{HCN} CH_3C(OH)(CN)CH_3$

14.13 (a) $C_6H_5-C(OMgCl)(CH_3)CH_2CH_3$ (b) $C_6H_5-C(OH)(CH_3)CH_2CH_3$ (c) $C_6H_5-C(OH)(CN)CH_3$

(d) $C_6H_5-CH_2C(OH)(OCH_3)CH_3$ (e) $C_6H_5-CH_2C(OCH_3)_2CH_3$

(f) $CH_3C(=O)CH_2CH_3 \xrightarrow{C_6H_5MgCl} C_6H_5-C(OMgCl)(CH_3)CH_2CH_3 \xrightarrow{H_2O \;\to\; MgOHCl} C_6H_5-C(OH)(CH_3)CH_2CH_3$

(g) $C_6H_5-CH_2C(=O)CH_3 \underset{CH_3OH}{\overset{CH_3OH}{\rightleftharpoons}} C_6H_5-CH_2C(OH)(OCH_3)CH_3 \underset{HCl,\ CH_3OH \;\leftarrow\; H_2O}{\overset{CH_3OH,\ HCl(g) \;\to\; H_2O}{\rightleftharpoons}} C_6H_5-CH_2C(OCH_3)_2CH_3$

14.14 (a) CH_3CH_2OH can hydrogen bond to itself (higher boiling point); both can hydrogen bond to water (equal solubility). (b) $CH_3CHOHCH_3$ can hydrogen bond to itself (higher boiling point); both can hydrogen bond to water (equal solubility).

(c) $H-\overset{O}{\overset{\|}{C}}-H$ is more polar (higher boiling point); $H-\overset{O}{\overset{\|}{C}}-H$ can hydrogen bond to water (higher solubility). **14.15** (a) ester (c) ketone (e) ketone (g) aldehyde (*Section 14.2*)

14.16 (a) $CH_3CH_2\overset{O}{\overset{\|}{C}}H$ (c) $C_6H_5-\overset{O}{\overset{\|}{C}}-C_6H_5$ (e) $CH_3\underset{}{\overset{Cl}{\overset{|}{C}}}HCH_2\overset{O}{\overset{\|}{C}}-H$ (g) Br–(cyclohexane ring)=O (*Sections 14.4, 14.9*)

14.17 (a) 3-methylbutanal (c) ethanal (e) 1,5-dichloro-3-pentanone (*Sections 14.4, 14.9*)

14.18 (a) Mix solution with sample; colorless solution forms silver mirror on sides of test tube in presence of aldehyde. $R-\overset{O}{\overset{\|}{C}}H + Ag(NH_3)_2^+ \rightarrow R-\overset{O}{\overset{\|}{C}}-O^- + NH_4^+ + Ag\downarrow$ (b) Mix solution with sample; blue solution decolorizes and forms reddish precipitate in presence of aldehyde. (*Section 14.6*)

14.19 (a) $CH_3\overset{O}{\overset{\|}{C}}-O^-$ (c) $H-\overset{O}{\overset{\|}{C}}-O^-$ (e) CH_3CH_2MgBr (g) $CH_3\overset{OH}{\overset{|}{C}}HCH_2CH_3$

(i) $C_6H_5-CH(OCH_2CH_3)_2$ (k) $C_6H_5-\overset{O}{\overset{\|}{C}}-O^-$ (m) N.R. (*Sections 14.6, 14.7, 14.10*)

14.20 (b) $C_6H_5-\overset{O}{\overset{\|}{C}}-H \underset{CH_3CH_2OH}{\overset{CH_3CH_2OH}{\rightleftharpoons}} C_6H_5-\underset{OH}{\overset{OCH_2CH_3}{C}}-H \underset{HCl;\ CH_3CH_2OH,\ H_2O}{\overset{CH_3CH_2OH,\ H_2O;\ HCl}{\rightleftharpoons}} C_6H_5-\underset{OCH_2CH_3}{\overset{OCH_2CH_3}{C}}-H$

(*Section 14.10*)

14.21 Propanol can form intermolecular hydrogen bonds; propanal cannot.

14.23 (a) CH_3OH (c) $H-\underset{OCH_3}{\overset{OH}{C}}-H$

Chapter 15

15.1 (a) $CH_3\overset{CH_3}{\overset{|}{C}}HCO_2H$ (b) CH_3CO_2H (c) $O_2N-C_6H_4-CO_2H$ (d) HCO_2H (e) $CH_3\overset{OH}{\overset{|}{C}}HCH_2CO_2H$

15.2 (a) *p*-chlorobenzoic acid (b) chloroethanoic acid (c) pentanoic acid

15.3 (a) $CH_3CH_2CO_2H$ has both the higher boiling point and solubility. (b) $CH_3CH_2CO_2H$ has both the higher boiling point and solubility. (c) CH_3CH_2OH has the higher boiling point; they have similar solubilities. (d) $C_6H_5-CO_2H$ has both the higher boiling point and solubility.

15.4 (a) $CH_3CH_2OH + H_2O \rightleftharpoons CH_3CH_2O^- + H_3O^+$ (b) $C_6H_5-OH + H_2O \rightleftharpoons C_6H_5-O^- + H_3O^+$

(c) $CH_3CO_2H + H_2O \rightleftharpoons CH_3CO_2^- + H_3O^+$ (d) $CH_3CH_2OH + OH^- \rightleftharpoons CH_3CH_2O^- + H_2O$

(e) $C_6H_5-OH + OH^- \rightleftharpoons C_6H_5-O^- + H_2O$ (f) $CH_3CO_2H + OH^- \rightleftharpoons CH_3CO_2^- + H_2O$

(g) $CH_3CH_2OH + HCO_3^- \rightleftharpoons CH_3CH_2O^- + H_2CO_3$ (h) C_6H_5—OH + $HCO_3^- \rightleftharpoons C_6H_5$—$O^- + H_2CO_3$

(i) $CH_3CO_2H + HCO_3^- \rightleftharpoons CH_3CO_2^- + H_2CO_3$ **15.5** (a) acetic anhydride (b) ethanamide or acetamide (c) acetyl chloride (d) ethyl acetate

15.6 (a) C_6H_5—C(=O)—OCH_2CH_3 + NaCl + H_2O (b) $CH_3CH_2C(=O)$—$NHCH_3$ + $CH_3\overset{+}{N}H_3Cl^-$

(c) $CH_3C(=O)$—OCH_3 + CH_3CO_2H (d) $CH_3C(=O)$—$NHCH_3$ + $CH_3C(=O)$—O^- + $CH_3NH_3^+$

(e) C_6H_5—C(=O)—NH_2 + CH_3CH_2OH (f) $CH_3C(=O)$—$O^-\overset{+}{N}H_3CH_2CH_3$ (g) $CH_3C(=O)$—$NHCH_2CH_3$ + H_2O

15.7 (a) H—C(=O)—$OCH_2CH_2CH_2CH_3$ (b) $CH_3CH_2CH_2C(=O)$—O—C_6H_5 (c) $CH_3C(=O)$—O—C_6H_{11}

15.8 (a) methyl formate (b) cyclopentyl propanoate (c) pentyl acetate

15.9 (a) $CH_3CO_2H + CH_3CH_2OH$ (b) C_6H_5—$CO_2H + CH_3OH$ (c) $CH_3CH_2CO_2H$ + C_6H_5—OH

15.10 $C_6H_4(OH)(CO_2H)$ + CH_3CO_2H
o-hydroxybenzoic acid

15.11 RO—P(=O)(OH)—OH + H_3PO_4 → RO—P(=O)(OH)—O—P(=O)(OH)—OH + H_2O

15.12 (a) $CH_3CH_2CH_2C(=O)$—NH_2 (b) Cl—C_6H_4—C(=O)—NH_2

15.13 (a) hexanamide (b) *p*-nitrobenzamide **15.14**

R—C(=O)—NH_2 hydrogen bonded to water: C=O···H—O—H; N—H···OH_2; N—H···OH_2; N···H—O—H

15.15 Amides have two hydrogens with which to form hydrogen bonds; carboxylic acids have only one.
15.16 In base-catalyzed hydrolysis, the carboxylate ion is produced because the free RCO_2H cannot exist in basic solution. In acid-catalyzed hydrolysis, the free RCO_2H forms.

15.17 (a) $CH_3C(=O)$—O^- + NH_3 (b) C_6H_5—C(=O)—OH + NH_4^+

15.18

	Common name	IUPAC name	
(a) $H-C(=O)-OH$	formic acid	methanoic acid	
(c) $CH_3CH_2CO_2H$	propionic acid	propanoic acid	
(e) $C_6H_5-CO_2H$	benzoic acid	benzoic acid	(*Section 15.3*)

15.19

	Common name	IUPAC name	
(a) $HC(=O)-OCH_2CH_3$	ethyl formate	ethylmethanoate	
(c) $CH_3CH_2C(=O)-OCH_2CH_3$	ethyl propionate	ethyl propanoate	
(e) $C_6H_5-C(=O)-OCH_2CH_3$	ethyl benzoate	same	(*Section 15.8*)

15.20

	Common name	
(a) $HC(=O)-NH_2$	formamide	
(c) $CH_3CH_2C(=O)-NH_2$	propanamide	(*Section 15.14*)

15.23 nylon (*Sections 15.12, 15.17*) **15.24** (a) $C_6H_5-CH_2CO_2H$ (c) $C_6H_5-C(=O)-NH_2$ (*Sections 15.3, 15.7, 15.14*)

15.25 (a) 2,4-dinitrobenzoic acid (c) chloroacetic acid (e) formamide (*Sections 15.3, 15.14*)

15.26 (a) $CH_3CO_2H + NaOH\ (aq) \rightarrow \underbrace{CH_3CO_2^-Na^+}_{\text{(soluble)}} + H_2O$; benzene does not react and is insoluble in water.

(d) $C_6H_5-OH + NaOH\ (aq) \rightarrow \underbrace{C_6H_5-O^-Na^+}_{\text{soluble}} + H_2O$ Cyclohexanol does not dissolve. (*Section 15.5*)

15.27 (a) $C_6H_5-C(=O)-OCH_2CH_3$ (b) $CH_3CH_2C(=O)-OCH_2CH_2CH_3$ (c) $CH_3C(=O)-\overset{-}{O}\ \overset{+}{N}H_3CH_3$

(d) $CH_3C(=O)-NHCH_3$ (e) $CH_3C(=O)-NHCH_3 + CH_3CH_2OH$

(f) $Cl\!\left(\!C(=O)(CH_2)_4C(=O)-NH(CH_2)_6NH\right)_{n}\!H + nHCl$ (g) $CH_3CH_2CO_2^- + (CH_3)_2NH$

(h) $CH_3CH_2CO_2H + (CH_3)_2\overset{+}{N}H_2$ (i) $CH_3CO_2H + CH_3OH$ (j) $CH_3CO_2^- + CH_3OH$

(k) $CH_3CO_2Na + CH_3OH$ (l) $(\overset{CH_3}{\underset{CO_2^-}{C}}\ \ CH_2)_n + n\ CH_3OH$ (m) $C_2H_5O-C_6H_4-NH_2 + CH_3CO_2^-$

(n) $CH_3CH_2OH + HO-C_6H_4-\overset{+}{N}H_3 + CH_3CO_2H$ (*Sections 15.5–15.7, 15.9–15.11, 15.16*)

15.29 I (*Section 15.5*)

Chapter 16

16.1 trisaccharide **16.2** a small polysaccharide **16.3** (a) aldotetrose (b) ketohexose (c) aldohexose (d) ketopentose **16.4** (c), (d) are chiral; (a), (b) are achiral. **16.5** (d)
16.6 CH_3, CH_3CH_2, OH, and H

16.7 (a) $CH_3\overset{*}{C}HBrCH_2CH_3$ (b) $CH_3\overset{*}{C}HOH\overset{O}{\overset{\|}{C}}H$ (c) $C_6H_5-\overset{*}{C}H(CH_3)CO_2H$ **16.8** $-13.52°$

16.9 no **16.10** (a)  CHO–CHOH–CH_2OH (b) 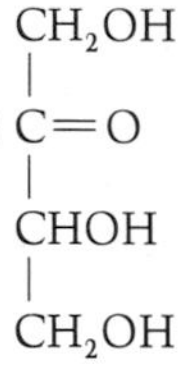 CH_2OH–C=O–CHOH–CH_2OH

16.11 No.

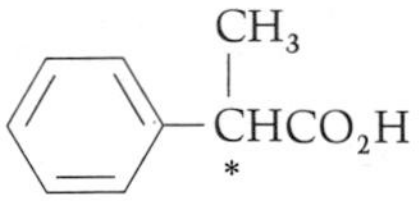

16.12

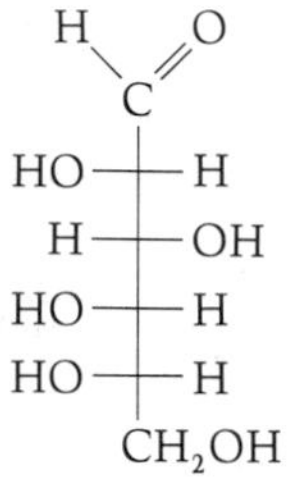

	CHO	
HO	—	H
H	—	OH
HO	—	H
HO	—	H
	CH_2OH	

16.13

	CHO	
HO	—	H
H	—	OH
H	—	OH
HO	—	H
	CH_2OH	

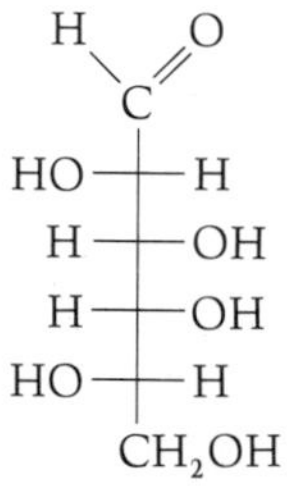

16.14 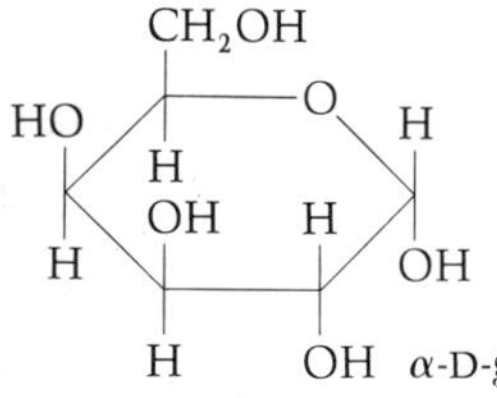 α-D-galactose 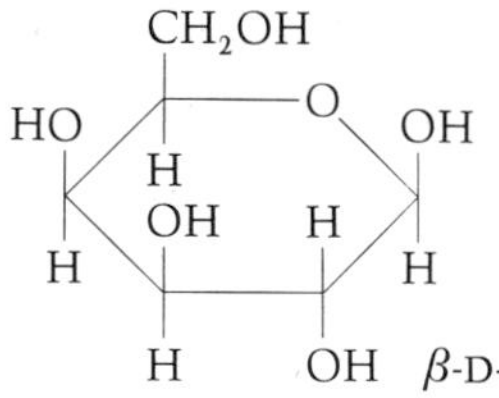 β-D-galactose

16.15 (a)

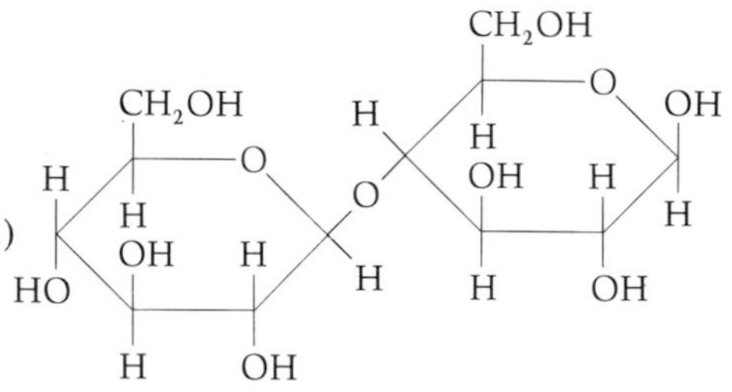

(b) 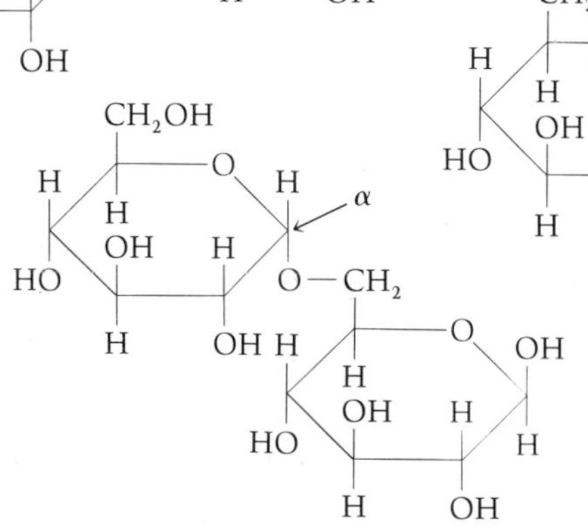 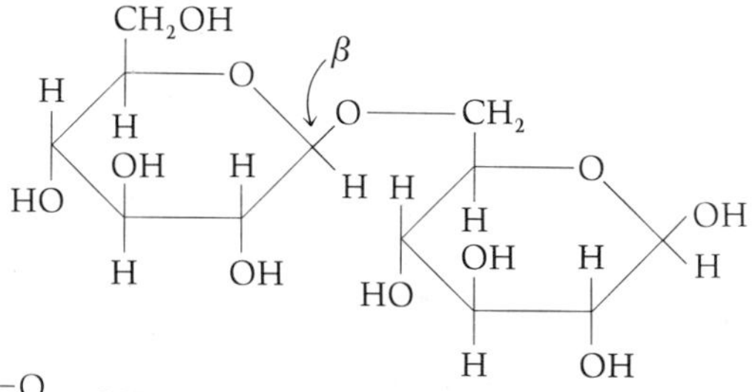 α β

16.16 (a) $C_6(H_2O)_6$ (c) $C_6(H_2O)_6$ (*Section 16.1*) **16.17** (a) and (b) are monosaccharides (c) disaccharide (*Section 16.1*) **16.18** pentasaccharide (*Section 16.1*)

16.20 (a)

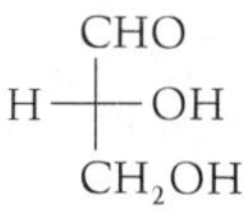

CHO
H—┼—OH
CH_2OH

(c)

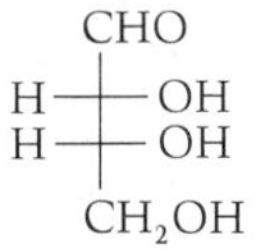

CHO
H—┼—OH
H—┼—OH
CH_2OH

(*Section 16.8*)

16.21 (a)

CH_2OH
H ring O OH
H
HO OH OH H
H H

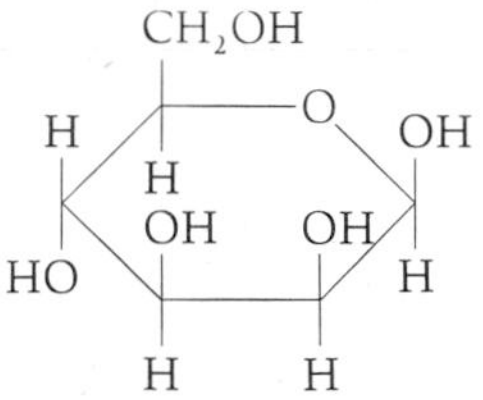

(*Section 16.10*)

16.22 (b), (d), and (f) (*Section 16.4*) **16.24** the highest numbered chiral carbon atom (*Section 16.9*)
16.25 (a) yes (c) no (*Sections 16.8, 16.9*)

16.26 (a)

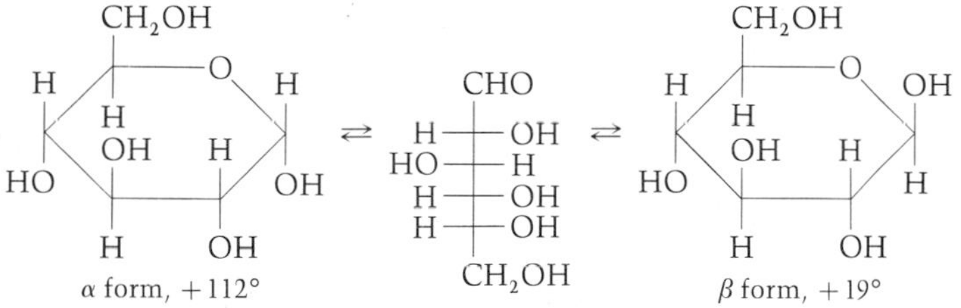

CH_2OH
H ring O H
H
HO OH H OH
H OH
α form, +112°

⇌

CHO
H—┼—OH
HO—┼—H
H—┼—OH
H—┼—OH
CH_2OH

⇌

CH_2OH
H ring O OH
H
HO OH H H
H OH
β form, +19°

(*Section 16.10*)

16.28 (b) Ring *B* is a furanose ring. (*Sections 16.10, 16.11*)

Chapter 17

17.1 $^{+}H_3N-\overset{R_1}{\overset{|}{C}H}-CO_2^- + {}^{+}H_3N-\overset{R_2}{\overset{|}{C}H}-CO_2^- + {}^{+}H_3N-\overset{R_3}{\overset{|}{C}H}-CO_2^- \rightarrow$

$$^{+}H_3N-\overset{R_1}{\overset{|}{C}H}-\overset{O}{\overset{\|}{C}}-NH-\overset{R_2}{\overset{|}{C}H}-\overset{O}{\overset{\|}{C}}-NH-\overset{R_3}{\overset{|}{C}H}-CO_2^- + 2H_2O$$

17.2 Catalyst, transport vehicle in body fluids, membrane transport vehicle, hormone receptor, structural material, protective material.

17.3 (a) 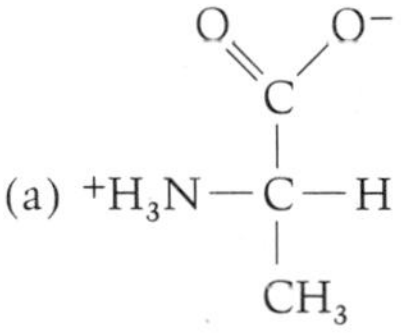 $^{+}H_3N-C(CO_2^-)(H)-CH_3$ (b) $^{+}H_3N-C(CO_2^-)(H)-CH_2OH$ (c) $^{+}H_3N-C(CO_2^-)(H)-H$ (d) 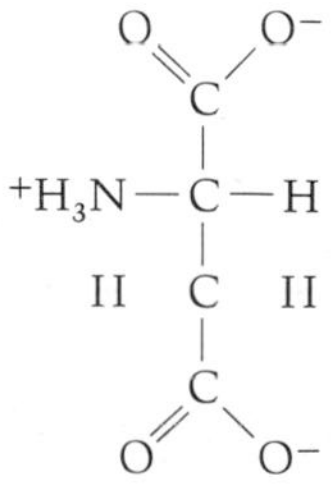 $^{+}H_3N-C(CO_2^-)(H)-CH_2-CO_2^-$

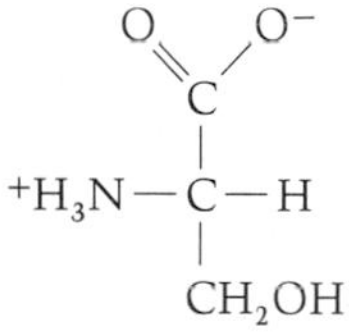

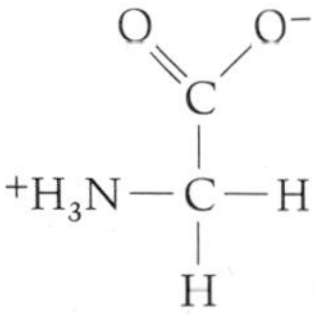

17.4 Proline; it is a substituted or secondary amine (—NHR, not —NH_2).

17.5

(a) 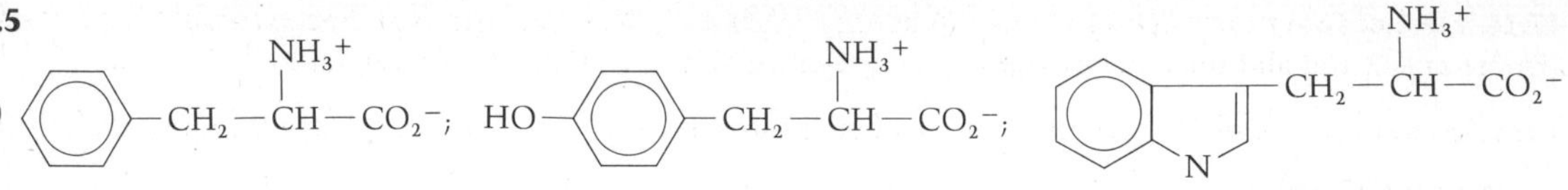 $C_6H_5-CH_2-CH(NH_3^+)-CO_2^-$; $HO-C_6H_4-CH_2-CH(NH_3^+)-CO_2^-$; (indolyl)$-CH_2-CH(NH_3^+)-CO_2^-$

phenylalanine; tyrosine; tryptophan

(b) $HO-CH_2-CH(NH_3^+)-CO_2^-$; $CH_3-CH(OH)-CH(NH_3^+)-CO_2^-$

serine; threonine

(c) $HS-CH_2-CH(NH_3^+)-CO_2^-$; $H_3CSCH_2CH_2CH(NH_3^+)-CO_2^-$

cysteine; methionine

17.6 ala-trp-gly-asp; ala-gly-trp-asp; ala-trp-asp-gly; ala-asp-gly-trp; ala-gly-asp-trp; ala-asp-trp-gly, and so on. (There are a total of 24 possible sequences, since the total number of sequences is given by $n!$, where n = the number of amino acid residues in the peptide. Here $n = 4! = 4 \cdot 3 \cdot 2 \cdot 1 = 24$.)

17.7 $^+NH_3-CH(CH_3)-C(=O)-NH-CH(CH_2CO_2^-)-C(=O)-NH-CH(CH_2\text{-indolyl})-C(=O)-NH-CH(H)-CO_2^-$ ala-asp-trp-gly

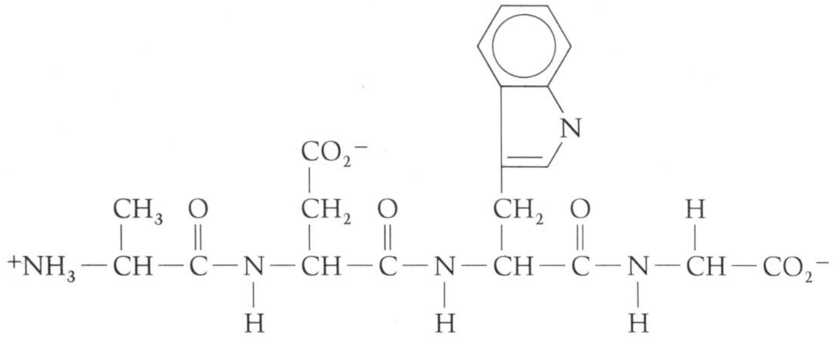

17.8 Certain kinds of fish are pickled without cooking. In this case strong salt and/or acid do the denaturing.

17.9 $^+H_3N-CH(CH(CH_3)CH_2CH_3)-C(=O)-NH-CH((CH_2)_4NH_3^+)-CO_2^-$ Net charge is +1 (*Section 17.5*)

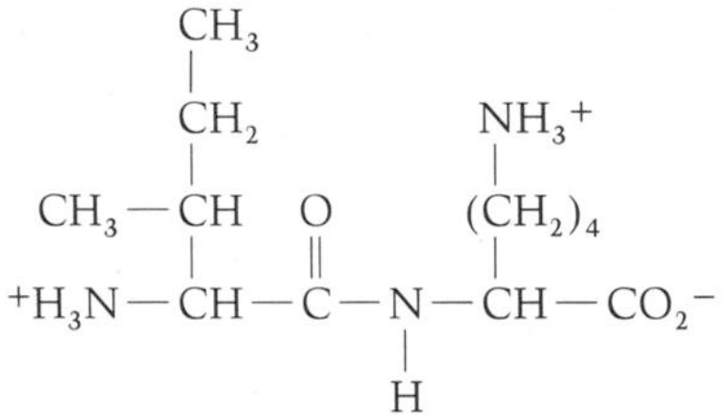

17.11 (a) $H_2N-C(=NH_2^+)-NH-(CH_2)_4-CH(NH_3^+)-CO_2H$; +2 (*Section 17.5*)

(b) 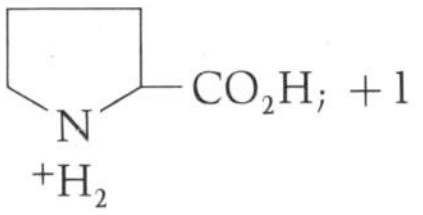 (pyrrolidine ring, $\overset{+}{N}H_2$)$-CO_2H$; +1

(c) 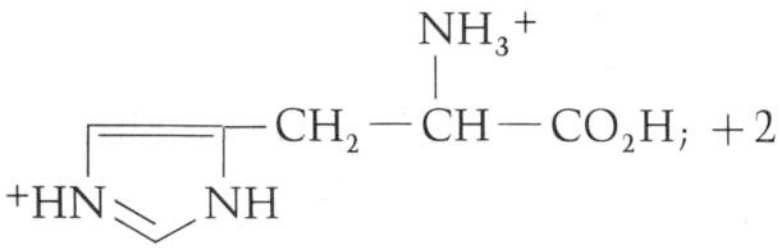 (imidazolium ring, ^+HN, NH)$-CH_2-CH(NH_3^+)-CO_2H$; +2

(d) $CH_3-CH_2-CH(CH_3)-CH(NH_3^+)-CO_2H$; +1

(e) $HO_2CCH_2CH_2CH(NH_3^+)-CO_2H$; +1

17.12 No, because their net charges would be the same (−2). Yes, because aspartic acid would have a net charge of −2 and alanine a net charge of −1. (*Section 17.5*) **17.14** Hydrogen bonding between peptide groups (>C=O···H—N<). In α-helical structure, the hydrogen bonded groups are four residues apart along the same chain and the chain is uniformly coiled. In β-structure, the hydrogen bonded groups are in separate polypeptides or in distinct regions of the same polypeptide, and the chains are stretched or extended rather than coiled in the hydrogen-bonded regions. (*Section 17.10*)
17.16 (a) serine (b) globular protein (c) globular protein at a pH greater than its isoelectric pH (*Sections 17.13, 17.14*) **17.19** by noncovalent bonds; denaturing agents such as heat, nonpolar solvents, acid, base, and so on (*Section 17.12*)

Chapter 18

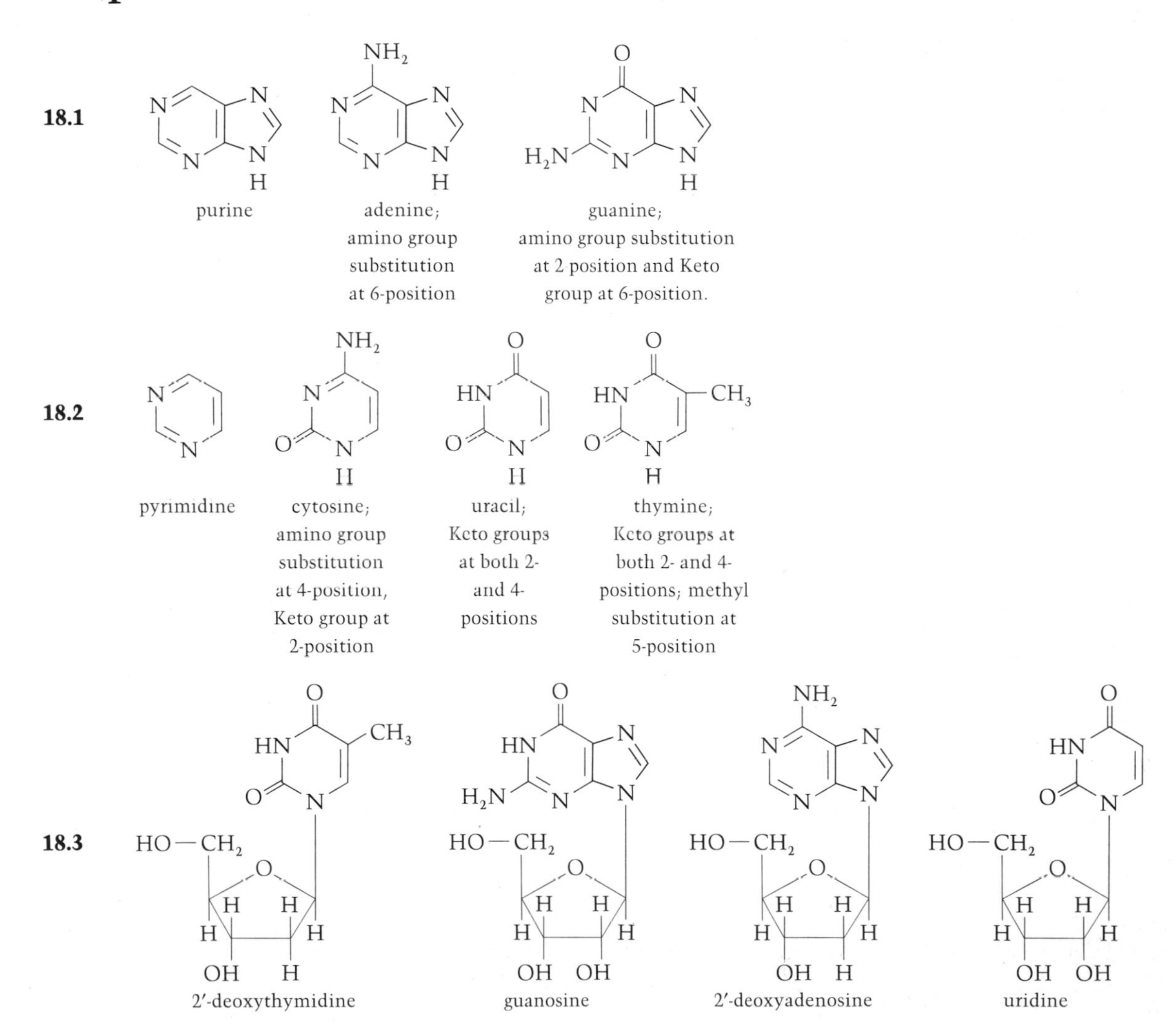

18.4

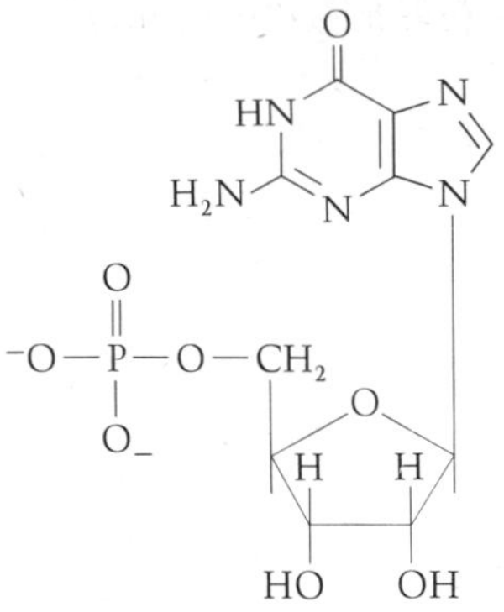

GMP

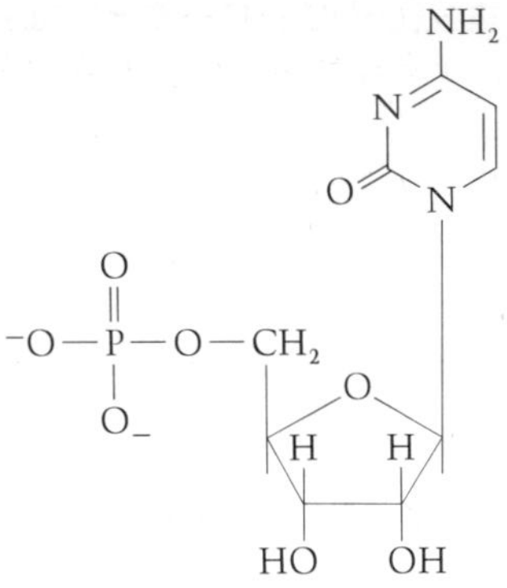

CMP

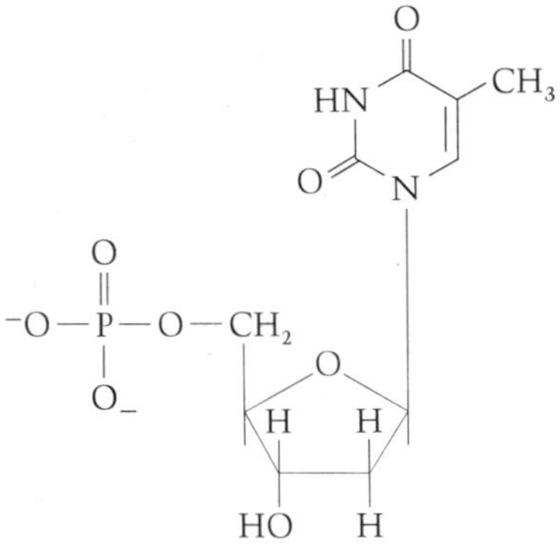

dTMP

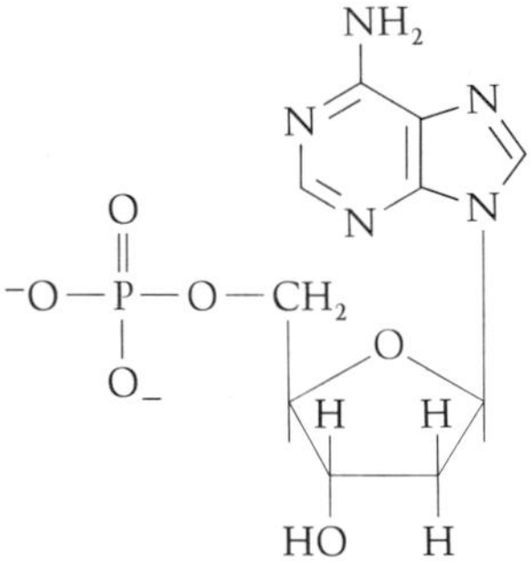

dAMP

18.5 5′ A—C—U 3′

A—U—C

C—A—U

C—U—A

U—A—C

U—C—A

(Total number is $n!$, where n = number of nucleotides in sequence. $n = 3$ and $3! = 6$.)

18.6 5′ A—C—U 3′

5′

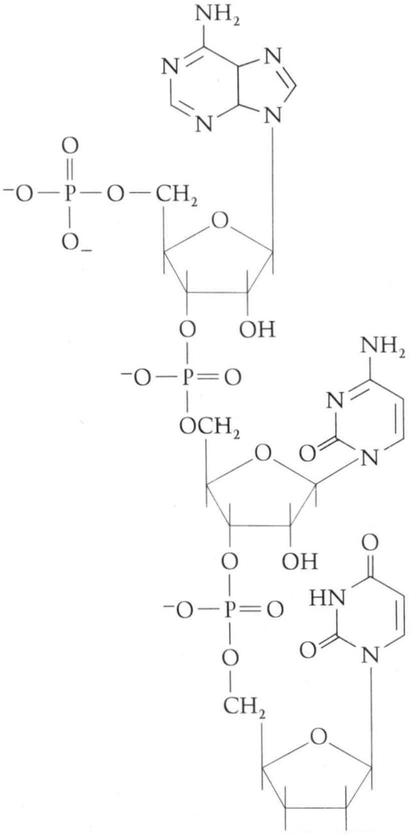

3′

18.7 5′—C—G—A—C—U—C—A—U—C—G—U—3′ **18.8** 31% thymine, 19% cytosine
18.9 phosphodiester bond; the purine and pyrimidine bases (*Section 18.4*)

18.11

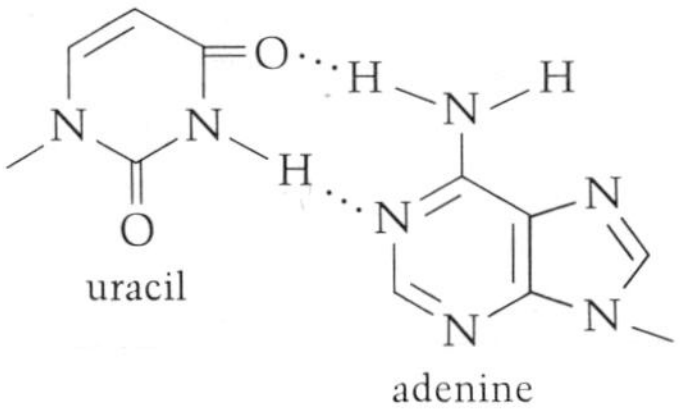

uracil

adenine

(*Section 18.6*)

18.13 Adenine is not very water-soluble because it is quite nonpolar; adenosine has the polar ribose sugar, which increases its water solubility; adenosine-5′-monophosphate has the very polar phosphate group, which makes it very water soluble.
18.15 T—G—C—T—A—A—C—G—A—T—T—C—A—T—G (*Section 18.6*)

Chapter 19

19.1

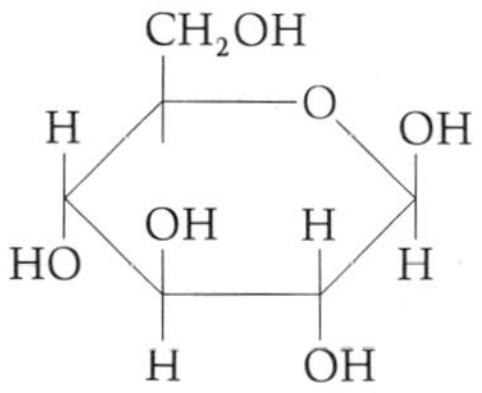

19.2

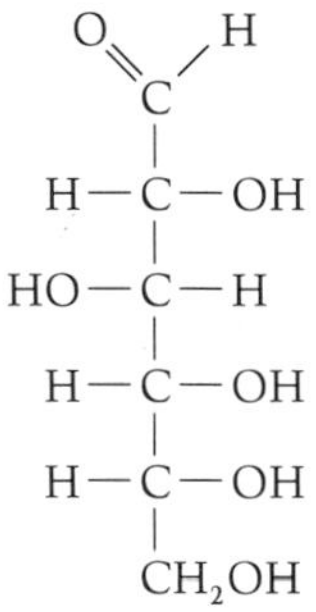

19.3

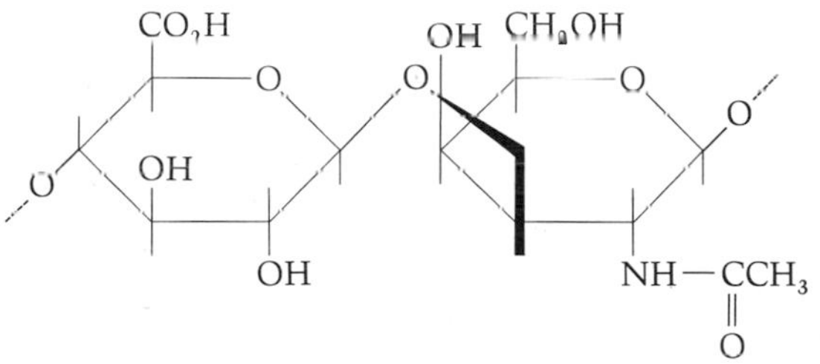

19.4 hydrogen bonding

19.5

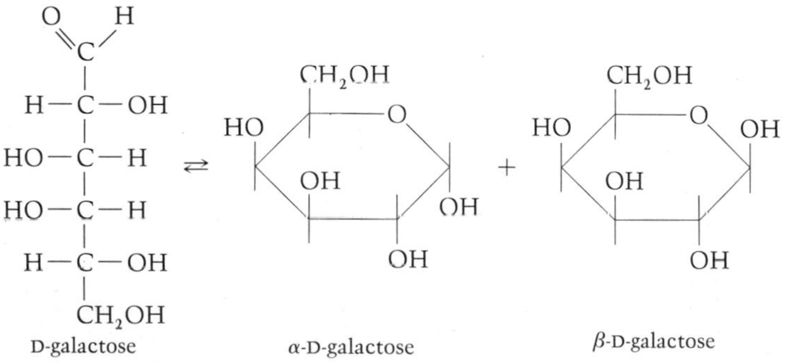

D-galactose ⇄ α-D-galactose + β-D-galactose

19.6 (structure: two pyranose rings, each with CH_2OH, ring O, OH groups, linked by O; terminal HO and OH)

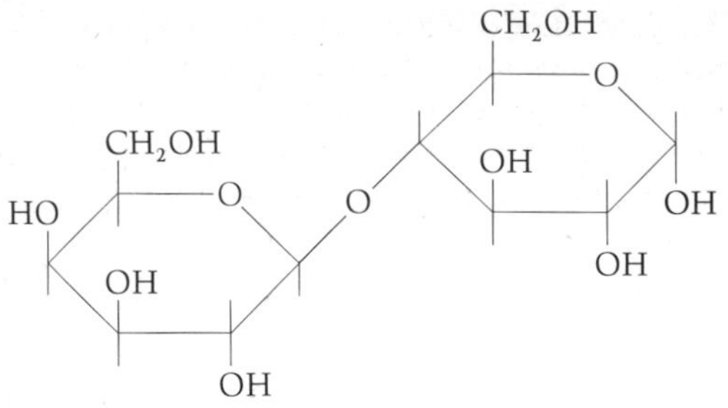

19.9 (structure: disaccharide with CH_2OH, ring O, OH groups, HO, linked by O) $+ H_2O \rightarrow 2$ (monosaccharide with CH_2OH, ring O, OH, HO, OH)

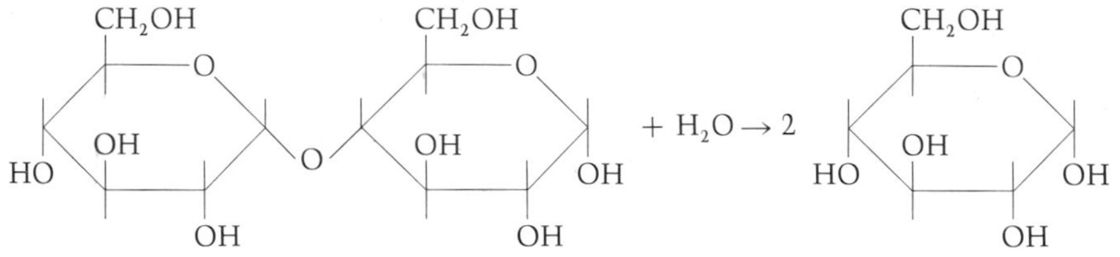

Chapter 20

20.1 For example: $CH_3CH_2C(=O)-OH$ (propionic acid); $CH_3CH_2CH_2CH_2CH_2CH_2CH_2C(=O)-OH$ (octanoic acid); $CH_3CH_2CH_2CH_2CH_2CH_2CH_2CH_2CH_2CH_2CH_2CH_2CH_2C(=O)-OH$ (myristic acid)

20.2

$CH_2OH-CH(OH)-CH_2OH + 2CH_3(CH_2)_4CH{=}CHCH_2CH{=}CH(CH_2)_7CO_2H + CH_3(CH_2)_{14}CO_2H$

$\rightarrow$

$$\begin{array}{l} CH_2-O-C(=O)-(CH_2)_7CH{=}CHCH_2CH{=}CH(CH_2)_4CH_3 \\ | \\ CH-O-C(=O)-(CH_2)_7CH{=}CHCH_2CH{=}CH(CH_2)_4CH_3 \\ | \\ CH_2-O-C(=O)-(CH_2)_{14}CH_3 \end{array}$$

20.3 Yes, there is a chiral carbon atom (the middle carbon of the glycerol backbone) when the fatty acids esterified at the two end carbon atoms are different. A compound containing one chiral carbon atom will exhibit optical activity.

20.4

$$\begin{array}{l} CH_2-O-C(=O)(CH_2)_7CH{=}CH(CH_2)_7CH_3 \\ | \\ CH-O-C(=O)-(CH_2)_7CH{=}CH(CH_2)_5CH_3 \\ | \\ CH_2-O-P(=O)(O^-)-O-CH_2CH_2NH_3^+ \end{array}$$

$$\begin{array}{l} CH_2-O-C(=O)-(CH_2)_7CH{=}CHCH_2CH{=}CH(CH_2)_4CH_3 \\ | \\ CH-O-C(=O)-(CH_2)_7CH{=}CH(CH_2)_7CH_3 \\ | \\ CH_2-O-P(=O)(O^-)-OCH_2-CH(NH_3^+)-CO_2^- \end{array}$$

20.5 A micelle is a group of associated molecules that arrange themselves into a globule with polar portions on the surface and nonpolar portions buried inside. A bilayer is a sheet of molecules two molecules in thickness. The molecules arrange themselves with their polar portions on the two surfaces of the sheet and their nonpolar portions buried inside. A monolayer is a sheet of molecules one molecule in thickness; all are oriented with polar portions together and nonpolar portions together.

20.7

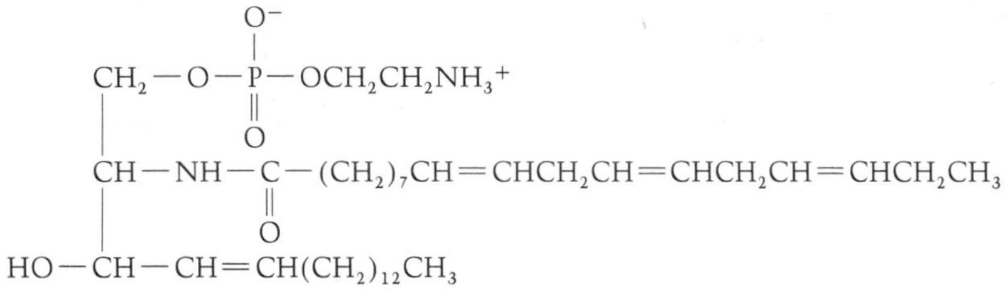

$$\begin{array}{l} \qquad\qquad\quad O^- \\ \qquad\qquad\quad | \\ CH_2-O-P-OCH_2CH_2NH_3^+ \\ | \qquad\qquad\ \| \\ | \qquad\qquad\ O \\ CH-NH-C-(CH_2)_7CH=CHCH_2CH=CHCH_2CH=CHCH_2CH_3 \\ | \qquad\qquad\ \| \\ | \qquad\qquad\ O \\ HO-CH-CH=CH(CH_2)_{12}CH_3 \end{array}$$

20.9 $CH_3(CH_2)_{14}\overset{O}{\overset{\|}{C}}O-$(cholesterol ring system)

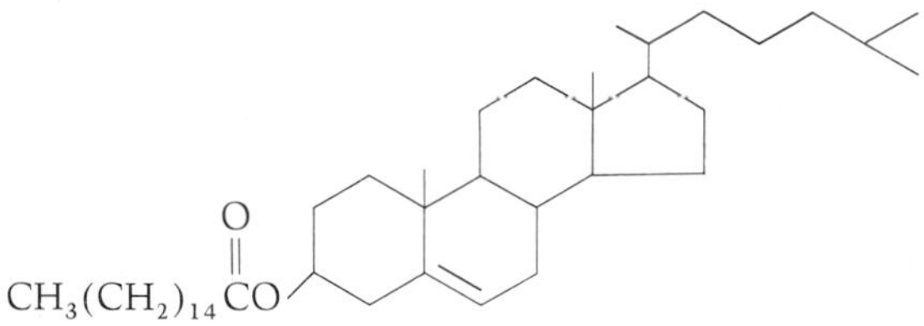

20.10 The ester group **20.11** Yes; complex.

20.14

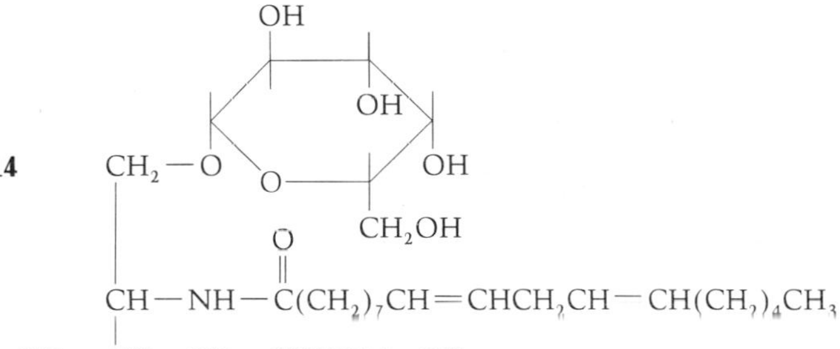

$$\begin{array}{l} CH_2-O-(\text{galactose ring: OH, OH, OH, }CH_2OH) \\ | \qquad\qquad\quad O \\ | \qquad\qquad\quad \| \\ CH-NH-C(CH_2)_7CH=CHCH_2CH=CH(CH_2)_4CH_3 \\ | \\ HO-CH-CH=CH(CH_2)_{12}CH_3 \end{array}$$

(*Section 20.5*)

20.16 phosphoglycerides, sphingomyelins, cerebrosides, cholesterol (*Sections 20.3, 20.4, 20.5*)

Chapter 21

21.1 (a) The spontaneous association or interaction of specific macromolecules leading to the formation of a functional structure (b) The specific association or interaction of one molecule (often a small molecule) with another (often a macromolecule) **21.2** Proteins deeply embedded in the hydrophobic core of the lipid bilayer will probably have a large percentage of nonpolar amino acids. **21.3** Polar surfaces—the proteins are probably bound by hydrogen bonds and ionic bonds. **21.4** Procaryotes have no nuclei—they evolved early before such specialized cell structures arose. Eucaryotes have true nuclei. **21.5** phenylalanine, tryptophan, isoleucine, leucine, and so on (see Table 17.2 for structural formulas) **21.6** lysine, arginine, histidine (see Table 17.2 for structural formulas) **21.7** H^+ donors: glutamic acid, aspartic acid, cysteine; H^+ acceptors: lysine, arginine, histidine **21.8** (a) about 10% of maximum rate (b) about 90% of maximum rate **21.9** An activator that builds up when the amino acid is in short supply might activate the enzyme and increase its reaction rate. **21.12** Bacterial cells do not have internal organelles and they are much smaller cells. They have only one chromosome rather than several or many. (*Sections 21.5, 21.6, 21.7*) **21.13** Rough e.r. has ribosomes on the surfaces of its membranes and is therefore involved in protein synthesis. Smooth e.r. has no ribosomes, rather it has smooth membranes; s.e.r. is involved in many metabolic processes, but not protein synthesis. (*Sections 21.6, 21.7*) **21.15** The ability of an enzyme to bind only molecules with particular structural features or properties and to catalyze only one or a very small number of different types of reactions. (*Section 21.8*)

Chapter 22

22.1 Water falling over a dam drives electric generators (Mechanical energy → electrical energy). Coal or oil are burned to heat water and make steam which drives the generators (Chemical → heat → electrical energy) Gasoline is burned in the internal combustion engine of your car to drive the pistons and eventually, in turn, the car's wheels. (Chemical → mechanical energy) **22.2** Negative; positive; zero. **22.3** Probably not. **22.4** Yes, because the animals they feed upon are herbivores.

22.5 $^{-}O-\overset{\overset{O}{\|}}{\underset{\underset{O^-}{|}}{P}}-O-\overset{\overset{O}{\|}}{\underset{\underset{O^-}{|}}{P}}-O-\overset{\overset{O}{\|}}{\underset{\underset{O^-}{|}}{P}}-O-CH_2$ (ribose ring with OH OH, attached at N to the guanine ring: HN, H_2N, N, N, N, =O), and so on

GTP

22.6 Much of the energy given off in biological reactions is used to make ATP. ATP hydrolysis is used in turn to supply energy to those biological reactions that require it.

22.7

Metabolic reaction	$\Delta G^\circ = +3.2$ kcal/mole
ATP hydrolysis	$\Delta G^\circ = -7.3$ kcal/mole
Net	$\Delta G^\circ = -4.1$ kcal/mole

Since the overall ΔG° is very negative, plenty of energy is available to make the metabolic reaction go, and indeed to make it go very far toward product formation.

22.8

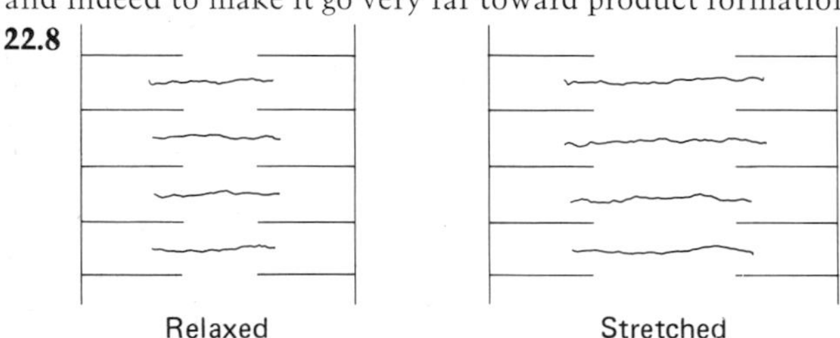

22.9 The formation of bonds between actin and myosin molecules and a conformational change in the myosin molecule that causes the thick and thin filaments to slide along each other, shortening the length of the sarcomere.

22.10 $R-\overset{\overset{O}{\|}}{C}-H + 2(H\cdot) \rightleftharpoons R-CH_2OH$ **22.11** $H_2O \rightleftharpoons \frac{1}{2}O_2 + 2(H\cdot)$

22.12 It would cause uncontrolled respiration and rapid breakdown of available foodstuff molecules so that the body's energy stores, mainly fat, would be broken down without regard for the actual need of the body for energy. **22.13** Ca^{2+} is necessary for the reaction between the thick filament myosin molecules and the thin filament actin molecules which leads to shortening or contraction of the sarcomeres, and in turn of the muscle fibers. When the Ca^{2+} concentration in the cytoplasm is low, contraction cannot occur and the muscle is relaxed. When the Ca^{2+} concentration in the cytoplasm rises, contraction begins, and when the concentration is lowered again, the muscle relaxes. (*Section 22.10*) **22.14** High Ca^{2+} concentrations would cause the heart muscle fibers to be somewhat contracted at all times—pumping of blood by the heart would be weak; if Ca^{2+} was high enough the muscle would be unable to relax at all and beating would stop. Low Ca^{2+} concentrations would cause very weak contractile function—weak pumping action would also occur here. (*Section 22.10*) **22.16** Phosphoenolpyruvate, 1,3-diphosphoglycerate, phosphocreatine, phosphoarginine. These are chosen because they have $\Delta G°'$s of hydrolysis that are more negative than the $\Delta G°$ of ATP formation from ADP and P_i ($\Delta G° = +7.3$ kcal/mol). (*Section 22.9*) **22.18** Flavoproteins **22.20** Because it is a small somewhat hydrophobic organic molecule that is lipid-soluble. (*Section 22.16*)

Chapter 23

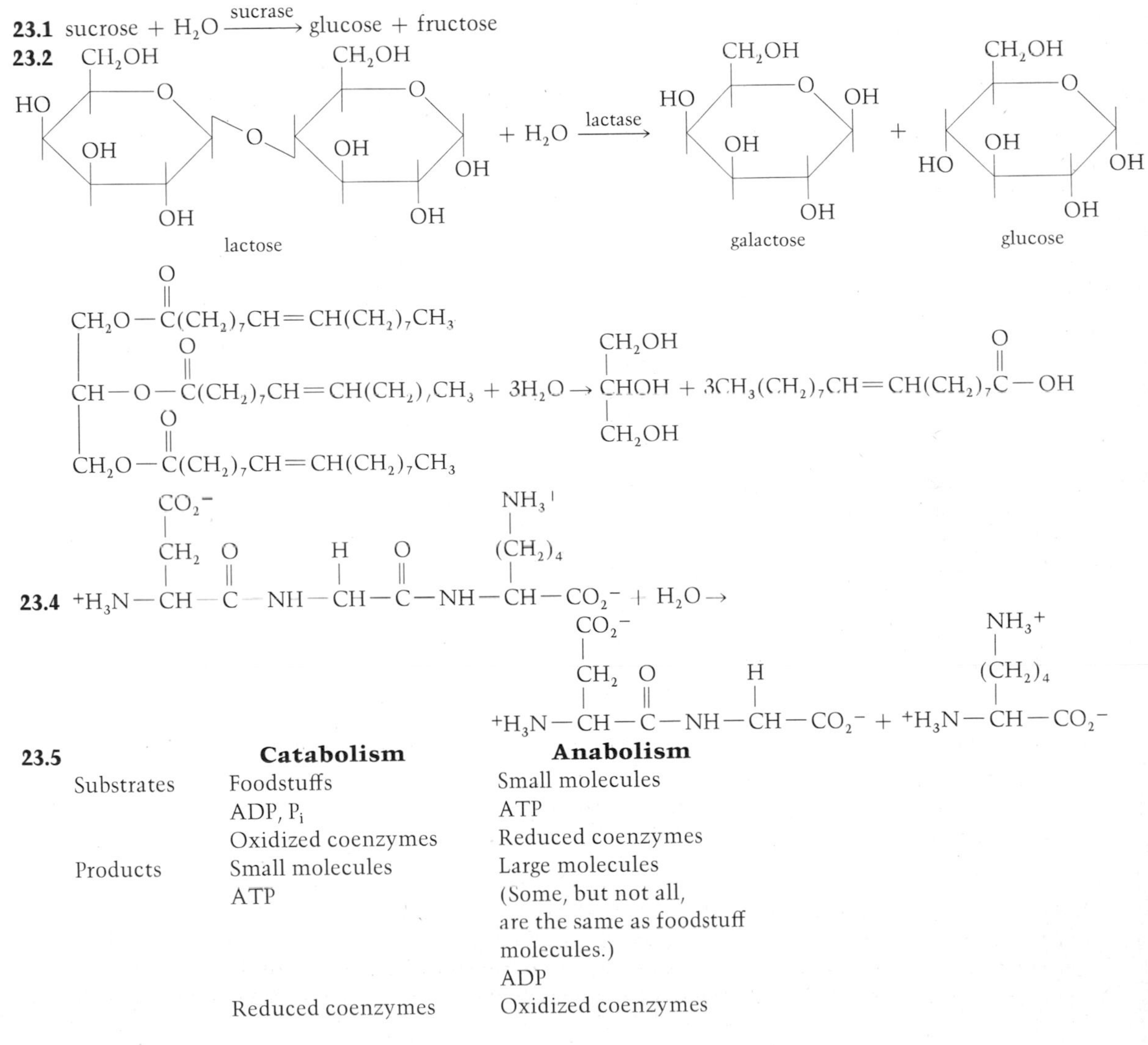

23.5

	Catabolism	**Anabolism**
Substrates	Foodstuffs ADP, P_i Oxidized coenzymes	Small molecules ATP Reduced coenzymes
Products	Small molecules ATP Reduced coenzymes	Large molecules (Some, but not all, are the same as foodstuff molecules.) ADP Oxidized coenzymes

23.6 $R{-}\underset{\text{OH}}{\overset{|}{C}}H{-}R' + NAD^+ \xrightarrow{\text{a dehydrogenase}} R{-}\overset{O}{\overset{\|}{C}}{-}R' + NADH + H^+$

23.7 $NADH + H^+ + \frac{1}{2}O_2 \longrightarrow NAD^+ + H_2O$

$3H^+ + 3ADP + 3P_i \longrightarrow 3ATP + 3H_2O$

23.8 Because it is used as a reactant but also formed again as a product; when a substance appears in the same amount on both sides of a chemical equation, it cancels out and does not appear in the final equation. **23.10** Acetyl CoA would build up or accumulate inside the mitochondria because of inadequate amounts of oxaloacetate for it to react with so as to undergo oxidation in the citric acid cycle. (*Section 23.9*)

23.12 isocitrate^{2-} + $NAD^+ \rightarrow$ α-ketoglutarate^{2-} + CO_2 + NADH + H^+
α-ketoglutarate + NAD^+ + HSCoA → succinyl-S-CoA + NADH + CO_2 (*Section 23.10*)

23.14 No, no ATP is made although one molecule of GTP is made when succinyl CoA is converted to succinate. Energy is stored in the form of reduced coenzymes NADH and FPH_2. (*Section 23.10*)

Chapter 24

24.1 (a) mitochondria (b) mitochondria (c) cytosol or cytoplasm (d) mitochondria

24.2 (a) pyruvate (b) lactate (c) CO_2 + ethanol

24.3 [glucose ring structure: CH_2OH, O, OH, HO, OH, OH] $+ 2NAD^+ + 2P_i^{2-} + 2ADP^{3-} \rightarrow 2CH_3\overset{O}{\overset{\|}{C}}CO_2^- + 2ATP^{4-} + 2H_2O + 2NADH + 2H^+$

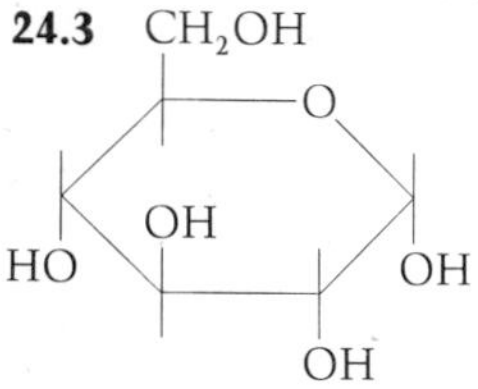

(6 C–glucose; 6 O–glucose; 12 H–glucose; 6 C–2 pyruvates; 6 O–2 pyruvates; 6 H–2 pyruvates; 6 H elsewhere (remember that P_i gives up $2H^+$)

24.4 UDP-glucose + galactose-1-phosphate →

UDP-galactose + glucose-1-phosphate

[structures: UDP-glucose (CH_2OH, O, OH, HO, OH; O—P(=O)(O⁻)—O—P(=O)(O⁻)—O—CH_2—ribose(OH, OH)—uracil (HN, O, N, O)); galactose-1-phosphate (CH_2OH, HO, O, OH, OH; O—P(=O)(O⁻)—O⁻); UDP-galactose; glucose-1-phosphate]

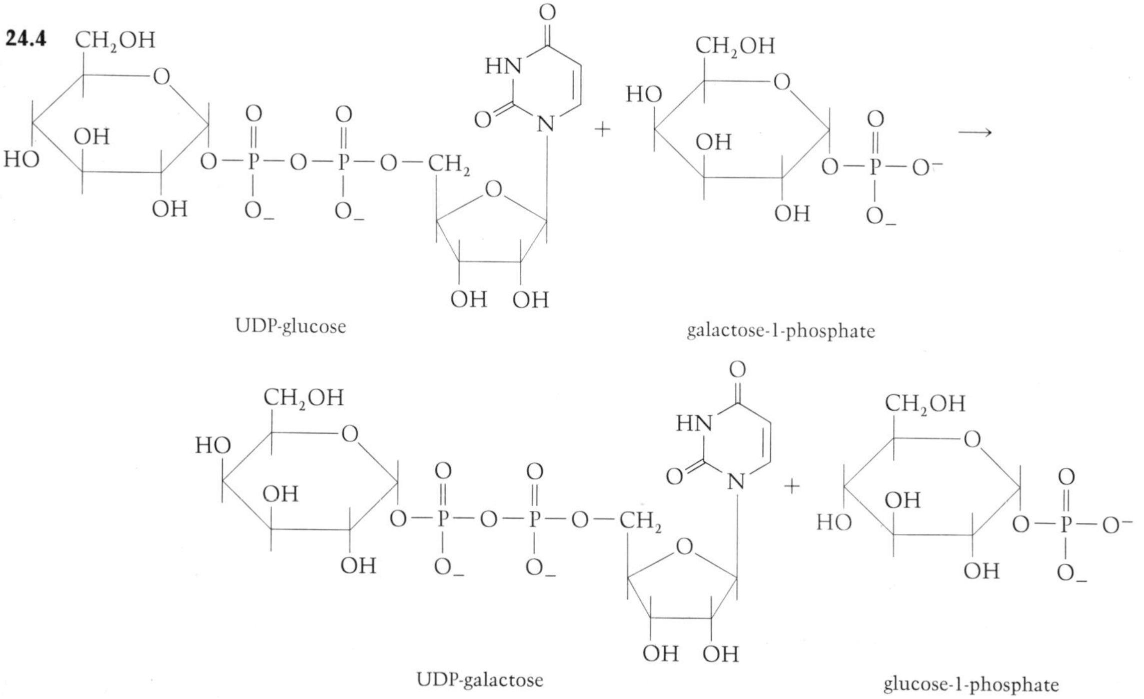

24.5 $4NADH + 2O_2 + 12ADP^{3-} + P_i^{2-} + 16H^+ \rightarrow 4NAD^+ + 16H_2O + 12ATP^{4-}$
Overall equation for aerobic glucose oxidation:
Glucose $+ 6O_2 + 38ADP^{3-} + 38P_i^{2-} + 38H^+ \rightarrow 6CO_2 + 38ATP^{4-} + 44H_2O$

24.6

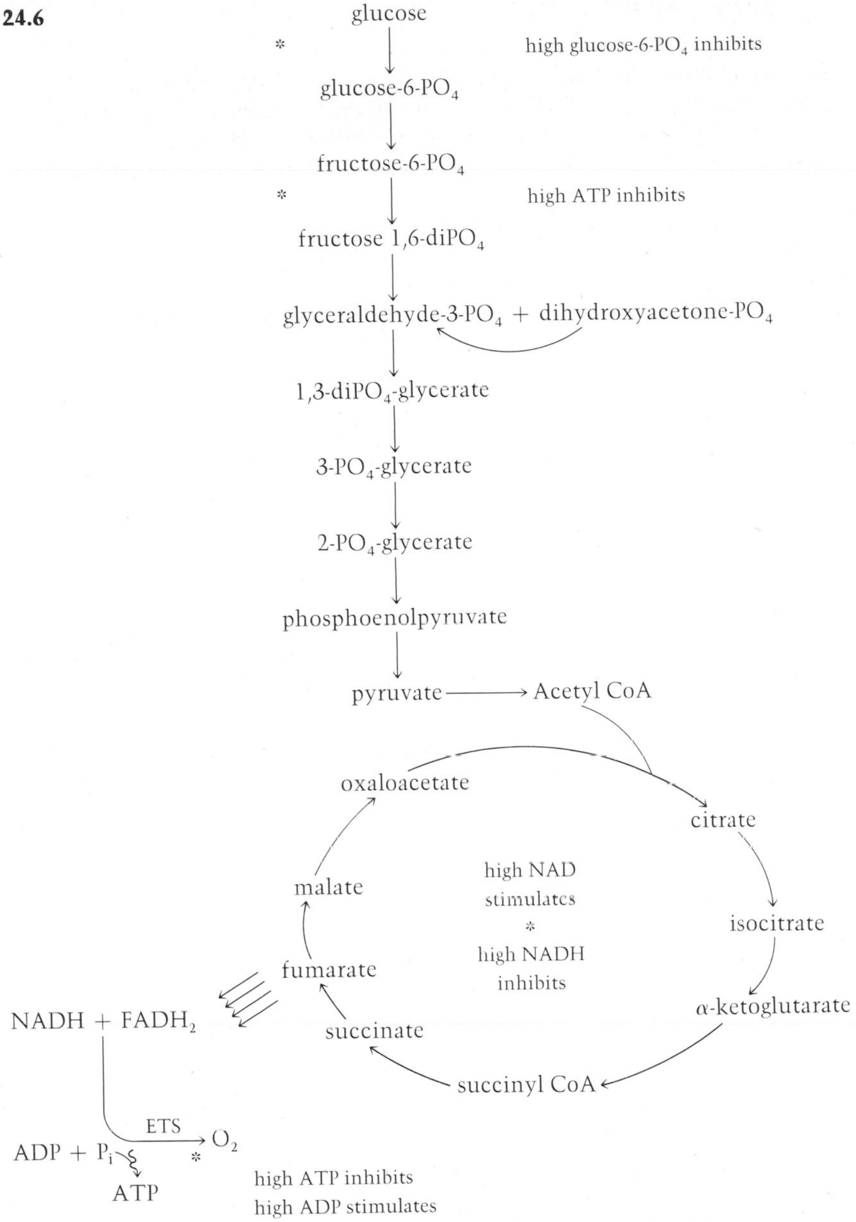

glucose
* ↓ high glucose-6-PO_4 inhibits
glucose-6-PO_4
↓
fructose-6-PO_4
* ↓ high ATP inhibits
fructose 1,6-diPO_4
↓
glyceraldehyde-3-PO_4 + dihydroxyacetone-PO_4
↓
1,3-diPO_4-glycerate
↓
3-PO_4-glycerate
↓
2-PO_4-glycerate
↓
phosphoenolpyruvate
↓
pyruvate ⟶ Acetyl CoA

oxaloacetate → citrate → isocitrate → α-ketoglutarate → succinyl CoA → succinate → fumarate → malate → oxaloacetate

high NAD stimulates
*
high NADH inhibits

NADH + $FADH_2$

ETS → O_2

ADP + P_i → ATP *

high ATP inhibits
high ADP stimulates

24.7 succinyl CoA + GDP + $P_i \rightarrow$ succinate + GTP + HSCoA
succinate + $FP_2 \rightarrow$ fumarate + FP_2H_2 fumarate + $H_2O \rightarrow$ L-malate
L-malate + $NAD^+ \rightarrow$ oxaloacetate + NADH + H^+ oxaloacetate + GTP → PEP + CO_2 + GDP
24.8 6; 4 **24.9** 2lactate + $2NAD^+ \rightarrow$ 2pyruvate + 2NADH + $2H^+$
$2H_2O$ + 2pyruvate + 2ATP + $2CO_2 \rightarrow$ 2oxaloacetate + 2ADP + $2P_i$ + $4H^+$
2oxaloacetate + 2GTP → 2PEP + $2CO_2$ + 2GDP 2PEP + $2H_2O \rightarrow$ 2-phosphoglycerate

2 2-phosphoglycerate ⟶ 2 3-phosphoglycerate
2 3-phosphoglycerate + 2ATP ⟶ 2 1,3-diphosphoglycerate + 2ADP
2 1,3-diphosphoglycerate + 2NADH + $2H^+$ ⟶ 2glyceraldehyde-3-phosphate + 2NAD + $2P_i$
glyceraldehyde-3-phosphate ⟶ dihydroxyacetone-phosphate
glyceraldehyde-3-phosphate + dihydroxyacetone phosphate ⟶ fructose-1,6-diphosphate
fructose-1,6-diphosphate + H_2O ⟶ fructose-6-phosphate + P_i
fructose-6-phosphate ⟶ glucose-6-phosphate   glucose-6-phosphate + H_2O ⟶ glucose + P_i

24.10 Insulin **24.11** Only target cells have the right receptors to bind hormones and mediate response to the hormone binding. **24.12** Manufacture of ATP (*Section 24.2*) **24.14** Reactions can occur, except for the one directly involved, but net ATP formation is not possible. (*Section 24.5*)
24.16 α-ketoglutarate dehydrogenase (*Section 24.8*) **24.19** DHAP ⟶ glyceraldehyde-3-phosphate
DHAP + glyceraldehyde-3-phosphate ⟶ fructose-1,6-diphosphate
and so on ⟶ glucose (*Section 24.11*)
24.20 The message is that oxaloacetate levels are too low. Accumulation of acetyl CoA allows acetyl CoA to activate pyruvate carboxylase, which catalyzes oxaloacetate formation. If the energy level (ATP level) in the cell is high enough and carbohydrate synthesis is needed, oxaloacetate will be channeled into gluconeogenesis. (*Section 24.14*)
24.22 glucose-6-phosphate ⟶ glucose-1-phosphate
glucose-1-phosphate + UTP ⟶ UDP − glucose + PP_i
$\underset{\text{glycogen}}{(\text{glucose})_n}$ + UDP-glucose ⟶ $\underset{\text{longer glycogen polymer}}{(\text{glucose})_{n+1}}$ + UDP (*Section 24.15*)

Chapter 25

25.1 Both involve ATP hydrolysis. However, in glucose catabolism ATP is used to phosphorylate glucose, yielding glucose-6-phosphate and ADP, while in fatty acid breakdown ATP hydrolysis drives the formation of a CoA ester of the fatty acid and ATP ⟶ AMP + PP_i.

25.2

$$\mathrm{CH_3CH_2CH_2CH_2CH_2CO_2^- + HSCoA + ATP \longrightarrow CH_3CH_2CH_2CH_2CH_2\overset{O}{\overset{\|}{C}}SCoA + AMP + PP_i}$$

$$\mathrm{CH_3CH_2CH_2CH_2CH_2\overset{O}{\overset{\|}{C}}SCoA + FP \longrightarrow CH_3CH_2CH_2CH{=}CH\overset{O}{\overset{\|}{C}}SCoA + FPH_2}$$

$$\mathrm{CH_3CH_2CH_2CH{=}CH\overset{O}{\overset{\|}{C}}SCoA + H_2O \longrightarrow CH_3CH_2CH_2\overset{OH}{\overset{|}{C}}HCH_2\overset{O}{\overset{\|}{C}}SCoA}$$

$$\mathrm{CH_3CH_2CH_2\overset{OH}{\overset{|}{C}}HCH_2\overset{O}{\overset{\|}{C}}SCoA + NAD^+ \longrightarrow CH_3CH_2CH_2\overset{O}{\overset{\|}{C}}CH_2\overset{O}{\overset{\|}{C}}SCoA + NADH + H^+}$$

$$\mathrm{CH_3CH_2CH_2\overset{O}{\overset{\|}{C}}CH_2\overset{O}{\overset{\|}{C}}SCoA + HSCoA \longrightarrow CH_3CH_2CH_2\overset{O}{\overset{\|}{C}}SCoA + CH_3\overset{O}{\overset{\|}{C}}SCoA}$$

$$\mathrm{CH_3CH_2CH_2\overset{O}{\overset{\|}{C}}SCoA + FP \longrightarrow CH_3CH{=}CH\overset{O}{\overset{\|}{C}}SCoA + FPH_2}$$

$$\mathrm{CH_3CH{=}CH\overset{O}{\overset{\|}{C}}SCoA + H_2O \longrightarrow CH_3\overset{OH}{\overset{|}{C}}HCH_2\overset{O}{\overset{\|}{C}}SCoA}$$

$$\mathrm{CH_3\overset{OH}{\overset{|}{C}}HCH_2\overset{O}{\overset{\|}{C}}SCoA + NAD^+ \longrightarrow CH_3\overset{O}{\overset{\|}{C}}CH_2\overset{O}{\overset{\|}{C}}SCoA + NADH + H^+}$$

$$\mathrm{CH_3\overset{O}{\overset{\|}{C}}CH_2\overset{O}{\overset{\|}{C}}SCoA + HSCoA \longrightarrow CH_3\overset{O}{\overset{\|}{C}}SCoA + CH_3\overset{O}{\overset{\|}{C}}SCoA}$$

25.3 146 ATPs
25.4 palmitate + $23O_2$ + 131ADP + $131P_i$ + $131H^+$ → $16CO_2$ + 130ATP + $154H_2O$ + AMP + PP_i
Since an ATP will be required to make ADP from the AMP appearing in the right-hand side of the equation (AMP + ATP → 2ADP), and PP_i will be broken down to yield $2P_i$(PP_i + H_2O → $2P_i$ + H^+) adding these equations to the above will give us the real overall equation and our expected yield of 129ATP: palmitate + $23O_2$ + 129ADP + $129P_i$ + $130H^+$ → $16CO_2$ + 129ATP + $153H_2O$
25.5 The molecular weight of the lipid is less than that of the carbohydrate, so that more molecules of hexanoic acid will be found per unit weight than in the same weight of glucose ($C_6H_{12}O_6$ has MW180, $C_6H_{12}O_2$ has MW116). **25.6** Because fatty acids bound to serum albumin cannot cross the blood-brain barrier, but the small ketone bodies are just dissolved in the blood plasma and readily diffuse across to the cells of the brain.
25.7

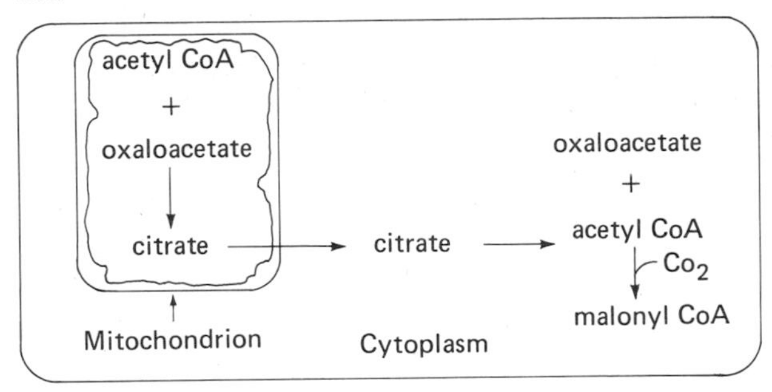

25.8 (a) 7 (b) formation of malonyl CoA

25.9

	F.A. oxidation	**F.A. synthesis**
cellular location of enzymes	mitochondria	cytoplasm
coenzymes	NAD^+ and FAD (of the flavoprotein dehydrogenase)	NADPH
	coenzyme A is carrier throughout process	acyl carrier protein carries growing chain; coenzyme A only used in initial stages
energetics	yields a lot of energy-rich reduced coenzymes and ATP	requires energy-rich reduced coenzymes and ATP
activators	none of great importance	citrate
inhibitors	high energy level, that is, high ATP and NADH.	long chain fatty acyl CoAs are feedback inhibitors
When does it occur?	When cell needs energy and fatty acids are available	When cell has excess energy and needs to store excess foodstuffs

25.10 In uncontrolled diabetes, cells such as muscle cannot burn glucose for energy. When the cells are unable to transport amino acids, this means that one more major class of food—proteins—is closed off as an energy source. Thus the dependence on fatty acid oxidation is made even greater. In addition it means that these cells will not be able to make new proteins at the normal rates, and less important proteins in the cells will be broken down faster to provide amino acids for synthesis of new protein.

25.11 Transport of fatty acids (*Section 25.2*)
25.13 $CH_3(CH_2)_{10}CO_2H + 17O_2 + 95ADP + 95P_i + 95H^+ \rightarrow 12CO_2 + 95ATP + 112H_2O$ (*Section 25.5*)
25.15 The cells of both experience carbohydrate starvation and must depend extensively on fatty acid oxidation for energy, and protein degradation to supply carbons for gluconeogenesis. Ketone body formation occurs as byproduct of liver's rapid oxidation of fatty acids, and ketoacidosis may result. (*Section 25.9*) **25.17** Those carbons bearing branches cannot undergo the hydration and ketone formation that occurs in beta-oxidation. (*Section 25.7*)

25.19 $^-O_2CCH{=}CHCO_2^- + H_2O \rightarrow {}^-O_2C{-}CH(OH){-}CH_2{-}CO_2^-$ (*Section 25.5*)
fumarate → L-malate

25.21 palmitic acid (*Section 25.13*) **25.23** acetyl CoA—from citrate exported from mitochondria malonyl CoA—from acetyl CoA by a carboxylation reaction NADPH—NADP is reduced to NADPH in certain catabolic pathways (*Sections 25.12, 25.13*)

25.25 $CH_3{-}CH(NH_3^+)CO_2^- + \alpha\text{-ketoglutarate} \rightarrow CH_3{-}C(=O)CO_2^- + \text{glutamate}$

$CH_3C(=O)CO_2^- + HSCoA + NAD^+ \rightarrow CH_3C(=O)SCoA + NADH + CO_2$

$CH_3C(=O)SCoA \xrightarrow{\text{citric acid cycle}} 2CO_2,\ 3NADH,\ FPH_2,$ and so on

$\text{reduced coenzymes} + O_2 \xrightarrow{\text{ETS}} H_2O + \text{oxidized coenzymes} + ATP$

$H_2O + \text{glutamate} + NAD^+ \rightarrow \alpha\text{-ketoglutarate} + NH_4^+ + NADH + H^+$ (*Sections 25.18, 25.20*)

25.27 $NH_4^+ + CO_2 + 3ATP^{4-} + \text{L-aspartate}^- + H_2O \rightarrow$
$NH_2C(=O)NH_2 + \text{fumarate}^{2-} + 2ADP^{3-} + AMP^- + 3P_i^{2-} + 3H^+$ (*Section 25.19*)

Chapter 26

26.1 5′—G—C—A—T—G—C—G—C—3′ **26.2** 1248; 2496
26.3 fatty acid + HSCoA + ATP → fatty acyl CoA + AMP + PP_i **26.4** This is an exergonic reaction that tends to pull the main reaction toward product formation. **26.5** They are "activated" by nature—since they are high energy nucleoside triphosphates. **26.6** UUU or UUC; AAA or GAA
26.7 1248 **26.8** (a) lys-phe-arg-arg-gly-ser (b) val-ser-ser-ser-ala-ser-pro (c) met-ala-lys-tyr (d) pro-arg-leu (*Section 26.7*) **26.10** Code words that indicate the end of a message. An incomplete polypeptide would be made whenever that *m*RNA message was translated. (*Section 26.13*)
26.12 AGG → arg, GGA → gly, GAG → glu (*Section 26.7*)
26.14 Because in the double-stranded DNA, every nucleotide in one chain is matched with its complementary nucleotide in the other. So the total adenine nucleotide content must equal the total thymine nucleotide content, and the same for guanine and cytosine nucleotides. (*Section 26.4*)
26.16 Once the amino acid is attached to a *t*RNA, the only information regarding the identity of that amino acid resides in the anticodon of the *t*RNA. If the wrong match is made when amino acyl *t*RNAs are made, a wrong translation will be made. (*Section 26.10*)

Chapter 27

27.1 Red cells carry O_2 and CO_2. White cells are part of the body's defense mechanism.
27.2 In lymphocytes **27.3** Since it decreases the H_2CO_3 and HCO_3^- concentration in the blood, the buffering capacity is reduced. **27.4** LDL and HDL **27.5** A high O_2 concentration shifts the equilibrium to the side of HbO_2 (Le Chatelier's principle) and allows O_2 to take the place of CO molecules on hemoglobin heme groups that have bound CO. CO can then be excreted from the body by the lungs.
27.6 $H_2PO_4^- + OH^- \rightarrow H_2O + HPO_4^{2-}$ **27.7** $HPO_4^{2-} + H^+ \rightarrow H_2PO_4^-$
27.8 Red bone marrow (*Section 27.1*) **27.10** Presence of lipid or carbohydrate (*Section 27.2*)
27.12 Erythrocytes have no organelles and are highly specialized and streamlined to perform one role—that of O_2 transport. Liver cells are packed with mitochondria, endoplasmic reticulum, and so on, and perform a wide variety of metabolic functions. (*Section 27.9*)
27.14

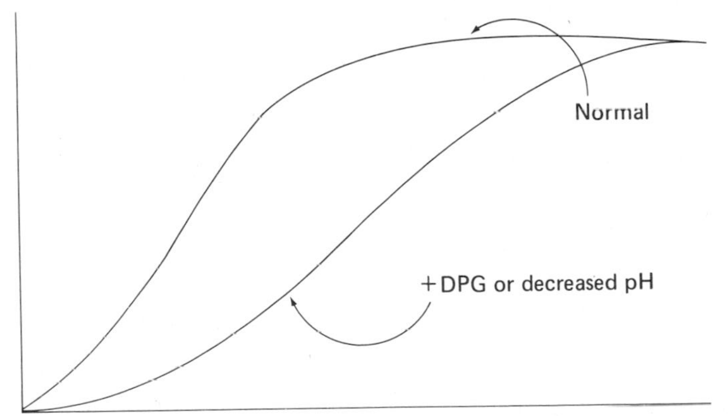

(*Section 27.10*)

27.16 In plasma, Na^+ is high ($\simeq$ 150 mEq/L) and K^+ is low ($\simeq$ 5 mEq/L); in cell fluid, Na^+ is very low (doesn't even appear on scale in Figure 27.4) and K^+ is high ($\simeq$ 150 mEq/L). (*Section 27.12*)

27.18 hydrocortisone, aldosterone } adrenal glands testosterone} testes estrogens, progestins } ovaries and placenta
(*Section 27.14*)

INDEX

ABCDEFGHIJ-M-8210